K.P. Joy □ A. Krishna □ C. Haldar (Eds)

Comparative Endocrinology and Reproduction

K.P. Joy □ A. Krishna □ C. Haldar (Eds)

Comparative Endocrinology and Reproduction

Springer-Verlag

Narosa Publishing House

EDITORS
K.P. Joy, A. Krishna and C. Haldar
Department of Zoology
Banaras Hindu University
Varanasi-221 005, INDIA

Exclusive distribution in Europe, North America, Canada, Mexico, South America and Japan by Springer-Verlag Berlin Heidelberg New York.

ISBN 3-540-63691-9 Berlin Heidelberg New York
ISBN 0-387-63691-9 New York Berlin Heidelberg
ISBN 81-7319-194-8 Narosa Publishing House, New Delhi

Printed in India.

Professor J.P. Thapliyal (M.Sc., Ph.D., FNA, FNASc)

Professor Jagdamba Prasad Thapliyal born on 1920 at Garhawal, North India, received M.Sc. (1945) and Ph. D (1950) degree from the Department of Zoology, B.H.U., Varanasi under Late Professor A.B. Mishra, the founder of the Department of Zoology, Banaras Hindu University, Varanasi. He worked with Professor Emil Witschi, U.S.A. as a Rockefeller Foundation Research Associate (1958–60). He joined the Department as Lecturer in 1948 and became Professor in 1968. He was Head of the Department during 1971–74 and 1978–80. He published over 175 research papers in highly reputed National and International journals and trained 24 Ph. D. students. His students have worked with many eminent scientists in prestigious laboratories in India and abroad. He served as Incharge of the Special Assistant Programme of University Grants Commission, New Delhi. He was elected as General President of Indian Society of General and Comparative Endocrinology 1977; Biological Section and National Academy of Science. University Grants Commission had selected him as National Lecturer in 1979–80. He became Fellows of the Indian National Science Academy Delhi (1980) and National Academy of Sciences, Allahabad (1978). Besides, he was President of several Professional Societies and member of editorial boards of reputed Journals and funding bodies.

Because of his notable contributions in avian endocrinology and the hypothesis on photoperiodism and avian thyroid function, he is commonly known as father of "Indian Avian Endocrinology" and is still continuing his scientific and academic pursuits on tropical avian endocrinology.

Professor A.G. Sathyanesan (M.A., M.Sc., Ph.D., D.Sc., F.N.A)

Prof. Appagi Gilbert Sathyanesan was born on October 3, 1926 at Neyyoor in Kannyakumari, Tamilnadu and had his early educataion in the erstwhile Travancore State. He took his Master's degree in Zoology from Annamalai University where he came under the sway of Prof. R.V. Seshiya, F.N.A., A devoted teacher and researcher who introduced him to endocrine research. Prof. Sathyanesan took his Ph.D. and D.Sc. from Banaras Hindu University. The theses were evaluated and appreciated by stalwarts in endocrinology and neuroendocrinology namely, Dr. Emil Witschi (U.S.A.), Dr. G.J. Van Oordt (Holland), Dr. J.M. Dodd (England), Dr. Howard Bern (U.S.A.), and Dr. M. Gabe (France). As a fellow of the Population Council of the Rockefeller Foundation, he had his post-doctoral research training at Columbia University, New York, Friday Harbor Marine Laboratory, Washington State and Wayne State University, Detroit. He taught Physiology for sometime in the Washington University, Seattle (U.S.A).

Prof. Sathyanesan started his career in 1951 as Lecturer in Zoology in the Andhra Christian College, Guntur (A.P.). Later he shifted to BHU, and worked as Lecturer in Zoology, Senior Scientist at the Centre of Experimental Medicine and Surgery, Institute of Medical Sciences and Professor of Zoology. After superannuation in 1986, he joined the Central Marine Fisheries Research Institute at Cochin as C.S.I.R. Emeritus Scientist. Prof. Sathyanesan is a recipient of several honours that included Hari Om Ashram Trust Award, Dr. Rafi Ahmad Kidwai Memorial Prize and Dr. Sunder Lal Hora Medal. He was elected as a Fellow of the Indian National Science Academy in 1982. He has published over 200 research papers in Neuroendocrinology and Reproductive Physiology of fishes and mammals in international journals and guided 15 students for Ph.D. He has been associated with many scientific societies, editorial boards of scientific journals and national committees on academic and educational subjects.

True to the ancient maximal teacher is known by his students', several students of Prof. Sathyanesan whom he guided to obtain Ph.D. degrees have achieved recognition in their respective fields. Prof. Sathyanesan's other interests include comparative religion and philosophy, environment, and service to society.

Professor C.J. Dominic (M.Sc, Ph.D. Banaras, Ph.D. Cantab., FNA, FNASc)

Born on 1st June 1930 in Kattor, Alleppey, Kerela, Prof. Candanat John Dominic obtained his first Ph.D. degree from Banaras Hindu University in 1959 for his thesis on Avian Endocrinology. Later, he earned his second Ph.D. degree in 1965 from the University of Cambridge for his pioneering work on mammalian reproductive pheromones with late Prof. Sir A.S. Parkes, FRS. Prof. Dominic was inducted to the faculty of Zoology Department, Banaras Hindu University in 1958, and continued till his superannuation in May 1995. His research contributions in the areas of mammalian reproductive pheromones, mammalian reproduction and fertility control, avian reproduction, hypothalamo-hypophysial system and bursa of Fabricius in birds have been recognized both nationally and internationally. He has published over 250 research papers and contributed several reviews/chapters in books and supervised 18 research students.

Prof. Dominic is a recipient of Gulbenkian fellowship from the University of Cambridge (1962–65) and Biomedical fellowship of the Population Council, New York. He was elected as a Fellow of the Zoological Society, Calcutta in 1975, the Indian National Science Academy, New Delhi in 1984, the National Academy of Sciences, Allahabad in 1987 and as a Foreign member of the Erlangen Physio-Medical Society, Germany in 1989. He has been associated with several national and international Scientific Societies, and served in the editorial boards of several national and international journals. Besides, he served in various expert committees constituted by the UGC, ICAR, ICMR, UPSC etc., Government of India.

Preface

Endocrinology and Reproductive physiology are mainstream subjects in modern Zoology curiculum in Universities and Colleges and are offered as general or specialized courses at undergraduate, graduate and post-graduate levels. There has been a growing demand from students and teachers for textbooks and reference books containing updates of advancements made in Endocrinology and Reproduction. The editing of the present volume was taken up to make available a reference-cum-textbook at affordable price to readers interested in comparative endocrinology and reproductive physiology. The subject matter of the book is treated under four sections: Hypothalamus and Pituitary, Gonads and Reproduction, Pineal Gland, and Hormones and Metabolism. The articles were prepared by experts in the subject and we owe to them our sincere gratitude for encouragement and co-operation, and devoting time and energy to make available their contributions timely.

Prof. J.P. Thapliyal, Prof. A.G. Sathyanesan and Prof. C.J. Dominic in whose honour the publication of the book was initiated, have been in the forefront of research in Endocrinology and Reproductive Physiology. Not only did they contribute immensely to the growth and development of the subject in Banaras Hindu University, but also were the source of inspiration and encouragement to their students to set up laboratories for comparative endocrinology and reproductive physiology research in other institutions. It is our solemn duty and great previlege to dedicate the book to commemorate their significant contributions and pioneering leadership in teaching and research in comparative endocrinology and reproductive physiology. That several scientists from abroad and India have contributed articles to the book speaks of their standing and reputation. The editors and contributors wish and prey to Almighty God for their prosperity and good health to continue their academic work in the years to come.

M/S Narosa Publishing House, New Delhi, owes our sincere thanks for accepting and undertaking publication of the book in right earnest.

EDITORS

Contents

A

Hypothalamus and Pituitory

Comparative Endocrinology and Reproduction
K.P. Joy, A. Krishna and C. Haldar (Eds)

1. Mammalian Gonadotropin Releasing Hormone (GnRH): An overview

E. Vijayan and J. Variyar
School of Life Sciences, Pondicherry University
Pondicherry-605 014, India

Introduction

Gonadotropin releasing hormone (GnRH) is the central neuropeptide which governs the synthesis and release of luteinizing hormone (LH). It is also a key hormone involved in sexual behaviour. It acts as the humoral link between the neural and endocrine system. The purification, sequencing and synthesis of the decapeptide by Guillemin and Schally can be considered as a milestone in discovering the role of the neuroendocrine system in reproduction (Matsuo *et al.*, 1971; Burgus *et al.*, 1972). It was through the relentless efforts of Howleg, Harris, McCann, Everett and Sawyer spanning a period of fifty years that we got an understanding of the role of the central nervous system (CNS) in the control of LH and follicle stimulating hormone (FSH). Regulation of the gonadotropes (which secrete LH and FSH) is brought about by this neuropeptide which is carried from the median eminence through the pituitary portal capillaries into the anterior pituitary via the hypophysial portal veins, where it stimulates release of the gonadotropins, LH and FSH. Although secretion of both LH and FSH is stimulated by GnRH and suppressed by gonadal steroids, regulation of these hormones diverges at certain stages of the estrus cycle. Specifically on late proestrus and early estrus, when serum LH declines to basal levels, FSH secretion remains high.

Structure

The structure of GnRH is

pGlu-His-Trp-Ser-Tyr-Gly-Leu-Arg-Pro-Gly-NH

as determined by Matsuo *et al.* (1971) and Burgus *et al.* (1972) from ovine and porcine hypothalamic tissues, respectively. The structure is conserved in all mammalian species that have been studied so far. The key to intrinsic activity on GnRH peptides as shown by Guillemin *et al.* (1972) is the His residue at position 2. The elucidation of the structure has led to the development of agonists and antagonists which have been used for fertility regulation (Matsuo *et al.*, 1971; Burgus *et al.*, 1971, 1972).

Origin of GnRH Neurons

GnRH neurons differentiate from the olfactory placode of the midgestation embryo. Following their differentiation, these neurons migrate through the nasal septum and enter the ventral forebrain with the central roots of the nerve terminalis and vomeronasal nerves (Wray *et al.*, 1989; Fukuda and Pfaff, 1989). The GnRH neurons emerge from the mitotic cycle on embryonic day 10.5 and commence expression of GnRH genes by embryonic day 12.5. The number of GnRH cells are almost similar to that in adults suggesting that this olfactory derived group gives rise to the full complement of GnRH cells in the adult mammalian CNS (Wray *et al.*, 1989a, b; Hoffman and Finch, 1986). Subsequently these cells penetrate the medial ventral forebrain and migrate caudally to reside in the septal preoptic and anterior hypothalamic areas (Fukuda and Pfaff, 1989) by embryonic day 16.5. The migration of the GnRH cells out of the vomeronasal organ (VNO) coincides with the emergence of axons from the epithelium of the VNO on embryonic day 12 (Cusclueri and Banniser, 1975). A population of glial progenitor cells also migrate from the olfactory and vomeronasal epithelia with the outgoing axons and differentiates into unsheathing cells (Chuah and Au, 1991; Marin-Padilla and Amieva, 1989; Mendoza *et al.*, 1982). Ultrastructural observations reveal that axon fascicles and their ensheathing cells form a physical channel through which the GnRH cells can migrate across the nasal septum and enter the forebrain (Fukuda *et al.*, 1989). Once in the forebrain, the GnRH cells are no longer associated with axonal fascicles and appear to intermingle with heterogenous population of neurons and glia. It is plausible that these neurons rely on chemical rather than physical cues in choosing their migratory pathway in the forebrain. The ability of the GnRH neurons to release the mature neuropeptide does not take place till these neurons have migrated well into the forebrain.

There are distinct subpopulations of GnRH neurons in the CNS of adult mammals. Even these subpopulations are not clustered together but appear as a loose network spread through many divisions. In most species, the GnRH cells form a loose continuum from the telencephalic diagonal band of Broca and more dorsal septal areas, to the bed nucleus of the stria terminalis and diencephalic areas (including the periventricular area) medial and lateral preoptic areas, anterior hypothalamus and retrochiasmatic zone medial to the optic tract (Barry, 1979; Steinberger and Hoffman, 1978; Silverman *et al.*, 1979; Kerry and Silverman, 1983).

Around 50-75% of all GnRH neurons in the septal, preoptic and hypothalamic areas are afferent to the median eminence. Cells that supply the median eminence innervation are frequently located immediately next to those which do not project into the median eminence (function unknown Silverman *et al.*, 1987; Merchenthaler *et al.*, 1989). There are studies which suggest that it is these neurons which play an important role during the pulsatile and preovulatory surge of LH and the neurons in the median eminence is proposed to have only a supplementary role (Koves and Molnar, 1986). GnRH pathways have also been traced to other circumventricular organs (sites within the CNS in which the ependymal lining of the ventricular space is specialized and where the capillaries forming the blood supply are fenestrated, thereby permitting access of blood borne substances into the nervous tissue). GnRH innervations are very dense in organum vasculosum lamina terminalis (OVLT) (Barry, 1979). GnRH neurons have been indentified to olfactory related structures like the nervus eterminalis and other structures extending from the corpus callosum (anterior hippocampal rudiment) to the medial division of the anterior olfactory nucleus in the hippocampus and in the Ammon's horn (Merchenthaler *et al.*, 1984). The question whether GnRH neurons present in olfactory regions of the mammalian CNS play any direct role in reproduction is still unanswered.

Cytology of GnRH neurons

Two types of cells have been recognized in the rat: those with a smooth contour and those with a more rugged outline which may be further subdivided by the number of spinous processes (Krisch, 1980; Wray and Hoffman, 1986; Jennes *et al.*, 1985; Liposits *et al.*, 1984; Wray and Gainer, 1987). GnRH neurons have a large centrally placed nucleus and a thin rim of cytoplasm which extends into the tapering cones of the principal dendrites. Nuclei contain one or two large nucleoli and the nuclear envelope is frequently indented. The cytoplasm contains many stacks of rough endoplasmic reticulum, one or more prominent Golgi stacks and neurosecretory granules (Kozlowski *et al.*, 1980; Witkin and Demasio, 1990). In the preoptic area, the GnRH like cells are fusiform in shape, with one or two major dendrite and a single axon (Lehman *et al.*, 1990). The surface of the perikarya and dendrites may be smooth or spiny (Silverman *et al.*, 1990). Using transmission electron microscope, Hokfelt *et al.* (1974, 1978) have found that the infundibular stem contains terminals that stain for GnRH. Hypophysiotropic parvicellular systems terminate in the palisade layer of the external zone of the median eminence. Simultaneous localization of GnRH and dopamine (DA) demonstrates the GnRH fibers preferentially in the lateral palisade zone of the rat median eminence in close relationship with DA containing fibers (McNeill and Sladek, 1978; Ibata *et al.*, 1979). McNeill and Sladek (1978) found a population of aminergic terminals closely associated with the portal vessels and a second population of aminergic terminals closely related to GnRH containing terminals. They proposed that dopaminergic terminals inhibit GnRH released from petidergic terminals through axo-axonic synapses (McNeil and Sladek, 1978). DA containing terminals have been found closely apposed to GnRH containing terminals in rat median eminence (Ajika, 1979). GnRH has been found in the arcuate nucleus of the rat by microdissection and RIA. Immunoreactive GnRH has been identified in arcuate neurons in the mouse and rat (Naik, 1975; Kozlowski *et al.*,

1975; Kawano and Daikoku, 1981). The medial preoptic area/zone (MPOA) which is sexually dimorphic and the medial zone of POA were found to have GnRH (Selamnoff *et al.*, 1980). Stimulation of the POA of proestrus female rats caused an increase in plasma LH concentration and ovulation (Cramer and Barraclough, 1971). The integrity of the posterior region of the POA is necessary for the demonstration of male sexual behaviour in the rat. It is to be expected that this region in rodent is related to ovulation and hence to the feedback relationships of estrogen on GnRH containing neurons and to sexual behaviour (Stumpf, 1970; Pfaff and Keiner, 1973; Pfaff *et al.*, 1976).

Synaptic features

GnRH is found in membrane bound granules within axon profiles and neurosecretory terminals (Kawano and Daikoku, 1981; Hatten *et al.*, 1984). Granules range from 40-130 nm. GnRH-GnRH synapses have been reported. It is posible that these synaptic connections and other complex interactions between GnRH neurons and those between GnRH neurons and their non-immunoreactive neighbours form a part of the substrate for the coordination of the GnRH signal. Among those factors involved in synaptic input are the various neurotransmitter containing neurons and peptidergic neurons which synapse with GnRH neurons and dendrites. Recently there have been studies indicating that there might be substances in the GnRH producing neurons itself. Delta sleep-inducing peptide and galanin are two neuropeptides that have been identified in the GnRH neurons of rat (Vallet *et al.*, 1990, 1991).

Regulation of GnRH release

Catecholamines

Noradrenergic innervation of the hypothalamus is extrinsic. Norepinephrine (NE) containing terminals have a wide but uneven distribution throughout the hypothalamus. Adrenergic cell bodies have been identified in the rostral ventrolateral medulla and rostro-medial part of the nucleus of the solitary tract in the rostro-medial medulla (Hokfelt *et al.*, 1974; Howe *et al.*, 1980). High concentrations of epinephrine are found in the dorsomedial, paraventricular, periventricular, arcuate and supraoptic nuclei (Gugten *et al.*, 1976). Using inhibitors of NE synthesis and alpha receptors, it was proposed that NE is involved in the pulsatile release of GnRH. The largest group of neurons containing DA are the tuberoinfundibular neurons (Fuxe and Hokfelt, 1966). Epinephrine and DA do not seem to have an important role to play in the pulsatile release of LH. In the absence of steroid feedback, it appears that noradrenergic neurons are involved in pulsatile LH release. The release of NE in the medial preoptic nucleus is pulsatile although there is a lack of correlation between the pulses of NE and LH (Jarry *et al.*, 1986).

In rats that have been ovariectomized and implanted with silastic capsules with estrogen, there is a surge of NE in the MPOA, arcuate and suprachiasmatic nuclei and also in the median eminence. In animals treated with estrogen and progestrone, there is an advance in the LH surge (Wise *et al.*, 1981). These observtions are in close agreement with findings in cycling rats. The action of estrogen on NE turnover could be mediated directly on the nordrenergic neurons since they contain estrogen receptors (Hertiage *et al.*, 1977). The steroid effect on NE could be through opiate or GABA containing interneurons that contain steroid receptors. No changes in the DA turnover rate were seen in the arcuate and medial preoptic nuclei and median eminence in ovariectomized rats treated with estrogen + progestrone. There is an increase in epinephrine containing neurons in the hypothalamus, simultaneous with the ovulatory LH surge. It has been concluded that NE and E containing neurons are involved in the stimulation of the ovulatory surge of LH in rodents although there is an increase in DA containing neurons (Krieg and Sawyer, 1976).

Serotonin

Serotonergic neurons may play facilitatory or inhibitory roles in LH secretion depending on the steroid enviornment. Direct electrochemical stimulation of 5-HT neuron in the dorsal raphe nucleus inhibited LH secretion in ovariectomized rats and blocked spontaneous ovulation in normal rats (Carrer and Taleisnik, 1972; Arendash and Gallo, 1978). The opposite actions of 5-HT on LH secretion may also be explained by the involvement of serotonergic pathways affecting GnRH secretion directly as well as indirectly. Both electrochemical stimulation of the medial raphe nucleus blocks the preovulatory release of LH suggesting that the medial raphe nucleus plays a facilitatory role (Morello and Taleisnik, 1985). It is likely that the actions of serotonin on LH secretion and ovulation are expressed via changes in the outflow of GnRH from neurosecretory neurons rather than direct action at the pituitary gland (Coen and Mackinnon, 1979).

Gamma amino butyric acid (GABA), histamine and glutamate

GABA and is agonist, muscimol inhibit pulsatile release of LH suggesting an inhibitory role for GABA in the regulation of GnRH/LH (Lambertz *et al.*, 1984; Fuchs *et al.*, 1984). Correlated with these inhibitory effects of GABA on both pulsatile and steroid induced surges of LH, changes in the activity of noradrenergic neurons have been reported. Contrasting to this is the finding that GABA has an excitatory role in regulation of GnRH when infused in the vicinity of the median eminence (Vijayan and McCann, 1978). It has been proposed that GABA neurons in the POA mediate the negative feedback action of estrogen through estrogen receptors on the neurons in this region. There is a decrease in GABA concentration prior to LH surge. It could also be that GABA acts directly on GnRH inhibition because some GABA containing terminals synapse with neurons containing GnRH (Vijayan and McCann, 1978; Jarry *et al.*, 1992). The role of histamine in the physiological regulation of LH secretion is still unclear. The existing data seem to suggest that it might play a role in GnRH regulation (Van Hirk *et al.*, 1989). Glutamate, an excitatory amino acid has been shown to stimulate LH secretion in rats. The relevance of glutamate for gonadotropin control may be related to its capacity to amplify the firing of GnRH neurons and consequently participate in the synchronization of episodic release of the peptide (Brann and Mahesh, 1991).

Other neuroactive substances

Substance P, a tachykinin, stimulates the release of GnRH (Vijayan and McCann, 1979). Atrial natriuretic peptide (ANP) in high concentration has been reported to stimulate the release of LH from pituitary cells in vitro and prolong the effect of GnRH (Harvath *et al.*, 1986). Galanin increases LH secretion from pituitary cells in culture and do potentiate their response to GnRH. Galanin system in rat is sexually dimorphic with a higher concentration in females (Lopez *et al.*, 1991). Neuropeptide Y also has a positive action on GnRH (Bauer *et al.*, 1991) but in castrated animals (both sexes) it inhibits GnRH action. Neuropeptide Y appears to stimulate LH release by a direct, steroid dependent action on pituitary gonadotropes (Kerkerian *et al.*, 1985). Prolactin has an inhibitory role on GnRH. Oxytocin and Vasopressin advances preovulatory LH surge (Azad *et al.*, 1990; Johnston *et al.*, 1990; Gibbs, 1986). Angiotensin II is capable of stimulating LH release upon intraventricular administration in rats (Steele *et al.*, 1983). Endothelins stimulate GnRH release (Krsmanovic *et al.*, 1991). Vasoactive intestinal peptide has been observed to have a direct stimulatory action on GnRH as observed from synaptosomal preparations (Vijayan *et al.*, 1981). Gastrin inhibitory peptide (GIP) has a positive action on LH (Ottlecz *et al.*, 1985).

Opioid peptides

Morphine and opioid peptides inhibit ovulatory surge of LH in rats (Bruni *et al.*, 1977). Opioid antagonists (naloxone) elevate plasma levels of LH (Cicero *et al.*, 1979). The functional relevance of endogenous opiates for LH and prolactin regulation is highly dependent on the endocrine condition of experimental animals (Piva *et al.*, 1986; Bhanot and Wilkinson, 1992). Gonadotropin responses to opiates are also dependent upon maturation and sexual differentiation of the hypothalamus and the onset of puberty (Valenca and Negro-Vilar, 1986). The hypothalamus seems to be the major site of opiate-LH interactions (Cicero *et al.*, 1977). In addition to a direct presynaptic inhibitory action on GnRH neurons, opiates are also believed to affect the gonadotropin secretion by interfering with catecholaminergic neurotransmission in the hypothalamus (Leadem *et al.*, 1985). Interleukin 1B(IL-1B) was also shown to inhibit LH surge (Rivier and Vale, 1990).

Mechanism of action

Understanding the basic mechanisms by which GnRH alters gonadotropic function has been a valuable resource for development of clinical and veterinary strategies. GnRH receptors are found in the anterior pituitary and brain. Through various binding and cloning studies using natural and synthetic analogs (agonists and antagonists) it has been found that the GnRH receptor is a 327 aminoacid residue with a molecular weight of 37683 (Tsutsumi *et al.*, 1992). It contains three N-linked glycosylation sites. Hydrophobicity analyses of the aminoacid sequence suggest the existence of seven highly hydrophobic regions which exhibit all features typical for transmembrane spanning regions. This feature is commonly found in G-protein linked receptors. Unusual features of the protein include the lack of a C-terminus intracellular tail, as well as substitutions of highly conserved Asp or Glu by Asn at the second transmembrane segment and substitution of Tyr by Ser next to the transmembrane spanning region (Kakar *et al.*, 1992).

The binding of GnRH to its receptor causes extracellular Ca^{2+} mobilization (Marian and Conn, 1983). Although this mobilization is similar to that in nervous and muscle tissues it is not identical (Conn *et al.*, 1983). The GnRH receptors are regulated according to the physiological conditions. It is seen to increase before preovulatory surge of LH and decrease following the surge. The number also decreases during pregnancy and lactation (Clayton *et al.*, 1980; Marian *et al.*, 1981). The number of GnRH receptors are not only regulated by hormones like GnRH, LH and FSH but also by steroids and protein products from the gonads (Young *et al.*, 1984; Wise *et al.*, 1984; Greg and Nett, 1989; Wang *et al.*, 1989). Studies have indicated that GnRH agonists and antagonists have different intracellular routes. Labelling experiments have indicated that GnRH is internalized and is routed to either Golgi cisternae or lysosomes. Using a GnRH agonist, it was found that patching, capping and internalization is not essential for GnRH action (Jennes *et al.*, 1984, 1986). It has been observed that activators of protein kinase-C (PKC) can stimulate synthesis of GnRH receptors. Inhibin has been proved to affect the basal levels of GnRH receptors not by decreasing the synthesis rates of receptors but by preventing upregulation of GnRH receptor synthesis by homologous hormone (Wang *et al.*, 1988). Activin stimulates GnRH receptor synthesis in cell culture through a mechanism distinct from homologous stimulation by GnRH (Bardin and Conn, 1992). The action of PKC on the GnRH receptor, either direct or indirect requires the receptor to be in an activated state but does not require receptor internalization (Huckle *et al.*, 1988). GnRH requires ionic calcium for many of its actions. Omission or chelation of Ca^{2+} in the extracellular medium inhibits depolarization (Marian and Conn, 1979).

Behavioural studies showed that GnRH can facilitate mounting and lordosis behaviour after peripheral, intracerebroventricular, or local intracerebral injection and that these effects are independent of an intact pituitary (Moss and McCann, 1973; Pfaff, 1975; Pfaff *et al.*, 1994). These studies pointed to the brain as an important site of the GnRH action. Subsequent electrophysiological studies showed that GnRH causes a slow depolarization of sympathetic ganglion cells in rat hippocampal neurons. These studies together with the finding that GnRH immunoreactive material is contained in presynaptic terminals strengthened the view that GnRH can function as a neurotransmitter in addition to its action on the anterior pituitary (Badr and Pelletier, 1987; Jennes *et al.*, 1988; Reubi *et al.*, 1987). Using ligand binding studies it has been shown that there are specific binding sites in a variety of brain regions. GnRH appears to be present in the cerebrospinal fluid, it is possible that this peptide reaches the hippocampus not through synaptic terminals but through the ventricular system. The GnRH receptors seem to be similar to those found in the pituitary. It has been proved that most GnRH analogues which bind to pituitary GnRH receptors also bind to the hippocampal GnRH receptor with a similar affinity (Jennes and Conn, 1992).

GnRH has a crucial role to play in the male and female reproduction. Its action as stated before leads to the release of LH and FSH which play a major role in male and female reproduction. In male, LH stimulates androgen secretion which is the central point in the development and functioning of the male reproductive system. In females, LH stimulates ovulation and corpus luteum formation. FSH stimulates spermatogenesis in the male and growth and maturation of follicles in female (Hsuch and Jones, 1981). Controlling both FSH and LH is by GnRH. In females there is a basal and ovulatory surge of LH meaning that the LH secretion is biphasic. The release of GnRH is pulsatile. Pulsatile release of gonadotropins have been studied in gonadectomized animals. The concentration of circulating gonadotropins are known to depend on the frequency of GnRH pulses discharged into the pituitary portal vessels; these in turn depend upon the endocrine status. The extensive studies on rhesus monkeys by Knobil (1980) show that while continuous infusion of GnRH fails to release LH, pulsatile infusion is effective, optimal results obtained with a pulse frequency of one per hour. In rats continuous infusion of GnRH stimulates LH release (Blake, 1976) but pulsatile infusion is more effective (Weick, 1981). Castration of either sex increases the magnitude of episodic GnRH release, resulting in a rise of circulating gonadotropins (Wildt *et al.*, 1981). Replacement with gonadal steroids reduces the pulse frequency of LH at first, thus depressing plasma LH and FSH but prolonged exposure to estrogen in ovariectomized female reverses the inhibition (Wise *et al.*, 1979, 1981). There are correlated changes in electrical activity in the medial basal hypothalamus. Dufy *et al.* (1979) recorded pronounced pulsatile multiunit event in the arcuate nucleus of ovariectomized animals immediately preceeding each LH pulse. Since GnRH promotes secretion of both LH and FSH and no specific FSH releasing hormone has been identified, the means of differential egulation of these two secretions is a mystery. A suggestion of an answer comes from the observation that LH/FSH ratios are influenced by changes in GnRH pulse frequency. LH secretion being favoured by higher frequencies and FSH by lower frequencies (Dufy *et al.*, 1979; Wildt *et al.*, 1981; Wise *et al.*, 1981).

Gonadal steroids also influence the phasic release under its influence. There is a rapid increase of GnRH content in the medial basal hypothalamus just before critical afternoon period of proestrus. There are multiple pulses of GnRH in the portal blood which follow the increased frequencies of pulsatile release of LH at intervals of 16-25 sec (Gallo, 1981). The magnitude of these pulses is governed initially by the responsiveness of the anterior pituitary after being primed by estrogen. As plasma LH concentration rises, the responsiveness is enhanced partly by the self-priming action of GnRH and probably also by increased exposure of gonadotropin to progestrone (Wise *et al.*, 1981). Knobil *et al.* (1980) have shown that in female monkeys bearing long term lesions of the arcuate nucleus, the month long infusion of the GnRH pulses once per hour sustained complete menstrual cycles. In long term ovariectomized females having similar hypothalamic lesions and similarly treated with GnRH, administration of estrogen invoked LH surges. No relation to the time of the day was observed. Nevertheless, the amount of GnRH in the portal blood does vary in rhesus monkey, being relatively high during the estrogen induced LH surge.

Ovulation occur a number of hours after the release of a LH surge from the anterior pituitary. The time between the LH surge and ovulation varies among species (Menaker, 1974). In rats, hamsters and mice studied under standard laboratory lighting conditions (LD 14 : 10), it has been established that ovulation occurs at a fixed time relative to light-drk cycle (Schwartz, 1969). Numerous studies have now clearly demonstrated that the proestrus LH surge is tightly controlled by a neural circadian clock (McCormack and Sridaran, 1978). Studies by Levine and Ramirez (1982) show an increased release of GnRH from the medial basal hypothalamus at the time of the preovulatory gonadotropin surge. Although numerous neurochemicals including catecholamines, acetylcholine, serotonin, corticotropin releasing factor and endorphins have been implicated in the neural regulation of the LH surge, the neurotransmitter that specifically conveys the circadian signal to GnRH neurons remains unknown. NE may be involved, since numerous pharmacologic sudies have demonstrated a stimulatory effect of NE on the LH surge and NE turnover rates measured in the hypothalamic nuclei involved in the regulation of GnRH secretion show a circadian rhythm on the day of proestrus. The increase in NE turnover precedes the LH surge (Kalra and Kalra, 1982).

GnRH and its analogues

The isolation and structure determination of GnRH have led to the synthesis of an extensive series of analogues, some of which are agonists and others have antagonistic properties. GnRH or its derivatives might be useful for the control of fertility by disrupting the menstrual cycle and preventing ovulatory surge. The replacement or deletion of one or more amino acids in GnRH might result in analogues possessing features requisite for a functional state i.e. the stimulation of LH and FSH release. Chronic treatment with agonists of GnRH elicits first an increase in gonadotropins followed by desenisitization of the pituitary to further agonist stimulation. The phenomena of desensitization and inhibition of sex steriod levels by GnRH agonists are being ued for treatment of true idiopathic precocious puberty. The initial high circulating levels of gonadotropins cause a reduction in the number of gonadal LH receptors and steroidogenesis decreases. The antagonists function by preventing the release of gonadotropin and not by defecting the pituitary of hormonal stores. Approximately 1000 compounds have been synthesized and evaluated for agonist and antagonist activity for potential use as investigative tools or contraceptive agents (Koch *et al.*, 1974). The ability of GnRH to stimulate reproducive functions (at pulsatile doses) or suppress them (at high doses) has been clinically applied for different purposes, including the induction of ovulation and spermatogenesis, contraception, and for the treatment of precocious puberty, endomeriosis, and steroid dependent tumors (i.e. prostate and breast cancer) as well as a range of veterinary purposes.

Acknowledgements

Financial support from the Department of Science and Technology, Government of India and the University Grants Commission is gratefully acknowledged.

References

Ajika, K. (1979) J. Anat **128:** 331–338.

Arendash, G.W. and Gallo, R.V. (1978) Endocrinology **102:** 1199–1206.

Azad, N., Duttner, L., Paloyan, E.B., Reda, D., Kristain, L., Emanuele, N.V. and Lawrence, A.M. (1990) Endocrinology **127:** 1928–1933.

Badr, M. and Pelletier, G. (1987) Synapse **1:** 567–571.

Bauer-Dantoin, A., McDonald, J.K. and Levine, J.E. (1991) Endocrinology **129:** 402–408.

Barry, J. (1979) Annu Rev Cytol **60:** 179–219.

Bardin, J.D. and Conn, P.M. (1992) Endocrinology **130:** 2101–2105.

Bhanot, R. and Wilkinson, M. (1992) Endocrinology **131:** 2891–2897.

Blake, C.A. (1976) Endocrinology **98:** 451–460.

Brann, D.W. and Mahesh, V.B. (1991) Endocrinology **128:** 1541–1547.

Bruni, J.F., Van Vugt, D., Marshall and Meites, J. (1977) Life Sci. **21:** 461–466.

Burgus, R., Butcher, M. and Ling, N. (1971) Comp. Tren. Acad. Sci. **273:** 1611–1613.

Burgu, R., Butches, M., Amoss, M., Ling, N., Mohanan, M.W., Rivier, J., Fellows, R., Blackwell, R., Vale, W., and Guillemin, R. (1972) Proc. Natl. Acad. Sci. USA **69:** 278–285.

Carrer, H.F. and Taleinik, S. (1972) Brain Res. **38:** 299–313.
Chuah, M. and Au, C.J. (1991) Neurosci. Res. **29:** 179–180.
Cicero, T.J., Badger, T.M., Wilcox, C.E., Bell, R.D. and Meyer, E.R. (1977) J. Pharmacol. Exp. Ther. **203:** 548–554.
Cicero, T.J., Schainker, B.A. and Meyer, E.R. (1979) Endocrinology **104:** 1286–1291.
Clayton, R.N., Solano, A.R., Garcia-Vila, A., Dufau, M.L. and Catt, K.J. (1980) Endocrinology **107:** 699–706.
Coen, C.W. and MacKinnon, P.C.B. (1979) J. Endocrinol **82:** 105–113.
Conn, P.M., Rogers, D.C. and Seay, S.G. (1983) Endocrinology **113:** 1592–1595.
Cramer, O.M. and Barraclough, C.A. (1971) Endocrinology **88:** 1175.
Cusclueri, A. and Bannister, L.H. (1975) J. Anat. **119:** 277–286.
Dufy, B., Dufy-Barbe, L., Vicent, J.D. and Knobil, E. (1979) J. Physiol. (Paris) **75:** 105–108.
Fuchs, E., Manky, T., Stock, K., Vijayan, E. and Wuttke, W. (1984) Neuroendocrinology **38:** 484–489.
Fukuda, M., and Pfaff, D.W., Nature (1989) **338:** 161–163.
Fukuda, M., Bick, D. and Pfaff, D.W. (1989) Mol. Brain Res. **6:** 311–326.
Fuxe, K. and Hokfelt, T. (1966) Endocrinol. Acta Physiol. Scand. **66:** 245–250.
Gallo, R.V. (1981) Biol. Reprod. **24:** 100–104.
Gibbs, D.M. (1986) Psychoneuroendocrinology **11:** 131–140.
Greg, D.W. and Nett, T.M. (1989) Biol. Reprod. **40:** 288–293.
Gugten, J., Palkovits, M., Wijnen, H. and Versteg, D. (1976) Brain Res. **107:** 171–175.
Harvath, J., Ertl, T. and Schally, A.V. (1986) Proc. Natl. Acad. Sci. USA **83:** 3444–3446.
Haton, G.L., Perlmutter, L.S., Salm, A.K. and Tweedle, C.D. (1984) Peptides **5:** 121–138.
Heritage, A., Grant, L. and Stumpf, W. (1977) J. Comp. Neurol. **176:** 607–630.
Hoffman, G.E. and Finch, C.E. (1986) Neurobiol. Aging **7:** 45–48.
Hokfelt, T., Fuxe, K. and Goldstein, M. (1974) Brain Res, **66:** 239–251.
Hokefelt, T., Elde, R. and Fuxe, K. (1978) In: The Hypothalamus (Reichiln, S., Baldessarini, R.J., Martin, J.B., Eds.) pp. 69–78, Raven Press, New York.
Howe, P., Costa, M., Furness, J. and Chalmers, J. (1980) Neuroscience **5:** 2229–2238.
Hsueh, A.J.W. and Jones, P.B.C. (1981) Endocr. Rev. **2:** 437–461.
Huckle, W.R., McArdle, C.A. and Conn, P. (1988) J. Biol. Chem. **263:** 3296–3302.
Ibata, Y., Watanbe, K., Kinoshita, H. (1979) Neurosci. Lett. **11:** 181–186.
Jarry, H., Hirsch, B., Leonharaf, S. and Wuttke, W. (1992) Neuroendocrinology **56:** 133–140.
Jarry, H., Sprenger, M. and Wuttke, W. (1986) Neuroendocrinology **44:** 422–428.
Jennes, L., Stumpf, W.E. and Conn, P.M. (1984) Peptides **5:** 215–220.
Jennes, L., Stumpf, W.E. and Sheedy, M.E. (1985) J. Comp. Neurol. **232:** 534–547.
Jennes, L., Coy, D. and Conn, P.M. (1986) Peptides **7:** 459–463.
Jennes, L., Dalati, B. and Conn, P.M. (1988) Brain Res. **452:** 156–164.
Jennes, L., Conn, P.M. (1992) In: Handbook of chemical Neuroanatomy (Bjorklund, A., Hokefett, T., Kuhar, M.J., Eds.) pp. 223–246, Elsevier Amsterdam.
Johnston, C.A., Lopez, F., Samson, W.K. and Negro Vilar, A. (1990) Neurosci. Lett. **12:** 256–258.
Kakar, S.S., Musgrove, L.C., Devor, D.C., Sellers, J.C. and Neill, J.D. (1992) Biochem. Biophys. Res. Commun. **189:** 289–295.
Kalra, P.S., and Kalara, S.P. (1982) Endocrinol. Rev. **4:** 311–351.
Kawano, H. and Daikoku, S. (1981) Neuroendocrinology **32:** 179–186.
Kerkerian, L., Guy, J., Lefevre, C. and Pellefier, G. (1985) Peptides **6:** 1201–1204.
Kerry, L.C. and Silvenman, A.J. (1983) In: Brain Peptides (Krieger, D.T., Brownstein, M., Martin, J. Eds.) pp. 687–709, John Wiley & Sons, New York.

Knobil, E. (1980) The Neuroendocrine control of the menstrual cycle. Rec. Prog. Horm. Res. **36:** 56–88.

Knobil, E., Plant, T.M., Wildt, L., Belchetz, P.E., Marshall, G. (1980) Science **207:** 1371–1373.

Koch, Y., Wilchek, M. and Fridkin, M. (1974) Biochem. Biophys. Res. Commun. **61:** 95.

Kozlowski, G.P., Chu, L., Hostetter, G. and Kerdelhue, B. (1980) **1:** 37–46.

Kozlowski, G.P., Nett, T.M. and Zimmerman, E.A. (1975) In: Anatomical Neuroendocrinology (Stumpf, W.E., Grant, L.D., Eds.) pp. 185–192. Karger, Basel.

Koves, K. and Molnar, J. (1986) Neuroendocrinology **44:** 172–183.

Krieg, R. and Sawyer, C. (1976) Endocrinology **99:** 411–419.

Krisch, B. (1980) Cell Tissue Res. **212:** 443–445.

Krsmanovic, L.Z., Stojitkovic, S.S., Balla, T., Al–Damluji, S., Weiner, R.J. and Catt, K.J. (1991) Proc. Natl. Acad. Sci. USA **88:** 1124–1128.

Lamberts, R., Mansky, T., Stock, K., Vijayan, E. and Wuttke, W. (1984) Exp. Brain Res. **52:** 356–362.

Leadem, C.A., Crowley, W.R., Simpkins, J.W. and Kalra S.P. (1985) Neuroendocrinology **40:** 497–500.

Lehman, M.N., Silverman, A.J., Witkin, J.W. and Millar, R.P. (1990) Brain Res. Bull. **20:** 211–218.

Levine, J. and Remirez, V.D. (1982) Endocrinology **111:** 1439–1448.

Liposits, Z., Setalo, G. and Flerko, B. (1984) Neuroscience, 13.

Lopez, F.J., Merchenthaler, I., Ching, M., Wisniewski, M.G. and Negro-Vilar, A. (1991) Proc. Natl. Acad. Sci. USA **88:** 4508–1512.

Marian, J., and Conn, P.M. (1979) Mol. Pharmacol. **16:** 196–201.

Marian, J., Cooper, R. and Conn, P.M. (1981) Mol. Pharmacol. **19:** 339–405.

Marian, J. and Conn, P.M. (1983) Endocrinology **112:** 104–112.

Matsuo, H., Bab, Y., Nair, R.M.G., Arimura, A. and Schally, A.V. (1971) Biochem. Biophys. Res. Commun. **43:** 1334–1339.

Marin Padilla, M. and Amieva, M.B.R. (1989) J. Comp. Neurol. **288:** 339–352.

McCormack, C.E. and Sridaran, R. (1978) J. Endocrinol. **76:** 135–144.

Mc Neill, T.H. and Sladek, J. (1978) Science **200:** 72–73.

Menaker, M. (1974) Circannual clocks Annual Biological Rhythms, (Pengelley, E.T., Eds.) pp. 507–520, Academic Press, New York.

Mendoza, A.S., Breiphol, W. and Miragall, F. (1982) J. Embryol. Exp. Morphol. **69:** 47–59.

Merchenthaler, L., Gorcs, T., Setalo, G., Petrusz, P. and Flerko, B. (1984) Cell Tissue Res. **237:** 15–19.

Merchenthaler, I., Setalo, G., Csontos, C., Petruz, P., Flerko, B. and Negro-Villar, A. (1989) Endocrinology **125:** 2812–2821.

Morello, H. and Taleisnik, S. (1985) Brain Res. **360:** 311–317.

Moss, R.L. and McCann, S.M. (1973) Science **181:** 177–179.

Naik, D.V. (1975) Cell Tissue Res. **157:** 423–430.

Ottlecz, A.W., Samson, K. and McCann, S.M. (1985) Peptides **6:** 115–119.

Pfaff, D.W. (1975) Science **182:** 1148–1149.

Pfaff, D.W. and Keiner, M. (1973) J. Comp. Neurol. **151:** 121–129.

Pfaff, D.W., Gerlach, J.L., McEwen, B.S., Ferin, M., Carmel, P. and Zimmerman, E.A. (1976) J. Comp. Neurol. **170:** 279–286.

Pfaff, D.W., Chwatz-Gibilin, S., McCarthy, M.M., Kow, L.M. (1994) In: The Physiology of Reproduction, Vol. 2 (Knobil, E., and Neill, J.D., Eds.), pp. 107–220, Raven Press, New York.

Piva, F., Limonta, P., Maggr, R. and Martini, L. (1986) Neuroendocrinology **42:** 504–512.

Reubi, J.C., Palacios, J.M., and Maurer, R. (1987) Neuroscience **21:** 847–856.

Rivier, C. and Vale, W. (1990) Endocrinology **127:** 849–856.
Schwartz, N.B. (1969) Rec. Prog. Horm. Res. **25:** 1–42.
Selmanoff, M.K., Wise, P.M. and Barraclough, C.A. (1980) Brain Res. **192:** 421–428.
Silverman, A.J., Krey, L.C. and Zimmerman, E.A. (1979) Biol. Reprod. **20:** 98–110.
Silverman, A.J., Jamandas J. and Renand, L.P. (1987) J. Neurosci. **7:** 2312–2319.
Silverman, A.J., Witkin, J.W. and Millar, R.P. (1990) J. Histochem. Cytochem. **38:** 803–809.
Steele, M.K., Gallo, R.C. and Ganong, W.F. (1983) A.M.J. Physiol. **245:** R805–R810.
Steinberger, L.A. and Hoffman, G.E. (1978) Neuroendocrinology **25:** 111–128.
Stumpf, W.E. (1970) Am. J. Anat. **129:** 207–216.
Tsutsumi, M., Zhon, W. and Millar, R.P. (1992) Mol. Endocrinol. **6:** 1163–1169.
Valenca, M.M. and Nego-Vilar, A. (1989) Neuroendocrinol. Lett. **8:** 165–172.
Vallet, P.G., Charnay, Y. and Bouas, C. (1990) J. Chem. Neuroanat. **3:** 207–211.
Vallet, P.G., Charnay, Y., Bouras, C. and Kis, J.Z. (1991) Neuroendocrinology **53:** 103.
Van Hirk, E.A., Hatterman, S.D., Moss, G.E., Rose, J.D. and Murdoch, W.J. (1989) J. Am. Sci. **67:** 1006–1012.
Vijayan, E. and McCann, S.M. (1978) Brain Res. **155:** 35–43.
Vijayan, E. and McCann, S.M. (1979) Endocrinology **105:** 64–78.
Vijayan, E., Sason, W.K., Said, S.J. and McCann, S.M. (1981) Regul. Pept. **2:** 253–264.
Wang, Q.F., Farnworth, P.G., Findlay, J.K., Burger, H.G. (1988) Endocrinology **123:** 2161–2166.
Wang, Q.F., Farnworth, P.G., Findlay, J.K. and Burger, H.G. (1989) Endocrinology **124:** 363–368.
Weick, R.F. (1981) Can. J. Physiol. Pharmacol. **59:** 779–785.
Wildt, L., Hausler, A. and Marshall, G. (1981) Endocrinology **109:** 376–385.
Wise, P.M., Rance, N., Baer, G.D.F., Barraclough, A.C. (1979) Endocrinology **104:** 940–947.
Wise, P.M., Camp-Crissman, P. and Barraclough, C.A. (1981) Biol. Reprod. **24:** 820–830.
Wise, P., Rance, N. and Barraclough, C. (1981) Endocrinology **108:** 2186–2193.
Wise, M.E., Nieman, D., Stewart, J. and Nett, T.M. (1984) Biol. Reprod. **31:** 1007–1013.
Witkin, J.W. and Demasio, K.A. (1990) Neuroscience **34:** 777–783.
Wray, S. and Hoffman, G. (1986) Neuroendocrinology **43:** 93–97, 513–525.
Wray, S. and Gainer, H. (1987) Neuroendocrinology **45:** 413–419.
Wray, S., Grant, P. and Gainer, H. (1989) Proc. Natl. Acad. Sci. (USA) **86:** 8132–8136.
Young, L.S., Naik, S.I., Clayton, R.N. (1984) Endocrinology **114:** 2113–2122.

Comparative Endocrinology and Reproduction
K.P. Joy, A. Krishna and C. Haldar (Eds)

2. Neural Control of Gonadotropin-Releasing Hormone Secretion: Emphasis on neuropeptide Y, galanin, nitric oxide and opioids

A. Sahu
Dept. of Cell Biology and Physiology, University of Pittsburgh School of Medicine, Pittsburgh, PA 15261, USA

Introduction

Reproduction is the primary goal of each species for perpetuation and its survival. For that reason gonadal maturation in adult to produce sperm in male and ovum in female is an absolute necessary and it is under the control of gonadotropin secretion. Both the gonadotropins, luteinizing hormone (LH) and follicle stimulating hormone (FSH), are necessary for spermatogenesis and ovarian folliculogenesis. In male, basal secretion of both LH and FSH is pulsatile and essential to maintain spermatogenesis. In female, however, reproductive cycle is under the control of cyclic release of gonadotropins and steroids. Most dramatic event in cyclic females is the preovulatory LH surge that is responsible for ovulation to occur in eight to ten hours later. It has been evident that gonadotropin-releasing hormone (GnRH), also called as the luteinizing hormone-releasing hormone (LHRH), is the primary hypothalamic signal for the gonadotropin secretion.

GnRH was isolated and characterized in early 1970's (Matsuo *et al.*, 1971; Schally *et al.*, 1971; Amoss *et al.*, 1971; Burgus *et al.*, 1972). Radioimmunoassay (RIA) for GnRH and other hormones were also available during this period. In last twenty years, several neuropeptides were isolated in the hypothalamus. Central administrations of these neuropeptides have either stimulatory or inhibitory influence on GnRH and LH secretion. Immunohistochemical studies at both light and electron microscopic labels also reveal the existence of synaptic contacts between some of these neuropeptidergic neurons as well as with the GnRH neurons. Steroid receptors are present in some of these neuropeptidergic and catecholaminergic neurons in the hypothalamus. Thus, overwhelming evidence accumulated in last twenty years suggests the presence of a hypothalamic neural circuitry that controls the basal pulsatile and the preovulatory LH surge. As has been depicted in Figure 1, GnRH is the monarch of this circuitry, whereas, others are secondary, but necessary to maintain intricate balance in this neural circuitry. Overall, GnRH neuronal activity is under the influence of both excitatory and inhibitory signals. Recent evidence suggests that GnRH may also modulate its own secretion by an ultra-short loop feedback mechanism (vide infra). We will discuss separately the role of excitatory and inhibitory signals on GnRH secretion. First, a brief account of the GnRH neuronal system is presented with reference to its anatomy and physiology in the context of basal pulsatile and preovulatory LH surge.

Gonadotropin-Releasing Hormone

Anatomay and physiology

Gonadotropin-releasing hormone synthesis occurs from original 92-amino-acid precursor protein that, upon cleavage and further processing, produces mature GnRH decapeptide and a 56 amino-acid gonadotropin-releasing hormone associated peptide (Seeburg and Adelman, 1984, Adelman *et al.*, 1986, Seeburg *et al.*, 1987). The GnRH neuron is oval or fusiform in shape, with a maximum diameter of 10–20 μm. Dendrite processes emerge from one or both poles of the cell and axons emerge either from the cell body or dendrite (Silverman *et al.*, 1994). During the developmental process, GnRH neurons differentiate from the olfactory placode and migrate to live in the septal, preoptic and anterior hypothalamus (Schwanzel-Fukuda and Pfaff, 1989; Wray *et al.*, 1989a, 1989b; Ronnekleiv and Resko, 1990; Schwanzel-Fukuda *et al.*, 1992). In mice, differentiation of the GnRH neurons in the olfactory placode starts on embryonic day 10.5 (E 10.5). Biochemical and functional maturation, and migrations to their ultimate destination in the hypothalamic regions are completed by E 16.5 (Schwanzel-Fukuda and Pfaff, 1989; Wray *et al.*, 1989a, 1989b). Migration of these neurons into the hypothalamus is absolute necessary for proper functioning of these neurons to control gonadotropin secretion. This is evident in an inherited hypogonadal (Kallman) syndrome in human, where GnRH neurons can not immigrate into the forebrain (Schwanzel-Fukuda *et al.*, 1989), and as a result GnRH secretion does not occur. In adult mammal, distribution of the GnRH neurons in the brain regions varies between species (Silverman *et al.*, 1994). Overall, GnRH neurons are loosely distributed in the areas of the organum vasculosum of the lamina terminalis (OVLT), diagonal brand of Broca (DBB), dorsal septum, bed nucleus of the stria terminalis (BNST), periventricular area, medial and lateral preoptic areas, anterior hypothalamus and in retrochiasmatic area. In rat, total number of GnRH-producing neurons ranges from 1200 to 1600 (Ronnekleiv *et al.*, 1989; Malik *et al.*, 1991, Silverman *et al.*, 1994). Most of the GnRH neurons project into the median eminence (Witkin *et al.*, 1982; Merchanthaler *et al.*, 1984, 1989; Silverman *et al.*, 1987). The median eminence (ME) is the common pathway by which most of the neuropeptides, including GnRH, and neurotransmitters are being released into the portal vessel. Recent evidence suggests that regulation of GnRH secretion by excitatory and inhibitory signals occurs at two sites: one at the level of perikarya and other at the level of axon terminals in the ME. Moreover, regulation at the perikarya and axon terminals may be independent of each other.

GnRH pulse generator

It is well established that GnRH release into the hypophyseal portal vessel is characteristically episodic (Carmel *et al.*, 1976). Pulsatile mode of GnRH release is essential to drive the secretion of gonadotropins in the pituitary (Belchetz *et al.*, 1978; Kalra and Kalra, 1983; Lincoln *et al.*, 1985; Kalra, 1986; Levine *et al.* 1991; Kalra, 1993). *In vivo* and *in vitro* studies suggest a pulsatile secretion of GnRH in the medial basal hypothalamus (Dluzen and Ramirez, 1986; Levine *et al.*, 1985; Levine and Ramirez, 1982). Although, the pulsatile secretion suggests a coordinated activity, but the mechanism(s) of this coordination for a synchronized secretion *in vivo* in a small dispersed population of GnRH cells is far from clear. The presence of a 'pulse-generator' with in the hypothalamus has been suggested (Knobil, 1980; Knobil and Hotchkiss, 1988; Wilson *et al.*, 1984; Kalra, 1986, 1993; Levine *et al.*, 1991). It is not clear whether the GnRH pulse generator is separate from, or resides within the GnRH neuronal network. However, recent observations are in favor of the view that the pulse generator may reside within the GnRH neuronal network. For examples, (1) an immortalized GnRH cell line (GT1) expressing preproGnRH mRNA (Mellon *et al.*, 1990), secretes GnRH in a pulsatile manner (Escalera *et al.* 1992; Wetsel *et al.*, 1992; Krsmanovic *et al.*, 1992) and this pulsality is autoregulated by GnRH itself (Krsmanovic *et al.*, 1993), (2) secretion of GnRH from the embryonic GnRH neurons not yet positioned in their postnatal environment is pulsatile (Burich *et al.*, 1994) and these neurons also contain a highly differentiated voltage- and ligand-activated channels (Kusano *et al.*, 1995), (3) GnRH inhibits its own secretion from MBH *in vitro* (DePaolo *et al.*, 1987), (4) GnRH agonist administration reduces GnRH release into the hypophyseal portal vessel in ovx rats (Valenca *et al.*, 1987), (5) the injection of relatively large doses of GnRH into the third ventricle increases (Ben-Jonathan *et al.*, 1974) and lower dose reduces plasma LH (Castro *et al.*, 1985), and (6) there is presence of GnRH-GnRH synaptic contacts (Leranth *et al.*, 1985; Silverman *et al.*, 1994) and intercellular bridges or passageways between contiguous pairs of GnRH neurons (Witkin *et al.*, 1995). The syncytial contact between GnRH neurons (Witkin *et al.*, 1995), although in very small number (2-7% in rat), is of particular interest. It has been evident that adult female hypogonadal mice shows complete recovery of reproductive function when they are grafted with normal fetal preoptic area tissue containing less than a dozen immunocytochemically detected GnRH neurons (Gibson *et al.*, 1984). Thus, GnRH-GnRH contacts in few cells may be sufficient to maintain pulsatile GnRH secretion necessary for gonadotropin secretion and reproduction.

Alternatively, a growing body of evidence suggests that pulsatile GnRH secretion is controlled by a neural circuitry comprising amines and neuropeptides. Recent immunohistochemical studies at the light- and electron-microscopic levels have observed synaptic contacts between GnRH and several other excitatory and inhibitory neuronal systems (see Silverman *et al.*, 1994). Furthermore, some of these neuronal systems (stimulatory and inhibitory) also have synaptic contacts between themselves. These findings alongwith the evidence presented below may suggest that an intact neuronal network comprised of amines and neuropeptides is necessary for the pulsatile release of GnRH and LH secretion.

GnRH neuronal activity during basal and preovulatory LH surge

In male rats, GnRH and LH secretion are pulsatile (Gallo, 1980; Ellis and Desjardins, 1982; Levine and Duffy, 1988), but the pattern of GnRH secretion to maintain the basal mode of LH secretion is not well characterized (Levine *et al.*, 1991; Turek and Van Cauter, 1994). In female rats, the pattern of LH secretion varies considerably during the estrus cycle (Kalra and Kalra, 1974, 1983, 1989; Freeman, 1994). The most important and spectacular phenomenon, as has been mentioned earlier, occurs on the day of proestrus with an abrupt change (increase) in GnRH secretion prior to and during the preovulatory LH surge. GnRH secretion is pulsatile in nature and each pulse of GnRH, in most cases, is intimately associated with a pulse of LH *in vivo* (Levine and Ramirez, 1982; Ramirez and Dluzen, 1987; Levine *et al.*, 1991; Turek and Van Cauter, 1994). Sometimes LH secretion *in vivo* is used as a tool for assessing *in vivo* GnRH pulsatility. In fact by assessing LH secretion from frequent blood sampling (2-3 min. interval), it has been suggested that the rate of GnRH secretion may accelerate from one episode per hour in the morning to 2 to 3 episodes per hour in the afternoon of proestrus (Gallo, 1981; Allen *et al.*, 1988). This is proved by the evidence that infusion of GnRH at 20–30 min. interval will result in LH secretion that is seen in the proestrus afternoon (Allen *et al.*, 1988).

On the day of proestrus in rat, an increase in GnRH levels in the MBH, particularly in the ME, occurs 1-2 hours before the LH hypersecretion (Kalra *et al.*, 1973; Kalra and Kalra, 1979; Wise *et al.*, 1981) followed by an increased release of GnRH into the hypophyseal portal system (Sarkar *et al.*, 1976; Ching, 1982) to induce the preovulatory LH surge. It is clear from several studies, however, that changes at the GnRH genomic level do not occur before the LH surge. For example, c-fos and c-jun, early markers of increased genomic activity, expression in the GnRH neurons begin in the middle of the ascending phase of the LH surge or after the LH surge (Lee *et al.*, 1990a, 1990b, 1992a, 1992b). GnRH mRNA levels also remain lower during the most part of the ascending phase of the LH surge and become elevated in association with the peak LH release or thereafter (Zoller and Young, 1988; Zoller *et al.*, 1988; Park *et al.*, 1990). Recently, Petersen *et al.* (1995) reported that in estrogen (E2)-primed ovx rats, GnRH mRNA levels in rostral POA/OVLT area were increased at 1200 h as compared to morning (0800 h) levels and then gradually decreased to a low level at 2000 h, suggesting an increase of GnRH mRNA, in a particular group of cells, before the LH surge. Surprisingly, in contrast to another report (Kim *et al.*, 1989), these authors did not see any further increase in GnRH mRNA after progesterone (p) administration. Overall, most of the evidence suggest that GnRH release increases before the LH surge but, activation of GnRH neurons at the level of the perikarya may occur in response to the release at the terminal. Seemingly, the regulation of GnRH secretion occurs at two different sites—one at the level of the perikarya and other at the level of nerve terminals in the ME.

GnRH-steroid interactions

In spite of the absence of steroid receptors, GnRH neuronal activity is, undoubtedly, modulated by gonadal steroids. The general belief is that E2 and testosterone (T) inhibit LH secretion by decreasing GnRH release, but there is very little evidence (Kalra and Kalra, 1989; Kalra, 1993). On the other hand, several studies suggest that gonadectomy decreases and E2 or T reverses this effect on GnRH levels and release (Kalra and Kalra, 1989, 1991). However, effects of steriod on GnRH mRNA levels are still in debate. For example, E2 has been reported to decrease (Zoller *et al.*, 1988; Toranzo *et al.*, 1989), increase (Roberts *et al.*, 1989; Rothfeld *et al.*, 1989; Rosie *et al.*, 1990), or has no effect (Kelly *et al.*, 1989; Marks *et al.*, 1994) on GnRH mRNA in ovx rats. Similarly, some studies have reported that P increases (Kim *et al.*, 1989), decreases (Toranzo *et al.*, 1989), or has no effect (Park *et al.*, 1990) on cellular levels of GnRH mRNA in ovx E2-treated rats. Discrepancies in these studies may be due to different hormone regimen, time of the day and also the hypothalamic areas being studied.

Excitatory Signals Modulating GnRH Secretion

Catecholamines as well as a large number of neuropeptides and amino acids are considered to be excitatory to the GnRH neuronal system (Fig. 1). Some of these excitatory signals, however, act as inhibitory to the GnRH neurons in the absence of gonadal steroids. Since gonadectomy is not physiological, the inhibitory effects of these excitatory signals may not be apparent in normal condition. Role of catecholamines in basal pulsatile and in preovulatory hypersecretion of GnRH and LH, is mostly stimulatory and has been reviewed extensively (Kalra and Kalra, 1983; Barraclough *et al.*, 1984; Ramirez *et al.*, 1984; Kalra, 1986; McCann and Krulich, 1989, McCann, 1991; Kordon *et al.*, 1994). Neurotensin, delta-sleep inducing peptide, angiotensin II and oxytocin are among the other neuropeptides that stimulate GnRH and LH secretion (Sahu and Kalra, 1987, see Kalra *et al.*, 1992; Kalra, 1993; Freeman, 1994). The role of neuropeptide Y (NPY), galanin

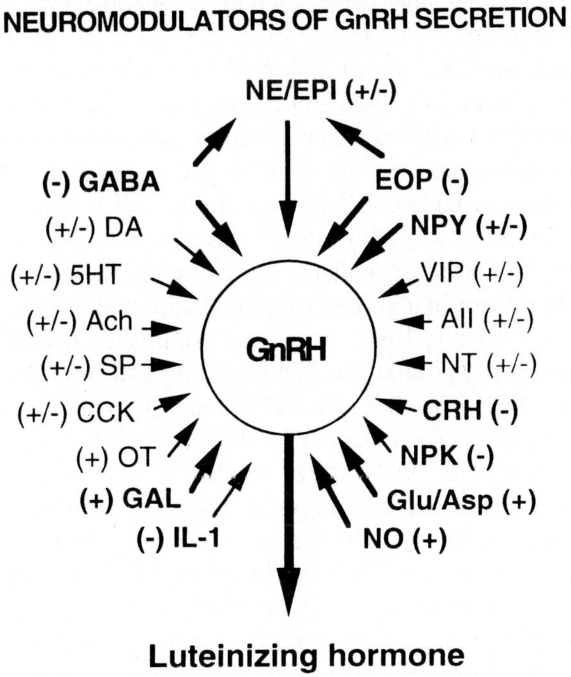

Fig. 1 Schematic presentation of the neuromodulators of GnRH secretion. GABA = gamma aminobytyric acid, DA = dopoamine, 5HT = serotonin, Ach = acetylecholine, SP = substance P, CCK = cholecystokinine, OT = oxytocin, GAL = galanin, IL-1 = interleukin-1, NE/EPI = norepinephrine/epinephrine, EOP = endogenous opioid peptides, NPY = neuropeptide Y, VIP = vesointestinal peptide, AII=angiotensin II, NT = Neurotensin, CRH = corticotropin-releasing hormone, NPK = neuropeptide K, Glu/Asp = glutamate/aspartate, NO = nitric oxide, (+) = stimulation, (–) = inhibition, (+/–) = stimulation/inhibition.

(GAL) and nitric oxide (NO) will be discussed in detail. The excitatory amino acids (EAA) are also a major component of the neural circuitry for GnRH regulation and their role will be briefly discussed alongwith NO.

Neuropeptide Y

Anatomy and physiology: Neuropeptide Y (NPY), a 36 amino acid peptide, is a member of the pancreatic polypeptide family. After its discovery in 1982 by Tatemoto, this peptide has been found to be involved in various physiological functions, including reproduction and feeding behavior (Kalra *et al.*, 1988, 1990, 1991; Sahu and Kalra, 1993). NPY is highly abundant in the hypothalamus and other brain areas involved in gonadotropin secretion. Hypothalamic nuclei receive NPY projection mainly from two groups of cell bodies—one resides in the brain stem where it is also colocalized with catecholamines and the other is localized in the hypothalamic arcuate nucleus (Everett and Hokfelt, 1989; Bai *et al.*, 1985; Sahu *et al.*, 1988b, 1990a). Experimental evidence suggests that NPY cell bodies in the ARC may be involved in regulation of GnRH and LH secretion. First demonstration that NPY is involved in GnRH and LH secretion came from the study showing that third cerebroventricular injection of NPY stimulated LH release in estradiol benzoate (EB) and P-primed ovx rats, but it inhibited LH release in ovx rats (Kalra and Crowley, 1984) indicating that the activity of NPY on LH release is steroid dependent. Subsequently, it was found that NPY stimulated LH release in a variety of animals, including rabbit and monkey (see Kalra and Crowley, 1992). It is now established that NPY stimulates LH release by acting both at the hypothalamic and pituitary levels. In the hypothalamus, NPY stimulates GnRH release (Crowley and Kalra, 1987; Khorram *et al.,* 1987; Sabatino *et al.,* 1989, 1990; Pau *et al.,* 1991, 1995; Woller and Terasawa, 1991, 1992; Woller *et al.*, 1992) by increasing intracellular Ca^{++} mobilization via postsynaptic Y1 NPY receptor mediated action (Crowley and Kalra, 1988; Kalra and Crowley, 1992). At the pituitary level, NPY potentiates the action of GnRH on LH release (Crowley *et al.*, 1987, 1990; Baur-Dantoin *et al.*, 1991, 1992, 1993a) by increasing GnRH binding sites (Parker *et al.,* 1991) and Ca^{++} entry through voltage-sensitive calcium channels (Crowley *et al.*, 1990). Progesterone up-regulates pituitary sensitivity to facilitatory action of NPY on the day of proestrus (Baur-Dantoin *et al.*, 1993b). While these findings clearly suggest an excitatory action on GnRH secretion by NPY, it also raises the question about how NPY stimulates GnRH neuronal system. Morphological communication in the form of synaptic contacts between NPY-containing axon terminals and GnRH-positive cell bodies in the septo-preoptic area have been reported (Guy *et al.,* 1988; Tsuruo *et al.,* 1990) and NPY injection into the POA increases GnRH mRNA (Li *et al.*, 1994) suggesting the possibility that NPY acts at the level of GnRH perikarya. At the level of ME, however, NPY seems to be acting through volume transmission modality (Fuxe and Agnati, 1991), because there is no direct contact between these two neuronal systems. On the other hand, recent evidence suggest that NPY may act on the GnRH neurons via some other neural systems. For example, anatomical evidence suggest the presence of direct appositions between NPY- and tyrosine-β hydroxylase-immunoreactive neurons in the ARC (Guy and Pelletier, 1988) and that NPY neurons innervate the galanin (Horvath *et al.,* 1993, 1996) and β-endorphin (Horvath *et al.,* 1992)-positive neurons in the hypothalamus. Furthermore, NPY and norepinephrine (NE) show similar effects on GnRH and LH secretion (see Kalra and Crowley, 1992), and they act synergistically to induce LH release in steroid-primed ovx rats (Allen *et al.*, 1987). Recently, we have also shown that prior injection with galantide, a galanin receptor antagonist, attenuated NPY-induced LH release in steroid-primed ovx rats (Sahu *et al.*, 1995; Horvath *et al.*, 1996). Since catecholamines, galanin and β-endorphin play an important role in GnRH secretion (See Kalra, 1986, 1993), these findings strongly argue in favor of the idea that NPY's action on GnRH secretion may, partially, be mediated via other neuronal systems such as galanin.

NPY and preovulatory GnRH/LH surge

From the physiological standpoint, following evidence suggest NPY as an obligatory signal for the genesis of the preovulatory GnRH and LH surges. First, like GnRH, NPY levels in the ME increase before the LH surge at proestrus (Fig. 2, Sahu *et al.*, 1989b) or that induced by ovarian steroids (Crowley *et al.*, 1985; Sahu *et al.*, 1990a) associated with an increased NPY and GnRH release in the MBH *in vivo* (Watanobe and Takebe, 1992). Second, preproNPY mRNA levels in the MBH increase long before the LH surge on the day of proestrus (Sahu *et al.*, 1995) or that induced by ovarian steroids in ovx rats (Fig. 2, Sahu *et al.*, 1994b). Third, hypophyseal portal concentration of NPY is significantly higher in the afternoon as compared to that in the morning of proestrus in prepubertal and adult rats, and this profile is similar to that of GnRH release (Sutton *et al.*, 1988a, 1988b). Fourth, immunoneutralization with NPY-antibody inhibits the steroid-induced LH surge in ovx rats (Sutton *et al.*, 1988a; Weherenberg *et al.*, 1989) as well as that occurs in the afternoon of first proestrus associated with a decreased GnRH release into the portal vessel (Minami *et al.*, 1990). Fifth, NPY antisense oligonucleotide injections into the lateral ventricle inhibits the steroid-induced LH surge in ovx rats (Kalra *et al.*, 1995) as well as blocks naloxone-, an opioid receptor antagonist, induced LH release in E2-primed ovx rats (Xu *et al.*, 1994, 1996). Thus, not only NPY synthesis and release are increased but results from the studies with oligonucleotide and antibody also suggest that increased NPY neurosecretion is obligatory in triggering GnRH/LH hypersecretion on the afternoon of proestrus. The neural trigger responsible for the antecedent changes in NPY neuronal activity leading to the GnRH/LH surge is not yet fully understood. However, our observations (1) that even after 48 hours of E2 exposure to ovx rats or increased E2 exposure during the period from the afternoon of diestrus II to the morning of proestrus did not increase hypothalamic NPY mRNA levels at 1000 h in the morning, and (2) that by 1200 h, mRNA levels were significantly increased in E2-primed ovx rats and in proestrus rats but not in the ovx or in the diestrus II rats (Sahu *et al.*, 1994b, 1995) are very intriguing. These results suggest that the neural clock (NC, Everett 1977), a timing device which times the preovulatory LH surge and located somewhere in the hypothalamus, possibly in the region comprised of the preoptico-tuberal, SCN and median preoptic nuclei (MPN) (see Kalra, 1993), is responsible for this antecedent increase in NPY mRNA and that estrogen facilitates the action of the NC. Furthermore, it was also apparent that P amplified the action of E2 resulting in further increase in NPY mRNA

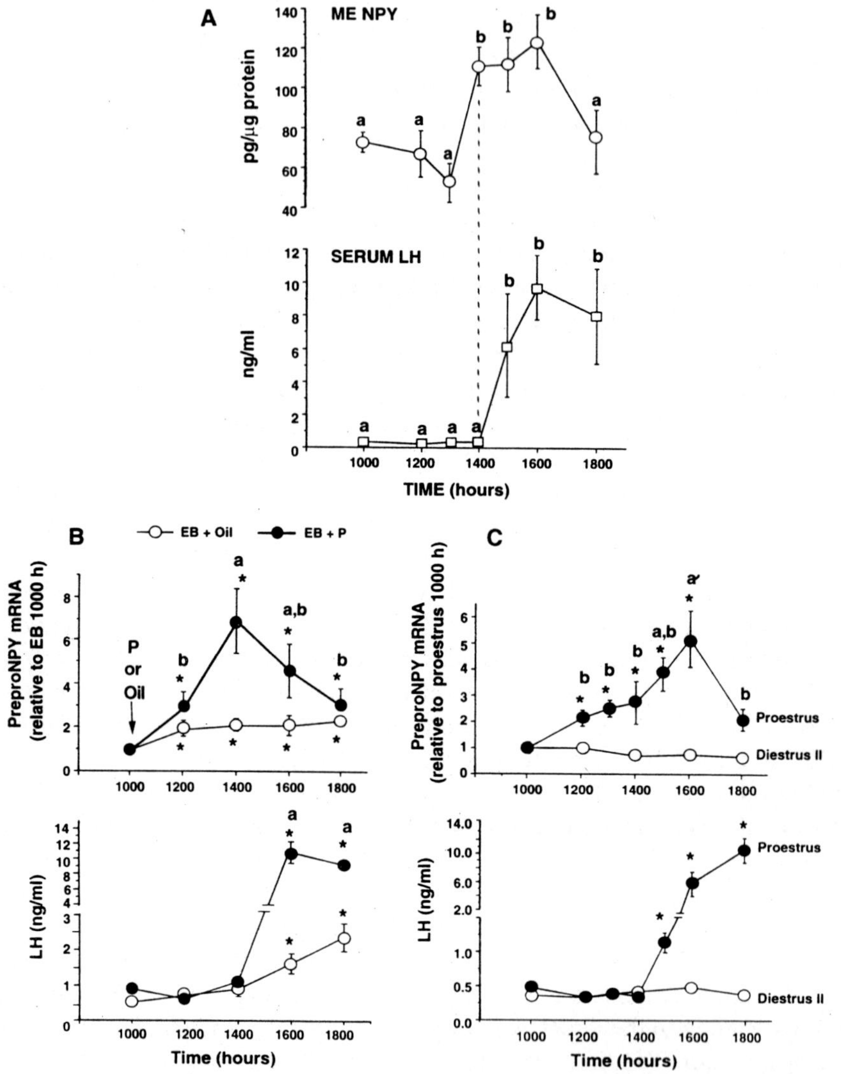

Fig. 2 Neuropeptide Y neuronal activity on the day of LH surge. A: Dynamic changes in NPY levels in the median eminence (ME) and serum LH levels on the day of proestrus. NPY levels in the ME increase abruptly before the rise in LH secretion and remain elevated during the course of the LH surge. B: preproNPY mRNA levels in the medial basal hypothalamus (MBH) on the day of the LH surge produced by ovarian steroids in ovx rats. Rats were injected with estradiol benzoate (EB) at 1000 h followed 48 h later by either oil or progesterone (P) injection. PreproNPY mRNA levels increase long before the spontaneous LH surge in EB-primed rats. P amplifies NPY mRNA levels in the EB-primed rats with a peak at 1400 h; this increase precedes the LH surge at 1600 h. C: PreproNPY mRNA levels in the MBH on the day of proestrus and diestrus II. Note that NPY gene expression is significantly increased long before the LH surge on proestrus and antecedent changes in NPY mRNA levels do not occur on the day of diestrus II. The dissimilar superscripts are significantly different from each other. *$p < 0.05$ vs 1000 h values (from Sahu *et al.*, 1989, 1994, 1995).

before the preovulatory LH surge (Sahu *et al.*, 1994b, 1995). Seemingly, increased P levels during the afternoon of proestrus not only enhances NPY synthesis and subsequent increase in NPY levels in the ME as well as NPY release in the portal vessel, but also facilitates the synergistic action between GnRH and NPY at the level of the pituitary (Bauer-Dantoin *et al.*, 1993) to induce a massive LH surge. Although, the mechanism(s) by which the NC sets the events in motion for the genesis of the GnRH and LH surges on the day of proestrus is not clear, one possibility may be the restraining of inhibitory tone in the hypothalamus. In support of this view there is experimental evidence which suggests that endogenous opioid peptide (EOP) system exerts a tonic inhibitory tone in the hypothalamus (see Kalra *et al.*, 1992; Kalra, 1993) resulting in low levels of LH release. In the morning between 1000 h and 1200 h at proestrus or in E-primed ovx rats, the NC restrains the inhibitory opioid tone and as a result excitatory neuronal systems, such as NPY neurons, become activated and induce hypersecretion of GnRH and LH in the afternoon. Our observations that the blockade of EOP tone by an opioid receptor antagonist, naloxone, increases NPY levels in the ME, ARC and MPOA in association with increased release of NPY and GnRH (Sahu *et al.*, 1990) and that naloxone induces hypothalamic preproNPY mRNA levels in E2-primed ovx rats (Xu *et al.*, 1996), strongly suggest a NC-opioid-NPY transmission line in genesis of the preovulatory GnRH and LH surges.

NPY and GnRH pulse generator network: Along with its physiological role during the GnRH/LH surge, the evidence that NPY release itself is pulsatile in nature in variety of animals including rat (Sahu *et al.*, 1992) and monkey (Woller *et al.*, 1992); and episodes of NPY either precede or coincide with GnRH episodes (Woller *et al.*, 1992); and that passive immunization with NPY-antibody in ovx animals decreases both the pulse frequency and amplitude of GnRH in monkey (Woller *et al.*, 1992), and that of LH in rabbit (Kaynard and Spies, 1991) and rat (Xu *et al.*, 1993) strongly suggest NPY as an important component of the hypothalamic GnRH pulse generating machinery.

NPY-steroid interaction: It is important to note that stimulatory action on GnRH and LH release of NPY is not only dependent on steroids, but NPY secretion, itself, is also modulated by gonadal steroids. We have shown that a sub-set of NPY cell bodies in the ARC concentrates E2 (Sar *et al.*, 1990) and that gonadal steroids stimulate NPY synthesis and release by a direct genomic action (Sahu *et al.*, 1987a, 1989a, 1992, 1994b, 1995). However, changes in the NPY neuronal activity with respect to the events occurring on the day of LH surge are of considerable importance.

NPY and aging: From the evidence presented above it is clear that NPY is an obligatory excitatory signal for GnRH and LH secretion. Thus we hypothesized that decreased LH levels in aging males, and attenuated or delayed LH surge for no LH surge in female rats during reproductive senescence (Wise *et al.*, 1991), could be due to an alteration in NPY neurosecretion. We have two lines of evidence to support this hypothesis. First, in middle-aged (MA) male rats with low levels of LH and T, hypothalamic NPY neurosecretion decreases (Sahu *et al.*, 1988a) and aging alters the responsiveness of hypothalmic NPY to T (Sahu *et al.*, 1989a). Recently, it has been evident that in MA regularly cycling female rats, showing attenuated and delayed LH surge, NPY levels in the ME show different pattern of changes on the day of proestrus (Sahu and Kalra, 1995). Whereas, in agreement with our earlier report (Fig. 2 Sahu *et al.*, 1989b), in young rats NPY levels in the ME increases significantly in the afternoon of proestrus before the LH surge, but in MA rats, NPY levels in the ME decrease in the afternoon as compared to morning levels and these levels are also significantly lower than in young rats. Similarly, in young rats, hypothalamic NPY gene expression increases prior to and during the LH surge in the afternoon of proestrus (Sahu *et al.*, 1995; Sahu and Kalra, 1997). In marked contrast, in MA rats NPY gene expression does not change between morning and afternoon hours on the day of proestrus (Sahu and Kalra, 1997). Since NPY is an obligatory signal for the GnRH and LH surges, these results suggest that diminution in NPY neuronal activity may be responsible for reduced GnRH and LH release in MA rats resulting in irregular cyclicity followed by acyclicity during reproductive senescence.

Galanin

Anatomy and physiology: Galanin (GAL), a 29 amino acid peptide originally isolated from porcine intestine (Tatemoto *et al.*, 1983), is widely distributed throughout the rat central nervous system with highest concentrations found in the hypothalmus (Skofitsch and Jacobowitz, 1985; Melander *et al.*, 1986). Galanin cell bodies are localized in the ARC, paraventricular nucleus (PVN), supraoptic, periventricular and preoptic area. Most of the GAL projections in the ME are from the ARC and paraventricular nuclei (Palkovits *et al.*, 1987; Merchenthaler *et al.*, 1993). Interestingly, galanin is also colocalized with GnRH in a significant number of perikarya in the MPOA (Coen *et al.*, 1990; Merchenthaler *et al.*, 1990). GAL receptors are found in the hypothalamic areas innervated by GAL and GnRH neurons (Skofitsch *et al.*, 1986; Melander *et al.*, 1992). Thus, anatomical evidence clearly suggests the possibility of interactions between Gal and GnRH neurons with respect to LH secretion. In 1987, Sahu *et al.*, first demonstrated that administration of GAL into the third cerebro-ventricle of ovx rats primed with ovarian steroids rapidly stimulated LH release in a dose dependent manner and recent study has confirmed this result (Sahu *et al.*, 1994a). A large body of evidence now attests to the view that GAL may be an important excitatory signal in the neural circuitry controlling basal pulsatile and surge related hypersecretion of GnRH and LH. Thus, GAL, like NPY, may act at two sites to stimulate LH release, one in the hypothalamus to release GnRH (Lopez and Negro-Vilar, 1990; Merchenthaler *et al.*, 1990; Sahu *et al.*, 1994a) via PGE2 involving alpha-adrenergic neurotransmission (Lopez and Negro-Vilar, 1990), and other at the level of pituitary to potentiate the action of GnRH by increasing GnRH binding sites (Lopez *et al.*, 1991; Coen *et al.*, 1990; Parker *et al.*, 1991).

GAL and preovulatory GnRH/LH surge: The observations that, on the day of proestrus, number of cells displaying coexpression of GAL and GnRH immunoreactivity rises sharply and GAL levels in the MBH increase in the morning (Merchenthaler *et al.*, 1993), and that GAL gene expression in GnRH perikarya in the preoptic area increases late in the afternoon (Marks *et al.*, 1993), suggest an increased GAL neuronal activity in association with LH surge. Furthermore, this increased GAL activity is also necessary for GnRH and LH hypersecretion because of the evidence that passive immunization of endogenous GAL with intravenous infusion of GAL-antibody attenuates the LH surge on proestrus (Lopes *et al.*, 1993) and that central injections of galantide, a galanin receptor antagonist (Bartfai *et al.*, 1991), completely blocks the LH surge on proestrus and that induced by ovarian steroids in ovx rats (Sahu *et al.*, 1994a). In addition, galantide also inhibits the basal GnRH release from the ME-ARC (Sahu *et àl.*, 1994a). These results suggest that initiation of GnRH hypersecretion by neural signals is mediated by GAL receptors in the hypothalamus.

GAL and GnRH pulse generator network: It is also apparent that GAL is a component of the hypothalamic GnRH pulse generator network. GAL is released into the portal vein in a pulsatile manner and GAL episodes are either in coincidence with those of LHRH or precede them (Lopez *et al.*, 1991). Central infusion with galantide (Sahu *et al.*, 1994a) or GAL-antibody (Xu *et al.*, 1994) suppresses LH levels as well as pulse amplitude and pulse frequency in ovx rats suggesting a role of GAL in pulsatile secretion of GnRH and LH. Furthermore, not only GAL and GnRH coexpress in the same cell but, GAL+ve synaptic-like buttons are also present on some GnRH+ve neurons in the MPOA (Merchenthaler *et al.*, 1990). Since, GAL neurons in the MPOA do concentrate E2 receptors (Bloch *et al.*, 1992; Horvath *et al.*, 1994) and GAL mRNA increases with estrogen therapy (Kaplan *et al.*, 1988), and the fact that GnRH neurons themselves do not appear to contain estradiol receptors (Shivers *et al.*, 1983), it is likely that some of the effects of steroids on GnRH neurons may, in part, be mediated by GAL neurons. Finally, GAL may regulate its own secretion by a negative ultrashort loop feedback mechanism at the level of the ARC-ME (Lopez *et al.*, 1992a).

Nitric oxide (NO)

Anatomy and physiology: Nitric oxide (NO), a free radical gas, has emerged from endothelium-derived relaxing factor (Palmer *et al.*, 1987; Ignarro *et al.*, 1987; Garthwaite *et al.*, 1988) to inter- and intracellular messenger molecule in various biological systems including central and peripheral nervous systems (Furchgott, 1990; Garthwaite, 1991; Moncada, 1992; Snyder, 1992; Snyder and Bredt, 1992; Vincent and Hope, 1992; Vincent, 1994). NO is synthesized from L-arginine by the enzyme NO synthase (NOS, Bredt and Snyder, 1990; Bredt *et al.*, 1991; Moncada, 1992). Two distinct types of NOS have been identified: an inducible Ca^{2+}-independent enzyme and a constitutive Ca^{2+}-calmodulin-dependent enzyme (Moncada *et al.*, 1991; Bredt and Snyder, 1990). Neuronal NOS is a constitutive isoform. Since NO is extremely labile and short lived with a half-life of about five seconds, most insight into NO in neuronal system have come from studies with NOS. NOS protein and mRNA have been localized in various brain regions including the hypothalamus. In the hypothalamus, NOS mRNA and NOS protein are localized in the paraventricular nucleus, supraoptic nucleus, OVLT, MPN, MPOA, lateral hypothalamus, amygdala, VMN and ME (Bredt *et al.*, 1990, Bredt *et al.*, 1991; Vincent and Kimura, 1992; Rettori *et al.*, 1993; Grossman *et al.*, 1993, 1994; Bhat *et al.*, 1995). Localization of NOS in the hypothalamus, especially in the areas innervated by GnRH neurons and other neuronal systems that modulate GnRH secretion, raised the possibility that NO may be involved in regulation of GnRH secretion.

NO and GnRH: Evidence accumulated in recent years clearly suggest that NO not only modulates GnRH secretion but it is also a major player in endocrine regulation of hypothalamopituitary axis. The fact that NE induces GnRH release by increasing prostaglandin E2 (PGE2) production (Ojeda *et al.*, 1979, 1986; Negro-Vilar *et al.*, 1986) and that NOS inhibitors decrease NE-induced PGE2 release from the MBH *in vitro* (Rettori *et al.*, 1992) strongly suggest an involvement of NO in GnRH secretion. This was supported by the findings that sodium nitroprusside (SNP), an NO donor (Kukovets *et al.*, 1991), stimulated GnRH release from the ME-ARC (Moretto *et al.*, 1993; Bonavera *et al.*, 1993; Rettori *et al.*, 1994a) or MBH explants (Rettori *et al.*, 1993; Bonavera *et al.*, 1993) as well as in GT1 GnRH cell line (Moretto *et al.*, 1993; Mahachoklertwattana *et al.*, 1994b). It was also apparent that, as in several other neuronal tissues (Bredt and Snyder, 1989; Moncada and Higgs, 1993; Schmidt *et al.*, 1993), SNP increased both inter- and intracellular cyclic guanosin monophosphate (cGMP) level (Moretto *et al.*, 1993; Canteros *et al.*, 1995) and which in turn stimulated GnRH release (Moretto *et al.*, 1993). However, an inhibitory effect of NO on GnRH release in GTI cells has also been reported (Sortino *et al.*, 1994). Furthermore, inhibition of GnRH release by either interleukin 1 alpha (Rettori *et al.*,

1994b) or ethanol (Canteros *et al.*, 1995) has been found to be due to blockade of the response of GnRH neuronal terminals to NO. Using NADPH-diaphorase reaction as the indicator of NOS Activity (Hope *et al.*, 1991; Bredt *et al.*, 1991; Dawson *et al.*, 1991), Okamura *et al.* (1994) examined colocalization of NOS and estrogen receptor (ER) in the hypothalamus and they found that 70% of the NOS positive neurons contained ER-immunoreactivity in the VMN. These authors also reported an increase in NOS positive neurons by estrogen in the VMN and MPOA of ovx rats. These findings raise the intriguing possibility that action of steroids on GnRH neurons may also be mediated by NOergic neurons. All these studies clearly suggest that NO is an important messenger molecule in regulation of GnRH secretion.

NO and GnRH/LH surge: Since GnRH release in pulsatile in nature and hypersecretion of GnRH is necessary for the preovulatory LH surge, thus the next question arises whether NO is involved in these phenomena. Bonavera *et al.* (1993) provided first evidence that subcutaneous administration of NOS inhibitors either as a single injection at 1000 h or as three injections at 1000, 1200 and 1400 h markedly suppressed the progesterone-induced LH surge in the afternoon in estrogen-primed ovx rats. They also showed that NOS inhibitor significantly diminished the N-methyl-D-asparate (NMDA)-induced GnRH release from the ME-ARC fragments. Excitatory amino acids (EAAs), such as glutamate and asparate, have been implicated to induce hypersecretion of GnRH both on proestrus and that induced by ovarian steroids in ovx rats (Arslan *et al.*, 1988; Bourguignon *et al.*, 1989; Lopez *et al.*, 1990; Donoso *et al.*, 1990; Brann and Mahesh, 1991, 1994). Specifically, activation of NMDA subtype of glutamate receptors in the hypothalmus plays an important role in hypersecretion of GnRH (Bourguignon *et al.*, 1989; Lopez *et al.*, 1992b; Brann and Mahesh, 1994). In addition, NMDA receptor activation results in NO efflux in variety of tissues (Garthwaite *et al.*, 1988, 1989; Garthwaite, 1991). Thus, studies of Bonavera *et al.* (1993) suggest an EAA-NO link in the generation of the GnRH and LH surges. Presence of NMDA receptor-1 in hypothalamic NOS+ve neurons (Bhat *et al.*, 1995), and the blockade of glutamic acid-induced GnRH release from the ME-ARC explants (Rettori *et al.*, 1994a) and that of NMDA-induced GnRH release from the GT1 cell line (Mahachoklertwattana *et al.*, 1994a) by NOS inhibitors strengthen this possibility.

NO, LH surge and aging: It has been recently evident that in young cycling rats, as compared to 1000 h, hypothalamic NOS mRNA levels on the day of proestrus increased significantly at 1200 h when LH levels were low basal and remained elevated during 1600 to 1800 h. In these rats LH surge occurred at 1600 h with peak values at 1800 h (Sahu, 1996). These findings suggest an enhanced NOS neuronal activity, and thereby an increased

NO production before the LH surge. This early increase in NO production could play an important role in stimulating other excitatory signals (GnRH, NPY, etc.) that are obligatory for the LH surge. The evidences are accumulating in favor of this idea (*vide infra*). Evidently, in middle-aged cycling rats, NOS mRNA levels in the hypothalamus do not change between morning and afternoon hours on the day of proestrus (Sahu, 1996). Thus, absence of antecedent increase in NO production could, in part, be responsible for dalayed and attenuated LH surge in middle-aged rats.

***NO and GnRH pulse generator network*:** NO has also been implicated in the control of pulsatile release of GnRH and LH secretion. Using GT1 cell line, Moretto *et al.* (1992) suggested NO as a possible synchronizing factor for GnRH release *in vitro*. Subsequently, Rettori *et al.* (1993) reported that microinjection of NOS inhibitor, NG-Monomethyl-L-arginine (NMMA), into the third cerebroventricle decreased plasma LH and abolished LH pulses in castrated male rats, suggesting NO is essential for the generation of pulsatile LH and GnRH release. The mechanisms by which NO is involved in pulsatile secretion of GnRH and LH are still not clear. If NO mediates or contributes to the synchronization of GnRH pulses then there should be a mechanism by which it turns on and off its own release as well as release of other hypothalamic signals controlling GnRH release. Recent observations that NO inhibits hypothalamic GnRH release by stimulating gama-aminobutyric acid (GABA) (Seilicovich *et al.*, 1995a) and that it also inhibits NE and dopamine release from the MBH (Seilicovich *et al.*, 1995b), however, definitely shed a light into the controlling mechanism for the pulsatile GnRH secretion. Accordingly, (1) NE stimulates NOergic neurons to produce NO, which diffuses to the GnRH nerve terminals at the level of ME-ARC where it stimulates GnRH release, (2) NO also diffuses to the noradrenergic terminals and inhibits NE secretion via an ultra-short-loop negative feedback mechanism, this termination of NE release by NO could be responsible to terminate GnRH release and (3) NO can also terminate the pulses of GnRH release by stimulating GABA release. It is also possible that extracellular NO may act at the NO redox modulatory site on the NMDA receptor to inhibit GnRH release (Lipton, 1993; Manzoni and Bockaet, 1993). Obviously, as eluded earlier, GnRH neuronal activity is modulated by several neurotransmitters (Fig. 1). Because of its gaseous nature it is quite likely that in addition to the NO-catecholaminergic-GnRH (Seilicovich *et al.*, 1995) or catecholamine-NOergic-GnRH (Reitori *et al.*, 1993) or EAA-NO-GnRH (Bonavera *et al.*, 1993; Rettori *et al.*, 1994a; Mahachoklertwattana *et al.*, 1994b) transmission line, NO also act through other neural systems. In line with this possibility SNP, a NO donor, has been found to stimulate NPY release from the MBH *in vitro* (Bonavera *et al.*, 1994). Since NPY stimulates GnRH release *in vivo* and *in vitro*, and NPY is an obligatory signal for the GnRH and LH surges and it is one of the component of

GnRH pulse generator (vide supra), this study strongly suggests that excitatory effects of NO on GnRH neurons are, at least in part, mediated by NPY neuronal system.

***NO and other neuropeptides*:** In addition to stimulation of GnRH release, NOergic neurons stimulate several other neuropeptides, such as corticotropin-releasing hormone (CRH, Karanth *et al.*, 1993), growth hormone-releasing hormone (Rettori *et al.*, 1994c), somatostatin (Aguila, 1994), prolactin-releasing factor (Rettori *et al.*, 1994d), and vessopressin (Ota *et al.*, 1993). Stimulation of CRH by NO is of particular interest. In view of the fact that CRH is inhibitory to GnRH neurons (Petraglia *et al.*, 1987; Gambaccini *et al.*, 1986; Nikolarakis *et al.*, 1986), increased CRH secretion by NO in certain circumstances, such as, under stress or disease condition, may be responsible for reduced and attenuated GnRH and LH secretion and, there by, inhibition of rcproduction (Christian 1971; Lee and McDonald 1985; Rivier *et al.*, 1986; Rivier and Vale, 1984).

NO also mediates sexual behavior in female rats (Mani *et al.*, 1994). These authors found that NOS inbibitor, NMMA, injection into the third cerebroventricle prevented pregesterone-facilitated lordosis behavior in E2-primed ovx rats. They also observed that SNP injection facilitated lordosis in E2-primed rats in the absence of P. These results along with the report that NO is a physiologic mediator of penile erection in rat (Burnett *et al.*, 1992), put this diffusible gas in the center stage to control reproduction. Thus, NO plays a major role in the GnRH and LH surges necessary for ovulation as well as in copulatory behavior for successful pregnancy for perpetuation and survival of the species.

Inhibitory Signals in GnRH Regulation

Endogenous opioid peptides (EOP)

Among the inhibitory signals that impinge on the GnRH system, the role of EOPs is being increasingly recognized. Three distinct classes of EOPs β-endorphins (β-END), dynorphin A (DYN), and enkephalins (ENK), and their receptors have been identified in the hypothalamic regions innervated by the GnRH network. Anatomical evidence of the existence of synaptic contacts between β-END and GnRH neurons in the POA (Chen *et al.*, 1989) and also close appositions of ENK and DYN fibers to GnRH neurons in the MPOA and ME (Khachaturian *et al.*, 1985) indicate a possible involvement of EOP in GnRH regulation. Indeed, experimental evidence suggests that hypothalamic EOPs inhibit LH release by decreasing GnRH release (see Kalra, 1986, 1993).

EOPs and preovulatory GnRH/LH surge: Except for a brief period in the afternoon of proestrus, the hypothalamic EOP network generally excerts a tonic inhibitory influence on GnRH and LH secretion throughout the estrus cycle (Gabriel *et al.*, 1983; Piva *et al.*, 1985; Kalra, 1986, 1993). Experimental evidence (1) that the opioid receptor antagonist, naloxone (NAL), stimulates GnRH and LH release, and the opiate agonist, morphine, inhibits spontaneous and P-induced GnRH/LH surge release (see Kalra, 1986, 1993), and (2) that NAL infusion advances the GnRH/LH surge on the day of proestrus (Allen and Kalra, 1986; Allen *et al.*, 1988) and that induced by E2 in ovx rats (Masotto *et al.*, 1990), support the prevailing idea that a restrain on this inhibitory opioid tone by the NC is necessary for the genesis of the GnRH and LH surges (Dyer and Bicknell, 1989). Additional observations, such as, (1) β-END secretion into the hypophyseal portal vessel decreases before the hypersecretion of GnRH (Sarkar and Yen, 1985), (2) steady state levels of proopiomelanocortin mRNA in the ARC decerease progressivley on the day of LH surge (Wise *et al.*, 1990; Bohler *et ai.*, 1991; Petersen *et al.*, 1993), and (3) opioid receptor bindings in the MBH also decreases in association with the P-induced LH surge in the E2-primed ovx rats (Jacobson and Kalra, 1989), strongly support restraining on hypothalamic opioid tone before the LH surge.

How the EOPs operate to exert an inhibitory effect on the GnRH neurons? The available morphological and experimental evidence suggests that the inhibitory effect of EOPs on the GnRH neurons may be a direct one or via inhibiting other neuronal systems that are excitatory to GnRH neurons. The evidence that NAL stimulates GnRH release from the MBH *in vitro* (Leaem *et al.*, 1985; Sahu *et al.*, 1990a) and the existence of morphological connections between EOPs and GnRH networks suggest a direct effect of EOPs on GnRH. On the other hand, the anatomical evidence of synaptic contacts between β-END and NPY (Horvath *et al.*, 1992), and between ENK and NPY (Li *et al.*, 1993) in the MBH as well as our evidence that restraint in opioid tone by NAL stimulates NPY synthesis and release (Sahu *et al.*, 1990a; Xu *et al.*, 1996; also see section on NPY), strongly suggest that the NPY neuronal system is functionally interposed between NC-EOP and GnRH network. Since activation of the NPY neurons occurs well before the onset of the GnRH/LH surge (Sahu *et al.*, 1989b, 1994b, 1995), and NAL stimulates NPY synthesis and release, we envision that on day proestrus, disinhibition of inhibitory opioid tone by the NC activates NPY synthesis and release resulting in GnRH hypersecretion in the afternoon. Furthermore, both NPY and β-END neurons concentrate estradiol (Sar *et al.*, 1990; Morell *et al.*, 1985) and their secretion is also modulated by ovarian steroid (Sahu *et al.*, 1987a, 1989b, 1994b, 1995; also see Kalra, 1993), suggesting again the possibility that these two interconnected neuropeptidergic neurons-one excitatory and other inbibitory, may constitute a mandatory link in the preovulatory GnRH surge.

Other Inhibitory signals

Besides the EOP's, tachykinins are known to have inhibitory effects on GnRH and LH secretion. There are at least five biologically active tachykinins present in the brain (Maggio, 1988): substance P (SP), neurokinin A (NKA), neuropeptide K(NPK) neuropeptide-gama and NKA (3–10). Among the tackykinins, NPK seems to be the most potent inhibitory signal for GnRH and LH release. In fact NPK is more potent in suppressing LH release than β-END (Leadem and Kalra, 1985; Sahu *et al.*, 1987b; Sahu and Kalra, 1992). NPK-containing neurons and fiber systems have been localized in those hypothalamic sites that are innervated by GnRH neuronal network (Valentino *et al.*, 1986). Sahu *et al.* (1987b) first demonstrated that central injection of NPK inhibited LH release in ovx rats. Recently, we have shown that central administration of NPK blocked the P-induced LH surge in E2-primed ovx rats through activation of NK-2 receptors (Sahu and Kalra, 1992). Furthermore, NK-2 receptor agonist inhibited basal and KCl-induced GnRH release *in vitro* (Sahu and Kalra, 1992). These results suggest that NPK is an important hypothalamic inhibitory neurotransmitter/neuromodulator in regulation of basal and cyclic GnRH and LH secretion.

Interleukin-1 (IL-1) and CRH are two other inhibitory signal that have been studied extensively with regard to their role in hypothalamic-pituitary-gonadal axis. Basically, CRH inhibits GnRH and LH secretion under the condition of interaction between immune and endocrine system such as the state of disease, inhibits GnRH/LH surge by decreasing GnRH release (see Kalra and Kalra, 1991; Kalra *et al.*, 1992).

Conclusions and Future Perspectives

An overview of the regulation of GnRH secretion and possible pathways during the generation of GnRH/LH surge is presented in Figure 3. Accordingly, sometime between 1000 h and noon of proestrus the NC restrains the hypothalamic inhibitory tone produced by the EOPs and other inhibitory signals (OIS), such as GABA and NPK. Disinhibition of this inhibitory tone results in excitation of the excitatory signals such as NPY, GAL, EAA, NE/EPI, NO and GnRH. It is possible that there may be a cascade of events occurring during this period starting with stimulation of one or more of these excitatory signals. The NC-EOPs-NPY-GnRH transmission line is already in existence and it suggests that disinhibition of opioid tone by NC stimulates NPY neurons which may initiate a chain of events including stimulation of GAL and GnRH secretion resulting the massive LH surge in the afternoon. Other important transmission lines such as EAA-NO-GnRH, NE-NO-GnRH and NO-NPY-GnRH are already in place suggesting a neural circuitry for the GnRH/LH surge. Disruption of the GnRH/LH surge by using antibody or antagonist to any of these excitatory signals strongly suggests the obligatory role of the circuitry. Establishment of transmission lines between these excitatory and inhibitory signals as well as the

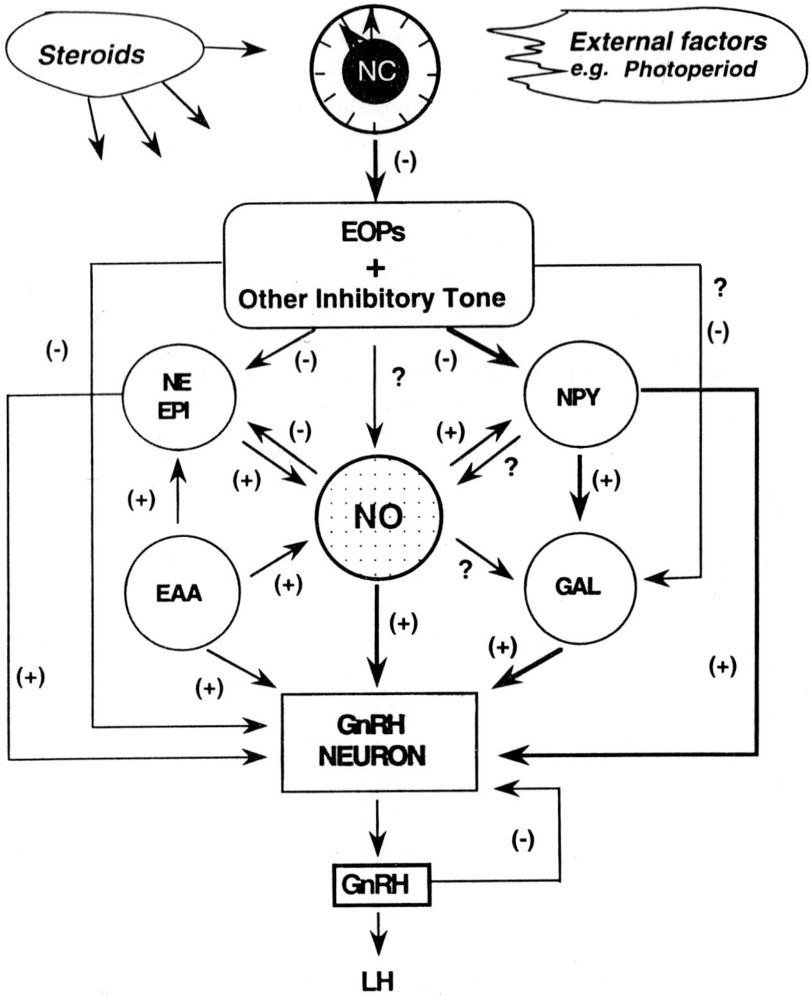

Fig. 3 An overview of the neural circuitry in control of GnRH secretion and possible pathways during the generation of the GnRH/LH surge. For abbreviation see text. (–), inhibition, (+), stimulation, and ?, unknown.

understanding of the mechanism(s) of operation of this circuitry, in my view, is the major task and a challenge to the scientific community. In this chapter, I have presented the role of major participant molecules in the neural circuitry that regulates the GnRH secretion. However, it is reminded to the readers that hypothalamic neuronal network controlling GnRH neuronal activity could be more complex than we understand today. There may be a lot of unidentified neural factors in the hypothalamus with potential role in GnRH regulation. For example, a decade or so ago, the concept of hypothalamic NPY and GAL that are now proved to be obligatory for GnRH secretion was inconceivable. At that time research was focused on catecholamines with regard to their role in GnRH secretion. Similarly, the profound effects of a diffusible gas, like NO, as a major neuronal messenger molecule was beyond imagination. It is true that NO is not unknown to the medical practitioner, but its role in reproduction as a whole—starting from GnRH secretion to penile erection and female copulation—was evidently not known until recently. Obviously, research on NO with regard to its role in GnRH secretion is just beginning and is still in its infancy. An explosion in NO research is expected and it is most likely that this gas will survive the test of time. Because of its diffusible nature, NO could be the way to explain many of the puzzling findings in reproductive neuroendocrinology. Recently, GTI cell line (immortalized GnRH neuronal cells) has been used extensively to further our understanding of the nature of GnRH regulation *in vivo*. Although this cell line behaves similar to the GnRH neurons in terms of its secretory activity and its regulation, it remains, however, to be explained why steroid receptors are present in GT1 cell line but not in the hypothalamic GnRH neurons. This could probably be because GT1 cell line is derived by genetically targetted tumorogenesis that may modify the normal characterisitics of GnRH neurons. Or it is possible that currently we lack the technique or method to measure steroid receptors in the small population of scattered GnRH neurons in the hypothalamus. Only time and further concerted research efforts will tell us. If the embryonic GnRH cells that are not yet migrated to the hypothalamus can secrete GnRH in pulsatile manner, then is it possible that GnRH neurons are carrying the pulse generator within themselves? If it is true, then why infusion of either NPY-antibody or GAL-antibody or GAL-antagonist disrupts the GnRH/LH pulsatility? However, at present, it is beyond doubt that a distinct neural circuitry controls GnRH neuronal activity.

Acknowledgement

This was supported partly by a grant from the National Institutes of Health (AG-10868).

References

Adelman, J.P., Mason, A.J., Hayflick, J.S., and Seeburg, P.H. (1986). Pro. Natl. Acad. Sci. USA. **83**: 179–183.

Aguila, M.C. (1994). Proc. Natl. Acad. Sci. USA **91**: 782–786.

Allen, L.G., and Kalra, S.P. (1986). Endocrinology **118**: 2375–2381.

Allen, L.G., Crowley, W.R., and Kalra, S.P. (1987). Endocrinology **121**: 1953–1959.

Allen, L.G., Hahn, E., Caton, D., and Kalra, S.P. (1988). Endocrinology **122**: 1004–1013.

Amoss, M., Burges, R., Blackwell, R., Vale, W., Fellows, R., and Guillemin, R. (1971). Biochem. Biophys. Res. Commun. **44**: 205–210.

Arslan, M., Pohl, C.R., and Plant, T.M. (1988). Neuroendocrinology **47**: 465–468.

Bai, F.L., Yamano, M., Shiotani, Y., Emson, P.C., Smith, A.D., Powell, J.F., and Tahyama, M. (1985). Brain Res. **331**: 172–175.

Barraclough, C.A., Wise, P.M., and Selmanoff, M.K. (1984). Recent Prog. Horm. Res. **40**: 487–529.

Bartfai, T., Bedecs, K., Land, T., Langel, U., Bertorelli, R., Girotti, P., Consolo, S., Xu, X., Wiesenfeld-Hallin, Z., Nilsson, S., Pieribone, V.A., and Hokfelt, T. (1991). Proc. Natl. Acad. Sci. USA **88**: 10961–10965.

Bauer-Dantoin, A.C., McDonald, J.K., and Levine, J.E. (1991). Endocrinology **129**: 402–408.

Bauer-Dantoin, A.C., McDonald, J.K., and Levine, J.E. (1992). Endocrinology **131**: 2946–2952.

Bauer-Dantoin, A.C., Knox, K.L., Schwartz, N.B., and Levine, J.E. (1993a). Endocrinology **113**: 2413–2417.

Bauer-Dantoin, A.C., Tabesh, B., Norgle, J.R., and Levine, J.E. (1993b). Endocrinology **133**: 2418–2423.

Belchetz, P.E., Plant, T.M., Nakai, Y., Keogh, EJ., and Knobil, E. (1978). Science **202**: 631–633.

Ben-Jonathan, N., Mical, R.S., and Porter, J.C. (1974). Endocrinology **95**: 18–25.

Bhat, G.K., Mahesh, V.B., Lamar, C.A., Ping, L., Aguan, K., and Brann, D.W. (1995). Neuroendocrinology **62**: 187–197.

Bloch, G.J., Kurth, S.M., Akesson, T.R., and Micevych, P.E. (1992). Brain Res. **595**: 301–308.

Bohler, H.C.L., Tracer, H., Merriam, G.R., and Peterson, S. (1991). Endocrinology **128**: 1265–1269.

Bonavera, J.J., Sahu, A., Kalra, P.S., and Kalra, S.P. (1993). Endocrinology **133**: 2481–2487.

Bonavera, J.J., Sahu, A., Kalra, P.S., Crowley, W.R., and Kalra, S.P. (1994). Soc. Neurosci. Abstr. **20**: 639.

Bourguignon, J.P., Gerarg, A., Mathieu, J., Simons, J., and Franchimont, P. (1989). Endocrinology, **125**: 1090–1096.

Brann, D.W., and Mahesh, V.B. (1991). Endocrinology **128**: 1541–1547.

Brann, D.W., and Mahesh, V.B. (1994). Front. Neuroendocrinol. **15**: 3–49.

Bredt, D.S., and Snyder, S.H. (1989). Proc. Natl. Acad. Sci. USA **86**: 9030–9033.

Bredt, D.S., and Snyder, S.H. (1990). Proc. Natl. Acad. Sci. USA **87**: 682–685.

Bredt, D.S., Hawang, P.M., and Snyder, S.H. (1990). Nature **347**: 768–770.

Bredt, D.S., Glatt, C.E., Hwang, P.M., Fotuhi, M, Dawson, T.M., and Snyder, S.H. (1991). Neuron **7**: 615–624.

Burich, A.J., Claude, P., and Terasawa, E. (1994). Soc.Neurosci. Abstr. **20**: 637.

Burgus, R., Butcher, M., Amoss, M., Ling, N., Monahan, M., Rivier, J., Fellows, R., Blackwell, R., Vale, W., and Guillemin, R. (1972). Proc. Natl. Acad. Sci. USA **69**: 278–282.

Burnett, A.L., Lowenstein, C.J., Bredt, D.S., Chang, T.S.K., and Snyder, S.H. (1992). Science **257**: 401–403.

Canteros, G., Rettori, V., Belova, Franchi, A., Genaro, A., Cebral, E., Faletti, A.N., Gimento, M., and McCann, S.M. (1995). Proc. Natl. Acad.Sci. USA **92**: 3416–3420.

Carmel, P.W., Araki, S., and Ferin, M. (1976). Endocrinology **99**: 243–248.

Castro, J.C.B., Khorram, O., and McCann, S.M. (1985). Proc. Soc. Exp. Biol. Med. **179**: 132–135.

Chen, W.P., Witkin, J.W., and Silverman, A.J. (1989). J. Comp. Neurol. **286**: 85–95.

Ching, M. (1982). Neuroendocrinology **34**: 279–285.

Christian, J. (1971). Biol. Reprod. **4**: 248–294.

Coen, C.W., Montagnese, C., and Opacka-Juffry, J. (1990). J. Neuroedocrinol. **2**: 107–111.

Crowley, W.R., and Kalra, S.P. (1988). Synapse **2**: 276–281.

Crowley, W.R., and Kalra, S.P. (1987). Neuroendocrinoology **46**: 97–103

Crowley, W.R., and Kalra, S.P. (1987). Neuroendocrinology **120**: 941–945.
Crowley, W.R., Shah, G.V., Carroll, B.L., Kennedy, D., Dockter, M.E., and Kalra, S.P. (1990). Endocrinology **127**: 1487–1497.
Crowley, W.R., Tessel, R.E., O'Donohue, T.L., Adler, B.A., and Kalra, S.P. (1985). Endocrinology **117**: 1151–1155.
Dawason, T.M., Bredt, D.S., Fotuhi, M., Hwang, P.M., and Snyder, S.H. (1991). Proc. Natl. Acad. Sci. USA **88**: 7797–7801.
DePaolo, L.V., King, R.A., and Carrillo, A.J. (1987). Endocrinology **120**: 272–279.
Dluzen, D.E., and Ramirez, V.D. (1986). Endocrinology **118**: 110–113.
Donoso, A.O., Lopez, F.J., and Negro-Vilar, A. (1990). Endocrinology **126**: 414–420.
Dyer, R.G., and Bicknell, R.J. (1989). (Eds) Brain Opioid Systems in Reproduction, pp. 1–357, Oxford University Press, Oxford, UK.
Ellis, G.B., and Desjardins, C. (1982). Endocrinology **110**: 1618–1627.
Escalera, G.M., Choi, A.L.H., and Weiner, R.I. (1992). Proc. Natl. Acad. Sci. USA **89**: 1852–1855.
Everett, J.W. (1977). J. Endocrinol. **75**: 1P–13P.
Everett, B.J., and Hokfelt, T. (1989). In: Neuropeptide Y (V. Mutt, T. Hokfelt, K. Fuxe and J.M. Lundberg, Eds), pp. 61–72, Alan R. Liss, Inc., New York
Freeman, M.E. (1994). In: The Physiology of Reproduction, **Vol. 2,** Second Edition, (E. Knobil and J.D. Neill, Eds), pp. 613–658.
Furchgott, R.F. (1990). Acta Physiol. Scand. **139**: 257–270.
Fuxe, K., and Agnati, L.F. (1991). In: Volume transmission in the Brain: Novel Mechanisms for Neurotransmision (K. Fuxe and L.F. Agaati, Eds), pp. 1–9, Raven Press, New York.
Gabriel, S.M., Simpkins, J.W., and Kalra, S.P. (1983). Endocrinology **113**: 1806–1811.
Gallo, R.V. (1980). Neuroendocrinology **30**: 122–131.
Gallo, R.V. (1981). Biol. Reprod **20**: 100–105.
Gambaccini, M., Yen, S.S.C., and Rasmussen, D.D. (1986). Neuroendocrinology **43**: 533–536.
Garthwaite, J. (1991). Trends. Neurosci. **14**: 60–67.
Garthwaite, J., Charles, S.L., and Chess-Williams, R. (1988). Nature **336**: 385–388.
Garthwaite, J., Garthwaite, G., Palmer, R.M.J., and Moncada, S. (1989). Eur. J. Pharmacol. **172**: 413–416.
Gibson, M.J., Krieger, D.T., Charlton, H., Zimmerman, E.A., Silverman, A.J., and Perlow, M.J. (1984). Science **225**: 949–951.
Grossman, A., Kabigting, E.B., Cadd, G., Rossmanith, W.G., Clifton, D.K., and Steiner, R.A. (1993). J. Endocrinol. (Suppl) **139**: 19 (Abst.).
Grossman, A., Rossmanith, W.G., Kabigting, E.B., Cadd, G., Clifton, D.K., and Steiner, R.A. (1994). J. Endocrinol. **140**: R5-R8.
Guy, J., Li, S., and Pelletier, G. (1988). Regulatory Peptides **23**: 209–216.
Guy, J., and Pelleier, G. (1988) Peptides **9**: 567–570.
Hope, B.T., Michael, G.J., Knigge, K.M., and Vincent, S.R. (1991). Proc. Natl. Acad. Sci. USA **88**: 2811–2814.
Horvath, T.L., Naftolin, F., Kalra, S.P., and Leranth, C. (1992). Endocrinology **131**: 2461–2467.
Horvath, T.L., Naftolin, F., Kalra, S.P., and Leranth, C. (1993). Soc. Neurosci. Abstr. **19**: 1693.
Horvath, T.L., Kalra, S.P., Leranth, C., and Naftolin, F., (1994). Brain Res. **675**: 321–324.
Horvath, T.L., Kalra, S.P., Naftolin, F., and Leranth, C. (1995). J. Neuroendocrinol. **7**: 579–588.
Horvath., T.L., Naftolin, F., Leranth, C., Sahu, A., and Kalra, S.P. (1996). Endocrinology **137**: 78–74.

Igaarro, L.J., Buga, G.M., Wood, K.S., Byrne, R.E., and Chaudhuri, G. (1987). Proc. Natl. Acad. Sci. USA **84**: 9265–9269.
Jacobson., W., and Kalra, S.P. (1989). Endocrinology **124**: 199–205.
Kalra, S.P. (1986). Frontiers Neuroendocrinol. **9**: 31–75.
Kalra, S.P. (1993). Endocr. Rev. **14**: 507–538.
Kalra, S.P., and Crowley, W.R. (1984). Life Sci. **35**: 1173–1186.
Kalra, S.P., and Crowley, W.R. (1992). Frontiers Neuroendocrinol. **13**: 1–46.
Kalra, S.P., and Kalra, P.S. (1983). Endocr. Rev. **4**: 311–351.
Kalra, S.P., and Kalra, P.S. (1989). Biol. Repord. **41**: 559–570.
Kalra, S.P., and Kalra, P.S. (1974). Endocrinology **95**: 1711–1718.
Kalra, S.P., and Kalra, P.S. (1979). Acta endocr. (Copenh.) **85**: 449–455.
Kalra, S.P., and Kalra, P.S. (1991). In: Brain Endocrinology (M. Motta, Ed), 2nd ed., pp. 177–216, Raven Press, New York.
Kalra, S.P, Krulich, L., and McCann, S.M. (1973). Neuroendocrinology **12B**: 321–333.
Kalra, S.P., Clark, J.T., Sahu, A., Dube, M.G., and Kalra, P.S. (1988). Synapse **2**: 254–257.
Kalra, S.P., Clark, J.T., Sahu, A., Kalra, P.S., and Crowley, W.R. (1989). In: Neuropeptide Y (V. Mutt, T. Hokfelt, K. Fuxe and J.M. Lundberg, Eds), pp. 229–241, Alan R. Liss, Inc., New York.
Kalra, S.P., Clark, J.T., Sahu, A., Kalra, P.S., and Crowley, W.R. (1990). Ann. N.Y. Acad. Sci. **611**: 273–283.
Kalra, S.P., Allen, L.G., Sahu, A., and Crowley, W.R. (1991). In: Brain Gut Peptides and Reproductive Function (C. Johnston and D. Barnes, Eds), pp. 153–178, CRC Press, Boca Raton, Florida.
Kalra, S.P., Sahu, A., and Kalra, P.S. (1992). In: Frontiers in Reproductive Physiology (D. Ghosh and J Sengupta, Eds), pp. 21–38, Wiley Eastern Limited, New Delhi.
Kalra, P.S., Bonavera, J.J., and Kalra, S.P. (1995). Regulatory Peptides **59**: 215–220.
Kaplan. L.M., Gabriel, S.M., Koenig, J.I., Sunday, M.E., Spindel, E.R., Martin, J.B., and Chin, W.W. (1988). Proc. Natl. Acad. Sci. USA **85**: 7408–7412.
Karanth, S., Lyson, K., and McCann, S.M. (1993). Proc. Natl. Acad. Sci. USA **90**: 3383–3387.
Kaynard, A.H., and Spies, H.G. (1991). Endocrinology **128**: 2769–2775.
Kelly, M.J., Garrett, J., Bosch, M.A., Roselli, C.E., Douglass, J., and Adelman, J.P. (1989). Neuroendocrinology **49**: 88–97.
Khachaturian, H., Lewis, M.E., Schafer, M.K.H., and Watson, S.J. (1985). Trends Neurosci. **8**: 111–119.
Khorram, O., Pau, K.Y.F., and Spies, H.G. (1987). Neuroendocrinology **45**: 290–297.
Kim, K., Lee, B.J., Park, Y., and Cho, W.K., (1989). Mol. Brain Res. **6**: 151–158.
Knobil, E. (1980). Recent Prog. Horm. Res. **36**: 53–88.
Knobil, E., and Hotchkiss, J. (1988). In: The Physiology of Reproduction (E. Knobil and J.D. Neill, Eds), pp. 1971–1994, Raven Press, New York.
Kordon, C., Drouva, S.V., Escalera, G.M., and Weiner, R.I. (1994). In: The Physiology of Reproduction, Volume 1 (E. Knobil and J.D. Neill, Eds), 2nd ed., pp. 1621–1681, Raven Press, New York.
Krsmanovic, L.Z., Stojilkovic, S.S., Merelli, F., Dufour, S.M., Virmani, M.A., and Catt, K.J. (1992). Proc. Natl. Acad. Sci.USA **89**: 8462–8466.
Krsmanovic, L.Z., Stojilkovic, S.S., Mertz, L.M., Tomic, M., and Catt, K.J. (1993). Proc. Natl. Acad. Sci. USA **90**: 3908–3912.
Kukovets, W.R., Holsmann, S., and Schmidt, K. (1991). Eur. Heart J. **12** (Suppl. E): 16-24.
Kusano, K., Fueshko, S., Gainer, H., and Wray, S. (1995). Proc. Natl. Acad. Sci. USA **92**: 3918–3922.
Leadem, C.A., Crowley, W.R., Simpkins, J.W., and Kalra, S.P. (1985). Neuroendocrinology **40**: 497–500.

Leadem, C.A., and Kalra, S.P. (1985). Endocrinology **117**: 684–689.
Lee, A., and McDonald, I. (1985) Oxf. Rev. Reprod. Biol. **7**: 261–304.
Lee, W-S., Smith, S.M., and Hoffman, G.E. (1990a). Proc. Natl. Acad. Sci. USA **87**: 5163–5167.
Lee, W-S., Smith, S.M. and Hoffman, G.E. (1990b). Endocrinology **127**: 2606.
Lee, W-S., Abbuel, R., Smith, S.M., and Hoffman, G.E. (1992a). Endocrinology **130**: 3101–3103.
Lee., W-S., Smith, S.M., and Hoffman, G.E. (1991b). J. Neuroendocrinol. **4**: 161–166.
Leranth, C., Segura, L.M.C., Palkovits, M., Maclusky, N.J., Shanabrough, M., and Naftolin, F. (1985). Brain Res. **345**: 332–336.
Levine, J.E., and Duffy, M.T. (1988). Endocrinology **122**: 2211–2221.
Levine, J.E., and Ramirez, V.D. (1982) Endocrinology **111**: 1439–1448.
Levine, J.E., Bethea, C.L., and Spies, H.G. (1958). Endocrinology **116**: 431–438.
Levine, J., Bauer-Dantoin, C., Besecke, L., Conaghan, L.A., Legan, S.J., Meredith, J.M., Strobl, F.J., Urban, J.H., Vogelsom, K.M., and Wolff, A.M. (1991). Recent Prog. Horm. Res. **47**: 97–153.
Li, S., Hisano, S., and Daikoku, S. (1993). Neuroendocrinology **57**. 306–313.
Li, S.Y., Hong, M., Fournier, A., Stpierre, S., and Pelletier, G. (1994). Mol. Brain Res. **26**: 69–73.
Lincoln, D.W., Fraser, H.M., Lincoln, C.A., Martin, G.B., and McNeilly, A.S. (1985). Recent Prog. Horm. Res. **41**: 369–411.
Lipton, S.A. (1993). Trends. Neurosci. **16**: 527–532.
Lopez, F.J., Donoso, A.O., and Negro-Vilar, A. (1990). Endocrinology **127**: 2431–2436.
Lopez, F.J., and Negro–Vilar, A. (1990). Endocrinology **126**: 1771–1776.
Lopez, F.J., Merchenthaler, I., Ching, M., Wisniewski, M.G., and Negro–Vilar, A. (1991). Proc. Natl. Acad. Sci. USA **88**: 4508–4512.
Lopez, F.J., Liposits, Z., and Merchenthaler, I. (1992a). Endocrinology **130**: 1499–1507.
Lopez, F.J., Donoso, A.O., and Negro-Vilar, A. (1992b), Endocrinology **130**: 1986–1992.
Lopez, F.J., Meade, E.H., and Negro-Vilar, A. (1993). Endocrinology **132**: 795–800.
Maggio, J.E. (1988). Annu. Rev. Neurosci. **11**: 13–28.
Malik, K.F., Silverman, A.J., and Morell, J.I. (1991). Anat. Rec. **231**: 457–466.
Mahachoklertwattana, P., Sanchez, J., Kaplan, S.L., and Grumbach, M.M. (1994). Endocrinology **134**: 1023–1030.
Mahachoklertwattana, P., Black, S.M., Kaplan, S.L., Bristow, J.D., and Grumbach, M.M. (1994). Endocrinology **135**: 1709–1712.
Mani, S.K., Allen, J.M.C., Rettori, V., McCann, S.M., O'Malley, B.W., and Clark, J.H. (1994). Proc. Natl. Acad. Sci. USA **91**: 6468–6472.
Manzoni, O., and Bockaet, J. (1993). J. Neurochem. **61**: 368–370.
Marks, D.L., Smith, M.A., Vrontakis, M., Clifton, D.K., and Steiner, R.A. (1993). Endocrinology **132**: 1936–1844.
Marks, D.L., Lent, K.L., Rosmanith, W.G., Clifton, D.K., and Steiner, R.A., (1994). Endocrinology **134**: 1991–1998.
Masotto, C., Sahu, A., Dube, M.G., and Kalra, S.P. (1990). Endocrinology **126**: 18–25.
Matsuo, H., Baba, Y., Nau, R.M.G., Arimura, A., and Schally, A.V. (1971). Biochem. Biophys. Res. Commun. **43**: 1334–1339.
McCann, S.M., and Krulich, L. (1989). In: De Groot: Endocrinology (G.M. Besser, Ed), 2nd ed., pp. 117–130, Saunders, Philadelphia.
McCann, S.M. (1991). In: Brain Endocrinology (M.Motta, Ed), 2nd ed., pp. 1–30, Raven Press, New York.
Melander, T., Hokfelt, T., and Rokaeus, A. (1986). J. Comp. Neurol. **248**: 475–517.
Melander, T., Kohler, C., Nilsson, S., Fisone, G., Bartfai, T., and Hokfelt, T. (1992). In:

Handbook of Chemical Neuroanatomy (A. Bjorklund, T. Hokfelt and M.J. Kuhar, Eds), **vol 2**, pp. 187–221, Elsevier, Amsterdam.

Mellon, P.L., Windle, J.J., Goldsmith, P.C., Padula, C.A., Roberts, J.L., and Weiner, R.I. (1990). Neuron **5**: 1–10.

Merchanthaler, I., Gores, T., Setalo, G., Petrusz, P., and Flerko, B. (1984). Cell Tissue Res. **237**: 15–29.

Merchanthaler, I., Setalo, G., Csontos, G., Csontos, C., Petrusz, P., Flerko, B., and Negro-Vilar, A. (1989). Endocrinology **125**: 2812–2821.

Merchenthaler, I., Lopez, F.J., and Negro-Vilar, A. (1990). Proc. Natl. Acad. Sci. USA **87**: 6336–6330.

Merchenthaler, I., Lopez, F.J., and Negro-Vilar, A. (1993). Prog. Neurobiol. **40**: 711–769.

Minami, S., Frautschy, S.A., Poltsky, P., Sutton, S.W., and Sarkar, D.K. (1990). Neuroendocrinology **51**: 112–115.

Moncada, S., Palmer, R.M.J., and Higgs, E.A. (1991). Pharmacol. Rev. **43**: 109–142.

Moncada, S. (1992). Acta Physiol. Scand. **145**: 201–227.

Moncada, S., and Higgs, A. (1993). N. Eng. J. Med. **329**: 2002–2012.

Morell, J., McGinty, F., and Pfaff, D.W. (1985). Neuroendocrinology **41**: 417–426.

Moretto, M., Lopez, F.J., and Negro-Vilar, A. (1993). Endocrinology **133**: 2399–2402.

Moretto, M., Lopez, F.J., Petrusz, P., and Negro-Vilar, A. (1992). Soc. Neurosci. Abstr. **18**: 915.

Negro-Vilar, A., Conte, D., and Valenca, M. (196). Endocrinology **119**: 2796–2802.

Nikolarakis, K.E., Almeida, O.F.X., and Hertz, A. (1986). Brain Res. **377**: 399–390.

Ojeda, S.R., Negro-Vilar, A., and McCann, S.M. (1979). Endocrinology **104**: 617–624.

Ojeda, S.R., Urbanski, H.F., Katz, K.H., Costa, M.E., and Conn, P.M. (1986). Proc. Natl. Acad. Sci. USA **83**: 4932–4936.

Oka, Y. (1992). Neurosci. Lett. **142**: 119–122.

Okamura, H., Yokosuka, M., McEwen, B.C., and Hayashi, S. (1994). Endocrinology **135**: 1705–1708.

Ota, M., Crofton, J.T., Festavan, G.H., and Share, L. (1993). Am. J. Physiol. **57**: 955–959.

Palkovits, M., Rokaeus, A., Antoni, F.A., and Kiss, A. (1987). Neuroendocrinology **46**: 417–423.

Palmer, R.M.J., Ferrige, A.G., and Moncada, S. (1987). Nature **327**: 524–526.

Park, O.K., Gugneja, S., and Mayo, K.E. (1990). Endocrinology **127**: 365–372.

Parker, S.L., Kalra, S.P., and Crowley, W.R. (1991). Endocrinology **128**: 2309–2316

Pau, K.Y.F., Kaynard, A.H., Hess, D.L., and Spies, H.G. (1991). Neuroendocrinology **53**: 396–403.

Pau, K.Y.F., Berria, M., Hess, D.L., and Spies, H.G. (1995). J. Neuroendocrinol. **7**: 63–67.

Petersen, S.L., Keller, M.L., Carder, S.A., and McCrone, S. (1993). J. Neuroendocrinol. **5**: 643–648.

Petersen, S.L., McCrone, S., Keller, M., and Shores, S. (1995). Endocrinology **136**: 3604-3610.

Petraglia, F., Sutton, S., Vale, W., and Plotsky, P. (1987). Endocrinology **120**: 1083–1088.

Piva, F., Maggi, R., Limonta, P., Motta, M., and Martini, L. (1985). Endocrinology **117**: 766–772.

Ramirez, V.D., and Dluzen, D.E. (1987). Biol. Reprod. **36**: 59–76.

Ramirez, V.D., Feder, H., and Sawyer, C.H. (1984) Frontiers Neuroendocrinol. **8**: 27–84.

Rettori,V., Kamat, A., and McCann, S.M. (1994a). Brain Res. Bull. **33**: 501–503.

Rettori, V., Belova, N., Kamat, A., Lyson, K., Gimento, M., and McCann, S.M. (1994b). Neuroimmunomodulation **1**: 86–91.

Rettori, V., Belova, N., Yu, W.H., Gimento, M., and McCann, S.M. (1994c). Neuroimmunomodulation **1**: 195–200.

Rettori, V., Belova, N., Gimento, M., and McCann, S.M. (1994d). Neuroimmunomodulation **1**: 116–120.
Rettori, V., Belova, N., Dees, W.L., Nyberg, C.L., Gimento, M., and McCann, S.M. (1993). Proc. Natl. Acad. Sci. USA **90**: 10130–10134.
Rettori, V., Gimento, M., Lyson, K., and McCann, S.M. (1992). Proc. Natl. Acad. Sci. USA **89**: 11543–11546.
Rivier, C., and Vale, W. (1984). Endocrinology **114**: 914–921.
Rivier, C., Rivier, J., and Vale, W. (1989). Science **231**: 607–609.
Roberts, J.L., Dutlow, C.M., Jakubowski, M., Blum, M., and Millar, R.P. (1989). Mol. Brain Res. **6**: 127–134.
Rosie, R., Thompson, E., and Fink, G. (1990). J. Endocrinol. **124**: 285–289.
Ronnekleiv, O.K., and Resko, J.A. (1990). Endocrinology **126**: 498–511.
Ronnekleiv, O.K., Naylor, B.R., Bond, C.T., and Adelman, J.P. (1989). Mol. Endocrinol. **3**: 363–371.
Rothfeld, J., Hejtmancik, J.E., Conn, P.M., and Pfaff, D.W. (1989). Mol. Brain Res. **6**: 121–125.
Sabatino, F.D., Collins, P., and McDonald, J.K. (1989). Endocrinology **124**: 2089–2098.
Sabatino, F.D., Collins, P., and McDonald, J.K. (1990). Neuroendocrinology **52**: 600–607.
Sahu, A. (1996). Soc. Neurosci. Abst., **22**: 1345.
Sahu, A., and Kalra, S.P. (1987). Life Sci. **40**: 1201–1206.
Sahu, A., and Kalra, S.P. (1992). Endocrinology **130**: 1571–1577.
Sahu, A., and Kalra, S.P. (1993). Trends Endocrinol. Metab. **4**: 217–224.
Sahu, A., and Kalra, S.P. (1995). 77th Ann. Mtg. Endocrine Society, Abst. # P3–680, pp. 638.
Sahu, A., and Kalra, S.P. (1997). 79th Ann. Mtg. Endocrine Society, Abst. # P2–381, pp. 380.
Sahu, A., Kalra, S.P., Crowley, W.R., O'Donohue, T.L., and Kalra, P.S. (1987). Endocrinology **120**: 1831–1836.
Sahu, A., Crowley, W.R., Tatemoto, K., Balasubramaniam, A., and Kalra, S.P. (1987). Peptides **8**: 921–926.
Sahu, A., Kalra, P.S., Crowley, W.R., and Kalra, S.P. (1988). Endocrinology **125**: 2199–2203.
Sahu, A., Kalra, S.P., Crowley, W.R., and Kalra, P.S. (1988). Brain Res. **457**: 376–378.
Sahu, A., Kalra, S.P., Crowley, W.R., and Kalra, P.S. (1989). Endocrinology **124**: 410–414.
Sahu, A., Jacobson, W., Crowley, W.R., and Kalra, S.P. (1989). J. Neuroendocrinol. **1**: 83–87.
Sahu, A., Crowley, W.R., and Kalra, S.P. (1990). Endocrinology **126**: 876–883.
Sahu, A., Kalra, S.P., Crowley, W.R., and Kalra, P.S. (1990). Brain Res. **515**: 287–291.
Sahu, A., Phelps, C.P., White, J.D., Crowley, W.R., Kalra, S.P., and Kalra, P.S. (1992). Endocrinology **130**: 3331–3336.
Sahu, A., Xu, B., and Kalra, S.P. (1994). Endocrinology **134**: 529–536.
Sahu, A., Crowley, W.R. and Kalra, S.P. (1994). Endocrinology **134**: 1018–1022.
Sahu, A., Crowley, W.R., and Kalra, S.P. (1995). J. Neuroendocrinol. **7**: 291–296.
Sar, M., Sahu, A., Crowley, W.R., and Kalra, S.P. (1990). Endocrinology **127**: 2752–2756.
Sarkar, D.K., and Yen, S.S.C. (1985). Endocrinology **116**: 2075–2079.
Sarkar, D.K., Chiappa, S.A., Fink, G., and Sherwood, N.M. (1976). Nature **264**: 461–463.
Schally, A.V., Arimura, A., Baba, Y., Nair, R.M.G., Matsuo, H., Redding, T.W., Debeljuk, L., and White, W.F. (1971). Biochem. Biophys. Res. Commun. **43**: 393–399
Schmidt, H.H.H.W., Lohman, S.M., and Walter, U. (1993). Biochimica Biophysica Acta **1178**: 153–175.
Schwanzel-Fukuda, M., and Pfaff, D.W. (1989). Nature **338**: 161–164.

Schwanzel-Fukuda, M., Bick, D., and Pfaff, D.W. (1989). Mol. Brain Res. **6**: 311–326.

Schwanzel-Fukuda, M., Jorgenson, K.L, Bergen, H.T., Weesner, G.D., and Pfaff, D.W. (1992). Endocr. Rev. **13**: 623–634.

Seeburg, P.H., and Adelman, J.P. (1984). Nature **311**: 666–668.

Seeburg, P.H., Mason, A.J., Stewart, T.A., and Nikolocs, K. (1987). Rec. Prog. Horm. Res. **43**: 69–98.

Seilicovich, A., Duvilanski, B.H., Pisera, D., Theas, S., Gimeno, M., Rettori, V., and McCann, S.M. (1995). Proc. Natl. Acad. Sci. USA **92**: 3421–3424.

Seilicovich, A., Lasaga, M., Befumo, M., Duvilanski, B.H., Diaz, M.D.C., Rettori, V., and McCann, S.M. (1995). Proc. Natl. Acad. Sci. USA **92**: 11299–11302.

Shivers, B.D., Harlan, R., Morrell, J., and Pfaff, D.W. (1984). Nature **304**: 345–347.

Silverman, A.J., Jhamandar, J., and Renaud, P. (1987). J. Neurosci. **7**: 2312–2319.

Silverman, A.J., Livne, I., and Witkin, J.W. (1994). In; The Physiology of Reproduction, **Vol. 1**, 2nd ed., (E. Knobil and J.D. Neil, Eds), pp. 1683–1709. Raven Press, Ltd., New York.

Skofitsch, G., and Jacobowitz, D.M. (1985). Peptides **6**: 509–546.

Skofitsch, G., Sills, M.A., and Jacobowitz, D.M. (1986). Peptides **7**: 1029–1042.

Snyder, S.H. (1992). Science **257**: 494–496.

Snyder, S., and Bredt, D.S. (1992). Sci. Am. **266**: 68–77.

Sortino, M.A., Aleppo, G., Scapagnini, U., and Canonico, P.L. (1994). Endocrinology **134**: 1782–1787.

Sutton, S., Toyama, T., Otto, S., and Poltsky, P. (1988a). Endocrinology **123**: 1208–1210.

Sutton, S., Mitsugi, N., Poltsky, P., and Sarkar, D.K. (1988b). Endocrinology **123**: 2152–2154.

Tatemoto, K., (1982). Proc. Natl. Acad. Sci. USA **79**: 5485–5489.

Tatemoto, K., Rokaeus, A., Jorwall, H., McDonald, T.J., and Mutt, V. (1983). FEBS Lett. **164**: 124–128.

Toranzo, D., Dupont, E., Simard, J., Labrie, C., Couet J., Labrie, F., and Pelletier, G. (1989). Mol. Endocrinol. **3**: 1748–1756.

Tsuruo, Y., Kwano, H., Kagotani, Y., Hisano, S., Daikoku, S., Chihara, K., Zang, T., and Yanaihara, N. (1990). Neurosci. Lett. **110**: 261–266.

Turek, F.W., and Van Cauter, E. (1994). In: The Physiology of Reproduction, **Vol. 2**., Second Edition (E. Knobil and J.D. Neill, Eds), pp. 487–540. Reven Press, Ltd., New York.

Valenca, M.V., Johnston, C.J., Ching, M., and Negro-Vilar, A. (1987). Endocrinology **121**: 2256–2259.

Valentino, K.L., Tatemoto, K., Hunter, J., and Barchas, J.D. (1986). Peptides **7**: 1043–1059.

Vincent, S.R. (1994). Prog. Neurobiol. **42**: 129–160.

Vincent, S.R., and Hope, B.T. (1992). Trends. Neurosci. **15**: 108–113.

Vincent, S.R., and Kimura, H. (1992). Neuroscience **46**: 755–784.

Watanobe, W., and Takebe, K. (1992). Neurosci. Lett. **146**: 57–59.

Wherenberg, W.B., Cordor, R., and Gaillard, R.C. (1989). Neuroendocrinology **46**: 680–682.

Wetsel, W., Valenca, M.M., Merchenthaler, I., Liposits, Z., Lopez, F.Z.,Weiner, R.I., Mellon, P.L., and Negro-Vilar, A. (1992). Proc. Natl. Acad. Sci. USA **89**: 4149–4153.

Wilson, R.C., Kesner, J.S., Kaufman, J., Uema, T., Akema, T., and Knobil, E. (1984). Neuroendocrinology **39**: 256–260.

Wise, P.M., Rance, N., Selmanoff, M.K., and Barraclough, C.A. (1981). Endocrinology **108**: 2179–2185.

Wise, P.M., Scarbrough, K., Larson, G.H., Lioyed, J.M., Weiland, N.G., and Chiu, S. (1991). Frontiers Neuroendocrinol. **12**: 323–356.

Wise, P.M., Scarbrough, K., Weiland, N.G., and Larson, G.H. (1990). Mol. Endocrinol. **4**: 886–892.

Witkin, J.W. (1990). Neuroscience **37**: 501–506.
Witkin, J.W., Paden, C.M., and Silverman, A.J. (1982). Neuroendocrinology **35**: 429–438.
Witkin, J.W., O'Sullivan, H., and Silverman, A.J. (1995). Endocrinology **136**: 4323–4330.
Woller, M.J., and Terasawa, E. (1991). Endocrinology **128**: 1144–1150.
Woller, M.J., and Terasawa, E. (1992). Neuroendocrinology **56**: 921–925.
Woller, M.J., McDonald, J.K., Reboussin, D.M., and Terasawa, E. (1992). Endocrinology **130**: 2333–2342.
Wray, S., Nieburgs, A., and Elkabes, S. (1989A). Dev. Brain Res. **46**: 309–318.
Wray, S., Grant, P., and Gainer, H. (1989B). Proc. Natl. Acad. Sci. USA **86**: 8132–8136.
Xu, B., Sahu, A., Crowley, W.R., Leranth, C., Horvath, T., and Kalra, S.P. (1993). Endocrinology **133**: 747–754.
Xu, B., Sahu, A., Hyde, J.F., and Kalra, S.P. (1994). Soc. Neurosci. Abstr. **20**: 638.
Xu, B., Sahu, A., Kalra, P.S., Crowley, W.R., and Kalra, S.P. (1996). Endocrinology **137**: 78–84.
Zoller, R.T., and Scott Young III, W. (1988). Endocrinology **123**: 1688–1689.
Zoller, R.T., Seeburg, P.H., and Scott Young III, W. (1988). Endocrinology **122**: 2570–2577.

Comparative Endocrinology and Reproduction
K.P. Joy, A. Krishna and C. Haldar (Eds)

3. Neuroendocrine Role of Opioid Peptides

E. Vijayan and P. Sreenivasan
School of Life Sciences, Pondicherry University
Pondicherry-605 014, India

Introduction

Opiates like morphine and related compounds were of therapeutic interest for medical scientists and laymen alike since a long time. The demonstration of pharmacologically relevant opiate receptor binding sites in brain and gut (Pert and Snyder, 1973; Simon *et al.*, 1973; Terenius, 1973) and subsequent isolation and characterization of opioid peptides paved the way for renewed interest in opiate research. The first of the endogenous peptide with opioid activity, the enkephalins, was isolated from brain tissue and sequenced by Hughes *et al.*, in 1975. The enkephalins joined the growing list of neuroactive peptides being found in the mammalian brain, a list almost non-existent 25 years ago but one which today contains more than 80 peptide sequences and is still being expanded rapidly. Opioid peptides are derived from three separate genes, and distributed throughout the central nervous system (Bradbury *et al.*, 1976; Cuello, 1983), anterior and intermediate lobes of pituitary (Li and Chung,1976) and peripheral tissues such as pancreas, antrum, placenta (Feurle *et al.*, 1980; Tanaki *et al.*, 1982), testis and male reproductive tract (Tsong *et al.*, 1982) and ovary (Shaha *et al.*, 1984). The multiplicity of opioid peptides and receptors and a variety of biological functions are attributed to opioid peptides.

Chemistry of Opioid Peptides

Currently three distinct families are known of endogenous opioid peptides, derived from three distinct precursor molecules: proopiomelanocortin, prodynorphin and proenkephalin coded by three separate genes.

Proopiomelanocortin (*POMC*): This is the common precursor for β-endorphin (β-END), adrenocorticotropin (ACTH), α-melanocyte stimulating hormone (α-MSH), β-lipotropin (β-LPH), β-MSH, α-LPH and γ-MSH. There are major inter-tissue differences in the processing, post-translational modification and storage of β-END and ACTH related species. Thus in the anterior lobe of the pituitary, ACTH is the major secretory product and β LPH and β-END are present in equimolar amounts. In the intermediate lobe of the pituitary, POMC is processed to yield α-MSH and β-END. No β-END is found in the posterior pituitary. In the brain β-END synthesising neurons are confined to hypothalamic arcuate nucleus (ARC). β-END and MSH are the major peptides of POMC processing. Immunoreactive β-END has been localized in testis and ovary also (Tsong *et al.*, 1982; Shaha *et al.*, 1984; Civelli *et al.*, 1984; Watson *et al.*, 1984; Holt, 1986).

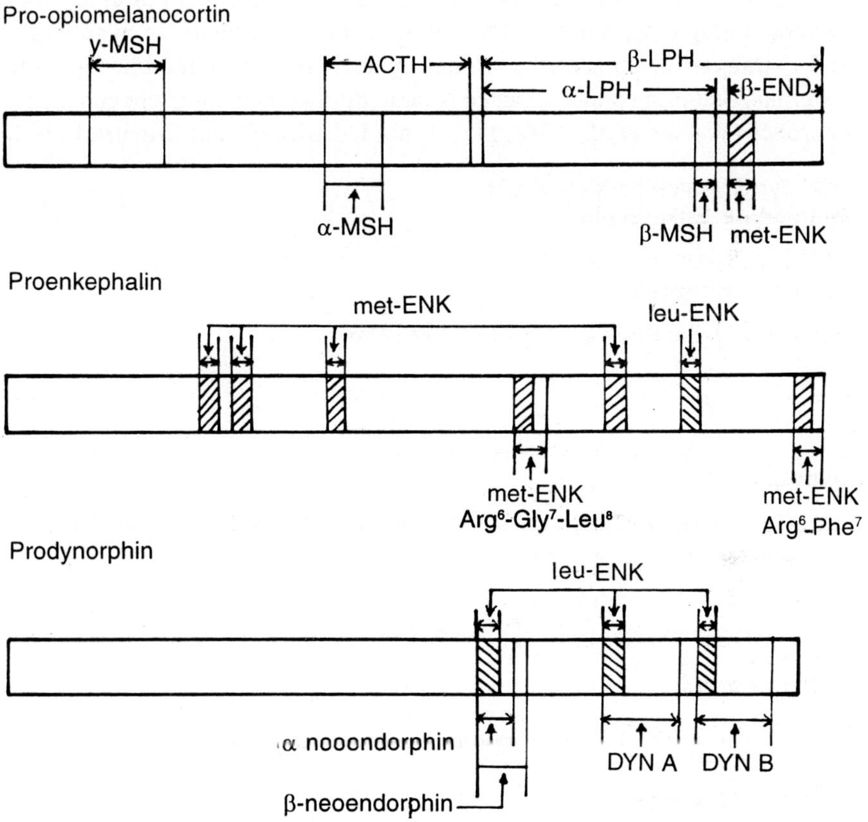

Fig. 1 Schematic representation of three opioid peptide gene products: Proopiomelanocortin (yielding β-End), proenkephalin (yielding met and leu-enkephalin) and prodynorphin (yielding dynorphin A and B, α- and β-neoendorphin).

Prodynorphin (Pro-DYN) or Pro-enkephalin B

The members of this family bear a N-terminal-located leucine-enkephalin (leu-ENK). They include dynorphin A (DYN-A, 17 amino acid residue), DYN 1-8, rimorphin (DYN B) and α- neoendorphin (α-NE). These peptides are localized in the anterior and neural lobes of the pituitary. In the brain dynorphin related peptides are found in many nuclei of hypothalamus such as suprachiasmatic nucleus (SCN), paraventricular nucleus (PVN), supraoptic nucleus (SON), arcuate nucleus (ARC), limbic system and midbrain (Watson *et al.*, 1984; Holt, 1986).

Pro-enkephalin A (Pro-ENK)

This is the precursor molecule for four copies of met-ENK, one leu-ENK, one hepta peptide (Tyr-Gly-Gly-Phe-Met-Arg-Phe) and one octapeptide (Tyr-Gly-Gly-Phe-Met-Arg-Phe-Leu). Met–ENK (Tyr-Gly-Gly-Phe-Met) and Leu-ENK (Tyr-Gly-Gly-Phe-Leu) is reported to be present in all pituitary lobes. Enkephalinergic neurones are present in the brain especially in hypothalamic and extrahypothalamic areas which are involved in endocrine regulation. External zone of median eminence is particularly abundant in enkephalinergic neurones (Watson *et al.*, 1984; Holt 1986; Udenfriend and Kilpatrick, 1984).

H2N-Tyr-Gly-Gly-Phe-Met-COOH
Methionine Enkephalin

H2N-Tyr-Gly-Gly-Phe-Leu-COOH
Leucine Enkephalin

H2N-Tyr-Gly-Gly-Phe-Leu-Arg-Arg-Ile-Arg-Pro
|
Lys
|
HOOC-Gln-Asn-Asp-Trp-Lys-Leu
Dynorphin A

H2H–Tyr-Gly-Gly-Phe-Met-Thr-ser-Glu-Lys-Ser-Gln-Thr-Pro-Leu-Val-Thr
|
Leu
|
HOOC-Glu-Gly-Lys-Lys-Tyr-Ala-Asn-Lys-Ile-Ile-Ala-Asn-Lys-Phe

Beta-Endorphin

Fig. 2 Amino acid sequences of opioid peptides.

Opioid Receptors

A natural corrollary of the heterogeneity of opioid peptides is the presence of multiple classes of opioid receptors. Pharmacological classification of opioid receptors by use of isolated tissue preparations and radioligand studies indicate that there are primarily four types of opioid receptors—μ, k, δ and ρ (Morley, 1983; Peterson *et al.*, 1984; Herz, 1987). In addition existence of receptor subtypes for μ and k types are also known (Herz, 1987).

Molecular properties of opioid receptors

Opioid receptors are localized on the plasma membrane of opioid-sensitive nerve cells. Ligand binding to the opioid receptor was shown to be stereoselective and sensitive to ionic strength, pH and temperature. Divalent and monovalent cations and guanosine triphosphate (GTP) affect differentially the binding of opioid agonists and antagonists (Blume, 1978, 1980; Sade, 1982). Occupation of receptor sites by opioid agonists and opioid peptides in intact neuroblastoma, glioma cells and cell membrane preparations leads to an inhibition of adenosine 3′, 5′ cyclic phosphate (cyclic AMP) production, via the inhibition of adenylatc cyclase activity (West and Miller, 1983). This opioid mediated inhibition of adenylate cyclase and thereby cAMP production is Na^+ and G protein dependent (Blume *et al.*, 1979). Intracisternal administration of Ca^{2+} is shown to antagonize morphine analgesia showing an adverse relationship between calcium and opiate activity (West and Miller, 1983). Opiates inhibit depolarization induced influx of calcium into nerve terminals and thereby reduce transmitter release. Opiates also inhibit uptake of Ca^{2+} into nerve terminals.

Studies indicate reduced protein kinase (phosphorylating enzymes) activity in microsomes from whole brain in response to opiate treatment. In vitro incorporation of ^{32}P into synaptic plasma membrane was also reduced following opiate treatment (West and Miller, 1983). Thus a reduction in calcium release, phosphorylation of proteins by reduced protein kinase activity and cAMP activity are the mechanisms by which opiates inhibit cellular mechanisms such as neurotransmitter release. The above responses may be acting alone or in a sequence. This is supported by the report that adenyl cyclase activity of the brain is stimulated by calmodulin, a Ca^{2+} dependent regulatory protein (Lynch *et al.*, 1977).

Physiological role of opioids

Opioids have been involved in various functions. The most well studied are their role in pain modulation and neuroendocrine regulation. Recent research indicates that opioids may be involved in paracrine regulation of cellular functions in mammalian reproductive organs.

Opioids in pain modulation

Relief from pain represents the most characteristic pharmacological effect of morphine. There are two principal sites of opioidergic modulation of pain; the brain and the spinal cord (Basbaum and Fields, 1984; Yaksh and Noneihed, 1985; Miller, 1986). The midbrain and pons are important sites of action of opioids on pain relief. Microinjection of minute amounts of morphine into the ventral periaqueductal grey (PAG) induces complete antinociception in experimental animals. In humans, local injection into the lower parts of the ventricular system results in full analgesic effect. The electrical stimulation of the above areas resulted in antinociception in animals and pain relief in man. This effect is partly antagonized by naloxone, an opioid receptor antagonist, indicating opioidergic mediation of the effects of stimulation. The receptor types involved are mainly μ and k with lesser density of δ. The endogenous opioids β-endorphin, dynorphin and enkephalins are present in these sites (Yaksh and Noneihed, 1985).

A high density of opioid receptors was found in the spinal cord, particularly in the grey matter. High receptor density was seen in the dorsal horn and specifically in the marginal cell zone and in the dorsal layer of the substantia gelatinosa. Physiological experiments showed that opiates specifically block the transmission of nociceptive stimuli in the dorsal horn of spinal cord when they are applied locally (Atweh and Kuhar, 1983). In addition to activating descending inhibitory influences on the spinal cord, opiates exert direct influences on the spinal mechanisms involved in nociception. Opioid receptors of the spinal cord are located presynaptically suggesting effect on neurotransmitter release. Consistent with this idea, opiates have been known to inhibit the release of Substance P in rat trigeminal nucleus (Jessen and Iverson, 1977). Substance P is the neurotransmitter that is released by some afferent C-fibers involved in nociception.

Opioids in neuroendocrine regulation

The observation that morphine inhibits the secretion of pituitary gonadotropins was reported more than 20 years prior to the discovery of endogenous opioids (Barraclough and Sawyer, 1955). Opioid peptides of the three classes derived from three distinct genes are now known to be influencing release of other hormones from the anterior and posterior pituitary.

Gonadotropin release

Most of the reports suggest that endogenous opioid peptides (EOPs) are inhibitory to luteinizing hormone (LH) release from the pituitary (Cicero *et al.*, 1979; Kalra *et al.*, 1989; Sreenivasan, 1994) although under certain experimental conditions they may also stimulate GnRH and LH release (Kalra *et al.*, 1986; Piva *et al.*, 1986). Opiate antagonists such as naloxone were found to be effective in stimulating LH release in intact male and cycling female rats (Cicero *et al.*, 1979; Piva *et al.*, 1985). In gonadectomised rats, intraventricular injection of β-END inhibited LH release. Continuous infusion of β-END for 3h markedly suppressed the amplitude and frequency of LH episodes in long-term ovariectomised rats. Dynorphin also had LH suppressive effect while met-ENK and leu-ENK produced divergent effects (Leadem and Kalra,1983).

Site of opioid action on gonadotropin release is the hypophysiotrophic areas in the hypothalamus. Naloxone implants as well as intracerebral injection of naloxone in the narrow medial zone extending caudally from medial preoptic area (MPOA) to the median eminence-arcuate nucleus (ME-ARC) readily stimulated LH release (Kalra, 1983). A number of reports suggest the involvement of a variety of neurotransmitters in the mediation of opioid inhibition of LH release. Studies by Kalra and co-workers (1984, 1989) demonstrated that concomitant with GnRH release, naloxone evoked release of noradrenaline and adrenaline from the hypothalamus. Diez-Guerra *et al.* (1986) showed that morphine attenuated the electrically stimulated release of noradrenaline from POA slices *in vitro*. Further, naloxone induced LH release was inhibited by prior blockade of post-synaptic adrenergic receptors (Kalra, 1981). These results suggested that endogenous opioids may communicate with GnRH system via the adrenergic neurons. Hypothalamic GABA-ergic system is known to be involved in the GnRH and LH release (Vijayan and McCann, 1978; Negro Vilar *et al.,* 1980; McCann *et al.,* 1981). Recent findings suggest that opioids may be exerting the inhibitory response on gonadotropin release via the GABA-ergic system (Masoto *et al.,* 1989). GABA-ergic system may affect gonadotropin secretion by influencing GnRH release from arcuate nucleus-median eminence (AN-ME). Activation of different types of GABA-ergic receptors, GABA-A and GABA-B, elicits either stimulation or inbibition of GnRH release (Masoto *et al.,* 1989). Activation of GABA-β receptors increases membrane conductance of potassium and generates a slow inhibitory postsynaptic potential. This type of receptor is selectively activated by high concentrations of GABA or by GABA-β agonist baclofen. Baclofen was reported to reduce the turnover rate of norepinephrine in estrogen-progesterone treated rats, an effect that is responsible for the inhibition of LH surge in these animals after baclofen treatment (Adler and Crowley, 1986). Thus the adrenergic system which probably interacts with GnRH fibers and terminals in the AN-ME area is modulated by GABA-ergic system through GABA-B receptors. Dopaminergic system is also reported to influence opioid peptide release. Dopamine stimulates human mediobasal hypothalamus (MBH) β-END release in vitro (Rasmussen *et al.,* 1987). Endogenous and exogenous dopaminergic stimulation appears to increase rat MBH POMC gene expression (Weihmuller *et al.,* 1991). In ovariectomised rats, MBH POMC and tyrosine hydroxylase (rate limiting enzyme in dopamine synthesis) mRNA levels are positively correlated within individual animals and in response to estrogen treatment (Ramussen *et al.,* 1990). Thus activation of this pathway may be one of the mechanisms by which increased prolactin levels suppress gonadotropin release (Sarkar and Yen, 1985; Foreman and Porter, 1981). From the above discussion, it can be concluded that modulation of gonadotropin secretion by opioids involves one or more neurotransmitter systems acting on the hypothalamus.

Prolactin

Predominant effect of both opiate alkaloids and opioid peptides in the human is to stimulate prolactin release (Delitalia *et al.,* 1983; Pende *et al.,* 1986). In the rat morphine could induce lactation. This effect is also shown by met-ENK, leu-ENK and β-endorphin (Bruni *et al.,* 1977). Naloxone is able to lower base-line levels of prolactin and suppress the release of prolactin in response to painful stress and suckling (Van Vugt and Meites, 1980). The great majority of reports on the effect of naloxone on serum prolactin in the human have shown no evidence for EOP modulation of prolactin release. The only clear cut evidence for a role of endogenous opioids in modulating prolactin release was shown in professional atheletes undergoing acute severe exercise (Moretti, 1983). Evidence accrued from studies in man and experimental animals suggest that opioid peptides/morphine modulate prolactin release via inhibition of tubero-infundibular dopaminergic pathway (Bentley and Wallis 1986; Bouloux *et al.,* 1986). The receptors involved are the hypothalamic μ and possibly k-receptors (Delitalia *et al.,* 1983).

Release of other pituitary hormones such as thyrotropin (TSH) and growth hormone are also influenced by opiates and opioid peptides. The acute effect of opioids in the rat is to depress circulating TSH, which is reversed by naloxone (Judd and Hedge, 1983). The effect of opioids on TSH release is thought to be mediated at the hyothalamic level (Lomax *et al.,* 1970; Muraki *et al.,* 1980) while a direct effect on the pituitary was also reported (May *et al.,* 1979). It is suggested that opioid control of TSH release is mediated via an inhibition of hypothalamic dopamine (Delitalia, 1981). In man, predominant effect of exogenous opioids is the significant elevation of circulating TSH (Grossman *et al.,* 1981). Both opiates and opioid peptides stimulate the release of growth hormone in the rat. This response is naloxone sensitive (Bruni *et al.,* 1977) as naloxone alone decreases circulating growth hormone levels. A great variety of neurotransmitters have been implicated in the mediation of opioid effect on growth hormone levels. Thus growth hormone releasing effect of opioids has been reported to be mediated by adrenergic system (Katakami *et al.,* 1981a; Arce *et al.,* 1991), GABA-ergic system (Katakami *et al.,* 1981b), histamine (Rivier *et al.,* 1977), cholinergic and dopaminergic pathways (Cocchi *et al.,* 1980). But none of these reports conclusively proved which particular transmitter was the sole mediator of opioid action.

In man opioids are not considered to be important modulators of growth hormone secretion. A variety opioid agonists elevate ACTH and corticosterone in the rat after acute administration (Haraez *et al.,* 1981). Chronic morphine treatment leads to the suppression of ACTH release. In man, both opioid peptides and opiates suppress pituitary-adrenal axis (Delitalia *et al.,* 1981). Physiological role of endogenous opioid tone appears to be involved in the ACTH response to stress, via an adrenergic pathway (Grossman *et al.,* 1982). Opioids have an inbibitory effect on oxytocin secretion. They act both upon the secretory nerve terminals in the neurohypophysis and upon the electrical discharge activity of oxytocin cell bodies in the hypothalamus (Leng and Russel, 1989).

Pathophysiology of Opioids

Opioid peptides of POMC class have been localized in reproductive organs of rat and mouse (Sharp *et al.,* 1980; Tsong *et al.,* 1982; Margioris *et al.,* 1983; De Bold *et al.,* 1988). In the rat testis, Leydig cells are the sites of synthesis of β-END and opioid receptors have been localized on Sertoli cells (Fabri *et al.,* 1985, 1988). These observations have suggested that POMC derived opioid peptides have local paracrine and autocrine effects on testicular function including suppression of Sertoli cell growth during the quiescent state of testicular development and modulation of rat androgen binding protein (rABP) and testosterone secretion in the adult (Fabri *et al.,*

1988; Gerendai *et al.*, 1986). Opioids have been recently shown to inhibit testicular enzymes: lactate dehydrogenase and sorbitol dehydrogenase activity in adult rats (Sreenivasan and Vijayan, 1996). This is significant since lactate and pyruvate secreted by Sertoli cells are the energy source for the developing spermatids (Jute *et al.*, 1983). β-END was also reported to be present in the rat, mice and human ovary (Wai-Sum, 1990). It has a possible role in the oocyte maturation, as the level of immunoreactive β-END is lowered following gonadotropin surge (Tam *et al.*, 1988). Further, β-END inhibited spontaneous germinal vesicle breakdown in isolated rat oocytes in culture (Wai-Sum, 1990). This inhibition could be reversed by naloxone, an opioid receptor antagonist.

A wealth of data and particularly the recent hypothesis that the opioid peptides act in the brain to mediate the effects of sex steroid hormones on both the secretion of gonadotropins and on the expression of sexual behaviour have unambiguously demonstrated that opioid peptides and opiates are involved in many neuroendocrine processes. The potential importance of opiates in the CNS and neuroendocrine function is a new area currently attracting considerable interest and research activity.

Acknowledgements

Financial support from the Department of Science and Technology, Govt. of India and the University Grants Commission is greatfully acknowledged.

References

Adler, B.A. and Crowley, W.R. (1986) Endocrinology 118, 91–95.

Arce, V.M., Cella, S.G., Locatelli, V. and Muller, E.E. (1991). J. Neuroendo. 3, 357–361.

Atweh, S.F. and Kuhar, M.J. (1983) Br. Med. Bull. 39, 47–52.

Barraclough, C.A. and Sawyer, C.H. (1955) Endocrinology, 57, 329–337.

Basbaum, A.I. and Fields, L. (1984) Ann. Rev. Neurosci. 7, 309–338.

Bentley, A.M. and Wallis, M., (1986) Jl. Endocrinol. 109, 313–320.

Blume, A.J. (1978) Proc. Natl. Acad. Sci. USA 75, 1713–1717.

Blume, A.J., Lichstein, D. and Boone, G. (1979) Proc. Natl. Acad. Sci. USA. 76, 5626–5630.

Blume, A.J. (1980) In: Receptors for Neurotransmitters and peptide hormones pp. 35, Raven Press, New York.

Bouloux, P.M.G., Grossman, A., Alloio, B., Delitalia, G. and Besser, G.M. (1986) Hor. Res. 23, 83–90.

Bradbury, A.F., Smyth, D.G. and Snell, C.R. (1976) Biochem. Biophys. Res. Cummun. 69: 950–956.

Bruni, J.F., Van Vugt, D.A., Marshall, S. and Meites, J. (1977) Life Sci. 21, 461–466.

Cicero, T.J., Shainker, B.A. and Meyer, E.R. (1979) Endocrinology 104, 1286–1291.

Civelli, O., Douglas, J. and Herbert, J. (1984) In: The Peptides vol. 6 (ed: S. Udenfriend and S. Meienhofer) pp. 70–90, Academic Press, London.

Cocchi, D., Casanueva, F., Betti, R., Zanardi, P., Motta, T. and Mantegazza, P. (1980) In: Neuroactive drugs in Endocrinology (ed: E.E. Miller) pp. 191–200, Elsevier/North Holland Biomedical Press, Amsterdem.

Cuello, C.A. (1983) Br. Med. Bull. 39, 11–16.
De Bold, C.R., Nicholson, W.E. and Orth, D.M. (1988) Endocrinology 122, 2648–2657.
Delitalia, G., Grossman, A. and Besser, G.M. (1981) Life Sci. 29, 1537–1544.
Delitalia, G., Grossman, A. and Besser, G.M. (1983) Neuroendocrinology 37, 275–279.
Diez Guerra, F.J., Augood, S., Emson, P.C., Dyer, R.G. (1986) Neuroendocrinology, 43, 89–91.
Fabbri, A., Tsai-Morris, L.H., Luna, S., Fraioli, F. and Dufau, M.L. (1985) Endocrinology 117, 2544–2546.
Fabbri, A., Knox, G., Buczko, E. and Dufau, M.L. (1988) Endocrinology 122, 749–755.
Feurle, G.E., Weber, U. and Helmstaedter, V. (1980) Life Sci. 27, 467–473.
Foreman, M.M. and Porter, J.C. (1981) Endocrinology 108, 800–804.
Gerendai, I., Shaha, C., Gunsalus, G.L. and Bardin, C.W. (1986) Endocrinology 118, 2039–2044.
Grossman, A., Stubbs, W.A., Gaillard, R.C., Delitalia, G., Rees, L.H. and Besser, G.M. (1981) Clin. Endocrinol. 14, 381–386.
Grossman, A., Gaillard, R.C., McCartney, P., Rees, L.H. and Besser, G.M. (1982) Clin. Endocrinol. 17, 279–286.
Haraez, J.L., Bloom, A.S., Wang, R.I.H. and Tsang, L.F. (1981), Neuroendocrinology 33, 170–175.
Herz, A. (1987) In: Trends in Medical Chemistry (eds. E. Mustchler and E. Winterfeld) pp. 337–349, Verlag Chemie, Weinheim.
Holt, V. (1986) Ann. Rev. Pharmacol. Toxicol. 26, 59–77.
Hughes, J., Smith, T.W., Kasterlitz, H.W., Fothergill, L.A., Morga, B.A. and Morris, H.R. (1975) Nature (London) 258, 577–579.
Jessel, J.M. and Iverson, L.L. (1977) Nature (London) 268, 549–551.
Judd, A.M. and Hedge, G.A. (1983) Endocrinology 113: 706–710.
Jute, N.H.P.M., Jansen, R., Grootegoed, J.A., Rommerts, F.F.G. and Van der Molen, H.J. (1983) J. Reprod. Fertl. 68, 219–226.
Kalra, S.P. (1981) Endocrinology 109, 1805–1810.
Kalra, S.P. (1983) In: Role of peptides and proteins in control of reproduction (eds. S.M. McCann and D.S. Dhindsa) pp. 63–87, Elsevier, New York.
Kalra, S.P. and Leadem, C.A. (1984) In: Opioid modulation of Bendocrine function (eds. G. Delitalia, M. Motta and M. Serio) pp. 171–184, Raven Press, New York.
Kalra, P.S., Crowley, W.R. and Kalra, S.P. (1986) Endocrinology 120, 178–185.
Kalra, S.P., Allen, L.G. and Kalra, P.S. (1989) In: Brain Opioids Systems in Reproduction (eds. R.G. Dyer and R.J. Bicknell) pp. 95–111, Oxford University Press, New York,
Katakami, H., Kato, Y., Matsushita, Y., Shimatsu, A. and Imura, H. (1981a) Endocrinology 109, 1033–1036.
Katakami, H., Kato, Y., Matsushita, Y., Hiroto, S., Shimatsu, A. and Imura, H. (1981b) Neuroendocrinology 33, 129–135.
Leadem, C.A. and Kalra, S.P. (1983) Neuroendocrinology 41, 342–352.
Leadem, C.A., Crowley, W.R., Simpkins, J.W. and Kalra, S.P. (1985) Neuroendocrinology 40, 497–500.
Leng, G. and Russel, J.A. (1989) In: Brain Opioid Peptides in Reproduction (eds. R.G. Dyer and R.J. Bicknell) pp. 231–256, Oxford University Press, New York.
Li, C.H. and Chung, D. (1976) Proc. Natl. Acad. Sci. USA 73, 1145–1148.
Lomax, P., Kokka, N. and George, R. (1970) Neuroendocrinology 6, 146–150.
Lynch, T.J., Tallant, E.A. and Cheung, W.X. (1977) Arch. Biochem. Biophys. 182, 124–133.
Margioris, A.N., Liotta, A.S., Vaudry, H., Bardin, C.W. and Krieger, D.T. (1983) Endocrinology 113, 663–667.
Masotto, C., Wisniewski, G. and Negro-Vilar, A. (1989) Endocrinology 125, 548–552.

May, P., Mittler, J., Manousian, A. and Ertel, N. (1979) Horm. Metab. Res. 11, 30–34.
McCann, S.M., Vijayan, E. and Negro-Vilar, A. (1981) In: GABA and Benzodiazepem receptors (eds. E. Costa, G. Chiara, G. Gessa) pp. 237–246, Raven Press, New York.
Millan, M.J. (1986) Pain 27, 303–347.
Moretti, C. (1983) Clin. Endocrinol. 18, 135–138.
Morley, J.S. (1983) Br. Med. Bull. 31, 5–10.
Muraki, T., Nakadate, T., Tokunaga, Y. and Kato, R. (1980) J. Endocrinol. 86, 357–361.
Negro-Vilar, A., Vijayan, E. and McCann, S.M. (1980) Brain Res. Bull. 5: 239–244.
Peterson, S.J., Robson, L.E. and Kosferlitz, H.W. (1984) In: The Peptides vol. 6 (eds. S. Udenfriend and J. Meienhofer) pp. 148–149, Academic Press, London.
Pende, A., Musso, N.R. Montaldi, M.L., Postorino, G., Arzese, M. and Devilla, L. (1986) Biomedicine and Pharmacotherapy 40, 178–182.
Pert, C.B. and Snyder, S.H. (1973) Science 179, 1011–1014.
Piva, F., Maggi, R., Limonta, R. Motta, M. and Martini, L. (1985) Endocrinology 11, 766–772.
Piva, F., Limonta, P., Maggi, R. and Martini, L. (1986) Neuroendocrinology 42, 504–512.
Rasmussen, D.D., Liu, J.H., Wolf, P.L. and Yen, S.S.C. (1987) Neuroendocrinology 45, 197–201.
Rasmussen, D.D., Jakubowski, M., Allen, D.L. and Roberts, J.L. (1990) Proc. Soc. Neurosci. Mtg. 16, 1202 (Abstr.).
Rivier, C., Brown, M. and Vale, W. (1977) Endocrinology 100, 751–754.
Sade, W., Pfiiffer, A. and Herz, A. (1982) J. Neurochem. 39, 659–667.
Sarkar, D.K. and Yen, S.S.C. (1985) Endocrinology 116, 2080–2085.
Sharp, B., Pekary, E.A., Meyer, N.V. and Hershman, J.M. (1980) Biochem. Biophys. Res. Comm. 95, 618–623.
Shaha, C., Margioris, A., Liotta, A.S, Krieger, D.T. and Bardin, C.W. (1984) Endocrinology 115, 378–384.
Simon, E.J., Hiller, J.M. and Edelmann, I. (1973) Proc. Natl. Acad. Sci. USA 70, 1947–1949.
Sreenivasan, P. (1994) Ph. D thesis, Pondicherry University, Pondicherry, India.
Sreenivasan, P. and Vijayan, E. (1996). Ind. J. Exp. Biol. 34, 363–365.
Tam, P.P.L., Ng, T.B. and Mao, K.R. (1988) In vitro Fertil. Embryo Transfer 5, 91–95.
Tanaki, I., Yoshikatsu, N., Kazuwa, N., Shogo, O., Masaki, N., Ohtsuki, H. and Imura, H. (1982) J. Clin. Endocrinol. Metab. 54, 392–396.
Terenius, L. (1973) Acta. Pharmacol. Toxicol. 32, 317–320.
Tsong, D.S., Phillips, D.M., Halmi, N., Liotta, A.S., Margioris, A. and Krieger, D.T. (1982a), Endocrinology 110, 2204–2206.
Tsong, S.D., Phillips, D.M., Halmi, N., Krieger, D. and Bardin, C.W. (1982b) Biol. Reprod. 27, 755–764.
Udenfriend, S. and Kilpatrick, L. (1984) In: The Peptides,Vol. 6 (eds. S. Udenfriend and J. Meienhofer) pp. 26–68, Academic Press, London.
Van Vugt, D.A. and Meites, J. (1980) Fed. Proc. 39, 2533–2538.
Vijayan, E. and McCann, S.M. (1978) Brain Res. 155, 35–39.
Wai-Sum, O. (1990) Mol. Cell. Endocrinology 68, 181–185.
Watson, S.J., Akil, H., Khachaturian, H., Young, E. and Lewis, M.E. (1984) In: Opioids past present and future (eds. J. Hughes, H.O.J. Collier and M.J. Rance) pp. 145–178 Taylor and Francis, London.
Weihmuller, F.B., O'dell, S.J., Cole, B.N. and Marshall, J.F. (1991) Brain Res. 549, 230–235.
West, R.E. and Miller, R.J. (1983) Br. Med. Bull 39, 53–58.
Yaksh, T.L. and Noneihed, R. (1985) Ann. Rev. Pharmacol. Toxicol. 25, 433–462.

Comparative Endocrinology and Reproduction
K.P. Joy, A. Krishna and C. Haldar (Eds)

4. The Hypothalamus of Teleosts

N.K. Subhedar, F.A. Khan, S.G. Saha, V.S. Burade and S. Sarkar
Department of Pharmaceutical Sciences, Nagpur University Campus, Nagpur-440 010, India

Although hypothalamus is an integral part of the neural machinery and cannot be treated as an isolated entity, in view of its unsurpassed importance as a nodal link integrating diverse signals, and as a seat of vegetative and neuroendocrine regulation, it has attracted neurobiologists over decades. While the teleostean hypothalamus conforms to the general vertebrate pattern, certain unique features like the presence of inferior lobes, horizontal commissure, invasive nature of lateral recesses of the third ventricle, absence of median eminence and occurrence of direct adenohypophysial innervation make it a structure calling for the special attention. While rostrally the hypothalamus is demarcated by the anterior commissure, the posterior recesses are generally considered as the caudal limits and lateral forebrain bundles constitute the lateral margins. The demarcation from thalamus is less clear; third ventricular sulcus is sometimes considered as the separating line. The preoptic area, tuberal area and inferior lobes constitute major anatomical subdivisions of the hypothalamus. In addition to some of the well defined nuclei like the nucleus preopticus (NPO) and nucleus lateralis tuberis (NLT), the teleostean hypothalamus consists of homogenous and poorly differentiated neuronal populations crowded along the border of the third ventricle or its recesses; the lateral hypothalamic areas are relatively devoid of neuronal perikarya. Cytoarchitectonic pattern of the hypothalamus has been studied in a limited number of teleosts (Peter *et al.*, 1975; Billard and Peter, 1982; Braford and Northcutt, 1983; Fernald and Shelton, 1985; Striedter,1990; Maler *et al.*, 1991; Rama Krishna and Subhedar, 1991) (Figs. 1 to 3).

The hypothalamus is the seat of neuroendocrine regulation. Unlike higher vertebrates, the median eminence is absent in teleosts and the hypothalamic neurons directly innervate the pituitary cell types. The teleostean hypothalamo-hypophysial system therefore offers a simple model to determine the anatomical relationship between the specific neurons, the bioactive substances present in them and the pituitary cell types on which they act. A study of this arrangement in fish is important since it provides clues for determining nature of neuroendocrine control that may be operative in other vertebrates.

Conventionally, the hypothalamic nuclei have been divided into the magnocellular (Gomori-positive) and parvicellular (Gomori-negative) groups. While the former consists of the NPO, the latter consists of several poorly defined groups, extending throughout the hypothalamus. It may be recalled that the discovery of NPO in the cyprinid *Phoxinus laevis* gave birth to the concept of neuroendocrine cell (Scharrer, 1928); an event that ushered in the science of neuroendocrinology. The NPO of teleosts, homologous with the nucleus supraopticus and nucleus paraventricularis of amniotes, attracts

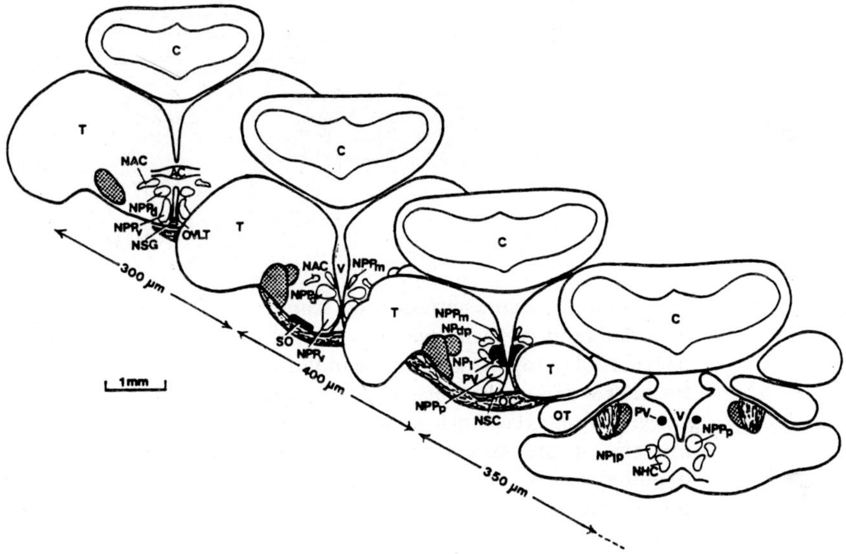

Fig. 1

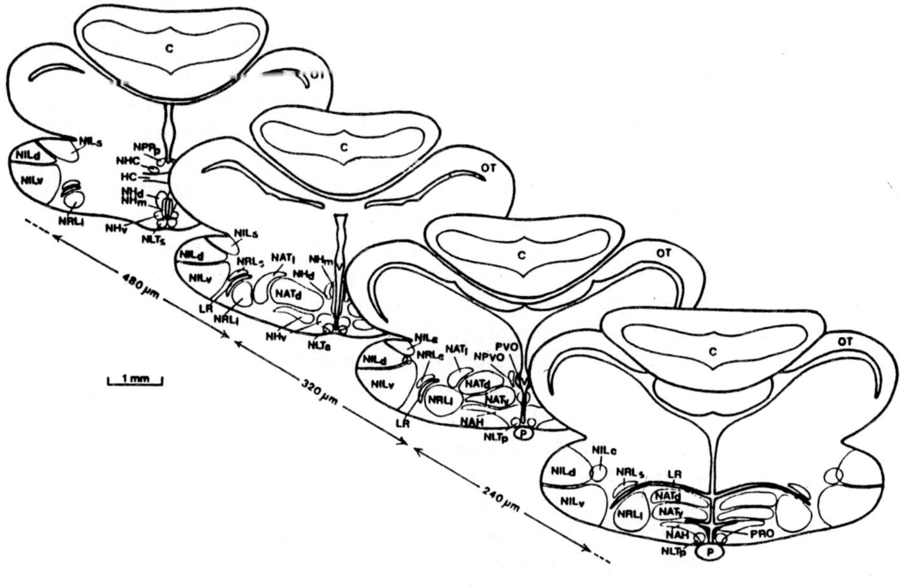

Fig. 2

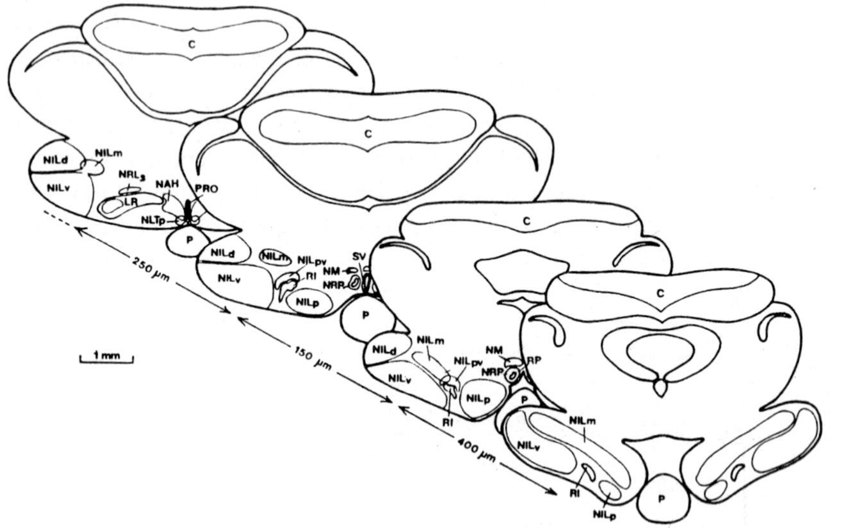

Figs. 1 to 3 **Diagrammatic representation of transverse sections through the hypothalamus of *C. batrachus*, showing magnocellular (dark) and parvocellular (open) neuronal complexes. Stippled area represents the lateral and medial forebrain bundles. AC: anterior commissure; C: cerebellum; HC: horizontal commissure; LR: lateral recess; NAC: nucleus of the anterior commissure; NAH: nucleus arcuatus hypothalamicus; NATd: nucleus anterior tuberis dorsalis; NATl: nucleus anterior tuberis lateralis; NATv: nucleus anterior tuberis ventralis; NHC: nucleus of the horizontal commissure; NHd: nucleus hypothalamicus dorsalis; NHm: nucleus hypothalamicus medialis; NHv: nucleus hypothalamicus ventralis; NILa: nucleus lobi inferioris anterior; NILc: nucleus lobi inferioris centralis; NILd: nucleus lobi inferioris dorsalis, NILm: nucleus lobi inferioris medialis; NILp: nucleus lobi inferioris posterior; NILpv: nucleus lobi inferioris periventricularis; NILv: nucleus lobi inferioris ventralis; NLTa: nucleus lateralis tuberis pars anterior; NLTp: nucleus lateralis tuberis pars posterior; NM: nucleus mamillaris; NPdp: nucleus preopticus dorsalis; NPL: nucleus preopticus lateralis; NPLp: nucleus postopticus lateralis; NPO: nucleus preopticus; NPPd: nucleus preopticus periventricularis dorsalis; NPPm: nucleus preopticus periventricularis medialis; NPPp: nucleus preopticus periventricularis posterior; NPPv: nucleus preopticus periventricularis ventralis; NPVO: nucleus of the paraventricular organ; NRLi: nucleus recessus lateralis inferior; NRLs: nucleus recessus lateralis superior; NRP: nucleus recessus posterioris; NSC: nucleus suprachiasmaticus; NSG: nucleus of the subventricular gray; OC: optic chiasma; OT: optic tectum; OVLT: organum vasculosum laminae terminalis; P: pituitary; PRO: posterior recess organ; PV: paraventricular division of the NPO; PVO: paraventricular organ; RI: recess of the inferior lobes; RP: recessus posterioris; SO: supraoptic division of the NPO: SV: saccus vasculosus; T: telencephalon; V: third ventricle (reproduced from Rama Krishna and Subhedar, 1991 by copyright permission of the Akademie Verlag Berlin).**

attention because of large size of somata, affinity for the selective neurosecretory stains like aldehyde fuchsin and the conspicuous efferent pathway to the pituitary gland. It is generally distinguishable into dorsal pars magnocellularis consisting of large cells and the ventral pars parvocellularis having smaller somata (Peter, 1977), although the pattern may vary (Subhedar *et al.*, 1990). Among the parvicellular goups, the NLT is distinctive because of large cells in the floor of the tuberal area that also project to the pituitary gland (Terlou and Ekengren, 1979; Rama Krishna and Subhedar, 1989). While other nuclear groups like the nucleus preopticus periventricularis, nucleus suprachiasmaticus and nucleus anterior tuberis are relatively well defined, remaining conglomerations of cells are poorly delineated and debate regarding their nomenclature and homology persists (Rama Krishna and Subhedar, 1991).

Neural Connections of the Hypothalamus

The hypothalamus occupies a strategic position with reference to other brain areas; it is closely associated with or serves as a passage for some of the major pathways of the brain. While the medial and lateral forebrain bundles constitute the bilateral trunk routes in the anterior hypothalamus, and serve as a conduit for ascending and descending fibers of the hypothalamus, the preoptico-hypophysial tract serves as an exit passage to the pituitary. One of the major commissural pathways of the hypothalamus is the anterior commissure; it is the component of the limbic system that serves for reciprocal innervation of the olfactory lobes. The optic chiasma carries visual signals to the preoptic area and the suprachiasmatic nucleus. Horizontal commissure has been identified as the forerunner of the supraoptic decussation of higher vertebrates that connects the two diencephalic halves.

While Levine and Dethier (1985) described olfactory projections to the nucleus preopticus in goldfish, Rooney *et al.* (1989) demonstrated the projections to the preoptic area and the nucleus posterior tuberis in *Gnathonemus petersii.* Electrical stimulation of the olfactory system triggered neurosecretory changes in the NPO (Kandel, 1964; Peter and Gorbman, 1968). In addition to these, ganglion cells of nervus terminalis (or its equivalent nucleus olfactoretinalis in fish with sessile olfactory bulbs) located in the olfactory system, are also found to be equipped with hypothalamic and also hypophysial connections (Halpern-Sebold and Schreibman, 1983; Rama Krishna *et al.*, 1992; Anglade *et al.*, 1993) (Fig. 4).

The retinal projections have been demonstrated to the nucleus suprachiasmaticus (Fernald, 1980; Springer and Mednick, 1985), nucleus preopticus periventricularis (Von Bartheld and Meyer, 1988) and NPO magnocellularis (Vanegas and Ito, 1983). In addition, neurons with retinopetal projections were observed in the preoptic area in *Novodon modestus* (Uchiyama, 1989), although these cells were absent in the goldfish (Uchiyama and Ito, 1984). Visual pathways leading from retinal ganglion cells through pretectal nuclei to the inferior lobes of the hypothalamus have been described (Wullimann and Meyer, 1990).

Several studies have shown that telencephalon projects fibers to the hypothalamus (see Airhart *et al.*, 1988). Some information on the connectivities between cerebellum and the inferior lobes is also available (Morita *et al.*, 1983; Ito and Yoshimoto, 1990). While spinal projections from the NPO neurons have been reported in some fish (Demski and Sloan, 1985; Prasada Rao *et al.,* 1987) stretch sensory projections from the ovary to the NPO have been described in the catfish (Subhedar *et al.*, 1987; 1996; Deshmkh *et al.,* 1992).

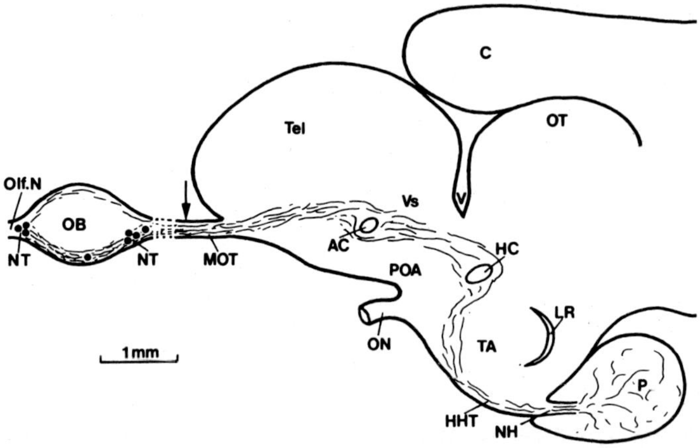

Fig. 4 Diagrammatic representation based on para-sagittal sections through the brain showing the organization of FMRF amide-immunoreactive neurons (black circles) of nervus terminalis (NT) and their processes innervating the pituitary gland (P) in *C. batrachus*. AC: anterior commissure; C: cerebellum; HC: horizontal commissure; HHT; hypothalamo-hypophysial tract; LR: lateral recess; MOT: medial olfactory tract; NH: neurohypophysis; OB: olfactory bulb; Olf. N: olfactory nerve; ON: optic nerve; OT: optic tectum; POA: preoptic area; TA: tuberal area; Tel: telencephalon; V: third ventricle; Vs, area ventralis telencephali pars supracommissuralis, (reproduced from Rama Krishna *et al.*, 1992 by copyright permission of the Academic Press, Inc.).

Although the preoptico-hypophysial pathway in teleosts is one of the best studied entities, it is only in the recent years that the distribution of the hypophysiotropic neurons was investigated using neuronal tracer injected via the pituitary gland (Rama Krishna and Subhedar, 1989; Anglade *et al.,* 1993; Prasada Rao *et al.,* 1993). In addition to the NPO and NLT, the fibers originating from the nucleus preopticus periventricularis, nucleus of the horizontal commissure, nucleus hypothalamicus ventralis, nucleus arcuatus hypothalamicus and some other groups were found to terminate in the pituitary (Rama Krishna and Subhedar, 1989) (Fig. 5).

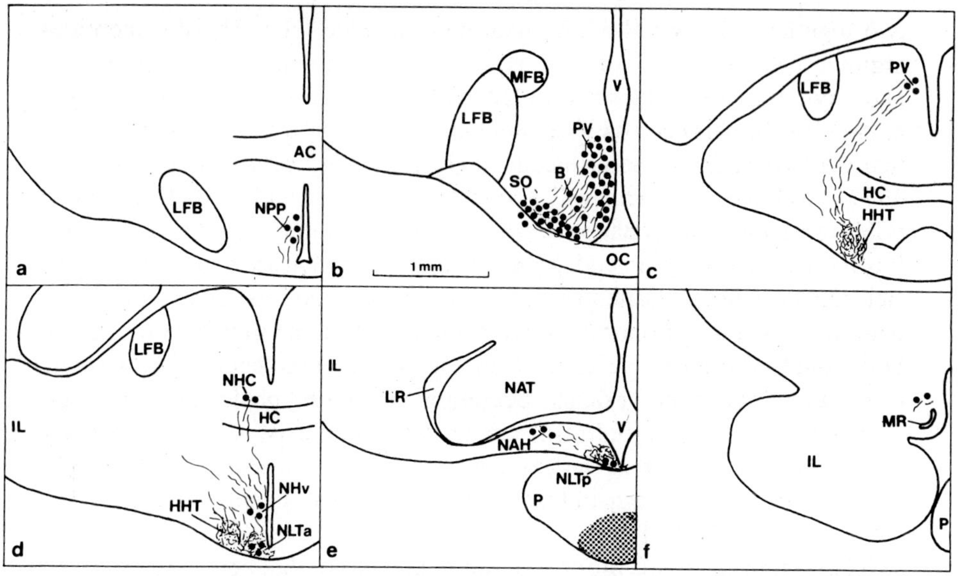

Fig. 5 A series of drawings that illustrate the distribution of retrogradely labelled neurons (black dots) and fibers in the hypothalamus of *Clarias batrachus* after neuronal tracer horseradish peroxidase administration in the pituitary gland. The drawings are arranged from rostral (a) to caudal (f). The stippled area represents the intrahypophysial site of HRP administration. Note the extensive labelling in the paraventricular group (PV), supraoptic group (SO) and intermediate bridge cells (B) of the nucleus preopticus. AC: anterior commissure; HC: horizontal commissure; HHT: hypothalamo-hypophysial tract; IL: inferior lobes; LFB: lateral forebrain bundles; LR: lateral recess; MFB: medial forebrain bundles; MR: mamillary recess; NAH: nucleus arcuatus hypothalamicus; NAT: nucleus anterior tuberis: NHC: nucleus of the horizontal commissure; NHv: nucleus hypothalamicus ventralis; NLTa: nucleus lateralis tuberis pars anterior; NLTp: nucleus lateralis tuberis pars posterior; NPP: nucleus preopticus periventricularis; OC: optic chiasma; P: pituitary; V: third ventricle (reproduced from Rama Krishna and Subhedar, 1989 by copyright permission of the Elsevier Science).

Endocrine functions of the hypothalamus

The teleostean hypothalamo-hypophysial system is unique among vertebrates, as there is a direct innervation of the pars distalis by the neurosecreory neurons of the hypothalamus and the typical median eminence of tetrapods is absent. The direct innervation from the hypothalamus might either play a role in the control of the pituitary endocrine cells or discharge their secretions in the capillaries (for review see Batten and Ingleton, 1987; Peter *et al.*, 1990). Earlier studies on the fish neuroendocrine system deal extensively with the nucleus preopticus-neurohypophysial system consisting of selectively stainable cells equipped with axonal projections to the pituitary gland (Sathyanesan and Jose, 1975). Within the pituitary gland, the neurohypophysial fibers interdigitate with the pars intermedia of the adenohypophysis and the two together are termed as the neurointermediate lobe. The nonapeptides arginine vasotocin and isotocin, equivalent of the mammalian vasopressin and oxytocin, are synthesized in the NPO, transported to the pituitary and released by the pars nervosa. The immunocytochemical techniques demonstrate that the two neurohormones occur in separate NPO cells (Goossens *et al.*, 1977); arginine vasotocin was found to colocalize with growth hormone (GH) releasing factor (Batten *et al.*, 1990a). Based on studies in mammals, it is inferred that isotocin and arginine vasotocin in fish are bound to cysteine rich carrier molecule called neurophysin for transportation and storage. The neurophysin and neurohormone are synthesized as one molecule in the cell body, and the hormone is cleaved from the precursor molecule by proteolytic enzymes in secretory vesicles. Secretion of neurohypophysial hormones, together with the neurophysin, occurs by exocytosis of the elementary neurosecretory granule at the axonal terminal (see Peter and Fryer, 1983). The neuroendocrine attributes of the neurohypophysial hormones are summarized in Table 1.

The importance of bioactive molecules like the neurotransmitters, neuromodulators and neurohormones in processing of neural signals cannot be overemphasized. Immunocytochemical studies have revealed hypothalamic fibers transporting these neuroactive substances and defined their precise relationship with the pituitary cell types (Moons *et al.*, 1988, 1989b).

Furthermore, deployment of two antibodies in proper sequence has revealed the relationship between two systems or the possible colocalizations of two bioactive substances in the same cell (Pickavance *et al.*, 1992). A bioactive substance in fiber terminals anywhere can have physiological significance, only if its specific receptors are also present in that area. Binding and autoradiographic studies reveal the presence of specific binding sites for sex steroids, several neuropeptides and/or neurotransmitters in brain and pituitary (Moons *et al.*, 1992). A wealth of information is available on the range of central factors, or their receptors, involved in the regulation of the pituitary and other physiological activities. For example, double immunocytochemical procedure using antibodies against estrogen receptors and tyrosine hydroxylase suggested that the nucleus preopticus pars anteroventralis may be mediating the negative feedback actions of estradiol on the gonadotropin secretion (Linard *et al.*, 1996). Some of the major findings in this area are summarized in the Table 1.

Additional neuroactive substances in teleostean hypothalamus: In addition to the above mentioned hypothalamic regulatory factors, an impressive catalogue of substances, mostly peptides, has been described in the hypothalamus of a variety of teleosts. Most of these are familiar substances already known to exist in the pituitary, gastrointestinal tract or elsewhere in the central nervous system. Although the physiological significance of these substances or of the neural circuitries in which they are present is yet to be determined, they are believed to perform neurotransmitter/neuroendocrine function. Organization of these substances is briefly considered here.

Table 1 Regulatory substances in the hypothalamus of teleosts*

Distribution of regulatory substances in the hypothalamus and pituitary	Nature of regulation
Dopamine and noradrenaline are present in NLT and PVO; fibers widely distributed (Sas *et al.*, 1990; Ekström *et al.*, 1990; Meek *et al.*, 1993). Dopamine cells in the preoptic area have been shown to innervate the GtH cells (Kah *et al.*, 1987). Dopamine fibers are associated with Gth cells (Peute *et al.*, 1987) and with the somatotrops (Wong *et al.*, 1993)	Dopamine inhibits GtH secretion in goldfish (Trudeau and Peter, 1995) but not in Atlantic croaker (Copeland and Thomas, 1989). It increases circulating growth hormone level (Chang *et al.*, 1985). MSH release may be under dopaminergic inhibitory control (Olivereau *et al.*, 1987). Norepinephrine stimulates GtH and inhibits GH release; modulates the release of GnRH (Peter *et al.*, 1990)
Serotonin is found in several CSF-contacting neurons (Meek and Joosten, 1989; Johnston *et al.*, 1990; Corio *et al.*, 1991; Khan and Thomas, 1993). Fibers also seen in the adenohypophysis (Groves and Batten, 1985)	Serotonin stimulates GtH II release (Somoza and Peter, 1991; Khan and Thomas, 1994); effect may be via GnRH neurons (Yu *et al.*, 1991; Khan and Thomas, 1993); may regulate prolactin cells (James and Wigham, 1984)
Neurohypophysial peptides isotocin and vasotocin are synthesized in the NPO and released in the pituitary; fibers also distribute in other areas of the brain (see Subhedar *et al.*, 1990). Binding sites for vasotocin were found in the NPO, tuberal hypothalamus and pituitary (Moons *et al.*, 1989a)	The peptides show hypophysiotropic actions (Moons *at al.*, 1988) and a role in osmoregulation. They increase pressure in the ventral aorta (Pang, 1977), cause contraction of smooth muscle in the ovary and oviduct (La Pointe, 1977), induce parturition in poeciliids (Venkatesh *et al.*, 1992) and influence the spawning behavior in killifish (Pickford *et al.*, 1980)
Corticotropin releasing factor is present in the NPO and nucleus preopticus periventricularis, also seen in the fibers in the vicinity of the ACTH cells (Olivereau and Olivereau, 1988)	Stimulates ACTH cells (Olivereau and Olivereau, 1991a; Jain *et al.*, 1994)
Somatostatin is distributed in the preoptic periventricular nucleus, anterior periventricular nucleus and the NLT and in the pituitary (Olivereau *et al.*, 1984; Sas and Maler, 1991); also seen in the neurons of the NPO (Power *et al.*,1996)	Decreases circulating GH level (Cook and Peter,1984; Sweeting and McKeown, 1986), reduces prolactin secretion *in vitro* (Williams and Wigham, 1994)
Growth hormone releasing hormone (GHRH), seen in the NPO and NLT neurons; the fibers are seen close to the GH cells (Moons *et al.*, 1988; Olivereau *et al.*, 1990)	A role of GHRH in promoting GH release is not yet fully elucidated (Olivereau *et al.*, 1990)
Gonadotropin releasing hormone (GnRH) is present in the NPO, NLT and several fibers in the tuberal region (Schreibman *et al.*, 1982; Subhedar and Rama Krishna,	Stimulates GtH cells (see Trudeau and Peter, 1995); also may play a modulating role in the control of oocyte meiosis at the level of the ovary (Habibi *et al.*, 1988)

(*Contd.*)

1988; Kah *et al.*, 1991). Fibers seen in the vicinity of the GtH cells (Peute *et al.*, 1987)	
Thyrotropin releasing hormone is present in the nucleus recessus lateralis; fibers widely distributed (Batten *et al.*, 1990c; Hamao *et al.*, 1990). Fibers numerous in the posterior pituitary.	No clear relationship with the thyroid stimulating hormone (TSH) cells, the peptide may be related to pars intermedia function, rather than TSH or prolactin releasing function (Batten *et al.*, 1990c).
Melanin concentrating hormone (MCH) is seen in the neurons of the NLT. Fibers project to the neurohypophysis and may innervate the cells of the pars intermedia (Batten and Baker, 1988)	MCH may act as a systemic hormone, a central neurotransmitter and a modulator of pituitary function. It may also act on ACTH cells (Batten and Baker, 1988).
GABA neurons are widely distributed in the hypothalamus; fibers are seen in the pituitary (Martinoli *et al.*, 1990). *Glutamate* is present in the hypothalamus and pituitary (Sloley *et al.*, 1991)	GABA stimulates GtH secretion (Trudeau *et al.*, 1993a). Stimulates and inhibits GtH secretion in gonadally regressed and developed Atlantic croaker, respectively (Khan and Thomas, 1995). Glutamate may be involved in the the stimulation of GtH II release (Trudeau *et al.*, 1993b) and inhibition of GH release (Trudeau *et al.*, 1996)

* See text for additional information.

Immunostaining of brain of goldfish, trout and eel revealed the presence of ACTH-like peptide in the ventral hypothalamus (Olivereau and Olivereau, 1990). In *Salmo gairdneri* also, ACTH was found to occur in certain cells of the NLT and fibers were seen projecting to the pituitary (Vallarino *et al.*, 1989). Immunoreactive α-MSH was detected in all regions of carp and trout brains, with highest concentration in the basal hypothalamus (Kishida *et al.*, 1988). The peptide cholecystokinin, as well as binding sites for the peptide were discovered in the hypothalamus (Batten *et al.*, 1990a; Moon *et al.*, 1992). While cholecystokinin as well as gastrin are believed to stimulate gonadotropic hormone (GtH) release in fish (Himick *et al.*, 1993), cholecystokinin reduced feeding in the goldfish (Himick and Peter, 1994a). The peptide bombesin has been described in the goldfish brain and pituitary (Himick and Peter, 1995); the peptide suppressed food intake in the goldfish while concomitantly elevating serum GH (Himick and Peter, 1994b). Galanin containing fibers were observed in the basal hypothalamus and seen to pass along the ACTH, GtH and GH cells in the pituitary (Batten *et al.*, 1990b; Olivereau and Olivereau, 1991b; Power *et al.*, 1996); galanin binding sites in the hypothalamus and pituitary were also studied (Moons *et al.*, 1991). Neuropeptide Y is extensively distributed in the hypothalamus and pituitary (Pickavance *et al.*, 1992; Subhedar *et al.*, 1996) and known to be involved in the regulation of GtH and GH secretion (Peng *et al.*, 1990; 1993a, b). Similarly neurotensin, calcitonin gene related peptide, and substance *P* (Batten *et al.*, 1990a), melanin-concentrating hormone gene-related peptide (Gröneveld *et al.*, 1995), FMRFamide (Rama Krishna and Subhedar, 1992) and histamine (Ekström *et al.*, 1995) have been reported to be present in fish hypothalamus and believed to directly or indirectly influence the hypophysial function.

Discovery of the endogenous opiate peptides and recognition of their importance in brain control, homeostasis and neuroendocrine regulation has been one of the most exciting episode in modern biology. While a great deal of information about the distribution and physiology of these peptides is available in mammals, researches in fish are just beginning to emerge. Enkephalin, β-endorphin and dynorphin have been described in the hypothalamus and the neuroendocrine system (Reaves and Hayward, 1979; Batten *et al.*, 1990a; Cepriano and Schreibman, 1993; Vecino *et al.*, 1995). α-as well as β-endorphins have been localized in the NLT (Vallarino *et al.*, 1989). The distribution of opioid binding sites has also been investigated (Bird *et al.*, 1988); the highest level of binding was found in the preoptic area (Rosenblum and Callard, 1988).

Circumventricular Organs of the Hypothalamus

The CSF-contacting neurons in the hypothalamus of teleosts can be encountered as isolated cells or as groups of cells constituting a well defined circumventricular organs (CVOs); these are generally considered organs of sensory/secretory function (Vigh-Teichmann and Vigh, 1983). Most of the CSF-contacting cells send apical terminals or cilia into the ventricular lumen. The organum vasculosum laminae terminalis (OVLT) of mammals is known to be involved in osmoregulation and reproduction; however, the function of this organ in fish is not known. The paraventricular organ (PVO) has been investigated in several teleosts and mainly reported to contain monoaminergic CSF-contacting cells. The dopaminergic CSF-contacting neurons that line the median, lateral and posterior recess of the third ventricle do not contain tyrosine hydroxylase immunoreactivity. This indicates uptake and not the synthesis of exogenous dopamine (Ekström *et al.*, 1992).

The CVOs might also serve as the sites for the secretion of peptides into the CSF. Many hypophysiotropic neurons in the preoptic region or the mediobasal hypothalamus have cytoplasmic processes contacting the ventricles (Kah *et al.*, 1993). While GnRH was found to be associated with OVLT (Kah *et al.*, 1986; Subhedar and Rama Krishna, 1988), NPY fibers communicate with the third ventricle (Subhedar *et al.*, 1996). In addition to above, the CSF contacting neurons in the hypothalamus have been shown to contain enkephalin (Vecino *et al.*, 1995), serotonin (Meek and Joosten, 1989) and somatostatin (Vigh-Teichmann *et al.*, 1983).

Vegetative Functions of the Hypothalamus

Studies on the vegetative functions of the hypothalamus are extremely limited. Lamb and Caprio (1993) have ascribed certain hypothalamic nuclei viz., nucleus centralis, nucleus diffusus and nucleus of the lateral recesses as the gustatory centers in the teleostean hypothalamus. Electrical stimulation of some of these nuclei produced complete feeding responses in bluegill (Demski and Knigge, 1971) and tilapia (Demski, 1973). Damage to the lateral hypothalamic regions of goldfish produced pronounced aphagia (Roberts and Savage, 1978).

Phylogenetic Considerations

The hypothalamic nuclei in teleosts, and also in amphibians, are poorly differentiated neuronal population crowded along the ventricular surface, rich in CSF-contacting extensions. The higher forms show specialization in terms of differentiated cluster-like accumulations of neurons, well developed intra-and internuclear neuropil and interposition of thick pad of neuropil between cell groups and the ependyma. The teleostean features, coupled with extensive lateral recesses of the third ventricle, emphasize the importance of the direct access to the CSF, either by individual neurons, or by way of well defined circumventricular organs. It may be recalled that in fish, as in some other vertebrates, the PVO is well developed, although the organ is strikingly absent in mammals. It may be recalled that the PVO has been considered evolutionarily a primitive feature (Butler and Northcutt, 1993). It seems that in fish, there might be greater dependence on the CSF as a medium of communication, while in higher forms, interneuronal contacts may have developed for more discrete processing of the signals.

While agnathans totally lacked direct contact between the hypothalamus and the pituitary, the anatomical substrate for hypophysial control perhaps evolved along two divergent lines in vertebrates. In teleosts the direct innervation of the hypophysial cell types emerged as the format for control, whereas, interposition of a portal system evolved in the tetrapods.

Acknowledgments

The work was supported by the grants from Indian Council of Agricultural Research (No. 4 (22)/92ASR (I) and 4(31)/92ASR(I)) and Department of Science and Technology (No. SP/SO/CO3/94), New Delhi.

References

Airhart, M.J., Shirk, J.O. and Kriebel, R.M. (1988). Brain Res. Bull. **20:** 503–514.

Anglade, I., Zandbergen, T. and Kah, O. (1993) Cell Tiss. Res. **273:** 345–355.

Batten, T.F.C. and Baker, B.I. (1988) Gen. Comp. Endocrinol. **70:** 193–205.

Batten, T.F.C., Cambré, M.L., Moons L. and Vandesande, F. (1990a) J. Comp. Neurol. **302:** 893–919.

Batten, T.F.C. and Ingleton, P.M. (1987) In: Fundamentals of comparative vertebrate endocrinology (I. Chester Jones P.M. Ingleton Phillips J.G., Eds), pp. 185–409, Plenum Press.

Batten, T.F.C., Moons, L., Cambré, M. and Vandesande F. (1990b) Neurosci. Lett. **111:** 12–17.

Batten, T.F.C., Moons, L., Cambré, M., Vandesande, F., Seki T. and Suzuki M (1990c) Gen. Comp. Endocrinol. **79:** 385–392.
Billard, R. and Peter, R.E. (1982) Reprod. Nutr. Dev. **22:** 1–25.
Bird, D.J., Jackson, M., Baker, B.I. and Buckingham, J.C. (1988) Gen. Comp. Endocrinol. **70:** 49–62.
Braford, M.R. (Jr.) and Northcutt, R.G. (1983) In: Fish neurobiology, Vol. 2, Higher, brain areas and functions (R.E. Davis and R.G., Northcutt, Eds), pp. 117–163, The University of Michigan Press, Ann. Arbor.
Butler, A.B. and Northcutt, R.G. (1993) J. Comp. Neurol. **328:** 527–546.
Cepriano, L.M. and Schreibman, M.P. (1993) Cell Tiss. Res. **271:** 87–92.
Chang, J.P., Marchant, T.A., Cook, A.F., Nahorniak, C.S. and Peter R.E. (1985) Neuroendocrinology **40:** 463–470.
Cook, A.F. and Peter, R.E. (1984) Gen. Comp. Endocrinol. **54:** 109–113.
Copeland, P.A. and Thomas, P. (1989) Gen. Comp. Endocrinol. **74:** 474–483.
Corio, M., Peute, J. and Steinbusch, H.W.M. (1991) J. Chem. Neuroanat. **4:** 79–95.
Demski, L.S. (1973) Comp. Biochem. Physiol. **44A:** 685–692.
Demski, L.S. and Knigge, K.M, (1971) J. Comp. Neurol. **143:** 1–16.
Demski, L.S. and Sloan, H.E. (1985) Neurosci. Lett. **55:** 283–288.
Deshmukh, M.K., Rama Krishna, N.S. and Subhedar, N. (1992) J. Exp. Zool. **263:** 231–234.
Ekström, P., Holmqvist, B.I. and Panula, P. (1995) J. Chem. Neuroanat. **8:** 75–85.
Ekström, P., Honkanen, T. and Borg, B. (1992) J. Chem. Neuroanat. **5:** 481–501.
Ekström, P., Honkanen, T. and Steinbusch, H.W.M. (1990) J. Chem. Neuroanat. **3:** 233–260.
Fernald, R.D. (1980) Vision Res. **20:** 1015–1019.
Fernald, R.D. and Shelton, L.C. (1985) J. Comp. Neurol. **161:** 125–142.
Goossens, N., Dierickx, K. and Vandesande, F. (1977) Gen. Comp. Endocrinol. **32:** 371–375.
Gröneveld, D., Balm, P.H.M. and Wendelaar Bonga, S.E. (1995) Neuroendocrinology **62:** 498–505.
Groves, D.J. and Batten, T.F.C. (1985) Cell Tiss. Res. **240:** 489–492.
Habibi, H.R., Kraak. G.V.D., Bulanski, E. and Peter, R.E. (1988) Am. J. Physiol. **24:** R268–R273,
Halpern-Sebold, L.R. and Schreibman, M.P. (1983) Cell Tiss. Res. **229:** 75–84.
Hamao, K., Inoue, K. and Yanagisawa, T. (1990) Gen. Comp. Endocrinol. **80:** 85–92.
Himick, B.A., Golosinski, A.A., Jonsson, A. and Peter, R.E. (1993) Gen. Comp. Endocrinol. **92:** 88–103.
Himick, B.A. and Peter, R.E. (1994a) Am. J. Physiol. **247:** R841-R851.
Himick, B.A. and Peter, R.E. (1994b) Physiol. Behav. **55:** 65–72.
Himick, B.A. and Peter, R.E. (1995) Neuroendocrinology **61:** 365–376.
Ito, H. and Yoshimoto, M. (1990) J. Comp. Neurol. **289:** 385–399.
Jain, M.R., Khan, F.A., Rama Krishna, N.S. and Subhedar, N. (1994). Neuroreport **5:** 2013–2016.
James, V.A. and Wigham, T. (1984) Gen. Comp. Endocrinol **56:** 231–239.
Johnston, S.A., Maler, L. and Tinner, B. (1990) J. Chem. Neuroanat. **3:** 429–465.
Kah, O., Breton, B., Dulka, J.G., Nuñez-Rodriguez, J., Peter, R.E., Corrigan A., Rivier, J.E. and Vale, W.W. (1986) Cell Tiss. Res. **244:** 327–337.
Kah, O., Anglade, I., Lepretre, E., Dubourg, P. and de Monbrison, D. (1993) Fish Physiol. Biochem. **11:** 85–98.
Kah, O., Dulka, J.G., Dubourg, P., Thibault, J. and Peter, R.E. (1987) Neuroendocrinology **45:** 451–458.

Kah, O., Zanuy, S., Mananos, E., Anglade, I., and Carrillo, M. (1991) Cell Tiss. Res. **266:** 129–136.
Kandel, E.R. (1964) J. Gen. Physiol. **47:** 691–717.
Khan, I.A. and Thomas, P. (1993) Gen. Comp. Endocrinol. **91:** 167–180.
Khan, I.A. and Thomas, P. (1994) J. Exp. Zool. **269:** 537–541.
Khan, I.A. and Thomas, P. (1995) In: Proceedings of the Fifth international Symposium on the Reproductive Physiology of the Fish (F.W. Goetz and P. Thomas, Eds), pp. 73, Austin, Texas.
Kishida, M., Baker, B.L. and Bird, D.J. (1988) Gen. Comp. Endocrinol. **71:** 229–242.
La Pointe, J.L. (1977) Amer. Zool. **17:** 763–773.
Lamb, C.F. and Caprio, J. (1993) J. Comp. Neurol. **337:** 400–418.
Levine, R.L. and Dethier, S. (1985) J. Comp. Neurol. **237:** 427–444.
Linard, B., Anglade, I., Corio, M., Navas, J.M., Pakdel, F., Saligaut, C. and Kah, O. (1996) Neuroendocrinology **63:** 156–165.
Maler, L., Sas, E., Johnston, S. and Ellis, W. (1991) J. Chem. Neuroanat. **4:** 1–38.
Martinolli, M.G., Dubourg, P., Geffard, M., Calas, M. and Kah, O. (1990) Cell Tiss. Res. **260:** 77–84.
Meek, J. and Joosten, H.W.J. (1989) J. Comp. Neurol. **281:** 206–224.
Meek, J., Joosten, H.W.J. and Hafmans, T.G.M. (1993) J. Comp. Neurol. **328:** 145–160.
Moons, L., Batten, T.F.C. and Vandesande, F. (1991) Neurosci. Lett. **123:** 49–52.
Moons, L., Batten, T.F.C. and Vandesande, F. (1992) Peptides **13:** 37–46.
Moons, L., Cambré, M., Batten, T.F.C. and Vandesande, F. (1989a) Neurosci. Lett. **100:** 11–16.
Moons, L., Cambré, M., Marivoet, S. and Batten, T.F.C. (1988) Gen. Comp. Endocrinol. **72:** 171–180.
Moons, L., Cambré. M., Ollevier, F. and Vandesande, F. (1989b) Gen. Comp. Endocrinol. **73:** 270–283.
Morita, Y., Murakami, T. and Ito, H. (1983) J. Comp. Neurol. **218:** 378–394.
Olivereau, M. and Olivereau, J.M. (1988) Peptides **9:** 13–21.
Olivereau, M. and Olivereau, J.M. (1990) Cell Tiss. Res. **262:** 115–123.
Olivereau, M. and Olivereau, J.M. (1991a) Cell Tiss. Res. **265:** 485–492.
Olivereau, M. and Olivereau, J.M. (1991b) Histochemistry **96:** 343–354.
Olivereau, M., Olivereau, J.M. and Lambert, J.F. (1987) Gen. Comp. Endocrinol. **68:** 12–18.
Olivereau, M., Olivereau, J. and Vandesande, F. (1990) Cell Tiss. Res. **259:** 73–80.
Olivereau, M., Ollevier, F., Vandesande, F. and Olivereau, J. (1984) Cell Tiss. Res. **238:** 289–296.
Pang, P.K.T. (1977) Amer. Zool. **17:** 739–749.
Peng, C., Huang, Y.P. and Peter, R.E. (1990) Neuroendocrinology **52:** 28–34.
Peng, C., Humphries, S., Peter, R.E., Rivier, J.E., Blomqvist, A.G., and Larhammar, D. (1993a) Gen. Comp. Endocrinol. **90:** 306–317.
Peng, C., Trudeau, V.L. and Peter, R.E. (1993b) J. Neuroendocrinol. **5:** 273–280.
Peter, R.E. (1977) Amer. Zool. **17:** 775–785.
Peter, R.E. and Fryer, J.N. (1983) In: Fish neurobiology, Vol 2, Higher brain areas and functions (R.E. Davis and R.G. Northcutt, Eds), pp. 165–201, The University of Michigan, Press.
Peter, R.E. and Gorbman, A. (1968) Neuroendocrinology **3:** 229–237.
Peter, R.E., Macey, M.J. and Gill, V.E. (1975) J. Comp. Neurol. **159:** 103–128.
Peter, R.E., Yu, K.L., Marchant, T.A. and Rosenblum, P.M. (1990) J. Exp. Zool. (Suppl) **4:** 84–89.
Peute, J., Schild, R.G., Schild, V.A., Buijs, R.M., Van Asselt, L.A.C. and Van Ordt, P.G.W.J. (1987) Gen. Comp. Endocrinol. **67:** 303–310.

Pickavance, L.C., Staines, W.A. and Fryer, J.N. (1992) J. Chem. Neuroanat. **5:** 221–233.

Pickford, G.E., Knight, W.R. and Knight, J.N. (1980) Rev. Can. Biol. **39:** 97–105.

Power, D.M., Canario, A.V.M. and Ingleton, P.M. (1996) Gen. Comp. Endocrinol. **101:** 264–274.

Prasada Rao, P.D., Jadhao, A.G. and Sharma, S.C. (1987) J. Comp. Neurol. **265:** 96–108.

Prasada Rao, P.D., Job, T.C. and Schreibman, M.P. (1993) Brain Behav. Evol. **42:** 24–38.

Rama Krishna, N.S. and Subhedar, N. (1989) Neurosci. Lett. **107:** 39–44.

Rama Krishna, N.S. and Subhedar, N. (1991) J. Hirnforsh. **32:** 289–308.

Rama Krishna, N.S. and Subhedar, N.K. (1992) Peptides, **13:** 183–191.

Rama Krishna, N.S., Subhedar, N. and Schreibman, M.P. (1992) Gen. Comp. Endocrinol. **85:** 111–117.

Reaves, T.A.Jr. and Hayward, J.N. (1979) Cell Tiss. Res. **200:** 147–151.

Roberts, M.G. and Savage, G.E. (1978) Brain Behav. Evol. **15:** 150–164.

Rooney, D.J., New, J.G., Szabo, T. and Ravaille-Vern, M. (1989) Cell Tiss. Res. **257:** 423–436.

Rosenblum, P.M. and Callard, I.P. (1988) J. Exp. Zool **245:** 244–255.

Sas, E. and Maler, L. (1991) J. Chem. Neuroanat. **4:** 155–186.

Sas, E., Maler, L. and Tinner, B. (1990) J. Comp. Neurol. **292:** 127–162.

Sathyanesan, A.G. and Jose, T.M. (1975) Acta Anat. **93:** 387–398.

Scharrer, E.A. (1928) Zeitschr. Vergl. Physiol. **7:** 1–38.

Schreibman, M.P., Margolis-Kazan, H., Halpern-Sebold, L., O'Neill, P.A. and Silverman, R.C. (1982) Brain Res. **302:** 180–183.

Sloley, B.D., Kah, O., Trudeau, V.L., Dulka, J.G. and Peter, R.E. (1991) J. Neurochem. **58:** 2254–2262.

Somoza, G.M. and Peter, R.E. (1991) Gen. Comp. Endocrinol. **82:** 1034–110.

Springer, A.D. and Mednick, A.S. (1985) J. Comp. Neurol. **236:** 179–196.

Striedter, C.F. (1990) Brain Behav. Evol. **36:** 329–354.

Subhedar, N., Cerdá, J. and Wallace, R.A. (1996) Cell Tiss. Res. **283:** 313–323.

Subhedar, N. and Rama Krishna, N.S. (1988) Gen. Comp. Endocrinol. **72:** 431–442.

Subhedar, N., Rama Krishna, N.S. and Deshmukh, M.K. (1987) Gen. Comp. Endocrinol. **68:** 357–368.

Subhedar, N., Deshmukh, M.K., Jain, M.R., Khan, F.A. and Rama Krishna, N.S. (1996) Brain Behav. Evol. **47:** 179–184.

Subhedar, N., Rama Krishna, N.S. and Prasada Rao, P.D. (1990) J. Hirnforsch. **31:** 25–40.

Sweeting, R.M. and McKeown, B.A. (1986) Can. J. Zool. **64:** 2062–2063.

Terlou, M. and Ekengren, B. (1979) Cell Tiss. Res. **197:** 1–21.

Trudeau, V.L. and Peter, R.E. (1995) In: Proceedings of the Fifth International Symposium on the Reproductive Physiology of the Fish (F.W. Goetz and P. Thomas, Eds) pp. 44 Austin, Texas.

Trudeau, V.L., Sloley, B.D. and Peter, R.E. (1993a) J. Neuroendocrinol. **5:** 129–136.

Trudeau, V.L., Sloley, B.D. and Peter, R.E. (1993b) Am. J. Physiol. **265:** R348-R355.

Trudeau, V.L., Sloley, B.D., Kah, O., Mons, N., Dulka, J.G. and Peter, R.E. (1996) Gen. Comp. Endocrinol. **103:** 129–137.

Uchiyama, H. (1989) Vis. Neurosci. **3:** 183–206.

Uchiyama, H. and Ito, H. (1984) Brain Res. **222:** 11–24.

Vallarino, M., Delbende, C., Bunel, D.T., Ottonello, I. and Vaudry, H. (1989) Peptides **10:** 1223–1230.

Vanegas, H. and Ito, H. (1983) Brain Res. **6:** 117–137.

Vecino, E., Perez, M.T.R. and Ekström, P. (1995) J. Chem. Neuroanat. **9:** 81–97.

Venkatesh, B., Tan, C.H. and Lam, T.J. (1992) Gen. Comp. Endocrinol. **87:** 28–32.

Vigh-Teichmann, I. and Vigh, B. (1983) Arch. Histol. Jpn. **46:** 427–486.

Vigh-Teichmann, I. and Vigh, B., Korf, H.W. and Oksche, A. (1983) Cell Tiss. Res. **233:** 319–334.
Von Bartheld, C.S. and Meyer, D.L. (1988) Cell Tiss. Res. **251:** 651–663.
Williams, A.J. and Wigham, T. (1994) Gen. Comp. Endocrinol. **93:** 398–405.
Wong, A.O.L., Chang, J.P. and Peter, R.E. (1993) Endocrinology **133:** 577–584.
Wullimann, M.F. and Meyer, D.L. (1990) Brain Behav. Evol. **36:** 14–29.
Yu, K.L., Rosenblum, P.M. and Peter, R.E. (1991) Gen. Comp. Endocrinol. **81:** 256–267.

Comparative Endocrinology and Reproduction
K.P. Joy, A. Krishna and C. Haldar (Eds)

5. Prolactin, ACTH and Growth Hormone-Secreting Cells in the Teleost Pituitary Gland: Environmental and hypothalamic control

M. Olivereau[1] and J.M. Olivereau[1]*
[1]Neuroendocrinology, 51 rue de Lancry, F-75010 Paris, France
*Université René Descartes, F-75005 Paris, France

Summary
Cells secreting prolactin, ACTH and growth hormone (GH) were identified in the pituitary pars distalis of teleost fish exposed to different experimental conditions (environmental changes, treatment with hormones or chemical inhibitors of their synthesis, catecholamines...). The identified functions were confirmed by immunocytochemistry using antisera raised against mammalian or fish PRL, ACTH and GH. Ultrastructural studies sometimes associated with immunocytochemistry gave more information on the processing of granules and their release during experimental stimulation or inhibition.

The synthesis and release of PRL, ACTH and GH are modulated by environmental factors (osmotic pressure, Na^+, Ca^{2+}, pH) and plasma levels of hormones secreted by the respective target glands, neurotransmitters, neuromodulators...). An excellent way to study the influence of these factors is to submit dispersed pituitary cells to some factors and to evaluate the production of PRL, ACTH and GH in the cells and their release in the perifusion or incubation medium by radioimmunoassay or incorporation of labeled amino acids. Results show that the synthesis of the three hormones is regulated by numerous stimulatory and inhibitory substances (electrolytes, dopamine, serotonin) and neuropeptides (CRF, TRH, GnRH, GHRH, SRIF...). As in other vertebrates, pituitary is progressively loosing its functional independence, admitted by the pioneers of comparative endocrinology, whereas the complexity of its hypothalamic control is unfolding rapidly.

This paper is dedicated to Professors C.J. Dominic, A.G. Sathyanesan and J.P. Thapliyal, eminent Comparative Endocrinologists who contributed to our knowledge of bird and fish endocrinology. Their warm and kind welcome at BHU is still present in my memory. I associate also with the souvenir of another colleague and friend, the late Prof. B.I. Sundararaj, Delhi.

Introduction

The pituitary gland or hypophysis cerebri is a small gland appending under the hypothalamus and attached to the brain by a pituitary stalk containing a wide or narrow infundibular recess. It is located just behind the optic chiasm in all vertebrates. In teleost fish, it extends back to the saccus vasculosus, is surrounded by the meningeal layers and enclosed in a shallow or deep depression of the cranial floor.

Pickford and Atz (1957*) were the first to collect all the data available so far on the "Physiology of the Pituitary Gland of Fish". The gland was considered as controlling several target glands (thyroid, interrenal homologous to the adrenocortical tissue, gonads) and some functions (growth, metabolism, ionic regulation) and involved in metamorphosis and migratory behavior. Hoar (1966*) and Ball and Baker (1969) reviewed recent data on fish pituitary histophysiology. It appeared evident that the release of each pituitary hormone was sensitive to the feedback, most often negative, from the secretion of its respective target gland. The functioning of the pituitary is no longer completely autonomous, but modulated by several neurotransmitters, mainly dopamine and serotonin, excitatory amino acids, NO, GABA and glutamate, various neuropeptides, etc. in mammals. Pituitary activity was also controlled by specific hypothalamic releasing or inhibitory factors whose effects were demonstrated *in vivo* and more easily *in vitro* in cultures or by perifusion of dispersed pituitary cells or glandular fragments. Cytophysiological studies tended to be replaced by radioimmunoassay of hormones released in the plasma or the incubation medium. Similar techniques were applied to teleost pituitary to evaluate the activity of a large number of hypothalamic factors. These various aspects of the functional study of teleost fish pituitary will be summarized, but limited to the three protidic hormones identified in the pars distalis, prolactin (PRL), adrenocorticotropin (ACTH) and growth hormone (GH). The literature is so abundant that an artificial selection was necessary. Some aspects such as the ontogeny of cell types and the circadian rhythms of their secretion will not be considered in this review.

Anatomy

The pituitary is composed of two different tissues, the adeno- and the neurohypophysis (NH), both of ectodermal origin. The adenohypophysis arises from the roof of the oral cavity (Rathke's pouch in mammals) and differentiates into a pars distalis (PD), itself divided into a rostral (RPD) and a proximal pars distalis (PPD), and an intermediate lobe (IL). The NH is an extension of the ventral hypothalamus and contains nerve fibers arising mostly from hypothalamic nuclei, glial cells, pituicytes and blood vessels. However, in teleosts IL and NH are not compact and separated as in other vertebrates, but intermingled forming a neurointermediate lobe (NIL) which may be as large as the PD. The functions of the IL are described in another article of the present book.

General Considerations

Several decades elapsed before the function of the various pituitary cell types could be identified. As a first step, cells were classified into acidophils and basophils after trichrome staining. These two heterogeneous groups corresponded to protidic and glycoprotidic cells. Later (1950–1970). Herlant's tetrachrome and histochemical reactions (Table 1) were applied to the pituitary of fish during migration, spawning, after adaptation to environments of a modified ionic composition, and treatment with hormones or their inhibitors. These experimental manipulations were not easily performed because fish pituitary hormones have structures different from those of mammalian hormones available at that tme (Fontaine 1969). Moreover, some target glands are not compact, but composed of scattered glandular islets impossible to remove: a complete thyroidectomy or adrenalectomy is impossible in almot all teleosts. Only chemical inhibitors could be used to suppress hormonal synthesis, and these inhibitors were tested in fish as soon as they became available. Various procedures led to the identification of cells secreting PRL, ACTH, thyrotropin (TSH), two gonadotropins (GTH I and GTH II) and GH in the pars distalis.

The immunocytochemical techniques first applied to fish pituitary used heterologous antisera raised against mammalian hormones and gave interesting but questionable results. More recently fish pituitary hormones were isolated and antisera prepared. They were not completely homologous in most cases as antisera were produced against hormones from one species and applied to another one. Immunocytochemistry allowed the identification of the hormones secreted by each cell type and fully confirmed results obtained with cytochemical staining. Indeed, cytophysiological studies furnished data not as obsolete as it was claimed sometimes, and the diagram of the sagittal section of an eel pituitary (Olivereau 1967*) is still valid.

Prolactin Cells

Long before cells could be considered as secreting a PRL-like hormone, trichrome stainings identified acidophils in the RPD, which secreted a protein. Actually, more than 300 papers deal with studies of this cell type. These cells show an erythrosinophilic staining similar to that of mammalian PRL cells (Olivereau and Herlant 1960*) and appear more active in fresh water (FW) than in sea water (SW) in eel and *Poecilia*. Simultaneously, Pickford and Phillips (1959*) note that hypophysectomized *Fundulus* die in FW but survive after an ovine PRL treatment. A correlation between erythrosinophilic cells and a PRL-like secretion was suspected (Ball and Olivereau 1964*).

However, the hypothesis of a "prolactin" in fish pituitary (Olivereau 1969) was not readily accepted by some scientists as PRL was mainly related to lactation in mammals. By using a heterologous antiserum to ovine or bovine PRL, the PRL-like secretion of the erythrosinophilic cells was demonstrated (Aler 1971, among others), but this was not yet a formal

Table 1 Adenohypophysial Cell Types in Teleosts

Adenohypophysis	Hormonal Secretion	Immunostaining with Antisera to	Acidophils Protidic	Basophils Glycoprotidic	Herlant's Tetrachrome	Lead Hematoxylin	Periodic Acid Schiff	Aldehyde-Fuchsin Alcian Blue
Pars Distalis, Rostral	PRL Prolactin	Fish PRL	+	0	Pink Erythrosin	0	0 or Gr	0
Pars Distalis, Rostral	ACTH Corticotropin	Human ACTH	+	0	Purple Alizarin blue	Black	0	0
Pars Distalis, Rostral	TSH Thyrotropin[1]	Human TSHb[2]	0	+	Blue Anilin blue	0	+	+
Pars Distalis, Proximal	GH Growth Hormone	Fish GH	+	0	Orange Orange G	0	0	0
Pars Distalis, Proximal	GTH-I Gonadotropin I	Salmon GTH-IB	0	+	Blue Anilin blue	0 or Gr[3]	+	+
Pars Distalis, Proximal	GTH II Gonadotropin II	Salmon GTH-IIB	0	+	Blue Anilin blue	0 or Gr[3]	+	+
Pars Intermedia	MSH Melanotropin	alpha-MSH	+	0	Light purple Alizarin blue	Black	0	
Pars Intermedia	SL Somatolactin	Cod SL	0	+	Blue or orange	0	+ or 0[4]	0 or +

1. TSH cells may be located in RPD (eel) and /or PPD (*Poecilia*) or between RPD and PPD.
2. Anti-TSHb labels TSH cells and GTH cells although the staining is often less intense than in TSH cells.
3. Gr: Only some large granules show an immunostaining of a variable intensity.
4. SL cells are PAS-negative in salmonidae.

proof. Therefore, it was necessary to isolate fish PRL. The first attempt (Farmer *et al.*, 1977) showed that the anti-tilapia PRL serum labeled PRL cells in different species (Nagahama *et al.*1981). Since then PRL has been identified in salmonids, carp, catfish, eel, sturgeon (Prunet and Auperin 1994, Sinha 1995) and cDNA cloned in carp (Chang *et al.*, 1992) and eel (Quérat *et al.*, 1994). Ultrastructural studies show that the immunoreaction is restricted to the secretory granules in goldfish (Dubourg *et al.*, 1986), *Poecilia* (Batten 1986a) and in 15 other species (Naito *et al.*, 1983).

Homology of amino acid sequences among fish PRLs reaches 60–75% and that with mammalian PRLs 20–40% only (Sinha 1995). Molecular weight is 20–23 kda. In fact, several species possess two isoforms of PRL (eel: Suzuki *et al.*, 1991, tilapia: Specker *et al.*, 1993, see below). A possible glycosylation of PRL molecule (Sinha 1995) as in reptiles has not yet been investigated in fish.

As regard PRL, cytophysiological observations stimulated biochemists to search for an unsuspected PRL-like hormone in fish pituitary, fish having no reason to secrete PRL. For other pituitary, following an inverse pathway, cytologists looked for cell types synthesizing hormones with a well known target gland or controlling growth and already identified in other vertebrates.

Cytology. Compact masses or small islets of PRL cells are intermingled with ACTH and some TSH cells in the RPD. They are spheroidal with abundant granules, often contacting the basal lamina between glandular and rostral neural ramifications. Elongated agranular (Agr) cells occur frequently among PRL cells. PRL cells are organized into follicular structures in Isospondyles (Salmonidae, Clupeidae) and Apods (Anguillidae) and unexpectedly in the ironfish, hybrid between funa and goldfish (Yamamoto *et al.*, 1982).

Follicular cells are elongated with a median or basal nucleus, and their apex may contact the follicular lumen containing a PAS+ colloid material. PRL follicles are surrounded by a connective membrane along which many Agr cells are observed. The triangular section of the cell body sends a long and thin process between PRL cells which ends to the lumen and often prevents the tip of PRL cells to contact the luminal fluid. Agr or stellate cells are bordered by a channel system communicating with extravascular spaces and revealed by the use of horseradish peroxidase in mullet (Abraham *et al.*, 1976).

Ultrastructure. In goldfish, spherical secretory granules (100–225 nm diameter) with an electron dense content are uniformly scattered in the cells or may tend to accumulate toward the basal lamina. The cytoplasm contains few short mitochondria, a small Golgi area, an endoplasmic reticulum with 2 or 3 perinuclear rows of cisternae, and numerous microtubules. Agr cells rarely contain a small granule of a low electron density, bundles of tonofilaments anchored on desmosomes. Desmosomes are observed between Agr cells (Olivereau *et al.*, 1986a).

In eels, secretory granules accumulate in the enlarged basal part of the cell. They are round or slightly oval (150–350 nm diameter) with a dense content, however two adjacent cells may have granules of different sizes. The endoplasmic reticulum is composed of 2–3 rows of cisternae around the nucleus or close to the plasma membrane and the Golgi complex is moderately developed. As in goldfish, exocytotic figures are scarce in FW fish, often located near the basal lamina. The apical cell area contains few granules, some mitochondria, numerous small clear vesicles and a microtubular system ending at the apical border. Centrioles are evident (9 triplets of hollow fibrils). From the pericentriolar structures or satellites arise two fibrous rootlets (1.5 μm long with 17 striae with a periodic structure (Olivereau *et al.*, 1981). Similar rhizoplast-like structures are described in *Alosa* (Betchaku and Douglas 1981) which possesses an orohypophyseal duct at juvenile stages. An apical cilium (2.5–3 μm long, 9 + 2 or 8 + 1 axonomes) is oriented toward the lumen. Few degenerating PRL cells seem to migrate into the lumen. Apically, the plasma membrane of PRL and adjacent Agr cells make several infoldings and junctional complexes (a zona occludens near the lumen followed by a zonula adherens and isolated or numerous desmosomes). These complexes are more numerous between adjacent Agr cells and rarely occur between adjacent PRL cells. Bundles of tonofilaments are anchored on desmosomes of Agr cells. These contain a subapical Golgi complex, numerous mitochondria, ribosomes, small and lucent vesicles and microtubules. Apices of PRL cells are almost always covered by Agr cells which may protrude into the lumen forming large pseudopodia or a few microvilli. Similar ultrastructures of PRL and associated Agr cells are described in the ironfish (Yamamoto *et al.*, 1982), *Engraulis japonica* (Olsson and Fujita 1976) and different salmonid species (see Olivereau *et al.*, 1981).

Responses to environmental changes: FW to SW and vice-versa

More than 250 papers describe responses of PRL cells to salinity changes in more than 30 species, with concordant results regarding an activity higher in FW than in SW and hypersaline lagoon (Abraham 1971). Marine teleosts transferred to brackish or FW (euryhaline species) show an enlarged RPD related to the PRL cell hypertrophy, but a mitotic activity is scarcely noted (mullet: Olivereau 1968*), an enlargement of the endoplasmic reticulum, an active Golgi area (Nagahama *et al.*, 1975) and more abundant exocytotic figures. Immunoreactive PRL is reduced in seabream transferred to brackish water (Mancera *et al.*, 1993). Migrating species entering the estuary (glass eel: Arakawa *et al.*, 1992) and during the upstream migration show stimulated PRL cells.

Conversely, FW teleosts adapted to SW show a gradual reduction in cell and nuclear size, number of secretory granules, smaller endoplasmic reticulum and Golgi complex, the nucleolemma becomes sinuous. These changes noted in 30–33‰ SW are accentuated in eels which withstand salinities of 50, 60 and 63‰. Granules mainly located in the supranuclear cytoplasm show a heterogeneous electron density and an undulated membrane and a reduced diameter (mean 130 nm). The follicular volume decreases, the lumen tends to enlarge and contains a dense material showing a typical hexagonal paracrystalline structure discernible in a few sites. Cilia, microvilli, microvesicles are scarce, rhizoplast-like structures are not identified (Olivereau *et al.*, 1981). A similar hypoactivity is noted in *Gambusia* kept in 30 and 45‰. SW (Dubourg *et al.*, 1980) and in RPD studied *in vitro* or after transplantation of the pituitary in *Poecilia* (Ball and Olivereau 1965*) and goldfish (Leatherland and Ensor 1973) in diluted SW. Agr cells appear larger in SW than in FW adapted fish according to the majority of authors.

Responses to salinity changes are not immediate and a new equilibrium is reached within 10–15 days. SW eels transferred to FW still show a low storage of granules although other biological parameters are typical of FW eels.

Hypoosmotic environments. Hypophysectomized eel, goldfish and salmonid survive in FW, cyprinodonts (*Poecilia, Gambusia* and some mugillidae) die within 5–15 days.

When intact fish of the first group are adapted to deionized water (DW) their survival was excellent although light and electron microscope studies never revealed stimulated PRL cells. Some indices of hypoactivity were noted: reduced cell height and nuclear size, underdevelopment of the endoplasmic reticulum and Golgi complex, smaller and less numerous granules and microvesicles, extreme scarcity of exocytoses (Olivereau *et al.*, 1981). Enlarged Agr cells possess pale granules close to the plasma membrane, abundant mitochondria, Golgi cisternae and numerous plasma membrane infoldings.

In goldfish adapted to DW for 20–40 days PRL cells never appeared stimulated. Cell parameters are not affected, but nuclei are often indented and the endoplasmic reticulum and Golgi area are poorly developed, exocytotic figures are scarce (Olivereau *et al.*, 1986a).

Species with a short survival in DW show considerably hypertrophied PRL cells, an almost complete loss of granules and numerous exocytotic figures and highly developed ergastoplasm and Golgi complex (Dubourg *et al.*, 1980). Cells are also stimulated in *Cirrhinus* (Joy 1986) and stickleback (Ogasawara and Yamada 1979).

What are the factors inducing these responses, low osmotic pressure, lack of Na^+ or Ca^{2+}? There is no clear answer. As hypophysectomized eels survive for 45 days at least in DW and can withstand salinity changes, it is evident that PRL does not play a major role in osmoregulation. However eel PRL cells react to environmental changes. Physiological changes of the osmolarity, *in vitro*, do not affect the percentage of release of total and newly synthesized PRL in *Oncorhynchus* (Kelley *et al.*, 1990), but in *Poecilia* PRL synthesis and release are inversely correlated to the external salinity (Batten and Ball 1977). A low Na^+ content of the medium increases eel PRL release *in vitro* (Baker and Ingleton 1975).

In DW addition of Na^+ greatly increases the survival of *Gambusia* and PRL cells retain a normal picture. Addition of Ca^{2+} is less effective on survival and normalization of cell ultrastructure (Dubourg *et al.*, 1983). In tilapia PRL cell activity is not influenced by ambient Na^+ or osmolarity, but by Ca^{2+} level as demonstrated in Ca-free SW, Mg^{2+} playing a minor role only (Wendelaar-Bonga and Van der Meij 1980). However PRL release appears inversely related to the osmotic pressure and independent from Ca^{2+} in the medium (Grau *et al.*, 1981) or also dependent from Ca^{2+} (Wendelaar-Bonga *et al.*, 1985). In goldfish in 1/3 Ca-free SW, PRL cells are slightly stimulated (Olivereau *et al.*, 1986a) and in eels kept in various environments free of or supplemented with Ca^{2+}, plasma osmolarity and Na^+ level seem to play a more important role than external or plasma Ca^{2+} in controlling PRL secretion (Olivereau and Olivereau 1983, Arakawa *et al.*, 1993).

Water acidification. In acid water (pH 4.5) ultrastructural studies show a stimulation of PRL cells and PRL synthesis *in vitro* and Na^+ is reduced in tilapia (Flik *et al.*, 1989). However cell stimulation appears independent from osmotic pressure, Na^+ and Ca^{2+} plasma levels (Wendelaar-Bonga *et al.*, 1988). In the acid tolerant mud-minnow PRL cells are more active at neutral pH than in acid water (Wendelaar-Bonga *et al.*, 1990).

The role(s) of PRL in teleost osmoregulation appears versatile according to the species. PRL binds to gill and kidney membranes in tilapia (Prunet and Auperin 1994) and affects gills, opercular membrane, mitochondira-rich cells, water permeability, kidney, etc., effects that will not be considered in this overview.

Two isoforms of PRL. Teleosts possess two isoforms of PRL, both localized in the same cells and the same granules in tilapia RPD. Two classes of PRL cells cannot be distinguished suggesting that a differential release of the 2 isoforms is not possible (Specker *et al.*, 1993). The role of the 2 PRL variants is still questionable. In eels, the 2 PRLs show a similar effect on Na^+ retention as in hypophysectomized *Fundulus* transferred from SW to FW (Suzuki *et al.*, 1991). In *Oreochromis niloticus* only PRL I reacts to the transfer from FW to brackish water, PRL II changes are undetectable in the gland. Injection of PRL I only induces ionic retention (Auperin *et al.*, 1994). In *O.niloticus* and *O. mossambicus,* PRLs are no longer detected after hypophysectomy. In pituitary and plasma of the 2 species in FW the ratio of the 2 PRLs is not the same, but both PRL decrease in hypertonic environment (Ayson *et al.*, 1993). When *O. mossambicus* is returned to FW the 2 PRLs are increased within 6h, then tiPRL 177 decreases and tiPRL 188 is still increasing for 3 days (Yada *et al.*, 1994). These results suggest a differential regulation of the 2 PRL secretion. During early development of tiapia PRL 177 mRNA and PRL 188 mRNA do not appear simultaneously at the time of hatching but are already involved in FW adaptation (Ayson *et al.*, 1994).

Hypothalamic control. PRL is under an *inhibitory* hypothalamic control evidenced by the stimulation of PRL cells in ectopic pituitaties (*Poecilia*: Olivereau and Ball 1966*, eel: Olivereau 1970a, *Gambusia*: Kah *et al.*, 1979) and *in vitro* (Batten *et al.*, 1983). In cultured trout, pituitary PRL release and labeling of PRLmRNA are increased (Yada *et al.*, 1991).

Dopamine (DA). DA is one of the inhibtiory factors reducing PRL cell activity in eels (Olivereau 1975a), PRL release in *Gillichthys,* tilapia (Nagahama *et al.*, 1975) and trout (Johnston and Wigham 1988) and PRL content in *Poecilia* (Wigham and Ball 1976). PRL cells are less active after injection of DOPA in tilapia (Zambrano *et al.*, 1974) and eel, and of bromocriptine (inhibitor of DA receptors) and stimulated after injection of 6-OHDA (partial destruction of aminergic fibers), reserpine (depletion of brain catecholamines) (Joy 1986), and pimozide (antagonist of DA receptors) (Olivereau 1978a), sulpiride and domperidone (antagonist of D2 receptors) (Olivereau *et al.*, 1988, Slowley *et al.*, 1991).

In the rostral NH, aminergic or B fibers were observed in the roach (Ekengren 1975), *Gambusia* (Kah *et al.*, 1978) and goldfish (Kaul and Vollrath 1974, Kah *et al.*, 1984, 1986), arising from various brain nuclei. Fibers may terminate on the cells (goldfish) or on the basal lamina (eel) (Knowles and Vollrath 1966a* and b*, Vollrath 1967*).

Somatostatin (SRIF). It is synthesized in various brain areas (see GH) and present in NH mainly close to the GH cells. Despite the scarcity of SRIF fibers close to the RPD (Kah *et al.*, 1982, Olivereau *et al.*, 1984a), *in vitro* studies show that SRIF inhibits PRL synthesis and release in *Poecilia* (Wigham and Batten 1984), tilapia (Grau *et al.*, 1982, 1987, Rivas *et al.*, 1986) and PRL cell activity (Batten and Wigham 1984). However, SRIF inhibition is maximal after 24 h and does not affect total PRL and PRLmRNA in trout pituitary (Le Goff *et al.*, 1992a). An antagonist of SRIF blocks the inhibitory effect of SRIF (Williams and Wigham 1994b).

Urotensin II. It is a partial analog of SRIF which inhibits PRL release (the 2 variants) by tilapia pituitary *in vitro* (Grau *et al.*, 1982, Rivas *et al.*, 1986).

Vasoactive Intestinal peptide (VIP) and *Peptide Histidine Isoleucine* (PHI). Both act as PRL releasing factors in mammals. When tested on tilapia RPD *in vitro* both neuropeptides inhibit PRL release in hypo- and hyperosmotic medium. VIP fibers are observed in the vicinity of the RPD and PPD. Opposite responses of PRL to VIP and PHI in tetrapods and fish are then suggested (Kelley *et al.*, 1988).

Stimulating factors: *Serotonin* (5-HT). The treatment with 5-hydroxytryptophan (5-HTP) a precursor of brain 5-HT, induces an increase in PRL cell nuclear area (eel: Olivereau and Olivereau 1979a, goldfish: Olivereau *et al.*, 1980) but cells remain granulated. 5-HT shows a similar but slower effect. Parachlorophenylalanine, a brain 5-HT depletor, reduces PRL cell height and nuclear area, granule density is increased, suggesting that PRL synthesis and release are reduced (Olivereau 1978a). *In vitro* inhibition of monoamine oxidase with pargyline increases PRL content in goldfish pituitary (Olcese *et al.*, 1979) and its release in trout. Tryptophan increases PRL content and cell nuclear area, and 5-HTP stimulates PRL release (James and Wigham 1984).

Estradiol (E2). In trout, E2 stimulates PRL synthesis and release (Williams and Wigham 1994a) or synthesis only (Wigham *et al.*, 1977) and its release in tilapia (Barry and Grau 1986). PRL cells are stimulated in E2-treated eels (Olivereau and Olivereau 1979b, c), showing a developed endoplasmic reticulum, a loss of granules and exocytotic figures along the basal lamina and between PRL and Agr cells (Olivereau *et al.*, 1986a). However, considering the paucity of E2 receptors in RPD compared to the high levels in PPD and IL and the absence of changes in PRL and PRLmRNA in E2-treated trout, a direct regulation of PRL by E2 cannot be demonstrated (Le Goff *et al.*, 1992b).

Thyrotropin-releasing hormone (TRH). It stimulates PRL secretion in mammals and PRL synthesis and release *in vitro* in tilapia (Batten and Wigham 1984, Wigham and Batten 1984) and eel (Hall and Chadwick 1983). However, this effect requires the presence of E2 (Barry and Grau 1986) and recently TRH was reported to reduce PRL synthesis and release in trout (Williams and Wigham 1994a) which possess TRH receptors on pituitary membrane (Schwartzentruber and Omeljaniuk (1995), an effect more marked on PRL II than on PRL I.

Adrenocorticotropic Hormone (ACTH)

Identification. Fish possess interrenal islets scattered in the lymphoid kidney or the wall of cardinal veins and a complete removal is impossible. When metopirone, an inhibitor of 11 β-hydroxylase inducing a sharp decrease of plasma cortisol level, became available, it was tested in eels and *Poecilia*. op'DDD destructing adrenocortical cells and aldactone blocking aldosterone synthesis were also used. These chemical adrenalectomies induce an ACTH hypersecretion allowing the identification of ACTH cells forming a layer along the rostral NH. Immunocytochemistry confirmed these data in many species since their original identification (Follénius and Dubois 1976, Olivereau *et al.*, 1976a, b).

Cytology. ACTH cells are elongated, perpendicular to the basal lamina and easily identified with the Herlant's tetrachrome and lead-hematoxylin (Olivereau 1964*). Metopirone induces a considerable cell and nuclear hypertrophy, a loss of granules apparently released along the basal lamina in eels (Olivereau 1965*) and *Poecilia* (Ball and Olivereau 1966*) and in eels treated with aldactone and op'DDD. Surgical adrenalectomy strongly stimulates ACTH cells (Olivereau and Olivereau 1968*). Conversely, cortisol-treated eels show much smaller ACTH cells with scarce granules (Olivereau 1972).

Ultrastructure. ACTH cells are globular in goldfish, elongated in eel and salmon. The spherical granules (125–170 nm diameter) are located along the plasma membrane. Mitochondria are short and few in number (Nagahama 1973, Olivereau *et al.*, 1986a).

Control of ACTH secretion. In addition to the negative feedback of corticosteroids on ACTH secretion, external osmolarity affects the ACTH cells *in vitro* (Batten *et al.*, 1983). In sea bream transferred to brackish water, the ACTH cells show larger nuclei and endoplasmic reticulum, a paler immunoreaction of larger granules and plasma cortisol level increases (Mancera *et al.*, 1993). In goldfish kept in DW and in 1/3 Ca-free SW, the ACTH cells possess enlarged nuclei, a developed endoplasmic reticulum, a large Golgi area containing numerous vesicles and small granules. Cells are not affected in 1/3 SW (Olivereau *et al.*, 1986a). The ACTH cells appear stimulated in *Gillichthys* treated with E2 (Nagahama *et al.*, 1975) and in precocious mature male parr and smolts of *Salmo salar* (Olivereau 1975b). Observations on radiothyroidectomized and thyroxine (T4)-treated eels suggest that the ACTH cells show reduced and increased activity, respectively (Olivereau 1968b*, c*).

Hypothalamic control. It was suspected as autotransplanted pituitaries possess fewer and less active ACTH cells than *in situ* gland (*Poecilia*: Olivereau and Ball 1966*, eel: Olivereau 1970a) and more or less atrophic interrenal cells (Ball *et al.*, 1965*, Olivereau 1971) although these are more active than in hypophysectomized fish.

The main neuropeptide controlling ACTH secretion is the corticotropin-releasing factor (CRF) identified in the preoptic nucleus (PON) and other brain regions (Bugnon *et al.*, 1983, Olivereau *et al.*, 1984b, 1988a, Batten 1986b, Coto-Montes 1994). CRF fibers end in the rostral NH facing the ACTH cells and in the IL. CRF may be colocalized with arginine vasotocin (AVT) (see Lederis *et al.*, 1994) and urotensin I (U I) and II (Yamada *et al.*, 1985). AVT, sauvagine, urotensin and 3 CRF/U I-like insect peptides stimulate ACTH release in fish (Lederis *et al.*, 1994). Injection of cortisol decreases the hypothalamic CRF. Long term-treated eels show a gradual and almost complete loss of CRF in the rostral NH. Reserpine which stimulates the pituitary-interrenal axis reduces CRF in the rostral NH only, not in the hypothalamus. CRF in the caudal NH is not clearly affected, suggesting a different regulation for CRF in the rostral and caudal NH (Olivereau and Olivereau 1988b). CRF accumulates in the proximal part of the pituitary stalk in hypophysectomized eels (Olivereau and Olivereau 1988c). Metopirone increases CRFmRNA (Lederis *et al.*, 1994), initially reduces and then increases (4 days and more) CRF in the PON. Opposite responses are noted in the rostral NH: CRF increases initially and then decreases. AVT slightly increases in the PON, not in the pituitary. CRF secreting neurons are enlarged and the percentage of perikarya containing both CRF and AVT increases. CRF and ACTH release seems to occur simultaneously. Caudal CRF is barely affected (Olivereau and Olivereau 1990a). An almost total surgical adrenalectomy induces a decrease of CRF immunostaining in enlarged perikarya of the PON, CRF fibers are almot unstained in the rostral NH only, suggesting a high release involved in the degranulation of ACTH cells (Olivereau 1991a).

Dopamine antagonists (domperidone, pimozide) stimulate the ACTH cells and reduce the amount of immunoreactive CRF in the rostral NH (Olivereau *et al.*, 1988). GABA inhibits the cortisol increase in stressed carp probably by an inhibitory action on ACTH release. MCH (melanin-concentrating hormone) inhibits basal and stress-induced ACTH release in trout (see: Olivereau and Olivereau next article).

Brain ACTH. An ACTH-like substance has been extracted from the brain of *Epinephelus* (Ng and Chan 1990). Neurons are identified by immunocytochemistry in the nucleus lateralis tuberis (NLT) in tench (König and Bonn 1990a), goldfish, eel and trout and a dense network of fibers runs in the basal telencephalon and hypothalamus (Olivereau and Olivereau 1990b). Brain ACTH is not affected by hypophysectomy showing that it does not arise from the pituitary, but is reduced in cortisol-treated eels. In metopirone-treated eels, one third of animals showed an increased immunostaining, suggesting an increased ACTH synthesis. Brain ACTH is not affected in other eels. Responses of ACTH in the brain and pituitary occur independently when cortisol synthesis is inhibited (Olivereau and Olivereau 1991b). The physiological role of brain ACTH remains to be elucidated.

Growth Hormone or Somatotropin

Identification. A growth-promoting factor was suspected as hypophysectomized teleosts do not grow (Ball and Baker 1969) unless they are treated with fish pituitary extracts. As anti-mammalian GH sera do not label teleost GH cells, it appeared necessary to isolate fish GH which may have a commercial interest. GH is now isolated from 10 species and is a single chain polypeptide of 196 amino acid residues with 2 conserved disulfide bonds (Chen *et al.,* 1994). By using an anti-ti-GH, labeling of GH cells confirmed the function of the orangeophilic cells of the PPD (Nagahama *et al.*, 1981). Only secretory granules are labeled (Ueda *et al.*, 1985).

Cytology. GH cells are ovoid, filled with oranger G-positive granules and react with neotetrazolium salt and the Barnett and Seligman technique demonstrating SH and S-S groups. They predominate in the PPD of immature fish but their percentage decreases during sexual maturation and GTH cell development. In salmonid, eel and many marine species GH cells are scattered in the PPD and may form an almost complete layer along the basal lamina. In cyprinodont fish, the PPD is composed of a dorsal part containing GH and TSH cells and a ventral part mainly occupied by GTH cells. GH cells are larger and appear more active in females than in males in species exhibiting a sexual dimorphism, the size of females being much larger than that of males (eel, *Poecilia*). In these two species, suppression of T4 synthesis by radiothyroidectomy or thiourea, and T4 administration do not clearly affect the picture of GH cells (Ball and Baker 1969).

Ultrastructure. GH cells possess abundant oval or quadrangular electron dense granules, a small Golgi area and some rough endoplasmic reticulum. Granules are of a variable size and two subpopulations can be discerned (*Gambusia*: Chambolle *et al.*, 1981, eel: Olivereau *et al.*, 1981). Agr cells extend between some GH cells (salmon: Nagahama *et al.*, 1977, eel: Olivereau *et al.*, 1986b), but are less common than in the RPD and rarely show the specialized junctions described among PRL cells.

Responses to environmental changes. In eels and coho salmon adapted to SW, the GH cells are often enlarged with larger granules, active Golgi area and dilated cisternae in the endoplasmic reticulum (Nagahama *et al.*, 1977). During amago salmon smoltification and transfer to SW plasma GH and pituitary GHmRNA are increased (Yada *et al.*, 1992). Plasma GH is higher in maturing parr of *Salmo salar* in 25‰ SW (Rydevik *et al.*, 1990). However ti-GH does not changes in SW (Ayson *et al.*, 1993, 1994) and the osmoregulatory capacity of hypophysectomized *O. niloticus* is not increased by ti-GH treatment (Auperin *et al.,* 1995). In *O. mossambicus* transferred to SW, GH increases in the plasma (Yada *et al.*, 1992) and the pituitary, the total volume of the GH cells in the PPD increases and ultrastructural studies confirm the stimulation of GH cells (Borski *et al.*, 1994). Opposite responses are observed in SW *O. mossambicus* returned to FW. GH treatment reduces the increase of plasma Na^+ after SW transfer in trout (Bolton *et al.*, 1987) in accordance with the first observation of Smith (1956).

Incubation of *Poecilia* pituitary in a medium of reduced osmolarity increases GH release, but Na^+ and Ca^{2+} changes have no effect (Batten *et al.*, 1983).

In DW-eels GH cells are larger and filled with granules (Olivereau 1967*). In goldfish kept in DW, nuclear and cell area increase, granules of an irregular shape are less abundant and exocytoses are often observed. The Golgi complex possesses numerous Golgi vesicles and granules in formation, the rough endoplasmic reticulum may fill degranulated cells (Olivereau *et al.*, 1986a). In eels, there is no biochemical indication for a GH effect on osmoregulation although the GH cells are stimulated in SW and DW.

Osmoregulatory actions of GH in salmonidae transferred to SW involve IGF-I mediation, gill Na^+-K^+-ATPase activity and mitochondria-rich cells. GH is acting through GH receptors in liver, gills, gut and kidney (Sakamoto *et al.*, 1993).

Effect of starvation. After prolonged starvation (eel, carp: Olivereau 1970a, *Boops*: Michéle 1973) GH cells are often completely degranulated with an enlarged ergastoplasm demonstrated by pyronine. Starved carp possess GH cells with a highly developed Golgi complex and endoplasmic reticulum, an activation reduced by refeeding (Gas 1975). Plasma GH level is 6 times higher in starved trout and a sustained exercise accentuates this increase, GH being 22 times higher (Barrett and McKeown 1988), plasma IGF-like level is low (Niu *et al* ., 1993). In starved sea bream free and total hepatic GH-binding and plasma IGF-I-like level are reduced and return to control values after refeeding (Perez-Sanchez *et al.*, 1994).

Reproduction. GH seems to stimulate ovarian steroidogenesis in sea trout (Singh and Thomas 1993) and potentiates the effect of GTH in goldfish (Van der Kraak *et al.*, 1990). GH may act directly on gonads which possess GH receptors.

Triiodothyronine (T3). GH increased peripheral conversion of T4 to T3 through the T4–5' deiodinase (De Luze and Leloup 1984). Iopanoic acid inhibits this conversion and GH is no longer efficient to allow survival of trout in SW. T3 is necessary for the action of GH in the development of osmoregulatory mechanisms in SW (Leloup and Lebel 1993).

Hypothalamic control. It appears evident in hypophysectomized *Poecilia* with an autografted pituitary. Fish start to grow compared to hypophysectomized ones, although moderately, and the GH cells are present in the graft (Ball *et al.*, 1965*). In eel and *Gambusia* pituitary, grafted GH cells are better preserved (Olivereau 1971, Chambolle *et al.*, 1981). However lesions of the periventricular PON induce an increase in plasma GH suggesting an inhibitory control (Cook and Peter 1983, Nishioka *et al.*, 1988). In culture trout pituitary releases high amount of GH for more than one week, the number of weakly labeled and partly degranulated cells with a dilated rough endoplasmic reticulum is increased, the GHmRNA remains stable (Yada *et al.*, 1991).

Dopamine. DA inhibits GH release in mammals. It stimulates GH release from goldfish pituitary through the D1 receptors (Wong *et al.*, 1992, 1993) an effect related to the entry of extracellular Ca^{2+} (Wong *et al.*, 1994). Apomorphine (DA agonist) increases basal and TRH stimulated GH release by carp pituitary (Lin *et al.*, 1993).

Somatostatin. SRIF is detected in many parts of the brain and pituitary (Kah *et al.*, 1982, Vigh-Teichmann *et al.*, 1983, Olivereau *et al.*, 1984a, b, Batten *et al.*, 1985, Grau *et al.*, 1985, König and Bonn 1990b, Bonn 1990). It is present under two forms: SRIF 14 and SRIF 28 (Conlon 1990)—both identified in salmon brain. SRIF 25 is present in fish pancreas only (Nozaki *et al.*, 1988). Both forms inhibit GH synthesis and/or release (Marchant and Peter 1985, Peter *et al.*, 1991). SRIF 22 from catfish pancreas has no effect. The use of ^{3}H-leucine shows that SRIF reduces GH synthesis and release in trout pituitary, GHmRNA is not affected. Addition of GRH to the culture overcomes the effect of SRIF in trout (Yada and Hirano 1992), tilapia (Fryer *et al.*, 1979) and *Poecilia* (Wigham and Batten 1984). SRIF reduces plasma GH during SW adaptation in coho salmon (Sweeting and McKeown 1986) and *in vitro* in tilapia pituitary in hyperosmotic medium (Helms *et al.*, 1987).

Growth Hormone-Releasing Hormone (GHRH). The anti-human pancreatic GHRH 1–44 serum reveals a GHRH-like system in the brain (sea bass: Marivoet *et al.*, 1988, trout: Luo and McKeown 1989, eel: Olivereau *et al.*, 1990) with perikarya in the PON and NLT. GHRH may coexist with AVT. The intrapituitary distribution of fibers varies according to the species. GHRH fibers are present in the RPD (seabass, sculpin, mullet) and in the PPD of all species, penetrating close to GH cells or running along the hypothalamo-hypophyseal tract to the caudal NH. In some species fibers enter the IL and seems to innervate directly the two IL cell types. According to its variable distribution in the gland, GHRH may differently modulate the activity of several cell types. A GHRH-like molecule has been isolated from salmon, carp and sturgeon brains. It contains 45 amino acids, perhaps 48 in catfish and 44 in human. Salmon possess a GHRH/PACAP (pituitary-adenylate cyclase activating peptide) gene (Sherwood *et al.*, 1994). Mammalian GHRH stimulates GH release in goldfish *in vivo,* not *in vitro*, fish GHRH is active *in vitro* (tilapia: Melamed *et al.*, 1995).

Gonadotropin-Releasing Hormone (GnRH). Eight different GnRH with 50–90% similarity have been isolated from fish and cyclostome brain and 6 of them detected in teleost brain (Sherwood *et al.*, 1994). GnRH stimulates directly GH release by goldfish pituitary *in vitro* (Marchant and Peter 1989). sGnRH is active *in vivo* and *in vitro* in tilapia (Melamed *et al.*, 1995). E2 stimulates basal (Le Gac *et al.*, 1993) and sGnRH stimulated GH release (Trudeau *et al.*, 1992).

Thyrotropin-Releasing Hormone. TRH has no effect on GH release *in vitro* (*Poecilia*: Batten and Wigham 1984, tilapia: Melamed *et al.*, 1995).

It is stimulating in eels (Hall and Chadwick 1983). *In vivo* TRH and TSHβ increase plasma GH although T4 is not affected by TRH. T3 increases GH synthesis (increased incorporation of ^{3}H-leucine, Melamed *et al.*, 1995). TRH stimulates GH release from PD fragments of mature and sexually regressed goldfish. *In vivo* treatment with E2 results in a 3-fold increase of the *in vitro* response to TRH, but testosterone is ineffective (Trudeau *et al.*, 1992). TRH is also active on perifused carp pituitary (Lin *et al.*, 1993).

Bombesin. Bombesin/gastrin-releasing peptide-like is detected by immunochemistry in goldfish NH. Bombesin stimulates GH release *in vitro* (Himick and Peter 1995).

Neuropeptide Y (NPY). It is present in brain and pituitary of goldfish (Pontet *et al.*, 1989) and stimulates GH release *in vitro* (Peter *et al.*, 1991), but it may also act indirectly through a presynaptic stimulation of GnRH release by Y2-like receptors which in turn stimulates GH release (Peng *et al.*, 1993). The stimulating effect of NPY is potentiated by testosterone and E2 (Peng *et al.*, 1994).

Urotensin II. It is a partial analog of SRIF with a structure similar to that of the central region of the SRIF 14 (Conlon 1990) but has no inhibitory effect on GH release in tilapia *in vitro* (Rivas *et al.*, 1986).

Insulinlike-Growth Factor I (IGF-I). It is present in many tissues, brain and plasma of trout (Chen *et al.*, 1994). GH seems to act on fish growth through IGF-I and IGF-I mRNA increases in concert with GH during adaptation to SW in organs involved in osmoregulation (gill, kidney) (Sakamoto and Hirano 1993). Injection of GH increases plasma IGF-I with a maximal value in coho salmon smolt (Moriyama *et al.*, 1994). Incubation of trout pituitary cells with IGF-I inbibits the basal secretion of GH (Perez-Sanchez *et al.*, 1992). Hepatic IGF-I mRNA increases in yearling coho salmon just after the GH plasma level in normally growing fish. In stunted fish despite an increased GH level, hepatic IGF-I mRNA is reduced suggesting hepatic GH resistance (Duan *et al.*, 1995). A role in osmoregulation and starvation is suggested in previous sections.

General Conclusions

The subdivisions of the PD, RPD and PPD, are analogous to the cephalic and caudal lobes of amphibian, reptile and bird PD, the PRL and ACTH cells being located in the cephalic lobe, the GH cells in the caudal one (Fontaine and Olivereau 1975).

The lack of a portal system and a typical median eminence in most teleosts need to be reexamined. Indeed, according to the old hypothesis of Diepen (1953) the rostral NH is homologous to the infundibulum of tetrapods. This internalized median eminence contains numerous neuropeptides (CRF, SRIF, GHRH, GnRH, TRH, galanin...) which react to the physiological conditions (Olivereau and Olivereau 1991a). These substances can be released directly on PD cells through synaptic contacts or close to the cells when fibers are ending on the basal lamina and adjacent blood vessels. A more primitive organization is apparent in agnathan in which hypothalamic factors reach the pituitary by diffusion. According to Gorbman (1995) another evolutionary pathway appears in holcephalian with a primitive portal system and in tetrapods with a typical median eminence. Teleosts may also be considered as an intermediate step as some species possess a hypophysial portal system (Sathyanesan 1972) and in all teleosts studied so far the rostral NH containing hypothalamic factors is homologous to an internal median eminence.

In mammals, neuropeptides are commonly reported in adenohypophysial cells in which they may have a paracrine or autocrine function. Such a localization is almost never observed in the PD of teleosts suggesting that these neuropeptides do not act through an autocrine pathway.

Fish only seem to synthesize somatolactin (SL) which belong to the PRL/GH family. SL cells are restricted to the IL. In the PD mammals possess mammosomatotropes containing both GH and PRL which may be located in the same granule. In teleosts, this type of cell has not been observed so far. In species with follicular PRL cells easily identified, immunocytochemistry never demonstrated the presence of GH.

The control of PRL, ACTH and GH secretion is rather similar to that described in other vertebrates, but aquatic animals appear more sensitive to environmental changes than terrestrial ones. The use of fish releasing factors instead of mammalian substances may modify some actual conclusion. Effects of artificially dissociated factors *in vitro* may not be similar to those observed *in vivo*. The multifactorial regulation of pituitary hormone release is of an increasing complexity in correlation to the constant discovery of new neuropeptides.

References

References with after the date are cited in Ball and Baker (1969)

Abraham, M. (1971). Gen. Comp. Endocrinol. **17**: 334–350.

Abraham, M., Kieselstein, M. and Lisson-Begon, S. (1976). Cell Tissue Res. **167**: 289–296.

Aler, G.M. (1971). Acta Zool. **52**: 145–156.

Arakawa, E., Kaneko, T., Tsukamoto, K. and Hirano, T. (1992). Zool. Sci. **9**: 1061–1066.

Arakawa, E., Hasegawa, S., Kaneko, T. and Hirano, T. (1993). J. Comp. Physiol. B **163**: 99–106.

Auperin, B., Rentier-Delrue, F., Martial, J.A. and Prunet, P. (1994). J. Mol. Endocrinol. **12**: 13–24.

Auperin, B., Leguen, I., Rentier-Delrue, F., Smal, J. and Prunet, P. (1995). Gen. Comp. Endocrinol. **97**: 145–159.

Ayson, F.G., Kaneko, T., Tagawa, M., Hasegawa, S., Grau, E.G., Nishioka, R.S., King, D.S., Bern, H.A. and Hirano, T. (1993). Gen. Comp. Endocrinol. **89**: 138–148.

Ayson, F. G., Kaneko, T., Hasegawa, S. and Hirano, T. (1994). Gen. Comp. Endocrinol. **95**: 143–152.

Baker, B.I. and Ingleton, P.M. (1975). J. Comp. Physiol. **100**: 269–283.
Ball, J.N. and Baker, B.I. (1969). Fish Physiol. **2**: 1–110.
Barrett, B.A. and McKeown, B.A. (1988). Can. J. Zool. **66**: 853–855.
Barry, T.P. and Grau, E.G. (1986). Gen. Comp. Endocrinol. **62**: 306–314.
Batten, T.F.C. (1986a). Gen. Comp. Endocrinol **63**: 139–154.
Batten, T.F.C. (1986b). Cell Tissue Res. **244**: 661–672.
Batten, T.F.C. and Ball, J.N. (1977). Cell Tissue Res. **185**: 129–145.
Batten, T.F.C. and Wigham, T. (1984). Cell Tissue Res. **237**: 595–603.
Batten, T.F.C., Young, G. and Ball, J.N. (1983). Gen. Comp. Endocrinol. **51**: 113–130.
Batten, T.F.C., Groves, D.J., Ball, J.N., Cambré, M.L., Verdonck, W.A., Ollevier, F. and Vandesande, F. (1985). Cell Tissue Res. **242**: 115–125.
Betchaku, T. and Douglas, W.W. (1981). Anat. Rec. **199**: 403–421.
Bolton, J.P., Collie, N.L., Kawauchi, H. and Hirano, T. (1987). J. Endocrinol. **112**: 63–68.
Bonn, U. (1990). Z. mikros-anat. Forsch. **104**: 497–508.
Borski, R.J., Yoshikawa, J.S.M., Madsen, S.S., Nishioka, R.S., Zabetian, C., Bern, H.A. and Grau, E.G. (1994). Gen. Comp. Endocrinol. **95**: 483–494.
Bugnon, C., Cardot, J., Gouget, A. and Fellmann, D. (1983). C.R. Acad. Sci. **296 s. III**: 711–716.
Chambolle, P., Kah, O., Olivereau, M., Dubourg, P. and Olivereau, J. (1981). Gen. Comp. Endocrinol. **43**: 85–95.
Chang, Y.S., Huang, F.L. and Lo, T.B. (1992). Gen. Comp. Endocrinol. **87**: 260–265.
Chen, T.T., Marsh, A., Shamblott, M., Chan, K.M., Tang, Y.L., Chen, C.M. and Yang, B.Y. (1994). Fish Physiol. **13**: 179–212.
Conlon, J.M. (1990). In: Progress in Comparative Endocrinology (A. Epple, C.G. Scanes and M.H. Stetson Eds) pp: 10–15, Wiley-Liss, New-York.
Cook, A.F. and Peter, R.E. (1983). Gen. Comp. Endocrinol. **51**: 175–182.
Coto-Montes, A., Garcia-Fernandez, J.M., del Brio, M.A. and Riera, P. (1994). Histol. Histopathol. **9**: 233–242.
De Luze, A. and Leloup, J. (1984). Gen. Comp. Endocrinol. **56**: 308–312.
Diepen, R. (1953). Anat. Anz. Erg. **100**: 111–122.
Duan, C.M., Plisetskaya, E.M. and Dickhoff, W.W. (1995). Endocrinology **136**: 446–452.
Dubourg, P., Chambolle, P., Olivereau, M. and Kah, O. (1980). Gen. Comp. Endocrinol. **41**: 554–562.
Dubourg, P., Chambolle, P., Kah, O., Maiza, S. and Olivereau, M. (1983). Gen. Comp. Endocrinol. **50**: 432–444.
Dubourg, P., Prunet, P., Gonnet, F. and Kah, O. (1986). C.R. Acad. Sci. **303: s.III**: 331–366.
Ekengren, B. (1975). Cell Tissue Res. **158**: 169–175.
Farmer, S.W., Papkoff, H., Bewley, T.A., Hayashida, T., Nishioka, R.S., Bern, H.A. and Li, C.H. (1977). Gen. Comp. Endocrinol. **31**: 60–71.
Flik, G., Van der Velden, J.A., Seegers, H.C.M., Kolar, Z. and Wendelaar-Bonga, S.E. (1989). Gen. Comp. Endocrinol. **75**: 39–45.
Follénius, E. and Dubois, M.P. (1976). Gen. Comp. Endocrinol. **28**: 339–349.
Fontaine, M. (1969). Coll. Internat. CNRS n° 177: 412 pages, CNRS Paris.
Fontaine, M. and Olivereau, M. (1975). Amer. Zool. **15** suppl.: 61–79.
Fryer, J.N., Nishioka, R.S. and Bern, H.A. (1979). Gen. Comp. Endocrinol. **39**: 244–246.
Gas, N. (1975). J. Microsc. Biol. Cell. **23**: 289–300.
Gorbman, A. (1995). Gen. Comp. Endocrinol. **97**: 171–178.
Grau, E.G., Nishioka, R.S. and Bern, H.A. (1981). Gen. Comp. Endocrinol. **45**: 406–408.
Grau, E.G., Nishioka, R.S. and Bern, H.A. (1982). Endocrinology. **110**: 910–915.
Grau, E.G., Nishioka, R.S., Young, G. and Bern, H.A. (1985). Gen. Comp. Endocrinol. **59**: 350–357.

Grau, E.G., Ford, C.A., Helms, L.M.H., Shimoda, S.K. and Cooke, I.M. (1987). Gen. Comp. Endocrinol. **65**: 12–18.

Hall, T.R. and Chadwick, A. (1983). IRSC Med. Sci. **11**: 1009–1010.

Helms, L.M.H., Grau, E.G., Shimoda, S.K., Nishioka, R.S. and Bern, H.A. (1987). Gen. Comp. Endocrinol. **65**: 48–55.

Himick, B.A. and Peter, R.E. (1995). Neuroendocrinology **61**: 365–376.

James, V.A. and Wigham, T. (1984). Gen. Comp. Endocrinol. **56**: 231–239.

Johnston, L.R. and Wigham, T. (1988). Gen. Comp. Endocrinol. **71**: 284–291.

Joy, K.P. (1986). Z. mikros-anat. Forsch. **100**: 500–505.

Kah, O., Chambolle, P. and Olivereau, M. (1978). C.R. Acad. Sci. **286 s.D**: 705–708.

Kah, O., Chambolle, P., Olivereau, M., Dubourg, P. and Surle've-Bazeille, J.E. (1979). Gen. Comp. Endocrinol. **38**: 253–263.

Kah, O., Chambolle, P., Dubourg, P. and Dubois, M.P. (1982). C.R. Acad. Sci. **294 sIII**: 519–524.

Kah, O., Dubourg, P., Chambolle, P. and Calas, A. (1984). Cell Tissue Res. **238**: 621–626.

Kah, O., Dubourg, P., Onteniente, B., Geffard, M. and Calas, A. (1986). Cell Tissue Res. **244**: 577–582.

Kaul, S. and Vollrath, L. (1974). Cell Tissue Res. **154**: 231–249.

Kelley, K.M., Nishioka, R.S. and Bern, H.A. (1988). Gen. Comp. Endocrinol. **72**: 97–106.

Kelley, K.M., Nishioka, R.S. and Bern, H.A. (1990). J. Exp. Zool. **254**: 72–78.

König, B. and Bonn, U. (1990a). Z. mikros-anat. Forsch. **104**: 327–336.

König, B. and Bonn, U. (1990b). J. Hirnforsch. **31**: 227–236.

Leatherland, J.F. and Ensor, D.M. (1973). Can. J. Zool. **51**: 225–235.

Lederis, K., Fryer, J.N., Okawara, Y., Schonrock, C. and Richter, D. (1994). Fish Physiol. **13**: 67–100.

Le Gac, F., Blaise, O., Fostier, A., Le Bail, P.Y., Loir, M., Mourot, B. and Weil, C. (1993). Fish Physiol. Biochem. **11**: 219–232.

Le Goff, P., Weil, C., Valotaire, Y., Gonnard, J.F. and Prunet, P. (1992a). J. Mol. Endocrinol. **9**: 137–146.

Le Goff, P., Salbert, G., Prunet, P., Saligaut, C., Björnsson, B.T., Haux. C. and Valotaire, Y. (1992b). Mol. Cell. Endocrinol. **90**: 133–140.

Leloup, J. and Lebel, J.M. (1993). Fish Physiol. Biochem. **11**: 165–173.

Lin, X. W., Lin, H.R. and Peter, R.E. (1993). Fish. Physiol. Biochem. **11**: 71–76.

Luo, D. and McKeown, B.A. (1989). Experientia **45**: 577–579.

Mancera, J.M., Fernandez-Llebrez, P., Grondona, J.M. and Perez-Figares, J.M. (1993). Gen. Comp. Endocrinol. **90**: 220–231.

Marchant, T.A. and Peter, R.E. (1985). Fish Culture B7–19.

Marchant, T.A. and Peter, R.E. (1989). Fish Physiol. Biochem. **7**: 133–139.

Marivoet, S., Moons, L. and Vandesande, F. (1988). Gen. Comp. Endocrinol. **72**: 72–79.

Melamed, P., Eliahu, N., Levavi-Sivan, B., Ofir, M., Farchi-Pisanty, O., Rentier-Delrue, F., Smal, J.,Yaron, Z. and Naor, Z. (1995). Gen. Comp. Endocrinol. **97**: 13–30.

Michéle, M. (1973). C.R. Soc. Biol. **167**: 1202–1206.

Moriyama, S., Swanson, P., Nishii, M., Takahashi, A., Kawauchi, H., Dickhoff, W.W. and Plisetskaya, E.M. (1994). Gen. Comp. Endocrinol. **96**: 149–161.

Nagahama, Y. (1973). Mem. Fac. Fish. Hokkaido Univ. **21**: 1–63.

Nagahama, Y., Nishioka, R.S., Bern, H.A. and Gunther, R. (1975). Gen. Comp. Endocrinol. **25**: 168–188.

Nagahama, Y., Clarke, W.C. and Hoar, W.S. (1977). Can. J. Zool. **55**: 183–198.

Nagahama, Y., Olivereau, M., Farmer, S.W., Nishioka, R.S and Bern, H.A. (1981). Gen. Comp. Endocrinol. **44**: 389–395.

Naito, N., Takahashi, A., Nakai, Y., Kawauchi, H. and Hirano, T. (1983). Gen. Comp. Endocrinol. **50**: 282–291.

Ng, T.B. and Chan, S.T.H. (1990). Comp. Biochem. Physiol. **95C**: 159–162.
Nishioka, R.S., Kelley, K.M. and Bern, H.A. (1988). Zool. Sci. **5**: 267–280.
Niu, P.D., Perez-Sanchez, J. and Le Bail, P.Y. (1993). Fish Physiol. Biochem. **11**: 318–391.
Nozaki, M., Miyata, K., Oota, Y., Gorbman, A. and Plisetskaya, E.M. (1988). Gen. Comp. Endocrinol. **69**: 267–280.
Ogsawara, T. and Yamada, J. (1979). Zool. Mag. **88**: 314–317.
Olcese, J.M., Hall, T.H., Figueroa, H.R. and de Vlaming, V.L. (1979). Gen. Comp. Endocrinol. **38**: 309–313.
Olivereau, M. (1969). Gen. Comp. Endocrinol. suppl. **2**: 32–41.
Olivereau, M. (1970a). Coll. Nat. Neuroendocrinol. n°927: 251–260 J. Benoit Ed, CNRS Paris.
Olivereau, M. (1970b). C.R. Acad. Sci. **270 s.D**: 2343–2346.
Olivereau, M. (1971). Acta Zool. **52**: 69–83.
Olivereau, M. (1972). Acta Zool. **53**: 179–194.
Olivereau, M. (1975a). Gen. Comp. Endocrinol. **26**: 550–561.
Olivereau, M. (1975b). Gen. Comp. Endocrinol. **27**: 9–27.
Olivereau, M. (1978a). Cell Tissue Res. **189**: 219–230.
Olivereau, M. (1978b). Cell Tissue Res. **191**: 93–99.
Olivereau, M. and Olivereau, J. (1979a). Cell Tissue Res. **196**: 397–408.
Olivereau, M. and Olivereau, J. (1979b). Gen. Comp. Endocrinol. **39**: 247–261.
Olivereau, M. and Olivereau, J. (1979c). Cell Tissue Res. **199**: 431–454.
Olivereau, M. and Olivereau, J. (1983). Cell Tissue Res. **229**: 243–252.
Olivereau, M. and Olivereau, J. (1988a). Peptides **9**:13–21.
Olivereau, M. and Olivereau, J. (1988b). Peptides **9**: 1261–1268.
Olivereau, M. and Olivereau, J. (1988c). Neurosci-Lett. **91**: 271–275.
Olivereau, M. and Olivereau, J. (1990a). Gen. Comp. Endocrinol. **80**: 199–215.
Olivereau, M. and Olivereau, J. M. (1990b). Cell Tissue Res. **262**: 115–123.
Olivereau, M. and Olivereau, J. (1991a). Gen. Comp. Endocrinol. **81**: 295–303.
Olivereau, M. and Olivereau, J.M. (1991b). Cell Tissue Res. **265**: 485–492.
Olivereau, M., Bugnon, C. and Fellmann, D. (1976a). C.R. Acad. Sci. **283 s.D**: 1321–1323.
Olivereau, M., Bugnon, C. and Fellmann, D. (1976b). C.R. Acad. Sci. **283 s.D**: 1441–1443.
Olivereau, M., Chambolle, P. and Dubourg, P. (1981). Bio. Cell **42**: 153–166.
Olivereau, M., Chambolle, P., Dubourg, P. and Olivereau. J. (1986a). Bio. Cell **57**: 77–88.
Olivereau, M., Dubourg, P., Chambolle, P. and Olivereau, J. (1986b). Cell Tissue Res. **246**: 425–437.
Olivereau, M., Olivereau, J.M. and Aimar, C. (1980). Cell Tissue Res. **207**: 377–385.
Olivereau, M., Olivereau, J.M. and Lambert, J.F. (1988). Gen. Comp. Endocrinol. **71**: 506–515.
Olivereau, M., Olivereau, J. and Vandesande, F. (1990). Cell Tissue Res. **259**: 73–80.
Olivereau, M., Ollevier, F., Vandesande, F. and Olivereau, J. (1984a). Cell Tissue Res. **238**: 289–296.
Olivereau, M., Ollevier, F., Vandesande, F. and Verdonck (1984b). Cell Tissue Res. **237**: 379–382.
Olsson, R. and Fujita, H. (1976). Cell Tissue Res. **172**: 185–194.
Peng, C., Chang, J.P., Yu, K.L., Wong, A.O.L., Van Goor, F., Peter, R.E. and Rivier, J.E. (1993). Endocrinology **132**: 1820–1829.
Peng, C., Gallin, W., Peter, R.E., Blomqvist, A.G. and Larhammar, D. (1994). Endocrinology **134**, 1095–1103.
Perez-Sanchez, J., Weil, C. and Le Bail, P.Y. (1992). J. Exp. Zool. **262**: 287–290.
Perez-Sanchez, J., Marti-Palanca, H. and Le Bail, P.Y. (1994). J. Fish Biol. **44**: 287–301.
Peter, R.E., Chang, J.P., Peng, C., Trudeau, V., Somoza, G., Nahorniak, C.S., Wong,

A. O.L., Helms, L.M.H., Rivier, J.E. and Vale, W.W. (1991). Gen. Comp. Endocrinol. **82**: 308–309.

Pontet, A., Danger, J.M., Dubourg, P., Pelletier, G., Vaudry, H., Calas, A. and Kah, O. (1989). Cell Tissue Res. **255**: 529–538.

Prunet, P. and Auperin, B. (1994). Fish Physiol. **13**: 367–392.

Quérat, B., Cardinaud, B., Hardy, A., Vidal, B. and d' Angelo, B. (1994). Mol. Cell Endocrinol. **102**: 151–160.

Rand-Weaver, M., Baker, B.I. and Kawauchi, H. (1991). Cell Tissue Res. **263**: 207–215.

Rivas, R.J., Nishioka, R.S. and Bern, H.A. (1986). Gen. Comp. Endocrinol. **63**: 245–251.

Rydevik, M., Borg, B., Haux, C., Kawauchi, H. and Björnsson, B.T. (1990). Gen. Comp. Endocrinol. **80**: 9–15.

Sakamoto, T. and Hirano, T. (1993). Proc. Nat. Acad. Sci. USA **90**: 1912–1916.

Sakamoto, T., McCormix, S.D. and Hirano, T. (1993). Fish Physol. Biochem. **11**: 155–164.

Sathyanesan, A.G. (1972). Acta Anat. **81**: 349–366.

Schwartzentruber, R.S. and Omeljaniuk, R.J. (1995). Gen. Comp. Endocrinol. **97**: 209–219.

Sherwood, N.M., Parker, D.B., McRory, J.E. and Lescheid, D.W. (1994). Fish Phyisol. **13**: 3–66.

Singh, H. and Thomas, P. (1993). Gen. Comp. Endocrinol. **89**: 341–353.

Sinha, Y.N. (1995). Endocr. Rev. **16**: 354–369.

Sloley, B.D., Trudeau, V.L., Dulka, J.G. and Peter, R.E. (1991). Can. J. Physiol. Pharmacol. **69**: 776–781.

Smith, D.C.W. (1956). Mem. Soc. Endocrinol. **5**: 83–101.

Specker, J.L., Kishida, M., Huang, L., King, D.S., Nagahama, Y., Ueda, H. and Anderson, T.R. (1993). Gen. Comp. Endocrinol. **89**: 28–38.

Suzuki, R., Yasuda, A., Konado, J., Kawauchi, H. and Hirano, T. (1991). Gen. Comp. Endocrinol. **81**: 391–402.

Sweeting, R.M. and McKeown, B.A. (1986). Can. J. Zool. **64**: 2062–2063.

Trudeau, V.L., Somoza, G.M., Nahorniak C.S. and Peter, R.E. (1992). Neuroendocrinology **56**: 483–490.

Ueda, H., Kagatva, H. and Fujimoto, S. (1985). Gen. Comp. Endocrinol. **59**: 149–154.

Van der Kraak, G., Rosenblum, P.M. and Peter, R.E. (1990). Gen. Comp. Endocrinol. **79**: 233–239.

Vigh-Teichmann, I., Vigh, B., Korf, H.W. and Oksche, A. (1983). Cell Tissue Res. **233**: 319–334.

Wendelaar-Bonga, S.E. and Van der Meij, J.C.A. (1980). Gen. Comp. Endocrinol. **40**: 391–401.

Wendelaar-Bonga, S.E., Balm, P.H.M. and Flik, G. (1988). Gen. Comp. Endocrinol. **72**: 1–12.

Wendelaar-Bonga, S.E., Flik, G., Löwik, C.W.G.M. and van Eys, G.J.J.M. (1985). Gen. Comp. Endocrinol. **57**: 352–359.

Wendelaar-Bonga, S.E., Flik, G., Van der Velden, J.A. and Kolar, Z. (1990). Gen. Comp. Endocrinol. **78**: 421–432.

Wigham, T. and Ball, J.N. (1976). J. Comp. Physiol. **110**: 135–143.

Wigham, T. and Batten, T.F.C. (1984). Gen. Comp. Endocrinol. **55**: 444–449.

Wigham, T., Nishioka, R.S. and Bern, H.A. (1977). Gen. Comp. Endocrinol. **32**: 120–131.

Williams, A.J. and Wigham, T. (1994a). Gen. Comp. Endocrinol. **93**: 388–397.

Williams, A.J. and Wigham, T. (1994b). Gen. Comp. Endocrinol. **93**: 398–405.

Wong, A.O.L., Chang, J.P. and Peter, R.E. (1992). Endocrinology **130**: 1201–1210.

Wong, A.O.L., Chang, J.P. and Peter, R.E. (1993). Endocrinology **133**: 577–584.

Wong, A.O.L., Van Goor, F. and Chang, J.P. (1994). Gen. Comp. Endocrinol. **94**: 316–328.

Yada, T. and Hirano, T. (1992). J. Comp. Physiol. B **162**: 575–580.
Yada, T., Urano, A. and Hirano, T. (1991). Endocrinology **129**: 1183–1192.
Yada, T., Kobayashi, T., Urano, A. and Hirano, T. (1992). J. Exp. Zool. **262**: 420–426.
Yada, T., Hirano, T. and Grau, E.G. (1994). Gen. Comp. Endocrinol. **93**: 214–223.
Yamada, C., Owada, K., Kobayashi, H. (1985). Zool. Sci. **2**: 813–816.
Yamamoto, T., Kataoka, K., Ochi, J. and Honma, Y. (1982). Arch. Histol. Jap. **45**: 355–364.
Zambrano, D., Clarke, W.C., Hawkins, E.F., Sage, M. and Bern, H.A. (1974). Neuroendocrinology **13**: 284–298.

Addendum. GH has been isolated and purified from fish pituitary and homologous radioimmunoassays and ELISA are now available: *Oncorhynchus keta* (Farbridge and Leatherland 1991) and *O. tshawytschal* (Le Bail *et al.*, 1991), gilthead sea bream (Le bail *et al.,* 1993), *Solea senegalensis* (Pendon *et al.*, 1994), goldfish (Mahmoud *et al.*, 1996), and European eel (Marchelidon *et al.*, 1996). African catfish GH is partially characterized (Berghman *et al.*, 1996). Amino acid sequences of these GHs show some differences, but even among cyprinid GHs, the number of cysteine residue is variable (Mahmoud *et al.*, 1996). GH and GHmRNA are precociously detected in larval sea bream (3 and 2 days after hatching, respectively) (Funkenstein and Cohen 1996).

Fish GH receptors are identified in liver membranes (eel: Hirano 1991, coho salmon: Gray *et al.*, 1992, tilapia: Ng *et al.*, 1992, *Channa*: Ng *et al.*, 1993, and other species: Ng *et al.*, 1994), the central servous system (trout: Perez-Sanchez *et al.*, 1991) and gonads (trout: Le Gac *et al.*, 1992, *Pleuronectes*: Ng *et al.*, 1993). Specific and saturable binding sites or GH are detected early in the growth period in the head region of *Sparus auratus* (Marti-Palanca and Perez-Sanchez 1994).

Injection of NMDA (N-methyl-D,L-aspartic acid), a glutamate agonist, inhibits GH release. E2 increases plasma GH and increases the inhibitory influence of NMDA in goldfish (Trudeau *et al.*, 1996). Calmodulin and arachidonic acid modulate GH release stimulated by dopamine and GnRH (Chang *et al.*, 1996). Studies on the production of fish GHs are stimulated by their application in fish farms and fish transfected with GH gene are now investigated (Gross *et al.*, 1992).

GHRH-like peptides are isolated from catfish (McRory *et al.*, 1995) and carp hypothalamus. cGHRH shows only 45% identity with mammalian GHRH and stimulates pituitary GH release in goldfish (Vaughan *et al.*, 1992). Immune serum against cGHRH stains several diencephalic nuclei, many pinealocytes projecting into the pineal lumen, few corticotropes and some central pars intermedia cells, as well as abundant diencephalic and telencephalic fibers (Rao *et al.*, 1996).

Additional roles of GH are now considered. GH binds in trout testis and ovary which also seem to produce a GHRH-like peptide (Le Gac *et al.*, 1992). GH potentiates the effect of E2 on vitellogenin synthesis in eel hepatocytes (Peyon *et al.*, 1996). Both GH and IGF-I increase salinity tolerance and gill Na^+ -K^+ -ATPase activity in Atlantic salmon, an effect which varies with the initial environmental salinity, and in the case of IGF-I, is modulated by cortisol (McCormick 1996). GH increases predation exposure (Jöhnsson *et al.*, 1996) and plays a complex role in dominance of rainbow trout (Jönsson and Björnsson 1994; Jöhnsson *et al.*, 1996). GH appears as a "long-term-survival hormone" (Björnsson 1997). In yellow eels starved for 3 months, the GH levels are increased in plasma and pituitary (Marchelidon *et al.*, 1996). In both fasted yellow and silver eels, the volume of the PPD mainly composed of GH cells is increased and the cross sectional area of GH cells increases by 90 to 108% after 3–4 years. The cell area is negatively and significantly correlated to the condition factor. Initially, GH cells show a dense immunostaining with the anti-eel GH. After 2 years and more, the general immunostaining is reduced and often restricted to a few group of granules mainly located in the basal pole, near the basal lamina. After a prolonged period of overstimulation, GH cells are unable to maintain their hormonal storage (Olivereau and Olivereau 1997). Starvation (15 days) or food restriction tend to increase plasma GH, but its level is not correlated to the specific growth rate in tilapia, growth being also controlled by triiodothyronine (Toguyeni *et al.*, 1996). Fasting (30–60 days) increases the IGF-binding protein I in the striped bass (Siharath *et al.*, 1996). As GH receptors (Gray *et al.*, 1992) and IGF-I levels are low in fasted fish, the increased circulating GH is prevented to express most of its peripheral effects. GH injection increases IGF-I plasma levels (Moriyama 1995). IGF-I which binds to its brain receptors (Leibush *et al.*, 1996) inhibits GH release from trout pituitary cells (Blaise *et al.*, 1995). The multifactorial control of GH secretion in teleost fish appears as complex as in mammals and fish GH targets are probably more diverse.

Berghman, L.R., Lescroart, O., Roelants, I., Ollevier, F., Kühn, E.R., Verhaert, P.D., De Loof, A., Van Leuven, F. and Vandesande, F. (1996). Comp. Biochem. Physiol. **113B**: 773–780.

Björnsson, B.T. (1997). Fish Physiol. Biochem. in press.

Blaise, O., Weil, C. and Le Bail, P.Y. (1995). Growth Regul. **5**: 142–150.

Chang, J.P., Abele, J.T., Van Goor, F., Wong, A.O.L. and Neuman, C.M. (1996). Gen. Comp. Endocrinol. **102**: 88–101.

Farbridge, K.J. and Leatherland, J.F. (1991). Gen. Comp. Endocrinol. **83**: 7–17.

Funkenstein, B. and Cohen, I. (1996). Growth Regul. **6**: 16–21.

Gray, E.S., Kelley, K.M., Law, S., Tsai, R., Young, G. and Bern, H.A. (1992). Gen. Comp. Endocrinol. **88**: 243–252.

Gross, M.L., Schneider, J.F., Moas, N., Alvarez, C., Myster, S.H., Lin, Z., Hallerman, E.M., Hackett, P.B., Guise, K.S., Faras, A.J. and Kapuscinski, A.R. (1992). Aquaculture **103**: 253–274.

Hirano, T. (1991). Gen. Comp. Endocrinol. **81**: 383–390.

Jöhnsson, J.I. and Björnsson, B.T. (1994). Anim. Behav. **48**: 177–186.

Jöhnsson, J.I., Jönsson, E. and Björnsson, B.T. (1996). Hormones Behav. **30**: 13–21.

Jönsson, E., Jöhnsson, J.I. and Björnsson, B.T. (1996). Proc. R. Soc. London B **263**: 647–651.

Le Bail, P.Y., Mourot, B., Zohar, Y. and Perez-Sanchez, J. (1993). Can. J. Zool. **71**: 1500–1505.

Le Bail, P.Y., Sumpter, J.P., Carragher, J.F., Mourot, B., Niu, P.D and Weil, C. (1991). Gen. Comp. Endocrinol. **83**: 75–85.
Le Gac, F., Ollitrault, M., Loir, M. and Le Bail, P.Y. (1992). Biol. Reprod. **46**: 949–957.
Leibush, B., Parrizas, M., Navarro, I., Lappova, Y., Maestro, M.A., Encinas, M., Plisetskaya, E.M. and Guttierez, J. (1996). Regul. Peptides **61**: 155–162.
Mahmoud, S.S., Moloney, M.M. and Habibi, H.R. (1996). Gen. Comp. Endocrinol. **101**: 139–144.
Marchelidon, J., Schmitz, M., Houdebine, L.M., Vidal, B., Le Belle, N. and Dufour, S. (1996). Gen. Comp. Endocrinol. **102**: 360–369.
Marti-Palanca, H. and Perez-Sanchez, J. (1994). Growth Regul. **4**: 14–19.
McCormick, S.D. (1996). Gen. Comp. Endocrinol. **101**: 3–11.
McRory, J.E., Parker, D.B., Ngamvongchon, S. and Sherwood, N.M. (1995). Mol. Cell. Endocrinol. **108**: 169–177.
Moriyama, S. (1995). Growth Regul. **5**: 164–167.
Ng, T.B., Leung, T.C., Cheng, C.H.K. and Woo, N.Y.S. (1992). Gen. Comp. Endocrinol. **86**: 111–118.
Ng, T.B., Cheng, C.H.K., Woo, N.Y.S. and Wright, G.M. (1993). Comp. Biochem. Physiol. **106A**: 483–488.
Ng, T.B., Kwan, L.H.Y. and Cheng, C.H.K. (1994). Comp. Biochem. Physiol. **107A**: 101–105.
Olivereau, M. and Olivereau, J.M. (1997). Fish Physiol. Biochem. in press.
Pendon, C., Martinez-Barbera, J.P., Perez-Sanchez, J.. Rodriguez, R.B., Grenett, H. and Valdivia M.M. (1994). Gene **145**: 237–240.
Perez-Sanchez, J., Smal, J. and Le Bail, P.Y. (1991). Growth Regul. **1**: 145–152.
Peyon, P., Baloche, S. and Burzawa-Gérard, E. (1996). Gen. Comp. Endocrinol. **102**: 263–273.
Rao, S.D., Prasada Rao, P.D. and Peter, R.E. (1996). Gen. Comp. Endocrinol. **102**: 210–220.
Siharath, K., Kelley, K.M. and Bern, H.A. (1996). Gen. Comp. Endocrinol. **120**: 307–316.
Toguyeni, A., Baroiller, J.F., Fostier, A., Le Bail, P.Y., Kühn, E.R., Mol, K.A. and Fauconneau, B. (1996). Gen. Comp. Endocrinol. **103**: 167–175.
Trudeau, V.L., Sloley, B.D., Kah, O., Mons, N., Dulka, J. G. and Peter, R.E. (1996). Gen. Comp. Endocrinol. **103**: 129–137.
Vaughan, J.M., Rivier, J., Spiess, J., Peng, G., Chang, J.P., Peter, R.E. and Vale, W. (1992). Neuroendocrinology **56**: 539–549.

Comparative Endocrinology and Reproduction
K.P. Joy, A. Krishna and C. Haldar (Eds)

6. The Intermediate Lobe of Teleost Fish: POMC-MSH and Somatolactin. Functions, Environmental Control and Enigma

J.M. Olivereau[1] and M. Olivereau*[1]
[1]Université René Descartes, F-75005 Paris, France
*Neuroendocrinology, 51 rue de Lancry, F-75010 Paris, France

Summary
The intermediate lobe (IL) of teleosts is ramified among neurohypophysial digitations and composed of two types stained with lead-hematoxylin and PAS in most species. They are immunolabeled with antisera to MSH or POMC and to somatolactin (SL). MSH cells react to a black (B) or white (W) background (Bg) in some species, not in others. They may be slightly stimulated in hypoosmotic environments or not, probably playing a minor role in osmoregulation. SL cells are strongly stimulated on a B-Bg and inactive on a W-Bg in species whose MSH cells are unresponsive. In some species, both cell types react to the Bg, making uncertain a general correlation between MSH cell activity and malanodispersion Hypothalamic control of MSH (MCH, dopamine, serotonin...) is briefly reviewed. SL seems to participate in ionoregulation and acid-base equilibrium. However the stimulation of SL cells of fish in deionized or freshwater compared to that in seawater does not occur in all species studied so far. The great variability of MSH and SL response to the same factor is an incitation to study more species to understand the role(s) and control of these two hormones. Experimental responses show such a high diversity that the precise functions of the large IL of teleosts remain enigmatic.

Introduction
The pituitary of almost all vertebrates, but birds, possesses an intermediate lobe (IL) which secretes a hormone controlling the dispersion of melanosomes in pigment cells. The name of intermediate appears justified in tetrapods because this lobe is located between the pars distalis (PD) and the neural lobe or neurohypophysis (NH). In teleost fish, neither the IL nor the NH shows a compact structure as in tetrapods, instead the IL is composed of elongated strands of cells exhibiting numerous digitations surrounded by equally numerous NH ramifications. These mixed lobes form a voluminous

neurointermediate lobe (NIL) compared to the rostral (RPD) and proximal (PPD) PD. IL is not sharply limited and some strands may extend along the hypothalamo-neurohypophysial tract in the PPD. In several species (*Mugil, Cottus,* sea bass, sculpin...) gonadotropic (GTH) cell islets are noted at the periphery of the NIL. This complex structure in teleosts makes impossible the removal of IL alone or IL and NH, easily performed in tetrapods. This typical organization of the NIL considerably increases the surface of contact between the endocrine cells (IL) and the NH which contains many neuropeptides and neurotransmitters. A pituitary cleft is not present.

In teleost fish as in other vertebrates, the NIL is caudally located in the pituitary. However, in large mature fish such as salmonids, the long axis of the pituitary is vertical, the NIL being ventral, whereas in small immature fish the longitudinal axis is horizontal with a caudal NIL. *Myoxocephalus* shows a still different anatomy: the axis is horizontal, but probably during the embryonic development a torsion of the gland results in a "rostral" PD located caudally and a NIL rostrally, the hypothalamo-hypophysial tract coursing anteriorly and not posteriorly as in other species (Olivereau and Callard 1985).

In tetrapods, IL is composed of a single cell type synthesizing a proopiomelanocortin (POMC) further processed into several POMC-related peptides including MSH (Smith and Funder 1988). In teleosts, in addition to cells secreting MSH initially called intermedin, chromatophore hormone (Pickford and Atz 1957*), then melanocyter-stimulating hormone or melanotropin, there is another acidophilic or basophilic cell type already identified in *Salmo salar* (Woodman 1939). These cells are PAS+ in many species, but unstained in salmonids (Olivereau and Olivereau, see the other article). They were recently identified as somatolactin (SL) secreting cells. As there is no evidence so far that MSH and SL cells are controlling each other, they will be considered separately.

MSH Cells

Hormonal secretion. MSH is a peptide corresponding to the 1-13 sequence of the POMC molecule (Eipper and Mains 1980) whose cleavage gives rise to MSH, ACTH 1-39, corticotropin-like intermediate peptide 18-39 (CLP) and endorphin (END). The NIL of goldfish releases a large number of POMC-derived peptides including 7 peaks of ACTH and 6 of MSH (Tran *et al.,* 1989). At least 3 forms of αMSH are identified in the pituitary, the plasma and *in vitro* in the incubation medium: desacetyl-, monoacetyl-, and diacetyl αMSH in cyprinids and trout, with a predominance of the diacetylated form in the pituitary and the monoacetylated form in the plasma (Follénius *et al.,* 1985, 1986). The release of the 3 forms is not related to their intracellular relative concentration (Lamers *et al.,* 1991). Gamma-MSH is neither present in the N-terminal peptide of salmon POMC (Kawauchi *et al.,* 1981) nor in the sequence of the POMCmRNA in the chum salmon (Kitahara *et al.,*1988), but CLIP (Kawauchi *et al., 1982),* βMSH and 2 END (Kawauchi *et al.*, 1980a, b) are identified. Trout NIL possesses 2 kinds of POMC with 2 distinct END (Rodrigues *et al.*, 1983). The 2 POMC genes are expressed in almost all IL cells (Salbert *et al.*, 1992).

Cytology. Abundant papers on MSH cells in various species are reviewed in Ball and Baker (1969). MSH cells are often columnar and perpendicular to the basal lamina (BL) which separates IL from the NH. Abundant granules surround the oval nucleus and seem to migrate toward the basal pole under stimulation. Blood vessels and capillaries are never observed in the IL (Jasinski 1961*, 1962*), but they are abundant in the NH, mainly along the BL. A mitotic activity is observed in salmonids, mainly in salmon collected in the ocean, but it is scarcely detected in eels, at least in animals collected during day-time.

MSH cells show no cytochemical staining. The lead-hematoxylin (PbH) technique (McConaill 1947) applied by Stahl (1958*) gives a typical blue-black staining in both MSH and ACTH cells (Olivereau 1964*, Baker 1967*, Ball and Baker 1969). In eels, PbH cell activity varies in parallel to the skin melanophore index (Olivereau 1971, Baker 1972).

Immunocytochemistry. This technique allows the demonstration of MSH synthesis in PbH cells in the stickleback (Follénius and Dubois 1974, 1977), tench (Romain *et al.*, 1974), *Boops* (Malo-Michele *et al.*, 1976), eel, trout (Olivereau *et al.*, 1976a, b), tilapia (Van Eys and van den Oetelaar 1981), *Aequidens pulcher* (Munro 1985), *Sparus auratus* (Quesada *et al.*, 1988). MSH cells are also immunostained with anti-ACTH serum (Malo-Michele *et al.*, 1976, Olivereau *et al.*, 1976a, b, Cambré *et al.*, 1986), anti-βEND (Follénius and Dubois 1978, Van Eys *et al.*, 1984, Naito *et al.*, 1984, Siegmund *et al.*, 1987) and the anti-NH^2- terminal peptide of salmon POMC (Naito *et al.*, 1984).

Ultrastructure. MSH cells are identified by their large granules (Knowles and Vollrath 1966*) which may vary under different conditions (Vollrath 1966*). MSH cells predominate and contain a heterogeneous population of pale and dense granules (Thornton and Howe 1974) or moderately electron dense granules (Olivereau *et al.*, 1981b). Their fixation is more delicate than that of the PAS granules and the membrane may not be adjacent to the dense core of the granule. A substructure may be apparent in the pale granules, consisting of bundles of filaments with regular cross-striations (Thornton and Howe 1974). A few large granules (up to 1.25 μm diameter) show a fibrillar structure and probably contain glycoproteins as they stain with PAS (Olivereau and Chambolle 1979). In goldfish in FW, MSH cells are more globular. The rough endoplasmic reticulum (RER) and Golgi area (GA) are poorly developed. The nucleus is so deeply indented that some cells seem to be binucleated. The cytoplasm is filled with electron lucent or pale granules, some darker granules and few mitochondria (Olivereau *et al.,* 1983).

Agranular cells (Agr). They are elongated between MSH and SL cells and less abundant than in the RPD. They contain few mitochondria and polymorphic or deeply indented nuclei. The GA is rarely discernible (Olivereau *et al.,* 1981b, 1983).

Innervation. Adjacent NH strands contain peptidergic and aminergic fibers identified by immunocytochemistry. Previous characterization according to the size, morphology and ultrastructure of the granules will not be discussed here. Carbocyanin tract tracing and/or immunocytochemistry allows the identification of peptidergic systems (arginine vasotocin/isotocin and dopamine) in salmon (Holmqvist and Ekström 1995), melanin-concentrating hormone (MCH) (Baker 1991) and many other neuropeptides. In agreement with the original observation of Follénius (1971) in the stickleback, nerve fibers incorporate ^{3}H-dopamine (Kah *et al.,* 1984), and react with antiserum to dopamine (Kah *et al.,* 1986), making synaptoid contacts with MSH and SL cells in goldfish. The presence and enigmatic role of corticotropin-releasing factor (CRF) in the NH and of thyrotropin-stimulating hormone (TRH) fibers among IL cells of the sea bass (Batten *et al.,* 1990b) are discussed later. ^{3}H-serotonin is concentrated in Agr cells of goldfish (Kah 1983), in unidentified cells of the IL in *Poecilia* (Groves and Batten 1985) and in the PAS cells of *Xiphophorus* (Margolis-Kazan *et al.,* 1985). ^{3}H-GABA is incorporated in nerve fibers in the stickleback mainly in the rostral NH close to the ACTH cells (Follénius 1972a) and inhibits cortisol release in stressed carp (Follénius 1977). However, it is not known if GABA acts on rostral ACTH cells and/or MSH cells, as MSH and END may act in concert as a corticotropic factor (Balm *et al.,* 1995). In goldfish, immunocytochemical and radioautographic studies show an uptake of ^{3}H-GABA in the NH and a localization in the anterior pituitary and NIL in apposition to MSH cells (Kah *et al.,* 1987). GABA is also identified in the NH of eel (Medina *et al.,* 1994).

Direct neuroglandular contacts are exceptionally observed in the eel in which the IL and NH are almost always separated by the BL and synaptic structures are not observed (Olivereau *et al.,* 1981b, 1983). Direct apposition occur frequently in goldfish and roach although synaptic structures are rarely observed (Bage *et al.,* 1975, Kah 1983).

Control of MSH Cells

Environmental factors: Background (Bg) adaptation. *Coecobarbus,* a blind-cave fish completely deprived of melanin, possesses a small IL with inactive cells (Olivereau and Herlant 1954*). Identification of MSH cell function was realized in eels and trout kept on different Bg (Baker 1964*, 1965*, 1967*, Olivereau 1971, 1972a, Thornton and Howe 1974), tilapia (Van Eys 1980a) and goldfish (Olivereau *et al.*, 1989), MSH cells being stimulated or inactive on a black (B) Bg or a white (W) Bg, respectively. On a B-Bg MSH cell show a hypertrophied RER from 12 to 35% of the cytoplasmic volume, a loss of secretory granules from 38 to 5% and the development of a system of microtubules at the secretory pole of the cells and in the GA which may play a role in transporting granules from the GA to the plasma membrane (Thornton and Howe 1974). These responses occur rapidly after transfer to a B-Bg and are reversed when fish are returned on a W-Bg. Plasma MSH levels are increased on a B-Bg (Bowley *et al.*, 1983, Baker *et al.*, 1984, Rodrigues and Sumpter 1984a). Bioassays (Baker 1972) and radioimmunoassays (Baker et al., 1984) show a variable increase in pituitary MSH content in eels on a W-Bg suggesting the persistence of some synthesis and a reduced release, but transfer on a B-Bg does not always induces a decrease in MSH content. In tilapia kept in complete darkness, MSH cell activity is reduced, but remain high in blinded fish unless they are also kept in darkness, suggesting that the pineal gland may be involved in these responses (Van Eys 1980a, Van Eys and Wendelaar-Bonga 1981). The amount of fluorescent catecholamines increases in eels adapted to a B-Bg (Fremberg and Meurling 1975).

However, in several species the MSH cells do not react to the Bg color: *Cichlasoma, Blennius* (Baker 1963*), *Poecilia* (Baker and Rance 1981, Ball and Batten 1981) and flounder, at least for the first 3 weeks in which plasma and pituitary MSH levels remain constant (Baker *et al.*, 1984). In other species the MSH and PAS cells react simultaneously to the Bg. In some species the MSH cells react less intensely than the PAS cells (see SL cell section).

END also responds to the Bg. Trout transferred from a grey Bg to a B-Bg and blinded trout show an increase of plasma βEND and αMSH. Conversely, transfer from a grey Bg to a W-Bg induces a gradual and prolonged decrease of both hormones. However, changes are not detected in αMSH, βMSH, ACTH and βEND content of the pituitary. Rapid adaptation to Bg occurs without changes in plasma MSH, but it reacts during long-term adaptation. Injection of anti αMSH serum does not affect the melanocyte response of trout transferred from a W-Bg to a B-Bg (Rodrigues and Sumpter 1984a, b).

An autocontrol of MSH is apparent: in tilapia kept on a B-Bg, the continuous infusion of αMSH increases the melanogenesis, but MSH cells are less active and comparable to MSH cells of untreated fish on a W-Bg. MSH-treated fish kept on a W-Bg show many disintegrating MSH cells, but SL cells are not influenced (Van Eys and Peters 1981).

Electrolytes. Responses of the MSH cells to external salinity are quite variable. A decrease of salinity (8 g NaC1/1) induces a marked stimulation (degranulation, hypertrophy, mitotic activity) of MSH cells without changes of the melanophore index in *Boops*; after 3h in 40 g NaCl/l, granule storage increases (Malo-Michele 1975). Immunolabeling of the MSH cell with anti-ACTH serum is reduced in *Sparus auratus* transferred to brackish water (Mancera *et al.,* 1993). The MSH cells are not affected in *Mugil* adapted to brackish water (Olivereau 1968*), goldfish adapted to 15 g NaCl/l (Harrisson *et al.,* 1974) and to deionized water (DW), but are stimulated in 1/3 Ca-free SW (exocytosis, RER slightly hypertrophied). The nuclear area is reduced in fish in DW supplemented with Ca^{2+}, not in DW supplemented with other ions (Olivereau *et al.,* 1981c). In eels kept in DW, FW, normal and concentrated SW (up to 63‰), MSH and Agr cells show minor ultrastructural changes (Olivereau *et al.,* 1983). DW-adaptation does not affect pituitary MSH content (Baker 1972).

Stress, corticosteroids and POMC. A neurogenic stress (noise) induces a simulation of the interrenal and MSH cells without clear response of the rostral ACTH cells in *Boops* (Malo-Michéle 1979, 1980a). MSH cells are also stimulated by an infectious disease of the skin, but the response is inhibited in fish kept in darkness (Malo-Michèle 1980b).

Metopirone, a drug inhibiting 11β-hydroxylase, blocks the cortisol synthesis, stimulates ACTH cells and induces a melanodispersion in eel, *Poecilia* and roach. The MSH cells are degranulated in eel and roach (Olivereau 1965*, Bage *et al.,* 1975). Cortisol-treated eels show a marked paling and a gradual atrophy of the ACTH and MSH cells (Olivereau 1972b).

In trout kept on a B-Bg stimulating MSH cells, plasma cortisol level is increased (Baker and Rance 1982). When interrenal tissue is incubated with ACTH, MSH ..., MSH does not seem to be a steroidogenic factor and does not potentiate the effect of ACTH (Rance and Baker 1981). Adaptation to a B-Bg facilitates the reponse to stresses (noise or osmotic stress) as shown by the rapid cortisol increase. This facilitation is reduced after NIL removal in eels and dexamethasone treatment. A complete blockade with dexamethasone is noted in W-Bg adapted trout. These data suggest that substances released by the NIL may modulate the response to the stress (Gilham and Baker 1985). The response may also vary with the nature (water temperature, netting of the fish) and the intensity of the stress in *Salmo trutta*. A thermal stress increases plasma cortisol without changes of MSH and END, but these two hormones increase when the stress is associated to confinement and manipulation of the fish (Sumpter *et al.,* 1985). Responses to stress of these trout are not influenced by the Bg (Pickering *et al.,* 1986). In rainbow trout, the N-terminal peptide of POMC is not correlated to the stress condition and does not react to the Bg color, but it affects the ACTH- and CRF-induced cortisol release (Takahashi *et al.,* 1990).

In *Tilapia,* the corticotropic function of the IL is mainly related to the diacetyl αMSH which stimulates *in vitro* cortisol production (Lamers *et al.*, 1991) Indeed the corticortropic activity would be related to the release of N-acetyl-βEND by the MSH cells. The simultaneous secretion of END and αMSH may be involved in some form of stress although αMSH decreases without END changes during confinement (Balm *et al.,* 1995). In superfusion of goldfish NIL, addition of TRH, CRF, urotensin I and sauvagine stimulates the release of POMC-related peptides (ACTH and MSH) (Tran *et al.*, 1989, 1990).

POMC, MSH and END are also detected in salmonid brain (Vallarino *et al.*, 1989, Vecino *et al.*, 1990, 1992). An opiate activity is detected in trout pituiary, in both PD and NIL (Hunter and Baker 1979). It does not react to metopirone, but is reduced after 2 weeks on a W-Bg without correlation with the MSH content (Carter and Baker 1980).

Hypothalamic control. *In vitro* MSH synthesis and release are detected in eel and trout pituitary, both IL cell types being degranulated (Baker 1964*, 1967*). In culture of tilapia NIL for 7 days, the MSH cells retain a normal ultrastructure with two kinds of granules (Rawdon 1976). In long-term grafted pituitaries, the MSH cells undergo an involution in *Poecilia* (Olivereau and Ball 1966*) but are better preserved in eels showing a dark spot close to the graft (Olivereau 1969a). Reserpine (see below) and metopirone induce a lower melanodispersion in grafted eels than in intact ones. Melanodispersion of grafted-eels is intermediate between that of intact and hypophysectomized fish (Olivereau 1972c). In the stickleback MSH cell diameter is reduced in the graft (Leatherland 1970), but both cell types appear active and revascularized after 4 months in *Gillichthys* (Nishioka *et al.*, 1973).

Inhibitory influences: Melanin-concentrating hormone (MCH). If MSH controls the melanodispersion, MCH controls its concentration. MCH is synthesized in ventral hypothalamic nuclei. Incorporation of ^{35}S-methionine in the hypothalamus decreases in trout on a B-Bg and increases on a W-Bg. MCH acts directly on chromatophores and also decreases MSH release in fish on a B-Bg and the MSH cells appear inactive. MCH inhbits basal and stress-induced ACTH release, observations which agree with the presence of MCH fibers in the NH, close to the ACTH cells and IL cells (review in Baker 1991).

Dopamine. An abundant dopaminergic innervation of the NIL seems to play an inhibitory effect on MSH release (Olivereau 1975). *Reserpine* which depletes brain catecholamines induces a rapid darkening in eel (Olivereau 1972a) and *Boops* (Malo-Michèle 1979), a cellular, nuclear and nucleolar hypertrophy and a degranulation of MSH cells. The response is reduced in eels treated for 20-25 days. Acid solvent has a small effect.

6-hydroxy-dopamine (6-OH-DA) induces degeneration of intrahypophysial aminergic fibers and a melanodispersion in *Gasterosteus* (Follénius 1972b). In eels, 6-OH-DA produces the same darkening and MSH cells are active (Fremberg and Olivereau 1973).

Dopamine antagonists (sulpirid, pimozide and domperidone) blocking DA receptors induce a rapid darkening and the hypertrophy and degranulation of MSH cells (Olivereau 1978a) in eels and a reduced immunolabeling of the cytoplasm in goldfish (Olivereau *et al.*, 1987). These antagonists reach the pituitary (Omeljaniuk *et al.*, 1987) and act through D2 receptors in goldfish (Omeljaniuk and Peter 1989) and trout (Omeljaniuk 1995) as studied by the reversible binding of ^{3}H-spiperone in both PD and NIL. Domperidone reverses the inhibitory effects of DA on αMSH release (Omeljaniuk *et al.*, 1989).

Injection of *DA* or *L-DOPA* reduces the activity of MSH cells and fish are paler (Olivereau 1975). DA is a powerful inhibitor of αMSH release but it does not modulate the relative abundance of various forms of MSH released by tilapia pituitary (Lamers *et al.*, 1991). Noradrenaline inhibits adenylyl cyclase activity in goldfish NIL (Deery 1975).

GABA which inhibits MSH release in frogs (Desrues *et al.*, 1995) does not affect adenylyl cyclase activity in NIL (Deery 1975). Studies of GABA on fish MSH are needed.

Interleukin I and bacterial endotoxin lipopolysaccharide (LPS) partly inhibit αMSH release in tilapia, LPS reduces the TRH-induced stimulation of MSH release (Balm *et al.*, 1993).

Stimulatory influences: Serotonin (5-HT). Injection of 5-HT induces a rapid darkening and a stimulation of the MSH cells in eels (Olivereau and Olivereau 1979). Conversely parachlorophenylalanine, a tryptophan hydroxylase inhibitor, reduces brain 5-HT and induces a skin paling, a decrease in cellular and nuclear area and a storage of granules in MSH cells (Olivereau 1978b). Injection of 5-hydroxytryptophan (5-HTP), a precursor of 5-HT synthesis, induces a rapid melanodispersion and the MSH cells are stimulated on both a dark or pale Bg. GA and RER are highly developed, secretory granules are scarce, but exocytoses are not observed (Olivereau and Chambolle 1979). Responses are similar in eels and goldfish (Olivereau *et al.*, 1980b).

CRF releases ACTH and βEND. It occurs in both rostral and caudal NH, along ACTH cells (RPD) and IL cells (Olivereau and Olivereau 1988). Treatments affecting the ACTH cells modify rostral CRF, but CRF changes in the NIL are not evident although MSH cells are stimulated (metopirone) or inhibited (cortisol). In *Sparus,* CRF present in the NIL only may be a releasing factor for MSH (Mancera and Fernandez-Llebrez 1995).

TRH induces a dose-related increase of αMSH release *in vitro,* response reduced by DA, but normalized by domperidone (Omeljaniuk *et al.*, 1989).

TRH strongly stimulates αMSH release, mainly that of the diacetylated form in tilapia (Lamers *et al.,* 1991).

Somatolactin (SL) Cells

SL cells are identified as Pb-H$^-$ and PAS$^+$ cells in a majority of species, unlabeled with antisera to POMC-releted peptides and labeled with anti-SL sera. They were called Ca-sensitive cells in species reacting to the external Ca^{2+} level.

Hormonal secretion. A single hormone seems to be secreted. Somatolactin is a protein which belongs to the growth hormone/prolactin (GH/PRL) family (Ono *et al.,* 1990, Wallis 1992). In *Paralichthys,* SL precursor possesses 231 amino acid residues and can be cleaved into a signal peptide (24 amino acids) and a mature protein (207 amino acids) with a single site of glycosylation (Ono *et al.,* 1990). In cod and salmon pituitaries, mature SL possesses 209 amino acids and the signal peptide 26. Cod has 2 potential sites of glycosylation, salmon SL is nonglycosylated (Takayama *et al.,* 1991a, Rand-Weaver *et al.,* 1991a). SL is also isolated from the red drum (Zhu and Thomas 1995). Structure of SL gene is known in chum salmon (Takayama et al.,1991b), lump fish and halibut (Iraqi *et al.,* 1993) and *Solea* (Pendon *et al.,* 1994). The Pit1/GH factor 1, involved in the gene expression of SL, is present in the nuclei of SL cells (Ono *et al.,* 1994, Ono and Kawauchi 1994).

Cytology. According to the species, the SL cells have different location and morphology. They are situated at the periphery of IL cell cords (goldfish) or more deeply, sending a thin proces towards the BL (eel, salmon). This location is not constant and stimulated cells show a thicker process directly abutting on the BL. The round nucleus is apically located and granules tend to concentrate in the basal portion of the cell.

Immunocytochemistry. Immunoreactive (ir) SL is localized in the PAS+ and Pb-H-cells of salmonids (Rand-Weaver *et al.,* 1991b, Kaneko and Hirano 1993, Kakizawa *et al.,* 1993, Olivereau and Rand-Weaver 1994a, b, Parhar and Iwata 1994), *Micropogonias* (Zhu and Thomas 1995) and goldfish. As SL belongs to the GH/PRL family, this explains the labeling of PAS cells with 2 anti-mammalian PRL in tilapia (Rawdon 1979), their weak labeling with anti-chum PRL in *Cottus* (Mukai and Oota 1995) and that of trout PRL cells with anti-chum SL (Kaneko *et al.,* 1993b), not in salmon with anti-cod SL (Olivereau and Rand-Weaver 1994a, b). With *in situ* hybridization SLmRNA is localized in the secretory granules indicating that synthesis and storage occur in the granules (Kaneko *et al.,* 1993b).

Ultrastructure. In FW eels, the SL cells rarely contact the BL. The thin layer of perinuclear cytoplasm contains some granules and few mitochondria.

Granules at the basal pole are rounded (70 to 300 nm with 2 peaks at 155 and 185 nm) and more electron dense than those of the MSH cells. Granules may be smaller and elongated close to the BL where exocytotic figures are occasionally noted. RER and GA are poorly developed (Olivereau *et al.*, 1981b). In goldfish, SL cells contact NH digitations. The SL cells close to the PPD appear more active than those in the caudal IL. The BL separating the NH from the IL is often interrupted so that peptidergic and aminergic fibers may surround the SL cells. In FW most SL cells show a picture of low activity with a polymorphic nucleus, sometimes deeply indented, granules concentrated over the secretory pole, a small GA, few mitochondria and 1 or 2 perinuclear rows of RER. Granules are small (small diameter 90-260 nm, mean 167 ± 4 nm) with a medium or high electron density. Exocytotic and mitotic figures were not observed (Olivereau *et al.*, 1983). Similar pictures were reported in tilapia (Van Eys 1980b, Wendelaar-Bonga *et al.*, 1984). SL cells of trout are elongated, with a thin process often ending on the BL and small granules (120-200 nm) (Kaneko *et al.*, 1993b).

Possible roles of somatolactin: Reproduction. Incubation of gonadal tissue of chum salmon with SL or GTH I shows that SL is much less effective than GTH I to increase the production of estradiol by ovarian follicles and testosterone and 11-ketotestosterone by testis, (Planas *et al.*, 1992). Plasma SL levels gradually increase during smoltification and sexual maturation, from the spring to November in correlation with the increase of estradiol in females and 11-ketotestosterone in males. The peak noted during spawning does not seem to be related to external salinity, being similar in fish kept in FW and in SW before spawning (Rand-Weaver *et al.*, 1992, Rand-Weaver and Swanson 1993). In rainbow trout, plasma SL levels show pronounced seasonal rhyhms with a peak in summer and a trough in winter, closely correlated with water temperature and out of phase with the natural daylight. Values are significantly higher in both mature males and females than in immature fish (Rand-Weaver *et al.*, 1995). In the red drum, plasma SL levels are low 10 days prior spawning, but higher than those in immature fish held under similar conditions (Zhu and Thomas 1995). No correlation between plasma SL and final gonadal maturation is detected in *O.keta*. In fish at the beginning of sexual development (Bering sea) SL values are high (up to 35 ng/ml) and Na^+ plasma levels unusual (230-240 mM/1). In mature fish collected in the bay and at a short distance from it, in the ocean and the river, maximal values (8-10 ng/ml) occur in the river after ovulation and spawning (Kakizawa *et al.*, 1995a).

Immunocytochemistry in *Oncorhynchus nerka, O. keta* and *O. tshawytscha* suggests that the SL cells are slightly stimulated at the beginning of the gonadal development in SW for *O. nerka* and *O. keta* and in FW for *O. tshawytscha* (spring) with a thicker process contacting the BL. The stimulation is more marked (hypertrophy, partial degranulation) in fish collected on the spawning ground. In *O.tshawytscha* the maximal hypertrophy occurs in fish of the spring run spending 5 months in FW before spawning than in fish of the fall run spending 5 weeks only in FW. In spent salmon, 2 months after spawning, the SL cells are smaller and less active (Olivereau and Rand-Weaver 1994a, b). SL may be involved in sexual maturation of salmonids.

In some species, GTH cells may be located at the periphery of the NIL. Gonadotropin-releasing hormone (GnRH) fibers present in the NH end in the vicinity of the GTH cells in the PD. In the platyfish, unusual localization of GTH and mammalian GnRH are detected in the granules of some IL cells (Margolis-Kazan *et al.,* 1981). Galanin is colocalized with SL in the IL (Magliulo-Cepriano *et al.,* 1993), although in other species galanin is never detected in adenohypophysial cells (Batten *et al.* 1990a, Olivereau and Olivereau 1991, Anglade *et al.,* 1994). In trout, immunocytochemistry reveals an identical distribution of mammalian, chicken II and salmon GnRH fibers in the pituitary, ending close to the GH and SL cells (Parhar and Iwata 1994). However the distribution of the various forms of GnRH shows specific differences in the brain and pituitary of species studied so far. The description of GnRH "close contacts with SL cells" needs to be verified at the ultrastructural level as neuroglandular contacts are exceptional in trout. The faint and atypical immunoreactivity of scarce SL cells is also surprising. A possible correlation between SL, GTH, GnRH and galanin needs to be reinvestigated.

Stress. In salmonids, plasma SL levels increased rapidly after a stress (low water volume, overpopulation, handling), a response faster (a few minutes) than that of cortisol (Rand-Weaver *et al.,* 1993). A similar response was noted after an acute stress or an exhaustive exercise (Kakizawa *et al.,* 1995b). However acute stresses do not affect plasma SL concentration in red drum and Atlantic croaker, both being fed or unfed, kept in a large or small tank, on either a light or dark Bg, or anesthetized with MS 222 (plasma sampled at different times) (Zhu and Thomas 1995) (see also electrolyte section).

Adaptation to the background. In *Cichlasoma* and *Blennius* adapted to a B-Bg for several weeks, PAS cells are enlarged with big nuclei and nucleoli and abundant granules. They appeared regressed in fish on a W-Bg (Baker 1963*). Acitvation of the PAS cells on a B-Bg under permanent light or natural photoperiod was reported in *Boops* and *Scorpaena,* the response of the MSH cells being much less intense (Malo-Michéle 1997a, b). Transfer to an illuminated W-Bg induces a regression of both cell types (Malo-Michéle *et al.,* 1976). *Tilapia* on a B-Bg shows stimulated PAS cells (large RER and GA, but no exocytoses). The maximal response occurs in B-Bg fish in complete darkness (14 days) and an inactive picture is noted on a W-Bg (Van Eys 1980b). The PAS cells of *Gambusia* are stimulated after a 45 day-adaptation on B-Bg as indicated by developed RER and GA, After 45 days in complete darkness, the stimulation is less intense (Maiza 1981). Adaptation to a B-Bg induces a skin darkening in *Poecilia* kept in SW, 1/3 SW and FW, an effect abolished by hypophysectomy, and a stimulation of the PAS cells (large RER and GA, partial degranulation) in contrast to unreactive MSH cells (Ball and Batten 1981). In the Atantic croaker and red drum, plasma SL concentrations are much higher (5 times) after adaptation to a dark Bg than on a light Bg for 30 and 42 days (Zhu and Thomas 1995).

The SL cells may react to the Bg (4h) (Knowles and Vollrath 1966*), but it seems that eel and goldfish SL cells do not clearly react to B-Bg color (Olivereau 1972a, Olivereau *et al.*, 1989, Baker 1972). In trout kept on a B-Bg or a W-Bg, plasma SL levels are not different after 7 and 14 days, but the final value on a W-Bg is lower than the initial one (Kakizawa *et al.*, 1995b). A great diversity in the response to the Bg color is evident.

Electrolytes. The PAS cells of *Boops* do not react to the gradual adaptation to diluted SW or to a sudden transfer into FW or hypersaline SW for 3 h (Malo-Michéle 1975). These 2 stressing conditions do not seem to affect SL secretion. Evident changes in the PAS cells of *Gambusia* adapted to SW or DW (Maiza 1981), *Poecilia* adapted to FW, 1/3 SW or SW (Ball and Batten 1981) and *Tilapia* adapted to low-Ca^{2+} or Ca^{2+} -enriched FW (Van Eys and Wendelaar-Bonga 1984) were not observed. Addition of Ca^{2+} to FW does not inhibit the stimulation of SL cells in *Tilapia* on a B-Bg and does not affect the stimulated SL cells of goldfish in acid water (Wendelaar-Bonga *et al.*, 1986).

Anguilla kept in DW react by a considerable stimulation of the SL cells: hypertrophy, mitotic activity, degranulation, "apparent migration" of the cells towards the NH increasing the contact area between the SL cells and BL, and gaps occur in the BL (Olivereau 1967*). The stimulation is still visible after 11 weeks: scarce granules concentrated at the enlarged basal pole, highly developed RER and GA. Responses are similar in FW- and SW-adapted eels transferred to DW: maximal hypertrophy within 3-4 weeks, degranulation, mitotic activity, presence of microtubules associated with small granules, but no exocytoses (Olivereau *et al.*, 1981b). These pictures suggest that SL synthesis and release are increased, resulting in a low hormonal storage. In order to test the influence of some electrolytes on the SL cells, eels are adapted to DW for 21-30 days then in DW+Ca^{2+} 2mM/1 for 2 days. The SL cells remain hypertrophied, but are more granulated and contacts with the BL are reduced. Eels kept in DW+Ca^{2+} (10, 21, 30 days) since the beginning of the experiment show an almost normal picture of the SL cells filled with ir-granules. In DW+Mg^{2+} (2 mM/1) SL cells remain hypertrophied with a slight immunostaining. After 1 month in Ca-enriched

FW (10 mM/1 as in SW), the number of SL cells increased. The labeling is often reduced in the smaller cells, suggesting a reduced SL synthesis and release. The number of ir-cells is higher than that of the PAS cells. After 1 injection of Ca^{2+}, the SL cells show an intense labeling although the PAS reaction is not clearly increased. External Ca^{2+} seems to reduce SL release but not its synthesis (Olivereau and Olivereau 1982b). The lack of parallel between SL-PAS positivity and immunostaining suggests that the degree of glycosylation may be modified, SL of cod exists in both glycosylated and non glycosylated forms (Rand-Weaver *et al.*, 1991b).

In SW-adapted eels, the nuclear area is gradually reduced in SL cells only (Olivereau and Olivereau 1982a). These have less lamellar RER, an inconspicuous GA, smaller granules mainly in the thin basal pole rarely contacting the BL and no microtubules. Mitotic figures are not detected in normal and concentrated SW-adapted eels. In 63g NaCl/l SW, the SL cells are small with few very small granules and a nuclear chromatin of a heterogeneous density. However in eels kept in 50 and 60‰ SW the SL cells show occasional exocytotic figures along the adjacent Agr cells (*Olivereau et al.*, 1981b). In 1/3 SW, SL cells activity is slightly reduced. In 1/3 Ca^{2+} -free SW, the number of ir-cells is almost twice that of the PAS+cells, their stimulation is never as high as that in DW and mitotic activity is not observed. In full strength Ca^{2+} -free SW, the SL cells are initially stimulated (2-8 days), enlarged and the increase in number is revealed with immunocytoshemistry only, not with the PAS staining. After 2 to 4 weeks, the SL cells are less stimulated and plasma electrolytes return to normal values, but Ca^{2+} remains reduced (Olivereau and Olivereau 1982c).

Japanese eels culture in Ca^{2+} low FW (0.43 mM/1) show large SL cells much more active than those of European eels in Paris FW (2 mM/1). Adaptation to natural SW (9mM/1) for 16 days induces a rapid reduction of SL cell activity which accentuates after 48 days. Natural yellow eels (1 mM/1) show smaller and less active SL cells than in cultured eels. After 20 days in SW, the SL cells were smaller, ir-granules accumulate in the narrow basal pole and the contact with the BL is almost undetectable (Olivereau 1969b).

Goldfish. After 2-3 weeks in DW, hypertrophied SL cells show highly developed RER and GA, few granules, a reduced immunostaining, several exocytotic figures, some mitoses and large desmosomes between Agr cells (Olivereau *et al.*, 1980a, 1983). After adaptation for 24 days in DW supplemented with Na^+ (0.35 mM/1), K^+ (0.05 mM/l) and Mg^{2+} (0.2 mM/l) (concentrations present in FW) the SL cells remain highly stimulated. In Ca^{2+}-supplemented DW (2 mM/l $CaCl_2$ or formate), the SL cells are smaller, well granulated, similar to those of FW goldfish (Olivereau *et al.*, 1981c). In buffered DW, $CaCl_2$ (2 mM/l) prevents SL cell stimulation, an excess of Mg^{2+} (2mM/l) has no effect. When calcium gluconate at 1 mM/l only is added after 20 days in buffered DW, the cell and nuclear areas are decreased within 2-3 days and 50% of cells are well granulated (Olivereau *et al.*, 1981a).

The SL cells show a low activity in goldfish in diluted SW (NaCl 7.5 and 11 g/l) and diluted Mg^{2+}-free SW. They are stimulated in 1/3 Ca^{2+}-Mg^{2+}-free SW and 1/3 Ca^{2+}-free SW: cellular and nuclear hypertrophy mainly in cells close to the PPD, degranulation, large RER, mitotic activity. Cells at the periphery of the NIL react moderately. The lack of Ca^{2+} only stimulates SL cells (Olivereau *et al.*, 1982). Electron microscopy confirms that the SL cells in FW and 1/3 SW are quite similar, without exocytoses. After 19 days in 1/3 Ca^{2+}-free SW, plasma Ca^{2+} levels are low, the SL cells are almost as stimulated as in DW and granules are close to the BL. Exocytotic figures are more frequent than in DW-adapted fish (2-3 in a single cell) mainly along the BL, but mitoses are not detected (Olivereau *et al.*, 1983).

Trout. In trout adapted to Ca^{2+}-rich FW (10 mM/l) or to 80‰ SW (Ca^{2+} 8mM/l) the nuclear area of the SL cells is reduced, SLmRNA decreases in Ca-rich FW after 10, but not 21 days and SL plasma level is not affected. When Ca-rich FW trout are returned to FW, SL cell nuclear area and SL levels seem to increase, mainly because the values continue to decrease in trout left in Ca-rich FW, and SLmRNA is higher after 21 days in FW. These results suggest a hypercalcemic action of SL (Kakizawa *et al.*, 1993) and perhaps an indirect hypocalcemic action of SL involved in hypercalcemic regulation (Kaneko and Hirano 1993).

Fundulus kept for 4.5 months in Ca^{2+}-free SW (0.55 mM/l) show a decrease of plasma Ca^{2+} and an increase in size and number of the PAS cells (Ball *et al.*, 1982).

Stickleback. PAS cells are more abundant in FW than in SW (Leatherland 1970).

In summary, it is evident that SL cells react either to the Bg or to the environmental composition, but not to both conditions. These discrepancies remain unexplained.

Acid water. The SL cells show similar cell and nuclear area in *Tilapia* kept in FW at pH 7.5 and 3.5 for 14 days whereas those of *Carassius* are rapidly stimulated at pH 3.5 (degranulation, development of GA and RER) (Wendelaar-Bonga *et al.*, 1984). However, in goldfish kept in Ca-rich FW (6 mM/l) for 4 weeks and subsequently submitted to a gradual lowering of the water pH, the SL cells are as stimulated as in FW (0.8 mM/1), suggesting that SL is not a hypercalcemic factor, but may be implicated in acid-base regulation (Wendelaar-Bonga *et al.*, 1986). But may pH 3.5 be considered as a physiological environment?

As water pH affects the SL cells and as pH of DW is 5.3, goldfish were kept in buffered DW ar pH 6.7 with imidazole. Two sets of experiments show that the nuclear hypertrophy and other indices of stimulation are similar at both pH (Olivereau *et al.*, 1981c).

Hypothalamic control. In primary monolayer cultures of IL from *Tilapia* for 7 days SL cells still possess small round to elongated electron-dense granules (Rawdon 1976). *In vitro* eel PAS cells become degranulated (Baker 1967*) Grafted pituitaries show a strong or moderate regression of the PAS cells in *Poecilia* (Olivereau and Ball 1966*) and eel (Olivereau 1969a). In "cobalt" variant of rainbow trout, the NIL separated from the hypothalamus is extremely reduced. The small number of the SL cells agrees with a very low level of plasma SL (Kaneko *et al.,* 1993a). These data suggest that a hypothalamic factor is necessary to maintain a normal functioning of the SL cells although its nature is not yet known.

Serotonin was demonstrated in the IL of some species (see introduction). However DA, DOPA, 5-HT and its precursor 5-HTP, DA and 5-HT antagonists affecting the MSH cells do not seem to control SL cell activity, at least in eel and goldfish.

General Discussion

There is no doubt that IL secretes several POMC-related hormones and SL which are synthesized in the MSH and SL cells, respectively according to immunocytological data. However the functions of these two hormones are not yet fully understood. In trout, eel and goldfish MSH controls melanodispersion and MSH pituitary content increase shortly after transfer to a W-Bg. MSH cells do not react markedly to ionic changes in water. The MSH cells of other species do not react to the Bg whereas the SL cells are strongly responsive. In *Tilapia* and *Boops,* both MSH and SL cells react to the Bg color and SL seems to participate in the control of melanodispersion. Other hormones such as ACTH, catecholamines and MCH affect the pigmentation and may interfere with the action of MSHs and/or SL.

There are some indications (eel, goldfish) suggesting that SL participates in calcium regulation and/or acid-base equilibrium. However no clear response is observed in *Gambusia, Poecilia, Tilapia* transferred to FW. At least in *Gambusia* with unresponsive SL cells in DW, the PRL cells become extremely stimulated. In contrast, the PRL cells do not react in eel and goldfish kept in DW (see other article), but the SL cells respond markedly. It seems then that two responses to a hypoosmotic medium are possible; either PRL or SL secretion. And indeed these two hormones belong to the same family. PRL and SL do not seem to be essential in fish to survive in FW after hypophysectomy. In other species, PRL is necessary but SL is ineffective as hypophysectomized *Poecilia* with grafted NIL do not survive in FW (Ball 1965*) Does SL play a role in reproduction and stress? It is probable, but knowing the great variability of responses according to the species, it seems important to study a larger number of species. Such a diversity of reactions is distressing! A hypothalamic control is suspected but its nature is still unknown. The probable interactions between MSH-related peptides or SL and the numerous neuropeptides present in NH fibers must be investigated in more species for a better understanding of IL functioning in teleosts.

References

*References with * after the date are cited in Ball and Baker (1969)
Anglade, I., Wang, Y.X., Jensen, J., Tramu, G., Kah, O. and Conlon, J.M. (1994). J.Comp. Neurol. **350:** 63–74.
Bage, G., Ekengren, B., Fernholm, B. and Fridberg, G. (1975). Acta Zool. **56:** 43–60.
Baker, B.I. (1972). Gen. Comp. Endocrinol. **19:** 515–521.
Baker, B.I. (1991). Internal. Review Cytology **126:** 1–48.
Baker, B.I. and Rance, T.A. (1982). J. Endocrinol. **79:** 135–140.
Baker, B.I., Wilson, J.F. and Bowley, T.J. (1984). Gen. Comp. Endocrinol. **55:** 142–149.
Ball, J.N. and Baker, B.I. (1969). Fish Physiology **2:** 1–110.
Ball, J.N. and Batten, T.F.C. (1981). Gen. Comp. Endocrinol. **44:** 233–248.
Ball, J.N., Uchiyama, M. and Pang, P.K.T. (1982). Gen. Comp. Endocrinol. **46:** 480–485.
Balm, P.H.M., Hovens, M.L.M. and Wendelaar-Bonga, S.E. (1995). Peptides **16:** 463–469.
Balm, P.H.M., Pepels, P., Van Lieshout, E. and Wendelaar-Bonga, S.E. (1993). Fish Physiol. Biochem. **11:** 125–130.
Batten, T.F.C., Moons, L., Cambre, M. and Vandesande, F. (1990a). Neurosci. Lett. **111:** 12–17
Batten, T.F.C., Moons, L., Cambre, M.L., Vandesande, F., Seki, T. and Suzuki, M. (1990b). Gen. Comp. Endocrinol. **79:** 385–392.
Bowley, T.J., Rance, T.A. and Baker, B.I. (1983). J. Endocrinol. **97:** 267–275.
Cabre, M.L., Verdonck, W., Ollevier, F., Vandesande F., Batten, T.F.C. and Kühn, E.R. (1986). Gen. Comp. Endocrinol. **61:** 368–375.
Carter, D.A. and Baker, B.I. (1980). Gen. Comp. Endcrinol. **41:** 225–232.
Deery, D.J. (1975). Gen. Comp. Endocrinol. **25:** 395–399.
Desrues, L., Vaudry, H., Lamacz, M. and Tonon, M.C. (1995). J. Mol. Endocrinol. **14:** 1–12.
Eipper, B. A. and Mains, R.E. (1980). Endocr. Rev. **1:** 1–27.
Follénius, E. (1971). C.R. Acad. Sci. Paris **273D:** 1039–1940.
Follénius, E. (1972a). C.R. Acad. Sci. Paris **275D:** 1435–1438.
Follénius, E. (1972b). Z.Zellforsch. **128:** 69–82.
Follénius, E. (1977). C.R. Acad. Sci. Paris, **285D:** 993–996.
Follénius, E. and Dubois, M.P. (1974). Gen. Comp. Endocrinol. **24:** 203–207.
Follénius, E. and Dubois, M.P. (1977). Front. Hormone Res. **4:** 46–50.
Follénius, E. and Dubois, M.P. (1978). Cell Tissue Res. **188:** 273–283.
Follénius, E., van Dorsselaer, A. and Meunier, A. (1985). Gen. Comp. Endocrinol. **57:** 198–207.
Follénius, E., van Dorsselaer, A. and Meunier, A. (1986). Gen. Comp. Endocrinol. **62:** 185–192.
Fremberg, M. and Meurling, P. (1975). Cell Tissue Res. **157:** 53–72.
Fremberg, M. and Olivereau, M. (1973). Acta Zool. **54:** 231–239.
Gilham, I.D. and Baker, B.I. (1985). J. Endocrinol. **105:** 99–105.
Groves, D.J. and Batten, T.F.C. (1985). Cell Tissue Res. **240:** 489–492.
Harrisson, F., Demal, J. and Remacle, C. (1974). Arch.Anat. Microsc. **63:** 299–306.
Holmqvist, B.I. and Ekstrom, P. (1995). J. Chem. Neuroanat. **8:** 125–145.
Hunter, C. and Baker, B.I. (1979). Gen. Comp. Endocrinol. **37:** 111–114.
Iraqi, F., Gong, Z. and Hew, C.L. (1993). Mol. Mar. Biol. Biotech. **2:** 96–103.
Kah, O. (1983). These Doct. Univ. Bordeaux I, pp. 120.
Kah, O., Dubourg, P., Chambolle, P. and Calas, A. (1984). Cell Tissue Res. **238:** 621–626.
Kah, O., Dubourg, P., Onteniente, B., Geffard, M. and Calas, A. (1986). Cell Tissue Res. **244:** 577–582.

Kah, O., Dubourg, P., Martinoli, M.G., Rabhi, M., Gonnet, F., Geffard, M. and Calas, A. (1987). Gen. Comp. Endocrinol. **67:** 324–332.
Kakizawa, S., Kaneko, T., Hasegawa, S. and Hirano, T. (1993). Gen. Comp. Endocrinol. **91:** 298–306.
Kakizawa, S., Kaneko, T., Ogasawara, T. and Hirano, T. (1995a). Fish Physiol. Biochem. **14:** 93–101.
Kakizawa, S., Kaneko, T., Hasegawa, S. and Hirano, T. (1995b). Gen. Comp. Endocrinol. **98:** 137–146.
Kaneko, T. and Hirano, T. (1993). J. Exp. Biol. **184:** 31–45.
Kaneko, T., Kakizawa, S. and Yada, T. (1993a). Gen. Comp. Endocrinol. **92:** 31–40.
Kaneko, T., Kakizawa, S., Yada, T. and Hirano, T. (1993b). Cell Tissue Res. **272:** 11–16.
Kawauchi, H., Adachi, Y. and Ishizuka, B. (1980a). Int. J. Peptide Protein Res. **16:** 79–82.
Kawauchi, H., Tsubokawa, M., Kanezawa, A. and Kitagawa, H. (1980b). Biochem. Biophys. Res. Communic. **92:** 1278–1288.
Kawauchi, H., Takahashi, A. and Abe, K.I. (1981). Int. J. Peptide Protein Res. **18:** 223–227.
Kawauchi, H., Takahashi, A. and Abe, K.I. (1982). Arch. Biochem. Biophys. **213:** 680–688.
Kitahara, N., Nishizawa, T., Iida, K., Odazaki, H., Andoh, T. and Soma, G.I. (1988). Comp. Biochem. Physiol. **91B:** 365–370.
Lamers, A.E., Balm, P.H.M., Haenen, H.E.M.G., Jenks, B.G. and Wendelaar-Bonga, S.E. (1991). J. Endocrinol. **129:** 179–187.
Leatherland, J.F. (1970). Z. Zellforsch. **104:** 337–334.
MacConaill, M.A. (1947). J. Anat. **81:** 371–372.
Magliulo-Cepriano, L., Schreibman, M.P. and Blum, V. (1993). Gen. Comp. Endocrinol. **92:** 269–280.
Maiza, S. (1981). DEA Biol. Physiol. Animale, University of Bordeaux I, pp. 52.
Malo-Michèle, M. (1975). Gen. Comp. Endocrinol. **27:** 1–8.
Malo-Michèle, M. (1977a). Bull. Soc. Zool. **102:** 19–24.
Malo-Michèle, M. (1977b). Gen. Comp. Endocrinol. **31:** 343-349.
Malo-Michèle, M. (1979). Gen. Comp. Endocrinol. **38:** 300–308.
Malo-Michèle, M. (1980a). Gen. Comp. Endocrinol. **42:** 101–104.
Malo-Michèle, M. (1980b). Gen. Comp. Endocrinol. **42:** 105–109.
Malo-Michèle, M., Bugnon, C. and Fellmann, D. (1976). C.R. Acad. Sci. Paris **283D:** 643–646.
Mancera, J.M. and Fernandez-Llebrez, P. (1995). Cell Tissue Res. **281:** 569–572.
Mancera, J.M., Fernandez-Llebrez, P. and Pérez Figares, J. M. (1993). Revista Espan.Fisoiol. **49:** 115–120.
Margolis-Kazan, H., Peute, J., Schreibman, M.P. and Halpern, L.R. (1981). J. Exp. Zool. **215:** 99–102.
Margolis-Kazan, H., Halpern-Sebold, S., and Schreibman, M.P. (1985). Cell Tissue Res. **240:** 311–314.
Medina, M., Repérant, J., Dufour, S., Ward, R., Le Belle, N. and Micelli, D. (1994). Anat. Embryol. **189:** 25-39.
Mukai, T. and Oota, Y. (1995). Zool. Sci. **12:** 91–97.
Munro, A.D. (1985). Gen. Comp. Endocrinol. **60:** 215–226.
Naito, N., Takahashi, A., Nakai, Y. and Kawauchi, H. (1984). Gen. Comp. Endocrinol. **56:** 185–192.
Nishioka, R.S., Nagahama, Y. and Bern, H.A. (1973). Boll. Zool. Biol. Mar. NS Sao Paolo **30:** 69–79.
Olivereau, M. (1969a). Z. Zellforsch. **98:** 74–87.
Olivereau. M. (1969b). Z. Zellforsch. **99:** 389–410.
Olivereau, M. (1971). C. R. Acad. Sci. Paris **272D:** 102–105.

Olivereau, M. (1972a). Z. Anat. Entwickl. Gesch. **137:** 30–46.
Olivereau, M. (1972b). Acta Zool. **53:** 179–194.
Olivereau, M. (1972c). Excerpta Med. Int. Congr. Ser. **341:** 136.
Olivereau, M. (1975). Folia Anat. Iugosl. **4:** 15–21.
Olivereau, M. (1978a). Cell Tissue Res. **189:** 231–239.
Olivereau, M. (1978b). Cell Tissue Res. **191:** 83–92.
Olivereau, M. and Callard, G.V. (1985). Gen. Comp. Endocrinol. **58:** 280–290.
Olivereau, M. and Chambolle, P. (1979). J. de Physiol. **75:** 109–112.
Olivereau, M. and Olivereau, J. (1979). Cell Tissue Res. **196:** 397–408.
Olivereau, M. and Olivereau, J. (1982a). Acta Zool. **63:** 165–170.
Olivereau, M. and Olivereau, J. (1982b). Cell Tissue Res. **222:** 231–241.
Olivereau, M. and Olivereau, J. (1982c). Cell Tissue Res. **225:** 487–496.
Olivereau, M. and Olivereau, J. (1988). Peptides **9:** 13–21.
Olivereau, M. and Olivereau, J.M. (1991). Histochemistry **96:** 343–354.
Olivereau, M. and Rand–Weaver, M. (1994a). Gen. Comp. Endocrinol. **93:** 28–35.
Olivereau, M. and Rand-Weaver, M. (1994b). Fish Physiol. Biochem. **13:** 141–151.
Olivereau, M., Aimar, C. and Olivereau, J.M. (1980a). Cell Tissue Res. **208:** 289–404.
Olivereau, M., Aimar, C. and Olivereau, J.M. (1981a). J. de Physiol. **77:** 895–989.
Olivereau, M., Bugnon, C. and Fellmann, D. (1976a). C.R. Acad. Sci. Paris **283D:** 1321–1323.
Olivereau, M., Bugnon, C. and Fellmann, D. (1976b). C.R. Acad. Sci. Paris **283D:** 1441–1443.
Olivereau, M., Chambolle, P. and Dubourg, P. (1981b). Cell Tissue Res. **219:** 9–26.
Olivereau, M., Chambolle, P. and Dubourg, P. (1983). Biol. Cell **48:** 53–64.
Olivereau, M., Olivereau, J.M. and Aimar, C. (1980b). Cell Tissue Res. **207:** 377–385.
Olivereau, M., Olivereau, J.M. and Aimar, C. (1981c). Cell Tissue Res. **214:** 23–31.
Olivereau, M., Olivereau, J.M. and Aimar, C. (1982). Cell Tissue Res. **225:** 497–505.
Olivereau, M., Olivereau, J.M. and Lambert, J.F. (1987). Gen. Comp. Endocrinol. **58:** 12–18.
Olivereau, M., Olivereau, J.M. and Chantrier, N. (1989). Gen. Comp. Endocrinol. **74:** 269.
Omeljaniuk, R.J. (1995). Can. J. Physiol. Pharmacol. **73:** 585–593.
Omeljaniuk, R.J. and Peter. R.E. (1989). Gen. Comp. Endocrinol. **74:** 57–67.
Omeljaniuk, R.J., Shih, S.H. and Peter, R.E. (1987). J. Endocrinol. **114:** 449–458.
Omeljaniuk, R.J., Tonon, M.C. and Peter, R.E. (1989). Gen. Comp. Endocrinol. **74:** 451-467.
Ono, M. and Kawauchi, H. (1994). Fish Physiol. **13:** 159–178.
Ono, M., Takayama, Y., Rand-Weaver, M., Sakata, S., Yasanaga, T., Noso, T. and Kawauchi. H. (1990). Proc. Nat. Acad. Sci. USA **87:** 4330-4334.
Ono, M., Harigai, T., Kaneko, T., Sato, Y., Ihara, S. and Kawauchi, H. (1994). Mol. Endocrinol. **8:** 109–115.
Parhar, I.S. and Iwata, M. (1994). Histochemistry **102:** 195–204.
Pendon, C., Martinez-Barbera, J.P. and Valdivia, M.M. (1994). Gene **147:** 227–230.
Pickering, A.D., Pottinger, T.G. and Sumpter, J.B. (1986). Gen. Comp. Endocrinol. **64:** 206–211.
Planas, J.V., Swanson, P., Rand-Weaver, M. and Dickhoff, W.W. (1992). Gen. Comp. Endocrinol. **87:** 1–5.
Quesada, J., Lozano, M.T., Ortega, A. and Agulleiro, B. (1988). Gen. Comp. Endocrinol. **72:** 209–225.
Rance, T. A. and Baker, B. I. (1981). Gen. Comp. Endocrinol. **45:** 497–503.
Rand-Weaver, M., Noso, T., Muramoto, K. and Kawauchi, H. (1991a). Biochem. **30:** 1509–1515.

Rand-Weaver, M., Baker, B.I. and Kawauchi, H. (1991b). Cell Tissue Res. **263:** 207–215.
Rand-Weaver, M., and Swanson, P. (1993). Fish Physiol. Biochem. **11:** 175–182.
Rand-Weaver, M., Pottinger, T.G. and Sumpter, J.P. (1993). J. Endocrinol. **138:** 509–516.
Rand-Weaver, M., Pottinger, T.G. and Sumpter, J.P. (1995). J. Endocrinol. **146:** 113–120.
Rand-Weaver, M., Swanson, P., Kawauchi, H. and Dickhoff, W. W. (1992). J. Endocrinol. **133:** 393–399.
Rawdon, B.B. (1976). Proc. Electron Micr. Soc. South Africa **6:** 23–24.
Rawdon, B.B. (1979), Gen. Comp. Endocrinol. **37:** 374–382.
Rodrigues, K.T. and Sumpter, J.P. (1984a). J. Endocrinol. **101:** 277–284.
Rodrigues, K.T. and Sumpter, J.P. (1984b). Gen. Comp. Endocrinol. **37:** 69–75.
Rodrigues, K.T., Jenks, B.G. and Sumpter, J.P. (1983). J. Endcorinol. **98:** 371–282.
Romain, R., Bugnon, C., Dessy, C. and Fellmann, D. (1974). C.R. Soc. Biol. **168:** 1245–1248.
Salbert, G., Chauveau, I., Bonnec, G., Valotaire, Y. and Jego, P. (1992). Mol. Endocrinol. **6:** 1605–1613.
Siegmund, I., Troncoso, S., Caorsi, C.E. and Gonzalez, B. (1987). Gen. Comp. Endocrinol. **67:** 348–355.
Smith, A.I. and Funder, J.W. (1988). Endocr. Rev. **9:** 159–179.
Sumpter, J.P., Pickering, A.D. and Pittinger, T.G. (1985). Gen.Comp. Endocrinol. **59:** 257–265.
Takahashi, A., Ogasawara, T., Kawauchi, H. and Hirano, T. (1990). Gen. Comp. Endocrinol. **77:** 98–106.
Takayama, Y., Ono, M., Rand-Weaver, M. and Kawauchi, H. (1991a). Gen. Comp. Endocrinol. **83:** 366–374.
Takayama, Y., Rand-Weaver, M., Kawauchi, H. and Ono, M. (1991b). Mol. Endocrinol. **5:** 778–786.
Thornton, V.F. and Howe, C. (1974). Cell Tissue Res. **151:** 103-115.
Tran, T.N., Fryer, J.N., Bennett, H.P.J., Tonon, M. C. and Vaudry, H. (1989). Peptides **10:** 835–842.
Tran, T.N., Fryer, J.N., Lderis, K. and Vaudry, H. (1990). Gen. Comp. Endocrinol. **78:** 351–360.
Vallarino, M., Delbende, C., Transhand Bunel, D., Ottonello, I. and Vaudry, H. (1989). Peptides **10:** 1223–1230.
Van Eys, G.J.J.M. (1980a). Cell Tissue Res. **208:** 99–110.
Van Eys, G.J.J.M. (1980b). Cell Tissue Res. **210:** 171–179.
Van Eys, G.J.J.M. and Peters, P.T.W. (1981). Cell Tissue Res. **217:** 361–372.
Van Eys, G.J.J.M. and van den Otelaar, P. (1981). Cell Tissue Res. **215:** 625–633.
Van Eys, G.J.J.M. and Wendelaar-Bonga, S.E. (1981). Cell Tissue Res. **220:** 561–571.
Van Eys, G.J.J.M. and Wendelaar-Bonga, S.E. (1984). Cell Tissue Res. **236:** 181–187.
Van Eys, G.J.J.M., Lowick, C.W.G.M., Cruijsen, P.M.J.M. and Wendelaar-Bonga, S.E. (1984). Comp. Biochem. Physiol. **77B:** 135–142.
Vecino, E. and Ekstrom, P. (1990). J. Comp. Neurol. **299:** 229–241.
Vecino, E., Pinuela, C., Arévalo, R., Lara, J. Alonso, J. R. and Aijon. J. (1992). J. Anat. **180:** 435–454.
Wallis, M. (1992). J. Mol. Edocrinol. **9:** 185–188.
Wendelaar-Bonga, S.E., van der Meij, J.C.A., van des Krabben, W.A.W.A. and Flick, G. (1984). Cell Tissue Res. **238:** 601–609.
Wendelaar-Bonga, S.E., Van Der Meij, J.C.A. and Flick, G. (1986). Cell Tissue Res. **243:** 609–617.
Woodman, A.S. (1939). J. Morphol. **65:** 411–437.
Zhu, Y. and Thomas, P. (1995). Gen. Comp. Endcorinol. **99:** 272–288.

Addendum. Plasma SL levels are significantly lowered in fasted trout with a two-fold difference after 16 weeks, and are significantly and negatively correlated to most metabolic parameters (Rand-Weaver *et al.*, 1995). In long-term starved eels (1-4 years) the SL cells appear hypoactive (smaller size and weak immunostaining) in the rostral half of the NIL and more variable in the caudal half. The SL cells never appear stimulated (Olivereau and Olivereau 1997). Immunoreactive cells are not detected in the sturgeon and the moray eels. Few SL cells are observed in *Lepisosteus* which also contain α-MSH and in *Amia* all cell's contain both SL and α-MSH (Dores *et al.*, 1996). Transfer from an illuminated to a black background without illumination induces a highly significant increase in plasma SL and a smaller increase in pituitary SL content. The MSH cells are simultaneously stimulated (Zhu and Thomas 1996). SL increase hepatic IGF-ImRNA althought it is less active than GH (Duan *et al.*, 1993) and does not increase IGF-I level (Moriyama *et al.*, 1994). SL may be involved in the acidobasic regulation in trout (Kakizwa *et al.*, 1996). Bacterial production of SL from *Solea* will facilitate biophysical and physiological studies (Pendon *et al.*, 1996).

Dores, R.M., Hoffman, N.E., Chilcutt-Ruth, T., Lancha, A., Brown, C., Marra, L. and Youson, J. (1996). Gen. Comp. Endocrinol. **102:** 79–87.

Duan, C, Duguay, S.J. and Plisetskaya, E.M. (1993). Fish Physiol. Biochem. **11:** 371–379.

Kakizawa, S., Kaneko, T. and Hirano, T. (1996). J. Exp. Biol. **199:** 1043–1052.

Moriyama, S., Swanson, P., Nishii, M., Takahashi, A., Kawauchi, H., Dickhoff, W.W. and Plisetskaya, E.M. (1994). Gen. Comp. Endocrinol. **96:** 149–161.

Olivereau, M. and Olivereau, J.M. (1997). Fish Physiol. Biochem. in press.

Pendon, C., Martinez-Barbera, J.P., Ortiz, M. and Valdivia, M.M. (1996). Protein Express. Purif. **7:** 389–394.

Rand-Weaver, M., Pottinger, T.G., Guest, A., Martin, P., Smal, J. and Sumpter, J.P. (1995). Netherl. J. Zool. **45:** 129–131.

Zhu, Y. and Thomas, P. (1996). Gen. Comp. Endocrinol. **101:** 21–31.

Comparative Endocrinology and Reproduction
K.P. Joy, A. Krishna and C. Haldar (Eds)

7. Neuroendocrine Integrative Mechanisms in the Control of Gonadotropin Secretion in Teleosts

H.J.Th. Goos[1], B. Senthilkumaran[2] and K.P. Joy[2]

[1]Department of Experimental Zoology, Research Group for Reproductive Endocrinology, University of Utrecht, Padualaan 8, 3584 CH Utrecht, The Netherlands

[2]Centre of Advanced Study in Zoology, Banaras Hindu University, Varanasi-221005, India

Introduction

In most teleosts, reproduction is a seasonal phenomenon and is characterized by the occurrence of well-defined annual reproductive cycles in which phases of gonadal growth, maturation, and spawning alternate regularly with a period of gonadal quiescence. The initiation, maintenance and successful termination of the reproductive activity in fishes, as in other vertebrates, are controlled by the participation and sequential interaction of a host of hormones in which gonadotropic hormones (GTHs) have a central position. Environmental cues are transduced to the brain and translated into neuroendocrine signals, which control the synthesis and release of the GTHs. They are released into circulation and regulate the gonads in their dual function of production of gametes and synthesis and release of gonadal hormones. The gonadal hormones constitute another set of endocrine regulators which, on one hand control gonadal activity and on the other control the activity of hypophysical GTH-producing cells.

In this chapter, the recent advances in the study of teleost gonadotrophs, GTH(s), and gonadotropin-releasing hormones (GnRHs), and the integrative control mechanisms of GTH secretion are reviewed briefly.

Gonadotropic Hormones and the Cells of Origin

In teleosts, the pituitary in general is constituted by three parts: the rostral pars distalis (RPD), the proximal pars distalis (PPD) and the neuro-intermediate lobe (NIL), (see also Olivereau and Olivereau, this volume). The rostral pars distalis mainly consists of prolactin-producing cells and adrenocorticortropic cells, the latter bordering the neurohypophysis; in a few species, like rainbow trout, some gonadotrophs intermingle with prolactin cells. The neuro-intermediate lobe is the homologue of the pars nervosa of higher vertebrates, in which the pars intermedia tissue penetrates. Hence, it contains the pars nervosa-characteristic vasotocin and isotocin-containing nerve endings, the intermediate lobe-characteristic melanotropic cells, and the recently discovered somatolactin-producing cells.

GTH is secreted by one or two basophilic cell type(s), the gonadotrophs. They are distributed in the PPD, together with somatotrophs and thyrotrophs. In their extensive review, Van Oordt and Peute (1983) described the origin and morphology of the gonadotrophs. In most teleosts, the adenohypophysis develops from a solid placode in the roof of the ectodermal stomodeum, Consequently, the adenohypophysis in virtually all teleosts does not contain a lumen. Exceptions occur in the pituitaries of salmonids, clupeids and apodes, where the rostral part of the adenohypophysis is composed of follicles, arising from a schizocoel, which is formed in the placode during adenohypophysial development. In most vertebrates, the gonadotrophs arise from the *Zona tuberalis* of the adenohypophysis. However, in teleost fish a *Zona tuberalis* can not be distinguished but the ventral part of the PPD may be its homologue and the first gonadotrophs differentiate in this area. Very often the gonadotrophs remain restricted to the ventral PPD, but they may migrate into other regions of the pars distalis as well.

As a rule the gonadotrophs are relatively large rounded cells. They are characterized by the presence of round or oval electron dense secretory granules with a diameter of 100-500 nm. Secretory granules may fuse to larger structures, the globules with a diameter of 0.5-3 μm (Sharp-Baker *et al.*, 1995). At least in two species, the African catfish, *Clarias gariepinus* (Zandbergen *et al.*, 1993) and the rainbow trout, *Salmo gairdneri* (Nozaki *et al.*, 1993), in contrast to the secretory granules, the globules do not contain the complete hormone and are considered a part of the hormone degradation pathway. In the African catfish, even larger fusion products are found, described as irregular membrane-bound masses, which are probably also involved in the degradation of GTH (Peute *et al.*, 1986). Moreover, the gonadotrophs are characterized by the presence of irregular cisternae of rough endoplasmic reticulum.

Functional relationships between the pituitary and gonads have been investigated by cytophysiological approaches, supported by immunocytochemical techniques at the light- and electron microscopical levels using homologous or heterologous antibodies against GTH, and more recently also by *in situ* hybridisation. Such investigations have provided detailed insights into the functional identification and the ultrastructure of the gonadotrophs, and their cyclic variations during the annual reproductive cycle and changes upon removal of the gonads, the latter being interpreted as compensatory hypertrophy (Peute *et al.*, 1984; 1986; Dubourg *et al.*, 1985; Schreibman, 1986; Nozaki *et al.*, 1990a, b; Naito *et al.*, 1993). For example, in the African catfish both the ultrastructural appearance of the gonadotrophs and the pituitary gonadotropin content showed cyclical changes, which paralleled gonadal development during the reproductive cycle. In summer, during the breeding period, the gonadotrophs were large and fully granulated, and pituitary gonadotropin reached maximum levels. After spawning, the resting period begins, with an increasing number of fusion products of secretory granules and globules, indicating a breakdown of stored hormone. This is followed by the appearance of residual bodies, cell shrinkage and a considerable drop in GTH content. In late winter and early spring, during gonadal recrudescence, the gonadotrophs redevelop; i.e., they increase in size and granulation, and at the same time the GTH content augments (Peute *et al.,* 1986). Similar changes were observed after castration and steroid replacement (De Leeuw *et al.,* 1986). Castration of male African catfish resulted in a decrease in pituitary GTH content and a degranulation of the gonadotrophs. Replacement therapy with aromatizable androgens or estrogens caused a restoration of both gonadotroph ultrastructure and pituitary GTH content. In contrast to mammals, there are many indications, as will be discussed later, that in teleosts GTH gene transcription and translation, and GTH storage are stimulated by aromatizable androgens or estrogens, rather than these steroids feed back negatively on the gonadotrophs.

Purification and characterisation of GTHs have been successfully accomplished in a number of teleosts. These studies demonstrate one (carp, *Cyprinus carpio,* Burzawa-Gerard, 1971; rainbow trout, *Salmo gairdneri,* Breton *et al.,* 1976; tilapia, *Tilapia mossambica,* Farmer and Papkoff, 1977; pike eel, *Muraenesox cinereus,* Huang *et al.,* 1981; carp, Kabayashi *et al.,* 1985; African catfish, *Clarias gariepinus,* Goos *et al.*, 1986, Koide *et al.*, 1992; spotted sea trout, *Cynoscion nebulosus,* Copeland and Thomas, 1992) or two (referred to as GTH I and GTH II, respectively) (chum salmon, *Onchorhynchus keta,* Suzuki *et al.*, 1988a, Swanson *et al.*, 1991; carp, Van Der Kraak *et al.*, 1992) GTHs in teleosts. In the Indian species, *Channa punctatus* (murrel) and *Catla catla* (carp), only one GTH was isolated (Banerjee *et al.,* 1989).

As in other vertebrates, GTHs in teleosts are hetero-dimeric glycoproteins. Amino acid analysis revealed large homology of GTH I and II with FSH and LH of higher vertebrates, respectively. In all cases where only one molecular form of GTH could be isolated, the hormone resembles GTH II mostly (references above). It has been suggested by Prat *et al.* (1996) to refer GTH I and II to FSH and LH, respectively. Because of the functional and structural homology with GTHs of other vertebrates this suggestion will be followed in this paper from now on. Further, since no studies on regulation of GTH-I (FSH) are available, the discussion is centred on GTH-II (LH).

The introduction of radioimmuno assays (RIAs) for the measurement of hormone levels in fishes has given further impetus to define more precisely the brain-pituitary and pituitary-gonadal relationships. Although data on blood levels of gonadal steroids in relation to the annual reproductive cycle or gametogenesis and spawning are available for many teleosts, studies relating changes in sex steroid levels with GTH levels are mostly confined to salmonids, cyprinids and catfish (Billard *et al.*, 1978; Crim and Idler, 1978; Peter, 1981; Fostier *et al.*, 1983; Kobayashi *et al.*, 1986; Schulz *et al.*, 1994). This is because sensitive and specific assays for GTHs in a wide range of teleosts are not available.

The levels of FSH and LH in chum salmon show distinct seasonal patterns; plasma levels of FSH increase during vitellogenesis and spermatogenesis and decline at spawning whereas the levels of LH increase rapidly at the time of spawning (Suzuki *et al., 1988b;* Weil *et al.*, 1995; Prat *et al.*, 1996). Accordingly, two types of gonadotrophs (FSH and LH cells) were described in this species; the FSH immunoreactive cells are present prior to puberty, and appear to be active prior to the onset of spermatogenesis and vitellogenesis, while immunoreactive LH cells are present in the mature (adult) fish and appear coincidentally with the onset of spermatogenesis and vitellogenesis (Nozaki *et al.*, 1990a, b). Although salmon FSH and LH are equipotent in terms of estradiol-17β production, they showed different potencies in enhancing 17α, 20β-dihydroxyprogesterone production and oocyte maturation, LH being more potent than FSH (Suzuki *et al.*, 1988c). In species where only one GTH has been described, it is structurally homologous to the LH and controls all reproductive events, such as the steroidogenesis, gametogenesis and spawning. In the African catfish only one GTH, which is homologous to the LH, was isolated and characterized, and was found to stimulate testicular steroid production *in vitro* (Koide *et al.*, 1992). In common carp, although LH- and FSH-like hormones were purified and characterized, they displayed similar biological actions (Van Der Kraak *et al.*, 1992). Thus, the current situation is that in some teleost species (salmonids) two gonadotropic hormones are found, which display specific biochemical, immunological and functional properties. In other species (cyprinids), although two biochemically distinguishable GTHs can be demonstrated, their immunological and functional features are overlapping. In a third category (catfish, eels) only one GTH has been found. In the African catfish, the available data concerning the absence of a FSH-like hormone have been interpreted in two ways (Schulz *et al.*, 1995). Either FSH would be indistinguishable from LH biochemically, immunologically as well as regarding its steroidogenic activity and would be produced by the same type of gonadotroph, or FSH would be devoid of steroidogenic activity and produced by a pituitary cell type that does not show the ultrastructural characteristics of gonadotrophs. Both interpretations are difficult to reconcile with the information available from other teleost species. Schulz *et al.*, (unpublished) studied the ontogenesis of the gonadotropic cells in the pituitary in relation to testicular development. They showed that LH is present early during ontogenesis, that it increased slightly with male sexual differentiation, but very strongly with the onset of spermatogenesis. They, therefore, proposed a third possibility, namely that a single gonadotropic hormone, the LH-like form, fulfils all functions depending on gonadotropic regulation in this species, and assume that a gene encoding the β-subunit of FSH, if present at all is silent in the African catfish.

The brain-pituitary-gonad axis is the prevailing neuroendocrine system that controls reproductive processes, Gonadotropic hormones occupy a pivotal position in this system as they control directly the two functions of the gonads. Hence, the regulation of gonadotropin secretion is of main importance to create the endocrine status within the organism, required for each step in the reproductive cycle. In teleosts it is well documented that environmental factors, notably, photoperiod and temperature, influence gonadal activity (Peter, 1981; Sundararaj, 1981; Borg *et al.*, 1987). However, the mechanism by which environmental factors synchronise and modulate reproduction is still not clearly understood.

Neuroendocrine Control of GTH Secretion

GTH secretion is controlled by several regulatory mechanisms: a dominant stimulatory brain mechanism identified as the gonadotropin-releasing hormone (GnRH) system (Peter, 1981; De Leeuw *et al.*, 1987) and inhibitory or stimulatory mechanisms, in which neurotransmitter substances, such as monoamines, amino acids, neuropeptide Y, and opioids act as signal molecules (Peter *et al.*, 1991). These neurotransmitter systems may be considered to be the links between the environment and the GnRH system that dominate the GTH control and may be the physiological substrate for fine-tuning the gonadotropic activity of the pituitary.

Gonadotropin-releasing hormones (GnRH)

To date, eleven molecular forms of GnRH have been isolated from brains of different vertebrate species. They have been named after the species in which first isolated (different mammalian species: mGnRH chicken: cGnRH-I and cGnRH-II; salmon: sGnRH; catfish: cfGnRH; seabream: sbGnRH; lamprey: lGnRH-I and IGnRH-III; dogfish: dfGnRH). Out of the eleven forms, eight have been extracted from the fish brain (Sherwood *et al.*, 1993). All forms are structurally related. They are decapeptides, in which the amino acid residues on positions 1, 2, 4, 9 and 10 are conserved, while the position 8 shows the most variation (Sherwood *et al.*, 1993). In most vertebrate species at least two molecular forms of GnRH are found. In teleosts cGnRH-II is almost universally present. In addition, either mGnRH, sGnRH or one or more species or genus-specific or unidentified forms are found. Thus, in addition to cGnRH-II, cfGnRH is present in two catfish species (Bogerd *et al.*, 1992; Ngamvongchon *et al.*, 1992); in salmonids and cyprinids, for example, sGnRH is the additional form and in seabream, sbGnRH and sGnRH (King and Millar, 1992; Powell *et al.*, 1994). Since cGnRH-II is the most wide spread GnRH among all vertebrates, it is considered to have arisen early in vertebrate evolution.

In several fish species, the cDNAs encoding the GnRH-preprohormones have been cloned and characterized (Bogerd *et al.,* 1994). They all consist of a domain encoding the signal peptide, GnRH, amide-donating glycine, a processing site and a GnRH-associated peptide (GAP), respectively. The GnRH domain is strongly conserved, while the GAP domain is not conserved at all, considering cDNAs from that encode different GnRHs. However, comparing the GAP domains in the cDNAs from different species, but encoding the same GnRH, it shows a high degree of homology (Bogerd *et al.,* 1994).

Which of the GnRHs, in a given species is/are involved in the control of GTH secretion? This question can not be solved by studying the GTH release inducing capacity of the GnRHs separately *in vivo* or *in vitro,* since all forms of GnRH, including the ones that are not native to a given species, are more or less active, due to the high degree of homology in molecular structure. Therefore, anatomical studies to elucidate the localisation of GnRH neurones and their connection with the pituitary are essential. In the past, the anatomical distribution of GnRHs has been demonstrated by homologous and heterologous immunocytochemistry, using antisera raised against mammalian LHRHs or fish GnRH (s-GnRH in most cases) (Joy *et al.,* 1994). These studies have demonstrated an extensive distribution of GnRH neurones in the fish brain extending throughout the olfactory systems, ventral telencephalon, preoptic area, and *nucleus lateralis tuberis* (NLT). Immunoreactive GnRH fibres have been shown to make close contacts with gonadotrophs (Peute *et al.,* 1987). Since the identity of the GnRHs was not known in many species studied, the results relied on cross-reactivities of the sometimes unspecific antibodies that have been used. More recently, in a number of fish species the GnRHs have been identified, so that homologous and more specific antibodies could be used. In cases where the GnRH genes have been cloned, *in situ* hybridisation could be applied. Results of these studies show that in teleost brains there are two main GnRH systems. One consists of GnRH neurones situated in the olfactory bulbs and the medial olfactory tract and, more caudally, close to the longitudinal brain midline in the ventral telencephalon (sometimes concentrated in the nucleus olfacto-retinalis), the ventral hypothalamus and the pituitary stalk. A variety of molecular forms of GnRH has been found in this system. In cyprinids, salmonids and marine perciforms sGnRH, in the eel *(Anguilla anguilla)* and sturgeon *(Acipencer baeri* and *A. transmontanus)* mGnRH, in two catfish species cfGnRH and in the seabream sGnRH and sbGnRH were detected (for review: Kah *et al.,* 1993). Axons of these neurones project into the proximal pars distalis of the pituitary and terminate on, or in the vicinity of, the gonadotrophs. However, projections have also been demonstrated in the telencephalon, the optic tectum, the cerebellum and the spinal cord. This indicates that GnRH from this system fulfils other functions, aside the control of gonadotropin release. A second

GnRH system is localised in the midbrain. In all species studied so far this system contains cGnRH-II (Powell *et al.,* 1994).

Because of the high degree of homology between GnRHs, immunocytochemical cross reactivity easily occurs and hampers the localisation and identification of "GnRH-like" neurones. Different strategies should be followed before definite conclusions can be drawn as shown by the studies on GnRH localisation in the African catfish (Zandbergen *et al.,* 1995). In the African catfish, the cDNAs encoding prepro-cfGnRH and prepro-cGnRH-II have been cloned and characterised (Bogerd *et al.,* 1994). This allowed the localisation of the GnRHs by means of *in situ* hybridisation. Moreover, the respective GAPs were expressed in *E. coli* and used to raise antibodies. Cell bodies expressing prepro-cfGnRH mRNA were found scattered in two lines lateral to the midline of the brain extending from the ventral telencephalon to the caudal hypothalamus, in the medial olfactory tract, *area ventralis telencephali pars ventralis,* in the area of the *nucleus preopticus periventricularis,* in the area of the *nucleus anterioris periventricularis* and in the medial basal hypothalamus. Prepro-cGnRH-II mRNA expressing neurones were restricted to the midbrain. By using the respective anti-GAPs in an immunocytochemical study, the *in situ* hybridisation results were confirmed. Numerous fibres from the cfGnRH neurones project into the proximal pars distalis of the pituitary where the gonadotrophs are situated. From the cGnRH-II neurones in the midbrain tegmentum, only a few descending fibres were observed along the border of the third ventricle but none were found in the pituitary. In goldfish and rainbow trout the situation appears to be different. Using the fluorescent retrograde tracer "DiI", Anglade *et al.,* (1993) showed not only the sGnRH neurones in the olfactory bulbs, the ventro-medial hypothalamus, and the preoptic area to be connected to the pituitary, but also the cGnRH-II secreting neurones in the midbrain tegmentum.

The totally different distributions of the two GnRH systems suggest different functions. Since the "species-specific" GnRHs (sGnRH, mGnRH, cfGnRH, sbGnRH) are present in the system which is directly connected to the pituitary, it is likely that they represent the functional gonadotropin releasing message. However, in a number of species (catfish, goldfish), cGnRH-II is also present in the pituitary, binds to the GnRH receptors and exhibits GnRH release activity. In the African catfish, cGnRH-II is even about a hundred times more active and shows a similar higher affinity to the GnRH receptor, compared to the cfGnRH. Schulz *et al.,* (1993) suggested that in the African catfish cfGnRH is concerned with low basal release of GTH which serves gonadal recrudescence, while the far more potent cGnRH-II brings about GTH surge which precedes ovulation. Others, however, propagate that the telencephalic and ventro-medial hypothalamic GnRH complex controls the hypophysial gonadotropic function, while cGnRH -II from the midbrain fulfils a neuromodulatory role.

The telencephalic/ventromedial hypothalamic GnRH system seems not to consist of only one functional type of GnRH neurones. Magliulo-Cepriano *et al.* (1994) demonstrated that variant forms of GnRH are present at defined stages of development in specific regions of the brain and pituitary and suggested that different forms of GnRH regulate different aspects of the reproductive system development and physiology. Precocious sexual maturation in the platyfish *(Xiphophorus maculatus)* induced by androgen treatment is mainly reflected by increased activity of GnRH neurones in the *nucleus olfacto-retinalis* (Schreibman *et al.*, 1986). Amano *et al.*, (1997) showed that in the masu salmon *(Onchorhychus masou)*, the sGnRH containing neurones in the ventral telencephalon are under negative control of gonadal steroid hormones, while the GnRH neurones of the preoptic area are positively influenced by the same hormones.

The GnRH-GTH-system shows profound seasonal variations with reference to GnRH levels in the brain (Yu *et al.*, 1987), GTH release-response (Weil *et al.*, 1978; Trudeau *et al.*, 1991; Peter *et al.*, 1991; Tharakan and Joy, 1996), and GnRH receptor binding capacity (Habibi *et al.*, 1989b).

GnRH binds to specific G-protein coupled receptors on the gonadotrophs. In African catfish, flounder and stickleback, a single type of high affinity/low capacity receptors has been demonstrated (Schulz *et al.*, 1993). The goldfish pituitary contains an additionally low affinity/high capacity class of GnRH receptors (GnRH-Rs) (Habibi *et al.*, 1987). Only the high affinity binding sites seem to control LH secretion (Habibi *et al.*, 1989a). The GnRH binding capacity may vary during the annual cycle (Habibi *et al.*, 1989b) but the underlying mechanism is not yet known. In the African catfish steroid hormones have been shown to down-regulate the GnRH binding capacity of the gonadotrophs, thus decreasing the sensitivity of the cells for the ligand (Habibi *et al.*, 1989b).

GnRH signal transduction has been studied in goldfish, African catfish and tilapia. As in mammals, findings about the involvement of cAMP in the action of GnRH are controversial. For tilapia as well for the African catfish it has been presumed that cAMP is a second messenger (Van Asselt *et al.*, 1991; Levavi-Sivan and Yaron, 1992). Pituitary cells were reported to accumulate cAMP as a response to GnRH stimulation and increased concentration of cAMP caused LH release. Chang *et al.*, (1992) analysed the role of cAMP in GnRH-induced LH release from goldfish pituitaries. Although elevated cAMP levels induced LH release in a dose-dependent way, the action of cAMP was supposed to be indirect one, based on the observations that GnRH did not induce cAMP production, the effects of GnRH and cAMP being additive and the cAMP-induced LH release being independent of calcium in contrast to the GnRH-induced release. Stimulation of catfish GnRH-Rs led to a dose dependent phosphoinositide turnover (Tensen *et al.*, 1997).

In the gonadotrophs of the African catfish and several other teleosts studied so far, calcium fulfils all requirements of a second messenger. In the absence of calcium or in the presence of voltage-gated calcium channel (VGCCs) blockers, the release of LH is impaired, whereas calcium or calcium ionophores could mimic the effect of GnRH (Van Asselt *et al.,* 1989a). Moreover, Bosma *et al.,* (1995) studied the role of calcium in single, fura-2-loaded gonadotrophs. In contrast to mammalian gonadotrophs, the catfish cells did not show spontaneous or GnRH-induced calcium oscillations, but GnRH evoked a biphasic spike-plateau response of the $[Ca^{2+}]_i$ with a dose dependent amplitude. In both goldfish and tilapia gonadotrophs, similar mechanisms play a role: calcium entrance via VGCCs is essential to GnRH-induced LH release (Jobin and Chang, 1992; Levavi-Sivan and Yaron, 1993).

There are indications that protein kinase C (PKC) is involved in GnRH-induced gonadotropin release. Activators of PKC could mimic the effect of GnRH by stimulating calcium entry through VGCCs, whereas inhibitors of PKC impaired the release of LH (Jobin and Chang, 1993).

In several membrane-bound receptor mediated processes, arachidonic acid (AA) plays a role as second messenger. In mammalian LH cells, such a role was demonstrated by Stojilkovic *et al.,* (1994). AA is an unsaturated fatty acid which is hydrolysed from membrane phospholipids by phospholipase A2 (PLA2). It can be metabolised to prostaglandins and thromboxines via the cyclo-oxygenase route and to leucotrienes via the lipoxygenase pathways, respectively. AA as well as its lipoxygenase, but not the cyclo-oxygenate metabolites, can act as second messengers. Addition of PLA2 to goldfish gonadotrophs increased the GnRH-induced LH release. Application of AA had the same effect. sGnRH, but not cGnRH-II (the two GnRHs native to goldfish) induced the production of AA and its lipoxygenase metabolites. Moreover, inhibition of PLA2 and lipoxygenase, but not cyclo-oxygenase, impaired the sGnRH-evoked release of LH (Chang *et al.,* 1995). Also in tilapia it was shown that AA may induce LH release (Levavi-Sivan and Yaron, 1993).

Recent studies on the transcription of the LH-β gene in tilapia showed that the steady-state level of its mRNA increased when pituitary cells were exposed to sGnRH in vitro. Activation of PKC by TPA had identical results. Increasing levels of cAMP, evoked by forskolin or IBMX resulted in higher LH-β mRNA. These results indicate that GnRH, mediated by cAMP stimulates transcription and translation of LH-β encoding gene (Melamed *et al.,* 1995).

Bioamines

In vertebrates, the central monoaminergic (MA) systems project at different levels of the hypothalamo-hypophyseal system and participate in the regulation of secretory activity of the pituitary gland. In fishes and other lower vertebrates, the hypothalamus contains the largest population of MA neurones (Parent *et al.,* 1984). The major aminergic centres have been identified by various techniques: the Falck-Hillarp paraformaldehyde fluorescence method and glyoxylic acid histofluorescence method, immunocytochemical techniques using specific antibodies against 5-HT (serotonin), DA (dopamine), NA (noradrenaline) and by enzymecytochemical methods for MA-synthesizing or degrading enzymes (tyrosine hydroxylase (TH), DA-β-hydroxylase (DBH), phenylethanolamine-*N*-methyltransferase (PNMT) and monoamine oxidase (MAO). These centres are in the preoptic area including the *nucleus preopticus* (NPO), *nucleus recessus lateralis* (NRL) and *nucleus recessus posterioris* (NRP) (see Joy, 1993). The question if the *nucleus lateralis tuberis* (NLT) is aminergic or not is still an open one. While a majority of investigations point out that it does not contain any MA-perikarya, studies at least in some species have demonstrated aminergic perikarya in this nucleus (Corio *et al.*, 1991; Joy *et al.*, 1992). Furthermore, the NLT neurones also showed strong MAO activity in some teleosts (Joy and Sathyanesan, 1976).

Anatomical studies have demonstrated that NPO and NLT, which are concerned with the synthesis of various hypophysiotropic hormones, receive a strong catecholaminergic and serotonergic innervation. It is indicated that the innervation of the NLT is most likely from the paraventricular organ (PVO) (Corio *et al.*, 1991; Joy *et al.*, 1992). The origin of fibres innervating the NPO is not clearly understood. Various fluorescent and immunocytochemical investigations have demonstrated the presence of MA-ergic fibres in the neurohypophysis and in all parts of the adenohypophysis (Joy, 1993). Immunocytochemical studies at the light- and electron miroscopical level have shown that the amines present in the pituitary are 5-HT, DA and NA (Kah and Chambolle, 1983; Corio *et al.*, 1991; Khan and Thomas, 1993). In the African catfish and goldfish, DA immunoreactivity was found in part of the type B fibres that abut on the gonadotrophs (Kah *et al.*, 1984b; Peute *et al.*, 1987). The origin of the MA fibres in the pituitary has been investigated in the eel and goldfish in some detail (Fremberg *et al.*, 1977; Fryer *et al.*, 1985). In the eel, the PD innervation may be from the PVO which contains both DA and 5-HT. The NA fibres may have an extrahypothalamic origin or may be from the PVO-accompanying cells. In the goldfish, the PVO is considered the possible origin of MA fibres in the pituitary (Fryer *et al.*, 1985). However, immunocytochemical studies have demonstrated the existence of a prominent preoptico-hypophysial DA-ergic pathway to the rostral and proximal PD (Hornby *et al.*, 1987; Kah *et al.*, 1987). Although our knowledge of anatomical distribution of MAs and some related enzymes are extensive, quantitative studies on their distribution are meagre in teleosts (Khan and Joy, 1987; 1988a; Manickam and Joy, 1989, 1990; Saligaut *et al.*, 1992a; Trudeau *et al.*, 1993a, b).

Dopamine as a GRIF

In goldfish, large lesions in the *nucleus lateralis tuberis* region as well as destruction of the pituitary stalk caused a dramatic rise in LH release

(Peter *et al.,* 1978). This points towards the existence of a gonadotropin release-inhibitory factor (GRIF). On the basis of studies on the localisation of TH (Kah *et al.,* 1978), lesioning experiments (Peter and Paulenco, 1980) and retrograde labelling (Anglade *et al.,* 1993) it was concluded that GRIF originates from the antero-ventral preoptic region. In a series of pharmacological studies (for review, see Peter *et al.,* 1991) the GRIF was identified as DA, having an inhibitory effect on spontaneous and GnRH-induced LH release. A similar mechanism has been demonstrated in a number of teleosts, such as African catfish (De Leeuw *et al.,* 1987). In the Atlantic croaker a dopaminergic inhibition of GTH release could not be demonstrated (Copeland and Thomas, 1989), which means that this mechanism is not general to all teleosts. From studies with specific D_1 and D_2 agonists and antagonists, Van Asselt *et al.,* (1988) and Omeljaniuk *et al.* (1989) concluded that DA acts via a high affinity D_2 type receptor on the gonadotropic cells of the African catfish and goldfish, respectively.

Monoamines and GnRH Interaction

The discovery of the combined action of GnRH and DA on gonadotropin release has had considerable impact on the development of techniques for the manipulation of reproduction of fish species that are of economic interest. Many fish species fail to reproduce spontaneously in captivity. Although oocyte and sperm development may not be influenced, final oocyte maturation, ovulation, spermiation and spawning behaviour are more or less blocked. These processes are under control of a surge of LH, which fails in many species to occur under conditions of captivity. It can be considered a stress response or in some cases, it has been overcome by ongoing domestication. In most species, however, the phenomenon hampered profitable farming. Treatment with GnRH (or more active analogues) caused an increase of LH release, but not sufficiently to induce a high percentage of oocyte maturation and ovulation. GnRH in combination with a dopamine antagonistic drug has been shown in many cases to be successful to induce the release of viable eggs. In a number of species a similar procedure has to be followed to obtain sperm. Anti-DA drugs generally used are the non-specific dopamine receptor antagonist pimozide or the more specific D2-antagonist domperidone. As GnRH preparation, superactive analogues, constituted by modification on position 6 (which slows down enzymatic degradation) are used (De Leeuw *et al.,* 1987; Peter *et al.,* 1991).

The mechanism of interaction between GnRH and DA has not been completely elucidated, but there are indications that more than one mechanism is involved. De Leeuw *et al.,* (1988, 1989) demonstrated in catfish and goldfish that DA may influence the number of GnRH receptors, thus interfering with the dynamics of the GTH cell's sensitivity for GnRH. However, also GnRH post receptor mechanisms may be involved, following the studies by Van Asselt *et al.,* (1989a, 1991) and Levavi-Sivan and Yaron

(1992) and Levavi-Sivan *et al.*, (1995) on African catfish and tilapia, respectively. DA had no effect on the GnRH- or forskolin-induced increase in cAMP levels. On the other hand GnRH- or cAMP-induced GTH release appeared to be Ca^{++} dependent. This indicates that mechanisms that are involved in the increase of the intercellular Ca^{++} concentration may be the target of DA action. DA also inhibits GnRH release in the telencephalon-preoptic region and in the pituitary (Yu and Peter, 1992), which explains part of the effects of DA antagonists *in vivo.* NA, the hydroxylated product of DA appeared to have opposite effects. It stimulates GnRH activity via α_1-like noradrenergic receptors in the preoptic region (Peter *et al.*, 1991). There is no information about the existence of NA fibres in the pituitary of teleosts.

5-HT has stimulatory effects on LH release in goldfish (Somoza and Peter, 1991) and in the Atlantic croaker (Khan and Thomas , 1992). The effects are likely to be exerted directly on the pituitary level via 5-HT_2 receptors. The distribution of 5-HT in the brain of teleost fishes has been studied in a variety of species (for references, see above). Kah and Chambolle (1983) reported the presence of 5-HT containing fibres in the pituitary. They suggest, however, that these fibres do not originate from neurones in the brain but rather are long protrusions from hypophysial cells. This implicates that 5-HT, acting on the gonadotrophs, originates from the pituitary itself or reaches the pituitary via the circulation or the cerebrospinal fluid. An indication for the latter may be the observation of liquor contacting 5-HT neurones in the paraventricular organ. Immunocytochemical studies by Khan and Thomas (1993) in the Atlantic croaker showed that the GnRH and 5-HT system are in close proximity, especially in the olfactory bulbs and the preoptic hypothalamic area, suggesting that the two systems interact in the control of GTH secretion.

Monoamines and the Integration of Environmental Signals: Photoperiod and temperature

The following paragraph summarizes the effect of environmental factors, such as photoperiod and temperature on the quantity and turnover rate in the brain of MA neurotransmitters. In goldfish *(Carassius auratus)* and murrel *(Channa punctatus),* both photoperiod and temperature have been shown to influence hypothalamic 5-HT and monoamine oxidase (MAO) activity in a season-dependent manner. In the goldfish, Olcese *et al.,* (1980) reported that the fish maintained in a short photoperiod did not show any day-night differences in hypothalamic 5-HT content and turnover (activity) but when transferred to a long photoperiod regime, the diurnal patterns in 5-HT content and turnover appeared. Likewise, exposure of the goldfish to long photoperiod regimes, but not to a short photoperiod regime, induced a day-night variation in hypothalamic MAO activity. In the murrel and catfish, *Heteropneustes fossilis*, hypothalamic and/or telencephalic 5-HT and MAO show day-night patterns exclusively during the gonadal preparatory phase (February and March) (Khan and Joy, 1988a; Senthilkumaran and Joy, 1993). The appearance of the day-night patterns coincides with the time when photoperiod reaches the threshold level to signal the hypothalamo-hypophyseal system to initiate GTH secretion. GTH is detectable in plasma of *H. fossilis* and *C. batrachus* from February onwards (unpublished). In the murrel (Khan and Joy, 1987, 1988b. 1990a) and *H. fossilis* (Senthilkumaran and Joy, 1994a), exposure to a long photoperiod regime in the gonadal preparatory phase caused a shift in the day-night patterns of hypothalamic 5-HT and/or MAO. Further, when the animals were exposed to differentially-altered photoperiod and temperature regimes in the gonad-quiescent phase (December), the day-night patterns in hypothalamic 5-HT and MAO, which are normally not present in this phase, appeared in the long photoperiod-maintained fish with or without changing the ambient temperature. Exposure to a high temperature regime with or without altering the photoperiod increased the DA and NA contents but the DA turnover was suppressed concomitantly with an elevation of the NA turnover.

There is considerable evidence to show that the pineal organ integrates changes in environmental photoperiod and temperature via the secretion of melatonin (Vivien-Roels, 1985). In mammals, it has been shown that the pineal organ and melatonin influence the activity of a variety of central neurotransmitters like 5-HT and CA (Glass, 1988). In teleosts also, pinealectomy and melatonin treatment have been shown to influence preferentially 5-HT and/or MAO activity in the hypothalamus (Khan and Joy, 1988b; Senthilkumaran and Joy, 1995b). Hypothalamic 5-HT and MAO are considered to be important components of photosexual mechanisms in fishes (De Vlaming and Olcese, 1981; Joy and Khan, 1991). It has been demonstrated in the murrel that photoperiodic influence on the daily patterns of 5-HT and MAO activity in the gonadal preparatory phase is mediated through and by the interaction of the pineal and lateral eyes (Khan and Joy, 1990b). In view of the high concentration of 5-HT in the hypothalamus and its projections into the pituitary, the 5-HT system forms an ideal link in the mediation of photoperiodic responses on pituitary-dependent physiological actions, including reproduction.

Seasonal/annual variations in hypothalamic MA (5-HT, DA, NA and/or A), and their degrading (MAO and/or COMT), or synthesizing (DBH and PNMT) enzymes have been reported in the murrel (Khan and Joy, 1990b), *C. batrachus* (Manickam and Joy, 1989; 1990) and *H. fossilis* (Senthilkumaran and Joy, 1993; 1995a). The results of these studies show that the contents and turnovers of 5-HT, NA and A are high during the recrudescent phase and low in the quiescent phase. The DA content and turnover show a seasonal pattern opposite to that of the other amines. The seasonal patterns of DA and NA in subtropical species are different from those reported in the goldfish (Trudeau *et al.,* 1993a,b) and rainbow trout (Saligaut *et al.,* 1992a). In *H.*

fossilis, photoperiod and temperature exert differential effects on hypothalamic CA and their enzymes (Senthilkumaran and Joy, 1995a). High temperature and long photoperiod stimulate DBH and PNMT activities, the former exerting a more pronounced effect. The progressive decline in the DA metabolism with a concomitant increase in NA/A activities during the breeding phase can be attributed to the increase in environmental photoperiod and temperature. Thus, the stimulatory effect of photoperiod and temperature on GTH activity is mediated partly through their differential actions on hypothalamic CA.

It is evident from the foregoing discussion that the environmental photoperiod and temperature modulate hypothalamic MA-ergic activity. The stimulatory influence of long photoperiod on GTH secretion via the 5-HT pathway was illustrated in *H. fossilis* (Joy and Senthilkumaran, 1995b). Long photoperiod (16L:8D for 30 days) elevated hypothalamic 5-HT content and turnover, plasma levels of GTH-II, E_2 and testosterone. The administration of para-chlorophenylalanine (pCPA, a 5-HT depletor) decreased significantly all these variables, the decrease being higher in long photoperiod-held fish. The administration of 5-hydroxytryptophan (the precursor of 5-HT) along with pCPA nullified the pCPA effect, implicating clearly the 5-HT mediation of long photoperiodic stimulation of LH secretion. Likewise, alpha-methylparatyrosine (alpha-MPT, a CA inhibitor) also inhibited plasma LH and E_2 levels and ovarian weight and vitellogenesis in normal photoperiod -held catfish, the effect being less pronounced than that of pCPA or melatonin (Senthilkumaran and Joy, 1995b). In short, the hypothalamic MA systems play a vital role in the neurondocrine integration of environmental photoperiod and temperature signals.

Neuropeptide Y (NPY), GABA and Other Signal Molecules

NPY, GABA, taurine, monosodium glutamate, bombesin-like peptides, etc., have been demonstrated in many studies to be involved in the control of GTH release (see Prasada Rao, this volume). However, their integrative role in the regulation of the sexual cycle in fish is not established yet.

Integrative feedback action of steroid hormones

Among endogenous factors, gonadal steroids are well known to exert a feedback regulation of GTH secretion (Billard and Peter, 1977; Bommelaer *et al.*, 1981; De Leeuw *et al.*, 1987; Goos, 1987; Van Oordt, 1987; Kobayashi and Stacey, 1990). One of the first observations indicating the existence of such a regulatory mechanism came from Severinghaus (1939), who noticed the appearance of hyperactive cells in the pituitary of the castrated male rat. Later, a correlation was found between the presence of these "castration cells " and elevated plasma gonadotropin levels. Both could be restored by gonadal steroid replacement. A number of investigations have shown that a comparable feedback system is present in fish. In the Indian catfish, *Heteropneustes fossilis,* androgens and estrogens have a negative feedback effect on GTH release (Sundararaj, 1981). Likewise, anti-estrogens such as clomiphene citrate and tamoxifen appeared to cause a rise in the plasma GTH levels as demonstrated in goldfish (Billard and Peter, 1977). Castration also resulted in increased plasma GTH levels in the rainbow trout, *Onchorhynchus mykiss* (Van Putten *et al.,* 1981) and the three spined stickleback, *Gasterosteus aculeatus* (Borg *et al.,* 1985). This was accompanied by a partial degranulation of GTH cells. Bommelaer *et al.,* (1981) showed that the increase in GTH levels could be prevented by testosterone and estradiol, and to some extent, by 11-ketotestosterone. In the African catfish, De Leeuw *et al.,* (1986) showed that the aromatizable androgens, androstenedione and testosterone, but not 11-oxygenated 11β-hydroxyandrostenedione, had the capacity to decrease castration-induced elevated GTH levels and to restore the ultrastructure of the gonadotrophs.

All these results point towards a negative feedback action of steroid hormones on gonadotropin secretion and are in accordance with the well known mammalian concept. It should be realised, however, that the castrated fish may not be the ideal model to study the feedback actions of steroid hormones. Ablation of the gonads not only removes the complete set of gonadal steroid hormones, but also other non-steroidal factors that are implicated in the functioning of the brain-pituitary axis. Thus, castration changes the entire hormonal status of the animal. Moreover, gonadotropin storage, as a result of GTH synthesis, release and break down, and GTH release are all different functions of the gonadotrophs and there is increasing evidence that different steroid hormones may control each of these functions differently. To make it even more complex, these influences may be stimulatory as well as inhibitory. In mammals, it is well established that estrogens are essential to provoke the ovulation inducing LH surge by a stimulation effect on LHRH pulse frequency and amplitude, LHRH receptor availability and LH gene transcription and translation. Also in fish positive effects of steroid hormones on the gonadotropic function of the pituitary have been demonstrated as can be illustrated by the following examples. Crim and Evans (1979), Gielen *et al.,* (1982) in rainbow trout, Dufour *et al.,* (1983) in the European eel *(Anguilla anguilla),* and De Leeuw *et al.,* (1986) in the African catfish reported a positive effect of testosterone or estradiol on LH synthesis and storage. Cavaco *et al.,* (1995) studied the effect of several androgens on LH storage and plasma levels in immature African catfish. They showed that testosterone, but not 11-ketotestosterone, enhances LH storage with no effect on LH release. In recrudescent goldfish, Trudeau *et al.,* (1991, 1993a) showed that testosterone stimulates the GnRH-induced LH release. Similar effects of androgens have also been reported for the rainbow trout during different stages of spermatogenesis (Weil and Marcuzzi, 1990). cDNAs encoding the GTHs in several fish species have been cloned now, enabling to study the effects of steroid hormones on GTH subunit gene expression. Testosterone and estradiol treatments have been shown to stimulate LH β-subunit gene expression in pituitary cells from juvenile rainbow trout (Trinh *et al.*, 1986; Xiong *et al.*, 1994) as well as LHβ mRNA levels in the European eel (Querat *et al.*, 1991). Further studies in juvenile and adult rainbow trout demonstrated that both testosterone and estradiol specifically stimulate LH-β gene expression. Liu *et al.*, (1995) showed that in salmonid fish, the promoter of the GTH II β-subunit gene contains estrogen response elements which can overrule the activity of silencer sequence, thus allowing estrogen stimulated-GTH II β-subunit expression under the influence of estrogens or aromatizable androgens.

In most of these studies, attention has been focussed on the effect of estradiol or aromatizable androgens. In the pituitary tissue of several teleost species, including the African catfish, a highly active aromatase enzyme has been demonstrated (De Leeuw *et al.*, 1985) so that a stimulation of pituitary aromatase activity or an increased availability of aromatizable androgens may contribute to the LH gene expression. However, investigations by Rebers *et al.*, (1997) demonstrated an increase in LH mRNAs after *in vivo* or *in vitro* treatment with 11-ketotestosterone in adult male African catfish, indicating that the effect is not limited to one of the LH subunits, aromatizable androgens or estrogens nor to fish in the juvenile stage. Comparable results were reported by Huggard *et al.* (1996) in goldfish.

Thus, different steroid hormones have either negative or positive feedback effects on gonadotropin synthesis, storage and release. To date, no clear concept can be presented as how positive and negative feedbacks by gonadal steroids integrate the functioning of the hypothalamo-pituitary complex to regulate pubertal development and to establish sexual cycles. From studies on GTH gene expression it is rather obvious that several gonadal steroid homones exert a positive feedback on GTH subunit synthesis rather than a negative one, which seems to be essentially different from the situation in mammals. A negative feedback action by steroid hormones seems to affect mainly the release of gonadotropins, but it should be realised that this effect has only been shown in castration and steroid replacement experiments, of which the physiological significance may need reconsideration. GnRH receptors, controlling to some extent the sensitivity of GTH cells for GnRH have been shown to be under the control of gonadal steroids.

A number of studies have shown that the action of steroids may occur at the hypothalamic level as well. Francis *et al.* (1992) and Amano *et al.* (1997) demonstrated a relation between gonadal steroid levels and the size and number of GnRH-producing neurones in the brain.

The integrative action of steroid hormones on GTH synthesis and release is not limited to their direct effects on GnRH neurones, GnRH receptor sensitivity, and GTH synthesis and release. Gonadal steroids have been shown also to affect the aminergic neuroendocrine systems that are involved in the control of GTH secretion. The telencephalon and hypothalamus are major centres of neuroendocrine regulation of reproduction. GnRH, MA-ergic and steroid-concentrating neurones (Kim *et al.*, 1978) have been demonstrated in these areas with overlapping or proximate distribution (Kah *et al.*, 1986). Reports in the catfish, *C. batrachus*, show that the hypothalamus is a major centre for MA distribution (Joy *et al.*, 1992), and the *nucleus lateralis tuberis* (NLT) , which is an important GnRH centre (Subhedar and Rama Krishna, 1988; Joy *et al.*, 1994) receives extensive MA-ergic innervation (Joy *et al.*, 1992). Furthermore, in catfishes, the posterior NLT contains MA perikarya (Corio *et al.*, 1991; Joy *et al.*, 1992). Nuclear E_2 receptor (NE_2R) binding studies show a high concentration of binding sites in the hypothalamus (unpublished data). These evidences strongly support the view that the hypothalamus is a major integrating centre of regulatory mechanisms concerned with the neuroendocrine control of GTH secretion. A large number of physiological studies have also indicated a functional interaction between GnRH and MAs in the mechanism of control of GTH secretion (see Peter *et al.*, 1991). Both CAs and 5-HT have been shown to modulate GnRH and GTH activity at the level of brain and/or pituitary (Peter *et al.*, 1986, 1991; De Leeuw *et al.*, 1987; Manickam and Joy, 1989, 1990; Somoza and Peter 1991; Khan and Thomas, 1992; Yu and Peter, 1992; Trudeau *et al.*, 1993a, b).

MA turnover rates and the amount of MAs available for pituitary regulation have been used as parameters to study the relation of gonadal steroids with hypothalamic MAs in the steroid-mediated feedback control of GTH secretion (De Leeuw *et al.*, 1987; Manickam and Joy, 1989, 1990; Van Asselt *et al.*, 1989; Khan and Joy; 1991; Saligaut *et al.*, 1992b; Trudeau *et al.*, 1993a, b; Senthilkumaran and Joy, 1994b, 1995b). Saligaut *et al.*, (1992a) correlated the pre-ovulatory drop in estradiol levels with the decrease in hypothalamic and hypophysial dopamine turn-over. Hypothalamic estradiol implants caused an increase in dopamine levels (Saligaut *et al.*, 1992b), which, however, could not be explained by an effect on tyrosine hydroxylase (Saligaut *et al.*, 1993).

E_2 has been shown to modify MA metabolism at various enzymic steps altering the amine tones (Manickam and Joy, 1989, 1990; Khan and Joy, 1991; Senthilkumaran and Joy, 1994b, 1995a). The E_2 effects have been demonstrated in ovariectomized and ovariectomized-E_2 replaced fish. The effect varied with seasons, there being no effect in the non-breeding season due to the quiescence of the hypothalamo-hypophyseal-ovarian system. In the breeding (prespawning) phase, OVX evoked biphasic responses on hypothalamic MA and related enzymes (MAO, DBH and PNMT) which could be correlated with the bimodal LH release peaking at week 4 (Senthilkumaran and Joy, 1994b, 1995a). NA and 5-HT and DBH and PNMT activities were high during the first phase (2, 3 and 4 weeks) and decreased in the second phase (week 6) of OVX. DA and MAO showed opposite trends. In 3-week ovariectomized fish, a low dose of E_2 stimulated/

restored MAO, DBH, PNMT, NA and A activites while a high dose depleted them. The DA content and turnover behaved in an opposite manner. Although these results suggest a strong functional correlation between differential MA activity and the GTH release during the E_2-mediated negative feedback, direct evidences of a specific involvement of the amines were still lacking. In a subsequent study in *H. fossilis*, the amine (5-HT, DA and NA) tones were altered by specific synthesis blockers and then restored by respective precursors in 3-week ovariectomized fish, and the GTH-II release and the amine levels were monitored (Senthilkumaran and Joy, 1996). Summarizing the results, it was found that both 5-HT and CA are involved simultaneously in the E_2-mediated negative feedback, with 5-HT and NA stimulating and DA inhibiting the GTH-II release.

In the goldfish, DA was reported to be exclusively degraded by the MAO mechanism (Sloley *et al.,* 1992). However, in the African catfish and rainbow trout, O-methylation has been demonstrated to have a singnificant role in the DA-methylation process (De Leeuw *et al.,* 1987; Timmers and Lambert, 1988, 1989; Saligaut *et al.,* 1990). Because both catecholamines (CAs) and catecholoestrogens (CEs) are degraded by COMT, a functional interaction between these two systems was suggested by Knuppen *et al.,* (1981) and Breuer (1981) in mammals. Since CEs are better competitive substrate than CA, the latter's methylation is impaired leading to an increase in their concentration and permitting prolonged neuroendocrine actions. The same hypothesis was evoked in the African catfish to explain the mechanism of negative feedback action of E_2 and aromatizable androgens (De Leeuw *et al.,* 1987; Timmers and Lambert; 1989; Van Asselt *et al.,* 1989b). In a recent study, hypothalamic COMT activity was significantly affected during the annual reproductive cycle, alterations in photoperiod and temperature, and OVX and E_2 replacement in *H. fossilis* (Joy and Senthilkumaran, 1998).

Conclusion

In this article we have collected what we considered to be important data in order to present the current view on the integration of the neuroendocrine signals that control the synthesis, storage and release of teleost gonadotropin(s). It turns out to be a complicated picture, incomplete and difficult to explain the gonadotropic function of the pituitary at every stage of pubertal maturation or reproductive cycle. Although today's research on reproductive endocrinology in fish is mainly limited to a few model species, the variations in neuroendocrine mechanisms that control gonadotropin secretion appear to be considerable. Molecular techniques become rapidly available now also for research on fish. It certainly will improve our knowledge on molecular mechanisms involved in the regulation of hormone synthesis, release and mode of action. Integrative processes, however, like the neuroendocrine control of reproduction at the level of the organism, and in relation with its environment are still poorly understood and should be given equal attention.

References

Amano, M., Ikuta, K., Kitamura, S., and Aida, K. (1997). Fish Physiol. Biochem. **17:** 63–70.

Anglade, I., Zandbergen, M.A., and Kah, O. (1993). Cell Tissue Res. **273**: 345–355.

Banerjee, P.P., Bhattacharya, S., and Nath, P. (1989). Gen. Comp. Endocrinol. **73**: 118–128.

Billard, R., Breton, B., Fostier, A., Jalabert, B., and Weil, C. (1978). In: Comparative Endocrinology (P.J. Gaillard and H.H. Boer, Eds.), pp. 37–48, Elsevier.North-Holland Biomedical Press, Amsterdam.

Billard, R., and Peter, R.E. (1977). Gen. Comp. Endocrinol. **32**: 213–220; erratum: **34**: 120–122.

Bogerd, J., Li, K.W., Jansson-Dommerholt, C., and Goos, H.J.Th. (1992). Biochem. Biophys. Res. Comm. **187**: 127–134.

Bogerd, J., Zandbergen, M.A., Andersson, E., and Goos, H.J.Th. (1994). Eur. J. Biochem. **222**: 541–549.

Bommelaer, M-C., Billard, R., and Breton, B. (1981). Reprod. Nutr. Dev. **21**: 989 –997.

Borg, B., Reschke, M., Peute, J., and Van Den Hurk, R. (1985). Acta Zool. (Stockholm) **66**: 47–54.

Borg, B., Peute, J., Reschke, M., and Van Den Hurk, R. (1987). Can. J. Zool. **65**: 14–19.

Bosma, P.T., Van Dijk, W., Van Haren, S.F., Kolk, S.M., Lescroart, O., Schulz, R.W., Terlou, M., and Goos, H.J.Th. (1995). In: Proc. 5th International Symp. Reproductive Physiology of Fish (F.W. Goetz and P. Thomas, Eds.), pp. 55–57, Fish Symp. Austin.

Breton, B., Jalabert, B., and Reinaud, P. (1976). Ann. Biol. Anim. Biochim. Biophys. **16**: 25–36.

Breuer, H., Schneider, H.T., Dobervauer, C., Grüder, S., and Ladosky, W. (1981). Exp. Brain Res. (Suppl.) **3**: 118–129.

Burzawa-Gérard, E. (1971). Biochemie 53. 545–552.

Cavaco, J.E.B., Schulz, R.W., Trudeau, V., Lambert, J.G.D., and Goos, H.J. Th. (1995). In: Proc. 5th International Symp. Reproductive Physiology of Fish (F.W. Goetz and P. Thomas, Eds.), p. 560, Fish Symp. Austin.

Chang, J.P., Wong, A.O.L., Van Der Kraak, G., and Van Goor, F. (1992). Gen. Comp. Endocrinol. **86**: 359–377.

Chang, J.P., Garofalo, R., and Neuman, C.M. (1995). Gen. Comp. Endocrinol. **199**: 339–354.

Copeland, P.A., and Thomas, P. (1989). Gen. Comp. Endocrinol. **74**: 474–483.

Copeland, P.A., and Thomas, P. (1992). Gen. Comp. Endocrinol. **88**: 100–110.

Corio, M., Peute, J., and Steinbusch, H.W.M. (1991). J. Chem. Neuroanat. **4**: 79–95.

Crim, L.E., and Idler, D.R. (1978). Ann. Biol. Anim. Biochim. Biophys. **18**: 1001–1005.

Crim, L.E., and Evans, D.M. (1979). Gen. Comp. Endocrinol. **37**: 192–196.

De Leeuw, R., Smit-Van Dijk, W., Zigterman, J.W.J., Van Der Loo, J.C.M., Lambert, J.G.D. and Goos, H.J. Th. (1985). Gen. Comp. Endocrinol. **60**: 171–177.

De Leeuw, R., Wurth Y.A., Zandbergen, M.A., Peute, J. and Goos, H.J.Th. (1986). Cell Tissue Res. **243**: 587–594.

De Leeuw, R., Goos, H.J. Th., and Van Oordt, P.G.W.J. (1987). Aquaculture **63**: 43–58.

De Leeuw, R., Van't Veer, C., Goos, H.J. Th., and Van Oordt, P.G.W.J.(1988). Gen.Comp. Endocrinol. **72**: 408–415.

De Leeuw, R., Habibi, H.R., Nahorniak, C.S., and Peter, R.E. (1989). J. Endocrinol. **121**: 239–247.
De Vlaming, V., and Olcese, J. (1981). In: The Pineal Gland Vol. II, Reproductive Effects (R.J. Reiter, Ed.),pp. 2–29, CRC Press, Florida.
Dubourg, P., Burzawa-Gèrard, E., Chambolle, P., and Kah, O. (1985). Gen. Comp. Endocrinol. **59**: 472–481.
Dufour, S., Delerue-Le Belle, N., and Fontaine, Y. A. (1983). Gen. Comp. Endocrinol. **52**: 190–197.
Farmer, S.W., and Papkoff, H. (1977). Life Sci. **20**: 1227–1232.
Fostier, A., Jalabert, B., Billard, R., Breton, B., and Zohar, Y. (1983). In: Fish Physiology Vol. 9A (W.S. Hoar, D.J. Randall and E.M. Donaldson, Eds.,) pp. 279–372, Academic Press, New York.
Francis, C.R., Jacobson, B., Wingfield, J.C., and Ferald, R.D. (1992). J. Neurobiol **23**: 1084–1093.
Fremberg, M., Van Veen, T., and Hartwig, H.G. (1977). Cell Tissue Res. **176**: 1–22.
Fryer, J.N., Boudreault-Chateauvert, C., and Kirby, R.P. (1985). J. Comp. Neurol. **242**: 475–484.
Gielen, J. Th., Goos, H.J. Th., Peute, J., Van Den Bosch, R.A. and Van Oordt, P.G.W.J. (1982). Cell Tissue Res. **225**: 45–56.
Glass, J.D. (1988). Pineal Res. Rev. **6**: 219–259.
Goos, H.J. Th. (1987). In: Proceedings of the Third International Symposium on Reproductive Physiology of Fish (D.R. Idler, L.W. Crim and J.M. Walsh, Eds.), pp. 16–20, Memorial University of Newfoundland, St. John's, Newfoundland, Canada.
Goos, H.J. Th., de Leeuw, R., Burzawa-Gérard, E., Terlou, M., and Richter, C.J.J. (1986). Gen Comp. Endocrinol. **63**: 162–170.
Goos, H.J. Th., Peute, J., Preker, D., Janssen, C., Zandbergen, T., and Schulz, R. (1991). In: Proceedings of the 4th International Symposium on the Reproductive Physiology of Fish (A.P. Scott, J.P. Sumpter, D.E. Kime and M.S. Rolfe, Eds.), pp. 46–50, The University of Sheffield, Sheffield, UK.
Habibi, H.R., Peter, R.E., Sokolowska, M., Rivier, J.E. and Vale, W.W. (1987). Biol. Reprod. **36**: 844–853.
Habibi, H.R., Marchant, T.A., Nahorniak, C.S., Van der Loo, H., Peter, R.E., Rivier, J.E., and Vale, W.W. (1989a). Biol. Reprod. **40**: 1152–1161.
Habibi, H.R., de Leeuw, R., Nahorniak, C.S., and Peter, R.E. (1989b). Fish Physiol. Biochem. **7**: 9–18.
Hornby, P.J., Piekut, D.T., and Demski, L.S. (1987). J. Comp. Neurol. **261**: 1–14.
Huang, F.L., Huang, C.J., Lin, S.H., Lo, T.B., and Papkoff, H. (1981). Int. J. Pept. Res. **18**: 69–78.
Huggard, D., Khakoo, Z., Kassam, G., Mahmoud, S.S., and Habibi, H. (1996). Biol. Reprod. **54**: 1184–1191.
Jobin, R.M., and Chang, J.P. (1992). Neuroendocrinology **55**: 6–16.
Jobin R.M., and Chang J.P. (1993). Fish Physiol. Biochem. **11**: 5–42.
Joy, K.P. (1993). Indian Rev. Life Sci. **13**: 83–119.
Joy, K.P., and Sathyanesan, A.G. (1976). J. Neural Transm. **39**: 47–61.
Joy, K.P., and Khan, I.A. (1991). J. Pineal Res. **11**: 12–22.
Joy, K.P., Manickam, P., and Sathyanesan, A.G. (1992). Biol. Struct. Morphogen. **4**: 58–67.
Joy, K.P., Kirubagaran, R., and Goos, H.J. Th. (1994). Internat. Conf. Neurobiology, Abstr. Cochin, India.
Joy, K.P., and Senthilkumaran, B. (1995). In: Reproductive Physiology of Fish (Eds.: F.W. Goetz and P. Thomas), pp. 67–69, Fish Symposium 95, Austin.
Joy, K.P., and Senthilkumaran, B. (1998). Comp. Biochem. Physiol. **119C**: 37–44.

Kah, O., and Chambolle, P. (1983). Cell Tissue Res. **234:** 319–333.
Kah, O., Chambolle, P., and Olivereau, M. (1978). C.R. Acad. Sci. (Paris) **286:** 705–708.
Kah, O., Chambolle, P., Dubourg, P., and Dubois, M.P. (1984a). Gen. Comp. Endocrinol. **53**: 107–115.
Kah, O., Chambolle, P., Thibault, J., and Geffard, M. (1984b). Neurosci. Lett. **48**: 293–298.
Kah, O., Breton, B., Dulka, J.G., Nuñez-Rodriguez, J., Peter, R.E., Corrigan, A., Rivier, J.E., and Vale, W.W. (1986). Cell Tissue Res. **242**: 27-337.
Kah, O., Dulka, J.G., Duburg, P., Thibault, J., and Peter, R.E. (1987). Neuroendocrinology **45**: 451–458.
Kah, O., Anglade, I., Lepretre, E., Dubourg, P. and Monbrison, D. de (1993). Fish Physiol. Biochem. **11**: 85–98.
Khan, I.A., and Joy, K.P. (1987). J. Interdiscipl. Cycle Res. **18**: 287–292.
Khan, I.A., and Joy, K.P. (1988a). Chronobiol. Int. **5**: 311–316.
Khan, I.A., and Joy, K.P. (1988b). Fish Physiol. Biochem. **5**: 187–190.
Khan, I.A., and Joy, K.P. (1990a). Fish Physiol. Biochem. **8**: 291–297.
Khan, I.A., and Joy, K.P. (1990b). J. Pineal Res. **8**: 277–287.
Khan, I.A., and Joy, K.P. (1991). Gen. Comp. Endocrinol. **81**: 426–432.
Khan, I.A., and Thomas, P. (1992). Gen. Comp. Endocrinol. **88**: 388–396.
Khan, I.A., and Thomas, P. (1993). Gen. Comp. Endocrinol. **91**: 167–180.
Kim, Y.S., Stumpf, W.E., Sar, M., and Martinez-Vargas, M.C. (1978). Am. Zool. **18**: 425–433.
King, J.A., and Millar, R.P. (1992). Trends Endocrinol. Metabol. **3**: 339–346.
Knuppen, R. (1981). Exp. Brain Res. (Suppl.) **3**: 108–117.
Kobayashi, M., Aida, K., Hanyu, I., and Ishii, S. (1985). Bull. Jap. Soc. Sci. Fish. **51**: 405–411.
Kobayashi, M., Aida, K., and Hanyu, I. (1986). Bull. Jap. Soc. Sci. Fish. **52**: 1153–1158.
Kobayashi, M., and Stacey, N.E. (1990). Zool. Sci. **7**: 715–721.
Koide, Y., Noso, T., Schouten, G., Peute, J., Zandbergen, M.A., Bogerd, J., Schulz, R.W., Kawauchi, H., and Goos, H.J.Th. (1992). Gen. Comp. Endocrinol. **87**: 327–341.
Levavi-Sivan, B. and Yaron, Z. (1992). Mol. Cell. Endocrinol. **85**: 175–182.
Levavi-Sivan, B., and Yaron, Z. (1993). Fish Physiol. Biochem. **11**: 51–59.
Levavi-Sivan, B., Ofir, M., and Yaron, Z. (1995). Mol. Cell. Endocrinol. **109**: 87–95.
Liu, D., Xiong, F., and Hew, C.L. (1995). Endocrinology **136**: 3486–3493.
Magliulo–Cepriano, L., Schreibman, M.P. and Blum, V. (1994). Gen. Comp. Endocrinol. **94**: 135–150.
Manickam, P., and Joy, K.P. (1989) Gen. Comp. Endocrinol. **75**: 437–445.
Manickam, P., and Joy, K.P. (1990). Gen. Comp. Endocrinol. **80**: 167–174.
Melamed, P., Gur, G., Levavi–Sivan, B., Elizur, A., and Yaron, Z. (1995). In: Proc. 5th International Symp. Reproductive Physiology of Fish (F.W. Goetz and P. Thomas, Eds.), p. 38, Fish Symp, Austin.
Naito, N., Suzuki, K., Nozaki, M., Swanson, P., Kawauchi, H., and Nakai, Y. (1993). Fish Physiol. Biochem. **11**: 241–246.
Ngamvongchon, S., Sherwood, N.M., Warby, C.M., and Rivier, J.E. (1992). Gen. Comp. Endocrinol. **87**: 266–274.
Noaki, M., Naito, N., Swanson, P., Miyata, K., Nakai, Y., Ooto, Y., Suzuki, K., and Kawauchi, H. (1990a). Gen Comp. Endocrinol. **77**: 348–357.
Nozaki, M., Naito, N., Swanson, P., Dickhoff, W. W., Nakai, Y., Suzuki, K., and Kawauchi, H. (1990b). Gen. Comp. Endocrinol. **77**: 358–367.
Nozaki, N., Suzuki, K., Nozaki, M., Swanson, P., Kawauchi, H., and Nakai, Y. (1993). Fish Physiol. Biochem. **11**: 241–246.
Olcese, J. M., Darr, C., Demuri, B., Hall, T. R. and de Vlaming, V. (1980). Comp. Biochem. Physiol. **66A:** 363–365.

Omeljaniuk, R.J., Habibi, H.R., and Peter, R.E. (1989). Gen Comp. Endocrinol. **74**: 392–399.
Parent, A., Poitras, D., and Dube, L. (1984). In: Handbook of Chemical Neuroanatomy Vol. 2 (A. Bjorklund and T. Hokfelt, Eds.), pp. 409–439, Elsevier, Amsterdam.
Peter, R.E., Crim, L.W., Goos, H.J.Th., and Crim, J. (1978). Gen. Comp. Endocrinol. **35**: 391–401.
Peter, R.E., and Paulenco, C.R. (1980). Neuroendocrinology **31**: 133–141.
Peter, R.E. (1981). Gen. Comp. Endocrinol. **45**: 294–305.
Peter, R.E., Chang, J.P., Nahorniak, C.S., Omeljaniuk, R.J., Sokolowska, M., Shih, S.H., and Billard, R. (1986). Recent Progr. Horm. Res. **42**: 513–548.
Peter, R.E., Trudeau, V.L., and Sloley, B.D. (1991). Bull Inst. Zool., Academia Sinica, Monograph **16**: 89–118.
Peute, J., de Leeuw, R., Goos, H.J. Th., and Van Oordt, P.G. W.J. (1984). Cell Tissue Res. **238**: 95–103.
Peute, J., Zandbergen, M.A., Goos, H.J. Th., de Leeuw, R., Pinkas, R., Viveen, W.J.A.R., and Van Oordt, P.G. W.J. (1986). Can. J. Zool. **64**: 1718–1726.
Peute, J., Schild, R.G., Schild, V.A., Buijs, R.M., Van Asselt, L.A.C., and Van Oordt, P.G.W.J. (1987). Gen. Comp. Emdocrinol. **67**: 303–310.
Powell, J.F.F., Zohar, Y., Elizur, A., Park, M., Fischer, W.H., Craig, A.G., Rivier, J.E., Lovejoy, D.A., and Sherwood, N.M. (1994). Proc. Natl. Acad. Sc. USA **91**: 12081–12085.
Prat, F., Sumpter, J.P., and Tyler, C.R. (1996). Biol. Reprod. **54**: 1375–1382.
Querat, B., Hardy, A., and Fontaine, Y.A. (1991). J. Mol. Endocrinol. **7**: 81–86.
Rebers, F.E.M., Bogerd, J., Tensen, C.P., Schulz, R.W., and Goos, H.J. Th. (1997). Fish Physiol. Biochem. **17**: 99–108.
Saligaut, C., Bailhache, T., Salbert, G., Breton, B., and Jego, P. (1990). Fish Physiol. Biochem. **8**: 199–205.
Saligaut, C., Salbert, G., Bailhache, T., Bennani, S., and Nego, P. (1992a). Gen. Comp Endrocrinol. **85**: 261–268.
Saligaut, C., Garnier, D-H., Bennani, S., Salbert, G., Bailhache, T., and Jego, P. (1992b). Gen. Comp. Endocrinol. **88**: 209–216.
Saliguat, C., Bennani, S., and Bailhache, T. (1993). Fish Physiol. Biochem. **11**: 139–144.
Schreibman, M.P. (1986). In: Vertebrate Endocrinology: Fundamentals and Biomedical Implications Vol. I, Morphological Considerations (P.K.T. Pang and M.P. Schreibman, Eds.), pp. 11–56, Academic Press, New York.
Schreibman, M.P., Margolis–Nunno, H., Halpern-Sebold, L.R., Goos, H.J. Th., and Perlman, P.W. (1986). Cell Tissue Res. **245**: 519–524.
Schulz, R.W., Bosma, P.T., Zandbergen, M.A., Van der Sanden, M.C.A., Van Dijk, W., Peute, J., Bogerd, J., and Goos, H.J. Th. (1993). Endocrinology **133**: 1569–1577.
Schulz, R.W., Sanden, M.C.A. van der, Bosma, P.T. and Goos, H.J. Th. (1994). J. Endocrinol. **140**: 265–273.
Schulz, R.W., Bogerd. J., Bosma, P.T., Rebers, F.E.M., Zandbergen, M.A., and Goos, H.J.Th. (1995). In: Proc. 5th International Symp. Reproductive Physiology of Fish (F.W. Goetz and P. Thomas, Eds.), pp. 2–6, Fish Symp, Austin.
Senthilkumaran, B., and Joy, K.P. (1993). Gen. Comp. Endocrinol. **90**: 372–382.
Senthilkumaran, B., and Joy, K.P. (1994a). Fish Physiol. Biochem. **13**: 301–307.
Senthilkumaran, B., and Joy, K.P. (1994b). J. Endocrinol. **142**: 193–203.
Senthilkumaran, B., and Joy, K.P. (1995a). Gen. Comp. Endocrinol. **97**: 121–134.
Senthilkumaran, B., and Joy, K.P. (1995b). Fish Physiol. Biochem., **14**: 471–480.
Senthilkumaran, B., and Joy, K.P. (1996). Gen. Comp. Endocrinol. **101**: 220–226.
Severinghaus, A.E. (1939). In: Sex and Internal Secretion (E. Allan, C.H. Danforth and E.A. Doisy, Eds.), pp. 1045–1087, Williams and Wilkins Co, Baltimore, USA.

Sharp–Baker, H.E., Peute, J., Diederen, J.H.B. and Brokken, L. (1995). Cell Tissue Res. **280**: 113–122.

Sherwood, N.M., Lovejoy, D.A. and Coe, I.R. (1993). Endocrine Rev. **14**: 241–254.

Sloley, B.D., Trudeau, V.L., and Peter, R.E. (1992). J. Exp. Zool. **263**: 398–405.

Somoza, G.M., and Peter, R.E. (1991). Gen. Comp. Endocrinol. **82**: 103–110.

Stojilkovic, S.S., Reinhart, J., and Catt, K. (1994). Endocrine Rev. **15**: 462–497.

Subhedar, N., and Rama Krishna, N.S. (1988). Gen. Comp. Endocrinol. **72**: 431–442.

Sundararaj, B.I. (1981). Aquaculture Development and Coordination Programme, Food and Agriculture Organisation of the United Nations, Rome. ADCP/REP/81/16, pp. 1–82.

Suzuki, K., Kawauchi, H., and Nagahama, Y. (1988a). Gen. Comp. Endocrinol. **71**: 292–301.

Suzuki, K., Kanamori, A., Nagahama, Y., and Kawauchi, H. (1988b). Gen. Comp. Endocrinol. **71**: 459–464.

Suzuki, K., Nagahama, Y., and Kawauchi, H. (1988c). Gen. Comp. Endocrinol. **71**: 452–458.

Swanson, P., Suzuki, K., Kawauchi, H., and Dickhoff, W.C. (1991). Biol. Reprod. **44**: 29–38.

Tharakan, B., and Joy, K.P. (1996). J. Fish Biol. **48**: 623–632.

Tensen, C.P., Okuzawa, K., Blomenrohr, M., Rebers, F.E.M., Leurs, R., Bogerd, J., Schulz, R.W., and Goos, H.J.Th. (1997). Eur. J. Biochem. **243**: 134–140.

Timmers, R.J.M., and Lambert, J.G.D. (1988). Comp. Biochem. Physiol. **91C**: 399–402.

Timmers, R.J.M., and Lambert, J.G.D. (1989). Fish Physiol. Biochem. **7**: 201–210.

Trinh, K.Y., Wang, N.C., Hew, C.L., and Crim, L.W. (1986). Eur. J. Biochem. **159**: 619–624.

Trudeau, V.L., Peter, R.E., and Sloley, B.D. (1991). Biol. Reprod. **44**: 951–960.

Trudeau, V.L., Sloley, B.D., Wong, A.O.L., and Peter, R.E. (1993a). Gen. Comp. Endocrinol. **89**: 39–50.

Trudeau, V.L., Sloley, B.D., and Peter, R.E. (1993b). Brain Res. **624**: 29–34.

Van Asselt, L.A.C., Goos, H.J. Th., Smit-Van Dijk, W., Speetjens, P.A.M., and Van Oordt, P.G.W.J. (1988). Aquaculture **72**: 369–378.

Van Asselt, L.A.C., Goos, H.J. Th., Smit-Van Dijk, W., and Braas, J. (1989a). Gen. Comp. Endocrinol. **76**: 46–52.

Van Asselt, L.A.C., Timmers, R.J.M., Goos, H.J. Th., Lambert, J.G.D., and Van Oordt, P.G.W.J. (1989b). Fish Physiol. Biochem. **6**: 207–212.

Van Asselt, L.A.C., Goos, H.J. Th., and Stoof, J.C. (1991). Gen. Comp. Endocrinol. **82**: 312–313.

Van Der Kraak, G., Suzuki, K., Peter, R.E., Itoh., and Kawauchi, H. (1992). Gen. Comp. Endocrinol. **85**: 217–229.

Van Oordt, P.G.W.J. (1987). In: Proceedings of the V Congress of European Icthyology (S. O. Kullander and B. Fernholm, Eds.), pp. 247–268, Swedish Museum of Natural History, Stockholm.

Van Oordt, P.G.W.J., and Peute, J. (1983). In: Fish Physiology Vol. 9A (W.S. Hoar, D.J. Randall and E.M. Donaldson, Eds.), pp. 137–186, Academic Press, New York.

Van Putten, L.J.A., Peute, J., Van Oordt, P.G.W.J., Goos, H.J. Th., and Breton, B. (1981). Cell Tissue Res. **218**: 439–448.

Vivien-Roels, B. (1985). In: The Pineal Gland, Current State of Pineal Research (B. Mess, C. Rúzsas, L. Tima and P. Pevet, Eds.). pp. 187–209, Elsevier, Amsterdam.

Weil, C., Billard, R., Breton, B., and Jalabert, B. (1978). Ann. Biol. Anim. Biochem. Biophys. **18**: 863–869.

Weil, C., and Marcuzzi, O. (1990). Gen. Comp. Endocrinol. **79**: 492–498.

Weil, C., Bougoussa, M, Gallais, C., Itoh, S., Sekine, S., and Valotaire, Y. (1995). Gen. Comp. Endocrinol. **100**: 327–333.

Xiong, F., Lui, D., Le Drean, Y., Elsholtz, H.P. and Hew, C.L. (1994). Mol. Endocrinol. **8**: 82–793.

Yu, K.L., Nahorniak, C.S., Peter, R.E., Corrigan, A., Rivier, J.E., and Vale, W.W. (1987). Gen. Comp. Endocrinol. **67**: 234–246.

Yu, K.L., and Peter, R.E. (1992). Gen. Comp. Endocrinol. **85**: 138–146.

Zandbergen, M. A., van den Branden, C.A.V., Schulz, R.W., Janssen-Dommerholt, J., Ruijter, J. M., Peute, J., and Goos, H.J. Th. (1993). Fish Physiol. Biochem. **11**: 255–263.

Zandbergen, M.A., Kah, O., Bogerd, J., Peute, J., and Goos, H.J.Th. (1995). Neuroendocrinology **62**: 571–578.

Comparative Endocrinology and Reproduction
K.P. Joy, A. Krishna and C. Haldar (Eds)

8. Regulation of Hypophysial Gonadotropin by Neuropeptides of the Brain and Gonadal Secretions in Teleosts

P.D. Prasada Rao
Department of Zoology, Nagpur University, Nagpur-440 010, India

Introduction

It has been known for a long time that hypothalamus regulates the secretion of pituitary hormones through various hormone-releasing and/or hormone-inhibiting factors or hormones. However, recent advances involving the application of high performance liquid chromatography, immunocytochemical and radioimmunoassay procedures have brought to light a wealth of new information that necessitated alterations in previous concepts about regulation of different hypophysial hormones. Steroidal as well as non-steroidal secretions of the gonads have also been known to play an important role in gonadotropin (GTH) regulation in teleosts. Gonadotropins are of importance in vertebrate physiology as they regulate various reproductive processes which constitute one of the most fundamental processes of life. The GTH physiology and its regulation occupy an important position in the breeding of major carps of India (*Labeo rohila, Cirrhinus mrigala* and *Catla catla*). These major carps which are extensively cultivated, do not breed in ordinary perennial tanks of India. As a result, there is a shortage of fish seed, and one has to depend upon induced breeding of fish, which mostly involves utilization of GTH for this purpose. Before going into the details of regulation of GTH, it is appropriate to understand the current status of knowledge about gonadotropins in teleosts.

Gonadotropins in Teleosts

Unlike the situation in mammals, initial studies did not reveal two different hormones comparable to follicle stimulating hormone (FSH) and luteinizing hormone (LH). Application of trichrome or tetrachrome technique was useful in identifying the basophils or cyanophils which included both the thyrotropin- and gonadotropin-secreting cells. Previous methods of functional identification of gonadotropin- and thyrotropin-secreting cell types involved application of classical staining techniques such as aldehyde fuchsin (AF) and periodic acid Schiff's reagent (PAS). These methods were helpful in delineating gonadotropin- and thyrotropin-secreting cells in the pituitary gland of mammals, but were not useful with the teleostean pituitaries. Consequently, for the identification of gonadotropin-secreting cells, one had to depend upon circumstantial or experimental evidence which very often involved prolonged and laborious experiments. For example, circumstantial evidence for the identification of gonadotropin-secreting cells stems from the study of annual cyclic changes in the pituitary gland in correlation with gonadal cyclic changes; the cells that underwent correlative seasonal cyclic changes such as degranulation and vacuolization in conjunction with the gonadal changes permitted the identification of gonadotropin-secreting cells (Prasada Rao, 1969).

Experiments such as gonadectomy or administration of gonadal steroids have also permitted the identification of GTH cells (Prasada Rao, 1972; Prasada Rao *et al.*, 1972). However, individual cells that occur isolated among numerous cells of other hormone-secreting nature often posed difficulties in their localization and resulted in ommission of such cells. Such inaccuracies or failures in identification of GTH cells were overcome by the initiation of application of immunocytochemical technique which is highly sensitive and reproducible, and the reaction product is demonstrable even when the antibody is applied at high dilutions. On account of these advantages, it became possible to demonstrate GTH cells in the teleostean hypophysis in an authentic manner.

Electron microscopic studies and previous histochemical studies have generated controversial data on the nature and types of GTH cells. The concept of occurrence of more than one GTH was first postulated by Idler (Review: Idler, 1982). A number of biochemical investigations have been conducted to purify fish gonadotropins, but the number and identity of gonadotorpins have remained controversial. A glycoprotein-rich gonadotropin with a molecular weight of 25,000–40,000, termed glycoprotein-rich "maturational gonadotropin", has been purified in several teleosts (Burzawa-Gerard, 1982). This gonadotropin was reported to stimulate almost all gonadal activities including gametogenesis and steroidogenesis. However, biochemical evidence is now available to show that teleosts possess two types of gonadotropins. Idler and his colleagues isolated two gonadotropins from pituitaries of different teleost species (Idler and Ng, 1979; Idler, 1982).

Subsequently, two distinct carbohydrate gonadotropins, designated as GTH-I and GTH-II were identified in chum salmon (Suzuki *et al.*, 1988; Kawauchi *et al.*, 1989). These two types of gonadotropins are distinctly different from each other in chemical characteristics, and structurally homologous to tetrapod FSH and LH, respectively. Each gonadotropin consists of α and β subunits. GTH I was shown to be predominant in the plasma and pituitary of vitellogenic females, whereas GTH II was predominant at the time of final oocyte maturation (Swanson *et al.*, 1991). Evidence is available for the presence of two types of GTH cells in juvenile African catfish; however, only GTH-II has been isolated so far from the catfish and only one cell type is present in the postpuberal catfish. GTH-I and II have been isolated in the common carp, and evidence was presented that both GTH-I and GTH-II can stimulate uptake of vitellogenin into goldfish oocytes (cf. Peter, 1993). However, it is only in the salmonids that radioimmunoassays with reasonable specificity fo GTH-I and GTH-II have been developed, whereas in other systems it has not been feasible to develop specific radioimmunoassays for GTH-I. It seems likely that some teleosts, such as the cyprinids, catfishes and eels may control gonadal function primarily with one, rather than with two gonadotropins, that is the GTH-II. It requires further studies to elucidate this aspect.

The Gonadotropin-Releasing Hormone Peptide

The gonadotropin-releasing hormone (GnRH) was originally referred to as luteinizing hormone-releasing hormone (LHRH), and it was isolated and synthesized in 1971 using porçine, ovine and bovine hypothalamic extracts (Amoss *et al.*, 1971; Matsuo *et al.*, 1971). Although it was thought to be a unique form, biochemical studies have revealed that GnRH molecule occurs as a family of structurally similar peptides (King and Millar, 1982; cf. Sherwood *et al.*, 1994). Discovery of the primary structure of mammalian LHRH (pGlu^1–His^2–Trp^3–Ser^4–Tyr^5–Gly^6; Leu^7–Arg^8–Pro^9–Gly^{10}–NH_2; LH-RH; for review, Vale *et al.*, 1981) was a milestone in neuroendocrine research. This has influenced research on the neuroendocrine regulation of gonadotropin secretion in the non-mammalian vertebrates. Todate, nine different forms of GnRH peptides have been isolated with a sequence similarity of 5% or greater (Fig. 1). In all the forms, aminoacids 1, 2, 4, 9 and 10 residues that are important for receptor binding and/or biological activity, are conserved (King and Millar, 1991). Among jawed vertebrates, only aminoacids 5, 7 and 8 are variable.

The fish GnRH was first isolated from chum salmon (cf. Sherwood *et al.*, 1994). The salmon GnRH (sGnRH) was not different from the mammalian GnRH (mGnRH); both have 10 aminoacids, 80% sequence similarity and the same N-terminal (pGlu) and C-terminal (NH_2) modifications. Following the first discovery of sGnRH structure, several distinct GnRH peptides have been isolated and sequenced from different species of fish. Among different fishes the identified GnRH forms include salmon GnRH, dogfish GnRH, catfish GnRH and chicken GnRH-II. So far GnRH bas been studied in 47 species of fish by one or more of the following methods: high-performance liquid chromatography (HPLC) combined with radio-immunoassay (RIA); aminoacid composition; aminoacid sequencing; cDNA or genomic analysis. The structure of GnRH gene has been presented for only two fish, *viz.*, Atlantic salmon (*Salmo salar*) (Klungland *et al.*, 1992) and sockeye salmon (*Oncorhynchus nerka*) (Coe *et al.*, 1992).

Recent studies revealed that more than one type of GnRH molecule exists in the same species. All the species which were examined for the presence of more GnRH molecules, revealed that chicken-II type (cGnRH-II) exists, but a second form is either salmon GnRH (sGnRH), mammalian type or catfish type depending on the species (Sherwood *et al.*, 1994; Kim *et al.*, 1995). Very recently, it was shown by Powell *et al.*, (1995) that there exists even three molecular forms of GnRH viz., sea bream GnRH (sbGnRH), cGnRH-II and s-GnRH in the brain of a cichlid and pumpkinseed fish.

	1 2 3 4 5 6 7 8 9 10
SALMON	pGLU-HIS-TRP-SER-TYR-GLY-TRP-LEU-PRO-GLY-NH_2
DOGFISH	pGLU-HIS-TRP-SER HIS GLY-TRP-LEU-PRO-GLY-NH_2
CATFISH	pGLU-HIS-TRP-SER HIS GLY LEU-ASN PRO-GLY-NH_2
SEABREAM	pGLU-HIS-TRP-SER-TYR-GLY LEU-SER PRO-GLY-NH_2
MAMMAL	pGLU-HIS-TRP-SER-TYR-GLY LEU-ARG PRO-GLY-NH_2
CHICKEN-I	pGLU-HIS-TRP-SER-TYR-GLY LEU-GLN PRO-GLY-NH_2
CHICKEN-II	pGLU-HIS-TRP-SER HIS GLY-TRP TYR PRO-GLY-NH_2
LAMPREY-III	pGLU-HIS-TRP-SER HIS-ASP TRP LYS PRO-GLY-NH_2
LAMPREY-I	pGLU-HIS TYR SER LEU-GLU TRP LYS PRO-GLY-NH_2

Fig. 1. Comparison of amino acid sequence of the nine identified GnRH peptides. Amino acids shown in boxes indicate residue changes with respect to the salmon GnRH.

They further mention that three forms of GnRH are widespread in the order Perciformes. However, studies are warranted on other species with a view to examine the possible existence of more than two molecular forms of GnRH in the same species.

Response of GTH Release to GnRH

Based upon the report that extracts of the fish hypothalamus stimulate GTH release from the sheep pituitary *in vitro* (Breton *et al.*, 1972) and other evidences, one might choose a mammalian pituitary bioassay for teleost GTH-releasing factor. With the help of *in vivo* and *in vitro* approaches it has been demonstrated that GnRH peptide stimulates the release of GTH in the goldfish (Marchant and Peter, 1989). It is well documented that GnRH and synthetic superactive analogues are effective in stimulating GTH release in *in vivo* and *in vitro* conditions in teleost fishes and for inducing gonadal development and ovulation (Peter, 1982).

The goldfish pituitary contains two classes of GnRH-binding sites; one is of high affinity/low capacity, and the other is low affinity/high capacity.

All GnRH forms are known to bind to these binding sites. Binding sites of the two classes undergo seasonal variation in number and are most abundant during late gonadal recrudescence. The potency of various GnRH forms to stimulate GTH release from the pituitary is well correlated with their binding to the high affinity/low capacity sites. The superactive analogue (D-Arg^6 Pro^9-NEt)–sGnRH was found to have the highest affinity and highest activity in the goldfish.

The post-receptor activity of GnRH effect on the release of GTH is mediated through a cascade of intracellular events spanning through the influx of Ca^{2+} phosphoinositide turnover, mobilization of Ca^{2+} from intracellular sources, formation of diacylglycerol, activation of protein kinase C, and the synthesis of arachidonic acid and its metabolites. Cyclic-AMP may participate in the mediation of the GnRH effect in tilapia as part of a parallel or interconnected transduction system, but perhaps not in he goldfish (cf. Yaron, 1995).

In teleosts, considerable evidence shows that GTH secretion is regulated by the dual effects of GnRH and dopamine which inhibits GnRH-stimulated GTH release (De Leeuw *et al.*, 1986). It is not known whether dopamine also influences the activities of brain GnRH neurons. An interaction between the central dopaminergic and GnRH peptidergic systems is suggested by the anatomical proximity of the preoptico-hypophysial dopaminergic and GnRH peptidergic pathways. In teleosts, instead of transport of neurohormones to the pars distalis by the hypothalamo-hypophysial blood portal system in most vertebrates, the neurosecretory fibres directly innervate the pituitary gland. Thus, measurement of GnRH levels in the pituitary in a teleost is homologous to measurement of GnRH levels in the median eminence in other vertebrates. Gonadotrophs are directly innervated by the neurosecretory fibres immunoreactive to GnRH (Kah *et al.*, 1986a) and dopamine (Kah *et al.*, 1986b). GTH release is stimulated by GnRH, while it is inhibited by dopamine through its action as a GTH-release inhibitory factor. However, the inhibitory action of dopamine on GTH secretion varies in potency among different species. For example, in the goldfish and the common carp the dopamine inhibition is very strong. However, it is less pronounced in the Chinese loach, salmonids and tilapia. In the catfish, *Heteropneustes fossilis,* it is relatively minor, and dopaminergic inhibition is entirely absent in *Micropogonias undulatus* (cf. Yaron, 1995). Studies conducted on cultures of dispersed goldfish pituitary cells using specific DA agonists and antagonists have led to the conclusion that dopaminergic inhibition of GTH release is conveyed by binding to D2 receptors on the gonadotrophs (Chang *et al.*, 1990).

The superactive mammalian GnRH agonist [D-Ala^6, Pro^9-NEt]-LHRH (LHRH-A) and salmon GnRH agonist [D-Arg^6, Trp^7, Leu^8, Pro^9-NEt]- LHRH (sGnRH-A) stimulate only a modest increase in serum GTH and are ineffective in inducing ovulation and spawning in these species. *In vitro* studies in the intracellular mechanism of DA inhibition in tilapia revealed that DA antagonizes the effect of GnRH by repressing the transduction of GnRH signal at a site distal to Ca^{2+} flux, but proximal to the formation of arachidonic acid. Dopamine is known to inhibit GTH release in response to cAMP agonists, but does not affect the formation of cAMP in response to GnRH.

Administration of the dopamine receptor antagonist pimozide (PIM) or domperidone (DOM) greatly potentiates the actions of LHRH-A and sGnRH-A on GTH release. Combined injections of PIM or DOM and LHRH-A or sGnRH-A are highly effective in inducing ovulation and spawning in these species. The combination of a GnRH analogue and a dopamine antagonist (domperidone) for induced ovulation and spawning of cultured fish is called the Linpe method. The Linpe method has been commercialized by Syndel Laboratories, Inc., Canada under the commercial name Ovaprim (cf. Lin and Peter, 1991). Currently it is marketed in our country by Glaxo, India.

Serotonin

Treatment of intact female goldfish with 5-HT resulted in an increase in serum GTH levels indicating that serotoninergic neurons have some direct or indirect influence over GTH cells (Peter *et al.*, 1985).

Immunocytochemical studies revealed that serotonin (5-HT)-containing neurons are peresent in the *nucleus recessus lateralis, nucleus recessus posterioris* and in the *nucleus dorsolateralis thalami*, located ventrolateral to the posterior commissure. The last group of cells extends posteriorly as far as the rostral midbrain tegmentum (Kah and Chambolle, 1983). Other studies revealed that serotonin stimulates GTH release in the teleostean fish (Somoza and Peter, 1991; Khan and Thomas, 1992).

Melatonin

Injection of melatonin during the late-light phase of the day-night cycle elicited a significant elevation in plasma GTH-II levels in the Atlantic Croaker (*Micropogonias undulatus*) with fully developed gonads. The stimulatory effect of melatonin was dose dependent (Khan and Thomas, 1996). However, no significant effect of melatonin was observed on basal or LHRHa-induced GTH-II release in regressed coraker.

Neuropeptide Y

Neuropeptide Y (NPY) is a 36-amino-acid peptide which was first isolated from porcine brain (Tatemoto *et al.*, 1982). Recently, NPY has also been characterized in the goldfish and the sequence deduced goldfish cDNA clones reveal that goldfish NPY (gNPY) has over 86% identity with human NPY (hNPY; Bloomqvist *et al.*, 1992). The dose response relationships of gNPY on GTH-II release were first tested using an *in vitro* perifusion system for pituitary fragments. It was found that gNPYstimulated GTH-II release in a dose-dependent manner (Peng *et al.*, 1993a). There also occurs interaction between gNPY and other GTH regulators such as GnRH. Prolonged infusion of sGnRH induced an initial peak release of GTH-II, followed by a second phase of sustained hormone release at a reduced level. The gNPY-induced GTH-II release was reduced by dopamine, a known inhibitor for GTH-II. Intraperitoneal injection of gNPY induced time- and dose-dependent increase in serum GTH-II levels (Peng *et al.*, 1993b).

Norepinephrine

Norepinephrine (NE) has been shown to have a direct stimulatory effect on release of maturational GTH from perifused fragments and dispersed cells of the pituitary gland. Chang and Peter (1983) found that systemic treatment with NE caused an increase in serum GTH levels in goldfish. In addition to a direct action on the pituitary to stimulate GTH, illustrated by the *in vitro* studies, NE has a stimulatory action on sGnRH release from the preoptic region of the goldfish (cf. Peter *et al.*, 1986, 1991) which could also have contributed to the enhanced GTH release.

γ-Aminobutyric acid (GABA)

The occurrence of neurons containing γ-aminobutyric acid (GABA) has been demonstrated in the forebrain (Martinoli *et al.*, 1990) and pituitary gland (Kah *et al.*, 1987). Furthermore, a close association between GABA-containing neurons and gonadotrophs has been indicated (Kah *et al.*, 1992). This suggests that GABA might have some involvement in the regulation of pituitary function. Recently, it has been demonstrated that GABA is involved in the regulation of GTH-II secretion in goldfish (Trudeau *et al.*, 1993a). This work indicates that GABA has a stimulatory effect on GTH secretion through a GABA-like receptor. However, the effect does not appear to be direct as dispersed pituitary cell cultures do not release GTH-II in response to GABA (Kah *et al.*, 1992). Further evidence indicates that in goldfish, GABA stimulates GnRH release from pituitary slices (Kah *et al.*, 1992) and that γ-vinyl GABA (GVG) reduces pituitary and hypothalamic dopamine turnover (cf. Trudeau *et al.*, 1993a). Treatment with GVG caused a significant increase in serum GTH concentrations (Sloley *et al.*, 1994).

Taurine

Intraperitoneal injection of taurine into goldfish undergoing gonadal recrudescence resulted in a significant increase in serum GTH concentrations at 30 min after administration. A similar effect was also reported after injection of taurine precursor, hypotaurine. Intracerebroventricular administration of taurine also significantly elevated serum GTH concentration (Sloley *et al.*, 1992).

Glutamate

Injection of monosodium glutamate (MSG) into sexually recrudescent goldfish resulted in an elevation of GTH 24 hr after administration, but not at other intervals. Fish treated with MSG for 48 hr before the administration of sGnRH analogue (sGnRHa) showed a potentiated serum GTH response. Fish in midstages of ovarian recrudescence were not responsive to sGnRHa alone; on the other hand, fish pretreated with MSG showed a marked elevation of serum GTH concentration following sGnRHa injection (Sloley *et al.*, 1992).

Endogenous Opioid Peptides

Endogenous opiod peptides (EOPs) have been implicated in the regulation of hypothalamic-pituitary-gonadal axis in several vertebrates. Immunoreactive EOP-like material has been demonstrate in brains of several teleost species (Dubois *et al.*, 1979; Vallarino, 1985). In addition, specific opioid receptors have been described in the brains of different teleosts (Pert *et al.*, 1974; Rosenblum and Callard, 1988). Since both GnRH and dopamine regulate the secretion of GTH from the teleost pituitary, the EOP system was shown to play an important role by modulating both the stimulatory (GnRH) and inhibitory (dopamine) regulators of GTH secretion (Rosenblum and Peter, 1987).

Mammalian GnRH-Antagonists

It has been known that sGnRH and cGnRH-II stimulate the release of GTH-II in goldfish. Modification of GnRH structure at positions 1,2,3 and 6 often results in an antagonistic in effect in this species. Interestingly, in a preliminary study on goldfish, a mammalian GnRH antagonist, [Ac-D(2) Nal^1, Cl-D-Phe^2, D(3)-$Pal^{3,6}$, Arg^5, D-Ala^{10}]-m GnRH (analogue L) weakly stimulated the release of GTH-II (Murthy *et al.*, 1994). In a goldfish pituitary fragment perifusion system, analogue L stimulated GTH-II in a dose dependent manner. Further, analogue L stimulated GTH-II release from the enzymatically dispersed goldfish pituitary cells, indicating a direct action of analogue L at the pituitary cell level. It was further found that analogue L specifically inhibits sGnRH and cGnRH-II actions on GTH-II release (cf. Murthy *et al.*, 1994). Similarly, *in vitro* studies using certain other mammlian GnRH-antagonists have also revealed the ability of stimulating the release of GTH from the pituitary fragments.

Superactive GnRH-Analogues

Several synthetic GnRH-analogues are available and their ability to stimulate the release of GTH in teleostean species has been investigated. Synthetic GnRH analogues in which the L-amino-acid position 6 is replaced by D-aminoacids, and the Gly at the carboxy terminalis replaced by ethylamide are more resistant to enzymatic degradation by the pituitary, kidney and the liver, conferring on the analogue a superior biological activity (cf. Yaron, 1995). The most resistant form of the several analogues tested in the sea bream is [D-Arg^6, Pro^9 -NEt]-sGnRH. *In vitro* studies using pituitary fragment perifusion system have shown that [D-Arg^6]-GnRH was more potent than sGnRH which was itself significantly more potent than [Arg^6]- GnRH. [D-Ala^6] and [D-Leu^6]-GnRH were as potent as sGnRH, whereas the corresponding L-aminoacid-containing analogues, [Ala^6]- and [Leu^6]-GnRH were significantly less potent. [D-Arg^6, Trp^7, Tyr^8]-GnRH and [D-Lys^6, Trp^7, Tyr^8]-GnRH had a high potency relative to sGnRH. Although [His^5]-GnRH had a lower potency, [His^5, D-Trp^6]-GnRH had a higher potency than sGnRH. These findings indicate that substitution of selected D-aminoacids in position 6 is essential for development of superactive analogues of GnRH. Generally, [D-Arg^6]-or [D-Lys^6]-GnRH in combination with constituent hydrophobic aminoacids of salmon GnRH and chicken GnRH-II provides analogues of highest potency in the goldfish (cf. Peter *et al.*, 1995).

Gonadal Secretions

Certain gonadal secretions such as inhibin and activin, as well as the gonadal steroids have been found to play a role in GTH release. The following account summarized the results obtained along this direction.

Inhibin and Activin

Crude aqueous extracts from goldfish ovaries exhibit stimulatory effect on basal as well as sGnRH-stimulated GTH-II release after long term treatment. Furthermore, the extracts of goldfish testis have also revealed an acute stimulatory action on pituitary GTH-II release in perifusion with pulsatile treatment for as short as 2 min eliciting a rapid response of GTH-II release (Ge and Peter, 1994). In mammals, a variety of factors have been identified in goandal preparations that can either stimulate or inhibit pituitary GTH release (Li *et al.*,1987). In mammals, it has been known that in addition to steroidal hormones, gonads also produce a variety of non-steroidal proteins and peptides that can feed back to regulate pituitary gonadotropin and activin. In sharp contrast to mammalian gonadal extracts which usually have inhibitory effects on mammalian gonadotropin secretion due to inhibin the gonadal extracts stimulate pituitary GTH-II. These actions are similar to the effects of porcine follicular fluid on goldfish GTH-II release. It was also found that immunoreactive inhibin and activin subunits are present in the goldfish gonads.

Gonadal steroids

The serum GTH-II response to LHRH-analogues was enhanced in male goldfish that were implanted with 17β-estradiol. Thus 17β-estradiol was shown to potentiate the serum GTH-II response to exogenous GnRH (Trudeau *et al.*, 1993b). Serial implantation of testosterone pellets, at 30-day intervals to day 75 or at 15-day intervals to day 105, stimulated the pituitary content of GTH to increase approximately 24-fold although no significant change was found in serum GTH levels in the Japanese silver eel, *Anguilla japonica* (Lin *et al.*, 1991).

Concluding Remarks

It is evident from the above that the two most important hypothalamic factors regulating the secretion of GTH release from the teleostean pituitary are GnRH, which has a stimulatory effect and dopamine, which antagonizes GnRH effect by inhibiting the release of GTH. However, such a generalization cannot be made since the role of dopamine is variable among different species of teleosts. Furthermore, various brain hormones and neurotransmitters are implicated in the regulation of GTH release from the pituitary gland. Besides GnRH and its various forms, neuropeptide Y, γ-aminobyric acid, taurine, glutamate, N-methyl-D-aspartate, dopamine, norepinephrine, serotonin, endogenous opioid peptides, sex steroids and there are indications that cholinergic nerves (Mikolajczyk *et al.*, 1993) may also participate in GTH regulation. It would be worthwhile to investigate and elucidate the precise role played by various peptides/transmitters in the regulation of GTH so as to permit evolving appropriate technologies to induce ovulation and spawning in fishes which are of value in farming. The studies available on the superactive GnRH-analogues and their role in GTH release are limited to *in vitro* studies. It would be of great value to extend the studies to demonstrate the role of superactive GnRH-analogues through *in vivo* studies.

Acknowledgments

The author wishes to express thanks to Mr. Sandeep P. Rao for typing the manuscript and Mr. Santosh M. Baby and Mr. D.V.N.S. Suresh for their help in preparation of references.

References

Amoss, M., Burgus, R., Blackwell, P., Vale, W., Fellows, R. and Guillemin, R. (1971). Biochem. Biophys. Res. Commun. **44:** 205–210.

Bloomqvist, A.G., Soderberg, C., Lundell, I., Milner, R.J. and Larhammar, D. (1992). Proc. Natl. Acad. Sci. USA. **89:** 2350–2354.

Breton, B., Weil, C., Jalabert, B. and Billard, R. (1972). C.R. Acad. Sci. Ser. **D274:** 2530–2533.

Burzawa-Gerard, (1982). Can. J. Fish. Aquat. Sci. **39:** 80–91.

Chang, J.P., Yu, K.L., Wong, A.O.L. and Peter, R.E. (1990). Neuroendocrinol. **51:** 661–674.

Chang, J.P. and Peter, R.E. (1983). Gen. Comp. Endocrinol. **55:** 89–95.

Coe, I.R., Grier, H.J. and Sherwood, N.M. (1992). J. Exp. Zool. **261:** 414–432.

De Leeuw, R., Goos, H.J. Th. and Van Oordt, P.G.W.J. (1986). Gen. Comp. Endocrinol. **63:** 171–177.

Dubois, M.P., Billard, R., Breton, B. and Peter, R.E. (1979). Gen. Comp. Endocrinol. **37:** 220–231.

Ge, W., and Peter, R.E. (1994). Zool. Sci. **11:** 717–724.

Idler, D.R. and Ng, T.B. (1979). Gen. Comp. Endocrinol. **38:** 421–440.

Idler, D.R. (1982). In: Proceedings of the International Symposium on the Reproductive Physiology of Fish (C.J.J. Ritcher and H.J.Th. Goos, Eds.), pp. 4–13. Pudoc, Wagheningen.

Kah, O., Breton, B., Dulka, J.G., Nunez-Rodriguez, J., Peter, R.E., Corrigan, A., Rivier, J.E. and Vale, W.W. (1986a). Cell Tissue Res. **244:** 327–337.

Kah, O., Dubourg, P., Onteniente, B., Geffard, M. and Calas, A. (1986b). Cell Tissue Res. **244:** 577–582.

Kah, O. and Chambolle, P. (1983). Cell Tissue Res. **234:** 319–333.

Kah, O., Dubourg, P., Martinoli, M.G., Rabhi, M., Gonnet, F., Geffard, M. and Calas, A. (1987). Gen. Comp. Endocrinol. **67:** 324–332.

Kah, O., Trudeau, V.L., Sloley, B.D., Chang, J.P., Dubourg, P., Yu, K.L. and Peter, R.E. (1992). Neuroendocrinol. **55:** 396–404.

Kawauchi, H., Suzuki, L., Itoh, H., Swanson, P., Naito, N., Nagahama, Y., Nozaki, M., Nakai, Y. and Itoh, S. (1989). Fish Physiol. Biochem. **7:** 29–38.

Khan, I.A. and Thomas, P. (1992). Gen. Comp. Endocrinol. **88:** 388–396.

Khan, I.A. and Thomas, P. (1996). Gen. Comp. Endocrinol. **104:** 231–242.

Kim, M.H., Oka, Y., Amano, M., Kobayashi, M., Okuzawa, K., Hasegawa, Y., Kawashima, S., Suzuki, Y. and Aida, K. (1995). J. Comp. Neurol. **356:** 72–82.

King, J.A. and Millar, R.P. (1982). J. Biol. Chem. **257:** 10729–10732.

King, J.A. and Millar, R.P. (1991). In: Vertebrate Endocrinology: Fundamentals and Biomedical Applications, Vol. IV (P.K.T. Pang and M.P. Schreibman, Eds.), pp. 1–31. Academic Press, New York.

Klungland, H., Lorens, J.B., Anderson, O., Kisen, G.O. and Alestrom, P. (1992). Mol. Cell. Endocrinol. **84:** 164–174.

Li, C.H., Ramasharma, K., Yamashiro, D., and Chung, D. (1987). Proc. Natl. Acad. Sci., USA, **84:** 959–962.

Lin, H.R. and Peter, R.E. (1991). In: Cyprinid Fishes; Systematics, Biology and Exploitation (I.J. Winfield and J.S. Nelson, Eds.), Fish and Fisheries, Series 3, 590–622.

Lin, H.R., Zhang, S.M., Van Der Kraak, G. and Peter, R.E. (1991). Aquaculture, **96:** 87–95.

Marchant, T.A. and Peter, R.E. (1989). Fish Physiol. Biochem. **7:** 133–139.

Martinoli, M.G., Dubourg, P., Geffard, M., Calas, A. and Kah, O. (1990). Cell Tissue Res. **260:** 77–84.

Matsuo, H., Baba, Y., Nair, R.M. G., Arimura, A. and Schally, A.V. (1971). Biochem. Biophys. Commun. **43:** 1334–1339.

Mikolajczyk, T., Weil, C. and Breton, B. (1993). Comp. Biochem. Physiol. **105:** 83–88.

Murthy, C.K., Turner, R.J., Wong, A.O.L., Prasada Rao, P.D., Rivier, J.G. and Peter, R.E. (1994). Neuroendocrinol. **59:** 561–571.

Peng, C., Chang, J.P., Yu, K.L., Wong, A.O.L., Van Goor, F., Peter, R.E. and Rivier, J.E. (1993a). Endocrinology, **132:** 1820–1829.

Peng, C., Humphries, S., Peter, R.E., Rivier, J.E., Bloomqvist, A.G. and Larhammar, D. (1993b). Gen. Comp. Endocrinol. **90:** 306–317.

Pert, C.B., Aposhian, D. and Snyder, S.M. (1974). Brain Res. **75:** 356–361.

Peter, R.E. (1982). Can. J. Fish Aquat. Sci. **39:** 48–55.

Peter, R.E. (1993). Fish Physiol. Biochem. **11:** 445–451.

Peter, R.E., Chang, J.P., Nahorniak, C.S., Omeljaniuk, R.J., Sokolowska, M., Shih, S.H. and Billard, R. (1986). Rec. Progr. Horm. Res. **42:** 513–548.

Peter, R.E., Nahorniak, C.S., Sokolowska, M., Chang, J.P., Rivier, J.E., Vale, W.W., King, J.A. and Millar, R.P.(1985). Gen. Comp. Endocrinol. **58:** 231–242.

Peter, R.E., Trudeau, V.L. and Sloley, B.D. (1991). Bull. Inst. Zool. Acad. Sinica, Monogr. **16:** 89–118.

Peter, R.E., Nahorniak, C.S., Bastos, V.I., Murthy, C.S., Prasada Rao, P.D., Milton, R.C. de L., Millar, R.P. and Rivier, J.R. (1995). 5th Int. Symp. Reprod. Physiol. pp. 52. The University of Texas, Austin.

Powell, J.F.F., Fischer, W.H., Parke, M., Craig, A.G., Rivier, J.E. and Sherwood, N.M. (1995). Reg. Peptides, **57:** 43–53.

Prasada Rao, P.D. (1969). Acta anat. (Basel) **73:** 281–303.

Prasada Rao, P.D. (1972). Z. mikrosk.-anat. Forsch. **86:** 21–38.

Prasada Rao, P.D., Betole, U.K. and Kondawar V.V. (1972). Acta Zool. (Stockholm) **53:** 135–145.

Rosenblum, P.M. and Callard, I.P. (1988). J. Exp. Zool. **245:** 244–255.

Rosenblum, P.M. and Peter, R.E. (1987). Gen. Comp. Endocrinol. **73:** 21–27.

Sherwood, N.M., Parker, D.B., McRory, J.E. and Lescheid, D.W. (1994). In: Fish Physiology, Vol. XIII. Molecular Endocrinology of Fish (N.M. Sherwood and C.L. Hew, Eds.), pp. 3–66. Acad. Press Inc., New York.

Sloley, B.D., Kah, D., Trudeau, V.L., Dulka, J.G. and Peter, R.E. (1992). J. Neurochem. **58:** 2254–2262

Sloley, B.D., Trudeau, V.L., D'Antonio, M. and Peter, R.E. (1994). Endocrinol. J. **2:** 385–391.

Somoza, G.M. and Peter, R.E. (1991). Gen. Comp. Endocrinol. **82:** 103–110.

Suzuki, K., Kawauchi, H. and Nagahama, Y. (1988). Gen. Comp. Endocrinol. **71:** 292–301.

Swanson, P., Suzuki, K., Kawauchi, H. and Dickhoff, W.W. (1991). Biol. Reprod. **44:** 29–38.

Tatemoto, K., Carlquist, M. and Mutt, V. (1982). Nature (London), **296:** 659–660.

Trudeau, V.L., Sloley, B.D. and Peter, R.E. (1993a). J. Neuroendocrinol. **5:** 129–136.

Trudeau, V.L., Sloley, B.D. and Peter, R.E. (1993b). Am. J. Physiol. **265:** R348–R355.

Vale, W.W., Rivier, C., Perrin, M.S. and Rivier, J. (1981). In: Neurosecretion and Brain Peptides (J.B. Martin, S. Reichlin and K.L. Bick, Eds.), pp. 609–625. Raven Press, New York.

Vallarino, M. (1985). Gen. Comp. Endocrinol. **60:** 63–69.

Yaron, Z. (1995). Aquaculture **129:** 49–73.

B

Gonads and Reproduction

Comparative Endocrinology and Reproduction
K.P. Joy, A. Krishna and C. Haldar (Eds)

9. Molecular Mechanism of Steroid Hormone Action

V.K. Moudgil
Department of Biological Sciences, and The Institute for Biochemistry and Biotechnology, Oakland University, Rochester, Michigan, 48309-4401, USA

Introduction

The physiololgical actions of steroid hormones (e.g. estrogens, progestins, androgens, gluco and mineralocorticoids) are mediated via their initial interaction with specific intracellular *receptor* proteins found abundantly in the hormone-responsive (target) cells. The tissues that lack functional receptors or are receptor-deficient, fail to respond to the circulating levels of hormones. Natural mutations in receptor genes are associated with congenital and acquired human diseases. For example, a deletion of androgen receptor (AR) gene results in androgen insensitivity characterized by a female sex phenotype and absence of sexual hair. Defects in AR also lead to adult onset spinal/Bulbar muscular atrophy that results in gynecomastia and neuropathy. Alterations in or absence of functional estrogen receptor (ER) in the cancerous uterine or breast tissues that were responsive to estrogen prior to the malignant transformation make the clinical management of these conditions very difficult. Because of their widely recognized role in endocrine physiology, the past two decades have witnessed advances in the identification, purification, characterization, immunology and molecular biology of steroid receptors (SRs). My laboratory has focused on the mechanism of progesterone action for the past two decades. This chapter reviews various biochemical and molecular aspects of SRs studied by others and the progesterone receptor (PR) and glucocorticoid receptor (GR) studied in this laboratory.

Steroid Receptors as Regulators of Gene Expression While the precise molecular mechanism of their action is not completely understood, SRs are known to act as regulators of gene expression. The structural and functional similarities between receptors for steroid hormones, thyroid hormones, vitamin D_3, retinoic acid and v-erb A oncogene have necessitated their coalition into the *steroid/thyroid* hormone receptor *superfamily* (O'Malley, 1990). Members of the superfamily are also known as ligand-activated *trans-acting* transcriptional factors. Recently, a number of proteins were discovered which possess distinct homologies in their primary structure with the members of the steroid/thyroid superfamily but have no known ligands or defined functions (O'Malley & Conneely, 1992). Detailed structural and functional analyses of these *orphan receptors* will aid in studying regulation and evolution of the members of the superfamily.

Two-Step Model of Steroid Hormone Action: Historical Aspects The pioneering studies in the laboratories of Elwood Jensen and Jack Gorski laid the foundation of research into the hormone-receptor interactions. Selective accumulation of radioisotopic estrogen in the uterus and reproductive tissues established the target tissue concept. By mid 1960s, initial characterization of specific [^{3}H] estradiol binding in the rat uterus was accomplished (Toft & Gorski, 1966). In 1968, a *two-step model* of steroid hormone action was proposed (Jensen *et al.*, 1968; Gorski *et al.*, 1968). Accordingly, SRs exist in the cytoplasm of target cells where they interacted with the incoming hormone to form SR-complexes (SRc). The hormone binding causes conformational changes in the SRc collectively known to represent a process called *activation or transformation*, which facilitates *translocation* of SRs into the nucleus. The sequence-specific interaction between the SRc and the target cell genome results in the alteration of gene expression.

Since 1984, the two-step model has undergone reevaluation and some of its basic principles have been modified. Immunocytochemical analysis and enucleation of hormone-free intact cells have demonstrated that unoccupied SRs are found in the nuclei or the nucleoplasts of the enucleated cells (King & Greene, 1984; Welshons *et al.*, 1984). It now appears that in a majority of hormone-responsive systems, SRs (with a possible exception of GR) exist as nuclear entities even in a hormone-free environment (Gasc *et al.*, 1984; Welshons *et al.*, 1985; Perrot-Applant *et al.*, 1985; Antakly & Eisen, 1984; Wikstrom *et al.*, 1987). Thus, in the modified model, activation/ transformation processes would represent intranuclear event(s).

Activation/Transformation of Steroid Receptors In order to understand the sequence of molecular events which follow hormone binding and lead to a cellular response, changes in structure and function of SRs have been studied *in vitro*. In Cell-free environment, the SRs are recovered in the cytosol (150,000 g) fraction. Incubation of cytosol aliquots at 0-4°C with hormonal ligands yields *non-transformed* complexes which sediment as large 7–9 S structures exhibiting low affinity for DNA or isolated target cell nuclei (Jensen & DeSombre, 1973; Milgrom *et al.*, 1973; Moudgil, 1988). Although transformation *in vitro* of chicken oviduct or calf uterine progesterone receptor (PR) can be accomplished in the absence of exogenous ligands, the extent of transformation is generally lower (Moudgil & Hurd, 1987; Moudgil *et al.*, 1981, 1985). We have previously demonstrated that the 9 S (untransformed) to 4 S (transformed) conversion of rat liver GR is absolutely dependent on the occupancy of steroid binding site by hormonal ligands (Moudgil *et al.*, 1986a). Transformation can be induced *in vitro* by increasing the ionic strength of the medium, by incubating the receptor preparations at elevated temperatures in the presence of hormone (thermal transformation), and by incubation of the cytosol at 0-4°C with 10 mM ATP and other agents (Milgrom *et al.*, 1973; John & Moudgil, 1979; Moudgil & John, 1980; Moudgil *et al.*, 1981, 1985, 1986a). The transformed 4-5 S receptor exhibits increased affinity for isolated nuclei, chromatin, DNA-cellulose, phosphocellulose, and ATP-Sepharose. These resins are used in end-point assays for measurement of SRc transformation. The SR transformation, considered essential for a hormonal response, has also been described in intact target cells under physiological conditions (Munck & Foley, 1979).

Involvement of Phosphorylation in SR Transformation Phosphatase inhibitors (molybodate, tungstate, and vanadate) selectively stabilize SRc and block their thermal transformation to a DNA or ATP-Sepharose binding form(s) (Nishigori & Toft, 1980). Increased rate of transformation of GR observed upon incubation of receptor preparations with calf intestinal alkaline phosphatase (Barnett *et al.*, 1980), suggested that transformation of SRs involves dephosphorylation of the receptor itself or of some regulatory components(s) (Leach *et al.*, 1979; Barnett *et al.*, 1980). However, the action of molybdate on SRs may be direct and not exclusively related to SR dephosphorylation (Murakami & Moudgil, 1981; Housley *et al.*, 1982). Our laboratory reported that molybdates and tungstate exert actions on transformed and nuclear SRc suggesting the direct action of the ions (Murakami & Moudgil, 1981).

Association of Steroid Receptors with Heat Shock Protein (hsps) The SRs extracted from target tissue cytosols exhibit sedimentation rates of 7-10 S and these large forms can be stabilized in the presence of sodium molybdate, which blocks transformation of the 7-9 S SRc to their 4-5 S DNA binding form (Toft & Nishigori, 1979; Leach *et al.*, 1979; Nishigori & Toft, 1980). The molybdate-stabilized PR was isolated initially from chicken oviduct in its native 8 S form (Puri *et al.*, 1982; Renoir *et al.*, 1982). The PR in these preparations was associated with a major 90 kDa protein which did not bind hormone. Further studies revealed the presence of two 8 S receptor forms: each associated with the 90 kDa protein (Dougherty & Toft, 1982). The 90 kDa protein was shown to be associated with other untransformed SRs (Joab *et al.*, 1984; Housley *et al.*, 1985; Riechl *et al.*, Mendel *et al.*, 1986) and was identified as a major heat shock protein (hsp90) (Schuh *et al.*, Mendel *et al.*, 1985; Sanchez *et al.*, 1985; Catelli *et al.*, 1985). Toft and co-workers have reported additional nonreceptor proteins (23-90 kDa) which co-purify specifically with unactivated chicken

PR (Smith *et al.*, 1990a, b; Kost *et al.,* 1989). Since SRs can be reconstituted with hsp90 and hsp70, association of hsps with SRs may represent a biological phenomenon (Smith *et al.*, 1990a, b). Results of a number of studies indicate that (a) hsp90 may be necessary for inducing a SR conformation that promotes steroid binding (Bresnick *et al.*, 1989), (b) hsp90 is found associated with actin filaments and tubulin-containing complexes (Sanchez *et al.*, 1988; Miyata & Yahara, 1991). Hsp90 may be involved in the regulation of SR function by capping its DNA binding site to prevent it from binding to hormone response elements (HREs) and thus block the transcription of regulatory genes (Baulieu, 1989; Picard *et al.,*1990).

Baulieu (1989) proposed that while a hormone agonist may induce the dissociation of the receptor oligomer to unmask the functional DNA binding domain, an antagonist such as RU486, may stabilize the 8S form of PR that is unable to elicit a hormonal response. Accordingly, transformation results from dissociation of the hsp90 from the SR to expose DNA binding site masked in the 8S structure. We observed that 8S to 4S transformation *in vitro* of calf uterine PR is impaired when its ligand binding site is occupied by RU486 (Moudgil & Hurd, 1987).

The Heteromeric Structure of Chicken PR The chicken oviduct has been the most widely studied system for PR analyses. Two hormone-binding isoforms of PR have been described and desiganted PR-A (79,000 kDa) and PR-B (108,000 kDa) based on their differential ionic properties (Schrader & O'Malley, 1972; Schrader *et al.*, 1981). The chicken PR has been cloned and the cDNA successfully expressed to yield a functional PR (Conneely *et al.*, 1986, 1987; Jeltsch *et al.*, 1986; Gronemeyer *et al.*, 1987; Conneely *et al.*, 1987). PR-A and PR-B are produced from a single mRNA transcript by initiation of translation at alternate sites (Conneely *et al.*, 1987, 1989; Gronemeyer *et al.*, 1987). PR-A is a truncated version of PR-B; the two proteins differ by an additional 128 amino acids located at the N-terminus of the B protein. Both proteins appear to functionally activate a progesterone-responsive target gene (Carson *et al.*, 1987; Tora *et al.*, 1988; Conneely *et al.*, 1989). The avian PR occurs in association with hsp90, hsp70 and other recently identified proteins, p54, p50 and p23 (Smith *et al.*, 1990a, b).The biological significance of association of hsp with SRs, and the generality of this association are still under investigation.

Heat Shock and Immunosuppression Laboratories of Lee Faber and Etienne Baulieu have characterized another non-hormone binding protein p59, that appears to be a part of the hetero oligomeric SRs (Callebaut *et al.*, 1992; Lebeau *et al.*, 1992; Massol *et al.*, 1992). However, p59 does not interact directly with the SRs but is bound to hsp90. The p59 cDNA has been cloned and sequenced and the deduced sequence of 458 amino acids appears unique as there is a 55% homology seen in the 100 N-terminal amino acids to peptidyl-propyl isomerase and immunosuppressant FK506 binding protein. The latter is known to interact with immunosuppressants FK506 and Rapamycin. Thus p59 is an immunophilin that can associate with hsp90 and has been called, hsp-binding immunophilin (HBI). The p59/HBI may be involved in alterations of conformation and trafficking of intracellular proteins such as SRs.

A question is normally raised as to how the largely cytoplasmic hsps interact with SRs which are either nuclear or interact with nuclear sites upon hormone binding. A number of explanations have been presented. In those cells where GR is present in the cytoplasm in the absence of hormone, its subsequent entry into the nucleus and binding to HREs upon hormone binding must involve a trafficking system. The SR may attach to the hsp heterocomplex which may serve as a transport particle for intracellular trafficking (Pratt & Scherrer, 1994). According to Pratt, SRs remain "docked" to hsp90-containing protein complex until binding of hormone triggers its dissociation from hsp90 component to allow its high affinity interaction with transcriptional machinery. Docking to hsp complex may provide regulation against transcriptional activity in the absence of hormone. Thus, unliganded GR may remain cytoplasmic prior to its hormone driven transport to the nuclear sites. Since hsp70 is not present in the cytoplasmic GR heteromer, its association may be facilitated by the nuclear environment.

Modulation of Receptor Structure and Function by Phosphorylation Early observation of Munck and Brinck-Johnsen (1968) established that specific glucocorticoid binding in rat thymus lymphocytes was energy-dependent, and that the loss or acquisition of hormone-binding capacity paralleled changes in the levels of intracellular ATP (Ishii *et al.*, 1972; Rees & Bell, 1975). Subsequently, direct effects of ATP were noted on hormone binding (John & Moudgil, 1979; Sando *et al.*, 1979) and transformation of SRs (John & Moudgil, 1979; Moudgil & John, 1980; Barnett *et al.*, 1980). Steroid binding and receptor transformation were reported to be influenced by the phosphorylation-dephosphorylation processes. Consequently, phosphorylation of SRs and members of the nuclear receptor superfamily has been seen in a variety of systems (Reviewed in, Moudgil, 1990). While the exact role of phosphorylation in SR function is still under examination, it is well established that SR are phosphoproteins (Singh & Moudgil, 1985; Moudgil, 1990).

What is the role of receptor phosphorylation in steroid hormone action? Although no unequivocal answers are available, covalent modification of receptor may enhances its ligand binding capacity (Denner *et al.*, 1987; Sullivan *et al.*, 1988a, b; Moudgil, 1990). The three major proteins of chicken PR (PR-A, PR-B, and hsp90) can be phosphorylated in the absence of progesterone, but hormone administration leads to increased phosphorylation of PR-A and PR-B proteins with a loss of hsp90 (Dougherty *et al.*, 1982, 1984; Puri *et al.*, 1984; Denner *et al.*, 1987; Sullivan *et al.*, 1988 a, b; Nakao & Moudgil, 1989). Hormone-dependent phosphorylation also leads to an apparent decrease in the mobility of PR-A protein upon SDS-PAGE, the effect being termed "upshift" (Logeat *et al.*, 1985; Horwitz *et al.*, 1985, 1986). The upshift of phosphorylated PR and the loss of hsp90 following hormone administration are progesterone specific (Nakao & Moudgil, 1989). This hormone-dependent phosphorylation has been observed predominantly on serine residues located in the amino terminus of chicken and human PRs (Sullivan *et al.*, 1988a, b; Sheridan *et al.*, 1989; Denner, 1990b). Work from Weigel's Laboratory indicates that both PR forms A and B contain all four of the phosphorylation sites identified, sites one and two are phosphorylated in the absence of hormone and undergo hyperphosphorylation upon hormone administration. There may be additional phosphorylation sites since the constitutive sites which turn over slowly or sites which represent nuclear function might have lower level of phosphorylation that is not easily detected.

The significance of a phosphorylation-dependent post-translational maturation step in steroid hormone action is not well understood (Sheridan *et al.*, 1989). SR phosphorylation may be involved with processing or nuclear down regulation of PR (Wei *et al.*, 1987), affect the interaction of receptors with transcription elements (Sheridan *et al.*, 1988), or induce sequence specific DNA binding of PR to progesterone responsive genes (Bailly *et al.*, 1986). Results of a recent study have suggested that cAMP-mediated phosphorylation regulates PR transcription activity *in vivo* (Denner *et al.*, 1990a).

SR Phosphorylation in Cell-free Systems In order to obtain clues as to the nature of enzymes which phosphorylate PR *in vivo,* phosphorylation *in vitro* of SRs has been accomplished by a number of protein kinases (PKs) (Reviewed in Auricchio, 1989; Moudgil, 1990; Orti *et al.*, 1992). The question as to whether SRs possess any intrinsic kinase activity is still a matter of debate. Receptors for progesterone, glucocorticoids, and estrogen (Reviewed in Moudgil, 1990) all have been reported to possess PK activities attributed either to copurified enzymes or considered intrinsic to the receptor protein. In some instances, the PK activity was separable from the SR (Denner *et al.*, 1987; Sanchez & Pratt, 1986; Hapgood *et al.*, 1986; Perisic *et al.*, 1987; Garcia *et al.*, 1986). The presence of PK in highly immunopurified preparations of rabbit PR is claimd to be neither an artifact of the purification procedure nor of the cell homogenization (Savouret *et al.*, 1989). Among the candidates for the physiological PK responsible for phosphorylation of SRs are type II casein kinase and cAMP-dependent PK.

Functional Domains of SRs and Interaction with DNA SRs exhibit common structural and functional properties (Hollenberg *et al.*, 1986; Miesfeld *et al.*, 1986; Greene *et al.*, 1986; Loosfelt *et al.*, 1986; Conneely *et al.*, 1986; Jeltsch *et al.*, 1986; Misrahi *et al.*, 1987; Arriza *et al*, 1987; Chang *et al.*, 1988; Tilley *et al.*, 1989). Typically, a SR is composed of three major domains, the N-terminal immunoreactive domain of variable length (Hollenberg *et al.*, 1987), the highly conserve middle DNA-binding domain of approximately 70 amino acids, and the C-terminal steroid-binding domain of around 250 amino acids (Yamamoto, 1985, Hollenberg *et al.*, 1987; Conneely *et al.*, 1987; Carson *et al.*, 1987; Evans, 1988). Zinc fingers are characteristically present in the DNA binding domain of SRs (Yamamoto, 1985). Binding of the steroid ligand appears to be a prerequisite for specific DNA binding and transcriptional activation *in vivo* (Carson-Jurica *et al.*, 1990; Yamamoto, 1985). Occupancy of the steroid binding site by the hormonal ligands confers on SRs the ability to act as *trans*-acting regulatory proteins capable of altering gene expression. The steroid-charged receptor complexes have been shown to interact specifically with promotor *enhancer* elements of target cell genes (Green & Chambon, 1986; Green *et al.*, 1986). The enhancer or the hormone-responsive DNA sequences are located in the 5'-flanking region (Payvar *et al.*, 1981; Pfahl, 1982; Govindan *et al.*, 1982). Far upstream from the initiation site, the DNA binding site(s) have been located within introns of the transcription unit (Payvar *et al.*, 1981; Moore *et al.*, 1985). Accordingly, the transformed SRc bind preferentially to specific DNA sequences in the regions upstream from the trancriptional start site of hormone-regulated genes (Mulvihill *et al.*, 1982; Compton *et al.*, 1983; Karin *et al.*, 1984; Groner *et al.*, 1984; Scheidereit & Beato, 1984; Jost *et al.*, 1984).

Receptor Pharmacology and Clinical Considerations SR presence has been correlated with successful hormonal therapy in the breast cancer patients. In recent years, qualitative analyses of the samples have become better predictors of the clinical management of breast and endometrial cancers. Presence of ER in the breast cancer biopsies is regarded as one of the most important factors predicting response to endocrine therapy. Patients with positive ER expression in the cancerous tumors have 40–80% chance of response to antiestrogenic treatment, although several variations in ER expression could lead to drug resistance. Tamoxifen, an antiestrogen drug widely used in the treatment of breast cancer in the post-menopausal women, interacts with ER to block its action but without interfering with receptor dimerization or its interaction with DNA. However, long-term tamoxifen use is associated with drug resistance. This resistance could either be due to loss or an emergence of cells expressing ER mRNA variants.

Receptor Agonist/Antagonists and Breast Cancer Tumors originating from cells which normally grow under the influence of certain hormones may retain responsiveness to them following malignant transformation.

Tumorigenesis may also be accompanied by a loss/alteration of genetic information which results into formation of spontaneous cell variants. Many breast cancer cells become hormone resistant and do not express PR or ER. A hormone responsive breast cancer may transform into a hormone-insensitive entity. The mechanisms underlying the evolution of hormone-sensitive tumors to hormonally insensitive states are not known (Horwitz, 1994). The success rate of endocrine therapy involving SR antagonists is significantly higher in hormone-dependent breast cancers. The antagonists of SRs bind them to generate inactive complexes which fail to induce gene expression in hormone-responsive cells. Antagonists may also exert agonist effects. The success in cloning of cDNA for ER, and delineating the structures of ER gene and the receptor protein has allowed analyses of polymorphic forms of ER gene and identification of alternate mRNA transcripts. Accordingly, ER structure and regulation in normal cells may be tissue-specific. Furthermore, malignant cells also express ER mutants (Encarnacion & Fuqua, 1994).

RU486 and Clinical Potential Until 1990, RU486 was the only steroid analog available with demonstrable antiprogestin activity. It is now widely recognized as the effective antiprogestin and antiglucocorticoid: the two activities are discernable at different steroid concentrations (Baulieu, 1989; Ulmann *et al.*, 1990). Earlier attempts to use RU486 as a chemical probe for investigating molecular mechanism of progesterone action resulted in some unexpected observations. The chicken PR, which is the widely used system for studying progesterone action at the receptor level, failed to exhibit any affinity for RU486 (Moudgil *et al.*, 1986b; Eliezer *et al.*, 1987). RU486 interacted in the oviduct cytosol with a molecule which is immunologically distinct from chicken PR (Eliezer *et al.*, 1987). Our laboratory has subsequently shown that the other recently synthesized antiprogestin, ZK98299, also did not bind chichen PR (Moudgil *et al.*, 1991). These observations now represent a general phenomenon regarding the lack of interaction between RU486 and the antiprogestins: two additional antiprogestins, ORG 31806 and ORG 31710 also failed to bind chicken PR specifically (Mizutani *et al.*, 1992). The primary structural analysis of chicken PR revealed that single amino acid substitution of cysteine 575 with glycine in the hormone binding domain of chicken PR allows the resultant mutant to bind RU486 (Benhamou *et al.*, 1992). The above findings have potential implications in cases where RU486 fails to induce abortion (nearly 1% women); such patients may possess receptors which bind progestins optimally but are resistant to antiprogestins because of a possible mutation at position 722 which is required for RU486 binding (Baulieu, 1989; Benhamou *et al.*, 1992).

Mechanism of RU486 Action Initial work from this laboratory and by others suggested that RU486-bound calf uterine PR had impaired ability to undergo 8S to 4S transformation (Moudgil & Hurd, 1987), suggesting that RU486 binding to PR stabilizes the 8S nontransformed heteromeric complex. The presence in this complex of hsp90 (or other receptor components) precludes an interaction of a PRc with its cognate HRE (Baulieu, 1989). The inability of RU486-bound PRc to induce a functional response suggested to us that differential conformational changes were induced in the receptor by the agonist and antagonist ligands (Moudgil *et al.*, 1989). This view was supported by the finding that transcriptional inactivation of SRs by RU486 involves alterations in the conformation of receptor in the C-terminus of the steroid binding domain (Allan *et al.*,1992). There is some evidence that agonist and the antagonist ligands may bind to different regions within or around the hormone binding domain (Vegeto *et al.*, 1992). Milgrom and colleagues have reported that receptors bound to RU486 form abortive complexes with hormone HREs (Guiocho-Mentel *et al.*, 1988). These observations are supported by the finding that the antihormone-bound receptors can undergo transformation and interact with DNA in a manner that is transcriptionally nonproductive (Pham *et al.*, 1991). Another clue to RU486 activity comes from a study demonstrating that RU486 bound human PR interacts with HREs in an altered form (El-Ashry *et al.*, 1989). Further, Gronemeyer and colleagues have shown the presence of two transcription activation functions (TAFs) associated with the N-terminal (TAF-1) and the hormone binding domain (TAF-2). While both TAFs are active in the presence of agonists, TAF-2 was seen to be inactive in the presence of RU486 (Meyer *et al.*, 1990). This indicates that RU486-charged PRc binds to HRE but that this union does not result into a functional response.

Expression of PR in breast tissue has been recognized as an indicator of estrogen action. Knowledge about PR regulation in breast cancer by PR antagonists and the anti-cancer potential of these and other steroids is limited (Horwitz *et al.*, 1985; Horwitz, 1992). Recently, we have observed that RU486 is able to induce a proliferative state in T47D human breast cancer cells which is detected as an increase in total cell number when cells are cultured in media containing charcoal-treated serum (Herman *et al.*, 1993).

A SR Antagonist May Also Exert Agonist Activity Both the RU486 and the antiestrogen 4-OH tamoxifen have demonstrable agonist and antagonist effects (Clarke & Sutherland, 1990; Horwitz, 1992). Such partial agonists induce the sequence specific DNA binding activity of their respective receptors, but not a hormone-dependent transcriptional activation function. Our preliminary studies and work of others (Bowden *et al.*, 1989; Herman *et al.*, 1993; Hissom & Moore, 1987, 1989) suggest that RU486 exerts both stimulatory and inhibitory effects on the growth of T47D cells.

Steroids, SRs and Tumor Suppressors The functions of tumor suppressor genes have been the subject of numerous investigations. Retinoblastoma susceptibility gene has been directly related to suppression of tumorigenicity (Takahashi *et al.*, 1988; Weinberg, 1991). The protein product of the retinoblastoma gene, pRB, is a nuclear phosphoprotein and appears to control transit of cycling cells from Gl to S phase. Its activity is regulated by phosphorylation/dephosphorylation. Another gene product, p53, was initially identified as a tumor antigen that interacts with simian virus 40T (SV40) antigen and the adenovirus EIB oncoprotein (Lane & Crawford, 1979). The biological activity of p53 is associated with suppression of cell growth. Inactivation of p53 results in breakdown of cell growth regulation and tumor progression (Chin *et al.*, 1992). It may be the most frequently mutated gene in human cancers (Oren, 1992). Loss of one allele and point mutations in the other of the p53 gene appear to trigger events in tumor progression (Chen *et al.*, 1990). The protein product of the p53 gene, is a nuclear phosphoprotein which acts as a transcription factor directly regulating the expression of factors involved in cell cycle control (Kastan *et al.*, 1992) and programmed cell death (Miyashita *et al.*, 1994; Miyashita & Reed, 1995, Oltvai *et al.*, 1993).

There is a paucity of information regarding the influence of steroid hormones on tumor suppressor function in hormone-responsive breast cancer cells. We have examined the effects of estradiol and progestin in regulating p53 in T47D cells. In these cells, the levels of ER appear to be relatively low whereas PR is over-expressed (Horwitz *et al.*, 1982). Results of our preliminary studies show that p53 is significantly increased upon estradiol treatment while it is downregulated by both R5020 and RU486 (Hurd *et al.*, 1995). Additional work is in progress to examine relationship between steroids, SRs and p53 expression and explore correlations with cell proliferation.

Summary, Conclusions and Future Directions

Every physiological function appears to be influenced by the actions of steroid hormones. From their critical involvement in the processes of reproduction and conception, steroids play the determinable role in puberty and early manifestations of adulthood. Ironically, in the face of overwhelming data on the action and biological aspects of steroid hormones, their mechanism of action still remains incompletely understood. The concept that steroid hormones are regulators of gene expression is well cemented. SRs, which mediate the actions of steroids. are now referred to as ligand-activated *trans*-acting transcriptional factors. A number of nonreceptor (heat shock) proteins, which appear to significantly impact steroid action, are target of numerous investigations. At the level of genome, a number of accessory proteins (factors) have been identified which appear to be required to potentiate the action of SR-HRE. In cases where intracellular levels of certain cellular factors (e.g. cAMP, dopamine) are increased, a SR may function in the absence of a hormonal ligand. Thus, the simplicity and beauty suggested in the physiological function of a steroid hormone have traded places with complex phenomena relating the molecular basis to mode of action.

Historically, it was nearly thirty years ago that Gorski and Toft (1996) first reported a detailed characterization of a receptor molecule for estradiol. Following the identification of ER, nature and characteristics of other SRs were revealed leading to the successful cloning of different SRs. A variety of experimental systems have been employed to express the cloned receptors. The diversity of expression systems available allows anatomical dissection of the different domains of receptors to correlate SR structure with specific functions. Immunological advances have made cellular localization of receptors possible without compromising the cell integrity. A clinician is able to launch a pathological analysis of a specimen with tissue size that is a fraction of what was required prior to the availability of anti-SR monoclonal antibodies.

The predictions made in 1960s by Jensen and others that detection of SRs in human biopsies has potential in the endocrine treatment of hormone-dependent tumors have been realized with encouraging results. Antihormones (e.g. tamoxifen), which work at receptor level, have been employed in the treatment of postmenopausal breast cancers and are now being tested in the prevention of cancers in the high risk healthy population. Unprecedented social, political and scientific debates about the current use of the antiprogestin-antiglucocorticoid RU486 for fertility regulation has helped to focus the power and potential of steroids. Future efforts, therefore, could be invested in unlocking the molecular mysteries of SRs. The homologies in the functional domains of the members of the superfamily suggest that simpler, more common mechanisms of action may be revealed.

Important questions related to hormone action physiology demand consideration in discussions of the cited work. For example, is the scope of hormone activity limited to its interaction with the receptor, or does this interaction induce conformational changes which expose receptor domains masked by cellular components? Is hormone binding consequential and detrimental to the SRs interaction with select sequences on the hormone-responsive genome? The work reviewed in this chapter included modifications of receptor structure that influence its cellular regulation providing knowledge about the flexibility of the target cell response. And what about the *orphan* receptors? The search for the biological ligands for the latter may proceed at a rate that is compatible with investigations on the growth factors and cellular signalling pathways which influence steroid hormone actions. Steroid binding proteins are being reported from bacteria and lower eukaryotic forms. Evolutionary thoughts weigh in support of the hypothesis that mediators of chemical signals evolved from a common primordial molecule. The basic principles which govern the structure and function of SRs also guide clinicians in examining the relationship between SR and pathological conditions which originate due to hormone imbalance absence or alterations in SRs. Finally, the earlier held belief that hormones are secretions of well defined structural entities, the endocrine glands, is continuously evolving. Growth factors, cytokines, and other signal transducing biomolecules are as much on the forefront of hormone research as are efforts invested in elucidating the mechanism of steroid hormone action. A hypothetical model depicting the sequence of events, which follow entry of a steroid into a target cell nucleus and culminate in transcription of a steroid responsive gene, is shown in figure 1.

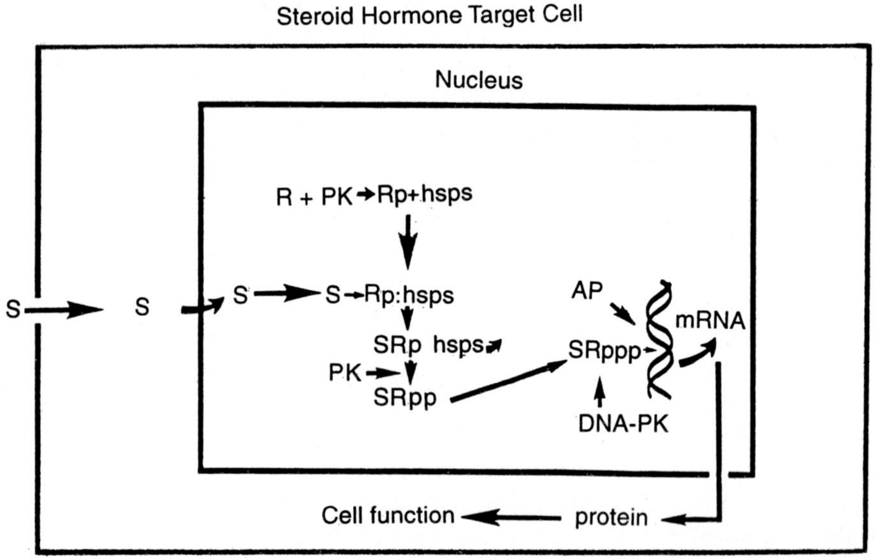

Fig. 1. Hypothetical model of steroid hormone action. S, steroid; R, receptor; Rp, phosphorylated receptor; hsps, heat shock proteins; AP, accessory nuclear factors; PK, protein kinase; DNA-PK, DNA dependent protein kinase.

Acknowledgments

The work in the authors laboratory is supported by the National Institutes of Health Grant DK-20893. The author gratefully acknowledges the effort of a number of past and current associates in the laboratory whose work has been discussed in this review. This chapter is dedicated to an inspiring and dedicated teacher, Professor C.J. Dominic. To this day, I utilize notes preserved from his lectures to understand the fundamental concepts in endocrinology, which remain unchanged amidst overwhelming recent technological advances.

References

Allan, G.F., Leng, X., Tsai, S.Y., Weigel, N.L., Edwards, D.P., Tsai, M.-J. and O'Malley, B.W. (1992). J. Biol. Chem., **267:** 19513–19520.

Antakly, T. and Eisen, H.J. (1984). Endocrinology **115:** 1984–1989.

Arriza, J.L., Weinberger, C., Carelli, G., Glaser, T.M., Handelin, B.L., Housman, D.E. and Evans, R.M. (1987). Science **237:** 268–275.

Auricchio, F. (1989). J. Steroid Biochem. **32:** 613–622.

Bailly, A., Le Page, C., Rauch, M. and Milgrom, E. (1986). EMBO J. **5:** 3235–3241.

Barnett, C.A., Schmidt, T.J. and Litwach, G. (1980). Biochemistry **19:** 5446–5455.

Baulieu, E.E. (1989). Science **245:** 1351–1357.

Benhamou, B., Garcia, T., Lerouge, T., Vergezac, A., Gofflo, D., Bigogne, C., Chambon, P. and Gronemeyer, H. (1992). Science **255:** 206–209.

Bowden, R.T., Hissom, J.R. and Moore, M.R. (1989). Endocrinology **124:** 2642–2644.

Bresnick, E.H., Dahlman, F.C., Sanchez, E.R. and Pratt, W.B. (1989). J. Biol. Chem. **264:** 4992–4997.

Callbaut, I., Renoir, J.M., Lebeau, M.-C., Massol, N., Burny, A., Baulieu, E.E. (1992). Proc. Natl. Acad. Sci. U.S.A. **89:** 6270–6274.

Carson-Jurica, M.A., Schrader, W.T. and O'Malley, B.W. (1990). Endocrine Rev. **11:** 201–220.

Carson, M.A., Tsai, M.J., Conneely, O.M., Maxwell, B.L., Clark, J.M., Dobson, A.D.W., Elbrecht, A., Toft, D.O., Schrader, W.T. and O'Malley B.W. (1987). Mol. Endocrinol. **1:** 791–801.

Catelli, M.G., Binart, N., Feramisco, J.R. and Helfman, D.M. (1985). Nucl. Acids Res. **13:** 6035–6047.

Change, C., Kokontis, J. and Liao, S. (1988). Proc. Natl. Acad. Sci. U.S.A. **85:** 7211–7215.

Chen, P.-L., Chen, Y., Bookstein, R. and Lee, W.-H. (1990). Science **250:** 1576–1580.

Chin, K.-V., Ueda, K., Pastan, I. and Gottesman, M.M. (1992). Science **255:** 459–462.

Clarke, C.L. and Sutherland. R.L. (1990). Endocrine Rev. **11:** 266–301.

Compton, J.G., Schrader, W.T., O'Malley, B.W. (1983). Proc. Natl. Acad. Sci. U.S.A. **80:** 16–20.

Conneely, O.M., Sullivan, W.P., Toft, D.O., Birnbaumer, M., Cook, R.G., Maxwell, B.L., Zarucki-Schulz, T., Greene, G.L., Schrader, W.T. and O'Malley, B.W. (1986). Science **233:** 767–770.

Conneely, O.M., Dobson, A.D.W., Tsai, M.J., Beattie, W.G., Toft, D.O., Huckaby, C.S., Zarucki, T., Schrader, W.T.and O'Mally, B.W. (1987a). Mol. Endocrinol. **1:** 517–525.

Conneely, O.M., Kettelberger, D.M., Tsai, M.J., Schrader, W.T. and O'Malley B.W. (1989). J. Biol. Chem. **264:** 14062–14064.

Denner, L.A., Bingman, III W.E., Greene, G.L., and Weigel, N.L. (1987). J. Steroid Biochem. **27:** 235–243.

Denner, L.A., Weigel, N.L., Maxwell, B.L., Schrader, W.T. and O'Malley, B.W. (1990a). Science **250:** 1740–1743.

Denner, L.A., Schrader, W.T., O'Malley, B.W. and Werigel, N.L. (1990b). J. Biol. Chem. **265:** 16548–16555.

Dougherty, J.J. and Toft, D.O. (1982). J. Biol. Chem. **257:** 3113–3119.

Dougherty, J.J., Puri, R.K. and Toft, D.O. (1984). J. Biol. Chem. **259:** 8004–8009.

Eliezer, N., Hurd, C. and Moudgil, V.K. (1987). Biochim. Biophys. Acta **926:** 34–39.

El-Ashry, D., Onate, S.A., Nordeen, S.K. and Edwards, D.P. (1989). Mol. Endocrinol. **3:** 1545–1558.

Encarnacion, C.A. and Fuqua, S.A.W. (1994). In: Steroid Hormone Receptors: Basic and Clinical Aspects. (Ed. V.K. Moudgil) pp. 427–446.

Evans, R.M. (1988). Science **240:** 889–895.

Garcia, T., Buchou, T., Renoir, J.M., Mester, J. and Baulieu, E.E. (1986). Biochemistry **25:** 7937–7942.

Gasc, J.M., Renoir, J.M., Radanyi, C., Joab, I., Tuohimaa, P. and Baulieu, E.E. (1984). J. Cell. Biol. **99:** 1193–1201.

Gorski, J., Toft, D.O., Shyamala, G., Smith, D. and Notides, A. (1968). Rec. Prog. Horm. Res. **24:** 45–80.

Govindan, M.V., Spiess, E. and Majors, J. (1982). Proc. Natl. Acad. Sci. U.S.A. **79:** 5157–5161.

Groner, B., Kennedy, N., Skroch, P., Hynes, N.E. and Ponta, H. (1984). Biochim. Biophys. Acta **781:** 1–6.

Green, S. and Chambon, P. (1986). Nature **324:** 615–617.

Green, S., Walter, P., Kumar, V., Krust, A., Bornet, J.M., Argos, P. and Cambon, P. (1986). Nature **320I:** 134–139.

Greene, G.L., Gilna, P., Waterfield, M., Baker, A., Hort, Y. and Shine, J. (1986). Science **231:** 1150–1154.

Gronemeyer, H., Turcotte, B., Quirin-Stricker, C., Bocquel, M.T., Meyer, M.E., Krozowski, Z., Jeltsch, J.M., Lerouge, T., Garnier, J.M. and Chambon, P. (1987). EMBO J. **6:** 3985–3994.

Guiochon-Mantel, A., Loosfelt, H., Ragot, T., Bailly, A., Atge, M., Misrahi, M., Perricaudet, M. and Milgrom, E. (1988). Nature **336:** 695–698.

Hapgood, J.P., Sabbatini, G.P. and Holt, C.V. (1986). Biochemistry **25:** 7529–7534.

Herman, M., Bhakta, A., Underwood, B., Kodali, S. and Moudgil, V.K. (1993). Annu. Meet Endocrine Soc., Las Vegas, NA, June 9–12.

Hissom, J.R. and Moore, M.R. (1987). Biochem. Biophys. Res. Commun. **145:** 706–711.

Hissom, J.R. Bowden, R.T. and Moor, M.R. (1987). Endocrinology **125:** 418–423.

Hollenberg, S.M., Weinberger, C., Ong, E.S., Cerelli, G., Oro, A., Lebo, R., Thompson E.B., Rosenfeld, M.G. and Evans, R.M. (1985). Nature **318:** 635–642.

Hollenberg, S.M., Giguree, V., Sequi, P. and Evans, R.M. (1987). Cell **49:** 39–46.

Horwitz, K.B., Mockus, M.B. and Lessey, B.A. (1982). Cell **28:** 633–642.

Horwitz, K.B. (1992). Endocrine Rev. **13:** 146–163.

Horwitz, K.B. (1994). J. Steroid Biochem. Mol. Biol. **49:** 295–302.

Horwitz, K.B. (1985). Endocrinology **116:** 2236–2245.

Horwitz, K.B., Francis, M.D. and Wei, L.L. (1985). DNA **4:** 451–460.

Horwitz, K.B., Wei, L.L. and Francis, M.D. (1986). J. Steroid Biochem. **24:** 109–117.

Horwits, K.B., Weil, L.L., Sediacek, S.M., D'arville, C.N.(1985). Rec. Prog. Horm. Res. **41:** 249–301.

Housley, P.R., Dahmer, M.K. and Pratt, W.B. (1982). J. Biol. Chem. **257:** 8615–8618.

Housley, P.R., Sanchez, E.R., Westphal, H.H., Beato, M. and Pratt, W.B. (1985). J. Biol. Chem. **260:** 13810–13817.

Hurd, C., Khattree, N., Alban, P., Nag, K., Jhanwar, S.C., Dinda, S. and Moudgil, V.K. (1995). J. Biol. Chem. **270:** 28507–28510.

Ishii, D.N., Pratt, W.B. and Aronow, L. (1972). Biochemistry **11:** 3896–3904.

Jeltsch, J.M., Krozowski, Z., Quirin-Stricker, C., Gronemeyer, H., Simpson, R.J., Garnier, J.M., Krust, A., Jacob, F. and Chambon, P. (1986). Proc. Natl. Acad. Sci. U.S.A. **83:** 5424–5428.

Jensen, E.V., Suzuki, T., Kawashima, T., Stumpf, W.E., Jungblut, P.W. and DeSombre, E.R. (1968). Proc. Natl. Acad. Sci. U.S.A. **59:** 632–638.

Jensen, E.V. and DeSombre, E.R. (1973). Science **182:** 126–134.

Joab, I., Radanyi, C., Renoir, J.M., Buchou, T., Catelli, M.-G., Bnart, N., Mester, J. and Baulieu, E.E. (1984). Nature **308:** 850–853.

John, J.K. and Moudgil, V.K. (1979). Biochem. Biophys. Res. Commun. **90:** 1242–1248.

Jost, J.P., Seldran, M. and Geiser, M. (1984). Proc. Natl. Acad. Sci. U.S.A. **81:** 429–433.

Kastan, M.B., Zhan, Q., El-Deiry, W.S., Carrier, F., Jacks, T., Walsh, W.V., Plunkett, B.S., Vogelstein, B. and Fornace, Jr., A.J. (1992). Cell **71:** 587–597.

Karin, M., Haslinger, A., Holtgreve, H., Richards, R.I., Krauter, P., Westphal, H.M. and Beato, M. (1984). Nature **308:** 513–519.

King, W.J. and Greene, G.L. (1984). Nature **307:** 745–747.

Kost, S.L., Smith, D.F., Sullivan, W.P., Welch, W.J. and Toft, D.O. (1989). Cell. Mol. Biol. **9:** 3829–3838.

Lane, D.P. and Crawford, L.V. (1979). Nature **278:** 261–263.

Leach, K.L., Dahmer, M.K., Hammond, N.D., Sando. J.J. and Pratt, W.B. (1979). J. Biol. Chem. **254:** 11884–11890.

Lebeau, M.-C., Massol, N., Herrick, J., Faber, L.E., Renoir, J.-M., Radanyi, C. and Baulieu, E.E. (1992). J. Biol. Che., **267:** 4284.

Logeat, F., LeCunff, M., Pamphile, R. and Milgrom, E. (1985). Biochem. Biophys. Res. Commun. **131:** 421–427.

Loosfelt, H., Atger, M., Misrahi, M., Guiochon-Mantel, A., Meriel, Logeat, F., Benarous, R. and Milgrom, E. (1986). Proc. Natl. Acad. Sci. U.S.A. **83:** 9045–9049.

Massol, N., Lebeau, M.-C., Renoir, J.M., Faber, L.E. and Baulieu, E.E. (1992). Biochem. Biophys. Res. Commun. **187:** 1330–1335.

Mendel, D.B., Bodwell, J.E., Gametchu, B., Harrison, R.W. and Munck, A. (1986). J. Biol. Chem. **261:** 3758–3763.

Meyer, M.-E., Pornon, A., Ji, J., Bocquel, M.-T., Chambon, P. and Gronemyer, H. (1990). EMBO J. **9:** 3923–3932.

Miesfeld, R., Rusconi, S., Gudowski, P.J., Maler, B.A., Okret, S., Wikstrom, A.C., Gustafsson, J.A. and Yamamoto, K.R. (1986). Cell **46:** 389–399.

Milgrom, E., Atger, M. and Baulieu, E.E. (1973). Biochemistry **12:** 5198–5205.

Misrahi, M., Atger, M. and Milgrom, E. (1987). Biochemistry **26:** 3975–3982.

Miyata, Y. and Yahara, I. (1991). J. Biol Chem. **266:** 8779–8783.

Miyashita, T., Harigai, M., Hanada, M. and Reed, J.C. (1994). Cancer Res. **54:** 3131–3135.

Miyashita, T. and Reed, J. (1995). Cell **80:** 293–299.

Mizutani, T., Bhakta, A., Kloosterboer, H.J. and Moudgil, V.K. (1992). J. Steroid Biochem. Mol. Biol. **42:** 695–704.

Moore, D.D., Marks, A.R., Buckley, D.I., Kapler, G., Payvar, F. and Goodman, H.M. (1985). Proc. Natl. Acad. Sci. U.S.A. **82:** 699–702.

Moudgil, V.K. (1988). Ed. Steroid Receptors in Health and Disease. Plenum Press, New York.

Moudgil, V.K. (1990). Biochim. Biophys. Acta. **1055:** 243 258.

Moudgil, V.K., Anter, M.J. and Hurd, C. (1989). J. Biol. Chem. **264:** 2204–2211.

Moudgil, V.K., Eessalu, T.E., Buchou, T., Renoir, J.M., Mester, J. and Baulieu, E.E. (1985). Endocrinology **116:** 1267–1274.

Moudgil, V.K, Kruczak, V.H., Eessalu, T.E., Paulose, C.S., Taylor, M.G. and Hansen, J.C. (1981). Eur. J. Biochem. **118:** 547–555.

Moudgil, V.K., Lombardo, G., Eessalu, T. and Eliezer, N. (1986a). J. Biochem. **99:** 1005–1016.

Moudgil, V.K., Lombardo, G., Hurd, C., Eliezer, N. and Agarwal, M.K. (1986b). Biochim. Biophys. Acta **889:** 192–199.

Moudgil, V.K. and Hurd, C. (1987). Biochemistry **26:** 4993–5001.

Moudgil, V.K. and John, J.K. (1980). Biochem. J. **190:** 799–808.

Moudgil, V.K., Nath, R., Bhakta, A. and Nakao, M. (1991). Biochim. Biophys. Acta **1094:** 185–192.

Mulvihill, E.R., Le Pennee, J.P. and Chambon, P. (1982). Cell **24:** 621–632

Munck, A. and Foley, R. (1979). Nature **278:** 752–754.

Munck, A. and Brink-Johnsen, T. (1968). J. Biol. Chem. **243:** 5556–5565.

Murakami, N. and Moudgil, V.K. (1981). Biochem. J. **198:** 447–455.

Nakao, M. and Moudgil, V.K. (1989). Biochem. Biophys. Res. Commun. **161:** 295–303.

Nishigori, H. and Toft, D.O. (1980). Biochemistry **19:** 77–83.

Oltvai, Z.N., Milliman, C.L. and Korsmeyer, S.J. (1993) Cell **74:** 609–619.
O'Malley, B.W. (1990). Mol. Endocrinol. **4:** 363–369.
O'Malley, B.W. and Conneely, O.M. (1992). Mol. Endocrinol. **6:** 1359–1361.
Oren, M. (1992). Faseb J. **6:** 3169–3176.
Orti, E., Bodwell, J.E. and Munck, A. (1992). Endo. Rev. **13:** 105–128.
Payvar, F., Wrange, O., Carlstedt-Duke, J., Okret, S., Gustafsson, J.A. and Yamamoto, K.R. (1981). Proc. Natl. Acad. Sci. U.S.A. **78:** 6628–6632.
Perisic, O., Radojcic, M. and Kanazir, D.T. (1987). J. Biol. Chem. **262:** 11688–11691.
Perrot-Applant, M., Logeat, F., Groyer-Picard, M.T. and Milgrom, E. (1985). Endocrinology **116:** 1473–1484.
Pfahl, M. (1982). Cell **31:** 475–482.
Pham, T.A., Elliston, J.F., Nawaz, Z., McDonnell, D.P., Tsai, M.-J. and O'Malley, B.W. (1991). Proc. Natl. Acad. Sci. U.S.A **88:** 3125–3129.
Picard, D., Kursheed, B., Grabedian, M.J., Fortin, M.G., Lindquist, S. and Yamamoto, K.R. (1990). Nature **348:** 166–168.
Pratt, W.B. and Scherrer, L.C. (1994). In Steroid Hormone Receptors: Basic and Clinical Aspects. (Ed. V.K. Moudgil) pp. 215–246, Birkhauser-Boston, Publishers.
Puri, R.K., Grandics, P., Dougherty, J.J. and Toft, D.O. (1982). J. Biol. Chem. **257:** 10831–10837.
Puri, R.K., Dougherty, J.J. and Toft, D.O. (1984). J. Steroid Biochem. **20:** 23–29.
Rees, A.M. and Bell, P.A. (1975). Biochim. Biophys. Acta **411:** 121–132.
Renoir, J.M., Yang, C.R., Formstecher, P., Lustenberger, P., Wolfson, A., Redeuilh, G., Mester, J., Richard-Foy, H. and Baulieu, E.E. (1982). Eur. J. Biochem. **127:** 71–79.
Renoir, J.M., Buchou, T. and Baulieu, E.E. (1986). Biochemistry **25:** 6405–6413.
Riehl, R.M., Sullivan, W.P., Vroman, B.T., Bauer, V.T., Pearson, G.R. and Toft, D.O. (1985). Biochemistry **24:** 6586–6591.
Rochefort, H., Bardon, S., Chalbos, D. and Vignon, F. (1984). J. Steroid Biochem. **20:** 105–110.
Sanchez, E.R., Toft, D.O., Schlesinger, M.J. and Pratt, W.B. (1985). J. Biol. Chem. **260:** 12398–12401.
Sanchez, E.R. and Pratt, W.B. (1986). Biochemistry **25:** 1378–1382.
Sanchez, E.R., Redmond, T., Scherrer, L.C., Bresnick, E.H., Welsh, M.J. and Pratt, W.B. (1988). Mol. Endocrinol. **2:** 756–760.
Sando, J.J., Hammond, N.D., Stratford, C.A. and Pratt, W.B. (1979). J. Biol. Chem. **254:** 4779–4789.
Savouret, J.F., Misrahi, M., Loosfelt, H., Atger, M., Bailly, A., Perrot-Applant, M., Vu Hai, M.T., Guiochon-Mentel, A., Jolivet, A., Lorenzo, F., Pichon, M.F., Bouchard, P. and Milgrom, E. (1989). Rec. Prog. Horm. Res. **45:** 65–120.
Scheidereit, C. and Beato, M. (1984). Proc. Natl. Acad. Sci. U.S.A. **81:** 3029–3033.
Schrader, W.T. and O'Malley, B.W. (1972). J. Biol. Chem. **247:** 51–59.
Schrader, W.T., Birnbaumer, M.E., Hughes, M.R., Weigel, N.L., Grody, W.W. and O'Malley, B.W. (1981). Rec. Prog. Horm. Res. **37:** 583–633.
Schuh, S.S., Yonemoto, W., Brugge, J., Bauer, V.J., Riehl, R.M., Sullivan, W.P. and Toft, D.O. (1985). J. Biol. Chem, **26:** 14292–14296.
Sheridan, P.L., Krett, N.L., Gordon, J.A. and Horwitz, K.B. (1988). Mol. Endocrinol. **2:** 1329–1342.
Sheridan, P.L., Evans, R.M. and Horwitz, K.B. (1989). J. Biol. Chem. **264:** 6520–6528.
Singh, V.B. and Moudgil, V.K. (1985). J. Biol. Chem. **260:** 3684–3690.
Smith, D.F., Faber, L.E. and Toft, D.O. (1990a). J. Biol. Chem. **265:** 3996–4003.
Smith, D.F., Schowalter, D.B., Kost, S.L. and Toft, D.O. (1990b). Mol. Endocrinol. **4:** 1704–1711.

Sullivan, W.P., Smith, D.F., Beito, T.G., Krco, C.J. and Toft, D.O. (1988a). J. Cell. Biochem. **36:** 103–119.

Sullivan, W.P., Madden, B.J., McCormick, D.J. and Toft, D.O. (1988b). J. Biol. Chem. **263:** 1417–14723.

Takahashi, R., Hashimoto, T., Xu, H.J., Xu, S.X., Matsui, T., Miki, T., Bigo-Marshall, H., Aaronson, S.A and Benedict, W. (1988). Proc. Natl. Acad. Sci., U.S.A. **88:** 5257–5261.

Tilley, W.D., Marcelli, M., Wilson, J.D. and McPhaul, M.J. (1989). Proc. Natl. Acad. Sci. U.S.A. **86:** 327–331.

Toft, D. and Gorski, J. (1966). Proc. Natl. Acad. Sci. U.S.A. **55:** 1574–1581.

Toft, D.O. and Nishigori, H. (1979). J. Steroid Biochem. **11:** 413–416.

Tora, L., Gronemeyer, H., Turcotte, B., Gaub, M.-P. and Chambon, P. (1988). Nature **333:** 185–188.

Ulmann, A., Teutsch, G. and Philibert, D. (1990). Sci. Amer. **262:** 42–48.

Vegeto, E., Allan, G.F., Schrader, W.T., Tsai, M.-J., McDonnell, D.P. and O'Melley, B.W. (1992). Cell **69:** 703–713.

Wei, L.L., Scheridan, P.L., Krett, N.L., Fransis, M.D., Toft, D.O., Edwards, D.P. and Horwitz, K.B. (1987). Biochemistry **26:** 6262–6272.

Welshons, W.V., Lieberman, M.E. and Gorski, J. (1984). Nature **307:** 747–749.

Welshons, W.V., Krummel, M. and Gorski, J. (1985). Endocrinology **117:** 2140–2147.

Weinberg, R.A. (1991). Science **254:** 1138–1146.

Wikstrom, A.C., Bakke, O., Okret, S., Bronnegard, M. and Gustafsson, J.A. (1987). Endocrinology **120:** 1232–1242.

Yamamoto, K.R. (1985). Ann. Rev. Genet. **19:** 209–252.

Comparative Endocrinology and Reproduction
K.P. Joy, A. Krishna and C. Haldar (Eds)

10. Gonadotropins and Gonadal Receptors

K. Muralidhar[1], K. Nirmala[2] and G. Arundhati[1]
[1]Laboratory for Biochemistry, Department of Zoology,
University of Delhi, Delhi-110 007, India
[2]Department of Biochemistry, Daulat Ram College,
University of Delhi-110 007, India

Introduction

Gonadotropins are tropic hormones required for growth and maintanance of differentiated functions of gonads in male and female vertebrates (Konig, 1993). The pituitary gland produces two gonadotropins in all mammals—follitropin (Follicle Stimulating Hormone, FSH) and lutropin (Luteinizing Hormone, LH). In addition, in certain mammals, the placenta synthesizes and secretes a gonadotropin—the chorionic gonadotropin (CG) (Wallis *et al.*, 1985). Another hormone involved in the regulation of gonadal activity is prolactin (PRL). Its more pronounced effects are on the mammary gland. Prolactin has a definite role to play in gonadal functions and hence can be grouped with FSH and LH for a physiological discussion of regulation of gonadal structure and function. Gonadotropins and prolactin have been subjected to extensive and intensive study on all aspects for over five decades now (Pierce and Parsons, 1981). A number of disorders in humans are directly attributed to hypo-or hypersecretion of these hormones. Measurement of CG forms the basis of all early pregnancy detection methods in humans. In the case of farm animals, the recently established embryo transfer biotechnology protocols include equine CG (mare serum gonadotropin, PMSG) or FSH for inducing superovulation (Donaldson *et al.*, 1986). An overview of gonadotropins including prolactin and their respective receptors is presented below. The discussion is restricted to mammals and covers only biochemical aspects.

Purification

Pituitary FSH and LH from humans in addition to CG have conventionally been isolated from urine, a readily available and cheap source (Bahl, 1969). Urine from post-menopausal women and pregnant women during first trimester have been the ready source material respectively for human menopausal gonadotropin, hMG (a mixture of LH and FSH) and hCG. Large quantities of urine are usually collected through volunteer donors and stored in carboys with thymol as preservative. All procedures involve an initial step of adsorption of glycoprotein on silicious material like kaolin at acid pH. Subsequent elution at alkaline pH results in concentration of the dilute urine sample into a crude hormone concentrate. All subsequent steps involve chromatography. Human chorionic gonadotropin can be easily isolated from crude urinary concentrates by direct chromatography on DEAE-Sephadex and stepwise elution with salt gradient. In the case of laboratory and farm animals, pituitaries collected from abattoirs by laboratories serve as a source material for purification of LH, FSH and PRL (Reichert, 1975). While traditional procedures have relied heavily on conventional biochemical techniques like salt and organic solvent fractionation or chromatography (Ellis, 1961), recombinant DNA techniques are being used in recent years for production of these hormones (Keene *et al.,* 1989). The use of rDNA techniques have yielded till today biologically active human CG, FSH and LH and bovine FSH (Simon *et al.*, 1988). However, bulk of the supply of these protein hormones is still obtained by conventional biochemical procedures (Reichert, 1975). Further, even in rDNA technologies the same biochemical techniques with few novel modifications are employed for the purification of the cloned and expressed gene products (Stewart *et al.,* 1988). Some characteristics of the rDNA technology based products are given in Table 1. It should be kept in mind that human LH, PRL and FSH are also prepared from human pituitaries (Ward *et al.*, 1973).

Several methods are described by various groups to isolate LH, FSH and PRL from various species either from the same batch of glands (Muralidhar *et al.,* 1994) or from separate batches by different protocols (Reichert, 1975). In general LH is obtained in good yields compared to FSH except

Table 1. Some aspects of production characteristics of recombinant DNA technology-derived gonadotropins and prolactin

Hormone	Expression system	Details of expression
hCG	C-127 cells	Stable expression, hormone biologically and immunologically active. Expression rate $=10^{-3}$ mIU/cell/24 h.
hCG	SF-9 cells	Stable expression, hormone active. Yield claimed high.
hFSH	CHO cells	Stable expression of biologically and immunologically active hormone dimer. Secretion rate = 500 ng/ml/10^6 cells/24 h Bioactivity = 1.1 ± 0.4 IU/10^6 cells/24 h.
oLH-β+hCG-α	CV-1 cells infected with vaccinia virus	Stable expression, generation of bioactive hetero-dimer which stimulates the production of testosterone in mouse Leydig cells.
bPRL	*E. coli*	Stable and efficient expression, product identified as 24 K protein, immunoreactive in western blots using specific antiserum.

in the case of pigs where yields of FSH are also high. The yields of LH and FSH in the case of cattle, for instance, are 80–120 and 3–5 mg, respectively, per kilogram of wet frozen pituitaries. An example of purification procedure for the isolation of LH and FSH from water buffaloes is given in Fig. 1.

Prolactin is relatively easier to purify. Both the salt fractionation and organic solvent fractionation schemes are applicable to yield pure and homogeneous preparations (Muralidhar *et al.*, 1987). The yields are also satisfactory. In the case of farm animals, anywhere between 300–500 mg of prolactin can be purified per kilogram of frozen pituitaries depending on the procedure. A set of procedures equally applicable to sheep, cattle, buffalo and pigs is depicted in Fig. 2. The advantage of this scheme is that the same batch of frozen glands can be used to purify LH, FSH and PRL. Each of them is obtained as a separate side fraction from which it can be further purified (Muralidhar *et al.*, 1994).

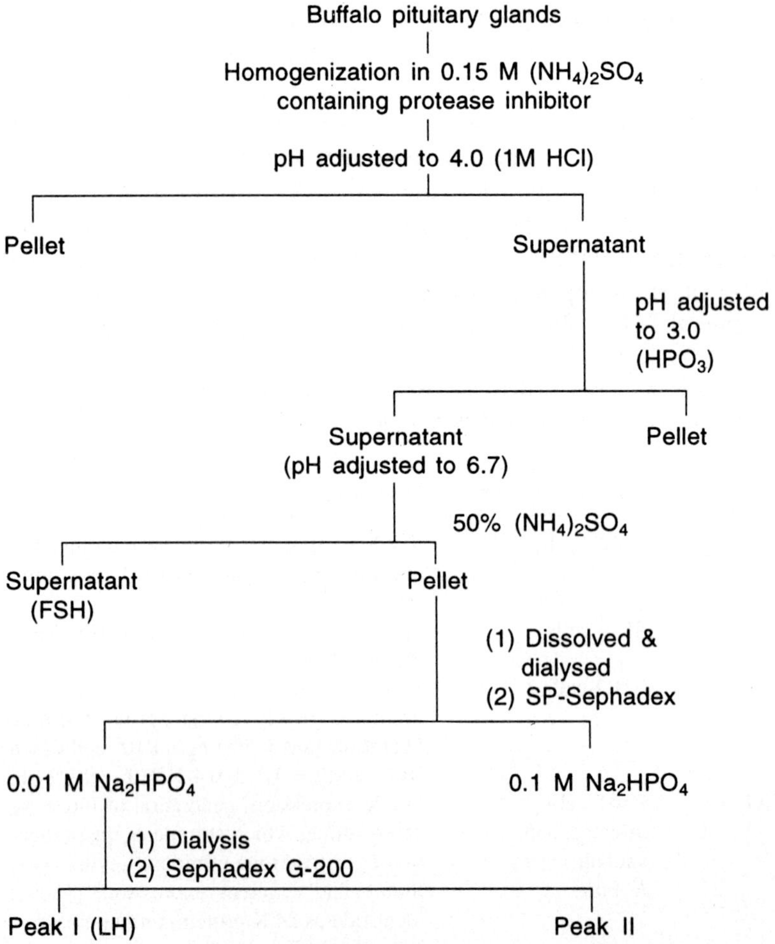

Fig. 1. A representative procedure for the isolation of LH and FSH from water buffalo pituitaries.

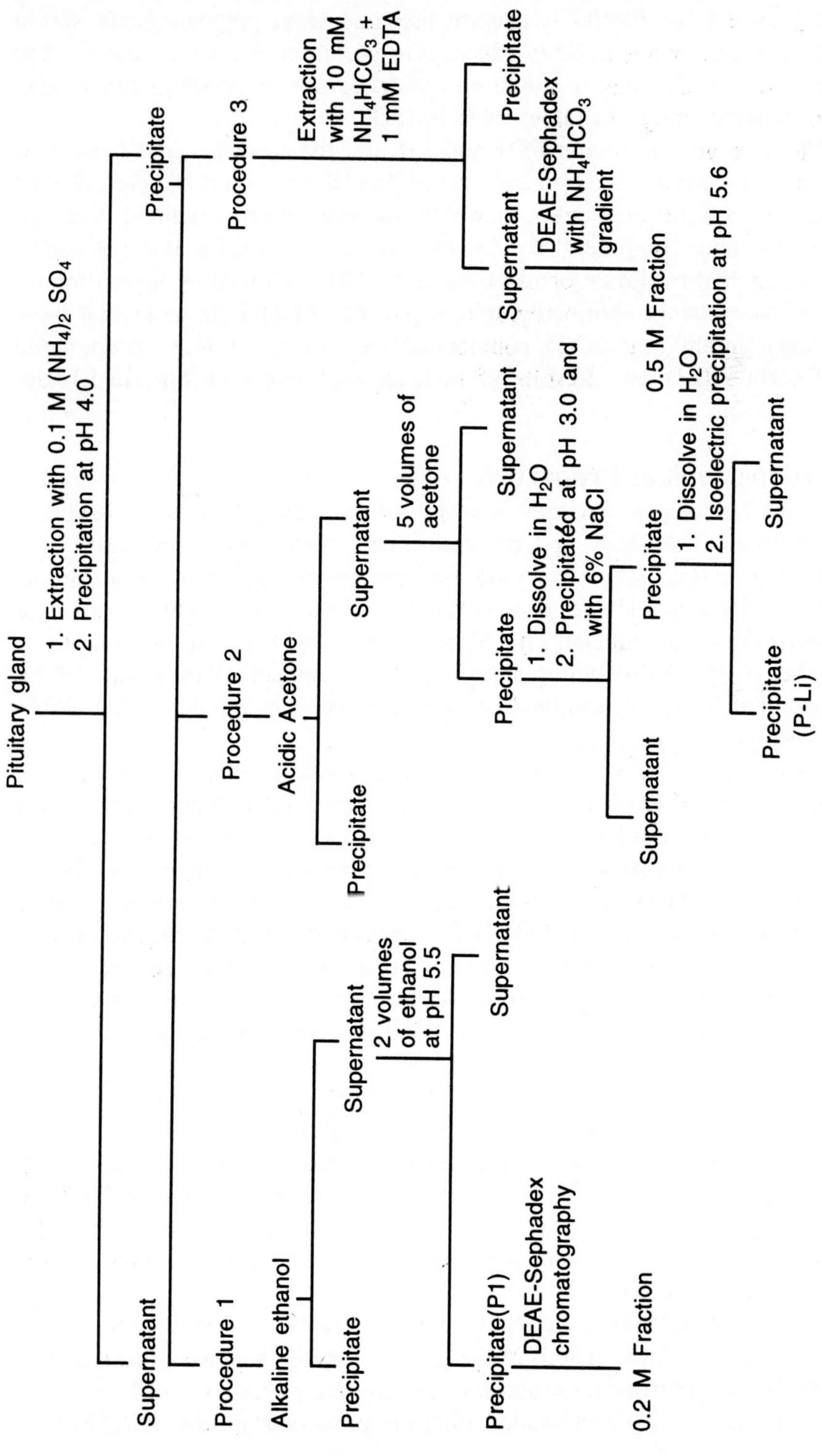

Fig. 2 Isolation protocols for Prolactin.

Equine CG or PMSG is usually prepared from pregnant mare serum collected from mares at 60 to 120 days of gestation. A good procedure was developed by Christakos and Bahl (1979) to obtain good quality PMSG from commercially available crude PMSG.

The two gonadotropins FSH and LH are glycoproteins and consist of two non-covalently linked subunits α and β and hence are called heterodimers. There are a number of ways in which subunits can be obtained from the intact hormone. In general the subunits are dissociated by acid pH and/or guanidine hydrochloride or 8M urea (e.g. hCG) followed by separation by using ion-exchange chromatography (e.g. cattle, sheep LH), molecular sieve chromatography (e.g. eCG), counter-current distribution (e.g. sheep, cattle and buffalo LH), etc. Results of such an experiment on buffalo LH are given in Fig. 3.

Physicochemical Properties

FSH and LH are glycoproteins. In many species, part of PRL is glycosylated. In addition to pituitary and placenta, many malignant trophoblastic cell lines and ectopic tumor cell lines secrete gonadotropins *in vitro* (Hussa, 1980). FSH and LH are 30,000 Da in molecular size. The amino acid composition and sequence of LH and FSH have been worked out for a number of species including sheep, cattle and human (Ward *et al.*, 1973). Tryptophan is absent in animal LH. Prolactin is a single polypeptide of 23–24K in size.

Being glycoproteins, FSH, LH and hCG have carbohydrate in addition to the protein backbone. The sugar content varies from 15 per cent in ovine LH to 45 per cent in PMSG. Both neutral sugars like mannose and galactose and amino sugars like galactosamine and glucosamine are present (Pierce and Parsons, 1981). In addition, sialic acid as N-acetylneuraminic acid is present in human CG and eCG. These sugars in the form of oligosaccharide chains are attached to specific asparagine and serine/threonine amino acid residues on the peptide backbone. The number, position and structure of oligosaccharides can vary with hormone and species. Some major physicochemical characteristics of LH, FSH, hCG and eCG are given in Table 2.

Prolactin was generally considered a simple protein for over 40 years. However, work of Lewis and others has established that part of PRL is indeed glycosylated (Lewis *et al.*, 1984) and sulfated (Kohli *et al.*, 1988). Some of its important physicochemical properties are listed in Table 3. Prolactin belongs to another group of protein hormones which includes growth hormone and placental lactogen. All the three share common structural features and receptors.

Microheterogeneity is a common structural feature of the three hormones FSH, LH and PRL (Keel and Grotjan, 1989). Varying degrees of deamidation, variations in N-terminal sequence due to limited proteolysis, differences in the structure of oligosaccharides, different post-translational modifications like sialylation, sulfation etc., all contribute to this phenomenon of microheterogeneity (Muralidhar *et al.,* 1992; de Medeiros *et al.,* 1993). The physiological significance is not very well understood. The different microheterogeneous forms react with different potencies in immuno- and bio-assays (Selvaraj and Moudgal, 1993; Ellis and Picciano, 1995).

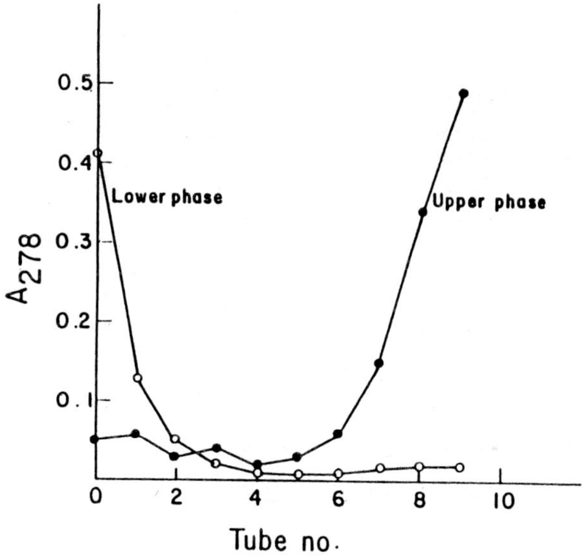

Fig. 3. Results of separation of α-subunit (O——O) from β-subunit (●——●) of buffalo LH by counter current distribution technique

Assays

The reproductive hormones LH, FSH and PRL can be assayed either by radio-receptor assays or radio-immunoassays. A number of such assay systems varying in sensitivity have been established (Behrman and Jaffe, 1978). Statistical methods of analysing the assay results to get potency estimates are available both for manual use and in the form of software for computer-aided use. In addition, conventional *in vivo* bioassays like the Steelman-Pohley assay for FSH and ventral prostate assay for LH are also used for potency estimates (Hari Paul Sharma *et al.,* 1994). Further, *in vitro* bioassays involving measurement of specific target cell biochemical response like cAMP or steroidogenesis are used for gonadotropins (Moyle and Ramachandran, 1973; Rao and Ramachandran, 1975). There is no satisfactory bioassay for PRL. Radio-receptor assay with hepatic tissue or mammary gland is used for PRL measurement (Kelly *et al.,* 1991). The pigeon crop-sac assay is not satisfactory. The Nb_2 lymphoma cell proliferation assay is reasonable but not specific to PRL action with regard to reproduction (Tanaka *et al.,* 1980).

Table 2. Some physicochemical characteristics of pituitary and placental gonadotropins

Property	Sheep LH	Sheep FSH	hCG	PMSG/eCG
Molecular weight	30,000	31,000	38,700	45,000–52,000
Stokes radius (Å)	30	32*	37	
Isoelectric point (pH)	7.3	4.6	4.3–6.6	
Sedimentation coefficient	2.74 (pH 8.4)	3.02 (pH 8.4)	3.40 (pH 10.0)	3.33 (pH 10.0)
($S_{20,w}$)	1.64 (pH 1.3)	1.54 (pH 1.3)	2.89 (pH 7.0)	3.08 (pH 7.2)
Carbohydrate content (%)	16	14	31	45
Number of amino acids				
α-subunit	96	96	96	96
β-subunit	119	118	145	145
N′-amino acid				
α-subunit	phe	phe	val	phe
β-subunit	ser	gly/asn	ser	ser

*Human FSH.

Potency estimates obtained from bio-assays and immuno-assays do not always match. Immunologically active and biologically inactive isoforms of these hormones are known to exist in circulation (Selvaraj and Moudgal, 1993; Ellis and Picciano, 1995).

Table 3. Some physiochemical characteristics of Prolactin*

Molecular weight	24,000
Stokes radius (Å)	24**
Isoelectric point (pH)	5.7
Sedimentation coefficient ($S_{20,w}$)	2.19
Total number of amino acids	198
N′-amino acid	Thr

*Sheep PRL; **Buffalo PRL.

Structure-Function Relationships

A number of biochemical studies using chemical modification, epitope-specific monoclonal antibodies and rDNA technique-derived fusion proteins (e.g. hFSH/hCG-β chimera combined with hCG-α) pointed to the topographic regions on the gonadotropin subunits involved in inter subunit interaction and receptor recognition (Nicoll *et al.*, 1986; Schwarz *et al.*, 1986; Kikuchi *et al.*, 1994). For example, the individual subunits of gonadotropins are biologically inactive *in vivo* or *in vitro* (Pierce and Parsons, 1981). However, the β-subunit, being hormone specific, has the minimal topography to recognise the receptors (Muralidhar and Moudgal, 1976). This can lead to a situation of the β-subunit being capable of inhibiting the response of target tissue to the intact hormone (Dighe *et al.*, 1979). A fusion protein composed of hCG-β and mouse antibody constant region (prepared by rDNA techniques) was shown to recognise receptors and cause complement mediated lysis (Johnson *et al.*, 1995).

The CMGCC sequence in the α-subunit (aa 28–32) and CAGYC in region in the β-subunit (Azuma *et al.*, 1990) of hCG appears to be involved in receptor recognition and biological activity.

The role of carbohydrate moeities in these hormones has also been delineated to some extent. For example, early studies established that sialic acid at the non-reducing end of the oligosaccharides of hCG confers longer plasma half-life and hence higher *in vivo* biopotency (Vaitukaitis *et al.*, 1976). In the case of the pituitary gonadotropins, the oligosaccharide at position 52 in the α-subunit is essential for signal transduction (Bishop *et al.*, 1994). Removal of either or both of the oligosaccharides at positions 7 and 24 increases metabolic clearance and hence decreases *in vivo* biopotency (Bishop *et al.*, 1995). A summary of their possible roles is given in Table 4.

Table 4. Functional contributions of sugar chains and sugarlinked sulfate residues in pituitary and placental gonadotropins

Sugar chains	Sulfate residues
Microheterogeneity	Microheterogeneity
Conformational stability	Resistance to glycosidases
Plasma half-life	Storage and secretion
Receptor recognition	
Biological activity	

Regulation of Secretion and Steady-State Circulatory Levels

Early physiological studies have given us an idea of the steady-state circulatory levels of FSH, LH and PRL under various natural and artificial states like puberty, menstrual cycle, pregnancy, aging, etc. These studies also resulted in a rich body of information on the various hormonal, non-hormonal and environmental factors which influence the secretory dynamics of these hormones (Campbell *et al.,* 1977; Muralidhar *et al.,* 1977; L 'Hermite *et al.*, 1978; Vijayan and McCann, 1978; Leiberman *et al.,* 1982; Clayton, 1989). To understand the mechanisms underlying steady-state levels in blood, both the biochemical details of biosynthesis and assembly of these hormones in the pituitary cells as well as molecular details of regulation of their gene expression have to be probed. Fortunately the advent of molecular biological techniques has permitted such studies. The current information is vast and exciting.

In humans, the FSH-β is coded by a single gene localised on chromosome 11 while LH-β and CG-β are coded by a gene, of which seven copies are present, and located on chromosome 19. The α- subunit of the human and bovine hormones is coded by a single gene composed of four exons separated by three introns (Ryan and Keutman, 1988). The ovine LH-β gene comprises three exons (Brown *et al.*, 1993).

The α and β-subunits of gonadotropins are translated as larger precursors on separate mRNAs (Chappel *et al.,* 1983). Simultaneous with the signal peptide cleavage the subunits undergo co-translational and subsequently post-translational modifications as in the case of hCG (Matzuk *et al.*, 1987, Peters *et al.,* 1989). The assembly of the two subunits to form whole hormone and the mechanism of packaging into secretory granules have been studied to some extent (Corless *et al.,* 1987; Matzuk and Boime, 1988; Ruddon *et al.,* 1989). O-glycosylation of hCG-β occurs after the assembly of the whole hormone. The genes for all the three hormones have been cloned and expressed (Boothby *et al.,* 1981; Fiddes and Talmadge, 1984; Policastro *et al.,* 1983; Reddy *et al.,* 1985; Rajendra Kumar *et al.,* 1992). The regulation of expression of these genes *in vivo* has been studied to some extent (Mercer, 1990; Rajendra Kumar and Low, 1993; Gharib *et al.,* 1990; Cecim *et al.,* 1995). It has been shown that progesterone and estradiol 17–β inhibit the transcription of genes coding for ovine FSH-α and β subunits by 70% (Phillips *et al.,* 1985). Progesterone was also shown to bring about shortening of poly A tails of the mRNAs for ovine LH-α and β subunits. This could possibly explain the low levels of LH during luteal phase when progesterone level is high in circulation (Jackson and Standart, 1990). It is possible that this is brought out by regulating poly A polymerase and/or some nuclease or by inhibiting the biosynthesis of proteins that bind to poly A tails protecting it from degradation. The effects of GnRH and cAMP on the expression of gonadotropin subunit gene expression has been studied to some extent (Mellon *et al.,* 1989). Based on studies using transgenic mice bearing human FSH-β gene, it was possible to show that species specific differences exist in the expression of FSH-β gene (Rajendra Kumar and Low, 1993). It was shown that hFSH-β gene expression was directly suppressible by dihydrotestosterone, a non-aromatisable androgen at the pituitary level (Rajendra Kumar and Low, 1993), Further proof was obtained when transgenic hypogonadal mice which are GnRH deficient were obtained and hFSH-β mRNA levels in pituitary cells from them in culture were shown to be inhibited by direct addition of DHT (Rajendra Kumar and Low, 1995). Even *in vitro* addition of activin to pituitary cells for example, was shown to activate FSH-β gene expression, increase primary transcript and mature mRNA levels (Weiss *et al.,* 1995). Estradiol regulates prolactin production in pituitaries by increasing its mRNA content (Maurer, 1980). Estradiol appears to bring about its effect without significantly influencing mammotroph proliferation (Lieberman *et al.,* 1982).

In situ hybridization studies indicate that FSH-β promoter is functional in testes. Some biological activity has also been ascribed to the α-subunit in testes (Erickson *et al.,* 1990).

Actions and Receptors

Physiological studies for over five decades have established the primary role of gonadotropins and prolactin in the growth, differentiation and maintenance of differentiated functions of gonads and accessory reproductive tissues in various mammalian species (Rani and Moudgal, 1983; Martin, 1985; Lakshmi Bhagat *et al.,* 1993). While the biological effects of FSH and LH appear to be well defined and restricted to gametogenesis and steroidogenesis, prolactin appears a mysterious hormone exhibiting more than hundred effects in various mammalian and non-mammalian animal groups. Further, while the biochemical details of FSH and LH actions are fairly well delineated, nothing is known about prolactin action. Studies in the late sixties and early seventies established the fact that all the three interact with membrane receptors in their target tissues (Moudgal *et al.,* 1971; Cuatrecausas, 1974). Initial studies demonstrated that cAMP is the second messenger mediating LH and FSH actions, at least, as far as steroidogenic effects are concerned (Wilson and Foster, 1985). In the case of PRL, no second messenger has as yet been discovered. The details of subsequent actions of cAMP have been worked out in some detail (Gilman, 1989; Krebs, 1989). While overwhelming evidence does not exist for the direct involvement of other second messenger systems like PI-Protein kinase, Ca^{2+}- Calmodulin, indirect involvement cannot be ruled out as the different second messenger systems do influence each other in subsequent steps in the cascade (Wurthner *et al.,* 1995). Recent work indicates that phospholipase A_2 (PLA_2) could be involved in LH action. Inhibitors of PLA_2 caused inhibition of LH stimulated testosterone (T) production in Leydig cells. They could inhibit cAMP stimulated T production indicating its position in the cascade as being later than cAMP production (Abyasekara *et al.,* 1990). By using specific inhibitors of A-kinase, C-kinase and Calmodulin kinase II, Morris and Richards (1995) showed that LH action during ovulation and luteinization involves induction of prostaglandin endoperoxide synthase-2 via A-kinase and C-kinase activation but not Calmodulin pathway. The regulation of steroidogenesis by LH in Leydig cells and ovarian cells and by FSH in Sertoli cells (i.e. action on aromatase activity) has been well worked out. The scheme fits into the general scheme of regulation of enzyme activity by covalent modification like phosphorylation (Krebs, 1989).

Membrane receptors to FSH and LH (hCG) are present in corpus luteum, Leydig cells, and in follicular granulosa cells at some stages of follicular growth (Griswold *et al.,* 1995). They belong to the general class of membrane receptors interacting with G-proteins and possess structural features characteristic of these, such as possesssing a long transmembrane hydrophobic region which spans the cell membrane seven times (Strade *et al.,* 1994). However, the extracellular domains are unusually long (Lewin, 1994). The cytoplasmic loop involved in the interaction with G-protein has been established (Okamato *et al.,* 1992). In the case of FSH-receptor (FSH-R), this spans residues 533–555 of the receptor (Grasso *et al.,* 1995a, b). Full length cDNAs representing FSH-R and LH-R have been cloned and protein sequence deduced (Sprengel *et al.,* 1990). LH-R has been purified even by biochemical procedures (Indrapichate *et al.,* 1992).

Receptors for prolactin have been described both in hepatic and mammary tissues and belong to the cytokine receptor family (Kelly *et al.,* 1991).

One of the unsolved problems in receptor physiology has been the stoichiometry between hormones and receptors. It has been assumed for more than 5 decades that this is 1:1. The only experimental work till date has been with PRL/GH receptors. The receptor is a single chain having the extracellular domain (ECD), transmembrane domain (TMD) and intracellular domain (ICD). The ECD is 200–250 aminoacids in length. Crystallographic and gel chromatographic evidence suggests a ratio of 2:1 for receptor: hormone in the case of hGH/bGH receptors (De Vos *et al.,* 1992). However, in the case of bPRL-R cloned and expressed in *E. coli*, stoichiometry comes to 1:1. Another puzzle that has set in due to recent work is that potency does not bear any relation to affinity to the receptor. For example, for the same receptor cloned and expressed as above, hGH had the highest affinity followed by bPRL and bGH when reconstituted in Nb_2 cell lines. However, the potencies remained same (Tchelet *et al.,* 1995).

Conclusion

The biochemical details of cellular biosynthesis and topological location of membrane receptor binding domain(s) for gonadotropins and PRL are to be worked out fully. Meanwhile, northern blot analysis has shown the presence of truncated receptors which are free floating as the sizes of mRNA were much shorter than the full size representative (Heckert *et al.,* 1992). More powerful techniques like gene-knock out experiments using embryonic stem (ES) cells (Ramirez-Solis and Bradley, 1994) have begun to yield very exciting results. This ES cell technology has ushered in the third approach to the study of gonadotropins or hormone action *in vivo*. The first was the surgical extirpation of the source, the second was the immuno-neutralization in circulation and now, the specific gene-knock out. For example, using this approach it was possible to show that tumorigenic effect of inhibin deficient rats could be overcome by simultaneous FSH deficiency (Rajendra Kumar, personal communication). Activin receptor gene-knock out resulted in new phenotypes with skeletal and facial abnormalities (Matzuk *et al.,* 1995) indicating new roles for activin. We are being ushered into a more exciting stage of gonadotropin research.

References

Abayasekara, D.R.E., Band, A.M., and Cooke, B.A., (1990) Mol. Cell Endocr. 70, 147–153.

Azuma, C., Miyai, K., Saji, F., Kamipura, S., Tokugawa, Y., Kimura, T., Ohashi, K., Koyama, M., Tijima, Y., Kashiwai, T., Hayashizaki, Y., and Tanizawa, O., (1990) Mol. Endocr. 5, 97–102.

Bahl, O.P., (1969) J. Biol. Chem. 244, 567–574.

Behrman, H.R., and Jaffe, B.L., (1978) Eds. "Methods in Hormone Radioimmunoassay", 2nd Edn. Academic Press, New York.

Bishop, L.A., Robertson, D.M., Cahir, N., and Schofield, P.R., (1994) Mol. Endocr. 8, 722–731

Bishop, L.A., Nguyen, T.V., and Schofield, P.R., (1995) Endocrinology 136, 2635–2640.

Brown, P., Mc Neilly, J.R., Wallace, R.M., Mc Nielly, A.S., Clark, A.J., (1993) Mol. Cell. Endocr. 93, 157–165.

Boothby, M.R., Ruddon, R.W., Anderson, C., Mc Williams, D., and Boime, I., (1981) J. Biol. Chem. 256, 5121–5127.

Campbell, G.A., Kurczy, M., Marshall, S., and Meites, J., (1977) Endocrinology 100, 580–587.

Cecim, M., Kerr, J., Bartke, A., (1995) Biol. Reprod. 52, 1144–1148.

Chappel, S.C., Ulloa-Aguirre, A., Coutifaris, C., (1983) Endocrine Rev. 4, 179–209.

Christakos, S., and Bahl, O.P., (1979) J. Biol. Chem. 254, 4253–4261.

Clayton, R.N., (1989) J. Endocr. 120, 11–19.
Corless, C.L., Matzuk, M.M., Ramabhadran, T.V., Krichevsky, A., and Boime, I., (1987) J. Cell Biol. 104, 1173–1181.
Cuatrecausas, P., (1974) Ann. Rev. Biochem. 43, 169.
De Vos, Abraham, M., Ultsch, M., and Kossiakoff, A. (1992). Science 255, 306–312.
Dighe, R.R., Muralidhar, K., and Moudgal, N.R., (1979) Biochem. J. (Lond) 180, 573–578.
Donaldson, L.E., Ward, D.N., and Glenn, S.D., (1986) Theriogenology 25, 747–757.
Ellis, S., (1961) Endocrinology 69, 554–570.
Ellis, L.A., and Picciano, M.F., (1995) Endocrinology 136, 2711–2720.
Erichson, L.D., Riza, S.A., Bergert, E.R., Charleswoth, M., Mc Cormik, D.J., Ryan, R.J., (1990) Endocrinology 126, 2555–2560.
Fiddes, J.C., and Talmadge, K., (1984) Rec. Prog. Horm. Res. 40, 43–78.
Gharib, S.D., Weirman, M.E., Shupnik, M.A., and Chin, W.W., (1990) Endocr. Rev. 11, 177–199.
Gilman, A.G., (1989) JAMA 262, 1819–1825.
Grasso, P., Leng, N., Reichert, L.E., Jr. (1995a) Mol. Cell Endocr. 108, 43–50.
Grasso, P., Leng, N., Reichert, L.E., Jr. (1995b) Mol. Cell Endocr. 110, 35–41.
Griswold, M.D., Heckert, L., and Linder, C., (1995) J. Steroid Biochem. Mol. Biol. 53, 215–218.
Hari Paul Sharma, Lakshmi Bhagat, and Muralidhar, K., (1994) Ind. J. Anim. Sci. 64, 959–961.
Heckert, L.L., and Griswold, M.D., (1992) Rec. Prog. Horm. Res. 48, 61–78.
Hussa, R.O., (1980) Endocrine Reviews 1, 268–293.
Indrapichate, K., Meehan, D., Lane, T.A., Chu, S.Y., Rao, Ch.V., Johnso, D., Chen, T.T., and Wimalasena, J., (1992) J. Biol. Chem. 46, 265–278.
Jackson, R.J., and Standart, N., (1990) Cell 62, 15–24.
Johnson, G.A., Hansen, T.R., Austin, K.J., van Kirk, E.A., Murdoch, W.J., (1995) Biol. Reprod. 52, 68–73.
Keel, B.A., and Grotjan, Jr, H.E., (1989) In "Microheterogeneity of Glycoprotein Hormones", CRC Press.
Keene, J.L., Matzuk, M.M., Otani, T., Fauser, B.C.J.M., Galway, B.A., Hseuh, A.J.W., and Boime, I., (1989) J. Biol. Chem. 264, 4769–4775.
Kelly, C.A., Kjiane, J., Vinay, M.C., and Edery, M., (1991) Endocr. Rev. 12, 235–351.
Kikuchi, T., Koyama, M., Miyai, K., Kimura, T., Nishikiori, N., Kimura, T., Azuma, C., Kusunoki, M., Saji, F., and Tanizawa, O., (1994) Mol. Cell Endocr. 102, 1–7.
Kohli, R., Chadha, N., Muralidhar, K., (1988) FEBS Letters 242, 139–143.
Konig, W., (1993) "Peptide and Protein Hormones" VCH Publishers, Weinheim, Germany.
Krebs, E.G., (1989) JAMA 262, 1815–1818.
Lakshmi Bhagat, Duraiswami, S., and Kumari, G.L., (1993) J. Biosciences 18, 59–72.
Lewin, B., (1994) Genes V Oxford University Press. New York
Lewis, U.J., Singh, R.N.P., Lewis, L.J., Seavery, B.K., and Sinha, Y.N., (1984) Proc. Natl. Acad. Sci. USA 81, 385–389.
L Hermite, M., Macleod, R.M., and Robyn, C., (1978) Acta Endocr. 89, 29–37.
Lieberman, M.E., Maurer, R.A., Claude, P., Gorski, J., (1982) Mol. Cell Endocr. 25, 277–294.
Martin, C.R., (1985) "Endocrine Physiology" Oxford University Press, London.
Matzuk, M.M., Kreiger, M., Corless, C.L., amd Boime, I., (1987) Proc. Natl. Acad. Sci. 84, 6345–6358.
Matzuk, M.M., and Boime, I., (1988) J. Cell. Biol. 106, 1049–1059
Matzuk, M.M., Rajendra Kumar, A.V., Bickenbach, J. R., Roop, D.R., Jaeisch, R., Bradley, A., (1995) Nature 374, 354–363.

Maurer, R.A., (1980) J. Biol. Chem. 255, 854–859.
deMedeiros, S.F., Amato, F., Matthews, C.D., and Norman, R.J., (1993) J. Endocr. 139, 519–532.
Mellon, P.L., Clegg, C.H., Correll, L.A., Mc Knight, G.S., (1989) Proc. Natl. Acad. Sci. USA 86, 4887–4891.
Mercer, J.E., (1990) Mol. Cell Endocr. 73, c63–c67.
Morris, J.K., and Richards, J.S., (1995) Endocrinology 136, 1549–1558.
Moudgal, N.R., Moyle, W.R., and Greep, R.O., (1971) J. Biol. Chem. 246, 4983–4986.
Moyle, W.R., and Ramachandran, J., (1973) Endocrinology 93, 127–134.
Muralidhar, K., and Moudgal, N.R., (1976) Biochem. J. (Lond) 160, 615–619.
Muralidhar, K., Rhoda Maneckjee, and Moudgal, N.R., (1977) Endocrinology 100, 1137–1142.
Muralidhar, K., Chadha, N., and Kohli, R., (1987) J. Biosciences 13, 201–214.
Muralidhar, K., Rajendra Kumar, T., and Hari Paul Sharma, (1992) Ind. J. Biochem. Biophys. 29, 168–172.
Muralidhar, K., Rajendra Kumar, T., Chadha, N., Khurana, S., Khanna, T., Hari Paul Sharma (1994) Ind. J. Exp. Biol. 32, 73–80.
Nicoll, C.S., Mayer, G.L., and Russell, S.M., (1986) Endocr. Rev. 7, 169–203.
Okamato, T., and Nishimoto, I., (1992) J. Biol. Chem. 267, 8342–8346.
Peters, B.P., Krzesicki, R.F., and Ruddon, R.W., (1989) Endocrinology 124, 1602–1612.
Phillips, C.L., Lin, L.W., Wu, J.C., Guzman, K., Misted, A., Miller, W.L., (1985) Mol. Endocrinol. 2, 641.
Pierce, J.G., and Parsons, T.F., (1981) Ann. Rev. Biochem. 50, 465–495.
Policastro, P., Ovitt, C.E., Hoshina, M., Fukuoka, H, Boothby, M.R., and Boime, I., (1983) J. Biol. Chem. 258, 11492–11499.
Rajendra Kumar, T., Fairchild-Huntress, V., and Low, M.J., (1992) Mol. Endocr. 6, 81–90.
Rajendra Kumar, T., and Low, M.J., (1993) Mol. Endocr. 7, 898–906.
Rajendra Kumar, T., and Low, M.J., (1995) Neuroendocrinology 61,628–637.
Ramirez-Solis, R., and Bradley, A., (1994) Current Opinion in Biotchnology 5, 528–533.
Rani, C.S.S., and Moudgal, N.R., (1983) In "Hormonal Proteins and Peptides" C.H. Li (Ed.) Academic Press, New York Vol. II, 135–184.
Rao, A.J., and Ramachandran, J., (1975) Life Sciences 17, 411–416.
Reddy, V.B., Beck, A.E., Garramone, A.J., Velluci, V., Lustbadder, J., and Bernsteine, E.G., (1985) Proc. Natl. Acad. Sci. USA 82, 3644–3648.
Reichert, Jr. L.E., (1975) Methods in Enzymol. 37, 360–380.
Ruddon, R.W., Krzesicki, R.F., Beebe, J.S., Loessel, L., Perini, F., and Peters, B.P., (1989) Endocrinology 124, 862–869.
Ryan, R.J., and Keutman, H.I., (1988) Rec. Prog. Horm. Res. 43, 383.
Schwarz, S., Berger, P., and Wick, G., (1986) Endocrinology 118, 189–197.
Selvaraj, N., and Moudgal, N.R., (1993) J. Reprod. Fert. 98, 611–616.
Simon, J., Danfarth, D.R., Hutchinson, J.S, and Hodgen, G.D., (1988) J. Amer. Med. Assoc. 259, 3290–3295.
Sprengel, R., Braun, T., Nikolics, F., Segaloff, D.L., and Seeburg, P.H., (1990) Mol. Endocrinol 4, 525–530.
Stewart, H.J., Jones, D.S.C., Pascall, J.C., Popkin, R.M., and Flint, A.P.F., (1988) J. Reprod. Fert. 83, 1–97.
Strade, L.D., Fong, T.M., Tota, M.R., and Underwood, D., (1994) Ann. Rev. Biochem. 63, 101–132.
Tanaka, T., Shin, R.P.Cl., Gont, P.W., Beer, C.T., Noble, R.L., and Friesen H.G., (1980) J. Clin. Endocrinol. Metab. 57 (5), 1058–1063.
Tchelet, A., Staten N.R., Creely, D.P., Krivi, G.G., and Gertler, A., (1995) J. Endocr. 144, 393–403.

Vaitukaitis, J.L., Ross, G.T., Braunstein, G.D., and Rayford, P.L., (1976) Rec. Prog. Horm. Res. 32, 289–331.

Vijayan, E., and McCann, S.M., (1978) Neuroendocrinology 25, 221–235.

Wallis, M., Howell, S.L., and Taylor, K.W., (1985) "Biochemistry of the Polypeptide Hormones" John Wiley & Sons, New York

Ward, D.N., Reichert, Jr, L.E., Liu–W.K., Nahm, H.S., Hsia, J., Lamkin, W.M., and Jones, N.S., (1973) Rec. Prog. Horm. Res. 29, 533–561.

Weiss, J., Guemdner, M.J., Halvorson, L.M., and Jameson, J.L., (1995) Endocrinology 136, 1885–1891.

Wilson, J.D., and Foster, D.W., (1985) Eds. William's "Textbook of Endocrinology", 7th Edn. W.B. Saunders Comp. Philadelphia.

Wurthner, J.U., Kistler, M., Kratzmeier, M., and Mukhopadhyaya, A.K., (1995) Endocrine 3, 579–584.

Comparative Endocrinology and Reproduction
K.P. Joy, A. Krishna and C. Haldar (Eds)

11. Evolution of Testis in Nonmammalian Vertebrates

S.K. Saidapur and B.A. Shanbhag
Department of Zoology, Karnatak University, Dharwad-580 003, India

Introduction

We feel greatly privileged to have been invited to write this article for the commemorative volume being prepared in honour of three eminent contempary endocrinologists of this country namely, Prof. J.P. Thapliyal, Prof. A.G. Sathynesan and Prof. C.J. Dominic. We have been associated with the three scientists in some way or the other and have derived considerable inspiration, guidance and timely help in shaping up our own scientific pursuits. This review is not meant to be an exhaustive literature survey of species specific data but it intends to highlight the salient comparative structural organization of the testis and the current concepts concerning the evolution of testicular structures in different groups of nonmammalian vertebrates. The literature cited therefore is highly representative and in no way it undermines the importance of numerous excellent publications that are not cited here.

1. General Considerations

Testicular histoarchitecture is well defined in reptiles, birds and mammals (amniotes), being distinctly divided into intratubular (or germinal) and intertubular (or interstitial) compartments that are separated by a basement membrane. On the inner side of the tubular compartment, upon the basement membrane lies a permanent germinal epithelium composed the germ cells and the Sertoli cells. The interstitium between the tubules consists of interstitial cells (Leydig cell), fibroblasts, and blood and lymph vessels. Anamniote chordates on the other hand, exhibit great diversity in their testicular morphology, both between and within the different classes (urochordates, cephalochordates, agnathans, elasmobranchs, fishes and amphibians). For instance, the unit structure of the testis in cyclostomes and elasmobranchs is believed to be an "ampulla" or "follicle". The teleost testis is described as being "lobular" or "tubular". Also, these terms are used interchangeably/ indiscriminately without regard to distinction, leading to confusion and controversy.

In recent years based on ultrastructural studies attempts have been made to redefine the above terminology and establish more clearly the similarities and differences in organization of the testicular components (germ cells and nongerminal cells) in different classes of chordates (Grier, 1981, 1992; Callard, 1991). In order to evolve a uniform terminology that is consistent with homology and that literally describes germinal compartments in cyclostomes and elasmobranchs the frequently and/or inaptly used terms "ampulla" (reviewed by Dodd, 1983) and "follicle" (Pratt, 1988) have now been dropped. For, anatomically, an 'ampulla' is defined as "a sac or dilated part of a tube or cannal...". The agnathan testis is ductless and therefore the term 'ampulla' is inappropriate; however, there exists a duct system in elasmobranch testis to which the saclike germinal compartments attach. A "follicle" is defined as a "small saclike cavity for secretion or excretion" and this term has been used in the past to describe testes in both cyclostomes and elasmobranchs. Recently, Callard (1991) proposed the term "spermatocyst" to describe the individual germinal compartments within the elasmobranch testis. The root word "spermato" means seed, germ or beginning and the word "cyst" describes the morphology. In fact, the literal meaning of "spermatocyst" best describes the homologous germinal compartments in testes of both cyclostomes and elasmobranchs. Therefore, the term spermatocyst instead of ampulla or follicle is used in the succeeding sections. Since, testes of agnathans and cartilaginous fishes contain numerous spermatocysts, they are aptly termed "polyspermatocystic".

Until recently there have been major inconsistencies in the literature regarding the teleost testis morphology and the homology of different cell types (Dodd and Sumpter, 1984). The teleost and amphibian testes vary considerably in their histoarchitecture as compared to that of reptiles, birds or mammals, in being either lobular or tubular. However, there are no established criteria to distinguish these testes types. Further, the terms 'lobule' and 'tubule' have been used interchangeably without regard to distinction. However, Grier (1981) classified the teleost testes into two major types based on the distribution of spermatogonia within the tubules. In the *unrestricted spermatogonial testis-type* the spermatogonia are distributed along the entire length of the tubule as in mammals. This testis type is typical of most teleosts. Tubules may form either anastomosing or a branching network. The second teleost testis-type is typical of the Atheriniformes. Spermatogonia are totally restricted to the distal terminus of the tubule immediately beneath the tunica albuginea. This category is now recognized as *restricted spermatogonial testis-type*. The other variants of the testes in teleosts may be grouped into one of these two testis-types. It is also pertinent to note here the definitions of 'lobular and 'tubular' testicular types proposed by Callard (1991). In the former, lobules terminate at the testis periphery, whereas in the latter, tubules form 'looped' structures that are common to all amniotes. The loops empty from both ends into the rete testis in mammals, but a rete testis has not been described in any anamniote (Grier, 1992).

2. Cyclostomes

The testes of both lampreys (Petromyzontidae) and hagfishes (Myxinoidea) contain numerous ovoidal or spherical germinal compartments. The unique feature of agnathan testis is that each germinal compartment develops from a primary spermatogonia and associated Sertoli cells. Initially, these cells are in direct contact with interstitial tissues and are not separated into germinal and interstitial compartments due to the absence of a basement membrane. The primary spermatogonia are thus not organized into a germinal epithelium. They are found singly or in clusters at the periphery of the testis in direct contact with the interstitium. However, in these primitive agnathans (e.g. *Petromyzon marinus* and *Eptatretus stouti*) also a basement membrane develops subsequently, around clusters of primary spermatogonia thereby marking the formation of secondary spermatogonia that enter meiosis. At this stage germinal and interstitial compartments are established and spermatocysts are produced. Since, numerous spermatocysts are produced, the testes of cyclostomes are termed "polyspermatocystic". Interestingly, both primary spermatogonia and Sertoli cells are randomly located in or connected to the basement membrane. The Sertoli cells may even be found in the center of the spermatocysts in the hagfish.

The pattern of spermatogenesis differs greatly between lampreys and hagfishes. In the former spermatogenesis is a single event since they breed once and die. All germ cells mature simultaneously and synchronously. The testes of lampreys lack stem cell spermatogonia. Whereas, the testes of hagfishes possess stem cell spermatogonia. The hagfishes do not commit all stem cells to spermatogenesis at one time. Hence, multiple spermatogenetic cycles can occur. The time of onset of spermatogenesis varies with the species. In both hagfish and lampreys, sperms are shed directly into the body cavity because testicular ducts are absent.

In *Lampetra fluviatilis* 'Leydig cells' were found to develop in the testis shortly before the development of secondary sexual structures (Larsen, 1973). In the testis of *E. stouti*, cells with ultrastructural features of the Leydig cells (smooth endoplasmic reticulum, mitochondria with tubular cristae) have been reported among spermatogenetic cysts (Tsuneki and Gorbman, 1977). *In vitro* incubation studies have shown that cyclostome testes produce very small quantities of sex steroids. However, little is known about the biological actions of sex steroids in cyclostomes which lack gonadal ducts (Gorbman, 1983).

3. Elasmobranchs

The testis of cartilaginous fishes is also polyspermatocystic as that of the agnathans. Based on the light microscopic studies Stanley (1966) used the term "spermatocyst" to denote a single Sertoli cell and its associated cohort of germ cells within a follicle in his description of testicular structure in a shark, *Scyliorhinus caniculus* and a ray, *Torpedo marmorata.* However, ultrastructural studies in *Squalus acanthias* showed that cytoplasmic processes from single Sertoli cell actually extend between its associated germ cells. Obviously, this could not be observed in light microscopic studies. Therefore, the cellular association between germ cells and Sertoli cells does not appear to conform to the definition of a cyst as given by Stanley (1966). Hence, such association between Sertoli cell and its clone of germ cells as seen in elasmobranchs has been defined as a "spermatoblast" and the follicular germinal compartment as a "spermatocyst" (Callard, 1991).

The histoarchitecture of testes of elasmobranchs differs strikingly from that of most vertebrates. The unit of testis structure is the spherical "spermatocyst" rather than the tubule. The mature testes are also unusual in having the zonate arrangement of the spermatocysts. The zonation pattern of spermatocysts differs with the species. Elasmobranch testis may be classified into three types based on the location and origin of propagation of spermatocysts (Pratt, 1988). In rajids, germinal zones are seen as discrete islands on the distolateral surface of the testis. The spermatocysts develop and move away from these islands. This type of testis is called "compound" owing to the presence of many discrete germinal zones. In the Carcarhinid-sphyrnid testis, spermatocysts develop from a germinal zone across the diameter of the testis, primarily in a dorsal to ventral direction. The testis is internally subdivided by interstitial tissue trabeculae radiating outwards from the germinal zone. This type of testis is termed "diametric". In the case of laminid-alopids, testis is subdivided into lobes with a germinal zone at the center of each lobe. Hence, the spermatocysts develop radially from these germinal zones and the term "radial" is applied to these testes.

Each concentric zone consists of spermatocysts at the same stage of spermatogenesis, and divisions (both mitotic and meiotic) of the germ cells are roughly synchronous within each spermatocyst. Each spermatogonium undergoes mitosis four times to give rise to sixteen secondary spermatogonia. At the same time, the Sertoli cell nuclei migrate from a central position to a peripheral one. The secondary spermatogonia then enter the prophase of the first meiotic division, becoming primary spermatocytes, and during this period spermatocysts increase in diameter. The duration of the secondary spermatocyte stage is short and it is followed by the production of spermatids and spermatozoa, respectively. The zones radiate from the site of their formation, and contain primary and secondary spermatogonia, primary and secondary spermatocytes, spermatids and spermatozoa, respectively, the latter being the widest. The spermatocysts remain closed until the end of spermatid development but later acquire ducts which open into the efferent duct system.

Leydig cells are found in some species but not in a few others studied. There is a good histochemical evidence to show that Leydig cells and Sertoli cells are capable of steroid synthesis. *In vitro* studies with labelled precursors also confirm the ability of testes for steroid production.

4. Teleosts

The literature pertaining to testicular structure and function in teleost fishes has been reviewed periodically (Hoar, 1969; Dodd, 1972; Lofts and bern, 1972; Guraya, 1976; Callard *et al.*, 1978; Hoar and Nagahama, 1978; Grier, 1981, 1992; Nagahama, 1983; Dodd and Sumpter, 1984). It is evident from these reviews that there has been a considerable confusion and controversy with regard to the morphology, distribution and homology of different cell types found in the fish testis. Therefore, it is widely believed that the teleost testis is a highly variable structure than that in the mammals. In mammals there is no confusion regarding the identity of intralobular Sertoli cells, intertubular Leydig cells and peritubular boundary cells. Whereas, in teleosts "lobule boundary cells" have been considered to be homologs of either Leydig or Sertoli cells depending upon the species studied and, based on light microscopic and histochemical studies. Further, the nonsteroidal interstitial tissue was generally considered to be fibroblasts or connective tissue, and no distinction was made between the boundary and interstitial cells. Myoid boundary cells circumscribe the mammalian tubule. Likewise, smooth muscles around the tubules have been described in the fish testis (Gresik, 1973).

The doctrine of the lobule boundary cell homology with the mammalian Leydig cells proposed by Marshall and Lofts (1956) dominated the literature for over two decades. The suggestion was based on the fact that sudanophilia, a feature considered to depict steroidogenic potentiality was found in the lobule boundary cells of *Esox lucius* (and later in several other species) rather than in the Leydig cells. Unfortunately, testis of *E. lucius* is atypical in not showing sudanophilia in their Leydig cells. Several other salmonids also exhibit similar phenomenon or at best poor sudanophilia of their Leydig cells. Interestingly, a reinvestigation at the ultrastructural level revealed that the sudanophilia in *E. lucius* is actually restricted to the intratubular Sertoli cells and that the so called Leydig cell homologs, the lobule boundary cells do not exhibit sudanophilia. Furthermore, it is now known that although the Leydig cells of salmonids lack sudanophilia, they possess 3β-hydroxysteroid dehydrogenase enzyme activity indicating their steroidogenic potential (Hurk *et al.*, 1978). The interstitial Leydig cells have also been reported subsequently in several other species of salmonids (references in Hoar and Nagahama, 1978; Grier *et al.*, 1980a, b; Grier, 1981) there by resolving the long-standing controversy with regard to the identity and homology of Sertoli, Leydig and lobule boundary cells. It is now evident that the so called 'lobule boundary cells' reported in several teleost testes reside immediately outside of the tubule basement membrane (Grier *et al.*, 1980a) and possess ultrastructural features (such as numerous microfilaments oriented parallel to the long axis of the cells, few to many pinocytotic vesicles along the plasmalemma, an elongated nucleus and rough endoplasmic reticulum) similar to myoid boundary cells in the mammalian testis (Dym and Fawcett, 1970). They lack characteristics of steroid producing cells. In mammals the boundary cells form a complete circumferential layer around the tubules. Whereas, in teleosts there may be regions that are not covered by the boundary cells. This permits the juxtaposition of basement membrane from neighbouring tubules or blood vessels or Leydig cells. The teleost testis lacks a lymphatic system as described in mammals (Dym and Fawcett, 1970) but it does possess an extravascular space containing blood plasma which leaks between blood vessel endothelial cells (Grier, 1981). Thus, the concept of homology between the so called 'lobule boundary cell' and Leydig or Sertoli cell is no longer held valid.

It is now clear that the teleost testis is tubular and it can be classified into two broad categories based on the distribution pattern of spermatogonia. In majority of teleosts, spermatogonia are distributed all along the length of the tubule as seen in mammals. This category is called *unrestricted spermatogonial testis type.* With the onset of spermatogenesis, the Sertoli cells form the borders of cysts in which synchronous spermatogenesis occurs. Sperms are released into the tubule lumen at spermiation. This is accompanied by the separation of Sertoli cell processes so that tubule and cyst lumen become continuous.

In the second category, the spermatogonia are exclusively restricted to the distal terminus of the tubule immediately beneath the tunica albuginea, in association with Sertoli cells. This type of teleost testis which is typical of Atheriniformes is called *restricted spermatogonial testis-type.* In these testes Sertoli cells reorganize to form cysts by the time spermatogonia transform into primary spermatocytes. The germ cell development then proceeds synchronously within the cyst. With the progress in spermatogenesis, the cysts move towards the efferent ducts. By the time spermatozoa are produced, the cysts are found near the efferent ducts. In this region spermiation takes place. It is interesting to note that during spermiation the sperms are shed into the efferent duct system and the Sertoli cells then transform themselves into efferent duct cells. The Sertoli cells hypertrophy and become secretory. Such efferent duct system is absent in teleosts with unrestricted spermatogonial testis type.

In addition to the formation of unencapsulated sperm bundles (spermatozeugmata) and synthesizing steroid hormones (at least in some species), Sertoli cells may phagocytize residual bodies cast off by developing spermatids and nourish germ cells via micropinocytosis.

Members of teleosts exhibit great diversity in their mode of reproduction. This includes oviparity, ovoviviparity and viviparity. While internal fertilization is a rule in live-bearing forms, it is practised by some oviparous species also. In the majority of egg laying fishes fertilization is external and it takes place in the aqueous medium which greatly differs in different habitat, ranging from freshwater to marine and, pelagic to benthic waters. Thus, mode of reproduction has resulted in the evolution of modifications of testis architecture, especially the germ cell-Sertoli cell interaction, sperm structure and physiology as well as production of sperm bundles that may be encapsulated (spermatophores) or unencapsulated (spermatozeugmata). Thus, spermiogenesis and sperm morphology show considerable species variation among teleosts. In addition, diverse mechanisms for transfer of sperms into the body cavity of female fishes which exhibit internal fertilization have also been evolved among different teleost species. The internal fertilization is associated by the evolution of elaborate courtship behaviour and modification of anal fin into a gonopodium which serves as an intromittant organ down which sperms slide into the female body cavity during copulation.

The interstitium of all teleosts may or may not exhibit sudanophilia but it does possess steroidogenic capability as in higher vertebrates (Nagahama, 1994). Previous reports that Leydig cells are absent in some species is now considered to be incorrect (Grier, 1981). However, there exists a diversity in the pattern of distribution of the Leydig cells in teleosts. In some teleosts, the Leydig cells are concentrated in areas other than in interstitium. In *Poecilia latipinna*, the Leydig cells are concentrated around the efferent duct (Hurk, 1973). In gobids, a glandular mass referred to as androgen gland or interstitial cell gland lies along the mesorchium (our unpublished observations) while in *Blennius* species, they are concentrated into a distinct accessory gland (Lofts, 1987). Also, fibroblasts and testis blood supply reside within the interstitium along with intercellular connective tissue fibers, mostly collagen.

5. Amphibians

The present day anurans, urodeles and apodans exhibit great differences between and/or within (e.g. urodeles) the different Orders in their testicular histoarchitecture. The salient features of testis in these three groups of amphibians are described below.

(a) Urodela

The urodelan testes are morphologically quite different from those of anurans. In some they are paired compact structures while in others (many salamanders and newts), each testis is composed of several individual lobes arranged in a linear chain and linked together with slender cords of tissue. The latter is also known to contain primordial germ cells. The number of lobes is variable. It also appears to increase (e.g. *Salamandra salamandra*) with age (references in Lofts, 1984). Each lobe consists of blindly ending saclike structures that are drained by efferent ductules. The latter runs into a larger collecting duct that links up with the vasa efferentia. In cross sections, the testicular units resemble seminiferous tubules but they differ from the latter in not having a permanent germinal epithelium. Therefore, they are termed 'lobules' (or 'ampullae'). The organization of lobules in relation to the main collecting duct varies with the species. In some urodeles (such as

Batrachoseps, Gyrinophilus and *Plethodon*) each testis has a centrally located collecting duct which runs longitudinally along the entire length of the gonad. In these testes, tubules are located radially around the collecting duct to which they are connected by their efferent ductules. Whereas, in *Necturus* and *Cryptobranchus* species, the longitudinal collecting duct runs superficially along the length of the testis near the mesorchium, the lobules are found radiating at this point so that in a transverse section they appear to be arranged like the spokes of an open fan. In the third category testicular lobules do not exhibit an orderly arrangement. Hence, there is no clearcut orientation of lobules towards the main collecting duct. In this type of testis, the lobules are more spherical and attached to the terminal apices of a highly branched efferent duct system (e.g. many salamanders). In transverse sections of such testes, the lobules appear more like sectioned tubules. In all the three types, the lobules are held together into a compact mass by connective tissue.

In urodeles also spermatogenesis is of cystic type. Cyst development begins when the primary spermatogonium becomes associated with several somatic cells. The primary spermatogonial divisions result in the production of secondary spermatogonial cysts, enclosed by the seminiferous lobules. Subsequent maturational divisions of these spermatogonia cause an increase in the size of the cysts by the development of spermatids that differentiate into spermatozoa. Throughout spermiogensis the developing spermatozoa remain enclosed by somatic cells and thus the integrity of the cysts remains intact. During the later stages the cysts become progressively distended and finally rupture, resulting in the separation of spermatozoa from the somatic cells. The free sperms reach the main collecting duct via lumen of the lobules. It is to be noted that once the cysts enter into the spermatogenetic process the germ cells present in each of these structures undergo the various developmental stages synchronously in a given seminiferous lobule. Therefore, each lobule will, in successive periods, contain in its maturing portion, spermatocysts, spermatids, and finally spermatozoa. The maturing portion will never at any one time contain all, or a combination of, these stages as in the seminiferous tubules of the amniotes (Pudney *et al.*, 1985).

The urodele testes unlike that of anurans, show a marked zonation in the distribution of the germ cell population along their length. Spermatogonial cysts are restricted to the cephalic region, and are succeeded by zones of increasing maturity as one progresses caudally. In species where the testes consist of many lobes, a similar gradient of development is seen in each of the individual lobes. The testicular areas occupied by the successive zones of spermatogenesis, spermatocytes, spermatids and mature spermatozoa, vary with the reproductive cycle, and can often be distinguished by the naked eye due to changes in the colour or opacity.

As in anurans, the Sertoli cells of urodeles are derived from the flattened, fibroblast-like follicle cells that surround the primary spermatogonia. During meiotic and maturation phases of spermatogenesis, these cells transform into conspicuous Sertoli cells with the necessary characteristics of steroid producing cells. Following spermiation, the Sertoli cells become heavily lipoidal and gradually degenerate as in anurans.

In most urodele males, spermatophores are produced. Whereas, in two primitive families, the Hynobiidae and Cryptobranchidae and some species of Sireniidae spermatozoa are discharged without being organized into a spermatophore. The form of the spermatophore varies from species to species. After the transmission of spermatophores into females, they are stored in the spermathecae located in the roof of the cloaca. Here, they are stored until ovulation. The duration of storage varies greatly from species to species, from a period of several months to as long as 2.5 years (e.g. *S. salamandra*).

As in certain teleosts there has been considerble confusion with regard to the occurrence and distribution of interstitial Leydig cells in the testes of urodeles. Early studies (references in Lofts, 1984) involving histological and histochemical techniques indicated that there exists no glandular steroidogenic tissue in the interlobular tissue of the urodele testis, instead it appeared that fibroblast-like cells of the lobule wall become steroidogenic in the testicular zone associated with spermiating and discharging germinal cysts. These cells were described as "lobule boundary cells" as in some teleosts. They are also termed "pericystic cells". The testicular zone containing glandular tissue appeared yellowish orange due to accumulation of lipids. The yellow zones were compared to mammalian corpus luteum by calling them a "varitable corps joune testiculaire". Recent electron microscopic studies have shown that in urodele testis as in some teleost fishes (see section IV) the so called lobule boundary cells are indicated as Sertoli cells. It is now clear that true Leydig cells also exist in the interlobular spaces and periodically undergo differentiation into a glandular tissue to assume steroid secretory role. The glandular tissue is not uniformly distributed throughout the testis, unlike in the anurans, but instead confined to the most spermatogenetically mature region where the lobules are discharging their seminal content. After discharge of lobular contents they become compressed into the glandular tissue which subsequently degenerate and become resorbed.

(b) Apoda

Studies on the histoarchitecture of caecilian testes are scanty (Seshachar, 1941; Wake, 1995). Each testis consists of variable number of oval shaped separate lobes. The lobes are connected to each other by a central duct which is associated with a series of transverse ducts that carry sperms to the kidney tubules. All the stages of spermatogenesis are found in each lobule. Hence, there is no caudo-cranial sequence so often seen in the urodeles. The Leydig cells are found in the interlobular areas.

(c) Anura

The testes are paired, compact and ovoid structures situated in close proximity to the kidneys. They are attached to the dorsal body wall by a short mesorchium through which run the vasa efferentia connecting the seminiferous tubules and the modified nephric tubules running to the Wolffian ducts. The testes are surrounded by an elastic fibrous coat. They consist of convoluted seminiferous tubules lined by a permanent germinal epithelium whose composition may vary seasonally or in accordance with the reproductive phase.

As in other anamniotes, spermatogenesis is of the cystic type i.e. germ cell proliferation occurs in coordinated clusters, each cluster being enclosed within a membranous capsule for much of its development. The capsule and its contained germ cells is termed a germinal cyst or nest or follicle. The cells present in a cyst are in the same stage of development and are derived from a single spermatogonium. In a cross section of a seminiferous tubule cell nests in different stages of spermatogenesis are encountered. Also, the sectioned testes display a remarkable uniformity in gametogenetic composition throughout the whole gonad (unlike in urodeles).

In the medullary cords of embryonic males, the primordial germ cells are surrounded by flattend cells of mesenchymal origin and in adult testis these are represented by the cells forming the wall of the germinal cysts. The nuclei of the so-called follicle cells are small, darkly stained, and generally crescent shaped. Each primary spermatogonium is completely enclosed by the follicle cells. The latter are placed in close proximity with the large and irregularly shaped nucleus of the primary spermatogonia. These follicle cells which later mature into secretory Sertoli cells do not possess ultrastructural features characteristic of steroidogenic cells in the intitial stages. However, with the progression of spermatogenesis these cells hypertrophy, acquire smooth endoplasmic reticulum, spherical mitochondria with tubular cristae, lipid dropletes and enzymes involved in steroidogenesis (Lofts, 1984; Saidapur, 1989). Following spermiation, Sertoli cells are detached from the basement membrane and are seen freely floating in the lumen of seminiferous tubules. They eventually degenerate.

The seminiferous tubules as in amniotes, are interspersed in an interstitial tissue composed of connective tissue, blood capillaries and large conspicuous Leydig cells. The Leydig cells exhibit pronounced or less drastic seasonal changes depending upon the species. Furthermore, Leydig cell and the spermatogenetic activities may or may not run parallel depending upon the species and its geographical location (Saidapur, 1989).

6. Reptiles and Birds

The basic architecture of reptilian and avian testis is similar to that in mammals (Lofts and Murton, 1973; Fox, 1977; Lofts and Massa, 1980; van Tienhoven, 1983; Licht, 1984; Lofts, 1987; de Reiver *et al.*, 1990). It consists of convoluted seminiferous tubules with permanent germinal epithelium. As germ cells mature, they move away from the basement membrane towards the central lumen of the tubule. The germ cells are not enclosed into germinal cysts. The spermatogenesis in reptiles and birds is thus noncystic type. However, the germ cells of a given generation produced from the same spermatogonium are interconnected by cyctoplasmic bridges to form heterkaryons.

It is not known whether cellular associations, and seminiferous epithelial cycles as seen in mammals also occur in the reptilian testes (Roosen-Runge, 1977; van Tienhoven, 1983) due to paucity of studies. On the other hand, several attempts have been made to understand the organization of the seminiferous epithelium in birds. Typical cellular associations have been defined in the mallard duck, the Japanese quail and the guinea fowl but not in the cockerel. In mallard duck, quail and guinea fowl eight stages of seminiferous epithelial cell cycle associated with ten developmental stages of spermatids have been described (Clermont, 1958; Yamamoto *et al.*, 1967). In cross sections of bird seminiferous tubules more than one cellular associations are often seen lying side by side. Since these associations are restricted to small areas of seminiferous tubules and due to overlapping between adjacent associations atypical associations are also formed. Furthermore, in the quail yet another type of atypical association has been described. This type is characterised by the presence of germ cells not belonging to neighbouring typical associations (Yamamoto *et al.*, 1967). Therefore, it was considered doubtful whether seminiferous epithelial cycles could be adequately defined in birds as that in mammalian testis (de Revier *et al.*, 1990). However, a recent study on the quail using semithin epoxyresin embedded sections of isolated seminiferous tubules has revealed a well defined seminiferous epithelial cycle with ten distinct stages (Lin *et al.*, 1990). The study thus showed that the various germ cells in the bird testis also form definite cellular associations, fundamentally, similar to the pattern seen in the mammalian testis.

Distinct Sertoli cells border the inner side of the basement membrane of the seminiferous tubules in both reptiles and birds and various stages of spermatogenesis are arranged in specific association with them. Electron microscopy has shown that the processes/folds from the Sertoli cells envelop the developing germ cells interspersed between them. It is suggested that this arrangement between the germ cells and Sertoli cell might serve to maintain integration of spermatogenetic processes. Further, the Sertoli cells in both these classes of vertebrates have ultrastructural, histochemical and biochemical features characteristic of steroid synthesizing/converting capacity (Guraya, 1976; Bhujle *et al.*, 1979a, b; Lofts and Massa, 1980; van Tienhoven, 1983).

Distinct Leydig cells are found in the interstitial area between the seminiferous tubules amidst closely packed connective tissue, blood and lymph spaces. However, the Leydig cells in some species of lizards, e.g.

Cnemidophorus (17 species), *Ameiva* and *Agama*, are concentrated beneath the tunica albuginea like a circumtesticular tunic (van Tienhoven, 1983; Lofts, 1987). In some bird species (e.g. barn owl), a large number of melanocytes are also encountered in the interstices imparting dark or gray colour to the postbreeding regressed testis. The reptilian and avian Leydig cells exhibit characteristics of steroid producing cells (Bhujle *et al.*, 1979a, b; Shanbhag and Nadkarni, 1982).

A close correlation between Leydig cell activity, plasma androgen levels and spermatogenetic activity is observed in most reptiles and in all the birds studied so far. However, in reptilian species exhibiting postnuptial pattern of spermatogensis, an uncoupling between peak spermatogenesis and Leydig cell activity is observed (e.g. most turtles and a few snakes from temperate regions; Licht, 1984; Lofts, 1987).

Rete testis is present in reptiles and birds (Lofts and Murton, 1973; Fox, 1977; Norris, 1987). It is derived from modified mesonephric tubules and connects seminiferous tubules to vasa efferentia which open into the epididymis.

7. Discussion

Although, studies on the testis of protochordates are scanty (Grier, 1992) in order to trace the evolution of testicular histoarchitecture in vertebrates, it is essential to succinctly consider the organization of testes in these primitive chordates. In *Branchiostoma* (*Amphioxus*) *lanceolatum* a metameric cephalochordate, a basement membrane delineates paired ovoidal, ventrolateral testes within each trunk metamere (Callard, 1991). Whereas, in the hermaphroditic, sessile urochordate *Styela plicata*, which lacks a metameric body structure, the basement membrane delineates a polylobed testis along the length of the body wall in association with the ovary (Grier, 1992). Although, gross testicular morphology differs between the Cephalochordata and Urochordata, both taxa possess clones of synchronously developing germ cells along the testicular periphery. Also, as in other chordates, germ cells within each clone are connected by intercellular bridges. The germ cells are not enclosed within spermatocysts unlike in the agnathans and elasmobranchs. Moreover, Sertoli cell processes grow between instead of surrounding the clones of germ cells holding these cells at the testicular periphery. Mature sperms are free within testicular lumen, often in direct juxtaposition to the surrounding cells. In the amphioxus, each ventrolateral testis was considered to represent a spermatocyst (Callard, 1991). However, Grier (1992) does not agree with this view based on his observations in *Branchiostoma caribbaeum*. The developing testis of *B. carribbaeum* contains sperm which had developed from preexisting clones, within the lumen and numerous interconnected, maturing germ cell clones at the periphery. Sertoli cell processes do not enclose single germ cell clones as in the fish and amphibians. Thus, testis structure in protochordates does not conform to the literal criteria for a spermatocyst. Therefore, testicular structure in protochordates is considered as "prespermatocystic" (Grier, 1992). Apparently, prespermatocystic type of testis represents a primitive organization that originated early in the chordate evolution. Later, this might have given rise to the polyspermatocystic testis found in cyclostomes and elasmobranchs.

Compartmentalization of the chordate testis into germinal and interstitial compartments by a basement membrane is a constancy that has survived half a billion years of evolution. Throughout the chordates, only two cell types reside within the testicular germinal compartment—the somatic Sertoli cells, and the germ cells that give rise to sperms. The constancy of the germinal epithelium and conserved Sertoli cell functions occur in all groups of chordates.

In all chordate testes, the germinal epithelium has a tripartite structure consisting of the acellular basement membrane, the Sertoli cells, and the germ cells. The basement membrane, which sequesters the Sertoli and germ cells within the germinal compartment, is of great importance in contributing to the morphological variations and possible evolutionary changes in the testicular structure in chordates. An epithelium, by definition covers or lines body surfaces and rests upon a basement membrane. The germinal epithelium lines the lumen of a germinal compartments, whether spermatocyst, lobule or tubule. However, the germinal epithelium of the testis is unique among epithelia since one of its cell types, the germ cell, divides meiotically to produce haploid sperms.

In anamniote vertebrates the spermatogenic unit of testis function is the spermatocyst, a "unit of evolution" in the chordate testis. In the cyclostomes and elasmobranchs, a spermatocyst is a spherical structure consisting of a basement membrane that surrounds a population of Sertoli cells and isogenetic germ cells, derived from a single spermatogonium. Since the testes in these taxa contain numerous spermatocysts they have been called the "polyspermatocystic". In polyspermatocystic type of testes primary spermatogonia do not form part of a tripartite germinal epithelium. A 'mesenteric type of testis' found in the hagfish, *E. stouti* is believed to the most primitive type of polyspermatocystic testis. Regions connecting spermatocysts are interconnected by tissue bridges that resemble the testicular mesentery. Every spermatocyst develops from a single spermatogonium and its associated Sertoli cells, but development is asynchronous. Only after the establishment of a tripartite structure of the germinal epithelium/ spermatocyst meiosis occurs. A second type of polyspermatocystic testis occurs in the lamprey *P. marinus* because all of the spermatocysts develop in synchrony. The lamprey testes are distinct organs as in higher chordates. Perhaps the "mesenteric" nature of the testis in *E. stouti* represents a "primitive" trait when compared to all other chordates, a gonadal characteristic that was evolutionarily lost within the agnatha. A third type, and the most evolutionarily advanced polyspermatocystic testis, occurs in Chondrichthyes which exhibits modifications that include a duct system and Sertoli cell and germ cell relationships that probably co-evolved along with internal fertilization.

In Osteichthyes and amphibia, the germinal epithelium resides within lobules or tubules. While spermatocysts of agnathans and Chondrichthyes are completely encompassed by a basement membrane, those of fish and amphibians are in contact with the basement membrane along one side only. Thus, the pattern of basement membrane formation seems to have played a key role in the evolution of testis types in the Chordata. Grier (1992) proposed the "Extra cellular Matrix Hypothesis" (EMH) to explain the possible mechanism of evolution of lobules and tubules from spherical spermatocysts. According to EMH, because Sertoli cells and primary spermatogonia in Agnatha and Chondrichthyes are not organized as a tripartite germinal epithelium, a major advance in the evolution of testis structure in the Vertebrata may have been the time-shifting of basement membrane gene activation. The earlier activation of these genes would result in Sertoli cells and their associated primary spermatogonia becoming permanently separated from the interstitium. That the extracellular matrix, the basement membrane, affects Sertoli cell organization and the structure of the germinal compartment in the chordate testis is now evident (Grier, 1992, 1993).

The evolution of elongated structures, lobules or tubules, from a polyspermatocystic testis could have involved a sequence of correlated events beginning with (1) the time-shifting of genes coding for basement membrane components, (2) the uncoupling of mitotic divisions so that Sertoli cells and germ cells would no longer be limited to a specific number, as occurs in elasmobranch spermatocysts, and (3) the evolution of Sertoli cell bipotentiality resulting in the formation of a duct system.

In elasmobranchs, germ cells and Sertoli cells stop dividing mitotically after the formation of spermtocysts. Then, germ cells enter meiosis. However, if Sertoli cells and germ cells could continue to multiply, the spherical spermatocysts, which are connected to a duct system can form elongated structures (lobules or tublues) due to lateral pressure exerted by the juxtaposed germinal compartments whose total volume increases progressively with mitotic division. Spermatocysts could then reside inside these lobules or tubules. The basement membrane would then define these larger structures instead of being restricted to surrounding individual spermatocysts as in the living Agnatha and Chondrichthyes. The synchrony between the synthesis of extracellular matrix and the mechanisms controlling Sertoli cell/germ cell divisions is thus an important step in the evolution of chordate testis.

In the testis of amniotes, tubules form loops beneath the tunica albuginea as do lobules in anamniotes. Conceptually, tubular type of testis can be derived from one that is lobular by joining blindly ending lobule termini, an event reflected in human testis development. The primary sex cords grow towards and terminate beneath the tunica albuginea. The termini of these join to form loops giving rise to testicular tubules. The evolution of tubular testis is thus reflected in the embryonic events that involve the formation of loops by the primary sex cords.

The extracellular matrix hypothesis proposed by Grier (1992) may be an oversimplification of the evolutionary process explaining the evolution of diverse testis types in higher chordates. Yet, it provides useful framework for progress in understanding the possible mechanism of evolution of testis structure.

Paradoxically, the evolution of germinal compartments in the chordate testis does not fit into a logical progression of types. For instance, one would expect that lower fish forms would possess lobular testis. Whereas lobular testes are found in perciform and atherinomorph fish. In most lower fish spermatocysts occur along the walls of germinal compartments that form anastomosing networks that often run parallel to, but do not terminate at, the testis periphery (Grier, 1993). Consequently, germinal compartments in the lower fish do not fulfil the criteria defined for a lobular testis. Since these compartments form loops at the testis periphery and they conform to the definition of a 'tubular' testis. However, unlike the tubules in the more advanced amniote testis, tubules in these lower fish form anastomosing network and such testes are termed "anastomosing tubular testis" to distinguish from the tubular testes of the Australian lungfish and amniotes (Grier, 1993). Such anastomosing tubular testes are possibly restriced to the lower fish.

Lofts (1984) categorised the amphibian testis into lobular (urodelan) and tubular (anuran) types because the latter possesses a permanent germinal epithelium, whereas the former does not. However, Grier (1992) considers both urodelan and anuran testis to be lobular types. In our opinion, more studies involving diverse species belonging to different genera are needed to resolve the true nature of anuran testis.

The testis of reptiles and birds are typically tubular with a permanent germinal epithelium. Testes in these taxa do not produce spermatocysts since Sertoli cells send processes between individual germ cells. The amniote tubular testis has been interpreted as being "postspermatocystic" in a progression of complexity in testicular diversity from the lower to the higher chordates (Grier, 1993).

The Sertoli cells are important components of germinal epithelia in all chordates. It is noteworthy that association between the germ cells and Sertoli cells and between neighbouring Sertoli cells has also progressively evolved during the course of evolution of chordate testis. It is the variation in the nature of association between the germ cells and Sertoli cells that has given rise to prespermatocystic (protochordates), spermatocystic (anamniote chordates) and postspermatocystic (amniote chordate) type of testes. Sertoli cells are responsible for the establishment of a "Sertoli cell barrier" for the first time in fishes. The barrier does not include the participation of blood vessel endothelial or boundary cells. Such barriers are not found in

Protochordata, Agantha and Chondrichthyes, due to the absence of tight junctions between the adjacent Sertoli cells. Whereas, in amniotes, well defined blood testis barriers are formed by the Sertoli cells in association with the blood vessel endothelial cells or boundary cells.

The phagocytic nature of the Sertoli cells is a mechanism evolved for removal of degenerating germ cells including residual sperm cells following the reproductive season in seasonal breeders, and residual bodies. This is a highly conserved phenomenon seen throughout the chordates.

Another important aspect of the Sertoli cell is its bipotentiality which has played a key role in the evolution of chordate testis. The first appearance of duct system in Chondrichthyes is believed to be the result of evolution of bipotentiality in Sertoli cells. Sertoli cell bipotentiality also occurs in fishes and amphibians. It appears that the putative Sertoli cells that remain in contact with spermatogonia, differentiate as Sertoli cells. Others, divide faster than spermatogonia and become dissociated from them form, cords of cells and differentiate as duct cells. The evolution of Sertoli cell bipotentiality thus must have preceeded the evolution of lobules and tubules because ducts are necessary to provide site for attachment from which germinal compartment elongation can produce a lobule or tubule. It is apparent that so far known functional morphology strongly suggests homology of Sertoli cells between all chordates.

In summary, the 'spermatocyst' is a unit of testicular evolution in chordates. Presumably, the prespermatocystic type of testis found in protochordates appears to have given rise to 'polyspermatocystic' type of testes encountered in the extant lower vertebrates, the cyclostomes and cartilagenous fishes. Cystic pattern of spermatogenesis is a rule in anamniote chordates. Lobular testis type found in many bony fishes and amphibians is a forerunner of tubular testes of amniotes. Evolution of a permanent tripartite germinal and, interstitial compartments could be explained at present by the extracellular matrix hypothesis of Grier (1992). The intricate association between germ cells and Sertoli cells and, Sertoli-Sertoli cells has become progressively complex during the evolution of testis in chordates. This also accounts for diversities in testicular organization among chordates. Yet, there occurs constancy of germinal epithelium and conservation of Sertoli cell functions vis-a-vis homology throughout the vertebrates.

Acknowledgement

The preparation of this paper was supported by a grant (SAP-II) from UGC, New Delhi.

References

Bhujle, B.V., Nadkarni, V.B. and Rao, M.A. (1979a). Gen. Comp. Endocrinol. **38:** 153–161.

Bhujle, B.V., Nadkarni, V.B. and Rao, M.A. (1979b). Histochem. J. **11:** 253–255.
Callard, G.V. (1991). In: Vetebrate Endocrinology: Fundamentals and Biomedical Implications, Vol 4, Part A (P.K.T. Pang and M. Schreibman, Eds.), pp. 303–341. Academic Press, New York.
Callard, I.P., Callard, G.V., Lance, V., Bolaffi, J.L. and Rosset, J.S. (1978). Biol. Reprod. **18:** 16–43.
Clermont,Y. (1958). Arch. Anat. Microsc. Morphol. Exp. **47:** 47–66.
Dood, J.M. (1972). Gen. Comp. Endocrinol. *Suppl.* **3:** 675–687.
Dodd, J.M. (1983). In: Fish Physiology, Vol. IXA (W.S. Hoar, D.J. Randall and E.M. Donaldson, Eds.), pp. 31–95. Academic press, New York.
Dodd, J.M. and Sumpter, J.P. (1984). In: Marshall's Physiology of Reproduction, Vol. 1 (G.E. Lamming, Ed.), pp. 1–126. Churchill Livingstone, London.
Dym, M. and Fawcett, D.W. (1970). Biol. Reprod. **3:** 308–326.
Fox, H. (1977). In: Biology of Reptilia,Vol. 6 (C. Gans and T.S. Parson, Eds.), pp. 1–122. Academic press, New York.
Gorbman, A. (1983). In: Fish Physiology, Vol. IXA (W.S. Hoar, D.J. Randall and E.M. Donaldson, Eds.), pp. 1–29. Academic Press, New york.
Gresik, E.W. (1973). Gen. Comp. Endocrinol. **21:** 210–213.
Grier, H.J. (1981). Am. Zool. **21:** 345–357.
Grier, H.J. (1992). J. Exp. Zool. **261:** 151–160.
Grier, H.J. (1993). In: The Sertoli Cell. (L.D. Russell and M.D. Griswold, Eds.), pp. 704–739, Cache River Press, Florida, USA.
Grier, H.J., Horner, J. and Mahesh, V.B. (1980a). Trans. Am. Microsc. Soc. **99:** 268–276.
Grier, H.J., Linton, J.R., Leatherland, J.F. and de Vlaming, V.L. (1980b). Am. J. Anat. **159:** 331–345.
Guraya, S.S. (1976). Int. Rev. Cytol. **47:** 99–136.
Hoar, W.S. (1969). In: Fish Physiology, Vol. 3 (W.S. Hoar and D.J. Randall, Eds.), pp. 1–72. Academic press, New York.
Hoar, W.S. and Nagahama Y. (1978). Ann. Bio. Anim. Biochem. Biophys. **18:** 893–898.
Hurk, R. van den (1973). Proc. Ken. Ned. Akad. Wet. Ser. C, **76:** 270–279.
Hurk, R. van den, Peute, J. and Vermeji, J.A.J. (1978). Cell Tissue Res. **186:** 309–325.
Larsen, L.O. (1973). Development in adult, freshwater river lampreys and its hormonal control. Starvation, sexual maturation and natural death. Thesis, University of Copenhagen.
Licht, P. (1984). In: Marshall's Physiology of Reproduction, Vol. 1 (G.E. Lamming, Ed.), pp. 206–282. Churchill Livingstone, London.
Lin, M., Jones, R.C. and Blackshaw, A.W. (1990). J. Reprod. Fertil. **88:** 481–490.
Lofts, B. (1984). In: Marshall's Physiology of Reproduction, Vol. 1 (G.E. Lamming, Ed.), pp. 127–205. Churchill Livingstone, London.
Lofts, B. (1987). In: Hormones and Reproduction in Fishes, Amphibians and Reptiles. (D.O. Norris and R.E. Jones, Eds.), pp. 283–325. Plenum Press, New York.
Lofts, B. and Bern, H.A. (1972). In: Perspectives in Endocrinology. (E.J.W. Barrington and C.B. Jorgensen, Eds.), pp. 37–125. Academic Press, New York.
Lofts, B. and Murton, R.K. (1973). In: Avian Biology, Vol. 3 (D.S. Farner and J.R. King, Eds.), pp. 1–109. Academic Press, New York.
Lofts, B. and Massa, R. (1980). In: Avian Endocrinology. (A. Epple and M.H. Steteson, Eds.), pp. 413–437. Academic Press, New York.
Marshall, A.J. and Lofts, B. (1956). Nature (Lond.), **177:** 704–705.
Nagahama, Y. (1983). In: Fish Physiology, Vol. IXA (W.S. Hoar, D.J. Randall and E.M. Donaldson, Eds.), pp. 223–275. Academic Press, New York.
Nagahama, Y. (1994). Int. J. Dev. Biol. **38:** 217–229.

Norris, D.O. (1987). In: Hormones and Reproduction in Fishes, Amphibians and Reptiles. (D.O. Norris and R.E. Jones, Eds.), pp. 327–354. Plenum Press, New York.

Pratt, H.L. (1988). Copeia, **3**: 719–729.

Pudney, J., Canick, J.A. and Callard, G.V. (1985). Brimleyana, **10:** 53–74.

Revier de, M.H., Courtens, J.L., Courat, M. and de Reveier, M. (1990). In: Marshall's Physiology of Reproduction, Vol. 2 (G.E. Lamming, Ed.) pp. 106–182. Churchill Livingstone, London.

Rossen-Runge, E.C. (1977). The Process of Spermatogenesis in Animals. pp. 102–132. Cambridge Univ. Press, London.

Saidapur, S.K. (1989). In : Reproductive Cycles of Indian Vertebrates. (S.K. Saidapur, Ed.), pp. 166–224. Allied Publ. Ltd., New Delhi.

Seshachar, B.R. (1941) Curr. Sci. **10:** 281–284.

Shanbhag, B.A. and Nadkarni, V.B. (1982). Curr. Sci. **51:** 407–409.

Stanley, H.P. (1966). Z. Zellforsch. **75:** 453–468.

Tienhoven van, A. (1983). Reproductive Physiology of Vertebrates. pp. 137–169. Cornell Univ. Press, London.

Tsuneki, K. and Gorbman, A. (1977). Acta Zool. (Stockholm), **58:** 17–25

Wake, M.H. (1995). J. Morph. **126:** 291–332.

Yamamoto, S., Tamate, H. and Itikawa, O. (1967). Tohoku. J. Agric. Res. **18:** 27–37.

Comparative Endocrinology and Reproduction
K.P. Joy, A. Krishna and C. Haldar (Eds)

12. Structure of Mammalian Epididymis

N.H. Gopal Dutt
Department of Studies in Zoology, University of Mysore,
Manasagangotri, Mysore-570 006, India

Introduction

The epididymis of the mammalian male duct system has attracted the attention of investigators because of its pivotal role in sperm maturation and as one as an extragonadal site which may help in the control of fertility without impairing libido and potentia. Further, the regional, anatomical and functional differences recorded in several mammalian species formed the basis for further studies on the subject (Hinton, 1980; Brooks, 1983; Abe *et al.*, 1983, 1984; Orgebin-Crist, 1986, Gopal Dutt and Suresh, 1988; Koe and Lee, 1993; Yeung *et al.*, 1994; Hermo *et al.*, 1994). A wealth of new information has accrued in recent years from scientists of different disciplines. (Turner and Reich, 1987; Goyal and Williams, 1991; Vieurula *et al.*, 1992; Parkkilla *et al.*, 1993; Tomomasa *et al.*, 1994). This necessiates periodic reviews and critical evaluation of the data gathered. The purpose of this chapter is to provide a general account on the structural aspects of epididymis in selected mammalian species.

Pre- and Postnatal Development

Embryologically, the mammalian epididymis is appropriated from the urinary system, the mesonephros or opisthonephros, to serve exclusively as an organ of reproduction (Romer, 1960). The mesonephros looses its urinary function, its peritoneal funnels and renal corpuscles degenerate, and its tubules which convey the sperm are called epididymal tubules and that part of the mesonephric duct involved in the epididymis is known as epididymal duct (Hyman, 1962). Complete exclusion of renal function in this manner provides the most advanced condition in mammals. Earlier investigators, who examined only scrotal mammals considered that those derivatives of the mesonephric duct which are adjacent to the testis as the epididymis and those which are distal to the testis as the vas deferens (Oudemans, 1982; Weber, 1898). In the primitive non-scrotal mammal, echidna (Djakiew and Jones, 1981), however, only the head (caput) of the epididymis is adjacent to the testis, the tail (cauda) and ductus deferens are separate and distal. Glover (1968) proposed that the proximal and distal regions of the derivatives of the testiconda mesonephric duct are homologous to the epididymis of scrotal mammals.

The cellular differentiation of epididymis occurs in foetal life (Alexander, 1972) and following dedifferentiation adult characteristics are acquired during prepubertal and pubertal life (Sun and Flickinger, 1979). In adult the epididymis still retains the features of nephric duct particularly with regard to the absorption of sodium and water (Brooks, 1983). Sun and Flickinger (1979) proposed a scheme of cellular differentiation, similar in some details to the 'two cell-option' model described in other developmental systems (Kauffman, 1973; Gardner and Rossant, 1976).

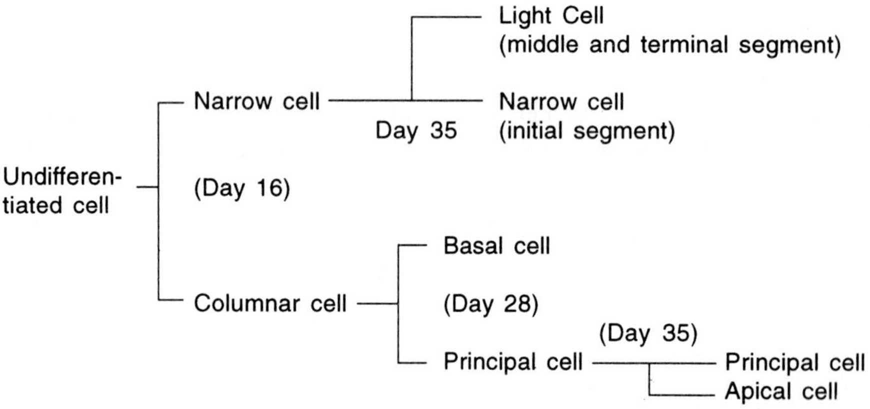

Proposed model of cellular differentiation in the rat epididymal epithelium (Sun and Flickinger, 1979)

According to this scheme, at a certain point, differentiating cells have the option of taking one of the two developmental pathways as illustrated above. In the rat, the postnatal development of the epididymis can be divided into three phases: (1) the undifferentiated period from 1 to 15 days; (2) the period of cellular differentiation from days 16 to 44; and (3) the period of growth extending from days 45 to 3 months. From birth up to day 5, the epididymis remains quiescent. Following cellular differentiation, narrow cells appear on day 16, the principal cells and basal cells on day 28. The narrow cells persist only in the initial segment, while in the middle and terminal segments they disappear by day 35, when light cells appear. Evidence suggests that the narrow cells are the precursors of light cells in the middle and terminal segments. The differentiation of the epididymal epithelium completes when the halo cells appear. Subsequently, Sun and Flickinger (1980) have studied by autoradiography the proliferating activity of cells in rat epididymis during postnatal development. Results have shown that the columnar cells appear to be precursors to the principal and basal cells. As expected, the terminally differentiated light cells are not mitotically active, where as columnar cells (undifferentiated) exhibit high level of proliferative activity. Cytologically the epididymis attains advanced stage of development even before the sperm arrival. It is noteworthy that in rat, rabbit and rehesus monkey (Leeson and Leeson, 1964; Alexander, 1972; Setty and Jehan, 1977), morphological differentiation of the caput epididymides preceds that of the cauda epididymides; the caput histological differentiation is completed several days before sperm entry, while in the cauda this occurs long after spermatozoa reach this segment (Setty and Jehan, 1977). Active proliferation of cells continues in the sexually mature phase of reproduction in the wild mouse (Suresh and Gopal Dutt, 1986).

Gross Anatomy

The epididymis is composed of a single, long, highly coiled duct closely applied to the testis and embedded in a variable amount of adipose tissue of the epididymal fat pad. Based on shape and position on the testis, the epididymis is conventionally divided into three regions viz., caput, corpus and cauda or tail. According to Glover and Nicander (1971) this terminology is misleading since it relates to the gross morphology of the epididymis but not to its microspcopic structure. They reported that the mammalian epididymis, histologically and functionally, can be divided into initial and middle segments where sperm maturation takes place and a terminal segment which subserves the function of sperm storage. These segments do not necessarily correspond to the traditional terms of caput, corpus and cauda regions described in several species.

Zonation

Variation in the distribution and morphology of cell types have been recognised all along the epididymal duct in a variety of mammals (Hamilton, 1975). Since the epididymal epithelium lacks uniformity in structure, zones have been recognised in a variety of mammalian species. Two histologically distinct zones have been located in the primitive mammal, echidna (Djakiew and Jones, 1981); five in the mouse (Takano, 1980); six in rat (Reid and Cleland, 1957) and tammar (Jones *et al.*, 1984), seven in guinea-pig (Hoffer and Greenberg, 1978); and eight in rabbit and human (Holstein, 1969). In addition various subzones have also been identified in the epididymis of rat (Reid and Cleland, 1957). Based on morphological differences, the goat epididymis is divided into five regions; region I and II and the proximal part of region III constitutes the head; the distal part of region III and region IV, the body; and region V, the tail (Goyal and Williams, 1991). In some species the epididymal zones may overlap with two functionally divergent zones. The zone VI of the epididymis of Siberian hamster described by Nagy *et al.* (1982) overlaps with two functionally distinct zones; one located in the distal portion of the middle segment and the other in the proximal portion of the terminal segment according to the divisions described by Glover and Nicander (1971).

The zones I, II and III constitute the caput, the zone IV forms the body and zone V, the tail in the epididymis of laboratory mouse (Takano, 1980).

Apart from these zones, Suresh and Gopal Dutt (1988) have described additionally zone VI in the tail of the epididymis of the wild mouse (Fig. 1). This zone, though generally resembles the ductus deferens in histological details, has been considered as a part of the epididymal duct due to its location in the tail (cauda) region. Nevertheless, there is a close resemblance between the laboratory mouse and wild mouse with regard to the distribution of PAS—positive substances and alkaline phosphatases (Suresh, 1987). Of the seven morphologically and histologically distinct zones identified in the guinea pig epididymis, the narrow zone I, which runs all along the entire length of the testis comprises efferent tubules.The narrow contour of zone I then gradually expands into a large globular region including zones II to VII of the epididymis. In most of the mammals, the caput epididymides corresponding to the zone I of the guinea pig, is club shaped, while the corpus and cauda are narrow and globose, respectively. In the Siberian hamster two distinct tubule populations are found in the zone VI; the larger tubules are laden with sperm and seminal fluid and the smaller tubules, which arise as diverticula of the larger ones, are filled with degenerating cells and cellular debris. These observations prompted Nagy *et al.* (1982) to conclude that the smaller tubules are specialised to serve as repository of dead cells and cell fragments. On the other hand, the most actively proliferating zone is only the zone III in the epididymis of golden Syrian hamster (Nagy and Edmonds, 1975). The human ductus epididymis is highly complex by the presence of multiple intra-epithelial cavities lined by principal and basal cells. The development of these pitted cavities are also androgen dependent (Nistal *et al.*, 1990a). Species specific regional specialisation is apparently a distinctive feature of mammalian epididymis.

Regional differences in relative proportions of the lumen, epithelial height and width of the musculature have been observed in different species. Some of these differences in the wild mouse are shown in Fig. 2 (Suresh and Gopal Dutt, 1988). There seems to be little variation in the morphological characteristics and histological features of the initial segment, while much variation is found in other segments. The initial segment in echidna, mouse, rat, hamster, guinea pig, cat, dog, rabbit, mole, pig, and bull is generally recognised by tall pseudostratified columnar epithelium, long densely arranged stereocilia in a tuft like form and relatively narrow tubules containing small number of sperms. (Nicander, 1957; Suzuki and Racey, 1976 and see Djakiew and Jones, 1981; Suresh, 1987). The tall epithelium of the initial segment is involved in fluid resorption (Nicander, 1970). The middle segment is characterised by slender, short and usually bent cilia, abundant pinocytotic vesicles and increase in the luminal spermatozoa (Glover and Nicander, 1971; Nicander and Glover, 1973). The terminal segment is a wide duct, lined by short epithelium with some pinocytotic vesicles, covered by thick muscular coat and its lumen contains high concentration of spermatozoa.

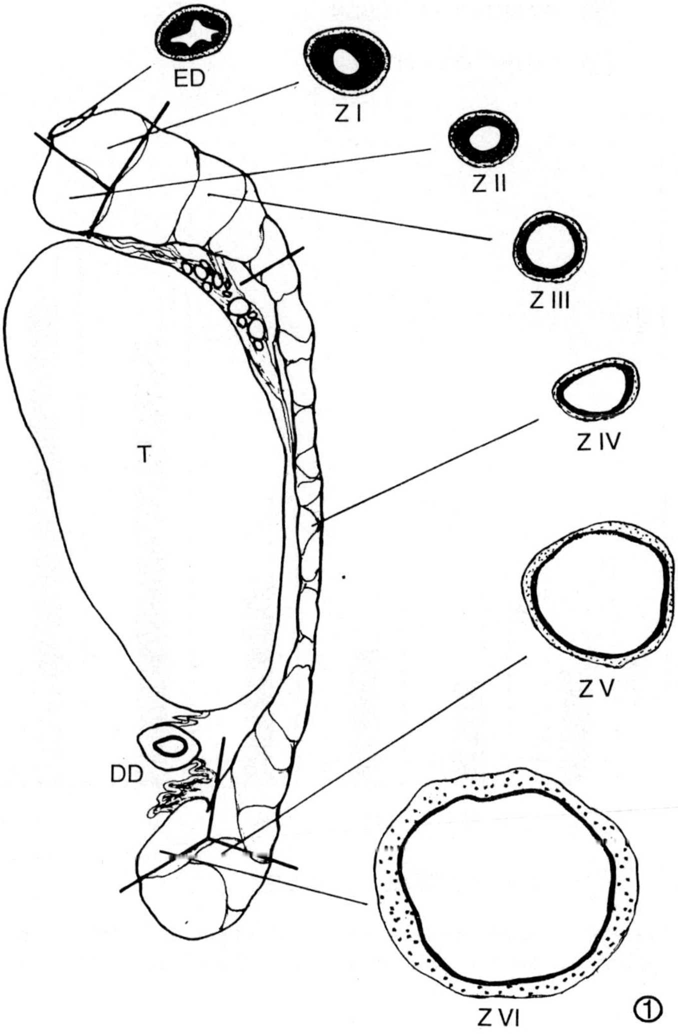

Fig. 1. Diagrammatic representation of testis-epididymis complex of the wild mouse (*M. platythrix*) depicting various zones and their relative proportions T: testis, Ed: efferent tubules, Z I to Z VI: zones of epididymis, DD: ductus deferens (from Suresh and Gopal Dutt, 1988).

Figures 3, 4 and 5 show histological details of the three principal divisions of the epididymis of the wild mouse.

Arterial Supply and Venous Drainage

The major blood supply to the epididymis is from the internal spermatic

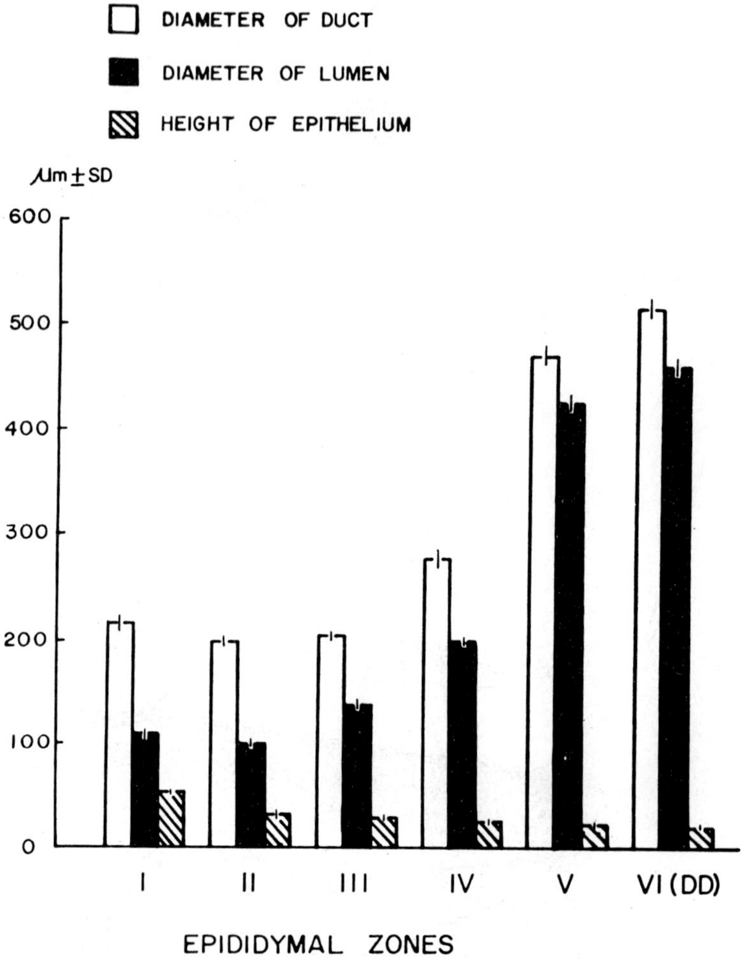

Fig. 2. Variation in the diameters of the duct, lumen and epithelial height of the epididymal zones of wild mouse (from Suresh and Gopal Dutt, 1988).

artery, which in sharp distinction from that of the testis especially in rat lacks convolutions. This branch divides into superior and inferior epididymal arteries whose distribution varies in different parts of the epididymis. The interductoral capillary network is tortuous and dense imparting reddish color to the caput in the rat but decrease in complexity distally in the epididymis (Kormano *et al.*, 1973; Hamilton, 1975). There is evidence that in the rat the rate of blood flow through this region is higher than elsewhere along the epididymis (Setchell *et al.*, 1964). In the goat epididymis, the head and tail regions are profoundly supplied with branches of epididymal arteries than in the body (Dhingra, 1978). Arterial and venous branches follow the pattern basically similar to those in the mouse, rat and rabbit (Chubb and Desjardins, 1982). These arteries anastomose all along the

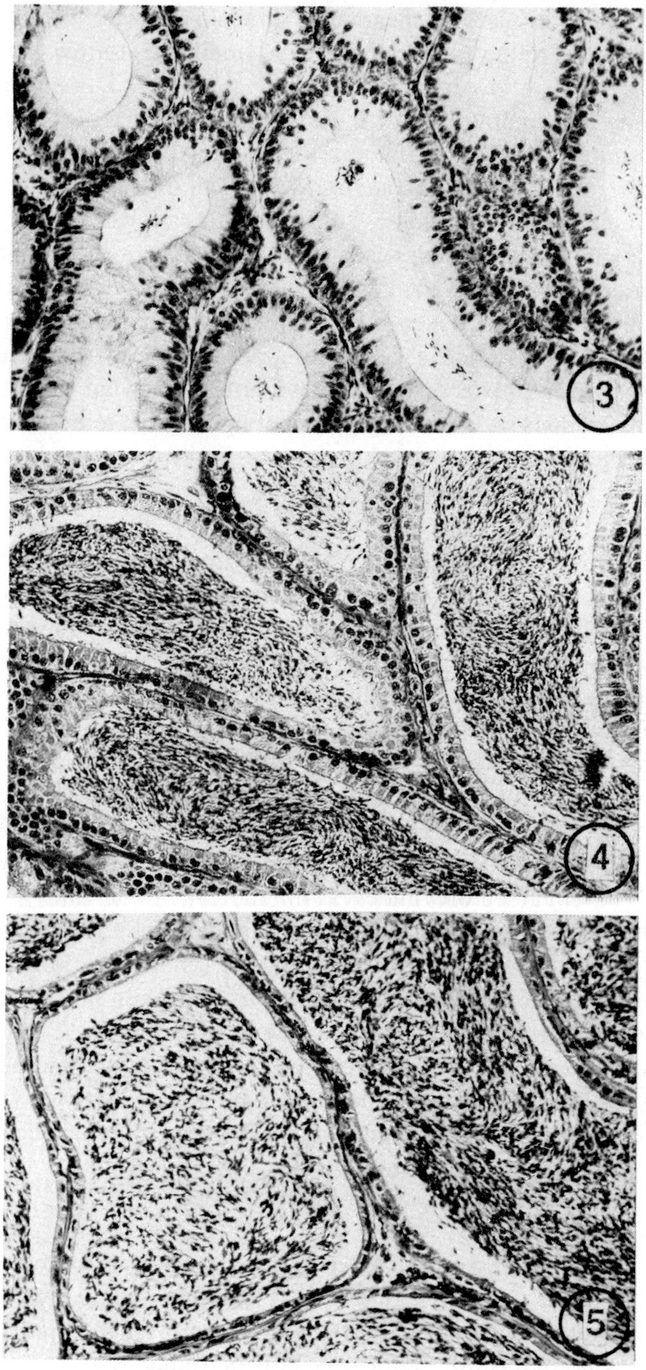

Figs. 3, 4 and 5. **Histological profile characteristic of caput, corpus and cauda of epididymides of the wild mouse × 200 (from Suresh and Gopal Dutt, 1986).**

epididymis and this arrangement may serve to enhance and stabilise the flow of the blood to the epididymis. Epididymal veins follow routes parallel to those of the epididymal arteries. As in the case of arteries, vein-vein anastomoses are common in all the abovementioned three species. In the rabbit, prominent vascular links join the testis and epididymis (Chubb and Desjardins, 1982). A similar but less conspicuous vascular link is found between the cauda epididymides and testis in rat (Kormano, 1967). The venous plexus that surrounds each artery typically in the rabbit serves to cool the arterial blood by a counter-current heat exchange mechanism (Chubb and Desjardins, 1982). Based on a study of the angioarchitecture of the epididymis of buffalo and sheep, Dhingra (1978, 1980) believes that the mini pampiniform plexus of the vein and arteries play an important role in the haemodynamics and thermorgulation of epididymis to prolong the fertile life of spermatozoa.

Innervation

The epididymis in most mammals receives sympathetic and parasympathetic innervation to the smooth muscles commencing at the mid corpus region. It completely lacks somatic innervation. The sympathetic pathway is through caudal mesenteric plexus → hypogastric nerve → middle spermatic nerve via the testicular artery and ductus deferens to the epididymis. The parasympathetic supply to the inferior spermatic nerves leading to the epididymis is controversial (Hodson, 1970). The intrinsic nerves to the epididymis are derived from the nerve trunk composed of both cholinergic and adrenergic fibres which follow and supply epididymal vessels (El Badawi and Schenk, 1967). These fibres vary in abundance in different epididymal regions. The caput epididymides as in the rat, rabbit and guinea-pig is poorly innervated, whereas in cat and dog both norepinephrine positive (adrenergic) and acetyl cholinesterase positive (cholinergic) fibres are found (Norberg *et al.*, 1967; El Badawi and Schenk, 1967). In most of the species, the corpus epididymides is rather poorly innervated. The peritubular cholinergic and adrenergic fibres progressively form intricate plexus as the muscular wall of the epididymis increase in thickness distally towards ductus deferens (Baumgarten and Holstein, 1968). The nerve fibres are both excitatory and inhibitory to the smooth muscles of the epididymis.

Musculature

The epididymis possesses characteristic distribution of musculature suited specifically to carry out their functions. Two morphologically distinct types of smooth muscle cells surround the entire length of the epididymal duct. The cells which constitute relatively thin muscular coat enables the slow movements at the poximal region and as the muscular coat increases in thickness proximo-distally, more powerful movements are induced in the distal region at the time of ejaculation (Holstien, 1967). Interestingly, the thick periductal muscle, which is characteristic of eutherian epididymis is already present in the primitive mammal, echidna (Djakiew and Jones, 1981). During the sexually inactive period in the soft furred field rat (Dechamma and Gopal Dutt, 1980), the epididymal muscle layer increases in thickness and its connective tissue is seen abundant. In marked contrast, in the active breeding season, it is characterised by a concomitent decrease in muscle layer and connective tissue component. Since the fluid transport takes place even though the flow from the testes is blocked by ligation of ductus efferentes (Macmillan and Auckland, 1960), it is concluded that the muscular layer is actively involved in the transport of fluid in the epididymis. This inference is further reinforced by the evidence that the luminal epithelium is lined with immotile cilia (Hamilton, 1975) and the transport takes place against increasing hydrostatic pressure form the testis to the caput and from the caput to the cauda epididymides (Johnson and Howards, 1976). The contractions of the smooth muscle layer lining the duct provide propelling force for transporting spermatozoa through epididymis (Baumgarten *et al.*, 1971). These contractions are peristaltic and the frequency of the electrical activity of the smooth muscle declines from the caput towards vas deferens (Risley, 1963; Talo *et al.*, 1979).

Blood-Epididymis Barrier

The development of a blood-epididymis barrier to provide a specialised microenvironment for sperm maturation is well documented. Evidence in support of this view is that the composition of luminal fluid is distinct from that of blood plasma (Setchell and Hinton, 1981). Ultrastructural studies on the epididymal duct in mammals have clearly established that various epithelial contacts such as tight junctions, gap junctions, zona occludens, and desmosomes do not simply allow the molecules to traverse the epithelium but they do seem to be regulated (Suzuki and Nagano, 1978a, b; Hinton, 1982). The blood-epididymis barrier performs a protective role by excluding the entry of immunoglobulins, many toxic metabolites and environmental agents into the lumen (see Hancock, 1981; Hinton, 1982).

Cytology

The epithelium lining the epididymis in mammals comprises of principal cells, basal cells (Fig. 6), narrow cells (Fig. 7), clear cells (Fig. 8), apical cells and 'halo cells' (intraepithelial lymphocytes/macrophages). These cell types show regional differences in distribution. For example, the principal and basal cells are of common occurrence in zones I to V, the apical cells in zones I and II, the less numerous 'halo cells' occur in zone II but rarely in zone III, and the clear cells in all the five zones of the wild mouse epididymis (Suresh, 1987).

Principal cells (Fig. 9)

The most numerous among the cell types so far described in all mammalian

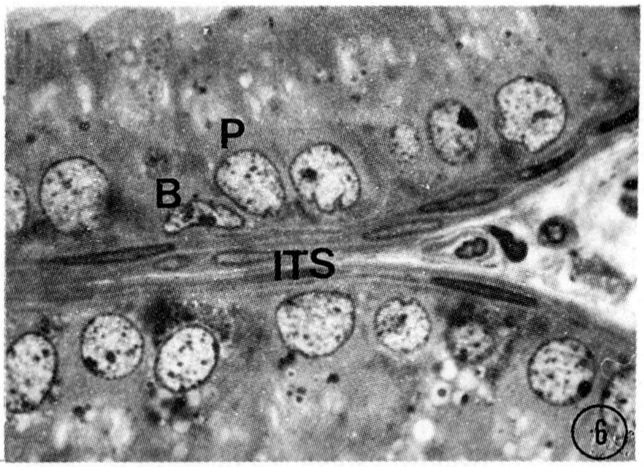

Fig. 6. Showing a basal cell (B) with its nucleus and its position in the epithelium along with principal cells (P).

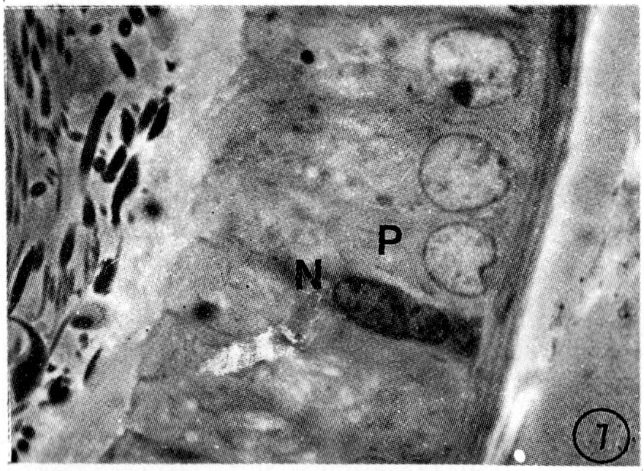

Fig. 7. A slender narrow cell (N) having densely stained elongated nucleus and cytoplasm wedged between the principal cells (P) × 1,250.

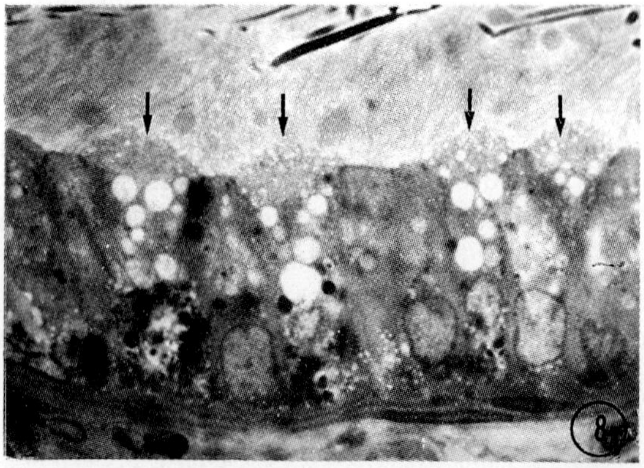

Fig. 8. Clear cells with distinct apical vacuoles and numerous large densely stained granules/lipid droplets. Arrows represent blebs at the apical surface and seretory material in the lumen × 1,000 (from Suresh, 1987).

species, are the principal cells, which have long since been the basis for classifying the zones of the duct system. The principal cells are present all along the entire length of the epididymis, though they vary in shape from tall columnar type extending from the basal lamina to the lumen in the proximal region to the squamous type found in the distal region. The principal cells thus decrease in height, increase in width from the proximal to distal segments. Quantitative study showed that among the various cell types the proportion of the principal cells is the highest in the central caput and lowest in the central corpus region of the ovine epididymis (Marengo and Amann, 1990). However, the number of principal cells per unit length of the basement membrane is same in all regions of the human epididymis (Regadera *et al.*, 1993). The principal cells have a single round or elliptical nucleus (Fig. 6) containing granular chromatin and one or more nucleoli. Multinucleate gaint cells associated with elevated androgen levels (Kuo and Gomez, 1981) and age related multinucleate principal cells have been observed in the human epididymis (Nistal *et al.*, 1990).

Ultrastructurally, the principal cells are characterized by well-developed stereocilia and long microvilli borne on the luminal surface. The stereocilia may differ regionally being tall, branched and irregular in the proximal region and short, dense and regular in the distal region. Several coated vesicles, pinocytotic invaginations and vesicles varying in diameter and content are usually observed in the apical region. Small coated vesicles are also often found in the viscinity of the Golgi body (Fig. 10). Regional specialization in the number of luminal surface invaginations and coated vesicles in the apical cytoplasm are not uncommon and those differences reflect on their function. The apical cytoplasm contains circular or oval multivesicular bodies (MVBs) (Fig. 11) and these vary in their size and content. The large MVBs may contain flocculent or amorphous material or a number of small vesicles. In addition, certain dense bodies as in the mouse epididymis contain myelin figures or granules of various sizes, foamy or dense globules in the cytoplasm (Abe *et al.,* 1983). These generally represent various stages of digestion of injested material (Jones *et al.*, 1979). The presence of microvilli, pinocytotic vesicles, coated vesicles and their ability to internalise fluid phase markers like ferritin, horse raddish peroxidase, thorotrast, etc. (Moore and Bedford, 1979; Flickinger, 1981) and/or receptor mediated markers such as alpha-2-microglobulin, transferin have provided impressive evidence for the absorptive function of the principal cells (Moniem and Glover, 1972; Moore and Bedford, 1979; Djakiew *et al.*, 1984; Robaire and Hermo, 1988). Evidence is available that certain exogenous proteins, colloids, and other substances are absorbed selectively by the principal cells.

The multivesicular bodies the and dense bodies which contain acid phosphatases hydrolyse these absorbed materials (Moniem and Glover, 1972). The principal cells of the caput region are more actively involved in this

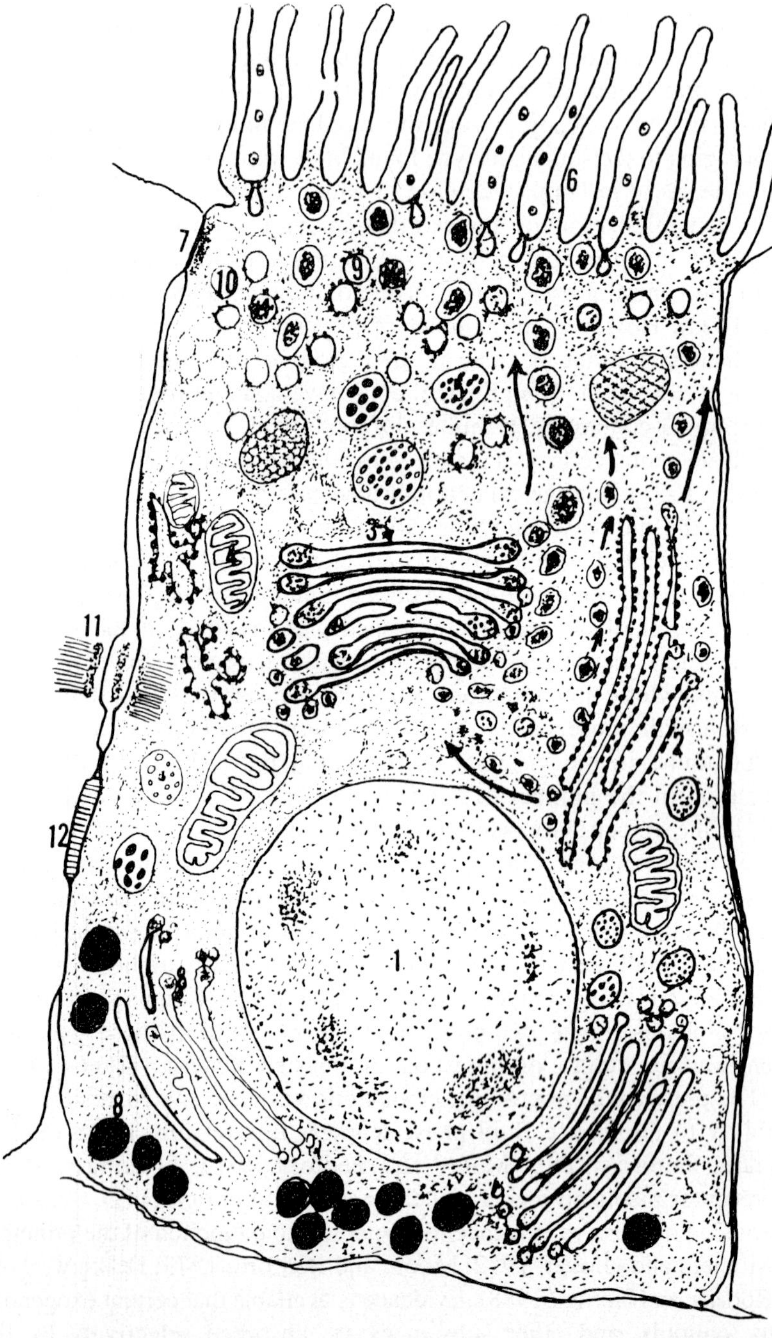

Fig. 9. A composite picture of a mammalian epididymal principal cell. 1. Nucleus, 2. Rough ER, 3. Golgi Complex, 4. Mitochondria, 5. Multivesicular body (MVB), 6. Microvilli, 7. Tight junction, 8. Lipid globules, 9. Mixed ER, 10. Smooth ER, 11. Desmosome, 12. Gap junction. Arrows indicate secretory pathways.

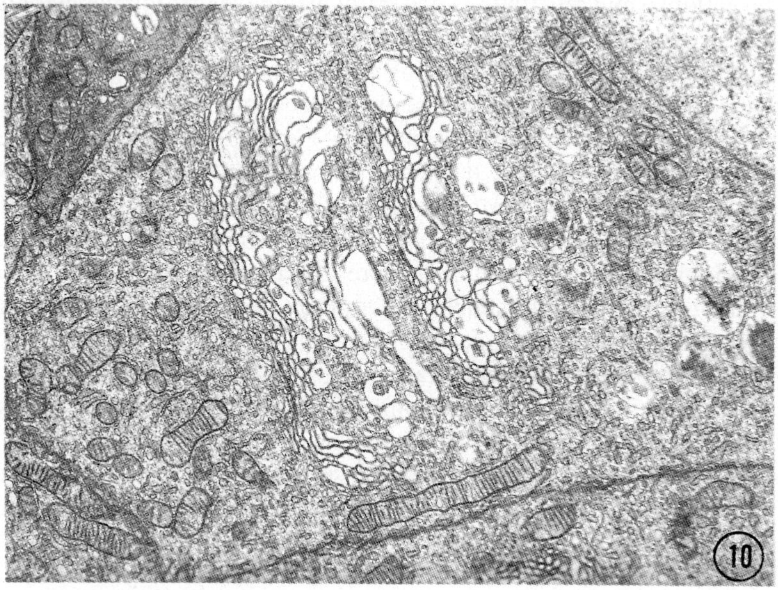

Fig. 10. Supranuclear cytoplasm of the principal cell with Golgi complex in the wild mouse × 10,000.

process than their counterparts in the caudal region. Rasweiler and Bedford (1982) have reported that in the rat cauda epididymides, the stercocilia of the principal cells serve as an important route for water absorption by a non-vascular mechanism.

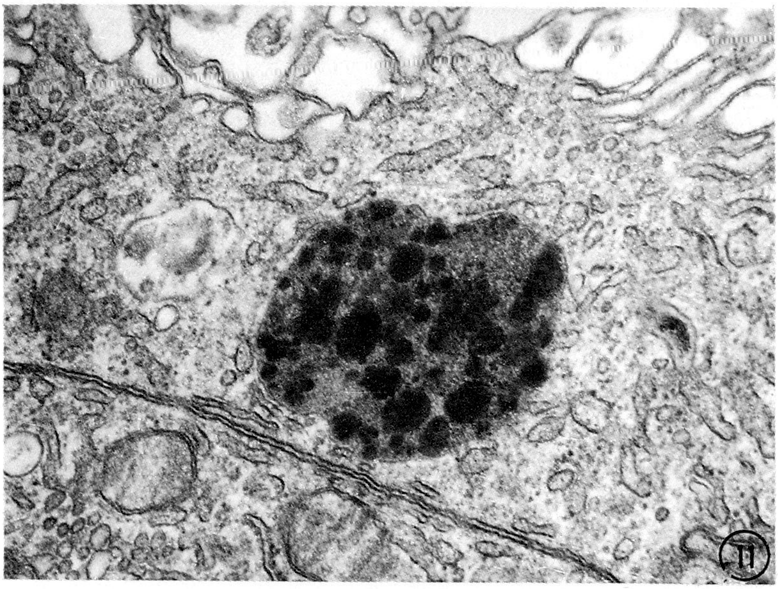

Fig. 11. The principal cell cytoplasm showing miultivesicular body × 40,000 (from Gopal Dutt and Suresh, 1988).

The endoplasmic reticulum is generally well developed in the cytoplasm. Rough ER may exist in differeznt forms. They may be found as flattened lamellae studded with ribosomes or as vesicles studded with ribosomes (Fig. 12). It is possible to find in some species mixed or heterogenous ER as in the adult male mouse described by Flickinger (1979). The RER is markedly well developed in the initial segment. In the mouse epididymis (Abe *et al.*, 1983), it has been shown that the marked development of RER is dependent upon testicular fluid in this segment since duct ligation regresses the rough endoplsmic reticulum (Moore and Bedford, 1979).

A major portion of supranuclear cytoplasm is occupied by an extensive Golgi complex, which is composed of parallely arranged stack of fenestrated cisternae of smooth surface membranes (Fig. 11). Some small sacs containing flocculent material different from pinocytotic vesicles are observed on the trans side of the Golgi appartus. Nicander and Ploen (1979) consider that these vacuoles may represent merocrine secretory vacuoles in all principal cells. Dense secretory granules are however seldom seen. Since the sacs of the Golgi are thick and rigid, they are referred to by Claude (1970) as rigid lamellae and these are of common occurrence in the principal cells of wild mouse (Gopal Dutt and Suresh, 1988) The lateral sides of the lamellae are swollen and appear to give rise to coated vesicles or autophagic vacuoles. In the ram, some fuzz coated vesicles are observed with open connectives to Golgi saccules or more rarely to the cisternae of the endoplasmic reticulum (Nicander, 1979). In the rabbit principal cells, small vesicles with opaque contents derived from the innermost Golgi cisternae move up to the cell apex to be released between the stereocilia (Nicander and Ploen, 1979). The mitochondria in the principal cells of the wild mouse are well developed with tubular cristae (Gopal Dutt and Suresh, 1988). Usually one end of these mitochondria is swollen enclosing a secretory droplet surrounded by a distinct vacuole. As the vacuole grows, the cristae of the mitochondria disappear. At least some of the large vacuoles filled with flocculent material seem to be derived from the mitochondria. Since the Golgi body is well developed and the RER and ribosomes are abundant, the principal cells are indeed capable of synthesizing protein. Using time course electronmicroscopic radioautography, Flickinger *et al.* (1984) have clearly demonstrated the passage of labelled proteins/glycoproteins from the RER to the Golgi body and then to coated vesicles which ultimately release their content into the lumen. Similarly, Nicander and Malmquist (1977) have provided ultrastructural evidence and pointed out that small vesicles, similar to those present in the Golgi area, are involved in the secretion of proteins in the initial segment of the epididymis. Gopal Dutt and Suresh (1988) have reported that the secretory product appears to be first formed in the RER which convey the same to the tiny vesicles produced in the Golgi region.These vesicles subsequently move to the cell apex where their contents seem to be released into the lumen (Fig. 9). Synthesis of secretory proteins in RER and their transport to the Golgi region, where they are packed and/or modified in accordance with similar pathways of protein secretion are known in other cell types (Jamieson and Palade, 1967).

Despite the presence of such morphological correlates of protein synthesis, the principal cells in mammals lack dense, homogenous protein granules characteristic of pancreas or pituitary cell types. Hoffer *et al.* (1973) have suggested an alternate pathway of protein secretion, where in the epididymal proteins after their synthesis in RER bypass Golgi body (Fig. 9). The major secretory products of rabbit epididymis are not proteins but glycerylphosphoryl choline, inositol and amino acids (Jones, 1978). The exact mechanism of secretion of such proteins into the lumen is perhaps through small vesiclels by exocytosis (Nicander and Ploen, 1979).

Based on the present knowledge of the epididymal physiology of mammals, the following secretory proteins are characterised in different species. They are: (1) forward motility protein in the bull (Acott and Hoskins, 1981); (2) sialoglycoprotein in the rat (Bayard e*t al.,* 1981), which is concerned with species specific sperm-egg recognition; (3) antiagglutinin in the ram and boar (Dacheux *et al.,* 1983); (4) quiescence factor in the bull (Acott and Carr, 1984); (5) acrosomal stabilising factor in the rabbit (Thomas *et al.,* 1984); (6) immobilin in the rat (Turner and Reich, 1987), which serves to immobolise the spermatozoa; and (7) cysteine proteinases (Cathepsins B and H) in rat involved in the intracellular degradation system (Tomomasa *et al.,* 1994); (8) glutathione peroxidase-like protein (MEP2H) in the caput epididymides of mouse (Rigaudiere *et al.,* 1992) and (9) 25-K Da protein in the mouse (Vierula e*t al.* 1992).

Secretion of steroids

The question, whether the epididymal epithelium is truely a site for steroidogenesis has now been resolved. It has been shown in a number of mammalian species including mouse, rat, rabbit and ram, that the epididymis is capable of synthesizing testosterone from acetate or cholesterol (Hamilton and Fawcett, 1970; Hamilton, 1971). A positive correlation seems to exist between the amount of smooth walled ER of the cell and the ability of the cells to synthesize cholesterol (Christensen and Fawcett, 1961). The principal cells of the terminal segment of rabbit epididymis contain large mitochondria with some tubular cristae and characteristic arrangement of smooth ER (Nicander and Ploen, 1979). In the wild mouse (Gopal Dutt and Suresh, 1988) the principal cells also contain smooth walled ER, well developed mitochondria with a complex system of tubular cristae aned lipid droplets (Fig. 13). Cytochemically, the epididymal epithelium is positive for $\Delta 5$-3β HSDH and 17β-HSDH in all the three regions. The reactions for these enzymes in the caudal region is however found to be high (Suresh and Gopal Dutt, 1984). Similar observations have been made in the epididymis of the long-tailed climbing mouse, *Vandeleuria oleracia* (Suresh and Gopal

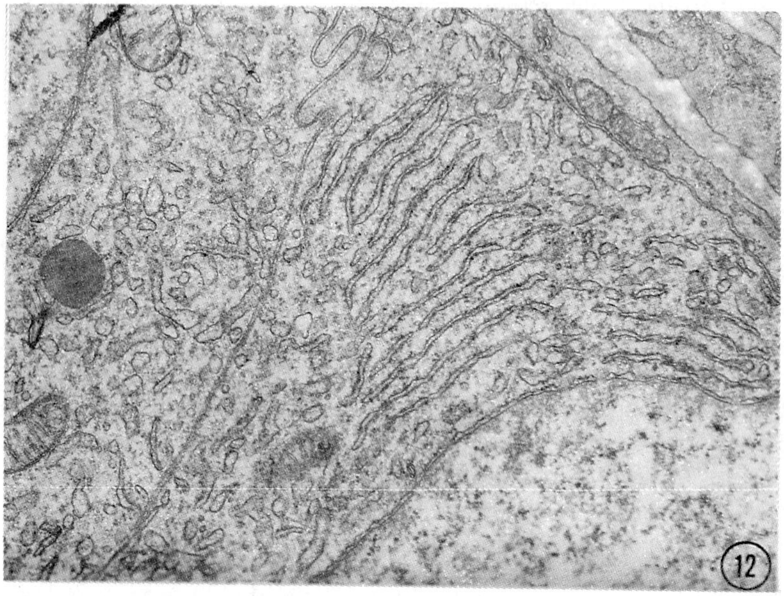

Fig. 12. Rough endoplasmic reticulum in the basal cytoplasm of a principal cell in the wild mouse × 12,000.

Dutt, 1985). In the rat and ram, the cauda epididymides synthesizes greater amount of testosterone compared to that of the caput epiididymides (Hamilton, 1972). The high enzyme activity in the cauda epididymides implies that

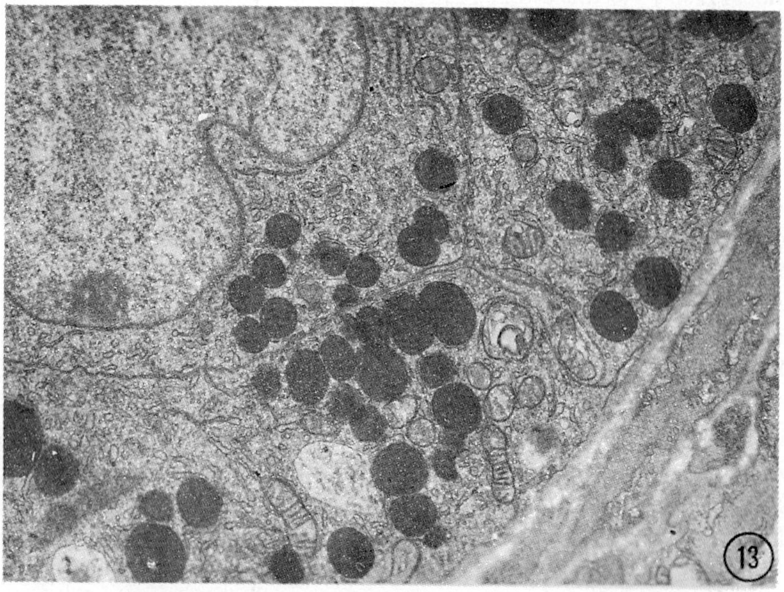

Fig. 13. Lipid droplets in the cytoplasm of a principal cell in the wild mouse × 15,000 (from Gopal Dutt and Suresh, 1988).

the sperms which occur in high concentration and reside for longer period here, are exposed to steroids probably for their maturation. It is not clear to what extent epididymal steroids are helpful in the maintenance of epididymal structure and function. Even if they do so, their role should be minimal since epididymis itself is dependent on circulating levels of androgens and/or molecules carried by the testicular fluid (Dinakar *et al.,* 1974; Rigaudiere *et al.,* 1992) for the maintenance of its normal structure and function.

Spermiophagy and phagocytosis of cytoplasmic droplet

All the sperms produced in the testes are never ejaculated and there is ample evidence that several of them disappear during their passage through excurrent duct (Amann and Almquist, 1962). One possibility in the elimination is by spermiophagy which has been reported commonly in the soft furred field rat (Guraya and Kaur, 1977; Dechamma and Gopal Dutt, 1980), but not in goat and bull (Goyal, 1985).

One of the important cytological events which occur during sperm maturation is the shedding of cytoplasmic droplets as the sperms pass through the epididymal duct. In bush tailed opossum, there is a clear-cut ultrastructural evidence that the principal cells phagocytise and digest the cytoplasmic droplets (Temple-Smith, 1984). This view gains support from the recent finding that the principal cells contain cysteine proteinases such as cathepsins B and H which are involved in the intracellular degradation system (Tomomasa *et al.,* 1994). The foregoing account lead to the conclusion that the principal cells, like Sertoli cells, are polyfunctional in mammals.

Basal cells (Fig. 14)

The basal cells vary considerably in size, shape and distribution along the epididymal duct. These cells, though occur sporadically in rat (Hamilton 1975) are more frequently seen in the wild mouse (Suresh, 1987). Rat basal cells isolated after unit gravity sedimentation seperation are slightly smaller than those from hamsters but generally contain more cytoplasmic organelles (Killan *et al.,* 1977). The basal cells are usually hemispherical, polygonal or pyramidal or round in shape and their nuclei are elongated and flattened against the basement membrane. Hamilton (1972) described that the shape of basal cells varies profoundly because of their close relationship with the principal cells. Certain projections from the plasma membrane of the basal cells interdigitate with the principal cells. Micropinocytotic folding on the plasma membrane are also not uncommon. The cytoplasm contains scattered spherical electron dense granules, although ultrastructural profiles do not suggest how the granules are formed. Mitochondria are sparsely distributed in the perinuclear area. The possibility that the basal ceells are involved in holocrine activity has been dispelled by ultrastructural studies (Hamilton, 1972). The available cytological evidence does not support the contention

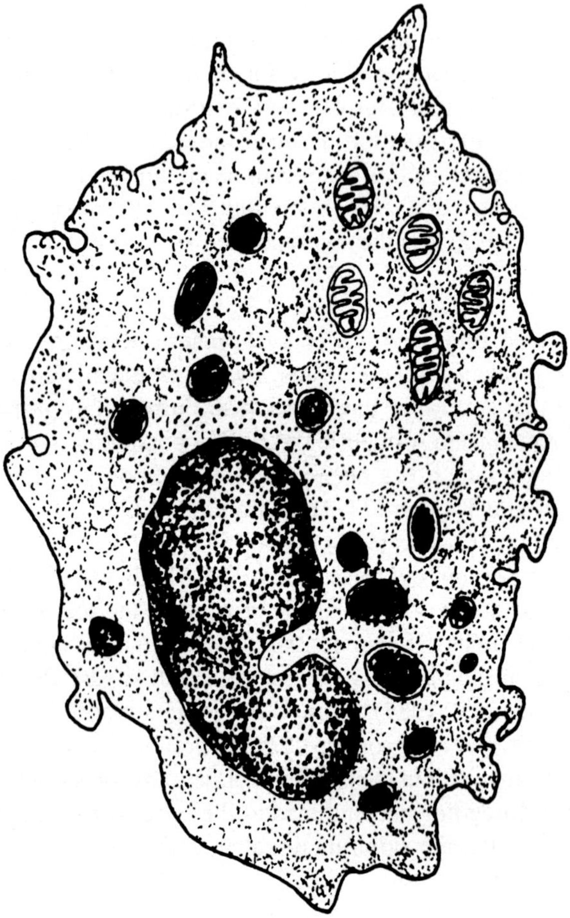

Fig. 14. A typical mammalian basal cell showing nucleus and sparse distribution of cell organelle.

that the basal cells are involved either in secretion or in absorption. This conclusion is drawn from the observations that the basal cells are numerically less, their location is far away from the lumen and they do not have complex cellular machinery for secretion or absorption. Recently, the glutathione-S-transferases (GSTs)—a family of isoenzymes that catalyse the conjugation of glutathione with various electrophilic compounds—have been localised in the cytoplasm of the basal cells of the corpus and proximal epididymides of rat. On the basis of this finding, Veri *et al.* (1993) have speculated that the basal cells are involved in the protection of epididymal epithelium from electrophilic attack. Two immunohistochemical types of basal cells are observed in the human epididymis—one is similar to the apical mitochondria rich cells (AMRC) and the other to the principal cells (Regadera *et al.*, 1993). The basal cells in the human epididymis are easily recognised by monoclonal antibodiy (mAB) 25Fq against mature tissue fixed macrophages

(Yeung *et al.*, 1994). These investigators have proposed a hypothesis that the basal cells are involved in a scavenging role in a local immune defence mechanism in which antigenic products derived from sperm degradation are phagocytised. The role of basal cells seems to be imperfectly understood, although these cells are present in all mammalian epididymides studied.

Narrow cells (Fig. 15)

The narrow cells can be easily distinguished from the neighbouring principal cells because of their deep staining as revealed by light-and electron microscopy. They appear as thin streaks of cells with their long axis perpendicular to basal lamina. The darkly staining slender or goblet shaped cells with apically located nuclei described in the epididymis of rat and guinea-pig are termed 'narrow columnar cells' (Clermont and Flonnery, 1970) or apical cells (Cohen *et al.*, 1976) both of which seem to correspond with the 'narrow cells' described by Sun and Flickinger (1979). Goyal and Williams (1991) believe that in rat epididymis the narrow cells are the same as AMRCs. The narrow cells, although not described in several mammalian species, are present only in the initial segment of the adult epididymis. The nuclei are elongated and centrally located containing granular chromatin and one or two perceptible nucleoli. The cytoplasm is meagre but in the apical region it is densely vacuolated. The microvilli are shorter and fewer as compared to those of the principal cells.

The mitochondria are rod-shaped with tubular cristae. Similar to the principal cells, cytoplasm contains multivesicular bodies. These ultrastructural characteristics, although occur in a low profile as compared to those in principal cells, suggest that the narrow cells play a subsidiary role in absorption or secretion, or both. This view, however, is conjectural and needs experimental proof.

Clear cells (Fig. 16)

The clear cells, also referred to as light cells or foamy cells by some workers, are remarkably striking in the study of mammalian epididymis. In rat, the early work of Kreth (1965) has shown two types of clear cells, whereas in the material examined by Hamilton (1975) only one type is seen. These cells have not been described at all in the epididymis of rabbit, stallion and bull (Nicander, 1957), mouse (Hamilton, 1975), mole (Suzuki and Racey, 1976) guinea-pig (Hoffer and Greenberg, 1978) and tammar (Jones *et al.*, 1984), monkey (Ramos and Dym, 1977) and human (Holstein, 1969). Surprisingly, the clear cells which were earlier reported to be absent in the mouse epididymis (Hamilton, 1975) are now located from the intermediate segment to the tail of the epididymis in the same species (Soranzo *et al.*, 1982). The clear cells are generally found in between the principal cells. They contain distinct ovoid nuclei placed slightly above basal position and contain one or more nucleoli and granular chromatin.

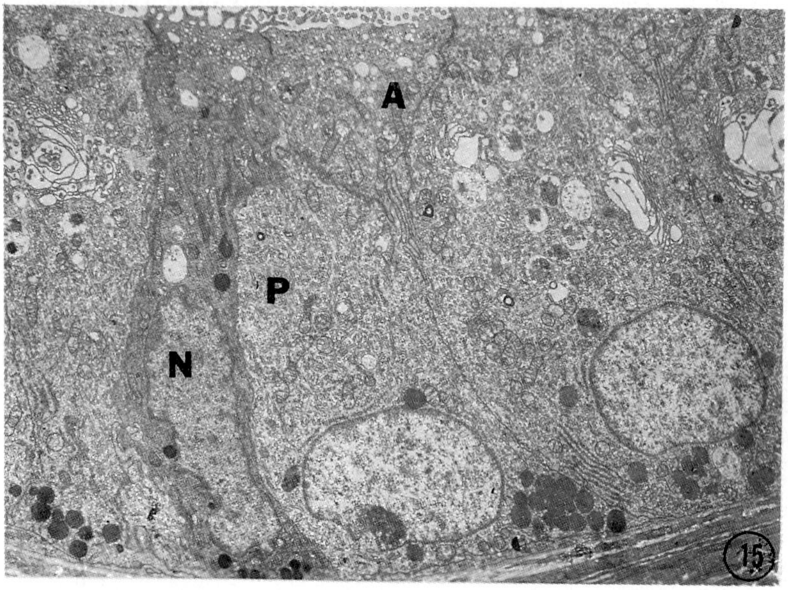

Fig. 15. **Low EM picture of the epididymal epithelium showing the narrow cell (N), principal cell (P) and apical cell (A) × 4,000 (from Gopal Dutt and Suresh, 1988).**

The clear cells have been implicted in holocrine secretory cycle (Martan and Risley, 1963) and as a principal source of glycerylphosphorylcholine

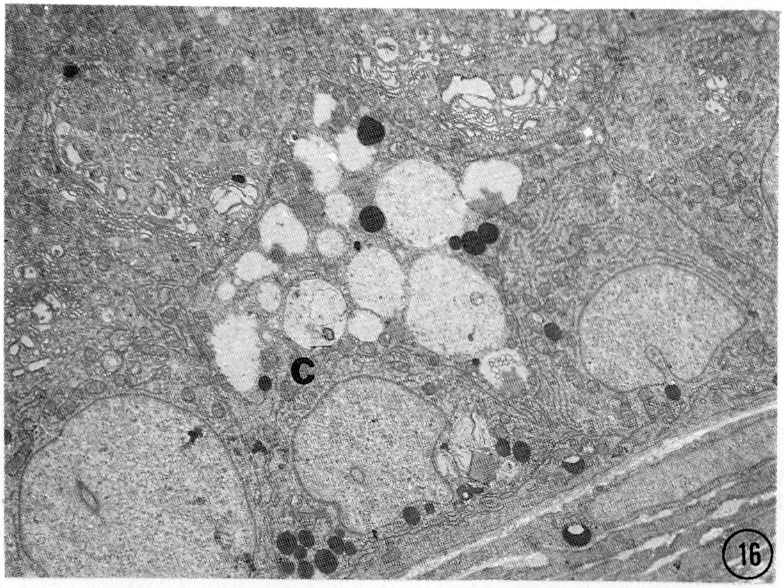

Fig. 16. **The apical cytoplasm of a clear cell (C) showing vacuoles of various sizes, mitochondria and nucleus × 9,600 (from Suresh, 1987).**

(GPC) (Scott *et al.,* 1963). These views have been subsequently refuted by Clermont and Flannery (1970), on the ground that the clear cells are absent in the head region of the epididymis of rat even though GPC is synthesized here more than in the tail region.

An impressive cellular feature of the clear cells is the presence of extensive cytoplasmic vacuolation which formed the basis for Reid and Cleland (1957) to divide these cells into different intergrading types. Using unphysiological concentration of sodium ion, Yeung and Cooper (1982) have demonstrated that the luminal Na^+ influences the size and density of the vacuoles and concluded that the sodium ion pumping could be one of the mechanisms by which concentration of vacuolar contents is achieved. The clear cells in the epididymis of the wild mouse (Suresh, 1987) are characterised by the appearance of prominent supranuclear vacuoles, pinocytotic pits near the cell surface and dense accumulation of lipid or glycolipid bodies in the basal cytoplasm. Electronmicroscopic study of clear cells in this species further revealed the presence of distinct blebs of secretion oozing out from their apical surface into the lumen. There is evidence in the case of rat that a glycoprotein secreted in the caput is partly taken up by the clear cells in the cauda epidymides (Lea *et al.,* 1978). These cells are also capable of absorbing both exogenous and endogenous proteins from the lumen. During absorption, the luminal content is taken up by micropinocytosis and transferred to supranuclear vacuoles (Moore and Bedford, 1979). Hermo *et al.* (1988) have reported that in the rat, the cytoplasmic droplets in the sperm first disintegrate into small particles and these are subsequently endocytosed and digested by the clear cells. Further in the rat epididymis, it has been reported that the immobilin protein which serves to immobilise spermatozoa is secreted by the principal cells of the initial segment and is absorbed by the distal caudal epithelial clear cells by endocytosis (Hermo *et al.*, 1994). The immobilin uptake by the clear cells continues despite the efferent duct ligation thereby indicating that the testicular factors are not essential for the stimulation of the clear cells.

Apical cells (Fig. 15)

Because of richness of mitochondria, the apical cells have been designated as apical mitochondria rich cells (AMRCs). They have been described in the initial segment of epididymis of rat (Reid and Cleland, 1957; Martan and Risley, 1963; Brown and Montesano,1980); hamster (Nicander and Glover, 1973), guinea-pig (Hoffer and Greenberg, 1978); mouse (Abou-Haila and Fain-Maurel, 1984); tammar (Jones *et al.,* 1984); *Loris* (Manjula, 1980); monkey (Ramos and Dym, 1977); and human (Parkkila e*t al.*, 1993). The AMCRs are located in between the principal cells towards the apical margin of the epithelial lining of the duct. Cytologically, the AMRCs generally resemble the adjacent principal cells except in the amount of cell organelles which is less. Similar to the principal cells, the AMRCs are acid phosphatase positive and alkaline phosphatase negative. However, immunohistochemically positive sites for oestradiol receptor protein (D5) and cytokeratin no 19 (Ks 4.62) are observed only in AMRCs but not in the principal cells (Palacios *et al.*, 1991).

The general cytological characteristics of AMRCs include apical location of nucleus, loosely lamellated scattered Golgi, extensive vacuolar system, vesicles containing electron dense granules and conspicuously punctate junctional areas, separating the principal cells. Since these cells lack microvilli, pinocytotic and coated vesicles and contain only less developed Golgi and RER, they cannot be considered to play a role either in absorption or secretion. Ultrastructural evidence seems to be inconclusive although, Palacios *et al.* (1991) have observed scant immunoreactivity for epithelial membrane antigen (EMA) and suggested that the AMRCs in the human epididymis are not markedly involved in glycoprotein secretion. On the other hand, the evidence in the mouse epididymis is striking in that the AMRCs are involved in the synthesis of SMAr4 glycoprotein which binds to the spermatozoon tail and prevents tail to tail spermatozoon aglutination (Feuchter *et al.*, 1987). The unique features of AMRCs are: (i) they are mitochondria rich (Farquhar and Palade, 1965; Whitear, 1975); (ii) they are carbonic anhydrase positive (Rosen and Friedly, 1973; Lonnerholm and Ridderstrale, 1974; Brown and Montesano, 1980); (iii) their plasma membrane contains elongated intramembrane particles; and (iv) their plasma membrane at the apical region contains characteristic unfoldings and/or microvilli (Brown and Montesano, 1980). The intense reaction for NADP, succinic dehydrogenase, and ATPase agrees with the ultrastructural finding of abundant mitochondria in the human ductus epididymides (Palacios *et al.*, 1991).

The abovementioned cellular features have also been described in other transporting epithelia such as those of the kidney collecting tubules of amphibians and mammals (Humbert *et al.*, 1975; Brown, 1978). Taking these findings into consideration and the presence of AMRC's only in the initial segment—a region of epididymis involved in fluid absorption and ion-transport, one can argue in favour of the AMRCs having an important function in trans-epithelial fluxes. Brown and Montesano (1980), and Parkkila *et al.* (1993) have suggested that the AMRCs in the human epididiymes contribute to the acidificatrion of the epididiymal fliuid that prevents premature sperm activation. In this respect it is important to consider that the mitochondrial aggregation observed in the apical cytoplasm might play a role in generating the ATP required for the transport of H^+ and Cl^- ions across the cell membrane (Goyal and Williams, 1991).

Halo cells (Fig. 17)

The halo cells, named by Reid and Cleland (1957) in the rat epididymis, contain perinulear empty cytoplasm in the light microscopic preparation. In well-fixed material for electronmicroscopy, however, Hamilton (1975) has demonstrated that there is discernible cytoplasm in the perinuclear area containing loosely scattered mitochondria, small Golgi bodies and sparse amount of rough ER. These cells vary in distribution in different zones in the epididymis of mammals. The halo cells occur on the apical side in zones III and IV in rat (Reid and Cleland, 1957), while in loris they are common in zones I and II of the epididymis (Manjula, 1980). They are occassionally seen in zones II and III in the epithelium of the epididymal duct in the wild mouse (Suresh, 1987). Reid and Cleland (1957) and later Hamilton (1972); Hofffer *et al.* (1973); Sun and Flickinger (1979); Goyal and Williams (1991) have pointed out that the 'halo cells' closely resemble the lymphocytes in their scanty and pale cytoplasm, hetrochromatic nulei, abundant RNP particles and glycogen. Other features, like lack of junctional complexes with other cells and blunt pseudopodia-like processes, indicate that the halo cell is a wandering leucocyte (Sun and Flickinger, 1979). Hamilton (1975) has reported that the lymphocytes transform into macrophages which are likely to take part in sperm resorption. Dym and

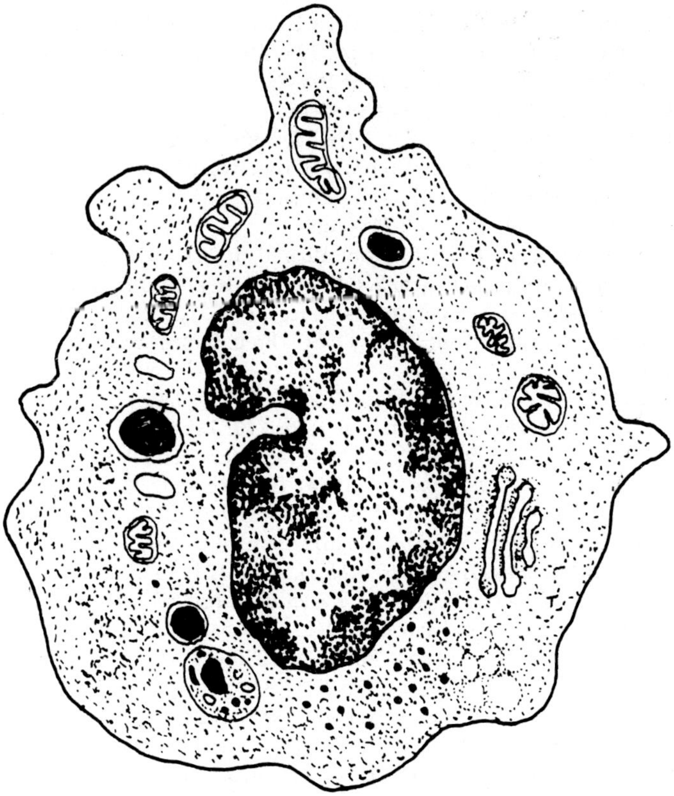

Fig. 17. A typical mammalian 'halo cell' with scanty pale cytoplasm, hetrochromatic nucleus and blunt pseudopodia-like process.

Romrell (1975) have suggested that these cells are immunocompetent and keep the epithelium of the male genital tract free from sperm antigens which are foreign to the body. There is no evidence so far regarding the actual passage of halo cells into the lumen. Sun and Flickinger (1979) believe that the halo cell does not appear to be a true epithelial cell of the epididymis.

References

Abe, K., Takano, H., and Ito, T. (1983). Arch. histol. Jap. **46:** 51–68.
Abe, K., Takano, H., and Ito, T. (1984). Arch. histol. Jap. **47:** 121–135.
Abou-Haila, A., and Fain-Maurel, M.A. (1984). Anat. Rec. **209:** 197–208.
Acott, T.S., and Carr, B.W. (1984). Biol. Reprod. **30:** 926–935.
Acott, T.S., and Hoskins, D.D. (1981). Biol. Reprod. **24:** 234–240.
Alexander, N.J. (1972). Am. J. Anat. **135:** 119–139.
Amann, R.P., and Almquist, J.D. (1962). J. Reprod. Fertil. **3:** 260–268.
Baumgarten, H.G., and Holstein, A.F. (1968). Z. Zellforsch. Mikr. Anat. **91:** 402–419.
Baumgarten, H.G., Holstein, A.F., and Rosengren, E. (1971). Z. Zellforsch. mikrosk. Anat. **120:** 37–79.
Bayard, F., Duguct, A., Mazzuca, M. and Faye, F.C. (1981). In: Reproductive Processes and Contraception. (K.W. Mekerns, Ed.), pp. 393–405. Plenum Press, New York.
Brooks, D.E. (1983). Aust J. Biol. Sci. **36:** 205–221.
Brown, D. (1978). J. Ultrastruct. Res. **63:** 35–40.
Brown, D., and Montesano, R. (1980). J. Cell Sci. **45:** 187–198.
Christensen, A.K. and Fawcett, D.W. (1961). J. Biophys. Biochem. Cytol. **9:** 653–661.
Chubb, C., and Dejardins, C. (1982). Am. J. Anat. **165:** 357–372.
Claude, A. (1970). J. Cell. Biol. **47:** 745–766.
Clermont, Y. and Flannery, J. (1970). Biol. Reprod. **3:** 282–292.
Cohen, J.P., Hoffer, A.P., and Rosen, S. (1976). Biol. Reprod. **14:** 339–346.
Dacheux, J.L., Paquignon, M., and Combarnous, Y. (1983). J. Reprod. Fertil. **67:** 181–189.
Dechamma, P.A. and Gopal Dutt, N.H. (1980). Proc. Indian Acad. Sci. (Anim. Sci.) **89:** 303–310.
Dhingra, L.D. (1978). J. Anat. Soc. India **27:** 111–114 (4 plates).
Dhingra, L.D. (1980). Haryana Agric. Uni. J. Res. **10:** 115–122.
Dinakar, N., Arora, R., and Prasad, M.R.N. (1974). J. Endocrinol. **60:** 399–408.
Djakiew, D., Byers, S.W., and Dym, M. (1984). Biol. Reprod. **31:** 1073–1075.
Djakiew, D., and Jones, R.C., (1981). J. Anat. **132:** 187–202.
Dym, M., and Romrell, L.J., (1975). J. Reprod. Fertil. **42:** 1–7.
El Badawi, A., and Schenk, E.A. (1967). Am. J. Anat. **121:** 1–13.
Farquhar, M.G., and Palade, G.E. (1965). J. Cell Biol. **26:** 263–291.
Feuchter, F.A., Green, H.F., and Tabet, A.J. (1987). Anat. Rec. **217:** 146–153.
Flickinger, C.J. (1979). Biol. Reprod. **20:** 1015–1030.
Flickinger, C.J. (1981). Biol. Reprod. **25:** 871–883.
Flickinger, C.J., Wilson, K.M. and Gray, H.D. (1984). Anat. Rec., **210:** 435–448.
Gardner, R.L., and Rossant, J. (1976). In: Embryogenesis in mammals: Ciba foundation Symposium 40 (New Series). (K. Elliot and M.O'connor. Eds.), pp. 5–25. Elsevier, New York.
Glover, T.D. (1968). Proceed. Internatal. Cong. Anim. Reprod. and Arti. Insem. (C. Thibault, Ed.) Volume I, pp. 273. Jouy-en-Josas, France.

Glover, T.D., and Nicander, B.L. (1971). J. Reprod. Fert. Suppl. **13:** 39–50.
Gopal Dutt, N.H. and Suresh, R. (1988). In: Regeneration and Development. (S. Inoue, Ed.), pp. 647–660. Okada Printing and Publishing Company, Maebashi, Japan.
Goyal, H.O. (1985). Am. J. Anat. **172:** 155–172.
Goyal, H.O. and Williams, C.S. (1991). Am. J. Anat. **190:** 349–369.
Guraya, S.S., and Kaur, R. (1977). Andrologia **9:** 265–269.
Hamilton, D.W. (1971). J. Reprod. Fert. Suppl. **13:** 89–97.
Hamilton, D.W. (1972). In: Reproductive Biology. (H. Balin and S. Glasser, Eds.), pp. 268–337. Excerpta Medica, Amsterdam.
Hamilton, D.W. (1975). In: Handbook of Physiology. Section VII: Endocrinology. (D.W. Hamilton and R.O. Greep, Eds.), pp. 259–301. American Physiological Society, Washington DC.
Hamilton, D.W., and Fawcett, D.W. (1970) Proc. Soc. Exp. Biol. **133:** 693.
Hancock, R.J.T. (1981). In: Oxford Reviews of Reproductive Biology. Vol. I. (G.A. Finn, Ed.), pp. 182–208. Clarendon Press, Oxford.
Hermo, L., Dworkin, J., and Oko, R. (1988). Am. J. Anat. **183:** 107–124.
Hermo, L., Barin, K., and Oko, R. (1994). Anat. Rec. **240:** 86–103.
Hinton, B.T. (1980). Invest. Urol. **81:** 1–10.
Hinton, B.T. (1982). Archiv. Androl. **9:** 75–83.
Hodson, N. (1970). In: The Testis I. (A.D. Johnson, W.R. Gomes and N.L. Vandemark, Eds.), pp. 47–99. Academic Press, New York.
Hoffer, A.P., and Greenberg, J. (1978). Anat. Rec. **190:** 659–678.
Hoffer, A.P., Hamilton, D.W., and Fawcett, D.W. (1973). Anat. Rec. **175:** 169–202.
Holstein, A.F. (1967) Z. Zellforsch. **76:** 498–510.
Holstein, A.F. (1969). Zwounglose A bhandl. Gcbiet Norm. Pathol. Anat. **20:** 1–91.
Humbert, F., Pricam, C., Perrelet, A., and Orci, L. (1975). J. Ultrastruct. Res. **52:** 13–20.
Hyman, L.H. (1962). In: Comparative Anatomy. pp. 338–427. The University of Chicago Press, Chicago.
Jamieson, J.D., and Palade, G.E. (1967). J. Cell Biol. **34:** 557–596.
Johnson, A.L., and Howards, S.S. (1976). Biol. Reprod. **14:** 371–376.
Jones, R. (1978). Comp. Biochem. Physiol. **61B:** 365.
Jones, R., Hamilton, D.W. and Fawcett, D.W. (1979). Amer. J. Anat. **156:** 373–400.
Jones, R.C., Hinds, L.A. and Tyndale. Boscos, C.H. (1984). Cell Tissue Res. **237:** 525–535.
Kauffman, S.A. (1973). Science **181:** 310–318.
Killan. G.J., Snyder, J., and Amann, R.P. (1977). Cell Tissue Res. **183:** 371–378.
Koe, T.H., and Lee S.H. (1993). Korean J. Anim. Sci. **35:** 377–383.
Kuo, T., and Gomez, L.G. (1981). Amer. Surg. Pathol. **5:** 483–490.
Kormano, M. (1967). Zeet Anat. Entwick. **126:** 138–153.
Kormano, M., Suoranta, H., and Reijinen, K. (1973). In: Advances in Bioscience. (G. Raspe, Ed.), pp. 73–82. Pergamon Press, Oxford.
Kreth, H.W. (1965). Z. Zellforsch. Mikroskop. Anat. **68:** 28–42.
Lea, O.A., Petrusz, P., and French, F.S.(1978). Int. J. Androl. (Suppl). **2:** 592–607.
Leeson, C.R., and Leeson, T.S. (1964). Anat. Anz. **114:** 159–167.
Lonnerholm, G., and Ridderstrale, Y. (1974). Acta. Physiol. Scand. **90:** 764–778.
Macmillan, E.W., and Auckland, J. (1960). J. Reprod. Fert. **1:** 139–149.
Manjula, A. (1980). Studies on the testis, epididymis and ductus deferens under normal and altered physiological conditions in the slender loris, *taris trdigradus gdekkerianus* (cabr). Ph.D. Thesis, Bangalore University, India.
Marengo, S.R., and Amann, R.P. (1990). Biol. Reprod. **42:** 167–179.
Martan, J., and Risley, P.L. (1963). Anat. Rec. **146:** 173–190.
Moniem, K.A., and Glover, T.D. (1972). J. Anat. **111:** 437–452.

Moore, H.D.M., and Bedford, J.M. (1979). Anat. Rec. **193:** 313–327.
Nagy, F., and Edmonds, R.H. (1975). Fertil. Steril. **26:** 460–468.
Nagy, F., Pendergrass, P.B. and Scott, J.N. (1982). J. Submicrosc. Cytol. **14:** 673–682.
Nicander, L. (1957). Acta. Morph. Nurl. Scand. **1:** 99–118.
Nicander, L. (1970). In: Morphological aspects of andrology. Vol. I. (A.F. Holstein and E. Horstmann. Eds.), pp. 121–124. Grosse Verlag, Berlin.
Nicander, L. (1979). Zbl. Vet. Med. C. Anat. Histol. Embryol. **8:** 318–330.
Nicander, L., and Glover, T.D. (1973). J. Anat. **114:** 347–364.
Nicander, L., and Malmquist, M. (1977). Cell Tissue Res. **184:** 487–490.
Nicander, L., and Ploen, L. (1979). Intl. J. Androl. **2:** 463–481.
Nistal, M., Iniguez. L., and Paniagua, R. (1990). J. Reprod. Fert. **89:** 655–661.
Nistal, M., Santamaria, L., and Paniagua, R. (1990a). Andrologia. **22:** 591–596.
Norberg, K.A., Risley, P.L., and Ungeredt, U. (1967). Z. Zellforsch. Mikr. Anat. **76:** 278–289.
Orgebin-Crist, M.C. (1986). Conntraception—fertilite—sexualite. **14:** 487–495.
Oudemans, J.T. (1982). Die accessorichen Geschlechtsdrusen der Siiugethiere. De Erven Loosjes. Harlem.
Palacios, J., Regadera, J., Nistal, M. and Paniagua, R. (1991). Anat. Rec. **231:** 82–88.
Parkkila, S., Parkkila, A.K., Kaunisto, K., Waheed, A., Sly, W.S., Rajaniemi, H. (1993). J. Histochem. Cytochem. **41:** 751–757.
Rasweiler, J.J., and Bedford, J.M. (1982). Biol. Reprod. **26:** 691–705.
Romos, A.S., and Dym, M. (1977). Am. J. Anat. **149:** 501–532.
Regadera, J., Cobo, P., Paniagua, R., Martizoz-Garua, F., Placios, J., Nistal, M. (1993). J. Anat. **183:** 507–514.
Reid, B.L., and Cleland, K.W. (1957). Aust. J. Zool. **5:** 223–246.
Rigaudiere, N., Ghyselinck, N.B., Faure, J., and Dufaure, J.P. (1992). Molecular Cell Ennodocrinol. **89:** 67–77.
Risley, P.L. (1963). In: Mechanism concerned with conception. (G.G. Hartman, Ed.), pp. 73–133. Macmillan, New York.
Robaire, B., and Hermo, L. (1988). In: The Physiology of Reproduction Vol. I. (E. Knobil, J.O. Neill, Eds.), pp. 999–1089.
Romer, A.S. (1960). The vertebrate Body. Saunders Company, London.
Rosen, S., and Friedley, N.J. (1973). Histochemistry **36:** 1–4.
Scott, T.W., Wales, R.G., Wallace, T.C. and White, I.G. (1963). J. Reprod. Fertil. **6:** 49–59.
Setchell, B.P., and Hinton, B.T. (1981). Prog. Reprod. Biol. **8:** 58–67.
Setchell, B.P., Waites, G.M.H., and Till, A.R. (1964). Nature **203:** 317–318.
Setty, B.S., and Jehan, Q. (1977). J. Reprod. Fertil. **49:** 317–322.
Soranzo, L., Dadoune, J.-P., and Fain-Maurel, M.-A. (1982). Reprod. Nutr. Develop. **22:** 999–1012.
Sun, E.L., and Flickinger, C.J. (1979) Am. J. Anat. **154:** 27–56.
Sun, E.L., and Flickinger, C.J. (1980). Am. J. Anat. **196:** 285–293.
Suresh, R., and Gopal Dutt, N.H. (1984). Z. mikrosk-anat. Forsch. **98:** 535–542.
Suresh, R., and Gopal Dutt, N.H. (1985). V Ann. Symp. Reprod. Biol. Comp. Endocrinology, Pondicherry (Abstract).
Suresh, R., and Gopal Dutt, N.H. (1986). Folia Morphol. **34:** 230–237.
Suresh, R. (1987). Contribution to the epididymal histophysiology in a wild mouse, *Mus platythrix,* Bennett. Ph.D. Thesis. University of Mysore, India.
Suresh, R., and Gopal Dutt, N.H. (1988). Proc. Nat. Semi. Rep. Biol. **74:** 67–68.
Suzuki, F., and Nagano, T. (1978a). Anat. Rec. **191:** 503–520.
Suzuki, F., and Nagano, T. (1978b). Dev. Biol. **63:** 321–334.
Suzuki, F., and Racey, P.A. (1976). J. Reprod. Fert. **47:** 47–54.

Takano, H. (1980). Acta Anat. Nippon. **55:** 573–587.
Talo, A., Jaakkola, U.M., and Markkula-Viitanen, M. (1979). J. Reprod. Fert. **57:** 423–429.
Temple-Smith, P.D., (1984). Biol. Repr. **30:** 707-720.
Thomas, T.S., Reynolds, A.I., and Coliphant, G. (1984). Biol. Reprod. **30:** 693–705.
Tomomasa, H., Waguri, S., Umeda, T., Koiso, K., Kominami, E., and Uchiyama, Y. (1994). J. Histochem. Cytochem. **42:** 417–425.
Turner, T.T., and Reich, G.H. (1987). Gam. Res. **18:** 267–278.
Vierula, M.E., Araki, Y., Rankin, T.L., Tulsiani, D.R.P., Orgenbin-Crist, M.C. (1992). Biol. Repro. **47:** 844–856.
Veri, J.P., Hermo, L., and Robaire, B. (1993). J. Androl. **14:** 23–44.
Weber, M. (1898). Studien uber Saugethiere. Part 2. Gustao Fisher Verlag, Jena.
Whitear, M. (1975). J. Zool. Lond. **175:** 107–149
Yeung, C.H. and Cooper, T.G. (1982). Cell Tissue Res. **226:** 407–425.
Yeung, C.H., Nashaw, D., Sorg, G., Oborpenning, F., Schulze, H., Nieschlag, E., and Cooper, T.G. (1994). Biol. Repr. **50:** 917–926.

Comparative Endocrinology and Reproduction
K.P. Joy, A. Krishna and C. Haldar (Eds)

13. Endocrine Regulation of Spermatogenesis in Amphibia

S. Kasinathan and V. Sriramulu*
Reader in Biology (Rtd), School of Life Sciences,
Pondicherry University, Pondicherry, India
*Professor and Head (Rtd), Department of Zoology,
Annamalai University, Annamalainagar-608 002, India

Introduction

Spermatogenesis in amphibians is an interdependent process and does not appear to be an isolated event. The impact of exogenous environmental factors in concert with the internal physiological rhythms of the animal play a significant role in exercising a decisive control over the strategy of reproduction in frogs. Most of the existing earlier reports in this field deal with the qualitative assessment of the spermatogenetic cycle but the publication of specific methods by Van Oordt (1956), for the quantitative assessment of spermatogenesis, provided a more specific way for evaluating the male sexual cycle. Testis, the principal male reproductive organ, is an extremely complex one with the interdependent exorcine and endocrine components existing side by side. The constituent cell types of the testis, which multiply at an astonishing rate, also appear to be extremely sensitive to environmental changes with a vital relationship which incidentally provides the greatest opportunity to study and to understand the functional physiology of this delicate system.

Spermatogenesis

The testis of *Rana tigrina* provides a clear picture of the continuous change in the histological characteristics of its cell types. Apart from this regression of the gonadal weight during the non-breeding season, increase in testicular weight during breeding season, enhanced tubular diameter, concomitant changes in secondary sexual characteristics like vocal sac, thump pad and colour change appear to be characteristic. *R. tigrina* has been studied in the laboratories located in Eastern and Southern zones of India and normal sexual cycle in this species, which is mostly in conformity, suggests that *R. tigrina* has a potentially continuous type of spermatogenesis. However, Basu (1969) has suggested that species exhibiting potentially continuous type of spermatogenesis should form a continuo-discontinuous variety since potentially continuous type also possesses the characteristics of potential discontinuity of spermatogenetic cycle. In *Rana cyanophlyctis* the average testis weight and tubular diameter do not exhibit any significant seasonal differences during breeding cycles. The seminiferous tubules contain sperm throughout the year, either embedded or scattered, suggesting the prevalence of a continuous spermatogenesis and capable of breeding throughout the season. In *Rana hexadactyla,* which is widely distributed in the Indian subcontinent, all stages of spermatogenesis occur in the seminiferous tubules throughout the year. Gopalakrishnan and Rajasekarasetty (1977) studied the spermatogenetic cycle of *Rana hexadactyla* from two regions in India (Mysore and Mangalore) which vary in their climatological conditions and inferred that the gonadal activity in *R. hexadactyla* is under the specific influence of the range of temperature and relative humidity with the pattern of rainfall of the particular locality.

Experimental Studies

Effects of Steroids on Spermatogenesis

The effects of steroids on anuran spermatogenesis has been reported on various frogs and toads, with conflicting reports of stimulation (Basu, 1962a,b) and inactivity (Lofts, 1974). The intensive and well planned work of Kasinathan and Basu (1977) on *Rana hexadactyla* has established the dose response of gonadal and adrenocortical steroids on spermatogenesis and incidentally provides a fundamental work for future studies. A variety of hormones, testosterone, androstenedione, 17-β-hydroxyprogesterone, progesterone, estradiol and estrone were separately homogenized and administered every day subcutaneously to each frog till their autopsy. Kasinathan and Basu (1977) have ascertained the minimal dose dependency of the hormones in eliciting spermatogenic inhibition, being specific for each hormone employed, commencing from the second week and with the rest of the stages at the end of the fourth week. Administration of testosterone, for one month, inhibited the cell nests of stage I, pycnosis of stage II, while stage IV appear reduced and spermatozoa occur in bundles attached to Sertoli cells. A month after the withdrawal of the hormone treatment initiation of testicular recovery were evident while higher doses of these hormones have an irreversible effect on the gonads. However, the phenomenon of 'rebound recovery' noticed in mammals after the cessation of testosterone treatment has not been observed in *R. hexadactyla* as restoration to normalcy with characteristic spermatogenesis is evident after the withdrawal of the hormone treatment.

Female sex steroids with a very low dosage normally inhibit spermatogenesis within a short period. However, with prolonged treatment, irrespective of the dosage administered, they cause complete atrophy and damage the spermatogenetic cells in experimental frogs, the damage being irreversible (Kasinathan and Basu, 1977). Synthetic female hormone, stilbesterol, inhibits spermatogenesis. Administration of a split dose (total 1.25 mg divided into five injections) has affected all stages of spermatogenesis. Additionally the overall effects of stilbesterol is evident in the disorganization of the seminiferous tubules thickening of the tubular wall and shrinkage of tubular lamina. However, spermatids and sperm bundles were found in abundance throughout the tubules in experimental frogs subjected to stilbesterol treatment (Kasinathan and Basu, 1974). Use of estradiol, implanted as pellets of 1 mg on the dorsal lymp sac, blocked spermatogenic activity akin to those reported for stilbesterol (Basu, 1968). Treatment with DOCA has resulted in a decrease in spermatogenetic stages other than Stages II and IV. Decrease in relative testicular weight concomitant with atrophy of interstitial tissue of steroid-treated frogs has also been observed.

In a series of experiments to ascertain the dose response study with testosterone, Kashinathan and Basu (1977) have observed an inhibition of spermatogenesis. In particular, the treatment has led to an inhibition of secondary spermatogonial cell-nests (Stage II) which suggests the blockade of cell division leading to degeneration of spermatogonia without interfering with advanced stages of spermatogenesis. Further, the interstitial tissue appeared highly reduced with a concomitant reduction in the relative testis weight.

Role of Gonadotrophins

Exogenous steroid administration suppresses gonadotrophin secretion (Van Oordt, 1963) and in turn inhibits spermatogenesis in treated frogs. Hypophysial mediation of spermatogenic blockade, due to, steroid treatment, is an established fact in several mammalian species. In poikilothermic animals pituitary feedback inhibition and suppression of spermatogenesis reflects two different target organs (Basu and Nandi, 1965). A study conducted on hypophysectomised *Rana hexadactyla* (Kasinathan and Basu, 1973) but supplemented with steroid hormones (PMSG, HCG, PMSG + HCG, testosterone + PMSG, testosterone + HCG, testosterone + PMSG + HCG) to ascertain the *modus operandi* of steroid activity. Two weeks after the hypophysectomy cells in spermatogenetic stages were reduced in experimental frogs while four weeks after all spermatogenetic stages were reduced excepting primary and secondary spermatogonia. Sperms occur scattered and disorganised and the interstitium is feebly developed. Only spermatogonia were found in hypophysectormised frogs treated with testosterone alone. Injection of PMSG to hypophysectomised frogs enhanced spermatogenic activity with pronounced delineation of tubular outline with well-developed sperms in peripheral tubules. Injection of HCG enhanced cell nests of different spermatogenetic stages and spermiation was evident. Treatment with PMSG + HCG enhanced spermatogenetic activity. Primary spermatocytes, sperms in bundles were noticed in all recipient frogs. Combined treatment with PMSG and testosterone has resulted in a greater spermatogenetic inhibition and sperms lie scattered.

Injection of HCG and testosterone has inhibited spermatogenetic stages excepting stages 0 and 1. Treatment with PMSG + HCG + testosterone has resulted in a greater spermatogenic inhibition while primary spermatogonia are found in abundance, secondary spermatogonia are poorly developed and the rest of the stages are missing.

Results indicate that hypophysectomised frogs were more sensitive to testosterone treatment than nonhypophysectomised ones as rarely stages beyond primary spermatogonia were found in the tubules. The action of testosterone on germinal epithelium appears to be a direct one on the testis due to its effects on the suppression of gonadotrophins. Spermatogenic arrest, due to exogenous androgen treatment, is effected via two pathways. First, the adenohypophysial feedback mechanism is interrupted and the second possible pathway is a direct effect on the germinal epithelium of the testis leading to the inhibition of spermatogenetic cells.

Effect of Steroids on Pituitary Cell Types

Kasinathan, Basu and Sriramulu (1977) also studied the influence of steroid (Stilbesterol, Testosterone, Oestradiol, DOCA and Cortisol) on the hypophysial cell types, their impact on spermatogenic activity in relation to the seasonal impact of pars distalis cell types, in *Rana hexadactyla.* A parallelism exists between the activity of B_3 cells and interstitial function, both under normal and experimental conditions. Van Oordt and Lofts (1963) have suggested that B_3 cells are the source of ICSH like hormone. When interrenal cells were also included in the studies during circenical cycle and under the influence of high temperature changes in B_3 cells run parallel in the different tissues (Van Kamenade and Van Oordt, 1968). In *Rana temporaria* activation of B_3 cells coincided with interrenal cells while the interstitial cells regressed. Since the influence of B_3 cells are more profound on the interrenal tissue than on interstitium of the testis, they are more corticotrophic than gonadotrophic in their function. The discrepancy whether the B_3 cells produce ICSH like hormone or ACTH finds an acceptable explanation in the studies on *R. hexadactyla* by Kasinathan and Basu (1973). By administering exogenous steroid hormones they found a decrease in inter-renal activity, regression of interstitium and inhibition of spermatogenesis in the testis. Hyperplasia and granulation of B_3 cells suggests that they are chiefly concerned with the production of ICSH like hormone and corticoids. Secretory activity of pars distalis runs parallel with the functional differentiation of the testis as spermiation and initial multiplication of primary spermatogonia occurs during the strong extrusion of B_3 cells. Impairment in the function of B_3 cells leads to an inhibition of gametogenetic activity. *In vitro* studies indicated that spermatogenic cells are affected directly by steroids (Basu, 1969) which suggest that steroids not only decrease FSH and LH output from the gonadotrophs but also alter their cytomorphological characteristics.

Testicular Steroidogenesis

Testicular homogenates of *Rana hexadactyla* and *Rana cyanophlyctis* when subjected to chromatographic analysis of steroids revealed the presence of progesterone and estradiol 17-β in these two species whereas, testosterone, estriol and estrone were absent (Kasinathan and Basu, 1973). The study indicates that progesterone is metabolized in the testicular tissue and testosterone is produced in small quantities as a metabolic product to meet the requirements. There is evidence to show that the biosynthetic pathway for the formation of C^{19} steroids from C^{21} steroids is similar to that operating in other vertebrates. Using radioactive C^{14} substances, Ozon (1966) found in the testes of *Pleurodeles waltilli* that testosterone is synthesized from progesterone. Hooli *et al.* (1973) have identified 3-β HSDH activity in Leydig cells of the amphibian testis, *Rana tigrina*, and the activity of G-6-PDH, DPNH and TPNH diaphosphorses were intensive in the Leydig cells. Saidapur and Nadkarni (1973) assessed the activity of these enzymes throughout the year to correlate the seasonal changes and reproductive cycle with the enzyme activity. Δ^5 3-β-HSDH activity occurred mainly in the Leydig and Sertoli cells suggesting their probable role in steroidogenesis. The occurrence of male hormone synthesizing systems in Leydig and Sertoli cells proved that the synthesis of testosterone is possible in the tubular lumen from the available precursors. In *Rana hexadactyla* Kasinathan and his collaborators (unpublished observations) observed enzyme activity confined to the Leydig and Sertoli cells suggesting their ability to oxidize Δ^5-3-β-hydroxysteroids to Δ^4 3-ketosteroids which further indicates that amphibian testis is capable of synthesizing steroid hormones.

Role of Fat Bodies

The digitated yellowish fatty processes at the caput of the testis of frogs are normally regarded as reserve nutritional substances (Dubois, 1972). It has been demonstrated subsequently that fat mobilization in *Bufo bufo* can only be effected through the pars distalis (Larsen, 1963; Van Dongen *et al.*, 1966). Detailed studies on *Rana esculenta* by Chieffi and Rastogi (1974) and Chieffi *et al.* (1975) suggested the importance of fat bodies in the normal functioning of the hypothalamo-hypophysial-gonadal axis. In the green frog, *Rana hexadactyla,* Kasinathan *et al.* (1978 and 1979) have found that the fat body controls spermatogenesis and steroidogenesis of the testis and mobilization of fatty substances is effected only by the hypophysial hormones. Further experiments with prostaglandins (PGs) and cyclic AMP and their biochemical estimations indicated that fat bodies appear to contribute cholesterol possibly for steroidogenesis. It was also evident that fat bodies probably contribute PGs and/or cAMP in the testicular tissue for the continued spermatogenesis and steroidogenesis. PGs appeared to be more involved in spermatogenesis while cAMP in steroidogenesis. Spectral analysis of the fat extract suggested the presence of PGF alpha in *Rana hexadactyla.* Thus fat bodies constitute an important link organ in the hypothalamo-hypophysial-gonadal axis.

Influence of Pituitary

Hypophysectomy leads to the atrophy of testis in several amphibians. Homoplastic transplantation of pituitary under the skin of hypophysectomised *Bufo arenarum* abolished testicular atrophy (Houssay *et al.*, 1929) but has virtually produced hypertrophy of the testis and stimulated the germinal epithelium considerably. Later workers adopted injection of pituitary material (pituitary extract, FSH, LH preparations, mammalian serum gonadotrophin, pregnant mare serum gonadotrophin) and reported an increase in spermatogenic activity (Burgos and Rufino, 1953; Van Oordt *et al.*, 1949; Basu and Mondal, 1961a, b). These studies underlie the importance of gonadotrophin for spermatogenesis emphasizing the axiomatic role of hypothalamo-hypophysial-gonadal axis for maintaining male sexual cycle in frogs. However, the importance of exogenous environmental influence in modulating the influence of the axis cannot be ruled out (Basu, 1969). In *Rana temporaria* pituitary gonadotrophin content varies with the season which in turn is in concert with the stage of spermatogenesis (Van Oordt, 1956) in this frog which exhibits a discontinuous type of spermatogenesis. Kasinathan and Basu (1972) have investigated the effects of exogenous injection of homoplastic pituitary extracts, collected from male frogs during specific seasons, summer and winter, on *Rana hexadactyla* which exhibits a continuous spermatogenic cycle.

Winter pituitary extracts when administrated to winter frogs induced precocious onset of spermatogenesis, spermiation and a reduction in the relative testicular weight. Summer pituitary extracts to winter frogs induced early stages of spermatogenesis and a decrease in relative testicular weights. Winter pituitary extracts to summer frogs accelerated spermatogenesis and spermiation, the tubules being devoid of spermatozoa, and a decrease in testicular weight. Summer pituitary extracts to summer frogs accelerated early stages of spermatogenesis upto Stage III. *Rana hexadactyla,* with a continuous spermatogenic cycle, the number of different stages vary and there is a wave in the cycle of spermatogenesis with a predominance of certain stages during a seasonal cycle. This suggests the prevalence of seasonal fluctuations in hypophysial gonadotrophin secretion despite the continuous sexual cycle. Winter pituitary extract exerts a lower stimulatory effect on spermatogenesis both in summer and winter frogs. On the other hand, summer pituitary extract on winter frogs shows a significant augmentation of all spermatogenic stages compared to the effects of winter pituitary extracts on winter frogs. Winter pituitary extract administrartion enhances only secondary spermatogonia in winter frogs but when administered to summer frogs induces not only secondary spermatogonia but also numerous primary and secondary spermatocytes. Injection of summer pituitary extract on winter and summer frogs has led to a statistically significant increase in all spermatogenetic stages compared to that of winter pituitary extract on winter and summer frogs. Even winter pituitary extract injection on summer and winter frogs accelerates spermatogenesis but the effect elicited is far less than that of summer pituitary extract administration. The results indicate that summer pituitary extract is more effective than that of winter pituitary extract. Further, spermatogenetic activity during summer is considerably higher than during winter. The sensitivity of the germinal epithelium to gonadotrophic hormone is lower in winter than during summer period and can be ascribed to low sensitivity of spermatogonia to gonadotrophic hormone. In *Triturus cristatus,* a species with continuous spermatogenic cycle, low winter temperature renders the germinal epithelium insensitive to gonadotrophic hormone (Galgano, 1941).

Effects of Gonadectomy

Kasinathan *et al.* (1979) studied the effects of partial gonadectomy (removal of left testis, removal of 1/2 left testis, removal of left testis and 1/2 right testis and removal of both testis and their transplantation to spleen) and their impact on the remaining testis in *Rana hexadactyla,* to ascertain whether compensatory hypertrophy depends on the size and volume of the other excised organ or the phenomenon is independent of such a factor. Removal of the left testis has significantly increased Stage II and III spermatocytes, well developed sperms and interstitium in the remaining right testis. Removal of one-half of the left testis has resulted in the occurrence of sparce interstitium, reduced tubular diameter, disorganized occurrence of sperms and a reduction in secondary spermatogonia in the remaining half of the testis while the intact right testis appears normal compared to controls. A marked reduction in spermatogenesis, spermatogonia and seminiferous tubules are noticed in frogs deprived of their entire left testis and one-half of the right testis. In transplanted testis with spleen, reduction in the spermatogenetic stages were marked along with copious development of connective tissue in the central part of the transplants. Also the sperms occur disorganized. Slight stimulation of spermatogenesis in unilaterally castrated frogs may be due to the feedback regulatory mechanism in the hypothalamo-hypophysial axis. Partial excision of the testis has induced only a mild stimulation of gonadotrophin secretion. The remaining one-half of the remaining testis has been strongly influenced by the injury to innervation and blood supply rather than gonadotrophin stimulation. Intensity of compensatory hypertrophy is not dependent on size or the volume of the excised testis but limited to the physiological capacity of the concerned animal. The degenerative change observed in transplanted gonadal tissue with spleen may be due to inadequate trophic condition than to subnormal gonadotrophic stimulation. Thyagaraja and Sarkar (1972–73) have also reported hypertrophy in the contralateral gonad in *Rana tigrina* since an increase in seminiferous tubules, spermatogenetic stages and interstitial cells were observed.

Pituitary Histoarchitecture

The pars distalis of amphibian hypophysis secretes various hormones and of them gonadotrophin is one which is essential for gonadal function and maintenance.

In *Rana hexadactyla* (Kasinathan *et al.,* 1977) the gonadotrophin secreting cells reveal a kind of regular wave of secretory activity and consequent reduction in size from November till February–March, followed by complete degranulation in the months of April–May till the end of August. The phase of regrowth and granule accumulation in the cytoplasm ensures from September until the secretory phase commences again. Further, a characteristic difference in the granulation of acidophils and basophils is evident. During the time of strong extrusion cells appear partly granulated whereas the period of gradual release is characterized by cell shrinkage. Basophils (B_3) loose most of their granules at the peak of reproductive activity while in the remaining period they are sparsely granulated. Basophils (B_2) after loosing orange granules accumulate new granules in April, probably storage of gonadotrophin due to negative feedback of testicular hormone (Van Oordt and Lofts, 1963).

Pars distalis cytology was studied on fatectomized, gonadectomized and steroid treated *Rana hexadactyla.* Abdominal fatectomized frogs showed hypertrophy of A_1 cells and degranulation of B_2 cells whereas A_2 cells population increased. Administration of fat extract resulted in the restoration of the pars distalis cell types to the pre-operative condition. Degranulation of B_2 cells might indicate the release of gonadotrophic hormone but the presence of this hormone did not stimulate spermatogenesis and on the contrary inhibited it (Kasinathan *et al.,* 1978). Injection of fat-body extract has not only restored the percentage of A_1 and B_2 cells to normal condition, but also stimulated spermatogenesis. It is presumed that A_1 and B_2 cells are probably the source of some fat mobilizing factor, the secretion of which in turn stimulates testicular function through the abdominal fat bodies in male *Rana hexadactyla.*

A_2 cells are normally either ACTH or LTH secretory in nature. LTH is necessary for the mobilization of cholesterol in the amphibian testis to synthesize male steroidal hormone by Sertoli cells (Bartke, 1971). In *Rana hexadactyla* A_2 cells increase in their count after gonadectomy, unilateral fatectomy, and fatectomized frogs treated with fat extracts. However, this cell type decrease when gonadectomized animals were treated with testosterone or in frogs subjected to bilateral fatectomy. The data on the behaviour pattern of A_2 cells in the above experiments is difficult to interpret as information on the interrenal tissues where the other probable hormone, ACTH of A_2 cells could have acted upon is lacking. But, it is clear that A_2 cells in some way or other are related to the testicular and fat body functions in addition to that of B_2 and B_3 gonadotrophs. Further critical study in this field will certainly reveal many exciting information on LTH functions in amphibia.

Role of Pineal Complex in Reproduction

Pineal glands and their impact in reproduction has attracted immense attention in mammals due to their anti-gonadal action. A few reports are available on amphibian pineal eye, investigated both from anatomical and functional angles. Rastogi *et al.* (1976) observed that the pineal complex together with the eyes play a greater role in reproduction in *Rana esculenta.* This species kept in constant darkness or with the presence of either eye or pineal complex results in the inhibition of spermatogonia and stimulation of later stages of spermatogenesis. On the other hand exposure to light for 24 hours regimen initiated massive spermatogonial multiplication and degeneration of secondary spermatocytes. Rastogi (1976) further showed that melatonin injections have no effect on spermatogenesis in *Rana esculenta.* Kasinathan *et al.* (1981) have studied the effects of light on spermatogenesis in *Rana hexadactyla* and reported that the photic cycle plays a co-ordinating role in male reproduction. Either an increase or decrease of the duration of light in normal frogs exhibit respective increase or decrease in spermatogenic stages while blinded frogs are non-responsive to photic stimuli. Further studies of Kasinathan and Anandan (1982) revealed that the red component of the white light of 320 ft candle illuminance and total irradiance of 74 W/ cm^2-nm^{-1} stimulate spermatogenesis with added effectiveness than other colours. With the available knowledge one can assume that light received by the photoreceptor organs stimulates hypothalamus to release the hormone which in turn stimulates the hypophysial-gonadal axis. The exact relationship in tropical frogs will provide an extremely exciting field for further future investigations.

Melatonin is known to be present in frogs. It is a pineal hormone that produces either progonadal or antigonadal effects in mammals depending on the time and mode of administration. Information on melatonin effects on the testis in toads indicated varied changes during the breeding and hibernating seasons (Chanda and Biswas, 1982). Studies on *Rana hexadactyla,* a continuous breeder, revealed that administration of melatonin at a dosage of 50mg/frog/day either in the morning or in the evening for a week inhibited spermatogenesis. However, when melatonin was administered both in the morning and evening or for a longer period, this inhibitory effect was lost. Therefore, the physiological action of melatonin seems to be: (1) inadequate secretion to disturb testicular function and (2) countering its own effect and making the pituitary-gonadal axis refractory. This is possible because the hormone is produced continuously to regulate the metabolism in melanocytes to maintain the skin colour of frogs (Kasinathan and Gregalatchoume, 1988).

Hypothalamus and Gonadotrophin-Releasing Hormones (GnRH)

Amphibian hypothalamo-hypophysial system represents a typical vertebrate pattern in its morphological organization. The hormone released from the hypothalamic region influences the hypophysis for the synthesis/release of the gonadotrophin for further activity in the testis. The GnRH is produced by the preoptic nucleus of the hypothalamus which is mediated through the portal vessels of the median eminence of the neurohypophysis for controlling the pars distalis secretion. In *Rana temporaria,* a GnRH area has been located in the periventricularis area of the ventral tubercinerum of the caudal hypothalamus both by conventional methods of aldehyde fuschin negative nerve cells and by ultrastructural studies (Dierickx, 1965, 1966; Dierickx *et al.,* 1972). In *Rana esculenta,* ultrastructural evidences show that several neurosecretory cells are present in the nucleus infundibularis ventralis located at the caudal hypothalamus. Evidently type Y cells are involved in GnRH synthesis to regulate pars distalis in turn.

Role of the central nervous system exercising a control on pars distalis through neurosecretory cells and the neurohypophysial relationship has been experimentally studied in considerable detail by Jorgensen and Larsen mainly on *Bufo bufo* (1960). The medical eminence and neurohypophysis were extripated and the pars distalis was implanted to be directly in contact with the axons, normally ending in neurohypophysis. In other toads the pars distalis was transplanted either in the eye muscle or under the oral mucosa. Well developed spermatogenesis was evident when pars distalis was transplanted under the hypothalamus while a degenerated condition was evident in individuals with pars distalis transplanted away from the hypothalamus. These experiments illustrate the hypothalamic control of gonadotrophin function possible with the presence of medial eminence.

Strategy of Reproduction in Amphibia

The mechanism of reproduction and the physiology of the male gonads in amphibia is complex compared to the higher vertebrates. Such complexity is pronounced due to the involvement of external environmental factors for the onset of reproductive processes. Temperature plays a greater role in amphibian spermatogenesis in species exhibiting potentially discontinuous type of spermatogenesis. Those exhibiting continuous and potentially continuous type of spermatogenesis temperature alone may not be the sole factor, as it is associated with humidity and rainfall to facilitate the entire process. Further, it has been noticed that tropical frogs can breed at any time in a year provided optimal environmental condition and nutrition are available. Very little information is available with regard to light impact on spermatogenesis in tropical frogs to draw any substantial conclusion. In this regard, the involvement of pineal cannot be completely ignored and one can only draw a questionable relationship of light-pineal complex-optic tract and their effects on higher nerve centres to stimulate hypothalamus. Probably, the type-Y cells of the nucleus—infundibularis ventralis of the hypothalamic region produces the release hormone which stimulates through median eminance the B_2 and B_3 cells of the pars distalis to produce FSH and LH-ICSH like hormones for testicular maintenance through gonadal fat bodies. Gonadal fat bodies appear to form an important link organ between the pars distalis and testis for spermatogenesis and steroidgenesis. Prostaglandins (PGF-alpha-2) and cAMP are also suspected to be implicated

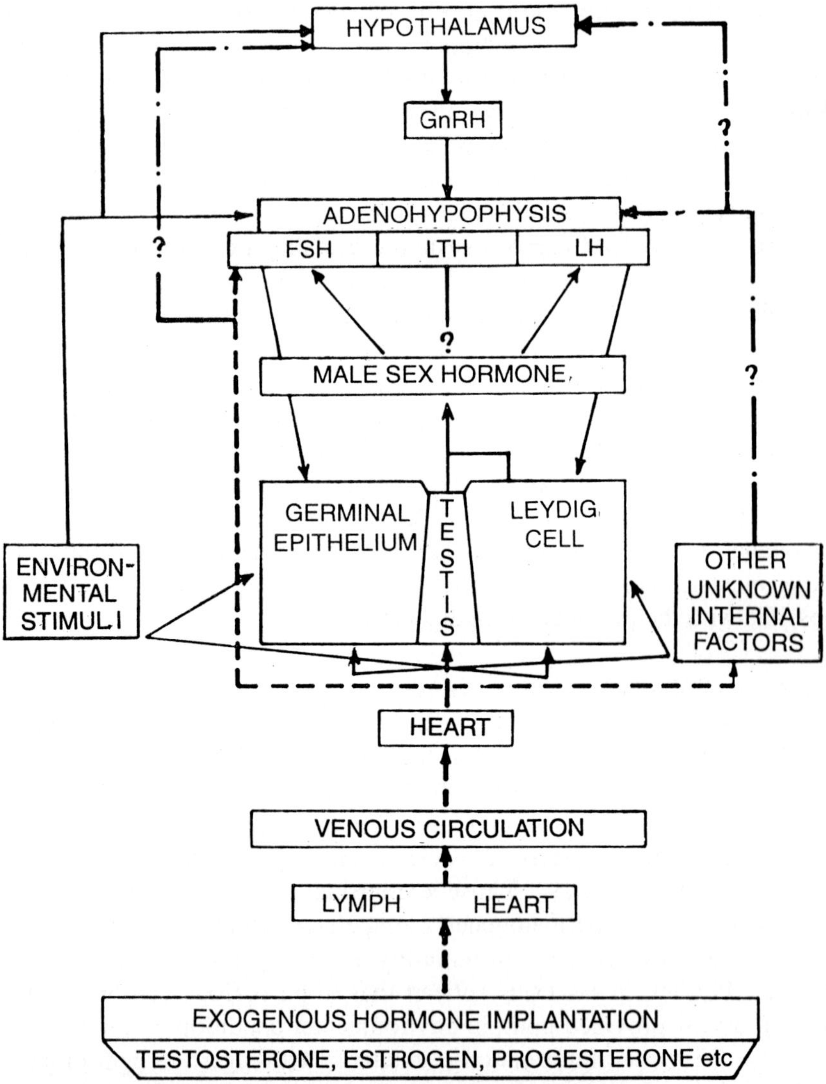

Fig. 1. Schematic representation of inter-relationships of hypothalamo-hypophysial-gonadal axis in relation to internal and external environmental factors in salientia. The probable pathway of action of exogenous steroid implants has been represented (adapted from Basu, 1969).

as a messenger of information relay. The steroid produced by the interstitium of amphibian testis appears to be mainly progesterone or estradiol. No testosterone could be located by chromatographic analysis. However, reports suggest that exogenous testosterone can maintain the thumb pad and vocal sac of male frogs as the male hormone is needed for these target organs. It is logical to assume that testosterone is converted from progesterone as and when needed either in the interstitium or Sertoli cells since the enzyme systems necessary for the conversion has been located in these places.

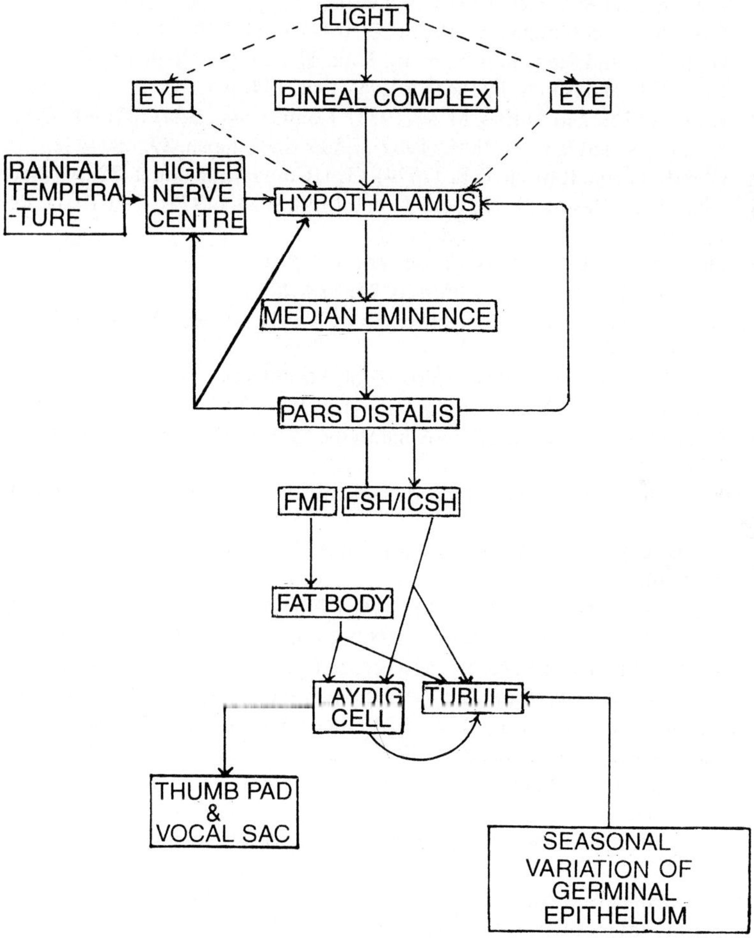

Fig. 2. Schematic representation of the control of male reproductive biology in frogs. It is drawn from the available information of all different sources and not strictly confined to the tropical frogs described here. Dotted lines indicate paucity of information. FMF, Fat Mobilising Factor.

The feedback mechanism is operative between the testis, hypothalamus and hypophysis. The male testicular and other associated accessory sex organs perform their functions in amphibia hand-in-glove with the external stimuli on the one hand and internal rhythm on the other (Figs. 1 and 2).

References

1. Bartke, A. (1971). J. Endocrinol. **49:** 317–324.
2. Basu, S.L. (1962a). Naturwiss. **49:** 188–189.
3. Basu, S.L. (1962b). Proc. Zool. Soc. **15:** 163–170.
4. Basu, S.L. 1968). J. Expt. Zool. **169:** 133–142.
5. Basu, S.L. (1969). Gen. Comp. Endocrinol. Suppl. **2:** 203–213.
6. Basu, S.L. and Mondal, A. (1961a). Naturwiss. **24:** 739–740.
7. Basu, S.L. and Mondal, A. (1961b). Folia Biol. (Warsaw). **9:** 135–142.
8. Basu, S.L. and Nandi, J. (1965) J. Exp. Zool. **159:** 93–112.
9. Burgos, M.H. and Rufino, M.A. (1953). Compt. Soc. Biol. **147:** 900–902.
10. Chanda, S. and Biswas, N.M. (1982). Endocrinol. Japan. **29:** 483–488.
11. Chieffi, G. and Rastogi, R.K. (1974). Gen. Comp. Endocrinol. **22:** Abstract 57.
12. Chieffi, G., Rastogi, R.K., Fela, L. and Milone, M. (1975). Cell Tissue Res. **161:** 157–165.
13. Dierickx, K. (1965). Z. Zellforsch. **66:** 504–518.
14. Dierickx, K. (1966). Z. Zellforsch. **74:** 188–203.
15. Dierickx, K., Druyts, A., Vandenberghe, M.P. and Goossens, W. (1972). Z. Zellforsch. **134:** 450–504.
16. Dubois, A.M. (1972). Rev. Suisse Zool. **34:** 499–581.
17. Galgano, M. (1941). Monit. Zool. Ital. **52:** 223–230.
18. Gopalakrishnan, M. and Rajasekharasetty, M.R. (1977). Ind. J. Exp. Biol. **15(12):** 1204–1296.
19. Hooli, M.A., Saidapur, S.K. and Nadkarni, V.B. (1973). Curr. Sci. **42(10):** 356–357.
20. Houssay, B., Giusti, L. and Lescano-Ganza Lez, J.M. (1929). Compt. Rend. Soc. Biol. **102:** 854–870.
21. Jorgensen, C.B. and Larson, L.O. (1960). Proc. Soc. Exp. Biol. N.Y. **103:** 685–688.
22. Kasinathan, S. and Anandan, V. (1982). Comp. Physiol. Ecol. **7(4):** 288–290.
23. Kasinathan, S. and Basu, S.L. (1972). Endocrinol. Japan. **19(1):** 59–67.
24. Kasinathan, S. and Basu, S.L. (1973), Proc. Indi. Acad. Sci. **77(5):** 190–195.
25. Kasinathan, S. and Basu, S.L. (1974). Ind. J. Exp. Biol. **12(5):** 450–451.
26. Kasinathan, S. and Basu, S.L. (1977). Ind. J. Exp. Biol. **15(5):** 377–378.
27. Kasinathan, S., Basu, S.L. and Sriramulu, V. (1977). Ind. J. Expt. Biol. **15:** 703–707.
28. Kasinathan, S., Basu, S.L. and Sriramulu, V. (1979). Comp. Physiol. Ecol. **4(1):** 7–10.
29. Kasinathan, S., Chandrababu, S., Anandan, V., Bindu Madhava Rao, P.V. and Basu, S.L. (1981). Comp. Physiol. Ecol. **6(1):** 10–12.
30. Kasinathan, S. and Gregalatchoumi, (1988). Endocrinol. Japan. **35(2):** 255–260.
31. Kasinathan, S., Gunasingh, A. and Basu, S.L. (1978). Biol. Zool. **45:** 15–22.
32. Kasinathan, S., Gunasingh, A. and Basu, S.L. (1979). Comp. Physiol. Ecol. **4(4):** 201–206.
33. Larson, L. O. (1963). Gen. Comp. Endocrinol. **3:** 713–714.
34. Lofts, B. (1974). In: Physiology of Amphibia (B. Lofts, Ed), pp. 107–218. Academic Press Inc. New York.
35. Ozon, R. (1966). Annl. Biol. Anim. Biochem. Biophys. **6(4):** 537–551.
36. Rastogi, R.K. (1976). Biol. Zool. **43:** 151–172.
37. Rastogi, R.K., Jela, L., Saxena, P.K. and Chieffi, G. (1976). J. Exptl. Zool. **196:** 151–166.
38. Saidapur, S.K. and Nadkarni, V.B. (1973). Gen. Comp. Endocrinol. **21:** 225–230.

39. Thygaraja, B.S. and Sarkar, H.B.D. (1972–73). J. Mysore Univ. Sci. B. **25:** 45–49.
40. Van Dongen, W.J., Jorgensen, C.B., Larsen, L.O., Rosenkilde, P., Lofts, B. and Van Oordt, P.G.W.J. (1966). Gen. Comp. Endocrinol. **6:** 491–518.
41. Van Kemenede, J.A.M. and Van Oordt, P.G.W.J. (1968). Z. Zellforsch. **91:** 96–111.
42. Van Oordt, P.G.W.J. (1956). Thesis. Utrecht Unit 1–116. G.W.Van der Well & Co.
43. Van Oordt, P.G.W.J. (1963). Cytologie Adenohypophyse Colloq. Intern. Cent. Natl. Rech. Sci. (Paris). **128:** 301–313.
44. Van Oordt, C.J., Crautzberg, F. and Spronk, N. (1949). Proc. K. Nedrl. Adad, Wetens (C). **52:** 535–537.
45. Van Oordt, P.G.W.J. and Lofts, B. (1963). J. Endocr. **27:** 137–146.

Comparative Endocrinology and Reproduction
K.P. Joy, A. Krishna and C. Haldar (Eds)

14. Biochemistry of Sperm Motility Initiation During Epididymal Maturation

G.C. Majumder, B.S. Jaiswal*, D. Nath, S. Banerjee, M. Barua, M. Sarkar, A.P.S. Rana, S. Mitra***, P. Dutta, T. Chatterjee, S. Misra[1] and A. Ghosh[1]**

Indian Institute of Chemical Biology, 4, Raja S.C. Mullick Road, Jadavpur, Calcutta - 700 032, India

[1]Department of Chemistry, Bose Institute, Calcutta - 700 009, India

Introduction

Mammalian testicular spermatozoa are incapable of fertilizing ova. They undergo a maturation process during transit through epididymis when they acquire the potential of forward progression and fertility. The epididymis is a single highly convoluted tubule that links the efferent ducts to the vas deference and it rests on the tunica albuginea of the testis. This tissue is made up of two major components: the epithelium and the lumen. The major epithelial cells are termed as "tall columnar" cells. They are the secretory cells being highly active in endocytosis/exocytosis. The epididymis is divided into three parts—the caput (The head), a narrow central portion—the corpus (The body) and the cauda (The tail). The anatomy of epididymis varies significantly in different species. The sperm from the rete testis, pass through the lumen of the caput tubules and during transport through the long epididymal tubule the male gametes are exposed to the epididymal plasma (EP): a secretory product of the epididymal epithelial cells. Spermatozoa acquire maturity during this transit and finally the mature cells are stored in the cauda-end of the lumen. The profiles of acquiring forward motility (FM) during the epididymal transport are markedly different in various species. In goat, most of the sperm cells acquire FM at the terminal stage of transit i.e. from proximal-cauda to distal-cauda stage. Cell proliferation in epididymis as well as the sperm maturation *in vivo* are dependent on the action of androgen (Brooks, 1981), although the biochemical basis of its action is largely unknown.

Present address: *Department of Membrane Research & Biophysics, Weizmann Institute of Science, 76 100 Rehovot, Israel.
**Diabetes Research Unit, Massachusettes General Hospital, MGH East, Bldg 149, 18th Street, Charlestown, MA 02129, USA.
***2445, Westervelt Avenue, Bronx, New York 10469, New York, USA.

This review will mostly cover the studies on major biochemical alterations of mammalian spermatozoa during epididymal maturation in intact animals and *in vitro* studies on the initiation of flagellar motility in the immature caput epididymal sperm, with special emphasis on goat epididymal sperm model that has been developed in our laboratory. Washed mature/immature epididymal sperm stick to the glass surface of haemocytometer counting chamber which is widely used for sperm motility assays and this phenomenon of sperm adhesion to glass may pose as an artifact in motility analysis (Stephens *et al.*, 1981; Roy and Majumder 1989). Recent studies from our laboratory have identified novel anti-sticking protein factors from epididymal plasma/blood serum that showed high affinity and specificity to inhibit sperm adhesion to glass (Roy and Majumder, 1989). In our studies on the goat sperm model, the anti-sticking factors (ASF) were added to the motility assays to eliminate the possibility of cell-sticking artifact in the motility estimations.

Biochemical Changes of Sperm During Epididymal Maturation

During the epididymal transit, spermatozoa undergo a variety of biochemical alterations (for review see Majumder *et al.*, 1990). Cell surface molecules regulate the functions of the mammalian cells by modulating cell-cell interactions, effector-receptor interactions, membrane permeability, transmembrane signalling, etc. (Yamada *et al.*, 1980). Sperm surface is believed to play a vital role in the regulation of its functions at all stages of the sperm life cycle-maturation/capacitation/acrosomal reaction/fertilization (Hammerstedt and Parks, 1987; Majumder *et al.*, 1990). This section reviews maturation dependent biochemical alterations of spermatozoa with special reference to cell surface.

Intrasperm alterations

Metabolism: In most species sperm metabolic rates increase and synthetic activities decrease during epididymal transit (Voglmayr *et al.*, 1967; Bedford, 1975; Dravland and Meizel, 1981). Glycolytic rates and respiration are several-fold higher in mature caudal or ejaculated spermatozoa than in testicular or caput cells from ram (Voglmayr *et al.*, 1967; Voglmayr, 1975; Inskeep and Hammerstedt, 1982, 1983), bull (Voglmayr and Amann, 1973, Hoskins *et al.*, 1975a), guineapig (Frenkel *et al.*, 1973) and boar (Dacheux *et al.*, 1979). The α-carnitine level also increases several-fold during epididymal passage in bull (Casillas, 1973) which acts as an endogenous substrate for the coincidental increasing level of respiration. Ram testicular spermatozoa convert only 60% as much glucose to lactate as do ejaculated cells, but convert more to CO_2 and synthesize more lipids, amino acids, inositol and carboxylic acids (Scott *et al.*, 1967; Voglymayr *et al.*, 1967; Voglmayr, 1975).

Adenosine 3′-5′ monophosphate (cyclic AMP) Cyclic AMP level increases in the spermatozoa of ram (Amann *et al.*, 1982), bovine (Hoskins *et al.*, 1974, 1975b), boar (Goh, 1983) and goat during epididymal transit. In goat, the epididymal maturation process was associated with a 4-fold rise in the intrasperm cyclic AMP level. The capacity of spermatozoa to respond to adenosine and its agonist increases markedly during epididymal passage (Vijayraghavan *et al.*, 1983b). The level of cAMP is determined by the relative rates of cyclic AMP synthesis and degradation i.e., by the relative activities of adenylate cyclase and its cyclic AMP phosphodiesterase (CPD), respectively. Adenylate cyclase activity of porcine sperm increases during epididymal transit, being maximal in corpus-sperm (Okamura *et al.*, 1985) whereas CPD activity of maturing sperm decrease in bovine (Stephens *et al.*, 1979), ram (Tash, 1976) and goat (Jaiswal and Majumder, 1996a). CPD activity of mature goat caudal sperm is one-third of that of immature caput-sperm. CPD may thus have an important role in epididymal sperm maturation (Jaiswal and Majumder, 1996a).

pH: The intrasperm pH has been found to increase 0.4 unit from caput (5.84 ± 0.1) to cauda (6.27 ± 0.05) in bovine (Vijayraghavan *et al.*, 1985) and 0.25 pH unit from caput (5.54 ± 0.02) to cauda (5.77 ± 0.01) in goat (Jaiswal, B.S and Majumder, G.C., unpublished observation) indicating thereby that intrasperm pH may have role in regulation of sperm motility initiation in epididymis. Biochemical basis of the maturation-dependent cellular alkalinization is still not clear. Alkaline intrasperm environment is believed to activate adenylate cyclase (Schackmann *et al.*, 1981; Okamura *et al.*, 1986; Rojas *et al.*, 1993) thereby causing increase of sperm cyclic AMP level. Alkaline pH (7.8–8.0) favours dynein microtubule sliding mechanism as shown in an invertebrate (sea urchin) sperm model (Shapiro *et al.*, 1985). It is thus possible that maturation-dependent intrasperm pH-shift may stimulate FM by activating the microtubules sliding process.

Sperm surface alterations

Sperm plasma membrane lipid composition Lipid, a major constituent of the plasma membrane (PM) plays an important role in determining the membrane structure. The PM lipids are thus believed to play an important role in the cellular recognition, species specificity and tissue immunity (Stubbs and Smith, 1984). Limited data are available on the lipid constituents of the maturing sperm membrane. PM lipids of the ram (Parks and Hammerstedt, 1985) and boar (Nikolopoulou *et al.*, 1985) spermatozoa have been shown to undergo marked changes during epididymal maturation. From our studies on the goat sperm model, we have developed an aqueous two-phase polymer method for the isolation of highly purified plasma membranes from maturing goat epididymal spermatozoa (Rana and Majumder,

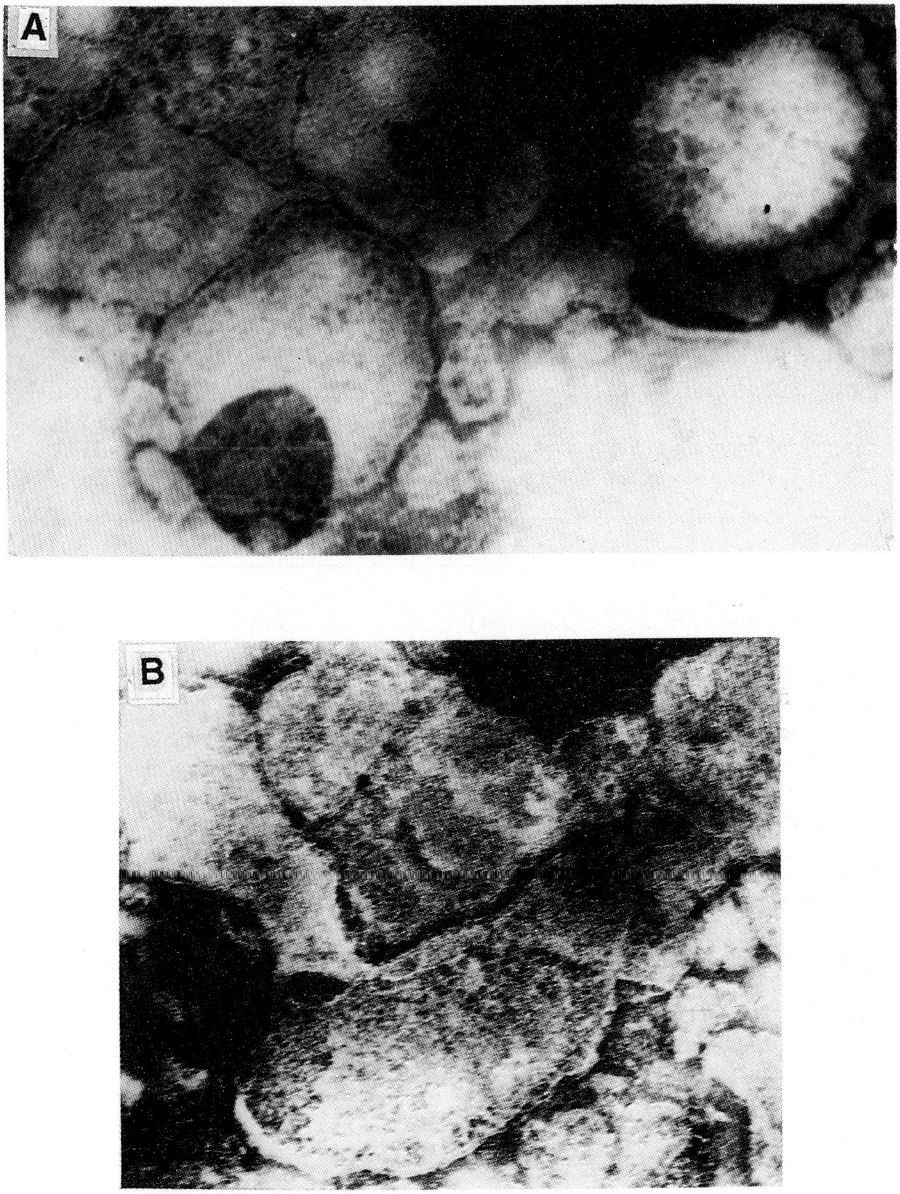

Fig. 1. Electron micrographs of isolated plasma membranes derived from the immature caput-epididymal spermatozoa (A: × 70,000) and the mature cauda-epididymal sperm (B: × 100,000). Membrane vesicles are seen widely scattered.

1987, 1989). The isolated PM showed high degree of purity as evidenced by marker enzyme analysis and electron microscopy (Fig. 1).

Phospholipid (approx. 75% w/w), neutral lipids (approx 15% w/w) and glycolipid (approx 10% w/w) represented the major sperm membrane lipids. The concentration of these lipids and their fatty acid composition underwent marked alteration during epididymal sperm maturation (Rana *et al.*, 1991). Table 1 shows the phospholipid(PL) composition of goat maturing sperm PM. The sperm PM is particularly rich in the ether lipids alkylacyl and alkenylacyl forms of phosphatidyl choline (PC) and phosphatidyl ethanolamine (PE). Of all the membrane phospholipids species diacyl PE decreased most strikingly (approx 65%) during the epididymal maturity of sperm. Maturation-dependent alterations of multiple neutral lipids including the novel wax ester and 1-0-alkyl-diacylglycerol (alkyl DAG) (Table 2) have been demonstrated for the first time using the goat model (Rana *et al.*, 1991, 1992). As speculated earlier (Rana *et al.*, 1991) the alkyl DAG may participate in cell signalling across the cell membrane. PM lipid composition

Table 1. Phospolipid composition of maturing sperm plasma membranes The data shown are mean of three experiments that are within ±10% of the mean

Phospholipids[a]	Caput	Corpus	Cauda
Total phosphatidyl ethanolamine (PE)	364.0	215.0	199.0
DiacylPE	84.7	41.6	28.4
AlkylacylPE	84.7	48.5	56.0
Alkenylacyl PE	195.0	125.0	112.0
Total phosphatidyl-choline (PC)	373.0	381.2	284.0
DiacylPC	50.8	125.0	28.4
Alkylacyl PC	89.1	55.0	57.0
AlkenylacylPC	237.2	200.0	200.0
Sphingomyelin	110.1	98.0	85.2

[a]Expressed as μg/mg protein. (Reprinted from Rana *et al.*, 1991, with permission of Elsevier Science—NL, Sara Burgerhartstraat 25, 1055 K V Amsterdam, The Netherlands).

Table 2: Neutral lipid composition of maturing sperm plasma membranes. The data shown are mean of three experiments, that are within ±12.0% of the mean

Neutral lipids[a]	Caput	Corpus	Cauda
Wax ester	5.9	7.4	3.3
Steryl ester	8.9	6.7	24.3
1-0-Alkyldiacylglycerol	9.0	7.1	5.3
Triacylglycerol	6.9	3.3	2.2
Fatty acid	11.6	11.0	6.0
Sterol	112.4	128.0	180.0

[a]Expressed as μg/mg of membrane protein. (Reprinted from Rana *et al.*, 1991, with permission of Elsevier Science—NL, Sara Burgerhartstraat 25, 1055 K V Amsterdam, The Netherlands).

of boar (Nikolopoulou *et al.*, 1985), ram (Parks and Hammerstedt, 1985) and goat (Rana *et al.*, 1991) showed some similarities and dissimilarities in respect of their alteration during sperm maturation. Although PC and PE were the major phospholipids in the sperm PM of all the species, marked differences were noted in their maturation-associated changes. The chemical nature of the glycolipids were different in boar (Nikolopoulou *et al.*, 1985) and goat (Rana *et al.*, 1991) sperm PM but both of them underwent marked decrease during maturation.

Membrane lipid asymmetry Little is known about the lipid asymmetry of sperm PM. Using phospholipase treatment, asymmetric distribution of phospholipids has been shown in the inner and outer membrane leaflets of ejaculated ram sperm (Hinkovska *et al.*, 1986). Phospholipid asymmetry in maturing sperm PM has been shown for the first time using goat model (Rana *et al.*, 1993).

The inner and outer membranes showed marked differences in the phospholipid composition at all stages of sperm maturation. The outer membrane was rich in PC and sphingomyelin whereas the inner leaflet was dominated by PE (Fig. 2). Profound increase in PC in the sperm outer membrane implicates that the outer membrane leaflet is rich in ether linked PC as sperm membrane PC has been shown to be rich in ether lipids (Rana *et al.*, 1991). Ether linkages retard the action of lipases on phospholipids and thus add to the stability of PM (Salivonchick *et al.*, 1980). Maturation-associated altered membrane leaflets may provide the infrastructure essential for the expression of sperm motility and fertility.

Membrane fluidity Several investigators have studied changes in sperm PM viscosity with the isolated membranes of specific domains of the sperm surface for determining regional lipid variation (Vijayasarathy *et al.*, 1982; Vijayasarathy and Balaram, 1982). Using pyrene and 1,6-diphenyl-1,3,5 hexatriene (DPH) as the lipophilic probes, it has been shown that the fluidity of the isolated goat sperm PM decreases significantly during the epididymal transit of sperm (Rana and Majumder, 1990). Arrhenious plots of the data on microviscosity at varying temperatures show that the maturing goat sperm membranes have two distinct lipid phase transitions in the temperature zones 19–25°C and 34–37°C (Fig. 3) (Rana and Majumder, 1995). The low temperature transitions for the immature caput- and mature cauda-sperm membranes were noted at 19–20°C and 24–25°C respectively, whereas both these membranes showed high temperature transitions at 36–37°C. The data implicate significant alteration of the sperm PM structure during epididymal maturation. It is well documented that the membrane microviscosity is greatly influenced by cholesterol/phospholipid and saturated/unsaturated fatty acid ratios of the lipids (Stubbs and Smith, 1984). It thus appears that the observed marked increases in the above-mentioned ratios of lipids and fattly acids during sperm maturation (Rana *et al.*, 1991), are primarily responsible for the maturation-dependent sperm membrane fluidity and phase transitions. The altered fluid state of the sperm PM may markedly influence membrane function with special reference to ion transport, masking and unmasking of cell-surface macromolecules, activities of PM-bound enzymes etc.

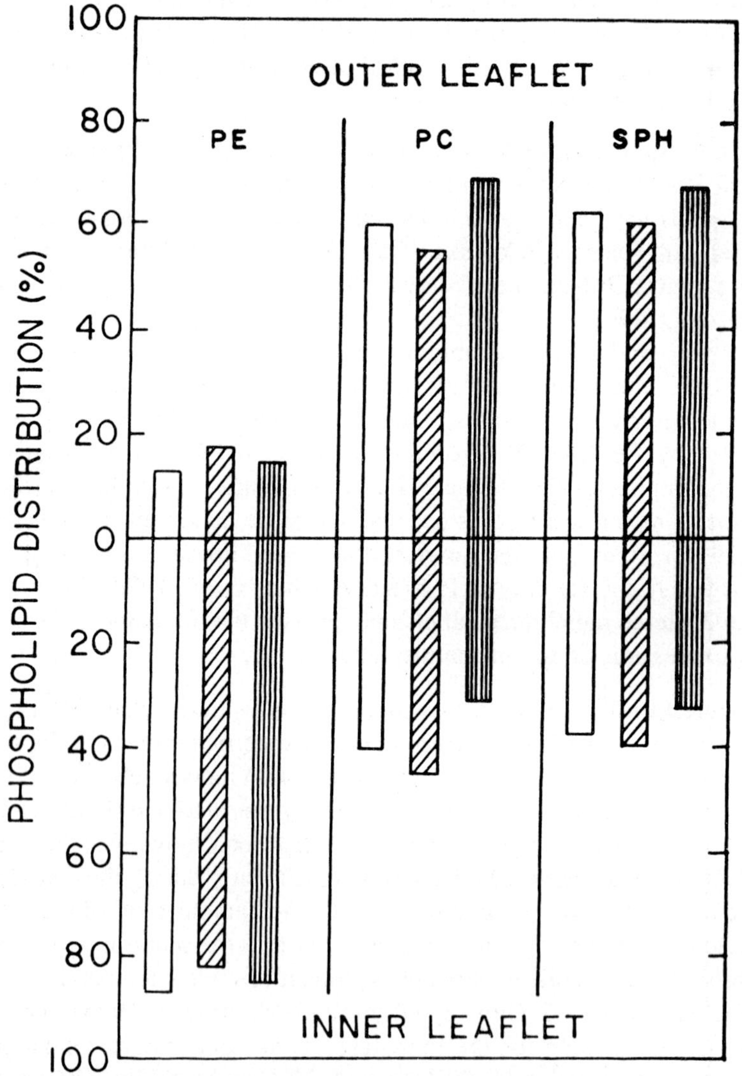

Fig. 2. **Distribution (%) of PLs in the inner/outer PM of maturing sperm. ▭ Caput sperm; ▨ Corpus sperm. ▥ Cauda sperm (reprinted from Rana *et al.*, 1993, with kind permission of Elsevier Science—NL, Sara Burgerhartstraat 25, 1055 KV Amsterdam, The Netherlands).**

Cell-surface protein phosphorylation/dephosphorylation mechanisms

Previous studies have provided several lines of evidences for the occurrence of cyclic AMP-dependent and independent protein kinases and phosphoprotein phosphatase (PPase) on the external surface of sperm of several species. The forward-motile spermatozoa are enriched in the activities of these ecto-enzymes (for review see Majumder *et al.*, 1990). The ecto-cyclic AMP-dependent protein kinase of goat sperm is incapable of phosphorylating membrane proteins and it consists of type-I and II kinases, these are enriched in forward-motile sperm (Majumder *et al.*, 1990; Dey and Majumder, 1990a) and Type-II kinase is maturation-specific (Dey and Majumder, 1990b).

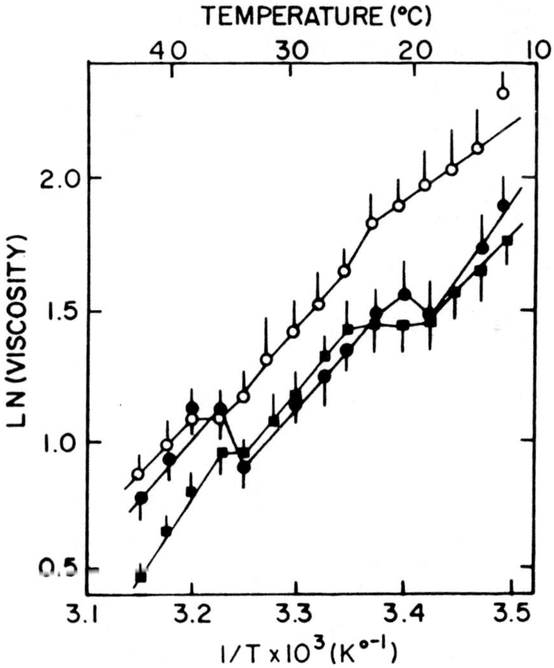

Fig. 3. Arrhenius plot of the microviscosity of the isolated plasma membranes from the maturing goat epididymal caput (●), corpus (■) cauda (○) spermatozoa. The polarization values were estimated using the temperature range 42 to 12°C. The data shown are the mean ± SEM of microviscosity values generated on three different membrane preparations from each sperm maturative stage (reprinted from Rana and Majumder, 1995, with kind permission of the Indian Academy of Sciences, Bangalore, India).

Our studies on the goat sperm model have shown that the ecto-cyclic AMP-independent protein kinase (ecto-CIK) of the intact cells causes phosphorylation of the endogenous membrane-bound phosphoproteins which undergo dephosphorylation by the ecto-PPase (Majumder *et al.*, 1990; Barua and Majumder, 1990). Ecto-protein phosphorylation profile of intact goat sperm by the endogenous ecto-CIK undergoes marked alteration during epididymal maturation (Mitra and Majumder, 1991). More recently the ecto-CIK has been purified from goat sperm and its enzymatic characteristics appear to be different from the other protein kinases reported earlier (Mitra *et al.*, 1994). The isolated CIK has markedly high substrate specificity for the phosphorylation of PM-bound phosphoproteins (Fig. 4). Figure 5 shows the time course of the sperm ecto-protein phsphorylation by the intact-cell CIK and the dephosphorylation of the ^{32}P-labelled proteins when the kinase reaction was arrested with EDTA that chelated Mg^{2+} required for the kinase activity. The dephosphorylated proteins were found to undergo rephosphorylation upon the addition of excess $MgCl_2$ that activated CIK. This study demonstrates conclusively the occurrence of a coupled enzyme system (ecto-CIK/PPase) on the sperm outer surface that regulates phosphorylation/dephosphorylation of outer cell-surface proteins of intact sperm. It thus appears that the ectoenzyme system may have an important role in modulating structure and function of sperm PM by the mechanism of protein phosphorylation/dephosphorylation.

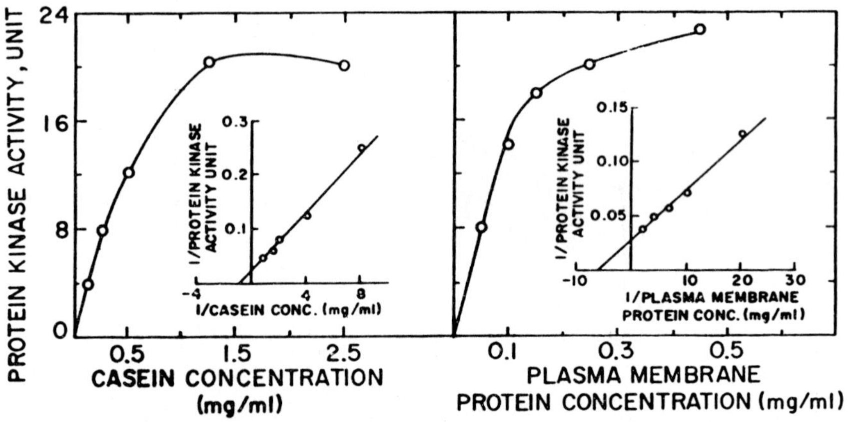

Fig. 4. Effect of various concentrations of: (a) casein and (b) plasma membrane proteins on the activity of isolated protein kinase. Standard assay conditions were used, except for the variation in the amount of casein and plasma membrane proteins. The inserts show the Lineweaver-Burk plot of these data. The data are represenatative of three such experiments (reprinted from Mitra *et al.*, 1994, with kind permission of National Research Council of Canada, Canada.)

Lectin binding sites Lectins, class of oligomeric proteins with high saccharide binding specificities, have been considered very popular probes. These probes are widely used to monitor the changes in PM macromolecules during physiologic maturation of spermatozoa. The differences in lectin mediated agglutination patterns studied so far in different species suggest that the alteration of sperm surface glycocomponents occur during the epididymal passage (Edelman *et al.*, 1971; Nicolson and Yanagimachi, 1972;

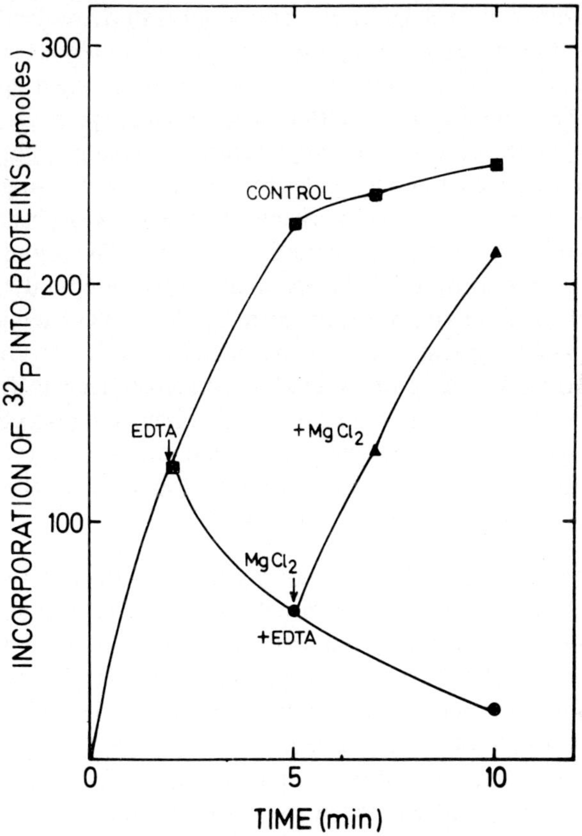

Fig. 5. Time course of phosphorylation (■), dephosphorylation (●) and rephosphorylation (Δ) of intact sperm ecto-proteins by the endogenous ecto-CIK and PPase (reprinted from Barua *et al.*, 1990, with kind permission of the Academic Press, Australia).

Nicolson *et al.*, 1977; Olson and Dango, 1981; Hammerstedt *et al.* 1982). Alterations in the distribution of the lectin receptors which reflect the sperm surface modification in the epididymal lumen have been visualized by the binding of the cell with lectin conjugated to fluorescein isothiocyanate (Lectin-FITC) in mouse, guinea pig, rat, ram and also in non human primates (Edelman *et al.*, 1971; Lewin *et al.*, 1979; Schwarz and Koehler, 1979; Olson and Dango, 1981; Voglmayr *et al.*, 1983). ConA binding to the PM of rabbit and ram sperm did not change during epididymal sperm maturation (Nicolson *et al.*, 1977; Curtens *et al.*, 1982) whereas in rat the population of ConA receptor decreased significantly (Olson and Dango, 1981). Similar observation have also made in rabbit with wheat germ agglutinin (WGA) Ricinus Communis agglutinin (RCA_2) (Nicolson *et al.*, 1977) as well as in guinea pig only with WGA (Enders *et al.*, 1983). In primates like monkey the ConA binding was first demonstrated in the distal corpus whereas WGA binding to the surface molecule was detected in caput epididymal sperm

(Fain-Mayrel *et al.*, 1984). In our laboratory a study on lectin binding sites reveals that the mature cauda epididymal sperm agglutinates more efficiently with ConA and WGA in comparison to immature caput sperm (Sarkar *et al*, 1991). The fluorescence of different epididymal sperm tagged with FITC-ConA conjugate did not show any significant change suggesting that the ConA specific saccharide molecules remain unaltered during the maturation process of sperm cell. The remarkable difference was observed in binding of FITC-WGA conjugate with epididymal sperm. The initial weak labelling of immature sperm by FITC-WGA conjugate with epididymal sperm. The initial weak labelling of immature sperm by FITC-WGA conjugate gradually increased as the sperm reach to the cauda epididymal lumen where the mature sperm gave strong fluoresence at the acrosomal cap. These differences of lectin binding sites of goat epididymal spermatozoa reflect the modification of cell surface carbohydrate moieties linked either with lipid or protein. In separate study the protein profile are also found to be altered during this maturation process (unpublished data).

Membrane antigens Much of the sperm transformation in the epididymis involves the acquisition of sperm coating antigens (Hunter, 1969; Barker and Amann, 1971) during epididymal transit. A rabbit globular glycoprotein component that is secreted by corpus epididymids (Thomas *et al.*, 1984, 1986), binds to the acrosomal regions, and can act to stabilize the acrosome. The association of other epididymal secretory proteins with the sperm have been reported in mouse (Murphy and Caroll, 1987; Cornwall *et al.*, 1990; Tracy *et al.*, 1992), human (Tezon *et al.*, 1985) and bovine (Acott & Hoskins, 1981). In boar, a maturation dependent surface antigen of mol. wt. 27 KDa is identified which is shown to be a proteolytic fragment of an epididymal protein. This antigen interacts with specific regions of sperm membrane during the process of maturation in the epididymis (Okamura *et al.*, 1992). In ram, another 18 K Dal sperm surface glycoantigen of epididymal origin have been identified as a maturation antigen (Mckinnon *et al.*, 1991). Studies from our laboratory have demonstrated the presence of multiple antigens in the highly purified goat epididymal sperm PM (Chatterjee and Majumder, 1989). As shown in Fig. 6, the profile of the goat sperm surface antigens are markedly altered during epididymal sperm maturation.

Sperm autoagglutination The maturing ram/boar epididymal spermatozoa derived from mid/distal caput and corpus epididymis have been shown to undergo autoagglutination, when incubated *in vitro* in Krebs Ringer bicarbonate medium at 37°C (Dacheux *et al.*, 1983). Biochemical basis of this maturation associated sperm agglutination event is largely unknown. Studies from our laboratory (Banerjee *et al.*, 1992) have shown that the goat distal-corpus epididymal spermatozoa specifically undergo head-to-head autoagglutination, when incubated *in vitro* in a chemically defined medium. This phenomenon of cell agglutination is accompanied by initiation of sperm forward motility (Banerjee *et al.*, 1992). Our recent studies have shown that the goat sperm autoagglutination event is dependent on Ca^{++} and it is inhibited specifically by the sugar D-galactose (Banerjee S. and Majumder G.C., 1994). The data implicate the occurrence of a galactose specific lectin on sperm surface that interacts with its receptor to cause the

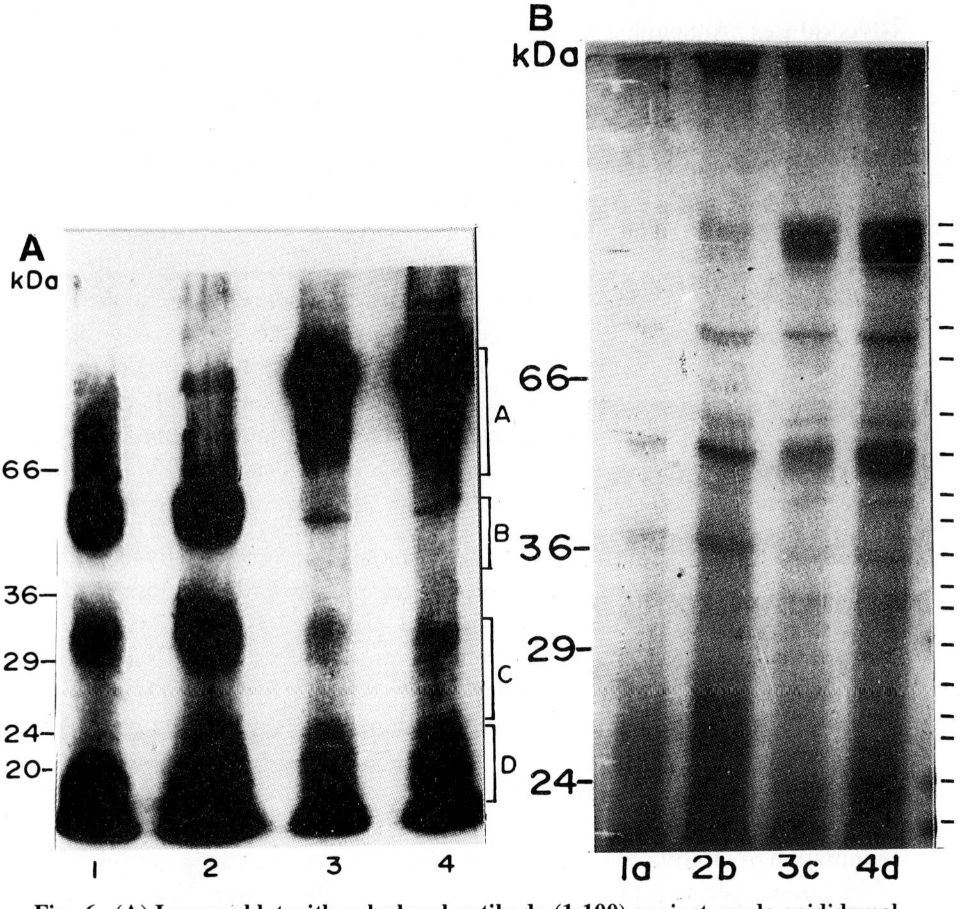

Fig. 6 **(A) Immunoblot with polyclonal antibody (1:100) against cauda epididymal goat sperm PM. The antisera was partially purified by 50% $(NH_4)_2SO_4$ precipitation and was extensively dialysed against 0.02 M PBS, pH 7.4. PM protein of caput (lane 1, 40 μg; *lane* 2, 80 μg) and cauda (lane 3, 40 μg; lane 4, 80 μg) epididymal sperm were electrophoresed in 10% SDS-PAGE and were immunoblotted. We have designated following region according to the approximate molecular weights of groups of polypeptides: A (96KDa, 82KDa, 78KDa, 68KDa); B (57KDa, 50KDa, 48KDA, 43KDa, 36KDa); C (34KDa, 32KDa, 31KDa, 28KDa, 26KDa, 25KDa); D (24KDa, 22KDa, 20KDa, 18KDa). (B) Coomassie blue stained of 10% SDS-PAGE of duplicate of gel (A). Lane la and 2b, caput epididymal sperm PM (40 μg and 80 μg). Lane 3c and 4d, cauda epididymal sperm PM (40 μg; 80 μg) (reprinted from Chatterjee and Majumder, 1989, with kind permission of the Academic Press Inc., USA.)**

maturation-dependent sperm aggregation phenomenon. The lectin and its receptor sites are greatly manipulated during epididymal sperm transit (Banerjee S. and Majumder G.C., unpublished results). The data suggest that sperm surface lectin-sugar interaction may play an important role in motility initiation in epididymis and in the subsequent events of the sperm life cycle.

Glycosidases Although glycosidases may play vital role in cellular regulation by modulating the sugar side chains of the biologically active cell surface glycoconjugates, limited data are available on the sperm PM-bound glycosidases. Tulsiani *et al.* (1989) have demonstrated a α-mannosidase in rat and human sperm membranes and the membrane associated glycosidase may mediate sperm oocyte interaction as shown in mouse model. Tulsiani *et al.* (1995) have recently shown that rat sperm PM-bound α-mannosidase undergoes proteolytic processing during the epididymal transit. Recent studies from our laboratory have shown the presence of a neutral β-galactosidase in goat epididymal sperm PM and the specific activity of the enzyme increase significantly (approx 2-fold) during epididymal maturation of sperm. The glycosidase isolated from goat cauda sperm PM markedly inhibits motility of the mature goat sperm (Dutta P. and Majumder G.C., Unpublished data). The isolated sperm β-galactosidase has the unusual characteristic of heat-stability.

Other changes Modification of sperm flagellar proteins by covalent carboxymethylation is involved in the regulation and development of mammalian sperm motility (Bardin and Gagnon, 1982). The protein carboxymethylation enzyme system includes protein carboxymethylase (PCM), protein methyl esterase and S-adenosyl-L-homocystein hydrolase. Maturation partially involves changes in sperm PCM and methyl acceptor proteins (MAP) (Bouchard *et al.*, 1980; Gagnon *et al.*, 1982). MAP capacity increases markedly as the capacity for motility develops during transit from the rat testis to the cauda epididymis (Bouchard *et al.*, 1980) but their physiological importance in initiation of motility in maturing sperm is not clear. The cauda spermatozoa of different species like rat (Olson and Dano, 1981), ram (Voglmayr and Dacheux, 1983), chimpanzee (Dacheux *et al.*, 1987), boar (Dacheux *et al.*, 1989), cynomolgous monkeys (Mahony *et al.*, 1994), appear to lack certain proteins which are initially detectable on caput spermatozoa. Thus, not only the expression of proteins but the loss of some surface components could also be correlated with the epididymal maturation.

In-Vitro Studies of Initiation of Sperm Forward Motility in Immature Epididymal Sperm

To gain an insight into the biochemical basis of the initiation of FM during epididymal transit of the male gametes, it is essential to develop an *in vitro* model system that permits induction of FM in the immature caput-epididymal sperm. Such an *in vitro* model was reported for the first time by Hoskins and his associates (Hoskins *et al.*, 1978; Vijayaraghavan *et al.*, 1985). These investigators have induced FM in the bovine caput-epididymal immature spermatozoa when incubated with theophylline; a cyclic AMP phosphodiesterase inhibitor and seminal plasma epididymal plasma (EP). Subsequently, several investigators have developed similar *in vitro* models for hamster (Kann and Serres, 1980; Yeung, 1984; Cornwall *et al.*, 1986) and goat (Jaiswal and Majumder, 1996b) caput sperm. Like the bovine model, the immotile goat caput sperm also acquire significant amount of FM upon incubation with theophylline and EP. In all these *in vitro* models, both theophylline and seminal plasma/EP are essential for FM initiation. Theophylline or EP alone is incapable of inducing FM in the caput sperm.

Role of forward motility protein Hoskins and his associates (Acott and Hoskin, 1978; Hoskins *et al.*, 1978) have shown that the active principle of bovine seminal plasma is a heat-stable protein designated as forward motility promoting protein (FMP). FMP is a 37 KDa glycoprotein that has been partially purified. FMP is essential for the induction of motility *in vitro* (Acott and Hoskins, 1978). It is produced in the epididymis (Brandt *et al.*, 1978) and is believed to be involved in the development of sperm motility as the sperm move from the caput to the cauda epididymis (Acott and Hoskins, 1978). As shown by Acott *et al.* (1983), there is bending in the neck part of bovine caput sperm and FMP is believed to activate motility in these cells by straightening the neck part of sperm. Little is known about the biochemical basis of the action of FMP.

Role of intrasperm level of cyclic AMP As mentioned above theophylline is essential for the *in vitro* initiation of FM in the caput sperm of all the species. Theophylline and caffiene: inhibitors of cyclic AMP phosphodiesterase have been shown to elevate intrasperm level of cyclic AMP (Hoskins and Casillas, 1975a; Garber and Kopf, 1980; Hoskins *et al.*, 1983). These data implicate that in the *in vitro* sperm models theophylline induces FM by elevating sperm cytosolic cyclic AMP level through its inhibitory action on the cyclic AMP phosphodiesterase. Theophylline activates FM in goat epididymal sperm of all stages of maturity. It is thus clear that cyclic AMP plays a vital role in sperm maturation. This view is supported by the finding that during *in vivo* maturation sperm level of cyclic AMP increases (Hoskins *et al.*, 1974; Amann *et al.*, 1982; Okamura *et al.*, 1983). As already reviewed (Majumder *et al.*, 1990) cyclic AMP is believed to activate a cyclic AMP-dependent protein kinase which in turn causes phosphorylation of several intrasperm phosphoproteins and one or more of these phosphoproteins may directly participate in the induction of flagellar motility.

Role of bicarbonate Exogenous bicarbonate has been shown to activate further the FMP-dependent *in vitro* FM initiation in the bovine caput sperm (Hoskins *et al.*, 1983; Vijayraghavan *et al.*, 1985). Bicarbonate is believed to activate motility by stimulating sperm adenylate cyclase activity, thereby enhancing intrasperm level of cyclic AMP: an activator of flagellar motility (Okamura *et al.*, 1986; Tajima *et al.*, 1987; Rojas *et al.* 1993).

Role of external pH Morisawa *et al.* (1993) suggested a close correlation among external pH, internal cyclic AMP and acquisition of sperm motility. A change in intracellular cyclic AMP level influenced by external pH (Miura *et al.*, 1992) through the regulation of adenyl cyclase and phosphodiesterase activity in sperm cells (Garber and Kopf, 1980) favoured sperm motility initiation in *Salmondae*. High external pH has been found to increase the internal pH and initiates the motility of boar caudal epididymal spermatozoa (Gatti *et al.*, 1993).

Acknowledgements

Research fellowships offered to B.S.J., S.B., M.B., M.S., A.P.S.R. and S.M. by the Council of Scientific and Industrial Research, New Delhi and to D.N. and A.P.S.R. by the Indian Council of Medical Research, New Delhi are gratefully acknowledged. We also thank Dr. J. Das, Director of Indian Institute of Chemical Biology for his interest in this study.

References

Acott T.S. and Hoskins D.D. (1978) *J. Biol. Chem.* **253:** 6744–6750.
Acott T.S., Katz D. and Hoskins D.D. (1983) *Biol. Reprod.* **29:** 389–399.
Amann R.P., Hay S.R. and Hammerstedt R.H. (1982) *Biol. Reprod.* **27:** 723–733.
Banerjee S., Dey N. and Majumder G.C. (1992) *Ind. J. Exp. Biol.* **30:** 1056–1061.
Banerjee S. (1994) Ph.D. Thesis, Jadavpur University.
Banerjee S. and Majumder G.C. (1994) In *Proceedings of the 16th International Congress of Biochemistry and Molecular Biology, Vol. II* Abs. p 364.
Bardin C.W. and Gagnon C. (1982) *Progress Clin. Biol. Res.* **87:** 217–234.
Barker L.D.S. and Amann R.P. (1971) *J. Reprod. Fertil.* **26:** 319–332.
Barua M., Halder S. and Majumder G.C. (1990) *Biochem. Int.* **20:** 1089–1096.
Bedford J.M. (1975) In: *Handbook of Physiology:* Hamilton D.W., Greep R.O. (eds.). *Am Physiol. Soc. Washington, D.C., Vol. 5,* p 303–317.
Bouchard P., Gagnon C., Phllips D.M. and Bardin C.W. (1980) *J. Cell. Biol.* **86:** 417–423.
Brooks D.E. (1981) *Physiol. Rev.* **61:** 515–555.
Brandt H., Acott T.S., Johnson D.J. and Hoskins D.D. (1978) *Biol. Reprod.* **19:** 830–835.
Cassillas E.R. (1973) *J. Biol. Chem.* **248:** 8227–8232.
Cassillas E.R., Elder C.M. and Hoskins D.D. (1980) *J. Reprod. Fertil.* **59:** 297–302.
Chatterjee T. and Majumder G.C. (1989) *Biochem. Biophys. Res. Commun.* **162:** 550–556.
Cornwell G.A., Smyth T.B., Vindivich D., Harter C., Robinson J. and Chang T.S.K. (1986) *Biol. Reprod.* **35:** 1065–1074.
Cornwell G.A., Vreeburg J.L., Holland M.K. and Orgebin Crist M.L., (1990) *Biol. Reprod.* **43:** 121–129.

Curtens J.L., Rozinek J. and Fóurnier-Delpech S. (1982) *Andrologia* **14:** 509–514.
Dacheux J.L., O'shea T. and Paquignon M. (1979) *J. Reprod. Fertil.* **55:** 287–296.
Dacheux J.L., Paquignon M. and Combarnous Y. (1983) *J. Reprod. Fertil.* **67:** 181–189.
Dacheux J.L., Chevvrien C. and Lanson Y. (1987) *Proc. Natl. Acad. Sci. USA*, **513:** 560–563.
Dacheux J.L., Dacheux F. and Paquignon M. (1989) *Biol. Reprod.* **40:** 635–651.
Dey C.S. and Majumder G.C. (1990a) *Biochem. Cell Biol.* **68:** 459–470.
Dey C.S. and Majumder G.C. (1990b) *Biochem. Int.* **21:** 659–665.
Dravland E. and Meizel S. (1981) *Gamete Res.* **4:** 515–523.
Dutta P. and Majumder G.C. (1993) *Biochem. Cell Biol.* **71:** 22–26.
Edelman G.M. and Millette C.F. (1971) *Proc. Natl. Acad. Sci. USA,* **68:** 2436–2440.
Enders G.C., Werb Z. and Friend D.S. (1983) *J. Cell Sci.* **60:** 303–329.
Fain-Maurel M.A., Dadoune J.P. and Regor J.F. (1984) *Anat. Res.* **208:** 375–382.
Frenkel G., Peterson R.N. and Freund M. (1973) *Proc. Soci. Expt. Med.* **144:** 420–425.
Gagnon C., Sherins R.J., Phillips D.M. and Bardin C.W. (1982) *New Eng. J. Med.* **306:** 821–825.
Garber D.L. and Kopf G.S. (1980) *Adv. Cyclic. Nucleotide Res.* **13:** 251–306.
Garber D.L., Lust W.D., First N.L. and Lardy H.A. (1971) *Biochemistry* **10:** 1825–1831.
Gatti J.L., Chevrier C., Paquignon M. and Dacheux J.L. (1993) *J. Reprod. Fertil.* **98:** 439–449.
Goh P. (1983) Ph.D. Thesis, University of Sydney, Australia.
Hall J.C. and Killian G.J. (1987) *Biol. Reprod.* **36:** 709–718.
Hammerstedt R.H., Hay S.R. and Amann R.P. (1982) *Biol. Reprod.* **27:** 745–754.
Hammerstedt R.H. and Parks J.E. (1987) *J. Reprod. Fertil.* **34:** (Suppl.) 133–149.
Hinkovska V.T., Dimitrov G.P. and Koumanov K.S. (1986) *Int. J. Biochem.* **18:** 1115–1121.
Hoskins D.D., Stephens D.J. and Hall M.L. (1974) *J. Reprod. Fertil.* **37:** 131–132.
Hoskins D.D. and Casillas E.R. (1975a) In: Singhal R.A., Thomas J.A. (eds.) *Advance in Sex hormone Research,* University Park Press, Baltimore, *Vol. I,* p 284.
Hoskins D.D., Munsterman D. and Hall M.L. (1975a) *Biol. Reprod.* **12:** 566–572.
Hoskins D.D., Hall M.L. and Musterman D. (1975b) *Biol. Reprod.* **13:** 163–176.
Hoskins D.D., Brandt H. and Acott T.S. (1978) *Fed. Proc.* **37:** 2534–2542.
Hoskins D.D., Acott T.S., Critchlow L.M. and Vijayraghavan S. (1983) *J. Submicros. Cytol.* **15:** 21–27.
Hunter R.H.F. (1969) *J. Reprod. Fertil.* **20:** 223–237.
Inskeep P.B. and Hammerstedt R.H. (1982) *Biol. Reprod.* **27:** 735–743.
Inskeep P.B. and Hammerstedt R.H. (1983) *J. Biochem. Biophys. Methods* **7:** 199–210.
Jaiswal B.S. and Majumder G.C. (1996a) *Biochem. Cell Biol.* **74:** 669–674.
Jaiswal B.S. and Majumder G.C. (1996b) *Int. J. Andrology* **19:** 97–102.
Kann M.L. and Serres C. (1980) *Reprod. Nutr. Dev.* **20:** 1739–1749.
Lewin L.M., Weissenberg R., Sobel J.S., Marcub Z. and Nebel L. (1979) *Arch. Androl.* **2:** 279–281.
Mandal M., Banerjee S. and Majumder G.C. (1989) *Biol. Reprod.* **41:** 983–989.
Mahony M.C., Fan Q., Cartwrigh S. and Hodgen G.D. (1994) *Am. J. Reprod. Immunol.* **31(2–3):** 133–140.
Majumder G.C., Dey C.S., Halder S. and Barua M. (1990) *Arch. Androl.* **24:** 287–303.
Mitra S., Nath D. and Majumder G.C. (1994) *Biochem. Cell Biol.* **72:** 218–226.
Mitra S. and Majumder G.C. (1991) *Biochem. Int.* **23:** 611–618.
McKinnon C.A., Weaver F.E., Yodev J.A., Fairbanks G. and Wolf D.E. (1991) *Mol. Reprod. Dev.* **29:** 200–207.
Miura T., Yamaguchi K., Takahashi H. and Nagahama Y. (1992) *J. Exp. Zoo.* **261:** 359–363.
Murphy G.J.P. and Caroll J. (1987) *Biochem. J.* **241:** 379–387.

Morisawa S., Ishida K., Okuno M. and Mrisawa M. (1993) *Mol. Reprod. Dev.* **34:** 420–426.
Nicolson G.L. and Yanagimachi R.A. (1972) *Science* **177:** 276–279.
Nicolson G.L., Usui N., Yanagimachi R., Yanagimachi H. and Smith J.R. (1977) *J. Cell Biol.* **74:** 950–962.
Nikolopoulou, M., Soucek D.K. and Vary, J.C. (1985) *Biochim. Biophys. Acta.* **815:** 486–498.
Okamura N. and Sugita Y. (1983) *J. Biol. Chem.* **258:** 13056–13062.
Okamura N., Tajima Y., Soejima A., Masuda H. and Sugita Y. (1985) *J. Biol. Chem.* **260:** 9699–9705.
Okamura N., Tajima Y., Ishikawa H., Yeshi S., Koiso K. and Sugita Y. (1986) *Fertil. Steril.* **45:** 265–272.
Okamura N., Dacheux F., Venien A., Onoe S., Huet J.C. and Dacheux J.L. (1992) *Biol. Reprod.* **47:** 1040–1052.
Olson G.E. and Danjo B.J. (1981) *Biol. Reprod.* **24:** 431–443.
Parks J.E. and Hammerstedt R.H. (1985) *Biol. Reprod.* **32:** 653–668.
Rana A.P.S. and Majumder G.C. (1987) *Prep. Biochem.* **17:** 261–281.
Rana A.P.S. and Majumder G.C. (1989) *Prep. Biochem.* **19:** 167–173.
Rana A.P.S. and Majumder G.C. (1990) *Biochem. Int.* **21:** 797–803.
Rana A.P.S., Majumder G.C., Misra S. and Ghosh A. (1991) *Biochem. Biophys. Acta.* **1061:** 185–196.
Rana A.P.S., Majumder G.C., Misra S. and Ghosh A. (1992) *Lipids* **27:** 75–78.
Rana A.P.S., Misra S., Majumder G.C. and Ghosh A. (1993) *Biochim. Biophys. Acta.* **1210:** 1–7.
Rana A.P.S. and Majumder G.C. (1995) *J. Biosci.* **20:** 525–530.
Rojas F.J., Patrizio P., Do J., Silber S., Asch R.H. and Rajas M.I. (1993) *Endocrinology* **133:** 3030–3033.
Roy N. and Majumdr G.C. (1989) *Biochim. Biophys. Acta.* **991:** 114–122.
Sarkar M., Majumder G.C. and Chatterjee T. (1991) *Biochim. Biophys. Acta.* **1070:** 198–204.
Schackmann R.W., Christen R. and Shapiro B.M. (1981) *Proc. Natl. Acad. Sci. USA*, **78:** 6066–6070.
Schwarz M.A. and Kochler J.K. (1979) *Biol. Reprod.* **21:** 1295–1307.
Scott T.W., Voglmayr J.K. and Setchell B.P. (1967) *Biochem. J.* **102:** 456–461.
Selivonchick D.P., Schmid R.C., Natarafon V. and Schmid H.H.O. (1980) *Biochim. Biophys. Acta.* **618:** 242–254.
Shapiro B.M., Schackmann R.W., Tombes R.M. and Kazazoglou T. (1985) *Current Topics Cell Regulation* **26:** 97–113.
Sosa M.A., Berbieri M.A. and Bertini F. (1991) *J. Reprod. Fert.* **93:** 279–285.
Stephens D.T., Wang J.L. and Hoskin D.D. (1979) *Biol. Reprod.* **20:** 483–491.
Stephens D.T., Acott T.S. and Hoskins D.D. (1981) *Biol. Reprod.* **25:** 483–491.
Stubbs C.D. and Smith A.D. (1984) *Biochim. Biophys. Acta* **779:** 89–137.
Tajima Y., Okamura N. and Sugita Y. (1987) *Biochem. Biophys. Acta.* **924:** 519–529.
Tash J.S. (1976) *J. Reprod. Fertil.* **47:** 63–72.
Tezon J., Romella G.E., Cameo M.S., Vazquez M.H. and Blaquier J.A. (1985) *Biol. Reprod.* **32:** 519–597.
Thomas T.S., Reynolds A.B. and Oliphant G. (1984) *Biol. Reprod.* **30:** 693–705.
Thomas T.S., Wilson L.S., Reynolds A.B. and Oliphant G. (1986) *Biol. Reprod.* **35:** 691–703.
Tracy L., Rankin K., James T., Holland M.K., Criswold M.D. and Orgebin Crist M.C. (1992) *Biol. Reprod.* **46:** 747–766.
Tulsiani D.R.P., Nagdas S.K., Skudlarek M.D., and Orgebin Crist M.C. (1995) *Dev. Biol.* **167(2):** 584–595.

Tulsiani D.R.P., Skudlarek M.D. and Orgebin Crist M.C. (1989) *J. Cell Biol.* **109:** 1257–1267.

Vijayasarathy S. and Balaram P. (1982) *Biochem. Biophys. Res. Commun.* **108:** 760–789.

Vijayasarathy S., Shivaji S. and Balaram P. (1982) *Biochem. Biophys. Res. Commun.* **108:** 585–591.

Vijayraghavan S., Critchlow L.M. and Hoskin D.D. (1983b) *Presented at the 5th International Conference on Cyclic Nucleotide and Protein Phosphorylation,* Milan, Italy 27 June-1 July.

Vijayraghavan S., Critchlow L.M., Hoskins D.D. (1985) *Biol. Reprod.* **32:** 489–500.

Voglmayr J.K., Scott T.W., Setchell B.P. and Waites G.M.H. (1967) *J. Reprod. Fertil.* **14:** 87–99.

Voglmayr J.K., and Amann R.P. (1973) *Acta Endocrinologia.* **73:** 196–208.

Voglmayr J.K. (1975) In: Hamilton D.W., Greep R.O. (eds.) *Handbook of Physiology, Am. Physiological Society Washington D.C., Vol. 5,* p 437–451.

Voglmayr J.K. and Dacheux J.L. (1983) *Biol. Reprod.* **28:** (Suppl. 1) 136 (Abstr.)

Voglmayr J.V., Fairbanks G. and Lewis R.G. (1983) *Biol. Reprod.* **29:** 767–775.

Yamada K.M., Olden K. and Halm L.H.E. (1980) In: *The Cell Surface: Mediator of Developmental Process* (Subtely S. and Wessels N.K. eds.) pp 43–77, Academic Press, New York.

Yeung C.H. (1984) *Gamete Res.* **9:** 99–114.

Comparative Endocrinology and Reproduction
K.P. Joy, A. Krishna and C. Haldar (Eds)

15. Male Contraceptive Agents

N.K. Lohiya and A.S. Ansari
Reproductive Physiology Section, Department of Zoology
University of Rajasthan, Jaipur-302 004, India

Introduction

There is an increasing demand that men should share not only the benefits but also the burden of contraception. The ideal contraceptive should be safe, effective, convenient to use, affordable, widely available and readily reversible. Unlike the contraceptive methods developed for women during the last three decades, no new male contraceptive has been developed. Currently men have a limited choice of available methods, namely, periodic abstinence, coitus interruptus, condoms and vasectomy. The reversible male methods have relatively higher failure rates. Vasectomy is by all means a permanent method of sterilization with a low rate of successful reversals. Although, studies have been conducted for new easier and reversible ways to block the vas, the silicone plug (Zhao *et al.*, 1992) and intravasal styrene maleic anhydride (SMA) (Guha *et al.*, 1993) injections, are in experimental/clinical stage. In this review we shall consider the risk and benefits associated with each of these approaches, the current status and likelihood of a successful outcome to this avenue of endeavour.

Antispermatogenic Agents

The antifertility effects of numerous compounds and classes of compounds have been well documented in both laboratory animals and the human male. Among these substances are the nitrofurans (Nelson and Steinberger, 1953), thiophenes (Steinberger *et al.*, 1956), dinitropyrroles (Patanelli and Nelson, 1964), bis-(dichloroacetyl) diamines (Drobeck and Coulston, 1962), chlorinated sugars (Ford and Waites, 1986), sulphasalazine (Ford and Waites, 1986), and other chemicals such as non alkylating insect chemosterilants, cytostatics, pesticides and fungicides. However, because of the concomitant toxicity further investigations on some of these agents as male contraceptives have been abandoned. It might, thus, well be feasible to reconsider only the highly effective compound.

Tolnidamine

The antispermatogenic activity of 3-indazole carboxylic acids was discovered during routine pharmacological screening (Corsi *et al.*, 1976). To date three active analogues are known viz., 4-chlorobenzyl (AF 1312/TS), lonidamine (AF 1890; 2, 4-dichlorobenzl) and tolnidamine (4-chloro-2-methylphenyl) or methyl-1H-indazole-3-carboxylic acid (Scorza Barcellona *et al.*, 1986). Tolnidamine is the most active and well studied compound as an antispermatogenic agent of the class.

The inhibitory activity of tolnidamine on the seminiferous epithelium was studied in rats using the doses ranging between 40 and 500 mg/kg (Heywood *et al.*, 1981; Scorza Barcellona *et al.*, 1986). In adolescent rhesus monkeys tolnidamine administration at 25 to 400 mg/kg/day for 28 days showed varied antifertility effects (Heywood *et al.*, 1981).

Further extensive studies on contraceptive efficacy and reversibility were carried out by the present authors. Adult male langur monkeys (*Presbytis entellus entellus* Dufresne) were treated orally with two dose regimens i.e., 50 mg/kg/twice a week and 50 mg/kg/alternate day for a period of 90 days and followed by a recovery period of 150 days. Treatments resulted in severe oligozoospermia with a significant decrease in sperm motility and vitality and increased percentage of sperm abnormality and immature germ cells. Sperm density did not return to normal level even after 150 days of recovery. At 90 days of treatment histological examination of testis showed varied degree of cellular damage both in germ cells and Sertoli cells. Diminished spermatogenesis particularly at spermatocyte stage was evident in most of the tubules. Depopulation of spermatogonia and Sertoli cells was evident. Disorganization and sloughing of immature germ cells were present. At 150 days of recovery, similar microscopic alterations were seen (Lohiya *et al.*, 1991a; Ansari *et al.*, 1992). Tolnidamine do not produce major alterations on sex hormones (Lobl *et al.*, 1981; Spitz *et al.*, 1985). In langurs no alterations in circulatory levels of testosterone were recorded (Lohiya *et al.*, 1991a, Ansari et al., 1992).

Heywood and associates (1981) carried out detailed toxicological investigations. Animals showed varied toxicological responses for tolnidamine. Rats exhibited no toxicological symptoms with daily oral doses ranging between 100 and 800 mg/kg for 4 weeks. While abnormal kidney functions, along with lessions in the histoarchitecture of kidney, were evident in adolescent rhesus monkeys after day 7 with 400 mg/kg dose and on day 28 with 200 mg/kg dose regimen. A reversible non-significant increase in the serum creatinine concentration at 50 mg/kg/twice a week and a significant but reversible elevation at 50 mg/kg/alternate day was observed in langur monkeys. Other haematological and blood/serum biochemical parameters were unaffected (Lohiya *et al.*, 1991a; Ansari *et al.*, 1992).

Post-Testicular Agents

The epididymis is regarded as an ideal extragonadal target site in the regulation of male reproduction. α-chlorohydrin and 6-deoxy-6-halosugars are two important compounds which are reported to have post-testicular action and are active antifertility agents. The 6-deoxy-6-halosugars induced infertility in rats. This compound exerts its effects at epididymal level and by interfering with metabolism of the sperm. Unfortunately it blocked glucose metabolism and thereby led to central neurotoxicity (Ford and Waites, 1986). α-Chlorohydrin induced epididymal spermatocoele and back pressure effects at the testis from high dose, together with frank toxicity in monkeys signalled its demise as candidate for human use (Prasad and Rajalakshmi, 1977).

Plant Products for Fertility Regulation

Fertility regulation with plants or plant preparations have been reported in the ancient literature of indigenous systems of medicine. A number of plant species have been tested for fertility regulation beginning about 50 years ago and were subsequently fortified by national and international agencies but no contraceptive of plant origin excluding steroids has yet emerged. This part of the review is being restricted to three plant products viz., gossypol from Gossypium plant, total glycosides of *Tripterygium wilfordii* and seed extracts of *Carica papaya* which are promising candidates for fertility regulation.

Gossypol

Gossypol is a yellowish polyphenolic compound, present in the seed, stem or root of the Gossypium plant. It is reported to possess antispermatogenic effect in rat, mice, hamsters, dog, monkeys and human beings (National Coordinating Group, 1978; Lohiya *et al.,* 1990). It is believed that gossypol exert its contraceptive effect during spermatogenesis including the post-testicular development and maturation of spermatozoa in the epididymis. Desquamation of the germinal epithelium occurred in the form of the multinucleated giant cells derived from spermatids (Hoffer *et al.,* 1988). The onset of gossypol induced infertility seems to be dose and duration dependent (Bardin *et al.,* 1981).

Clinical trials have been carried out at a number of centres in the People's Republic of China. The first trial was reported by Qian *et al.* (1972) using the higher doses ranged from 60 to 70 mg/day for a period of 5 to 6 weeks. Reversible antifertility effects along with no serious side effects were observed. After the conduction of this study several experiments were carried out with low dose level (Liu *et al.,* 1981; Coutinho and Melo, 1988; Xu *et al.,* 1988).

Untoward side effects figured during these studies were transient and disappeared during the recovery phase (Bajaj and Madan, 1983). The blood testosterone and LH levels were not significantly affected by gossypol administration (National Coordinating Group, 1978; Xue, 1981; Coutinho,1982). In some subjects circulating levels of FSH were increased (Zhang *et al.,* 1983; Gu *et al.,* 1985). The most serious side effect of gossypol administration was hypokalemia. It was reported in 66 cases out of 8806 treated men (Qian *et al.*, 1980; Qian, 1985; Liu and Lyle, 1987; Waites, 1988). Hypokalemia, potassium deficiency in the body, ultimately leads to muscular paralysis and other physiological complications.

To overcome the hypokalemic symptoms, supplementation of potassium salt was advised by some scientists (Qian *et al.*, 1980), while others did not agree with this view (Liu *et al.*, 1988). Toxicological screening has shown that rats, guinea pigs and hamsters are unsuitable as models for studies on hypokalemic paralysis (Qian, 1985). Therefore, expressing the serious concern about hypokalemia and other related side effects, WHO (1984) advised scientists to defer clinical trials and carry out toxicological investigations in suitable animal model with the possibility of extrapolation of findings in human situation.

After a critical analysis of drawbacks of gossypol, a number of studies were planned in langur monkey (Lohiya *et al.*, 1990; Kumar, 1991; Sharma, 1991). Highly purified gossypol acetic acid was orally administered at 5, 7.5 and 10 mg/animal/day alone and also in combination with potassium chloride at 0.25, 0.50 and 0.75 mg/animal/day dose regimens, respectively. The treatment caused a dose and duration dependent suppression in sperm motility along with increased abnormality and gradual decrease in the sperm count leading to oligo, severe oligo/azoospermia at the end of medication, respectively. Serum potassium concentration decreased significantly in gossypol alone groups, while values were comparable to pretreatment in combination groups except at 5 mg/day dose regimen where serum potassium levels were below normal. Serum GOT and GPT values were elevated significantly in both the groups. No other untoward side effects were observed. All the gossypol induced effects returned to normalcy. It was concluded from these studies that langur monkey is the only animal model which exhibited reversible infertility and the hypokalemic effects of the gossypol can be counteracted with supplementation of potassium salt.

Recently an international multicentre phase II randomized clinical trial is proposed and initiated by South to South Cooperation in Reproductive Health, Salvador, Brazil to investigate the efficacy, safety and degree of reversibility. The whole study consists of the dose regimen at 15 mg racemic gossypol tablet/day orally for 12 weeks during suppression phase followed by 7.5 mg/day or 10 mg/day for 40 weeks during maintenance phase (Ladipo, 1993). The future of the gossypol as an ideal male contraceptive will be decided after completion of this study.

Trypterygium wilfordii

Tripterygium wilfordii Hoof f. (family Celastraceae) has been used as a medicinal herb in Chinese traditional medicine in the treatment of rheumatoid arthritis, chronic nephritis, tuberculosis and other pulmonary diseases without serious side effects (Anonymous, 1978). However, long term medication of this plant leads to a decrease in testicular weight or menstrual disturbance in some patients (Jia, 1985). In the late 1970's the active constituents i.e., the multiglycosides of *Tripterygium wilfordii* (GTW) were extracted from the root xylem (Qian, 1987a). Based on these informations, preliminary experiments were conducted in mice, rats and human beings. In hybrid mice, when both male and female animals were fed simultaneously a laboratory chow containing GTW for 4.5 months, the body weight growth, birth rate and litter size were decreased (Zheng *et al.*, 1983). In wistar rats fed a laboratory diet containing 30 mg GTW/kg body wt./day for 35 or 80 days, marked seminiferous tubule damages were seen together with a decrease in the serum testosterone level in the 80 day group (Zheng *et al.*, 1985). Qian *et al.* (1986) and Qian (1987b) carried out further studies on the effect of GTW on fertility. In an experiment with wistar rats, GTW was given through gastric gavage at a dose of 10 mg/kg/day, 6 times a week, the fertility of the rats began to decrease by the end of 4 week of medication and at the end of the 8 week all animals were infertile with a significant decrease in the density and vitality of cauda epididymal sperms. The fertility recovered following 4 weeks of the withdrawal of the treatment. Similar results were also obtained by other investigators (Ding *et al.*, 1992; Feng *et al.*, 1992; Lan *et al.*, 1992). In addition, Lan *et al.* (1992) observed reduced viability of cultured rat Leydig and Sertoli cells by compounds purified from the extract of GTW. On the basis of these observations, it is concluded that GTW damages mainly the epididymal spermatozoa and to a less extent, the spermatogenic cells.

The WHO Task Force initiated the Bioassay directed compound isolation studies and the most active compound in the multiglycoside extract was found to be triptolide to carry out preclinical safety studies. A protocol for the assessment of testicular toxicity has been developed by Task Force scientists and this study was scheduled to start in 1995. The mutagenicity study will be conducted in accordance with standard protocols required by drug regulatory authorities and will be carried out at a commercial toxicology facility (WHO, 1995). The experiments with GTW have also been initiated in langur monkeys to find out the contraceptive efficacy, reversibility and side effects of this important plant product.

Carica papaya

The papaya (*Carica papaya* L., Family; Caricaceae) is a large herbaceous plant cultivated throughout the tropical world and in the warmest parts of the subtropics. Oral administration of aqueous suspension of ripe papaya seed powder (20 mg/animal/day) for 8 weeks produced sterility in 40% of male rats without affecting genital organs (Das, 1980).

After publication of the findings of Das (1980), experiments were conducted by Chinoy and associates (Chinoy *et al.*, 1985, 1994, 1995) using aqueous extracts. Marked inhibition of fertility in male rodents was observed following either oral or intramuscular administration of crude aqueous extract at varying dose regimens ranging between 0.5 and 20 mg/kg/day and concluded that it caused functional sterility in treated animals (Chinoy *et al.*, 1985, 1994, 1995).

In order to explore the possibility of developing an orally effective male contraceptive from *C. papaya* seeds, an extensive study was undertaken by Lohiya and co-workers with various crude extracts, their fractions and isolated compounds in rats. Administration of crude ethanolic (Lohiya *et al.*, 1992), aqueous (Lohiya *et al.*, 1994a) (10 and 50 mg/animal/day for 30, 60 and 90 days) and chloroform (5 mg/animal/day) extracts (Lohiya and Goyal, 1992), chromatographic fractions of crude chloroform extract, i.e., chloroform, ethyl acetate and methanol fractions (Lohiya *et al.*, 1994b) at 2 and 5 mg/animal/day dose regimens and isolated compounds (Goyal, 1991) from chloroform fraction viz., linoleic acid , linolenic acid and n-hentriacontanol at the dose regimens of 5 mg/animal/day for the period of 20, 40 and 60 days, respectively, curtailed fertility of male rats to varying degree. Prime effects were observed on cauda epididymal sperm motility associated with morphological defects and reduced sperm count. Testicular sperm count also reduced in all the treatment groups. Fertility test showed dose and duration dependent reduction. Sterile matings were observed in crude aqueous (50 mg/animal/day) and chloroform extract at 60 and 40 days of medication, respectively, chloroform fraction (5 mg/animal/day) at 20 days and isolated compound n-hentriacontanol at 60 days of treatment. While crude extract of ethanol, chromatographic fractions (ethyl acetate and methanol) and isolated compounds (linoleic and linolenic acids) exhibited moderate effects. Body and reproductive organ weights along with clinical parameters did not show any significant variation. Increase in the weights of testes and epididymis, however, was recorded in chloroform fraction treated animals. All the altered parameters returned to normalcy within 60 days of withdrawal of treatment.

Further extensive studies are required in different animal species including non-human primates with varying dose regimens to explore the possibility of this important plant product for development of a male pill.

Vaccines for Fertility Regulation

Three important approaches (anti-GnRH, oFSH and oLH) have been identified for development of a vaccine for male.

Anti-GnRH vaccine

Talwar and associates (1984) developed monoclonal antibodies against GnRH which blocked fertility in female primates. It is also usable in males to regulate fertility (Shaha *et al.*, 1986). Although, this vaccine renders male animals infertile, androgen supplementation is required to restore libido and extratesticular needs of the body. Appropriate combination dose regimens have been worked out to restore libido without return of fertility (Talwar *et al.*, 1991, 1992).

Immunization of male animals with this vaccine led to a decrease in serum testosterone levels with a reduction in prostatic size (Jayashankar *et al.*, 1989; Giri *et al.*, 1991). Based on these observations, South to South Cooperation in Reproductive Health developed a vaccine for immunotherapy of reproductive health disorders (precocious puberty, carcinoma of prostate, endometriosis, etc.). A phase I/II clinical trials were conducted using this vaccine in India, Austria and Mexico. Administration of 400 mg anti-LHRH vaccine plus flutamide supplement led to a drastic decline of testosterone to castrate levels with concurrent decline in the PSA values. The response lasted up to 2 years. In other patients, however, the antibody response was variable requiring frequent boosters (Ladipo, 1993; Talwar *et al.*, 1993). Additional studies are required to determine the appropriate dose necessary for provoking effective antibody response. If successful this LHRH vaccine may be used as a chemical prostatectomy.

Anti-oLH vaccine

Immunization of rats and rabbits with ovine LH results in suppression of spermatogenesis, cessation of Leydig cell function and loss of libido (Prasad and Rajalakshmi, 1977). However, different results were obtained by Talaat and Lawrence (1971). Passive immunization with LH antisera obtained from rabbits was more effective than active immunization (Lawrence and Hassouna, 1972). The inhibition of gonadal function with LH antiserum was reversible in hamsters (Moudgal, 1974). Active immunization of dogs with LH antiserum caused similar effects. Testosterone concentrations were markedly reduced but libido was not affected (Lunnen *et al.*, 1974). Further studies were not initiated with alone LH antiserum (Raj *et al.*, 1991) but it was proposed that it could be used as a good experimental tool to alter testicular androgen synthesis and to study the changes in the physiology of the epididymis in relation to sperm maturation (Prasad and Rajalakshmi, 1977).

Anti-oFSH vaccine

A vaccine against FSH has been developed by utilizing ovine FSH (oFSH) as the immunogen (Murty *et al.*, 1979). Subsequent studies either by passive or active immunization using oFSH in alhydrogel, an adjuvent acceptable for use in human, has been used for regulation of testicular function (Moudgal, 1981; Moudgal *et al.*, 1991). Based on the above findings, safety and efficacy trials were carried out with the oFSH vaccine in monkeys. Primary immunization was done by administering 1.0, 0.3, 0.1 and 0.01 mg of oFSH per monkey on day 1, 20, 40 and 70 of immunization. The booster injections were given every after 90–100 days. Sperm count was reduced to oligospermia by 120–150 days. However, controlled mating studies showed that all animals were infertile. Circulating levels of testosterone and libido remained unaffected. No other side effects were observed during treatment schedule (Moudgal *et al.*, 1991, 1994). Phase I clinical trails have been completed.

Hormonal Agents

Non-steroidal agents

Inhibin
Inhibin regulates the secretion of FSH by negative feedback (Setchell and Main, 1974). Besides in the gonads, it is synthesized in a number of organs such as prostate, epididymis, uterus, adrenal, placenta, pituitary and gastrointestinal tract. Action of inhibin on FSH release seems to be selective but not specific. Some preparations of inhibin can suppress prolactin, growth hormone, oxytocin and LH (Sheth, 1994). Immunocytochemical localization and/or biosynthesis of both inhibin and FSH in a number of reproductive organs, strong correlations in pathophysiological conditions and involvement in a classic negative feedback loop proved that both are twin hormones representing two facets of the same coin. Both inhibin and FSH together plays their role in endocrine processes and are important local growth regulatory factors acting by paracrine, autocrine and intracrine action. Moreover, inhibin is antagonistic to biological actions of FSH (Sheth *et al.,* 1994).

Prostatic inhibin peptide (PIP) is a secretary protein which is synthesized in human prostate. *In vitro* experiment with antiserum to PIP indicated that it is localized in the post acrosomal region of human spermatozoa and also impaired the sperm-egg interaction (Bandivdekar *et al.,* 1987). Passive immunization of adult male rats (Vanage *et al.,* 1990), hamsters and marmosets (Vanage *et al.,* 1992) with the rabbit anti-PIP indicated that it is a safe effective and reversible method of contraception. It acts at both testicular as well as epididymal level (Vanage *et al.,* 1992). Further research is required in different animal species and in non-human primate species to use inhibin as a contraceptive.

LHRH analogues
Luteinizing hormone-releasing hormone (LHRH) or gonadotropin-releasing hormone (GnRH) is the hypothalamic peptide that stimulates the release of LH and FSH from the anterior pituitary (Stewart, 1981). The two classes of analogues are: agonists, which are equal to or several hundred times more potent than LHRH itself and antagonists, which inhibits the expression of naturally occurring LHRH (Schally *et al.,* 1980).

LHRH agonists
Chronic administration of GnRH agonists to animals and human induced paradoxical antifertility effects (Smith *et al.,* 1979; Belanger *et al.,* 1980; Anand Kumar *et al.,* 1986; Lohiya *et al.,* 1991b, c). However, controversy exists regarding attainment of azoospermia. Azoospermia could not be induced in non-human primates by daily administration of a number of GnRH agonist

(Anand Kumar *et al.,* 1986; Lohiya *et al.,* 1991b). While, with an alternate and more effective mode of delivery system i.e. osmotic minipump, Akhtar *et al.* (1983) administered buserelin and observed azoospermia in all treated monkeys. Testosterone was administered by implantation of silastic capsules. Moreover, Ravindranath *et al.* (1989) and Lohiya *et al.* (1991c) reported complete azoospermia, respectively, with buserelin in bonnet monkeys and with leuprolide in langur monkeys.

Clinical studies have also been conducted with LHRH agonists aiming towards the development of a male contraceptive (Swerdloff and Bhasin, 1985; Swerdloff *et al.,* 1985; Pavlou, 1991). However, all men complained of decreased libido and potency. In subsequent studies, TE was added to this regimen (Doelle *et al.,* 1983; Rabin *et al.,* 1984). In spite of initial enthusiasm, oligospermia was achieved while in some sperm counts did not drop below 20 million/ml. These results prompted further studies using higher doses of agonists, mode of administration and/or other analogues of testosterone (Pavlou *et al.*, 1991. Rajalakshmi, 1994) but, successful induction of azoospermia was only in limited clinical trials.

LHRH antagonists

In view of the inability of LHRH agonists to induce uniform azoospermia and the requirement of androgen supplementation, attention was focussed on the use of antagonists for male fertility regulation. LHRH antagonists inhibit gonadotrophin secretion by competing with endogenous LHRH for binding to receptors on pituitary gonadotrophs (Pavlou, 1991) and inhibit both pituitary and testicular functions (Akhtar *et al.,* 1985; Adams *et al.,* 1986).

Chronic administration of LHRH antagonists to rodents and primates invariably suppress gonadal function in a rapid and consistent fashion (Pavlou, 1991; Rajalakshmi, 1994) and also exerted histamine-like side effects (Karten and Rivier, 1986). Therefore, efforts were made for synthesis of LHRH antagonists with relatively free of side effects. However, the use of these antagonists in inducing uniform azoospermia without lowering the circulating levels of testosterone could not be achieved in non-human primates (Akhtar *et al.,* 1985) and human males (Pavlou, 1991) emphasizing the requirement of androgen supplementation.

Combination treatments of LHRH antagonists with testosterone were then carried out both in non-human primates and humans. The treatment with Nal-Glu antagonist with two doses of testosterone buciclate (TB) incompletely suppressed spermatogenesis in cynomolgus monkeys (Weinbauer *et al.,* 1988). However, these investigators obtained sustained inhibition of spermatogenesis with another antagonist when TB was administered 6 weeks after the treatment of the analogue (Weinbauer *et al.,* 1989). Contrary, Bremner and associates (1991) used LHRH antagonist, Deterelix with simultaneous testosterone release from s.c. implanted silastic capsules in cynomolgus monkeys. They concluded that the level of testosterone replacement is not the important factor in inducing azoospermia, but maintenance of testosterone levels may help in preventing adverse effects on behaviour, lipids and bone mass. Similar conclusion was also drawn in clinical trial using the Nal-Glu antagonist (Pavlou *et al.,* 1991).

From these studies it was concluded that LHRH antagonists are more potent in the suppression of gonadotrophin secretion and sperm production than are LHRH agonists. However, the cost of synthesis of LHRH antagonists is very high for contraceptive use in developing countries (Waites, 1994).

Steroidal agents

Androgens

This approach utilizes the same methodological scheme as the female pill. The main advantage is that it combines spermatogenic-suppressing properties with androgen substitution in single compound. Systematic study for the use of this approach as a male contraceptive was first reported by Reddy and Rao (1972). They intramuscularly injected 25 mg/day testosterone propionate (TP) for a period of 60 days to 7 normal adult males. Azoospermia was induced in all treated subjects with no adverse side-effects. A complete recovery of sperm counts following cessation of treatment was observed.

In view of the inherent drawback of daily injection of TP, several studies have been conducted with TE at varying intervals and in different dosage regimens (Patanelli, 1978; Matsumoto, 1988). In most of these studies TE was given at a dose of 100–200 mg every week for up to 6 months. Approximately 50% of men became azoospermic and the rest remained oligospermic, while few men did not respond. However, the first multicentre efficacy study for hormonal male contraceptive using TE was completed under the auspices of WHO (1990). Only one pregnancy occurred during 1486 months of efficacy phase. After the testosterone had been stopped, the estimated median time from azoospermia to recovery was 3.7 months. A striking observation in this study was that men in the 3 Chinese centres achieved azoospermia more frequently than men in the 7 Caucasian centres (91% vs 60%). Therefore, a second study was initiated to assess whether TE-induced severe oligozoospermia (sperm counts < 3 million/ml) might also be sufficient to confer a contraceptive effects. Of the 356 volunteers who completed the suppression phase, 349 (98. 8%) entered into efficacy phase when sperm count obtained was 3 million/ml or less. The efficacy data thus obtained showed that the overall pregnancy rate in the men whose sperm counts remained at 3 million/ml was 1.4%, consisting of pregnancy rates of zero per cent in the men rendered azoospermic by the treatment, and 8.1% in the men whose sperm counts were reduced to levels between 0.1 million/ml and 3 million/ml, respectively (WHO, 1995).

In the search for a long acting androgen ester, 19-nortestosterone hexyloxyphenyl propionate (19-NT) was identified as a potential steroid for male contraception (Schürmeyer *et al.*, 1984). Besides its property, 19-NT has a stronger pituitary suppressing effect than does testosterone. In a study involving 12 volunteers, 19-NT was injected at three week interval, half were rendered azoospermic and half severely oligozoospermic. Since no side effects were observed (Knuth *et al.*, 1985), this steroid should be further considered if other new options should fail.

In view of the shortcoming of the existing testosterone esters, scientists have foreseened the requirement of long acting testosterone esters for substitution as well as alone treatments.

WHO has synthesized a long acting testosterone ester (trans-4-n-butylcyclohexyl-carboxylate; 20 Aet-1 or testosterone buciclate; TB) and reported to maintain the androgen level to four months by a single i.m. injection to castrated animals (WHO, 1984; Weinbauer *et al.*, 1986; Rajalakshmi and Ramakrishnan, 1989).

A clinical study carried out at the WHO collaborating centre in Münster, Germany, to assess the sperm suppressing ability of TB given as single injections at two dose levels (600 mg and 1200 mg) in normal men. Although serum testosterone levels remained within the physiological range in both dose groups, no reduction of sperm concentrations to oligospermic levels was observed in any of the four men in 600 mg dose groups. By contrast, three of the eight men in the 1200 mg dose group, whose LH and FSH levels were maximally suppressed, exhibited a corresponding suppression of spermatogenesis to azoospermia. The remaining 5 men in the high dose group had LH and FSH levels within the lower physiological range and were not suppressed to azoospermia (WHO, 1995).

As some evidence of possible genotoxicity observed, while *in vitro* and *in vivo* toxicity studies, including Ames tests, were conducted (WHO, 1995) which necessitated carrying out additional studies with TB before further clinical trials would be conducted.

Danazol and danazol-testosterone combination

Danazol is a heterocyclic steroid related chemically to 17α-ethinyltestosterone (ethisterone). The major pharmacologic action of danazol is its antigonadotropic effect on the pituitary (Krey *et al.*, 1981). But for such a diversity of clinical use, inhibition of gonadal steroidogenesis and a direct action on target tissues has been claimed (Reyniak and Lauersen, 1982; Lohiya and Sharma, 1986c).

FSH and LH suppressing properties and a direct action on target tissues of danazol have been explored by several investigators for its possible use in contraception. Danazol when administered alone was able to suppress testicular function and lower the sperm count inconsistently with impaired libido (Sherins *et al.*, 1971). But when administered together with testosterone it was able to induce severe oligospermia or complete azoospermia without change in libido. No serious side effects were observed (Lohiya and

Sharma, 1985). Since the rate of azoospermia was not significantly higher than that achieved with testosterone alone and the daily steroid load was quite high and liver toxicity had been reported by female users (for the treatment of endometriosis), studies were not continued further (Nieschlag *et al.,* 1992).

Antiandrogens

Two groups of antiandrogens are known; one group has only antiandrogenic properties called pure antiandrogen e.g., cyproterone, flutamide and the other group is both antiandrogenic and antigonadotrophic and the latter activity may be ascribed to their progestational properties e.g., cyproterone acetate and chloromadinone acetate. In this part only the activity of cyproterone acetate is discussed.

Cyproterone acetate (CA), 1,2-α-methylene-6-chloropregn-4, 6-diene-17α-ol-3, 20-diene-17α-acetate, is a potent androgen antagonist (Neumann *et al.,* 1970) and acts by competitive inhibition of endogenously or exogenously administered androgens. CA inhibits the uptake of androgens by the accessory organs by the formation of a nuclear-DHT protein complex possibly by inhibiting an initial interaction of androgens with cytoplasmic proteins or at the nuclear acceptor level (Fang and Liao, 1971). CA competitively inhibits binding of DHT by the epididymal cytosol (Ritzens *et al.,* 1971).

Animal studies and administration to sexual delinquents showed that CA may also be used for fertility regulation in the male (Prasad and Rajalakshmi, 1977; Neumann and Schenck, 1980). Prasad *et al.* (1970) and Rajalakshmi *et al.* (1971) suggested the use of low doses of CA for control of fertility in the human male. The administration of 30 mg/day of CA in human males resulted in oligo- or cryptozoospermia with reduced sperm motility and decreased plasma testosterone levels. Libido and potency were not disturbed (Petry *et al.,* 1972). Decreasing the orally administered dose of CA to 5 mg/day for a period of 20 weeks resulted in oligospermia and a gradual decrease in the motility of spermatozoa with an increase in the percentage of abnormal and immature forms (Roy *et al.,* 1976). From these studies it is clear that azoospermia can only be achieved in a small number of subjects treated with CA in doses of 5 to 20 mg/day. Moreover, reporting of pregnancies and a marked decrease in the circulatory levels of testosterone, impaired accessory glands function, libido and potency with some other side effects. Combining of this agent with long acting testosterone esters was suggested to counteract the testosterone lowering effects and simultaneous maintenance of accessory glands function, libido and potency (WHO, 1978).

Several combination studies of CA with testosterone enanthate (TE) have been conducted in rabbits (Lohiya and Sharma, 1984, 1986a; Sharma *et al.,* 1988; Sharma and Lohiya, 1989) and langur monkeys (Lohiya and Sharma, 1983, 1986b; Lohiya *et al.,* 1987). Though the results were promising for developing a reversible male contraceptive, however, extensive clinical trials have not been undertaken (Nieschlag *et al.,* 1992).

Progestins and progestin-androgen combination

Progestins are potent inhibitors of pituitary gonadotrophin secretion with negligible direct effects on the testis. Used alone, progestins suppress spermatogenesis but cause impotence (Frick *et al.,* 1981). Therefore, androgen replacement is necessary to maintain sexual function and to avert consequences of androgen deprivation (Nieschlag and Behre, 1990). Most of the progestins-androgen combinations have been unable to induce azoospermia (Schearer *et al.,* 1978), however, monthly injections of DMPA and TE induced azoospermia in 19/20 subjects (Pangkahila, 1991). Therefore, current status of DMPA-androgen combination has been made available in the present review.

The DMPA, commonly known as depo provera, is a synthetic hormone and has a close structural similarity to natural progesterone. DMPA possesses both antiandrogenic, antigonadotrophic properties as well as sperm suppressing properties (Mac Leod, 1965). The treatment with DMPA in combination with testosterone has been used for the regulation of male fertility in various animal species and men (Paulsen, 1986; WHO, 1990).

Studies in the 1970s with progestogen-androgen hormonal regimens established their safety and relative efficacy in Caucasian men but rarely achieved azoopermia in more than 50% of men (Schearer *et al.,* 1978). However, Pangkahila (1991) demonstrated that three injections at monthly intervals of DMPA (200 mg or 100 mg) and TE (250 mg or 100 mg) caused suppression of spermatogenesis to azoopermia in 19 out of 20 Indonesian men. Studies in Indonesian subjects indicate that such responses may be ethnically variable. Consequently, the comparative efficacy of two androgens, TE and 19-NT when each combined with DMPA for a treatment period of six months was assessed. The azoospermia rate reached more than 97% in both treatment groups (WHO, 1993).

The WHO Task Force has undertaken a Phase I dose finding study at two centres in Indonesia and Sweden to compare the gonadotrophin-suppressing and spermatogenesis-inhibiting of a combination regimen of a long acting progestogen levonorgestrel butanoate (HRP 002) and TE (WHO, 1995). A major drawback of the progestin-androgen combinations assessed so far is that the two compounds need to be injected at different intervals. Moreover, present complicated treatment regimen would be impractical to use on a large scale for family planning.

References

Adams, L.A., Bremner, W.J., Nestor, Jr.J.J., Vickery, B.H. and Steiner, R.A. (1986). J. Clin. Endocrinol. Metab. **62**: 58–63.

Akhtar, F. B., Marshall, G.R., Wickings, E.J. and Nieschlag, E. (1983). J. Clin. Endocrinol. Metab. **56**: 534–540.

Akhtar, F.B., Weinbauer, G.F. and Nieschlag, E. (1985). J. Endocr. **104**: 345–354.

Anand Kumar, T.C., Rajalakshmi, M., Mokkapati, S., Sharma, R.S. and David, G.F.X. (1986).

In: Gynaecology and Obstetrics (L. Ludwig and K. Thomsen, Eds.), pp. 722–726. Springer-Verlag, Berlin.

Anonymous (1978). Encyclopedia of Chinese Medicines. Hong Kong and Shanghai; Vol. 2, pp. 2460–2470. Commercial Press. Jiang Su Medical School, ed.

Ansari, A.S., Lohiya, N.K., Scorza Barcellona, P., Srivastava, S. and Goyal, R.B. (1992). Reprod. Contracept. **3**: 31–38.

Bajaj, J.S. and Madan, R. (1983). In: Proceedings of the international symposium on research on the regulation of human fertility, pp. 729–769. Stockholm, Sweden.

Bandivdekar, A.H., Gopalkrishnan, K. and Sheth, A.R. (1987). Adv. Contracept. **3**: 1–12.

Bardin, C.W., Sundaram, K.S. and Chang, C.S. (1981). Paper presented at the workshop on gossypol, programme for applied research on fertility regulation, Chicago.

Belanger, A., Labrie, F., Lemay, A., Caron, S. and Raynaud, J.P. (1980). J. Steroid Biochem. **13**: 123–126.

Bremner, W.J., Bagatell, C.J. and Steiner, R.A. (1991). J. Clin. Endocrinol. Metabol. **73**: 465–469.

Chinoy, N.J., D'Souza, J.M. and Padman, P. (1994). Reprod. Toxicol. **8**: 75–79.

Chinoy, N.J., D'Souza, J.M. and Padman, P. (1995). Phytother. Res. **9**: 30–36.

Chinoy, N.J., Geetha Ranga, M., Rao, M.V., Verma, R.J., Sam, M.G., Patel, K.G. and D'Souza, J.M. (1985). In: Methods for the regulation of male fertility (T.C. Anand Kumar and G.M.H. Waites, Eds.), pp. 95–106. Indian Council of Medical Research, New Delhi.

Corsi, G., Palazzo, G., Germani, C., Scorza Barcellona, P. and Silvestrini, B. (1976). J. Med. Chem. **19**: 778–783.

Coutinho, E.M. (1982). Arch. Androl. 9: 37–38.

Coutinho, E.M. and Melo, J.F. (1988). Contraception **37**: 137–152.

Das, R.P. (1980). Indian J. Exp. Biol. **18**: 408–409.

Ding, J.H., Tao, J.Z., Ji, L. and Qian, S.Z. (1992). In: Proceedings of the symposium on epididymis and male fertility (P.Y.D. Wong and T.G. Cooper, Eds.), pp. 3–10. Springer-Verlag, Hong Kong.

Doelle, G.C., Alexander, A.N., Evans, R.M., Linde, R., Rivier, J., Vale, W. and Rabin, D. (1993). J. Androl. **4**: 298–302.

Drobeck, H.P. and Coulston, F. (1962). Exp. Mol. Pathol. **1**: 251–274

Fang, S. and Liao, S. (1971). J. Biol. Chem. **246**: 16–24.

Feng, J., Yu, N., Xu, S., Qian, S.Z. and Wang, Y. (1992). In: Proceedings of the symposium on epididymis and male fertility (P.Y.D. Wong and T.G. Cooper, Eds.), pp. 3–9. Springer-Verlag, Hong Kong.

Ford, W.C.L. and Waites, G.M.H. (1986). In: Male contraception: advances and future prospects (G.I. Zatuchni, A. Goldsmith, J.M. Speiler and J.J. Sciarra, Eds.), pp. 89–106. Harper and Row, Philadelphia.

Frick, J., Danner C., Joos, H., Kunit, G. and Luukkainen, T. (1981). J. Androl. **2**: 331–338.

Giri, D.K., Jayaraman, S., Neelaram, G.S., Jayashankar, R. and Talwar, G.P. (1991). Exp. Mol. Pathol. **54**: 255–264.

Goyal, R.B. (1991). Ph.D. thesis, University of Rajasthan, Jaipur.

Gu, Z.P., Guang, Y., Sang, G.W., Wang, W.C., Wang, Y.X., Zhao, S.J., Cheng, Z.X., Ge, J.L. and Liu, X.H. (1985). In: Advances in fertility regulation in the male (S.Z. Qian and G.M.H. Waites, Eds.), pp. 42–44. The People's Medical Publishing House, Beijing.

Guha, S.K., Singh, S., Anand, S., Ansari, S., Kumar, S. and Koul, V. (1993). Contraception **48**: 367–375.

Heywood, R., James, R.W., Scorza Barcellona, P., Campana, A. and Cioli, V. (1981). Chemotherapy **27** (Suppl, 2): 91–97.

Hoffer, A.P., Agarwal, A., Meltzer, P., Naqvi, R. and Matlin, S.A. (1988). Contraception **37**: 301–331.

Jayashankar, R., Chaudhury, M.K., Singh, O., Alam, A and Talwar, G.P. (1989). Prostate **14**: 3–11.
Jia, L. (1985). Clin. Pharmac. Bull. **20**: 101–105.
Karten, M.J. and Rivier, J.E. (1986). Endocr. Res. **7**: 44–66.
Knuth, V.A., Behre, H.M., Belkien, L., Bents, H., Nieschlag, E. (1985). Fertil. Steril. **44**: 814–821.
Krey, L.C., Robbins, R. and Mc Ewen, B.S. (1981). Fertil. Steril. **35**: 467–472.
Kumar, M. (1991). Ph.D. thesis, University of Rajasthan, Jaipur.
Ladipo, O.A. (1993). Biennial Report 1991–1993. South to South Cooperation in Reproductive Health, Salvador, Bahia, Brazil.
Lan, Z.J., Gu, Z.P., Lu, R.F. and Zhuang, L.Z. (1992). Contraception **45**: 249–261.
Lawrence, K.A. and Hassouna, H. (1972). Biol. Reprod. **6**: 422–432.
Liu, G.Z., Hinton-Chiu, K., Cao, J., Zhu, C. and Li, B. (1988). Contraception **37**: 111–117.
Liu, G.Z. and Lyle, K.C. (1987). Fertil. Steril. **48**: 462–465.
Liu, Z.Q., Liu, G.Z., Hei, L.S., Zhang, R.A. and Yu, Z. (1981). In: Recent advances in fertility regulation (C.F. Chang, D. Griffin and A. Woolamn, Eds.), pp. 160-163. Beijing.
Lobl, T.J., Bardin, C.W., Gunsalus, G.L. and Musto, N.A. (1981). Chemotherapy **27** (Suppl. 2): 61–76.
Lohiya, N.K., Ansari, A.S. and Scorza Barcellona, P. (1991a). Contraception **43**: 485–496.
Lohiya, N.K. and Goyal, R.B. (1992). Indian J. Exp. Biol. **30**: 1051–1055.
Lohiya, N.K., Goyal, R.B., Jayaprakash, D., Ansari, A.S. and Sharma, S. (1994a). Planta Med. **60:** 400–404.
Lohiya, N.K., Goyal, R.B., Jayaprakash, D., Ansari, A.S., Srivastava, S. and Singh, P. (1994b). In: Current concepts in fertility regulation and reproduction (C.P. Puri and P.F.A. Van Look, Eds.), pp. 177–192. Wiley Eastern Limited, New Delhi.
Lohiya, N.K., Goyal, R.B., Jayaprakash, D., Sharma, S., Kumar, M. and Ansari, A.S. (1992). Int. J. Pharmacog. **30**: 308–320.
Lohiya, N.K., Jayaprakash, D., Sharma, R.S. and Sharma, O.P. (1991c). In: Perspectives in primate reproductive biology (N.R. Moudgal, K. Yoshinaga, A. J. Rao and P.R. Adiga, Eds.), pp. 353–362. Wiley Eastern Limited, New Delhi.
Lohiya, N.K., Sharma, K., Jayaprakash, D., Ansari, A.S., Kumar, M. and Sharma, S. (1991b). Contraception **43**: 187–200.
Lohiya, N.K., Sharma, K., Kumar, M. and Sharma, S. (1990). Contraception **41**: 519–532.
Lohiya, N.K. and Sharma, O.P. (1983). Contraception **28**: 575–586.
Lohiya, N.K. and Sharma, O.P. (1984). Contra. Deliv. Syst. **5**: 75–81.
Lohiya, N.K. and Sharma, O.P. (1985). In: Male fertility and its regulation (T.J. Lobl and E.S.E. Hafez, Eds.), pp. 343–349. MTP Press Ltd., Lancaster.
Lohiya, N.K. and Sharma, O.P. (1986a). Adv. Contra. Deliv. Syst. **2**: 64–72.
Lohiya, N.K., Sharma, O.P., Sharma, R.C. and Sharma, R.S. (1987). Biomed. Biochem. Acta **46**: 259–266.
Lohiya, N.K. and Sharma, R.C. (1986b). Adv. Contra. Deliv. Syst. **2**: 53–57.
Lohiya, N.K. and Sharma, R.C. (1986c). In: Endocrinology (C.P. Puri and T.C. Anand Kumar, Eds.), pp. 94–109. Endocrine Society of India, Bombay.
Lunnen, J.E., Faulkner, L.C., Hopwood, M.L. and Pickett, B.W. (1974). Biol. Reprod. **6**: 42–55.
Mac Leod, J. (1965). In: Agents affecting fertility (C.R. Austin and J.S. Perry, Eds.), pp. 93–123. Little Brown, London.
Matsumoto, A.M. (1988). Fertil. Steril. **50**: 324–328.
Moudgal, N.R. (1974). In: MTP international review of science (A.C. Guyton and D.F. Horrobin, Eds.), pp. 33–62. Butterworths, London.
Moudgal, N.R. (1981). Arch. Androl. **7**: 117–125.

Moudgal, N.R., Aravindan, G.R. and Suresh, R. (1994). In: Current concepts in fertility regulation and reproduction (C.P. Puri and P.F.A. Van Look, Eds.), pp. 107–115. Wiley Eastern Limited. New Delhi.

Moudgal, N.R., Ravindranath, N., Murthy, G.S., Dighe, R.R. and Aravindan, G.R. (1991). In: Perspectives in primate reproductive biology (N.R. Moudgal, K. Yoshinaga, A.J. Rao and P.R. Adiga, Eds.), pp. 297–306. Wiley Eastern Limited, New Delhi.

Murty, G.S.R.C., Sheela Rani, C.S., Moudgal, N.R. and Prasad, M.R.N. (1979). J. Reprod. Fert. Suppl. **26**: 147–163.

National Coordinating Group on Male Fertility Agents (1978). Chinese Med. J. **NS4**: 417–428.

Nelson, W.O. and Steinberger, E. (1953). Fed. Proc. **12**: 103–104.

Neumann, F. and Schenck, B. (1980). In: Regulation of male fertility (G.R. Cunningham, W.B. Schill and E.S.E. Hafez, Eds.), pp. 93–104. Martinus Nijhoff, The Hague.

Neumann, F., Von Bersworelt-Wallrabe, R., Elger, W., Steinbeck, H., Hahn, J.D. and Kramer, M. (1970). Rec. Prog. Horm. Res. **26**: 337–410.

Nieschlag, E. and Behre, H.M. (1990). In: Testosterone - action, deficiency and substitution (E. Nieschlag and H.M.Behre, Eds.), pp. 92–114. Springer-Verlag, Heidelberg.

Nieschlag, E., Behre, H.M. and Weinbauer, G.F. (1992). In: Spermatogenesis-fertilization-contraception (E. Nieschlag and U.F. Habenicht, Eds.), pp. 477–501. Springer-Verlag, Berlin.

Pangkahila, W. (1991). Int. J. Androl. **44**: 248–256.

Patanelli, D.J. (Ed.) (1978). Hormonal control of male fertility. US Department of Health, Education and Welfare, National Institute of Health, Bethesda.

Patanelli, D.J. and Nelson, W.O. (1964). Rec. Prog. Horm. Res. **20**: 491–543.

Paulsen, C.A. (1986). In: Male contraception: advances and future prospects (G.I. Zatuchni, A. Goldsmith, J.M. Spieler and J.J. Sciarra, Eds.), pp. 300–303. Harper and Row, Philadelphia.

Pavlou, S.N. (1991). In: Perspectives in primate reproductive biology (N.R. Moudgal, K. Yoshinaga, A.J. Rao and P.R. Adiga, Eds.), pp. 283–296. Wiley Eastern Limited, New Delhi.

Pavlou, S.N., Brewer, K. and Farley, M.G. (1991). J. Clin. Endocrinol. Metab. **73**; 1360–1369.

Petry, R., Mauss, J., Rusch-Stroomann, J.G. and Vermeulen, A. (1972). Horm. Metab. Res. **4**: 386–388.

Prasad, M.R.N. and Rajalakshmi, M. (1977). In: International review of physiology, vol. 13, reproductive physiology II (R.O. Greep, Ed.), pp. 153–199. University Park Press, Baltimore.

Prasad, M.R.N., Singh, S.P. and Rajalakshmi, M. (1970). Contraception **2**: 165–168.

Qian, S.Z. (1985). In: Methods for the regulation of male fertility (T.C. Anand Kumar and G.M.H. Waites, Eds.), pp. 58–71. Indian Council of Medical Research, New Delhi.

Qian, S.Z. (1987a). Contraception **36**: 335–345.

Qian, S.Z. (1987b). In: Fertility regulation today and tomorrow (E. Diczfalusy and M. Bygdeman, Eds.), pp. 217–232. Raven Press, New York.

Qian, S.Z., Hu, J.H., Ho, L.X., Sun, M.X., Huang, Y.Z. and Fang, J.H. (1972). In: Clinical pharmacology and therapeutics (P. Turner, Ed.), pp. 489–492. Macmillan, New York.

Qian, S.Z., Jiang, G.W., Wu, X.Y., Xu, Y., Li, Y.Q. and Zhou, Z.H. (1980). Chin. Med. J. **93**: 477–482.

Qian, S.Z., Zhong, C.Q. and Xu, Y. (1986). Contraception **33**: 105–110.

Rabin, D., Evans, R.M., Alexander, A.N., Doelle, G.C., Rivier, J., Vale, W., Liddle, G. (1984). J. Androl. **5**: 176–180.

Raj, H.G.M., Kotagi, S.G., Letellier, R., Nath, N. and Raj, S.G. (1991). In: Perspectives in primate reproductive biology (N.R. Moudgal, K.Yoshinaga, A.J. Rao and P.R. Adiga, Eds.), pp. 307–316. Wiley Eastern Limited, New Delhi.

Rajalakshmi, M. (1994). In: Current concepts in fertility regulation and reproduction (C.P. Puri and P.F.A. Van Look, Eds.), pp. 117–132. Wiley Eastern Limited, New Delhi.

Rajalakshmi, M. and Ramakrishnan, P.R. (1989). Contraception **40**: 399–412.

Rajalakshmi, M., Singh, S.P. and Prasad, M.R.N. (1971). Contraception **3**: 335–346.

Ravindranath, N., Rao, A.J., Prasad, M.R.N. and Moudgal, N.R. (1989). In: Proceedings of the national symposium on neuroendocrine regulation in fertility control. pp. 129–133. North Eastern Hill University, Shillong, India.

Reddy, P.R.K. and Rao, J.M. (1972). Contraception **5**: 295–301.

Reyniak, J.V. and Lauersen, N.H. (1982). Fertil. Steril. **37**: 475–477.

Ritzen, B.M., Nayfeh, S.N., French, F.S. and Dobbins, M.C. (1971). Endocrinology **89**: 143–151.

Roy, S., Chatterjee, S., Prasad, M.R.N., Poddar, A.K. and Pandey, D.C. (1976). Contraception **14**: 403–420.

Schally, A.V., Coy, D.H. and Arimura, A. (1980). Int. J. Gynaecol. Obstet. **18**: 318–323.

Schearer, S.B., Alvarez-Sanchez, F., Anselmo, G., Bremner, P., Coutinho, E., Lathen-Faundes, A., Frick, J., Heinild, B., Johansson, E.D.B. (1987). Int. J. Androl. Suppl. **2**: 680–712.

Schürmeyer, Th., Knuth, U.A., Belkien, L. and Nieschlag, E. (1984). Lancet **1**: 417–420.

Scorza Barcellona, P., Cioli, V., De Martino, C., Bardin, C.W. and Spitz, I.M. (1986). In: Male Contraception: advances and future prospects (G.I. Zatuchni, A. Goldsmith, J.M. Speiler and J.J. Sciarra, Eds.), pp. 237–251. Harper and Row, Philadelphia.

Setchell, B.P. and Main, S.J. (1974). Biblphy. Reprod. **24**: 245–367.

Shaha, C., Kaul, S., Kini, M., Chaudhury, M. K., Anand, R., Das, C. and Talwar, G.P. (1986). In: Immunological approaches to contraception and promotion of fertility (G.P. Talwar, Ed.), pp. 143–150. Plenum Press, New York.

Sharma, R.C. and Lohiya, N.K. (1989). Arch. Biol. **100**: 141–156.

Sharma, R.C., Lohiya, N.K. and Ansari, A.S. (1988). Iugoslav. Physiol. Pharmacol. Acta **24**: 291–303.

Sharma, S. (1991). Ph.D. thesis, University of Rajasthan, Jaipur.

Sherins, R.J., Gandy, H.M., Thorslund, T.W. and Paulsen, C.A. (1971). J. Clin. Endocrinol. **32**: 522–531.

Sheth, A.R., Garde, S.V., Joseph, R.J. and Patel, D. (1994). In: Recombinant and synthetic vaccines (G.P. Talwar, K.V.S. Rao and V.S. Chauhan, Eds.), pp. 275–282. Narosa Publishing House, New Delhi.

Smith, R., Donald, R.A., Espiner, E.A., Stromach, S.G. and Edwards, I.A. (1979). J. Clin. Endocrinol. Metab. **48**: 167–170.

Spitz, I.M., Gunsalus, G.L., Mather, J.P., Thau, R. and Bardin, C.W. (1985). J. Androl. **6**: 171–178.

Steinberger, E., Nelson, W.O. and Boccabella, A. (1956). Anat. Rec. **125**: 312–317.

Stewart, J.M. (1981). In: LHRH peptides as female and male contraceptives (G.I. Zatuchni, J.D. Shelton and J.J. Sciarra, Eds.), pp. 1–12. Harper and Row, Philadelphia.

Swerdloff, R.S. and Bhasin, S. (1985). In: Advances in fertility regulation in the male (S.Z. Quin and G.M.H. Waites, Eds.), pp. 58–66. The People's Medical Publishing House, Beijing, China.

Swerdloff, R.S., Handelsman, D.J. and Bhasin, S. (1985). J. Steroid Biochem. **23**: 855–861.

Talaat, M. and Lawrence, K.A. (1971). Fertil. Steril. **22**: 113–118.

Talwar, G.P., Pal, R., Arunan, K., Singh, O., Sad, S., Suri, A., Shaha, C., Gupta, S.K. and Upadhyay, S.N. (1992). In: Contraception-newer pharmacological agents, devices and delivery systems (R. Sitruk-Ware and C.W. Bardin, Eds.), pp. 161–192. Marcel Dekken Inc., New York.

Talwar, G.P., Singh, O. and Pal, R. (1991). In: Perspectives in primate reproductive biology (N.R. Moudgal, K. Yoshinaga, A.J. Rao and P.R. Adiga, Eds.), pp. 269–276. Wiley Eastern Limited, New Delhi.

Talwar, G.P., Singh, O., Pal, R., Chatterjee, N., Arunan, K., Sad S., Raghunathan, V., Sahai, P., Gupta, H. and Salunke, D. (1993). In: Immunology of reproduction (R.K. Naz, Ed.), pp. 235–249. CRC Press, New York.

Talwar, G.P., Singh, V., Singh, O., Das, C., Gupta, S.K. and Singh, G. (1984). In: Hormone receptors in growth and reproduction (B.B. Saxena, K.J. Catt, L. Birnbaumer and L. Martini, Eds.), pp. 351–359. Raven Press, New York.

Vanage, G.R., Garde, S.V., Sheth, A.R. and Gopalkrishnan, K. (1992). Int. J. Androl. **15**: 114–126.

Vanage, G.R., Gopalkrishnan, K. and Sheth, A.R. (1990). Mole. Reprod. Develop. **25**: 227–236.

Waites, G.M.H. (1988). In: Research in human reproduction, biennial report, 1986–87 (E. Diczfalusy, P.D. Griffin and J. Khanna, Eds.), pp. 119–223. WHO Special Programme of Research, Development and Research Training in Human Reproduction, Geneva.

Waites, G.M.H. (1994). In: Current concepts in fertility regulation and reproduction (C.P. Puri and P.F.A. Van Look, Eds.), pp. 93–105. Wiley Eastern Limited, New Delhi.

Weinbauer, G.F., Göckeler, E., Nieschlag, E. (1988). J. Clin. Endocrinol. Metab. **67**: 284–290.

Weinbauer, G.F., Khursid, S., Fingscheidt, U., Nieschlag, E. (1989). J Endocr. **123**: 303–310.

Weinbauer, G.F., Marshall, G.R. and Nieschlag, E. (1986). Acta Endocrinol. **113**: 128–132.

WHO (1978). Annual Report, Special Programme of Research, Development and Research Training in Human Reproduction, World Health Organization, Geneva.

WHO (1984). Annual Report, Special Programme of Research, Development and Research Training in Human Reproduction, World Health Organization, Geneva.

WHO (1993). Annual technical report 1992, Special Programme of Research, Development and Research Training in Human Reproduction, World Health Organization, Geneva.

WHO (1995). Annual technical report 1994, Special Programme of Research, Development and Research Training in Human Reproduction, World Health Organization, Geneva.

World Health Organization Task Force on Method for the Regulation of Male Fertility (1990). Lancet **336**. 955–959.

Xue, S.P. (1981). In: Recent advances in fertility regulation (C.F. Chang and D. Griffin, Eds.), pp. 122–146. S.A. Atar Press, Geneva.

Xu, D., Cai, W.J., Zhu, B.L., Dong, C.J., Zheng, Z.C. and Gao, Z.Q. (1988). Contraception **37**: 129–136.

Zhao, S.C., Lian, Y.H., Yu, R.C. and Zhang, S.P. (1992). Int. J. Androl. **15**: 465–467.

Zheng, J.R., Fang, J.L., Xu, L.F., Gao, J.W., Guo, H.Z., Li, Z.R., Sun, H.Z. (1985). Acta Acad. Med. Sinica **7**: 1–5.

Zheng, J.R., Liu, J.H., Hsu, L.F., Gao, J.W. and Jiang, B.L. (1983). Acta Acad. Med. Sinica **5**: 73–78.

Comparative Endocrinology and Reproduction
K.P. Joy, A. Krishna and C. Haldar (Eds)

16. Male Infertility: An Overview

Ashok Agarwal and Rakesh K. Sharma
Andrology Research and Clinical Laboratories, Department of Urology,
The Cleveland Clinic Foundation, Cleveland, Ohio 44195, USA

Introduction

Infertility is generally defined as the failure to achieve a desired pregnancy within 1 year of unprotected intercourse. By this definition, approximately 15% of couples attempting their first pregnancy are infertile. The definition is not based on biology, but it is a convenient clinical yardstick. Obviously, misinformation about the menstrual cycle and timing of intercourse can significantly impair fertility. There is no absolute requirement that a couple need to fail to achieve pregnancy for 1 year before fertility evaluation. In individuals with a risk factors or in older couples with a good understanding of the principles of fertility, a work-up may be appropriate sooner than 1 year. Data accumulated over the past few decades indicate that a male factor alone is responsible for infertility in about 30% of infertile couples, a female factor alone in about 50% and both male and female factors is involved in about 20% (Overstreet and Katz, 1987; MacLeod and Gold, 1951).

Epidemiology of Infertility

Infertility affects about one in six couples (15%). The incidence appears to be increasing because many women are delaying childbearing to pursue education or career goals. With increasing maternal age, not only does monthly fecundity decrease but exposure to diseases that ultimately decrease fertility (e.g., sexually transmitted diseases and endometriosis) also increases. In the USA alone, an estimated 8 to 10 million couples of reproductive age may have problems of conception.

Normal Male Reproductive Physiology

The human male reproductive system can be divided into three major components: the brain and hypothalamus, the anterior pituitary gland, and the testis. The system is triggered by pulsatile secretion of gonadotrophin releasing hormone (GnRH) secreted by the hypothalamus. In the normal male, GnRH is secreted with the frequency of about one pulse every 2 hours, although this normal frequency varies widely. Under the influence of GnRH, the anterior pituitary gland secretes luteinizing hormone (LH), which stimulates the secretion of testosterone by the Leydig cells and follicle stimulating hormone (FSH) which in concert with testosterone stimulates the development of spermatozoa along the seminiferous tubules.

The normal testicle ranges from 15 to 25 ml in volume. Approximately 90% of the testicular volume is filled by the seminiferous tubules; the remainder is composed of Leydig cells and other supportive structures. Therefore, the actual size of the testicle reflects the quantity of germinal epithelium. Because of this relationship, small testes indicate a small quantity of seminiferous tubular epithelium and probably a reduced sperm count. However, this relationship may be totally separate from the quantity of Leydig cells and their function, which may be completely normal even in small testes.

In general, LH secretion from the anterior pituitary is controlled by a negative feedback of testosterone from the gonads, which acts on the hypothalamus to decrease the frequency of GnRH secretion and to modulate the output of LH directly from the pituitary. In addition, estradiol acts on both the pituitary gland and the hypothalamus to decrease the amplitude of pulsatile release of LH. However, the control of FSH secretion is less clearly understood. It may involve not only the gonadal feedback of steroids (testosterone and estradiol) but also the gonadal peptide hormones, including inhibin and activin. Both of these hormones are secreted by the seminiferous tubules (from Sertoli cells) and are thought to selectively inhibit and increase FSH releases, respectively. Their exact physiological roles remain to be determined.

Normal male spermatogenesis requires about 75 days for an immature germ cell to develop from a primary spermatogonia into a mature spermatozoa. This process requires both high levels of testosterone under the influence of LH and stimulation of the seminiferous tubular epithelium under the influence of FSH. The exact relationship between these two factors remains to be clarified.

Male fertility requires the production of a sufficient number of mature and motile sperm that can undergo capacitation and acrosome reaction to bind and penetrate the zona pellucida of the oocyte, which subsequently divides and forms a viable fetus. A defect in any of these qualities, such as sperm concentration, maturation, motility, capacitation, acrosome reaction, binding, and penetration of the zona pellucida, could lead to defective sperm function, causing male infertility. Infertility attributed to the first two qualities, sperm production and maturation, although easy to diagnose, is difficult to treat without invasive procedures. However, several factors have been investigated with varying degree of success for enhancing the fertilizing capacity of subfertile sperm attributed to other factors; namely motility, capacitation, acrosome reaction, binding and penetration of the zona pellucida.

Laboratory Evaluation of Male Infertility

Infertility is a unique problem in medicine because it involves evaluating both the man and the woman as a couple and not just an individual patient. Recent progress in our understanding of the cell biology of spermatozoa provides a framework for rationally appraising the sperm functions required for fertility and the capability of contemporary tests to reveal the integrity of these functions. If wisely applied and interpreted, currently available assays have the potential to provide not only the diagnostic information regarding the cause of infertility, but also the prognostic (predictive) information on the likelihood of future fertility.

Basic Semen Evaluation

In many cases, the basic evaluation of sperm numbers, movement, and morphology provides adequate information for assessing the contribution of the male to a couple's infertility (Overstreet and Katz, 1987). Sperm function tests are used for assessing the fertility potential in men with consistently marginal semen quality. For men with normal semen quality, sperm function tests are seldom indicated early in the workup, although they prove useful later in establishing the diagnosis of unexplained infertility. Likewise, when there are obvious abnormalities of semen quality, sperm function tests are not required to establish the contribution of a male factor to infertility. In these cases, the test are often used in screening candidates for assisted reproductive techniques or as an aid in counselling patients who may be considering donor artificial insemination or adoption.

Although the sperm count is often used as an indicator of male infertility sperm number per ejaculate is, in itself, a poor predictor of fertility, and low sperm counts are seldom the cause of infertility. Our current standards for the "normal range" of sperm concentrations in semen are derived from the classical studies of Macleod (MacLeod and Gold, 1951). He evaluated semen samples from 1000 men in infertile marriages and compared their sperm count with 1,000 men whose wives were pregnant at the time of semen evaluation. The only clear difference between the fertile and infertile men was in the subgroup of men with sperm counts less than 20 million/mL: these men were found more frequently in the infertile population. These data are the basis of our current minimum values for a "normal" sperm count. A subsequent series of 15,000 semen evaluation by Macleod (MacLeod and Wang, 1979) showed that only about 20% of men in infertile marriages have sperm counts below 20 million/mL, suggesting that factors other than inadequate sperm numbers cause male infertility in many cases. Another line of evidence indicates that low numbers of sperm are adequate for fertility, providing that the sperm are functional cells. This situation is best illustrated by the fertility of hypogonadotrophic, hypogonadal men who are treated with gonadotrophins. This information is consistent with the premiss that most childless men are infertile, not because they produce inadequate numbers of sperm, but because the sperm cells they do produce are dysfunctional.

Sperm Motility as a Measure of Sperm Function

In basic semen evaluation, assessments of sperm motility is limited to the measurements of the percentage of motile cells (which reflects sperm viability) and assessment of sperm progression. Flagellar activity is an important function of the sperm cell and is required for migration through the female reproductive tract and for penetrating the oocyte. Abnormalities of flagellar movement are undoubtedly associated with some cases of male infertility. Computer-assisted semen analysis (CASA) can measure a range of sperm movement characteristics (Boyers et al., 1989). *In vitro* tests of sperm-cervical mucus interaction may be performed using slide preparations or capillary tubes. Measured end points generally include the number of sperm that penetrate the mucus, the depth of mucus penetration, and sperm survival in the mucus over time (World Health Organization, 1992). Quantitative standards for mucus penetration have proven difficult to establish because of the biologic variation encountered in normal semen and cervical mucus. Appropriate controls are important for interpreting these tests.

In conclusion, sperm flagellar activity is a manifestation of sperm function that can be directly measured. Sperm with abnormal morphology have associated abnormalities in motility, and in semen from infertile men, even sperm with normal morphology are less vigorous than normal sperm from fertile donors. However, sperm rapidly leaves the semen after ejaculation, and seminal plasma is not their natural environment. The sperm that succeed in gaining entry to the female tract are thus a selected population (Hanson and Overstreet, 1981). Therefore, assessments of sperm motility in semen may not provide information that can be directly related to subsequent sperm function in the female. Nevertheless, sperm motility characteristics in semen may still be useful for predicting male fertility because they provide indirect information on testicular and epididymal function.

Assessing Sperm-Oocyte Interaction

Fertilization *in vivo* requires sperm migration to the oviduct, sperm survival and capacitation in the female tract, reaction of the sperm acrosome in close proximity to the oocyte, penetration through the vestments surrounding the oocyte, and fusion with the plasma membrane of the oocyte. Many of these functions can now be assessed in the laboratory, either directly or indirectly.

Capacitation is described as changes within the sperm cell and on its surface that transform it from a potentially fertile cell to a cell fully competent to fertilize (Yanagimachi, 1988). The cell biology of capacitation is not well understood, even in animal models. However, two functional sequelae of capacitation have been identified in human sperm; hyperactivated motility and acrosome reaction. Hyperactivated motility is thought to be involved in sperm migration through the oviduct and penetration through the investing layers of the oocyte (Katz et al., 1989). The onset of hyperactivation involves changes in the flagellar movements of the sperm, and hence, alterations in its swimming trajectory. The sperm appears to be less progressive after hyperactivation, although its power output increases greatly. There is evidence that sperm from infertile men do not hyperactivate under capacitating conditions which normally promote hyperactivation in sperm from fertile men (Morales et al., 1988). Using CASA technology, sperm populations can be assessed for evidence of hyperactivation (Mack et al., 1989). However, the clinical utility of these measurements has not been shown, and this approach to sperm function assessment is still viewed as a research tool.

The acrosome reaction involves the fusion of sperm head membranes, the release of enzymes from the sperm head, and the exposure of previously sequestered sperm surfaces (Yanagimachi, 1988). The acrosome reaction is essential for penetrating the zona pellucida of the oocyte and for fusing with its plasma membrane. It is almost certain that the functional acrosome reaction takes place very near the oocyte and that the "trigger" for the reaction may be associated with oocyte investments (Yanagimachi, 1988). Thus, a functional sperm must be capable of resisting a premature acrosome reaction and be responsive to the biological inducer at the site of fertilization.

For many years, human acrosome reaction could only be observed with electron microscopy, but fluorescein conjugated lectins (Cross et al., 1986), monoclonal antibodies (Wolf, 1989), and other reagents can now be used to identify acrosome reaction with light microscopy. Because dead sperm may lose their acrosomes during degeneration, a viability stain should be included in acrosome reaction assays to distinguish 'functional" acrosome reactions from the degenerative ones. The rates of spontaneous acrosome reactions in semen are low and little information on sperm function is obtained by assessing the reaction.

One of the oldest sperm function tests is the sperm penetration assay (SPA) using zona-free hamster oocytes. Because the zona pellucida is the principle block to interspecies fertilization in this species, hamster oocytes will fuse with human sperm after removal of the zona (Yanagimachi, 1984). Sperm-oocyte fusion requires capacitation and an acrosome reaction, so the SPA is a bioassay for these sperm functions. The SPA is also thought to measure the specific sperm functions required for fusion with oolemma. A number of unrecognized biological factor may affect SPA results. Nevertheless, the SPA provides information on sperm function that is not fully duplicated by any other test or set of tests. Although normal ranges have been proposed for various characteristics of the SPA, there is still no good evidence that any outcomes other than positive (some eggs penetrated) and negative (no eggs penetrated) can be interpreted (Aitken, 1986). The test is most effectively employed in a laboratory with established protocols and in which tests results can be interpreted in the context of an internal database.

Recent interest has been considerably increased in the application of the sperm-zona binding assay for assessing sperm function (Burkman et al., 1988). Sperm binding to the zona is probably an essential step in the human fertilization process. However, under natural *in vivo* conditions, only a few sperm bind to the zona before fertilization, and the frequency of collisions between the sperm and zona must be low. It is, therefore, difficult to evaluate the biological meaning of in vitro assays involving thousands of sperm-oocyte collisions over a relatively long period of time. As in other sperm function tests, these assays may prove to be useful clinically as an integrated test of sperm quality although not revealing any specific functional abnormality.

Conclusions

Advances are being made in both the diagnosis and the treatment of male infertility. Despite the availability of many tests to evaluate male fertility, standard semen analysis remains one of the most important test. It is clear that patients with suboptimal semen have a reduced chance of inducing pregnancy; however, semen analysis does not absolutely differentiate patients who are fertile from those who are not. We can expect that more specific tests of sperm function will be developed.

Major improvements in the treatments of men with idiopathic infertility await a more precise understanding of the cause of this condition, and it is likely that multiple causes will be found. Only after these causes are defined can a specific therapy be developed. With continued work in this field and continued development of the diagnostic and treatment modalities, the urologist and gynecologist will be able to treat the infertile couple more successfully.

References

Aitken, R.J. (1986) Int J Androl, Suppl 6, Copenhagen.

Boyers, S.P., Davis, R.O. and Katz, D.F. (1989) Curr Probl Obstet Gynecol Fertil **12:** 165–200.

Burkman, L.J., Coddington, C.C., Franken, D.R., Kruger, T.F., Rosenwaks, Z. and Hodgen, G.D. (1988) Fertil Steril **49:** 688–697.

Cross, N.L., Morales, P., Overstreet, J.W. and Hanson, F.W. (1986) Gamete Res **15:** 213–226.

Hanson, F.W. and Overstreet, J.W. (1981) Am J Obstet Gynecol **140:** 173–177.

Katz, D.F., Drobnis, E.Z. and Overstreet, J.W. (1989) Gamete Res **22:** 443–469.

Mack, S.O., Tash, J.S. and Wolf, D.P. (1989) Biol Reprod **40:** 1162–1169.

MacLeod, J. and Gold, R.Z. (1951) J Urol **66:** 436–449.

MacLeod, J. and Wang, Y. (1979) Fertil Steril **31:** 103–116.

Morales, P., Overstreet, J.W. and Katz, D.F. (1988) J Reprod Fertil **83:** 119–128.

Overstreet, J.W. and Katz, D.F. (1987) Urol Clin of North Am **14:** 441–449.
Wolf, D.P. (1989) Am J Reprod Immunol **20:** 106–113.
World Health Organization (1992) Cambridge; Cambridge University Press.
Yanagimachi, R. (1984) Gamete Res **10:** 187–232.
Yanagimachi, R. (1988) Physiology of Reproduction vol. 1, 135–185.

Comparative Endocrinology and Reproduction
K.P. Joy, A. Krishna and C. Haldar (Eds)

17. Antispermatogenic Agents

S.K. Singh
Centre of Advanced Study in Zoology,
Banaras Hindu University, Varanasi-221 005, India

Introduction

Suppression of spermatogenesis by chemical substances has so far not been very successful approach for control of male fertility, but such studies have contributed significantly to our understanding of the male reproductive physiology, in particular the testicular physiology. On the other hand, antispermatogenic substances may have a viable future, at the very least, in the control of mammalian pests, e.g. rodents. It is pertinent to mention here that rodents destroy about 20% of the harvested crops of the world during storage and do much damage to buildings and equipments, thereby causing billions of dollars damage in a year (cf. Gao and Short, 1993).

Spermatogenesis

Before discussing the chemical substances which are shown to suppress spermatogenesis, it is desirable to give a brief overview of the process and the factors which regulate it.

Spermatogenesis is an intricate process in which primitive stem cells, type A spermatogonia, undergo proliferation and transformation to give rise to spermatozoa. The duration of this process varies among different mammalian species. For example, it requires approximately 51-53 days in rat, 34.5 days in mouse, 37-42 days in nonhuman primates and 74 days in men. The spermatogenic process is divided into proliferative, meiotic and spermiogenic phases (Fig. 1). In the proliferative phase, spermatogonial cells undergo a number of mitotic divisions, and proliferating spermatogonia then enter the meiotic phase of spermatogenesis as diploid spermatocytes. During meiosis, haploid spermatids are produced which in the final phase of spermatogenesis-spermiogenesis-differentiate to become spermatozoa. In the seminiferous epithelium, cells in these developmental phases are arranged in defined associations or stages. Along the seminiferous tubules, these stages follow each other in a regular fashion and give rise to the wave of the seminiferous epithelium.

It is believed that the spermatogenic process is regulated by mainly luteinizing hormone (LH) and follicle stimulating hormone (FSH), and the synthesis and release of these two hormones are, in turn, triggered by gonadotropin releasing hormone (GnRH) from the hypothalamus. FSH acts directly on the seminiferous epithelium whereas LH exerts its effects on the spermatogenic process through testosterone which is produced by Leydig cells under the influence of LH. Thus, LH, testosterone and FSH are the key hormones that regulate spermatogenic process and act synergistically to promote germ cell development.

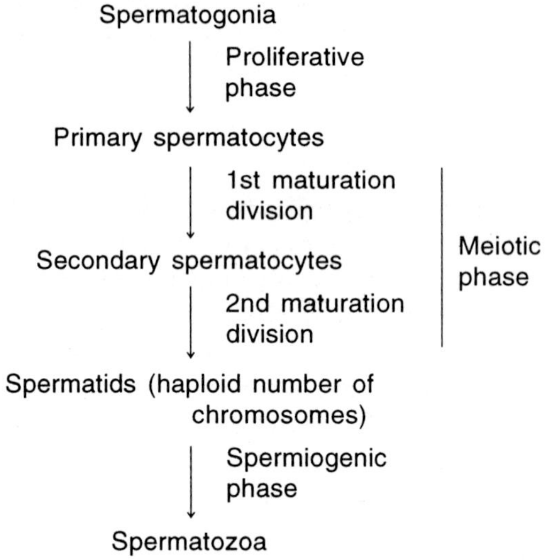

Fig. 1. Diagrammatic representation of spermatogenesis.

Compounds Suppressing Spermatogenesis

A large number of compounds are available which have been shown to induce antispermatogenic effects in several mammalian species (for reviews see, Fox and Fox, 1967; Jackson, 1966, 1970, 1973, 1975; Gomes, 1970; Patanelli, 1975; Neumann *et al.*, 1976; Setchell, 1978; Mann and Lutwak-Mann, 1981; Neumann, 1984; Wu, 1988; Comhaire, 1994). These compounds include steroids (e.g., androgens, antiandrogens, oestrogens, progestogens and anabolics), luteinizing hormone-releasing hormone (LHRH) analogues (agonists and antagonists), chemotherapeutics and antibiotics (e.g., thiophene, dinitrofuran, furadantin, furacin, actinomycin, chlorambucil, neomycin, gentamycin and trimethoprim), antimitotic and alkylating agents (e.g., cholchicine, vinblastine, busulphan, cyclophosphamide and methane sulphonic acid ester), nonsteroidal antiinflammatory drugs (e.g., indomethacin), antihypertensives (e.g., reserpine), heavy metals (e.g., cadmium and cobalt), tranquilizers (e.g., chlorpromazine), antimetabolites (e.g., 5-thio-D-glucose), plant products (e.g., gossypol)), prostaglandins (e.g., PG $F_{2\alpha}$), indazole-3-carboxylic acids, pipecolinomethylhydroxyindane maleate (PMHI) and variety of other substances.

In general, the action of any drug or compound on the testis is manifested in two ways: by suppression of spermatogenesis through inhibition of cell development or destruction of cells at one or more stages without any apparent influence on the endocrine function of the testis, or by inhibition of both germinal and Leydig cell function through suppression of the secretion of pituitary gonadotropins.

It is not possible to discuss all the compounds which have been shown to suppress spermatogenesis in a number of mammalian species. Thus, the present article describes the antispermatogenic effects induced by four compounds, viz. gossypol tetra-acetic acid (gossypol), prostaglandin (PG $F_{2\alpha}$), 1-(2, 4-dichlorobenzyl)-1H-indazole-3-carboxylic acid (DICA) and 1-p-chlorobenzyl-1H-indazole-3-carboxylic acid (AF 1312/TS) in the Parkes (P) strain laboratory mouse.

1. Gossypol tetra-acetic acid (Gossypol)

Gossypol is a yellow polyphenolic compound present in the seed of the cotton plant genus *Gossypium.* The discovery of this compound as a male fertility regulating agent by Chinese scientists is regarded as a major lead in the search for a male contraceptive agent (National Co-ordinating Group, 1978). This led several investigators around the world to confirm and extend the above information in a number of mammalian species including men (cf. Prasad and Diczfalusy, 1982; Liu and Segal, 1985; Rath, 1995). However, there is no consistent opinion among authors on the effect of gossypol. For example, in rat some authors (Chang *et al.*, 1982; Giridharan *et al.,* 1987) have reported no alterations in the histoarchitecture of the testis after gossypol treatment, while others (Hoffer, 1983; Rovan *et al.*, 1983; Wang *et al.*, 1985; Heywood *et al.*, 1986; Kalla *et al.*, 1986; Hoffer *et al.*, 1988; Kainz *et al.*, 1988; Kaur *et al.,* 1988; Srivastava *et al.*, 1989; Udoh *et al.*, 1992) have reported non-uniform regressive changes in the seminiferous tubules in testes of the treated rats. Likewise, in mice Hahn *et al.* (1981) have reported that gossypol treatment has no adverse effect on fertility. According to Coulson *et al.* (1980) and Hunt and Mittwoch (1984), in mice administration of gossypol causes a reduction in both the number of spermatozoa in the epididymis and the weight of the accessory reproductive organs. On the other hand, Wang *et al.* (1979) reported no adverse effect of gossypol on accessory glands in mice.

In P mice, gossypol treatment (20 or 40 mg/kg body weight/day, for 21 days) has no adverse effect on the body weight or on the weight of the testis; however, the same treatment causes a significant reduction in the weight of the seminal vesicle. In control mice, testis showed normal histological features (Fig. 2). By contrast, marked regressive changes were observed in the testes of the treated mice (Figs. 3 to 5). However, the effect of the gossypol treatment on the testis was not uniform; both regressed and normal seminiferous tubules were observed in the same section of the organ. In the testes of mice treated with 20 mg/kg body weight of gossypol,

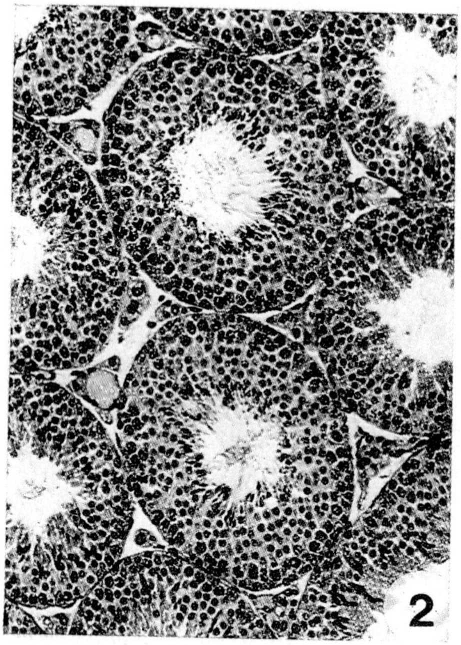

Fig. 2. Control. Note the normal appearance of the seminiferous tubules.

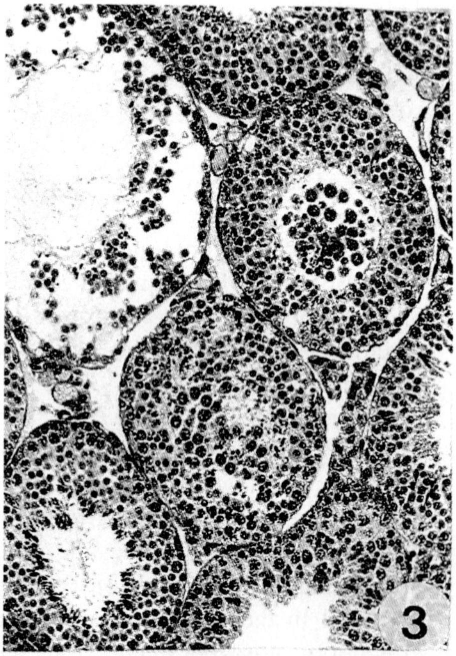

Fig. 3. After gossypol treatment, 20 mg/kg body wt/day, for 21 days. Note the presence of regressed and normal seminiferous tubules; the regressed tubules show germ cell exfoliation and epithelial disorganization.

there was exfoliation of germ cells from the epithelium into the lumen of the tubules; the tubular epithelium was also disorganized and showed much depletion of germ cells (Fig. 3). In some testes, the regressed seminiferous tubules were lined with only a thin layer of Sertoli cells, spermatogonia, few spermatocytes and rarely spermatids; multinucleated giant cells containing 2 to 4 nuclei of early spermatids were common in such tubules (Fig. 4). At a dosage of 40 mg/kg body weight, regressive changes in the affected seminiferous tubules were further pronounced; the tubules showed intraepithelial vacuolation and degeneration of germinal elements and nuclei of the Sertoli cells were sometimes exfoliated from the basal portion to the luminal portion of the tubules (Fig. 5). Normal characteristic features were, however, still observed in most of the seminiferous tubules in testes of the treated mice. When quantitatively analysed, the frequency of regressed seminiferous tubules was significantly higher in testes of the treated mice than in controls.

The mechanism of the antispermatogenic action of gossypol is not properly understood though it is suggested that it may cause disruption of testosterone secretion (Lin *et al.*, 1981), inhibition of germ cell differentiation (National Co-ordinating Group, 1978; Hadley *et al.*, 1981; Xue, 1981), or breakdown of the blood-testis barrier (Pelletier and Friend, 1980). Studies of Ahluwalia *et al.* (1986) indicate that antispermatogenic effect of gossypol is mediated locally within the germinal epithelium and that it does not involve hypothalamic-hypophyseal-gonadal axis. The significant decrease in the weight of the seminal vesicle in P mice treated with gossypol suggests that the treatment interferes with secretion of androgen (Singh and Rath, 1990). In mouse i*n vitro* studies, Donaldson *et al.* (1985) have also reported an inhibitory action of gossypol on the release of testosterone by the Leydig cells.

2. Prostaglandin ($PGF_{2\alpha}$)

In rat, prostaglandins (PGs) treatment is shown to depress testicular weight and interfere with spermatogenesis (Ericcson, 1972; Memon, 1973; Saksena *et al.*, 1974; Tso and Lacy, 1975; Kimball *et al.*, 1978; Chinoy *et al.*, 1980; Dev and Mangat, 1980). The antispermatogenic effects of PGs have also been reported in rabbit (Saksena *et al.*, 1978) and in Swiss-Webster strain of mouse (Abbatiello *et al.*, 1975, 1976). Furthermore, PGs treatment causes a reduction in the level of fructose in the rat seminal vesicle (Chinoy *et al.*, 1980; Dev and Mangat, 1980).

In P mice, PG $F_{2\alpha}$ treatment (3 mg/kg body weight/day, for 21 days) has no adverse effect on body weight. However, the same treatment causes a significant reduction in the weights of the testis, epididymis, seminal vesicle and Cowper's gland in treated mice as compared with controls; the weights of these organs, however, return to control levels by 56 days of drug withdrawal. The treatment also induced marked regressive changes in the

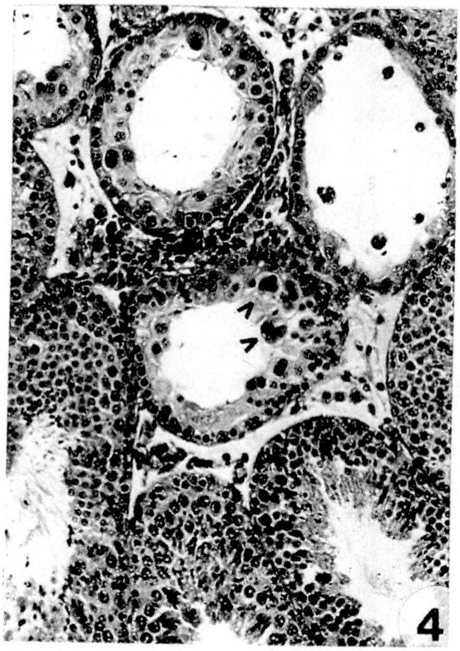

Fig. 4. After same treatment as shown in Fig. 3. Note the occurrence of giant cells (arrowheads) in the regressed seminiferous tubules; the regressed tubules are lined with only a thin layer of Sertoli cells, spermatogonia, few spermatocytes and rarely spermatids. Normal seminiferous tubules are also seen.

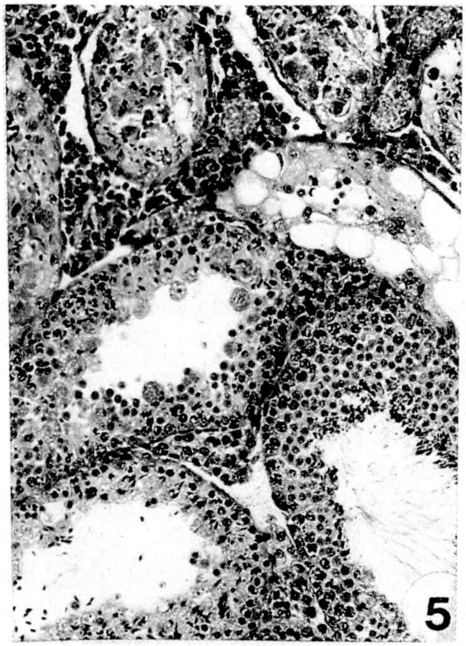

Fig. 5. After gossypol treatment, 40 mg/kg body wt/day, for 21 days. Note the marked degeneration of germinal elements in the regressed seminiferous tubules and the lumen of such tubules is filled with debris of germ cells and spermatozoa. A normal tubule is also seen at bottom on the right side.

Figs. 2 to 5. Sections of the testis of the laboratory mouse × 140.

seminiferous tubules leading to the suppression of spermatogenesis. The germinal epithelium was thin, and the tubules were devoid of spermatozoa and contained only Sertoli cells, spermatogonia, spermatocytes and occasionally spermatids. Several multinucleated giant cells as observed in testes of the gossypol-treated-mice were also noticed. The treatment, however, did not affect the levels of RNA, DNA and protein in the testis. The levels of sialic acid in both caput and cauda regions of the epididymis and that of fructose in the seminal vesicle were significantly decreased in treated mice as compared with controls. The alterations induced in the reproductive organs in P mice by PG $F_{2\alpha}$ treatment were, however, reversible and returned to pretreatment state by 56 days of drug withdrawal.

It is suggested that PGs cause suppression of spermatogenesis by acting directly on the seminiferous tubules or through interference with the secretion of androgens by the Leydig cells (cf. Abbatiello *et al.*, 1976). Suppression of spermatogenesis in P mice after PG $F_{2\alpha}$ treatment appears to be due to the deficiency of androgen, since there are significant reductions in the weights of the accessory sex organs, and in the levels of sialic acid in the epididymis and fructose in seminal vesicle in treated mice as compared with controls (Singh and Dominic, 1986).

3. 1-(2, 4-dichlorobenzyl)-1H-indazole-3-carboxylic acid (DICA)

The compound 1-(2, 4-dichlorobenzyl)-1H-indazole-3-carboxylic acid (DICA) is reported to induce antispermatogenic effects in rat and monkey (Coulston *et al.*, 1975; Lobl, 1979; Lobl *et al.*, 1979). By contrast, the testis of the musk shrew, a nonscrotal insectivore, is shown to be insensitive to DICA treatment (Singh and Dominic, 1983).

In P mice, DICA treatment (50 mg/kg body weight/day, for 28 days) has no adverse effect on body weight or weights of the epididymis, seminal vesicle, Cowper's gland and preputial gland. However, the same treatment causes a significant reduction in the weight of the testis though the testicular weight return to control level by 56 days of drug withdrawal. The treatment had also marked deleterious effects on the histoarchitecture of the testis. There was depletion of germ cells in the seminiferous tubules and the germinal epithelium consisted mainly of Sertoli cells, spermatogonia, spermatocytes and occasionally spermatids; some tubules were lined with only a ring of Sertoli cells and spermatogonia. Multinucleated giant cells as observed in the testes of gossypol—and $PGF_{2\alpha}$-treated mice were also observed. The levels of RNA and protein in the testis were significantly depressed in treated mice as compared with controls, whereas the level of testicular DNA was not affected by the treatment. The treatment had no effect on the level of fructose in the seminal vesicle. The alterations induced in the reproductive organs were, however, reversible and returned to pretreatment state by 56 days of drug withdrawal.

It is suggested that DICA acts directly on the testis and suppresses the spermatogenesis without interfering with the hypothalamo-hypophyseal-gonadal axis; the Sertoli cells appear to be the primary locus of DICA action in the testis (Buthala and Lobl, 1979; Lobl, 1979; Lobl *et al.*, 1979). The absence of alterations in the weights of the epididymis and accessory sex glands and in the level of fructose in the seminal vesicle in treated mice suggests that DICA treatment suppresses spermatogenesis probably without interfering with the pituitary-gonadal axis (Singh and Dominic, 1984).

4. 1-p-chlorobenzyl-1H-indazole-3-carboxylic acid (AF 1312/TS)

In rat, 1-p-chlorobenzyl-1H-indazole-3-carboxylic acid (AF 1312/TS) treatment depresses testicular weight and interferes with spermatogenesis without affecting the Leydig cells (Burberi *et al.*, 1975; DeMartino *et al.* 1975; Silvestrini *et al.*, 1975); the antispermatogenic effect induced by AF 1312/TS treatment is reported to be reversible (Coulston *et al.*, 1975). The compound also suppresses spermatogenesis in monkey (Coulston *et al.*, 1975)

In P mice, AF 1312/TS treatment (500 mg/kg body weight/day, for 28 days) does not affect the body weight or weights of the epididymis, seminal vesicle, Cowper's gland and preputial gland. However, the same treatment causes a significant decrease in the weight of the testis which is fully reversible by 56 days of drug withdrawal. The treatment also induced marked atrophic changes in the seminiferous tubules in the testis; the tubules showed only a thin layer of Sertoli cells and spermatogonia. The levels of RNA and protein in the testes of treated mice were significantly decreased as compared with controls; the level of testicular DNA, however, remained unaffected. The level of fructose in the seminal vesicle was not affected by the treatment. These alterations were, however, reversible and returned to normal levels by 56 days after withdrawal of the treatment.

It is reported that AF 1312/TS acts selectively on the testis and suppresses spermatogenesis without competing with vitamins and nutritive factors in the gonad (Silvestrini *et al.*, 1975). Studies of DeMartino *et al.* (1975) suggest that Sertoli cells are the primary locus of AF 1312/TS action in the testis and that the antispermatogenic effects induced by the drug could be due to the impairment of the regulative role of these cells in the maintenance of spermatogenesis. The suppression of spermatogenesis unaccompanied by any alterations in the weight of the accessory sex organs and in the level of fructose in the seminal vesicle in AF 1312/TS-treated P mice suggests that the drug acts selectively on the seminiferous tubules without interfering with androgen secretion by the Leydig cells (Singh and Dominic, 1985).

General Considerations and Future Prospects

In P mice, treatments with gossypol, prostaglandin, DICA and AF 1312/TS induce marked regressive changes in the seminiferous tubules in the testis.

The focal nature of damage in the seminiferous tubules after gossypol treatment as observed in P mice is also described in several studies using antispermatogenic agents (cf. Shemi *et al.,* 1989). The reason for the non-uniform response of the testis to these agents remains unclear. In general, the regressive changes observed in the seminiferous tubules in testes of P mice after treatments with all the four agents include exfoliation of germ cells, occurrence of multinucleated giant cells, degeneration of germinal elements and formation of vacuoles in the epithelium. Similar changes have also been described in mouse or in other mammalian species following treatments with several antispermatogenic agents (cf. Patanelli and Nelson, 1964; Singh, 1980; Sinha Hikim and Chakraborty, 1986; Rath, 1995). The occurrence of giant cells in the testis is considered to be an expression of germ cell degeneration. In P mice, giant cells were formed by fusion of mainly round nuclei of early spermatids. Germ cells in the seminiferous tubules are connected to each other by intercellular bridges (Dym and Fawcett, 1971). It has been suggested that giant cells are formed as a result of the fusion of spermatids due to alterations in the intercellular bridges (cf. Singh and Abe, 1987). It appears that spermatids and spermatocytes are the most susceptible germ cells in the seminiferous tubules to these treatments as these cells disappeared more readily from the epithelium by exfoliation, giant cell formation or degeneration. The spermatogonial cells, on the other hand, appear resistant and are able to proliferate as evident from the restoration of spermatogenic activity in the seminiferous tubules after withdrawal of the treatments.

The compounds described above are not suitable for use in human contraception, though there is still some hope with that of gossypol. However, it is possible that these compounds may prove useful in the control of rodent population.

Acknowledgements

I am greatly indebted to my revered teacher, Professor C.J. Dominic, FNA, for invaluable help and encouragement.

References

Abbatiello, E.R., Kaminsky, M. and Weisbroth, S. (1975). Int. J. Fertil. **20:** 177–182.

Abbatiello, E.R., Kaminsky, M. and Weisbroth, S. (1976). Int. J. Fertil. **21:** 82–88.

Ahluwalia, B., Hypolite, F. and Anderson, W. (1986). J. Androl. **7:** 254–263.

Buthala, D.A. and Lobl, T.J. (1979). Cytobios **25:** 23–28.

Burberi, S., Catanese, B., Cioli, V., Scorza Barcellona, P. and Silvestrini, B. (1975). Exp. Mol. Pathol. **23:** 308–320.

Chang, C.C., Gu, Z. and Tsong, Y.Y. (1982). Int. J. Fertil. **27:** 213–218.

Chinoy, N.J., Sharma, J.D., Seethalakshmi, L. and Sanjeevan, A.G. (1980). Int. J. Fertil. **25:** 267–274.

Comhaire, F.H. (1994). Hum. Reprod. **9:** 586–590.
Coulson, P.B., Snell, R.L. and Parise, C. (1980). Int. J. Androl. **3:** 507–518.
Coulston, F., Dougherty, W.J., LeFevre, R., Abraham, R. and Silvestrini, B. (1975). Exp. Mol. Pathol. **23:** 357–366.
DeMartino, C., Stefanini, M., Agrestini, A., Cocchia, D., Morelli, M. and Scorza Barcellona, P. (1975). Exp. Mol. Pathol. **23:** 321–356.
Dev, N.K. and Mangat, H.K. (1980). Curr. Sci. **49:** 94–96.
Donaldson, A., Sufi, S.B. and Jeffcoate, S.L. (1985). Contraception **31:** 165–171.
Dym, M. and Fawcett, D.W. (1971). Biol. Reprod. **4:** 195–215.
Ericcson, R.J. (1972). Adv. Biosci. **9:** 737–742.
Fox, B.W. and Fox, M. (1967). Pharmacol. Rev. **19:** 21–57.
Gao, Y. and Short, R.V. (1993). J. Reprod. Fertil. **97:** 39–49.
Giridharan, N., Sesikeran, B., Bamji, M.S. and Madhyastha, M.N. (1987). Contraception **35:** 89–100.
Gomes, W.R. (1970). In: The Testis (A.D. Johnson, W.R. Gomes and N.L. VanDemark, Eds), Vol. 3, pp. 483–554, Academic Press, New York.
Hadley, M.A., Lin, Y.C. and Dym, M. (1981). J. Androl. **2:** 190–199.
Hahn, D.W., Rusticus, C., Probst, A., Homm, R. and Johnson, A.N. (1981). Contraception **24:** 97–105.
Heywood, R., Lloyd, G.K., Majeed, S.K. and Gopinath, C. (1986). Toxicology **40:** 279–284.
Hoffer, A.P. (1983). Biol. Reprod. **28:** 1007–1020.
Hoffer, A.P., Agarwal, A., Meltzer, P., Naqvi, R. and Matlin, S.A. (1988). Contraception **37:** 301–331.
Hunt, S. and Mittwoch, V. (1984). J. Reprod. Fertil. **70:** 341–345.
Jackson, H. (1966). Antifertility compounds in the male and female. Charles C. Thomas, Springfield, Illinois.
Jackson, H. (1970). Antispermatogenic agents. Brit. Med. Bull. **20:** 107–114.
Jackson, H. (1973). Chemical methods of male contraception. Am. Scient. **61:** 188–193.
Jackson, H. (1975). Clin. Endocrinol. Metab. **4:** 643–663.
Kainz, V., Frick, J., Kainz, P., Kalla, N.R., Rovan, E. and Adam, H. (1988). Int. J. Androl. **11:** 533–546.
Kalla, N.R., Rovan, E., Ranga, A., Bhatia, S. and Frick, J. (1986). Contraception **34:** 177–190.
Kaur, P., Khullar, M., Mittal, S. and Rai, U.C. (1988). Indian J. Med. Res. **87:** 368–376.
Kimball, F.A., Frielink, R.D. and Porteus, S.E. (1978). Fertil. Steril. **29:** 103–108.
Lin, T., Murono, E.P., Osterman, J., Nankin, H.R. and Coulson, P.B. (1981). Fertil. Steril. **35:** 563–566.
Liu, G. and Segal, S. (1985). In: Gossypol: a potential contraceptive for men (S. Segal, Ed), pp. 1–7, Plenum Press, New York.
Lobl, T.J. (1979). Arch. Androl. **2:** 353–363.
Lobl, T.J., Forbes, A.D., Kirton, K.T. and Wilks, J.W. (1979). Arch. Androl. **3:** 67–77.
Mann, T. and Lutwak-Mann, C. (1981). Male Reproductive Function and Semen: Themes and Trends in Physiology, Biochemistry and Investigative Andrology, Springer-Verlag, Berlin.
Memon, G.N. (1973). Contraception **8:** 361–370.
National Coordinating Group on Male Antifertility Agents (1978). Chin. Med. J. **4:** 417–428.
Neumann, F. (1984). Arch. Toxicol. Suppl. **7:** 109–117.
Neumann, F., Diallo, F.A., Hasan, S.H., Schenck, B. and Traore, I. (1976). Andrologia **8:** 203–235.

Patanelli, D.J. (1975). In: Handbook of Physiology (D.W. Hamilton and R.O. Greep, Eds), Vol. V, Sect. 7, pp. 245–258, American Physiological Society, Washington, D.C.
Patanelli, D.J. and Nelson, W.O. (1964). Rec. Progr. Horm. Res. **20:** 491–513.
Pelletier, R.M. and Friend, D.S. (1980). J. Cell Biol. **87:** 1125 (abstract).
Prasad, M.R.N. and Diczfalusy, E. (1982). Int. J. Androl. Suppl. **5:** 53–70.
Rath, S.K. (1995). Ph.D. Thesis, Banaras Hindu University, Varanasi.
Rovan, E., Weinbauer, G.F., Frick, J. and Adam, H. (1983). Urol. Res. **11:** 75–85.
Saksena, S.K., Hunt, D.M. and Lau, I.F. (1978). Int. J. Androl. Suppl. 2 (abstract).
Saksena, S.K., Lau, I.F. and Bartke, A. (1974). Endrinology **95:** 311–314.
Setchell, B.P. (1978). The Mammalian Testis, pp. 393–400. Cornell University Press, Ithaca, New York.
Shemi, D., Sod-Moriah, U.A., Abraham, M., Friedlander, M., Potashnik, G. and Kaplanski, J. (1989). Andrologia **21:** 229–236.
Silvestrini, B., Burberi, S., Catanese, B., Cioli, V., Coulston, F., Lisciani, R. and Scorza Barcellona, P. (1975). Exp. Mol. Pathol. **23:** 288–307.
Singh, S.K. (1980). Ph.D. Thesis, Banaras Hindu University, Varanasi.
Singh, S.K. and Abe, K. (1987). Arch. histol. jap. **50:** 579–585.
Singh, S.K. and Dominic, C.J. (1983). Arch. Androl. **10:** 91–92.
Singh, S.K. and Dominic, C J. (1984). Exp. Clin. Endocrinol. **83:** 291–296.
Singh, S.K. and Dominic, C.J. (1985). Exp. Clin. Endocrinol. **85:** 297–302.
Singh, S.K. and Dominic, C.J. (1986). Exp. Clin. Endocrinol. **88:** 309–315.
Singh, S.K. and Rath, S.K. (1990). Arch. Histol. Cytol. **53:** 393–396.
Sinha Hikim, A.P. and Chakraborty, J. (1986). J. Androl. **7:** 277–284.
Srivastava, A., Gupta, G. and Setty, B.S. (1989). Contraception **39:** 337–355.
Tso, E.C.F. and Lacy, D. (1975). J. Reprod. Fertil. **44:** 545–550.
Udoh, P., Patil, D.R. and Deshpande, M.K. (1992). Contraception **45:** 493–509.
Wang, Y., Shi, X.-q. and Sun, Y.-b. (1985). Contraception **32:** 651–660.
Wang, Y.E., Luo, Y.G. and Tang, X.C. (1979). Acta Pharmacol. Sin. **14:** 663.
Wu, F.C.W. (1988). Clin. Endocrinol. **29:** 443–465.
Xue, S.P. (1981). In: Recent Advances in Fertility Regulation (C.F. Chang, D. Griffin and A. Woolman, Eds), pp. 122–147, Atar, Geneva.

Comparative Endocrinology and Reproduction
K.P. Joy, A. Krishna and C. Haldar (Eds)

18. Folliculogenesis and Oocyte Development in Mammals (Livestock Animals)

R. van den Hurk, M.M. Bevers* and S.J. Dieleman*
Department of Functional Morphology and *Department of Herd Health and Reproduction, Faculty of Veterinary Medicine, University of Utrecht, Utrecht, The Netherlands

Introduction

In temperate zones, ewe and goat normally exhibit oestrous cycles during a restricted period of the year (September-February in northern latitudes) in contrast with cow and sow, which are year-round cyclic (Cupps *et al.*, 1969). Ovarian cyclicity is dependent on daylight, temperature and nutritional conditions, which explains diffferences between species kept in temporate or (sub) tropical areas (Terrill, 1974).

Length of the oestrous cycle is usually about 17 days in ewes and about 21 days in sows, goats and cows, strain differences especially existing in sheep. As in all non-primate mammals, these livestock animals have an oestrous cycle that can be subdivided into a luteal and a follicular phase. In the luteal phase, one (e.g. in cows) or more (e.g. in pigs) corpora lutea are manifest. The follicular phase lasts for about 4 to 5 days in pigs and ruminants (Baird *et al.,* 1976; Foxcroft and Hunter, 1985; Spicer and Echternkamp, 1986). In the follicular phase, usually one (cow, ewe, goat) or more (sow) follicles grow to the preovulatory stage (Gougeon, 1993; Zeleznik, 1993; Greenwald and Roy, 1994). Under optimal hormonal conditions this or these follicles mature(s) and then ovulate(s), the oocyte released being ready for potential fertilization. In the luteal phase, maturation and ovulation of follicles is blocked by the negative feedback action on the hypothalamus of progesterone produced by one or more corpora lutea leading to suppressed gonadotropin output. The current views on the process of follicular development and oocyte development in these live-stock animals with long cycles summarized in the next paragraphs with particular emphasis on preantral follicle development, selection and dominance of antral follicles.

Formation of Preantral Follicles

Ontogeny

In all the mammals, the ovary develops as a thickening, called the genital ridge, along the ventral side of the mesonephros. During the first half of gestation, germ cells, originating in the yolk sac, migrate to and arrive at the genital ridge (Byskov and Hoyer, 1988). Within the indifferent gonad, both germ cells and somatic cells undergo mitotic proliferation.

In ovaries, the germ cells then become arranged in cortical cords. In these cords, groups of oogonia, connected with each other through intercellular bridges, are surrounded by somatic cells (Rüsse and Sinowatz, 1991; Fig. 1). After completion of the mitotic process (generally before birth, but in the pig up to the age of 35 days; Fulka *et al.,* 1972), these oogonia start meiosis. The female germ cells then become arrested in the dictyate stage of the prophase of the first meiotic division, and are called oocytes. At that time they are sloughed from the cortical cords and become enclosed within follicles. Each of these follicles consists of an oocyte surrounded by a single layer of a few (less than 10 min its cross-section) flattened somatic (= granulosa) cells, which on their turn are surrounded by a basement membrane. These follicles are called *primordial follicles,* and are formed during fetal life in some species (e.g. pigs and ruminants), but in others (e.g. rodents and rabbits) during the early neonatal period (Marion and Gier, 1971; Hirshfield, 1991; Rüsse and Sinowatz, 1991).

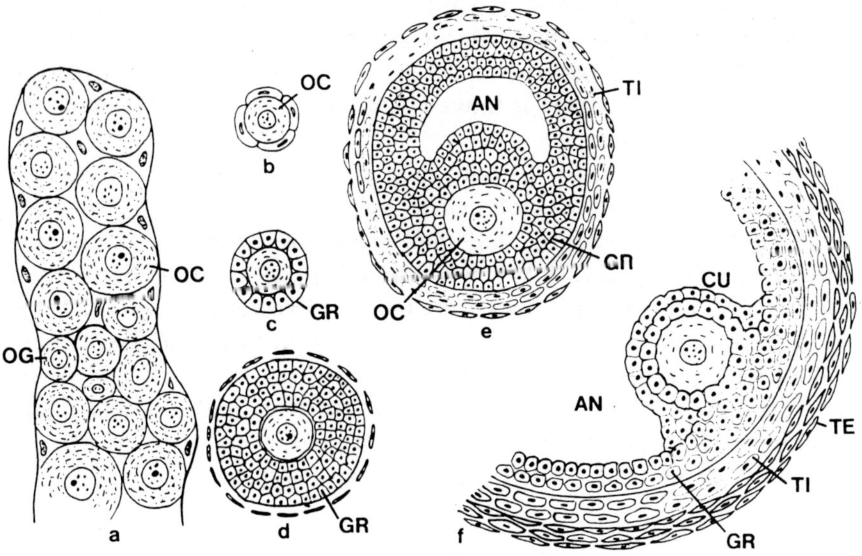

Fig. 1. Schematic presentation of follicular development in the cow. (a) cortical cord, containing oogonia (OG) and oocytes (OC), (b) a primordial follicle, (c) a primary follicle, (d) a secondary follicle, (e) a small antral follicle, and (f) part of a large antral (preovulatory) follicle. GR granulosa, TI theca interna, TE theca externa, AN antrum, CU cumulus.

Once the population of primordial follicles has been established, follicles gradually and continuously leave the resting pool to begin growth. The granulosa cells of these activated follicles become first cuboidal, and then proliferate, forming a multilaminar epithelium (Cahill and Mauleon, 1981; Rüsse and Sinowatz, 1991; van den Hurk *et al.,* 1995). Follicles with a single layer of cuboidal granulosa cells are called *primary follicles,* whereas those with more than one layer of cuboidal granulosa cells are called *secondary follicles* (Fig. 1). In the bovine, we recently observed, apart from cuboidal cells, also more elongated, vimentin-positive granulosa cells, in both primary and secondary follicles. In secondary follicles, they were located near the basement membrane and the oocyte (van den Hurk *et al.,* 1995). Primordial, primary and secondary follicles together form the population of *preantral follicles* .

Diameters increase from 30 to 40 μm for primary follicles to 5–8 mm (ewes; Mariana *et al.*, 1991), and 9–11 mm (sows; Hunter and Wiesak, 1990) and 15–20 mm (cows; Ireland, 1987) for follicles prior to ovulation. In livestock animals,preantral follicles, i.e. secondary follicles, may form up to 10 layers of granulosa cells and may reach a maximum diameter of 150–650 μm, dependent of the species and the authors who measured these structures (Erickson, 1966a; Rüsse and Sinowatz, 1991; Morbeck *et al.,* 1992). The cuboidal to polygonal granulosa cells contain abundant rough endoplasmic reticulum (RER) and mitochondria (Rüsse and Sinowatz, 1991), indicative of protein synthesis. During the preantral stage, the oocyte starts to grow, while a glycoprotein coat, the *zona pellucida* (secreted by the oocyte; Takagi *et al.,* 1989), is formed between the oocyte and the innermost layer of the granulosa.

The theca first consists of only long elongated connective tissue cells that are arranged parallel to the basement membrane underneath the granulosa. At the end of the preantral stage and especially during the antral stage, hormone producing large epitheloid cells and an extensive capillary network additionally constitute the theca. Like the blood vessels, the epitheloid cells are distributed throughout the innermost layer of the theca, the *theca interna,* and are ultrastructuraly characterized by the presence of smooth endoplasmic reticulum, mitochondria with tubular cristae and lipid droplets in their cytoplasm. The outermost theca layer, *theca externa* contains abundant connective tissue with arterioles that form the capillary network in the theca interna.

Dynamics

Calculations based on the number of granulosa cells in follicles of various size classes and the time required to double the number of cells within a follicle showed that in cows, sheeps and pigs, growth rates of a follicle varied with its size (Morbeck *et al.,* 1992). The final stages of bovine development occur comparatively rapidly (Scaramuzzi *et al.,* 1980; Lussier *et al.,* 1987), whereas the rate of growth of (preantral and early antral) follicles < 0.5 mm in diameter is estimated to be much slower (Lussier *et al.*, 1987). Although the time required for antral follicle development appeared to be similar for cows, pigs and sheep, preantral follicle development differed among the three species. It appeared considerably longer in sheep (130 days to grow from 60 to 200 μm; Cahill and Nauleon, 1980) than in cows and pigs (Morbeck *et al.*, 1992). The estimated time required for the growth of a primary follicle to reach the antral stage in pigs is 84 days, and another 21 days (the span of one oestrous cyle) to reach ovulatory status. In cows, growth of an activated primordial follicle to ovulatory size probably lasts somewhat shorter (80–100 days; Britt, 1991), since preantral follicle development proceeds faster than in pigs (Morbeck *et al.*, 1992) and antral follicles are able to grow over 10 mm within 2 days during the follicular phase (van den Hurk *et al.*, 1994). The different types of follicles and an estimation of the duration of the follicular growth from the primordial up to the preovulatory stage in bovine ovaries is shown in Fig. 2.

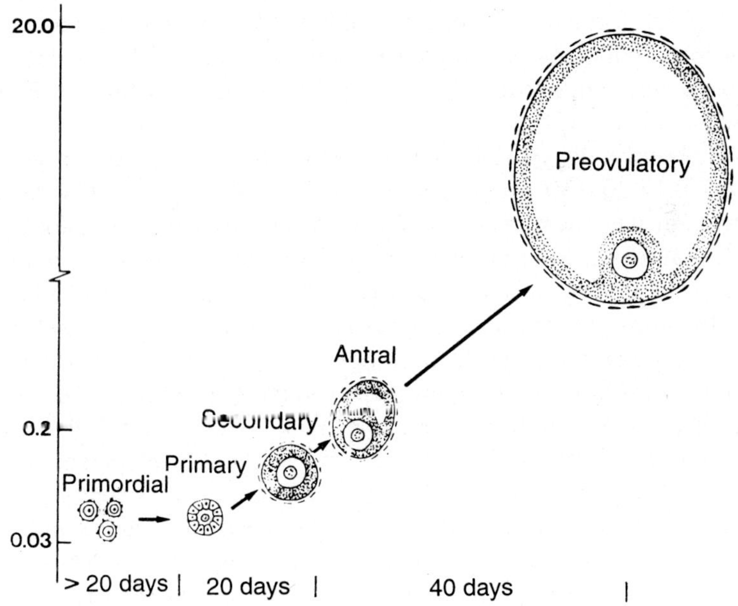

Fig. 2. Folliculogenesis in the bovine with respect to follicular diameters and estimated duration of follicular growth from the primordial to the preovulatory stage (modified from Gougeon, 1993, based on the data on bovine follicular growth by Lussier *et al.*, 1987).

All types of preantral follicles in ruminants already develop during the fetal period in livestock animals (Oxender *et al.*, 1979; Rüsse and Sinowatz, 1991; van den Hurk *et al.*, 1995) and in variable numbers. In the cow, total numbers of follicles varied from 18,000 to 200,000 per individual at the end of the fetal period, while the limits of this variation within the population sampled between birth and 24 months of age were zero (sterile) and 700,000 follicles (Ericsson, 1966b). In sheep, goat, and pig, maximum numbers of germ cells neonatally found are about 1 million 24,000 and 509,000, respectively (Rüsse and Sinowatz, 1991). The vast majority of these follicles degenerates and relatively few (less or far less than 1% of the available oocytes at birth may end ovulation during the reproductive lifespan (Saumande, 1991).

Oocytes

Oocytes of preantral follicles initially contain a few cell organelles, resulting in an electron-lucid appearance of their cytoplasm (Rüsse and Sinowitz, 1991). In growing oocytes, small vesicles arrange themselves into Golgi-fields underneath the oolemma. However, compared to small antral follicles, the number of cell, organelles in oocytes of preantral follicles is still scarce, especially when the amount of polyribosomes, RER cisternae, and cortical granules are considered. Furthermore, microvilli and the zona pellucida are formed in secondary follicles. Together with granulosa cell protrusions contacting the oolemma (Salustri *et al.*, 1993; van den Hurk *et al.*, 1995), microvilli of the oocytes penetrate the zona pellucida (Rüsse and Sinowatz, 1991).

In the bovine, oocyte diameters increase from 20 to 30 μm in primordial follicles till 120–150 μm in antral follicles (Höflinger, 1948). Maximum oocyte diameters (measured without zona pellucida) in other livestock animals are 132 μm (sow), 145 μm (goat) and 150 μm (sheep) (Rüsse and Sinowatz, 1991). In laboratory species (Sorensen and Wassarman, 1976), the oocyte generally stops growing and acquires competence to resume meiosis at the end of the preantral stage. In livestock species, however, this takes place just during the following stage, the (early) antral stage, of follicle development (Höflinger, 1948; Cahill and Mauleon, 1980; Motlik and Fulka, 1986).

Activation of primordial follicles and further development up to the early antral stage may begin at any stage of the female's reproductive life, from early infancy, throughout puberty, and during pregnancy until the end of the reproductive period (Gougeon, 1993). Nucleoplasmic RNA synthesis could be the first event in the initiation of oocyte growth in response to an inducative stimulus from the surrounding follicle cells (Lintern-Moore and Moore, 1979) or the depletion of an inhibitor. The RNA synthesis is high in growing oocytes and decreases drastically when the oocytes reaches full size (Oakberg, 1967; Moore *et al.*, 1974) which, in contrast to laboratory animals, is at the antral stage in livestock animals (Motlik and Fulka, 1986; Crozet *et al.*, 1986). In small laboratory animals some of the genes expressed during the primordial follicle stage have been demonstrated. These include proteins composing the zona pellucida and oocyte specific signalling factors, like c-kit tyrosine kinase receptor and c-kit ligand (for review, Richards, 1993). Besides, granulosa cells of preantral follicles express genes encoding of receptors for FSH (follicle stimulating hormone), receptors for NGF (nerve growth factor) and MIS (Mullerian inhibiting substance). Unfortunately, the precise role of each these gene products is unknown. Anyway, they are in support for the opinions that: (i) granulosa cells of the preantral follicles may bind FSH, (ii) the oocyte of the preantral follicles is responsible for the formation of the zona pellucida, and (iii) various peptides are involved in preantral follicle development.

Hormonal regulation

Every day, 2–3 follicles in sheep (Cahill and Mauleon, 1980) or six in cows (Scaramuzzi *et al.,* 1980) start to grow from the pool of primordial follicles. So far, information on the compounds that initiate the activation of the primordial follicles in mammals is scarce. In this respect, data concerning the role of FSH are confusing (Gougeon, 1993). Based on studies with small laboratory animals and monkeys, it is suggested that FSH might trigger the very first transformation of flattened into cuboidal granulosa cells. Thereafter, however, gonadotropins seem to play a minor role in this process, since in these animals (Richards, 1980; Gougeon, 1993) and in the ewe (Dufour *et al.,* 1979; Greenwald and Terranova, 1988) development of small antral follicles from preantral follicles seems relative impervious to the absence of gonadotropins after hypophysectomy, proliferative activity of granulosa cells from preantral and small antral follicles is not related to cyclic hormonal changes (Gougeon, 1986) and granulosa cells can divide in the absence of gonadotropins (Mulheron and Schomberg, 1993; Gong *et al.,* 1993). In contrast, under *in vitro* circumstances, preantral follicle development in men (Roy and Treacy, 1993) and small laboratory animals (Tonetta and DiZerega, 1989; Gore-Langton and Daniel, 1990; Roy and Greenwald, 1989; Nayudu and Osborn, 1992) appeared to depend on FSH action, which includes stimulation of granulosa cell proliferation and differentiation, antrum formation and reduction of the percentage of atretic follicles.

In small bovine preantral follicles (e.g. primary and early secondary follicles), cultured under serum-free in vitro circumstances, FSH also acts as a mitogen, resulting in growth of the follicles (Hulshof *et al.,* 1995). The autoradiographic demonstration of FSH binding sites in granulosa cells of bovine primary and secondary follicles (Wandji *et al.,* 1992a) is in accordance with the involvement of FSH in bovine preantral follicle development, although the FSH binding sites show a low density as compared with early antral follicles. Hirao *et al.* (1994) were able to stimulate antrum formation and growth of oocytes in late porcine preantral follicles, cultured *in vitro*, by addition of FSH and oestradiol. When oocytes were isolated from these latter follicles and further cultured, fully grown oocytes acquired the capacity to resume and complete the meiosis, and were subsequently penetrated by spermatozoa.

Oestradiol is generally produced only in granulosa cells, but in pigs also in theca cells (Evans *et al.*, 1981; Foxcroft and Hunter, 1985). It is a well-known mitogenic steroid in granulosa cells and increases the sensitivity of the follicle to further gonadotropin stimulation (Tonetta and DiZerega, 1989). Like other intragonadal factors, oestradiol acts to independently regulate cellular function or to attenuate or accentuate the actions of gonadotropins (Hsueh *et al.*, 1984). Although in rats (Tonetta and DiZerega, 1989) and in pigs (Hirao *et al.*, 1994) oestradiol synergizes with FSH to enhance granulosa cell proliferation and antrum formation in preantral follicles, it evokes only enlargement of granulosa cells in *in vitro* cultured bovine primary and small secondary follicles (Hulshof *et al.*, 1995). Unfortunately, it is unclear, whether this enlargement reflects differentiation on or dedifferentiation of granulosa cells.

Among the intragonadal factors, growth factors also seem to play a key role in folliculogenesis by controlling the proliferation and differentiation of granulosa cells. In this respect EGF (epidermal growth factor), bFGF (basic fibroblastic growth factor), TGFs (transforming growth factors) and IGF-1 (insuline-like growth factor-1) have been shown to influence porcine granulosa cell proliferation and differentiation *in vitro* (for review, Mulheron and Schomberg, 1993). EGF also appeared to be a mitogen for bovine granulosa cells *in vitro* (Gospodarowitz *et al.*, 1977), while IGFs have been shown to influence (antral) granulosa cell proliferation and differentiation in ovine (Monniaux and Pisselet, 1992) and in bovine (Schams *et al.*, 1988; Gong *et al.*, 1993) granulosa cells, probably by the amplification of FSH action. However, in contrast to antral granulosa cells, EGF binding in bovine preantral granulosa cells is low, while IGF binding is virtualy absent (Wandji *et al.*, 1992b), suggesting a role for EGF and IGFs only in antral follicle development. On the other hand, the latter authors demonstrated high bFGF binding in granulosa cells of preantral (i.e. primary and secondary) follicles in fetal and neonatal bovine ovaries. This indicates, that granulosa cells are the major sites of bFGF reception during preantral follicle development and that bFGF may be involved in the regulation of granulosa cell function in the preantral stage in cattle. Indeed, Nuttinck *et al.* (1993) recently demonstrated a stimulating effect of bFGF on the proliferative activity of bovine preantral granulosa cells *in vitro.*

Other proteins mentioned to be involved in the paracrine or autocrine regulation of preantral follicle development are activin and inhibin (Robertson *et al.*, 1988). Hulshof (1995) recently demonstrated the growth promoting effect of activin on bovine primary and early secondary follicles, cultured *in vitro,* while both activin and activin receptor could be immuno-histochemically demonstrated both in granulosa cells and oocytes of especially the preantral follicles. Furthermore, mRNAs for activins and inhibins are confined to granulosa cells of cows (Rodgers *et al.*, 1989; Torney *et al.*, 1989) as it is in rats (Meunier *et al.*, 1988) and primate (Schwall *et al.*, 1990). These findings may point to granulosa cells as sources of activin and to an autocrine action of activin on granulosa cells of bovine preantral follicles. The specific α-

subunits inhibin could first be visualized in antral follicles > 2 mm in diameter (van den Hurk and Dijkstra, 1992), which is in agreement with the presence of substantial amounts of α-subunit mRNA in only antral follicles in the bovine (Hopko-Ireland, and Ireland, 1994) and other mammals (Mather *et al.,* 1992) except for the ovine (Tisdall *et al.,* 1995).

Hulshof (1995) recently found that, apart from activin and FSH, vasoactive intestinal protein (VIP) appeared associated with preantral follicle growth, since it stimulates the development of isolated bovine primary and early secondary follicles *in vitro.* We previously observed, that formation of primary follicles is accompanied by the outgrowth of vasoactive intestinal protein (VIP) or neuropeptide Y (NPY) containing nerve vessels (Hulshof *et al.,* 1994) and vimentin containing blood vessels (van den Hurk, unpublished) towards the activated follicle. Whether chemotaxic signals coming from a follicle, e.g. follicular growth factors like VEGF, bFGF (both formed by bovine granulosa cells; Neufeld *et al.,* 1987; Wandji *et al.,* 1992b; Koos, 1993) and perhaps NGF (formed by gonadotropin stimulated antral follicles; Mattioli *et al.,* 1995) may function on as chemotaxic signals involved in angiogenesis and nerve outgrowth, or gradual and continuous outgrowth of the circulation and nerve system initiate recruitment of accidently reached primordial follicles is still unclear.

Research in the field of preantral follicle development is hard but intriguing, and will be indispensable to understand the mechanism that underlies recruitment of antral follicles from primordial follicles.

Formation of Antral Follicles

The formation of a fluid filled cavity within the granulosa, the follicular *antrum,* is characteristic for *antral follicles.* Antrum formation is brought about by both decreased adhesiveness of adjacent granulosa cells and filling of the intergranulosa cell space with fluid (Schweitzer *et al.,* 1981). Secretion of proteoglycans by granulosa cells evokes the influx of fluid as a result of increased osmotic pressure. This process is enhanced after vasodilation and increased vascular permeability during further antral follicle development. Due to the accumulation of fluid, antral follicles may enormously increase in size (to more than 20 mm in cows). The antral follicular fluid may act as an important source of regulatory or modulatory substances derived from blood or secretions of the folicular cells (e.g. gonadotropins, steroid hormones, growth factors, regulatory peptides, lipoproteins, proteoglycans, and enzymes; for review, Guraya 1985).

Apart from basal growth, formation of preovulatory follicles from antral follicles, emphasis selection and maturation of preovulatory follicles. The process of basal growth, in which early antral follicles enter into the pool of developing follicles is often referred to as *recruitment* of follicles in literature (Driancourt, 1991; Fortune, 1994). However, the actual recruitment of follicles is the development of early antral follicles from the pool of primordial follicles (Zeleznik, 1993). Hence, one may talk about recruitment of primordial follicles and recruitment of early antral follicles. *Selection* of follicles includes the removal of excess number of recruited antral follicles by atresia in order to reach the number of ovulations typical of the breed. The follicle(s) surviving the selection process and destined to ovulate is/are usually called *dominant* (Hodgen, 1982; Driancourt, 1991).

In dominant antral follicles, the oocytes are surrounded by the closely associated granulosa cells, referred to as *cumulus cells,* forming a compact cumulus cell-oocyte complex (COC). This complex is a peninsula, embedded in follicular fluid and connected with the *mural granulosa cells* that surround the follicular antrum via a small stalk consisting of antral granulosa cells (Fig. 1). Mural granulosa cells located in different regions of the follicle acquire different regions of the follicle acquire different morphologic and functional properties (for review, Salustri *et al.,* 1993; van den Hurk *et al.,* 1995). The more peripherally located layers of mural granulosa cells (the *peripheral granulosa cells)* generally have an elongated appearance and stop proliferating. The more antral located layers of mural granulosa cells (the *antral granulosa cells*) and the cumulus cells have a cuboidal to the polyhedral appearance and continue to proliferate. They also show limited steroidogenic activity and have fewer LH receptors. Interestingly, we had recently autoradiographically demonstrated in the bovine that, at the preovulatory LH surge, also the antral granulosa cells possess large numbers of LH receptors, whereas they remain absent in cumulus cells (Fig. 3; van den Hurk and Bevers, unpublished data). Peripheral granulosa cells are exposed to molecules produced by theca cells and to components of the basal membrane, whereas the cumulus cells are exposed to products coming from the oocyte. Molecules of small size that penetrate the granulosa cells, or are generated in granulosa cells after endocrine or paracrine stimulation, can be efficiently spread between mural granulosa cells, cumulus cells and the oocyte through an extensive network of gap junctions, thus coordinating follicular development (Salustri *et al.*, 1993).

In some species (e.g. cow, sheep and horse), formation of dominant follicles from recruited antral follicles occur at regular intervals (waves; Robertson, 1977; Fortune, 1994), whatever the reproductive stage, whereas in others (e.g. pigs, primates and rats) dominant follicles develop only during the follicular phase of the oestrous cycle (Spicer and Echternkamp, 1986; Fortune, 1994). In all species, the dominant follicles(s) present mature(s) and then the ovulates(s) after the generation of an adequate preovulatory LH surge at the end of the follicular phase (Fortune, 1994), i.e. usually about 24–48 h before ovulation (Tonetta and DiZerega, 1989).

Maturation of these dominant follicles (now often called *Graaffian or preovulatory follicles)* includes a.o. meiotic resumption of the oocyte, expansion of the COC, and the initiation of luteinization of the granulosa, marked by increase of the size of the granulosa cells and folding of the basal membrane (Salustri *et al.,* 1993; Eppig, 1993; Foxcroft and Hunter, 1985).

Ovulation is the release of an oocyte due to follicular rupture. The most important structural changes leading to breaching of the follicular wall and subsequent ovulation are dissolution of the extracellular matrix and dissociation of thecal collagen (Espey, 1967).

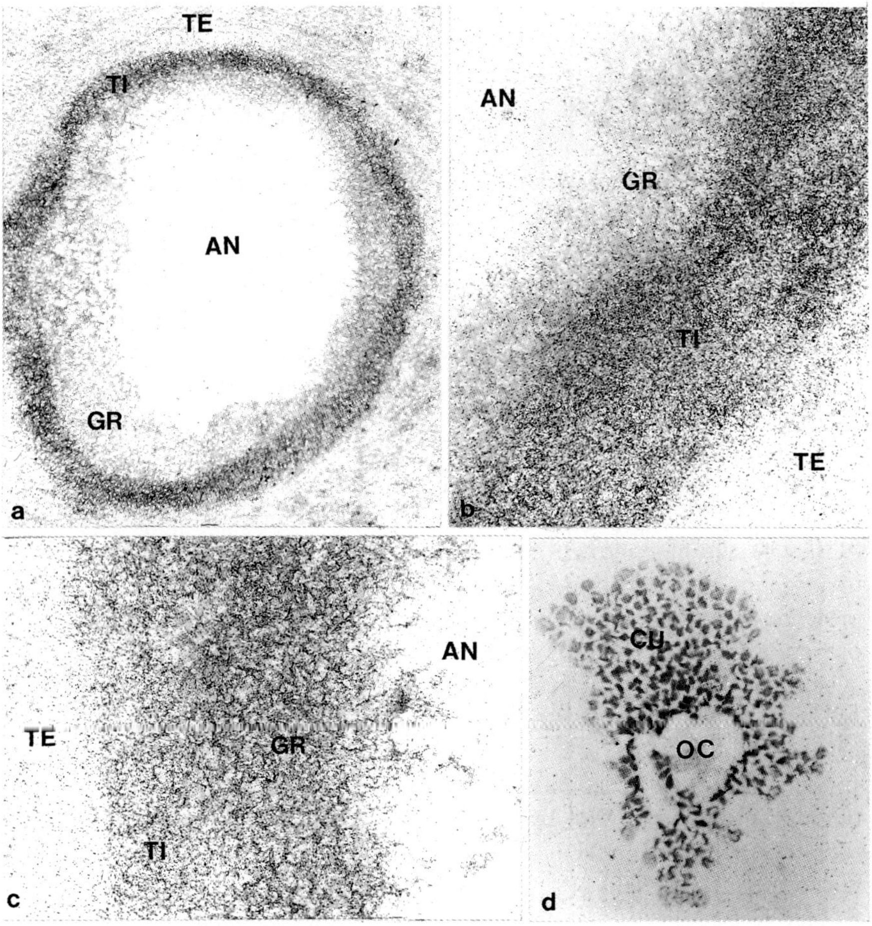

Fig. 3. Autoradiographic demonstration of I125 labeled hCG. LH binding to antral bovine follicles. During the luteal stage of the oestrous cycle, only the theca interna (TI) of small antral follicles is labeled (a, × 40; b, × 200). At the LH-surge, both the theca interna and the granulosa are labeled (c, × 200), whereas the cumulus (CU) is unlabeled (d, × 100).

Selection and Dominance

In general, the processes of selection and establishing of dominance of the ovulatory follicles take place within 48–60 h of onset of luteal regression during the oestrous cycle in sheep and cattle.

At the end of the luteal phase, one to five follicles commence to grow and become gonadotropin-dependent out of the pool of 20–30 recruited antral follicles (cow 2–5 mm; sheep 2–3 mm) which are gonadotropin-responsive. Dependent of the species or breed, one (cow; ewe) or a few (ewe, sow) of these emerging follicles are then selected. Subsequently, the selected follicle(s) establish(es) dominance over the rest of the follicles, increasingly producing oestradiol leading to a switch in the pattern of hypothalamic gonadotropin-releasing hormone (GnRH) secretion. The resultant massive release of GnRH triggers the preovulatory LH surge (for review, Karsch *et al.*, 1992) at which stage the dominant follicle has acquired all characteristics necessary to undergo final follicular maturation leading to ovulation.

Selection is a complex multifactorial process resulting in establishment of dominance. Although gonadotropins play a key-role, selection can be considered to be the outcome of gradual shifts of the balance in a follicle of the interaction between endocrine regulatory mechanisms, local paracrine and autocrine regulation and intra-ovarian modulation of endocrine signals. In the gonadotropin-dependent antral follicle, androgen production by the theca interna, necessary as precursor for granulosa aromatization into oestradiol, is stimulated by LH action which is attenuated by thecal EGF and granulosa activin but enhanced by granulosa inhibin. FSH is necessary to promote aromatization in the membrane granulosa and this signal is amplified by locally produced IGF-I and activin but counteracted by thecal EGF. The action of growth factors in turn is dependent on the presence of their receptors and the bio-availability of these factors is affected by binding proteins. In some cases the binding proteins are modied by specific proteases preventing the sequestration of the growth factors. The role of these complex systems in folliculogenesis has been reviewed for ruminants (Monget and Monniaux, 1995; Spicer and Echternkamp, 1995) and pigs (Hammond *et al.*, 1993).

Selection may be accomplished in two ways: (1) the increasing production of inhibin and oestradiol by the follicle in the process of being selected reduces the concentration of circulating FSH below a threshold level necesssary to maintain other follicles (sheep; Driancourt, 1991) and (2) the follicle to be selected produces factors that either attenuate FSH sensitivity or inhibit aromatizing activity of other follicles (cow; Campbell *et al.*, 1995). Further, the growing follicle will produce increasing amounts of follistatin a peptide inhibiting FSH activity possibly by binding activin which enhances FSH secretion (for review, Findlay *et al.*, 1992). The selected follicle is able to continue maturation until the preovulatory LH signal in spite of decreasing FSH concentrations. This is achieved by the higher oestradiol production interacting with FSH to stimulate granulosa IGF-I production which in turn amplifies the granulosa sensitivity to FSH. Further, the increment of intrafollicular IGF-I increases the number of receptors for LH necessary for maintaining dominance and proper response to start final maturation.

Follicular Maturation

It is generally accepted that the resumption of meiosis in mammalian oocytes is triggered by the preovulatory LH peak. Of the antral follicles, only those which have achieved an adequate degree of maturation i.e., the preovulatory follicles will respond to LH with a series of morphological and functional changes including the resumption of meiosis of the oocyte. There are no receptors for LH on the oolemma and LH acts on the somatic cells, which in turn transduce some regulatory signals to the oocyte. The sequence of events starting on the plasma membrane of granulosa cells with the binding of LH to its receptor is finally reflected in the oocyte, but the signals used for this intercellular communication are still unknown (review Mattioli, 1992). In this respect, it should be mentioned that quite a lot of research data originate from *in vitro* maturation studies (review Mattioli, 1992). However, since *in vitro* maturation studies are performed with oocytes from small and medium sized follicles, i.e. non-preovulatory follicles, it is dangerous to assume that *in vitro* maturation of oocytes reflects the *in vivo* process. Coincident with the LH-surge, thecal androstenedione production is immediately inhibited, depleting the membrana granulosa of its precursor for oestradiol, which causes a decrease of oestradiol in the microenvironment of the oocyte (Dieleman *et al.*, 1983a, b). The membrana granulosa, however, is capable of aromatization until halfway through the maturation process at 14 h after the LH-surge (Dieleman and Blankenstein, 1984). Shortly before ovulation, progesterone synthesis becomes dominant (Dieleman and Bevers 1993; Dieleman and Blankenstein, 1985). The morphology of the follicular wall changes throughout the process of maturation; cell density decreases and mitotic indices also decline after an initial high level during the first period of maturation (de Loos *et al.*, 1991). Shortly before ovulation, luteinization occurs at which time the follicular wall shows breakdown of the basal membrane and the invasion of eosinophilic granulocytes.

During the period between the LH-surge and ovulation, the oocyte undergoes a series of profound changes in both the nucleus and the cytoplasm. Nuclear maturation, lasting about 24 h in the cow and sheep, about 44 h in pig (Mattioli, 1992) and presumably 36 h in horse (King *et al.*, 1987), comprises several steps, which can be recognized in the process of nuclear progression from dictyate stage to metaphase II: (i) condensation of chromosomes into distinct bivalents, (ii) disappearance of the nucleolus, (iii) dissolution of the nuclear membrane referred to as germinal vesicle breakdown (GVBD), (iv) alignment of chromosomes on the metaphase I spindle, (v) separation of homologous chromosomes and extrusion of the first polar body, and finally (vi) rearrest of meiosis at metaphase II stage. Completion of the meiotic cycle is triggered by the activation stimulus provided by sperm penetration. In the cow and the sheep, GVBD is observed to occur 4 to 8 h after the LH-surge, while in the pig it takes about 24 h. Nuclear maturation is controlled by a cascade of events in the cytoplasm. The transition from the G2-phase to the M-phase of the cell cycle is dependent on the presence of an active maturation promoting factor (MPF). An initial activity increase of this factor is responsible for the GVBD and the metaphase plate assembly. When the activity declines, the nucleus proceeds through the anaphase and telophase stages and the second activity increase stabilized by a cytostatic factor maintains the oocyte in the M-II stage (review Parrish *et al.,* 1992). In cow oocytes, collected after the LH-surge (Kastrop *et al.,* 1992), marked changes in protein synthesis patterns were observed exclusively after GVBD, whereas oocytes that remained in the germinal vesicle stage showed consistent protein synthesis patterns. Similarly, protein phosphorylation patterns exhibit striking differences around GVBD with some proteins undergoing extensive phosphorylation before GVBD. In ungulate oocytes, an early phase of transcriptional activity in the cumulus cells surrounding the oocyte followed by a period of protein synthesis in the oocyte is an absolute requirement for GVBD (Kastrop *et al.,* 1991). In addition to the control and direction of meiotic progression, the cytoplasm has to change in order to acquire the conditions required to decondense penetrated spermatozoa and to form male pronuclei after fertilization. During maturation, an extensive redistribution of intracellular organelles takes place; mitochondria migrate to occupy a peri-nuclear position and the cortical granules migrate outwards to line up along the oolemma (Kruip *et al.,* 1983; review Cran, 1989). This migration is an important requisite for the oocyte to operate an effective block to polyspermy in case of fertilization. Concomitantly, a decrease in the development of the Golgi complexes and a clustering of the SER is seen (Kruip *et al.,* 1983).

Before and at the time of the LH-surge, the oocyte is surrounded by a compact cumulus investment. From the innermost cumulus cells, numerous projections penetrate the zona pellucida. These projections end in invaginations of the oolemma, where gap junctions are found. The perivitelline space is small or occluded. Already within the first two hours after the LH-surge, the perivitelline space is enlarged (Kruip *et al.,* 1983; Hyttel *et al.,* 1986), however, a considerable proportion of the junctions between cumulus cells and the oocyte are maintained. The final disruption of these junctions coincides more or less with GVBD. At the end of the maturation period, the oocyte with the enlarged perivielline space is surrounded by an expanded and clear cumulus cell mass (Hyttel *et al.,* 1986).

Follicular Atresia

As aforementioned, selection of dominant follicles includes atresia of the remainder antral follicles. The pattern of atresia is initialy characterized by degenerative changes in the granulosa, whereas the oocyte and zona pellucida are almost unchanged (Greenwald and Terranova, 1984). This degenerative process is an example of *apoptosis* (programmed cell death) and is marked by condensation of nuclear chromatin into crescantic caps at the periphery of the nucleus, nuclear desintegration and an overall reduction in nuclear size, in concert with cytoplasmic condensation (reduction in volume), compaction of cell organelles but mitochondria sparing (Hurwitz and Adashi, 1993). In contrast with antral follicles, preantral follicles predominately regress via *necrosis,* which is marked by membrane disruption, cell swelling and rupture. The necrotic changes start in the oocyte, whereas alterations in the granulosa are secondary.

Hence, atresia can occur at any stage of follicular development. It is, however, not equally prevalent across all stages of follicular development. Generally spoken, little atresia occurs in preantral and early antral follicles (Hirshfield, 1991). Atresia may be caused by hypoxia (Hirshfield, 1991), disruption of ionic gradients (Zeleznik *et al.,* 1989), complement activation (Farookhi, 1981) and inadequate differentiation of follicular components during the preantral stage (Richards, 1980). Besides, various inadequate hormonal signals, such as insufficient or excessive gonadotropin, oestrogen or androgen, and intraovarian or extraovarian proteinous factors like activin, inhibin, GnRH-like substanc(s), FSH-binding inhibitor, cytokines (like tumor necrosis factor-α) and growth factors (like EGF, TGFα and basic FGF) are considered to promote atresia (Gougeon, 1993). It is assumed that only those follicles that reach a critical stage in their development coincidently with appropriate changes in serum gonadotropins, steroid levels and possibly with other endocrine, paracrine, autocrine and intracrine factors will be saved from atresia and reach ovulation.

References

Baird, D.T., Land, R.B., Scaramuzzi, R.J., Wheeler, A.G. (1976). J. Endocr. **69:** 275–286.

Britt, J.H. (1991). Bovine practitioner **24:** 39–43.

Byskov, A.G. and Hoyer, P.E. (1988). In: The physiology of reproduction (E. Knobel and J. Neill, Eds), pp 265–302. Raven Press, New York.

Cahill, L.P. and Mauleon, P. (1980). J. Reprod. Fertil. **58:** 321–328.

Cahill, L.P. and Mauleon, P. (1981) J. Reprod. Fertil. **61:** 201–206.

Campbell, B.K., Scaramuzzi, R.J. and Webb, R. (1995). In: Reproduction in domestic ruminants III (R.J. Scaramuzzi, C.D. Nancarrow and C. Doberska, Eds.), pp 335–350. The Dorset Press, UK.

Cran, D.G. (1989). J. Reprod. Fert. Suppl. **38:** 49–62.

Crozet, N., Kanka, J., Motlik, J. and Fulka, J. (1986). Gamete Res. **14:** 65–73.

Cupps, P.T., Anderson, L.L. and Cole, H.H. (1969). In: Reproduction in domestic animals (H.H. Cole and P.T. Cupps, Eds), pp 217–250. Acad. Press, New York.

Dieleman, S.J. and Bevers, M.M. (1993). Molec. Reprod. Dev. **36:** 271–273.

Dieleman, S.J. and Blankenstein, D.M. (1984). J. Reprod. Fertil. **72:** 487–494.

Dieleman, S.J. and Blankenstein, D.M. (1985). J. Reprod. Fertil. **75:** 609–615.

Dieleman, S.J., Kruip, Th.AM., Fontijne, P., Jong, W.H.R. de and Weyden, G.V. van der (1983a). J. Endocrinol. **97:** 31–42.

Dieleman, S.J., Bevers, M.M., Poortman, J. and Tol, H.Y.M. van (1983b). J.Reprod. Fertil. **69:** 641–649.
Dufour, J., Cahill, L.P. and Mauleon, P. (1979). J. Reprod. Fertil. **57:** 301–309.
Driancourt, M.A. (1971). Theriogenology **35:** 55–79.
Eppig, J.J. (1993). In: The ovary (E.Y. Adashi and P.C.K. Leung, Eds), pp 185–208. Raven Press, New York.
Erickson, B.H. (1966a). J. Reprod. Fertil. **10:** 97–105.
Erickson, B.H. (1966b). J. Anim. Sci. **25:** 800–805.
Espey, L.L. (1967). Endocrinol. **81:** 267–276.
Evans, G., Dobias, M., King, G.J. and Armstrong, D.T. (1981). Biol. Reprod. **25:** 673–682.
Farookhi, R. (1981). In: Dynamics of ovarian function (N.B. Schwartz and M. Heinzicker, Eds), pp 13–23. Raven Press, New York.
Findlay, J.K., Robertson, D.M., Clarke, I.J., Klein, R., Doughton, B.W., Xiao, S., Russell, D.L. and Shukovski, L. (1992). In: Clinical trends and basic research in animal reproduction (S.J. Dieleman, B. Colenbrander, P. Booman and T. van der Lende, Eds), pp 319–328. Elsevier, Amsterdam.
Fortune, J.E. (1994). Biol. Reprod. **50:** 225–232.
Foxcroft, G.R. and Hunter, M.G. (1985). J. Reprod. Fertil. Suppl. **33:** 1–19.
Fulka, J., Kopecny, V. and Trebichavsky, W. (1972). Biol. Reprod. **6:** 46–50.
Gong, J.G., McBride, D., Bramley, T.A. and Webb, R. (1993). J. Endocrinol. **139:** 67–75.
Gore-Langton, R.E. and Daniel, S.A. (1990). Biol. Reprod. **43:** 65–72.
Gospodarowitz, D., Ill, C.R. and Birdwell, C.R. (1977). Endocrinol. **100:** 1108–1120.
Gougeon, A. (1986). Hum. Reprod. **1:** 81–87.
Gougeon, A. (1993). In: The ovary (E.Y. Adashi and P.C.K. Leung, Eds), pp 21–39. Raven Press, New York.
Greenwald, G.S. and Roy, S.K. (1994). In: The physiology of reproduction (E. Knobel and J. Neill, Eds), pp 629–725. Raven Press, New York.
Greenwald, G.S., Terranova, P.F. (1988). In: The physiology of reproduction (E. Knobel and J. Neill, Eds), pp 384–445. Raven Press, New York.
Guraya, S.S. (1985). Biology of ovarian follicles in mammals. Springer Verlag, Berlin.
Hammond, J.M., Samaras, S.E., Grimes, R., Leighton, J., Barber, J., Canning, S.F. and Guthrie, H.D. (1993). In: Control of pig reproduction IV (G.R. Foxcroft, M.G. Hunter and C. Doberska, Eds), pp 117–125. The Dorset Press, UK.
Hirao, Y., Nagai, T., Kubo, M., Miyano, T., Miyaki, M. and Kato, S. (1994). J. Reprod. Fertil. **100:** 333–339.
Hirshfield, A.N. (1991). Int. Rev. Cytol. **124:** 43–101.
Hodgen, G.D. (1982). Fertil. Steril. **38:** 640–647.
Höflinger, H. (1948). Acta Anat. (Basel) Suppl. **5:** 1–196.
Hopko-Ireland, J.L. and Ireland, J.J. (1994). Biol. Reprod. **50:** 492–501.
Hsueh, A., Adashi, E., Jones, P. and Welsh, T. (1984). Endocr. Rev. **5:** 76–127.
Hulshof, S.C.J., Dijkstra, G., Beek, E.N., van der, Bevers, M.M., Figueiredo, J.R., Beckers, J.F., Hurk, R. van den, (1994). Biol. Reprod. **50:** 553–560.
Hulshof, S.C.J., Figueiredo, J.R., Beckers, J.F., Bevers, M.M., Donk, J.A. van der and Hurk R. van den (1995). Theriogenology **44:** 217–226.
Hulshof, S.C.J. (1995). Ph.D. Thesis, University of Utrecht, The Netherlands.
Hunter, M.G. and Wiesak, T. (1990). J. Reprod. Fertil. Suppl. **40:** 163–177.
Hurk, R. van den and Dijkstra, G. (1992). Vet. Quart. **14:** 148–151.
Hurk, R. van den, Dijkstra, G. , Hulshof, S.C.J., Vos, P.L.A.M. (1994). J. Reprod. Fertil. **100:** 137–142.
Hurk, R. van den, Dijkstra, G., Mil, F.N. van, Hulshof, S.C.J., Ingh, T.S.G.A.M. van den (1995). Mol. Reprod. Develop. **41:** 459–467.

Hurwitz, A. and Adashi, E.Y. (1993). In: The ovary (E.Y. Adashi and P.C.K. Leung, Eds), pp 473–486. Raven Press, New York.

Hyttel, P., Callesen, H. and Greve, T. (1986). J. Reprod. Fert. **76:** 645–656.

Ireland, J.J. (1987). J. Reprod. Fertil. **84:** 39–54.

Karsch, F.J., Moenter, S.M. and Caraty, A. (1992). In: Clinical trends and basic research in animal reproduction (S.J. Dieleman, B. Colenbrander, P. Booman and T. van der Lende, Eds), pp 329–342. Elsevier, Amsterdam.

Kastrop, P.M.M., Hulshof, S.C.J., Bevers, M.M., Destree, O.H.J. and Kruip, Th. A.M. (1991). Mol. Reprod. Dev. **28:** 249–254.

Kastrop, P.M.M., Bevers, M.M., Destree, O.H.J., Dieleman. S.J. and Kruip, Th. A.M. (1992). Anim. Reprod. Sci. **29:** 133–143.

King, W.A., Bezard, J., Bousquet, D., Palmer, E. and Betteridge, K.J. (1987). Genome **29:** 679–682.

Koos, R.D. (1993). In: The ovary (E.Y. Adashi and O.C.K. Leung, Eds), pp 433–453. Raven Press, New York.

Kruip, Th. A.M., Cran, D.G., Beneden, Th.H. van and Dieleman, S.J. (1983). Gamete Res. **8:** 29–47.

Lintern-Moore, S. and Moore, G.P.M. (1979). Biol. Reprod. **20:** 773–778.

Loos, F.A.M.de, Bevers, M.M., Dieleman, S.J. and Kruip, Th. A.M. (1991). Theriogenology **35:** 527–535.

Lussier, J.P., Matton, P. and Dufour, J.J. (1987). J. Reprod. Fertil. **81:** 301–307.

Mariana, J.C., Monniaux, D., Driancourt, M.A., Mauleon, P. (1991). In: Reproduction in domestic animals (P.T. Cupps, Ed), pp 119–171. Acad. Press Inc., San Diego.

Marion, G.B. and Gier, H.T. (1971). J. Anim. Sci. **32** *(Suppl 1)*: 24–27.

Mather, J. P., Woodruft, T.K. and Krummen, L.A. (1992). Proc. Soc. Exper. Biol. Med. **201:** 1–15.

Mattioli, M. (1992). In: Embryonic development and manipulation in animal production (A. Lauria and F. Gandolfi, Eds), pp 17–29. Portland Press, London.

Matioli, M., Barboni, B., Gioia, L. and Lucidi, P. (1995). J. Reprod. Fertil. Abstr. Series **15:** p29.

Meunier, H., Cajander, S.B., Roberts, U.J., Rivier, C., Sawchenko, P.E., Hsueh, A.J.W. and Vale, W. (1988). Mol. Endocrinol. **2:** 1352–1363.

Monget, P. and Monniaux, D. (1995). In: Reproduction in domestic ruminants III (R.J. Scaramuzzi, C.D. Nancarrow and C. Doberska, Eds), pp 321–333. The Dorset Press, UK.

Monniaux, D. and Pisselet, C. (1992). Biol. Reprod. **46:** 109–119.

Moore, G.M.P., Lintern-Moore, S., Peters, H. and Faber, M. (1974). J. Cell. Biol. **60:** 416–422.

Morbeck, D.E., Esbenshade, K.L., Flowers, W.L., Britt, J.H. (1992). Biol. Reprod. **47:** 485–491.

Motlik, J. and Fulka, J. (1986). Theriogenology **25:** 87–96.

Mulheron, G.W., and Schomberg, D.W. (1993). In: The ovary (E.Y. Adashi and P.C.K. Leung, Eds), pp 337–362. Raven Press, New York.

Nayudu, P.L., Osborn, S.M. (1992). J. Reprod. Fertil. **95:** 349–362.

Neufeld, G., Ferrara, N., Schweigerer, L., Mitchell, R. and Gospodarowitz, D. (1987). Endocrinol. **121:** 597–603.

Nuttinck, F., Massip, A. and Dessy, F. (1993). J. Reprod. Fertil. Abstr. Series **12:** p34.

Oakberg, E.F. (1967). Arch. Anat. Microsc. Morphol. Exp. **56***(Suppl 3–4)***:** 171–184.

Oxender, W.D., Colenbrander, B., Wiel, D.F.M. van de, Wensing, C.J.B. (1979). Biol. Reprod. **21:** 715–721.

Parrish, J.J., Kim, C.I. and Bae, I.H. (1992). Theriogenology **38:** 277–296.

Richards, J.S. (1980). Physiol. Rev. **60:** 51–89.
Richards, J.C. (1993). In: The ovary (E.Y. Adashi and P.C.K.Leung, Eds), pp 93–111. Raven Press, New York.
Robertson, H.A. (1977). In: Reproduction in domestic animals (H.H. Cole and P.T. Cupps, Eds), pp 475–498. Acad. Press, New York.
Robertson, D.M., Hayward, S. and Irby, D. (1988). Mol. Cell. Endocrinol. **58:** 1–8.
Rodgers, R.J., Stuchbery, S.J., Findlay, J.K. (1989). Mol. Cell. Endocrinol. **62:** 95–101.
Roy, S.K. and Greenwald, G.S. (1989). J. Reprod. Fertil. **87:** 103–114.
Roy, S.K. and Treacy, B.J. (1993). Fertil. Steril. **59:** 783–790.
Rüsse, I. and Sinowatz, F. (1991). In: Lehrbuch der Embryologie der Haustiere (I. Rüsse and F. Sinowatz, Eds), pp 42–92. Verlag Paul Parey, Berlin und Hamburg.
Salustri, A., Hascall, V.C., Camaioni, A. and Yanagishita, M. (1993). In: The ovary (E.Y. Adashi and P.C.K. Leung, Eds), pp 209–226. Raven Press, New York.
Saumande, J. (1991). Rec. Vét. **167:** 205–218.
Scaramuzzi, R.J., Turnball, K.E. and Nancarrow, C.D. (1980). Aust. J. Biol. Sci. **33:** 63–69.
Schams, D., Koll, R. and Li, H. (1988). J. Endocrinol. **116:** 97–100.
Schwall, R.H., Mason, A.J., Wilcox, J.N. and Basset, D.T. (1990). Mol. Endocrinol. **4:** 75–90.
Schweitzer, M., Jackson, J.C., Ryan, J. (1981). Biol. Reprod. **24:** 332–340.
Sorensen, R. and Wassarman, P. (1976). Dev. Biol. **50:** 531–536.
Spicer, L.J. and Echternkamp, S.E. (1986). J. Anim. Sci. **62:** 428–451.
Spicer, L.J. and Echternkamp, S.E. (1995). Dom. Anim. Endocrinol. **12:** 223–245.
Takagi, J., Araki, L., Dobashi, M., Imai, Y., Hiroi, M., Tonosaki, A. and Sendo, F. (1989). Biol. Reprod. **40:** 1095–1102.
Terrill, C.E. (1974). In: Reproduction in farm animals (E.S.E. Hafez, Ed), pp 265–274. Lea and Febiger, Philadelphia.
Tisdall, D.J., Smith, P., Leeuwenberg, B. and McNatty, K.P. (1995). J. Reprod. Fertil. Abstr. Series **15:** p12.
Torney, A.H., Hodgron, Y.M., Forage, R., Kretser, D.M. de (1989). J. Reprod. Fertil. **86:** 391–399.
Tonetta, S.A., and DiZerega, G.S. (1989). Endocr. Rev. **10:** 205–229.
Wandji, S.A., Pelletier, G. and Sirard, M.A. (1992a). Biol. Reprod. **47:** 814–822.
Wandji, S.A., Pelletier, G. and Sirard, M.A. (1992a). Biol. Reprod. **47:** 807–813.
Zeleznik, A.J. (1993). In: The ovary (E.Y. Adashi and P.C.K. Leung, Eds), pp 41–55. Raven Press, New York.
Zeleznik, A.J., Ihrig, L.L. and Basset, S.G. (1989). Endocrinology **125:** 2218–2220.

Comparative Endocrinology and Reproduction
K.P. Joy, A. Krishna and C. Haldar (Eds)

19. Regulation of Follicular Development: Beyond Gonadotropins

Shyamal K. Roy
Division of Reproductive Endocrinology
Leland J. and Dorothy H. Olson Centre for Women's Health
Departments of OB/GYN, and Physiology and Biophysics
University of Nebraska Medical Centre
600 S. 42 nd Street, Omaha, NE 68198–4515, USA

Introduction

Development of ovarian follicles from the primordial to preovulatory stage is a complex process requiring a consolidated action of a variety of endocrine, autocrine and paracrine factors. Evidence has accumulated to provide some insights in the regulatory mechanisms of follicular development at antral stages (Greenwald and Roy, 1994) and to some extent, at the preantral stages (Roy, 1994; Greenwald and Roy, 1994). FSH is unquestionably the primary gonadotropin for early follicular development. Development of both granulosa and thecal cells is essential for full follicular development. Nevertheless, the involvement of a variety of growth factors and cytokines in follicular growth and maturation has become increasingly clear. Intraovarian growth regulators can be categorized into (1) proliferative, and (2) differentiation-inducing factors. An exquisitely time and regionalized action of these two types of factors determines the differential success of follicle development. The species difference in intraovarian growth regulators is as diverse as the factors themselves (Tonetta and DiZerega, 1989); nevertheless, IGF-I, FGF and EGF/TGF-α constitute the main mitogenic factors in most of the species (Roy, 1994; Greenwald and Roy, 1994; Tsafriri and Adashi, 1994). On the contrary, a variety of cytokines exert powerful differentiation-inducing stimulus. Some factors clearly mediate the action of gonadotropins while for the others, the relation is not so clear.

The role of growth factors in granulosa cell development has been studied extensively using mature cells *in vitro* (Adashi, 1985; Greenwald and Roy, 1994; Tsafriri and Adashi, 1994); however, information about preantral follicles is obtained by culturing intact follicles (Roy and Greenwald, 1994; Greenwald and Roy, 1994; Boland and Gosden,1994; Eppig and O'Brien, 1995) *in vitro*. The article is dedicated to the ovarian factors that have distinct roles in follicular cell activities and have a clear-cut relationship with gonadotropins.

Fibroblast Growth Factors and Follicular Cell Functions

The doubling time (3 folds), life span and lipid content (Gospodarowicz, Ill and Birdwell, 1977; Gospodarowicz and Bialecki, 1978) of bovine granulosa cells were enhanced significantly when cultured at low density in presence of bFGF, suggesting that cellular proliferation was favored over differentiation. The mitogenic effect of FGF, was also noted for hamster (Roy and Greenwald, 1987), rabbit, guinea-pig and human granulosa cells in culture (Gospodarowicz and Bialecki, 1979). FGF-induced reduction in granulosa cell apoptosis in the rat had been reported (Tilly *et al.,* 1992). Although theca is the exclusive site of bFGF gene expression (Koos and Olson, 1989), bFGF mRNA expression in granulosa cells of porcine and murine dominant follicles had been demonstrated by RT-PCR and *in situ* hybridization (Guthridge *et al.*, 1989).

A substantial research has led to the understanding of the influence of bFGF on steroidogenic and other physiological activities of granulosa cells. Basic FGF inhibited estrogen production and LH receptor induction (Baird and Hsueh, 1986) via cyclic AMP (Oury and Darbon, 1988), and significantly attenuated basal as well as FSH-stimulated inhibin-α and -β subunit mRNA expression (LaPolt *et al.*, 1990) in estrogen-primed rat granulosa cells *in vitro.* On the other hand, LH-induced estradiol and progesterone production in FSH-primed rat granulosa cells (Yamoto *et al.,* 1993), and plasminogen activator mRNA expression and oocyte maturation in the rat ovary (LaPolt *et al.,* 1990) were stimulated by bFGF. Basic FGF, however, did not influence steroidogenesis in hamster preantral follicles *in vitro* (Roy and Greenwald, 1987). Because bFGF is also a potent angiogenic facetor (Koos, 1989), its primary role in the ovary may be more than just mitogenic. Whether FGF peptide in follicular cells is induced by gonadotropins is unclear, nevertheless, all these lines of evidence clearly indicate an intraovarian regulatory role of bFGF.

Studies, albeit limited, have elucidated the role of bFGF on theca-interstitial functions. Basic FGF strongly attenuated androstenedione production by rat theca-interstitial cells in culture (Hurwitz, *et al.,* 1990), and inhibited forskolin and cyclic-AMP-induced CYP-17α activity, 17α-hydroxy-progesterone and dehydro-epiandrostenedione production but induced 3β-hydroxy-steroid dehydrogenase activity in cultured human theca cells (McAllister *et al.,* 1994). Recently, Spicer and Stewart (1996) documented that in bovine thecal cell culture, bFGF increased cell number in bovine thecal cells but inhibited steroidogenesis and IGF-I binding sites.

Epidermal Growth Factor Family and Follicular Functions

The role of EGF on follicular cell proliferation and differentiation (Roy, 1994; Roy and Greenwald, 1994; Tsafriri and Adashi, 1994) have been thoroughly investigated. EGF belongs to the family of mitogens that includes

TGF-α, heparin-binding EGF (HB-EGF), amphiregulin and heregulin peptides. The actions of EGF/TGF-α on the ovary have been reviewed recently (May and Schomberg, 1989). While TGF-α mRNA is present in rat granulosa cells (Yeh, Lee and Anderson, 1993), both EGF peptide and mRNA are expressed in hamster preantral and small antral follicles (Roy and Greenwald, 1990; Roy and Harris, 1994); EGF immunoreactivity dropped in granulosa cells as follicles reached the preovulatory stage (Roy and Greenwald, 1990). An EGF-like immunoreactivity has been identified in porcine follicular fluid (Hsu *et al.,* 1987) and in human granulosa cells (Stewart *et al.*, 1992, Maruo *et al.*, 1993). Granulosa cells and oocyte of porcine small antral follicles express EGF mRNA; however, EGF receptor message is present in all types of follicular cells (Singh, *et al.,* 1995). Evidence suggests that follicular EGF expression is primarily regulated by FSH. Removal of the pitutary in adult hamsters (Roy and Greenwald, 1990) or culturing hamster preantral follicles in the presence of antisense EGF oligonucleotides (Roy and Harris, 1994) resulted in a fall in immunoreactive EGF and the effect was reversed by FSH replacement (Roy and Greenwalds, 1990). Moreover, EGF binding to porcine granulosa cells increased in the presence of FSH—an effect that could be estradiol mediated (Fujinaga *et al.*, 1994). Exposure of rat granulosa cells to estradiol or testosterone resulted in a 2.5 folds increase in EGF binding without affecting the binding affinity (Liang *et al.*, 1990). These lines of evidence suggest that FSH regulates follicular growth by controlling cellular EGF and its receptor expression, and at least part of the FSH action may be mediated by follicular steroids, particularly estradiol.

The mitogenic action of EGF on granulosa cells is well documented. EGF stimulates DNA synthesis in porcine granulosa cells (May *et al.*, 1988) and in hamster preantral follicles (Roy and Greenwald, 1987); the latter was inhibited by an anti-EGF antibody (Roy and Greenwald, 1991). Interestingly, both FSH- and cyclic-AMP-induced DNA synthesis in hamster preantral follicles was also significantly attenuated by the EGF antibody (Roy and Greenwald, 1991) or antisense EGF oligonucleotides (Roy and Harrris, 1994). Therefore, FSH uses a novel pathway, at least in hamster granulosa cells, to regulate cell proliferation. EGF induced DNA synthesis in porcine granulosa cells from primary and secondary follicles of 3-day and 50–60-day old pigs, respectively, whereas FSH was effective only on the secondary follicles (Morbeck *et al.*, 1993). These results indicate that follicular response to EGF, at least in the pig, appears prior to a frank response to FSH. A gradual maturation of follicular cells may occur under the influence of local EGF. Whether secondary or tertiary follicles, which respond to FSH, are the source of the local EGF is unclear. EGF action on granulosa cells involves autophosphorylation of its receptors and activation of MAP-kinase for intracellular signaling (Keel *et al.,* 1995; Roy and Greenwald, 1996). Interestingly, activation of follicular MAP-kinase occurred when cyclic hamsters were exposed to exogenous PMSG (Roy, 1995; Roy and Greenwald, 1996). These lines of evidence suggest that intracellular signal transduction pathway(s) for EGF do exist in granulosa cells. Although a link between EGF and MAP kinase has been established for other cell types, whether granulosa cells possess an identical pathway is not yet known. Because granulosa cells, particularly of the preantral follicles, undergo crucial changes during follicular development through stages, a thorough understanding of the developmental changes in intracellular signal transduction pathways and the cross-talk between them is warranted.

[^{125} I]-EGF binds to rat thecal and granulosa cells (Chabot *et al.*, 1986; Feng *et al.*, 1987) and to porcine granulosa cells (Feng *et al.*, 1987; Buck and Schomberg, 1988). A number of pituitary hormones influence EGF receptor levels in granulosa cells. EGF receptor content of rat (Feng *et al.*, 1987) and porcine granulosa cells (Hattori, *et al.*, 1994) in culture was up-regulated by FSH but a decrease occured with further cell differentiation (Buck and Schomberg, 1988); however, a continuous induction of EGF receptor in cultured porcine granulosa cells had been documented (Fujinaga, *et al.*, 1992). The reason for such an apparent discrepancy is not clear, however, the stage of the estrous cycle, hence, the endogenous gonadotropin environment might be responsible for altered cell behavior. Interestingly, human growth hormone (hGH) significantly induced EGF binding sites in porcine granulosa cells in culture; FSH had an additive effect (Hattori *et al.*, 1994). These results suggest that (1) granulosa cells are target for EGF action, (2) more critical studies on the spatio-temporal induction of EGF receptors in granulosa cells under controlled hormonal environment is warranted to understand the mechanism of EGF regulation of folliculogenesis.

The physiological role of EGF on follicular development has been evaluated using hamster and mouse preantral follicles *in vitro.* Hamster follicles at the primary and secondary stages grow in response to EGF (Roy S.K., unpublished). EGF improved murine cumulus expansion (Buccione *et al.*, 1990), oocyte maturation (Downs, 1989) and viability of mouse follicles growing *in vitro* without affecting cellular DNA synthesis (Boland and Gosde, 1994). Removal of the oocyte inhibited the EGF- induced cumulus expansion; however, a reversal was noted when mature oocytes were co-cultured with the cumulus cells (Buccione *et al.*, 1990). In a recent study, Eppig and O'Brien (1996) elegantly demonstrated that EGF supported the development of primary follicles from the primordial pool in neonatal mouse ovaries maintained in an organ culture; FSH alone had no significant influence but synergized with EGF for follicular development. Although a sizable number oocytes could be fertilized and develop into preimplantation embryos *in vitro,* a limited number developed into live young upon transfer to foster mothers. In any case, this study represents a significant step toward understanding the physiological role of EGF in gonadal differentiation. It is noteworthy that FSH, although mandatory for *in vivo* development of antral follicles in all species, including mouse (Greenwald and Roy, 1994), had no influence *in vitro* on mouse follicles. These results clearly indicate that EGF is indeed an important physiological regulator of follicular development.

Besides mitogenesis, EGF modulates many differentiated-related functions, such as steroidogenesis and peptide production, of granulosa cells (Greenwald and Roy, 1994; Tsafriri and Adashi, 1994). EGF stimulated, in a dose-dependent manner, progesterone production by estradiol-primed rat granulosa cells *in vitro* but had no effect on estradiol biosynthesis (Jones and Hsueh, 1982); the increase in progesterone production occurred due to EGF-induced activation of 20α-hydroxy steroid dehydrogenase and 3β-ol-dehydrogenase, and increased production of pregnenolone. EGF stimulated the synthesis of P450 side chain cleavage and adrenodoxin (Trzeciak *et al.*, 1987), and progesterone production in hamster preantral follicles *in vitro* (Roy and Greenwald, 1987; Roy and Harris, 1994). However, EGF inhibited FSH-induced aromatase activity in rat (Adashi *et al.*, 1987) and FSH-induced aromatase gene transcripts in human (Steinkampf *et al.*, 1988) granulosa cells. More recently, Ghersevich *et al.* (1994) demonstrated that EGF also inhibited FSH-induction of 17β-hydroxy steroid dehydrogenase type 1 gene expression and enzyme activity, thus blocking another avenue of estradiol biosynthesis. These data clearly suggest that EGF critically modulates granulosa cell activities. One of the hallmark of differentiated granulosa cells is their ability to produce estradiol. EGF blocks this cytodifferentiation and directs cells toward proliferation.

TGF-α, which functions via EGF receptors (Fisher and Lakshmanan, 1990) had also been identified in follicular cells (Singh *et al.*, 1995); however, a distinct difference in granulosa cell response to EGF and TGF-α-had been documented (Roy and Greenwald, 1991). Although Kudlow *et al.* (rat: 1987) and Skinner and Coffey (bovine: 1988) suggested that TGF-α peptide and mRNA are theca-interstitial cell specific, Yeh *et al.* (1993), using a reverse transcription polymerase chain reaction (RT-PCR) amplification, demonstrated that rat granulosa cells expressed TGF-α gene. TGF-α strongly suppressed FSH-induced aromatase activity and estradiol production in rat granulosa cells *in vitro* (Adashi *et al.*, 1987). In contrast to the inhibitory actions of TGF-α and EGF on steroidogenesis of granulosa cells in the adults, recent studies suggest that TGF-α induces aromatase activity in prepubertal porcine granulosa cells (Gangrade *et al.*, 1991). Amphiregulin gene transcripts had been detected in human ovary (Johnson *et al.*, 1991), however, no distinct role of this peptide on follicular development or follicular cell functions has yet been demonstrated. All these lines of evidence underscore the complexity of growth factor actions on follicular cells—cells exposed to cyclic gonadotropins and ovarian steroids, such as in the adults, behave quite differently than cells in prepubertal animals.

Insulin-Like Growth Factors and Follicular Development

Evidence has accumulated to suggest that ovaries of many species possess an IGF system complete with the peptide, receptors and binding proteins (Adashi *et al.*, 1985; Adashi, 1994). IGF, have a variety of effects on follicular cells and perhaps it is the most widely studied growth factor in the ovary (Tsafriri and Adashi, 1994). Nevertheless, there are enough species difference to suggest that it is one of the intraovarian growth factors that control follicular activities. Immunoreactive IGF-I had been localized in the oocytes, granulosa and theca interna cells of developing rat follicles (Hansson *et al.*, 1988), and in porcine (Hsu and Hammond, 1987) and ovine (Monget *et al.*, 1989) follicular fluid whereas IGF-I message is expressed in granulosa cells of developing preantral and antral rat follicles (Oliver *et al.*, 1989) and increased with follicular development in the pig (Samaras *et al.*, 1993). FSH either alone or in combination with estradiol or cyclic AMP augmented IGF-I production in porcine (Hsu and Hammond, 1987) and IGF-I receptor mRNA expression in rat (Handez, 1995) granulosa cells. In contrast, neither granulosa nor thecal cells of human dominant follicle express IGF-I mRNA or immunoreactivity (el-Roeiy *et al.*, 1993) and human small antral follicles expressed both IGF-I and II mRNA but did not produce immunoreactive IGF-I (el-Roeiy *et al.*, 1993). These results suggest that despite a broad spectrum action of IGF-I on granulosa cells of different species, its role as a ubiquitous growth factor across species is not unequivocal.

In contrast to IGF-I gene, IGF-II mRNA is expressed in theca-interstitial cells of rat (Hernandez *et al.*, 1990) and in human granulosa cells (Geisthovel *et al.*, 1989). The presence of high affinity and low capacity IGF-I binding sites had been demonstrated in ovine (Monget *et al.*, 1989), porcine (Baranao and Hammond, 1984; Veldhuis, *et al.*, 1985) and murine (Davoren *et al.*, 1986; Adashi *et al.*, 1990) granulosa cells by radioreceptor assay. FSH stimulated IGF-I receptor contents in rat granulosa cells (Adashi *et al.*, 1988). IGF-II receptor gene, on the other hand, is expressed in both granulosa and thecal cells and appears to be constitutively expressed (el-Roiey *et al.*, 1993; Hernandez, 1995). IGF-I and IGF-II receptor messenger RNA have been detected in human fetal ovaries by RT-PCR at 10, 15, 19 and 22 weeks of gestation (Shifren *et al.*, 1993).

IGF-I stimulated DNA synthesis in porcine (Hammond and English, 1988), hamster (Roy and Greenwald, 1987), ovine (Monniaux and Pisselet, 1992), rat (Bley *et al.*, 1992) and human (Olsson *et al.*, 1990) granulosa cells, antral follicle formation in sheep and cattle (Campbell, *et al.*, 1995), and oxytocin gene expression and protein secretion by bovine granulosa-lutein cells in culture (Holtorf *et al.*, 1989; Meidan *et al.*, 1992). IGF-I profoundly influenced *in vitro* production of progesterone and estradiol by granulosa cells from a variety of species (Gong *et al.*, 1994; Gaetje, 1994; Wood, 1994; Spicer *et al.*, 1993; Erickson *et al.*, 1991; Schams *et al.*, 1988; Adashi, 1994; Devoto *et al.*, 1995).

In human granulosa cells; IGF-I stimulates basal and FSH-induced estradiol production (Erickson *et al.*, 1989) by inducing aromatase synthesis (Steinkampf *et al.*, 1988)—the action involved tyrosine phosphorylation (Costrici *et al.*, 1994). A role of IGF-I in human follicular and oocyte development had been proposed (Roussi *et al.*, 1989). However, a recent study using Laron-type dwarfism (IGF-I deficiency secondary to GH receptor abnormality) challenges the role of IGF-I in human follicular development (Dor *et al.*, 1992). Despite a very low levels of serum and follicular fluid IGF-I, and high levels of IGF-I binding protein, these patients have normal follicular development, and show high levels of serum estradiol in response to exogenous gonadotropins; however, granulosa cells of these patients do not produce estradiol in response to IGF-I. Similarly, monkey granulosa cells obtained from small antral and preovulatory vollicles did not respond *in vitro* to IGF-I for 48 h in terms of progesterone accumulation and aromatase activity; however, a response was noted following longer period of culture, such as 4–6 days (Shaw *et al.*, 1993), suggesting that cell characteristics had altered significantly during long-term culture.

Unlike other intraovarian growth factors, IGF action seems to be regulated by a family of endogenous binding proteins, called IGFBP, synthesized by both granulosa and thecal-interstitial cells (Adashi, 1994). There are at least 6 different IGFBP (1 through 6) with very high affinity with IGFs (Rutanen and Pekonen, 1990; Sara and Hall, 1990; Clemmons, 1990). In the pig, IGFBP-2 mRNA expression decreases as follicles grow in size (Samaras *et al.*, 1993). IGFBP-1, in a dose dependent manner, attenuated IGF-I-stimulated *in vitro* estradiol production by human granulosa cells and inhibited binding of [^{125} I] IGF-I to the membrane receptors (Iwashita *et al.*, 1994). IGFBP-3 in bovine follicular fluid decreased with increased estradiol production (Echternkamp *et al.*, 1994). IGF-I- induced progesterone production by cultured porcine granulosa cells (Samaras and Hammond, 1995) and inhibition of apototic DNA fragmentation in cultured rat preovulatory follicles (Chun *et al.*, 1994) were reversed by concurrent administration of IGFBP-3. All these lines of evidence indicate that IGFBP expression may be associated with cessation of follicular development. IGF-I, by itself, stimulates IGFBP-5 synthesis by estrogen-primed rat granulosa cells in culture (Adashi, 1994). Much work is needed to identify the physiological role(s) of different IGFBPs in follicular development.

Transforming Growth Factor-β and Follicular Functions

While follicular cells of most species use more than one mitogenic growth factor for growth, transforming growth factor-β (TGF) may be the most powerful cell differentiation factor having a universal appeal (Roy, 1994). TGF-β exists in three isoforms, such as TGF-β1-3, in mammals (Roberts and Sporn, 1988) and belongs to the family that includes mullerian inhibiting hormone, inhibin and activin (Massague, 1990). While TGF-β1 peptide and message are localized mainly in thecal cells (for ref. see Roy, 1994), both TGF-β2 peptide and message have been detected in granulosa cells of hamster (Roy *et al.*, 1992; Roy, S.K., unpublished) and rat (Mulheron and Schomberg, 1990) but not in the pig (Mulheron *et al.*, 1992). Evidence suggests that despite expression of TGF-β1 mRNA, porcine granulosa cells do not synthesize TGF-β1 protein (May *et al.*, 1996). However, earlier data from the same laboratory demonstrated TGF-β1 protein production by porcine granulosa and thecal cells cultured separately (May *et al.*, 1994). The apparent discrepancy between these two findings is not clear. Bovine theca cells, on the other hand, secrete only TGF-β1 (Gangrade *et al.*, 1993). In the hamster, TGF-β2 protein expression in granulosa cells is up-regulated by FSH while LH stimulates thecal TGF-β1 protein (Roy *et al.*, 1992). The granulosa-specific expression of TGF-β2 in the hamster helped us to study granulosa cells lineage in the ovary (Roy and Hughes, 1994). We observed that during early ovarian development, pluripotent cells that were apposed to germ cells expressed TGF- β2 protein and those cells eventually differentiated to form granulosa cells; cells elsewhere were negative (Roy and Hughes, 1994). Cellular expression of TGF-β2 increased with further differentiation, and granulosa cells proliferate to form prinordial and primary follicles. TGF-β1 expression occurs in the interstitial cells by postnatal day 13 when numerous large preantral follicles are visible (Roy and Hughes, 1994). These results suggest strongly that TGF-β may play a key role in selecting primary somatic cells, i.e., granulosa cells, for the follicles.

Despite the well-known differentiation-induction role, TGF-β significantly attenuated a FSH-induced progesterone production, and induced DNA synthesis in porcine (Kubota *et al.*, 1995) and rat (Bendell and Dorringotn, 1991) granulosa cells in culture. However, a critically designed study on hamster preantral follicles *in vitro* demonstrated that the apparent mitogenic action of TGF-β might result from the subsequent induction of endogenous mitogen, such as EGF (Roy, 1993a, b). TGF-β arrests cells in the "G1/S" boundary and lengthens the G1 phase of the cell cycle, hence synchronizes cells at the G1/S boundary (Shipley *et al.*, 1985). In dynamic cell cultures, such as intact preantral follicles, a fall in effective concentration of TGF-β will eventually shift the cell cycle in favor of the mitogenic factor(s), resulting in an entry of more cells in the 'S' phase. TGF-β induction of EGF receptors in cultured rat granulosa cells had been documented (Feng, *et al.*, 1986). Likewise, it is equally possible that rat granulosa cells may be unique to TGF-β action.

TGF-β stimulates FSH-induced aromatase activity, LH receptors and progesterone production in rat granulosa cells *in vitro* (Dorrington, *et al.*, 1988; Dodson and Schomberg, 1987). Magoffin *et al.* (1989) reported that TGF-β induced cytodifferentiation of rat theca-interstitial cells in culture that was associated with a reduction in androgen production. A recent report indicates that the TGF-$\beta 1$ *in vitro* induced, by a non-competitive mechanism, P450scc and 3β-ol-dehydrogenase mRNA expression in rat theca-interstitial cells by 96 h; however, it directly inhibited 17α-hydroxylase activity and a concurrent reduction in androgen production (Fournet, *et al.*, 1996). Interestingly, TGF-β1 enhanced LH-induced CYP17α mRNA levels within 2 days followed by a total decline by 4th day of culture (Fournet *et al.*, 1996). Whether the rise in mRNA indicates an immediaqte action of TGF-β1 is unclear. Collectively, these data indicate that differentiation of follicular cells is critically modulated by locally produced TGF-β. EGF, FGF and TGF-β significantly inhibit forskolin and dibutyril-cAMP-induced 17α-hydroxylase activity, 17α-OH progesterone and dehydroepiandrosterone production but stimulated progesterone production via 3β-ol-dehydrogenase in human theca interna cells (McAllister *et al.*, 1994). Depending on the developmental status, follicular cells will respond differently to TGF-β for follicular development.

TGF-β action on follicular cells is also determined by the presence of its receptors. Recently, we documented that hamster granulosa cells possess two forms of ovarian TGF-β type II receptors (TβRII): (1) a membrane bound form that recognized the ligand and (2) an abundant cytoplasmic form that showed limited interaction with the ligand; and both forms were up-regulated by FSH and estradiol but progesterone strongly attenuated estradiol-induced receptor expression (Roy and Kole, 1995). Moreover, the cytoplasmic form the receptor maintained an inverse relationship with the membrane anchored form throughout the estrous cycle. Our results clearly indicate that a spatio-temporal expression of growth factor ligand and receptor system is crucial for successful follicular development.

Ovarian Cytokines

Besides local growth factors, a variety of classical cytokines influenced follicular cell functions, particularly steroidogenesis, ovulation, luteinization and apoptotic processes (Adashi, 1991). Among the various cytokines, interleukins (IL)-1, -3, -6 and -10, and TNF-α are well studied (see for ref., Tsafriri and Adashi, 1994; Greenwald and Roy, 1994; Roy, 1994; Simon and Polan, 1994). Human follicular fluid cells express both IL-1mRNA and protein, which increases during the luteal phase of the cycle (Polan, *et al.*, 1994); significant amounts, however, may come from the macrophage contamination. IL-1β and IL-receptor type I (IL-1R tl) mRNA expressions were detected in human granulosa-lutein cells (Piquette *et al.*, 1994). In mice while IL-1α, IL-1β and IL-1R tl are present in thecal cells, oocytes possess only IL-1R tl (Simon *et al.*, 1994). During ovulation, granulosa cells expressed the full complement of the IL-1 system (Polan *et al.*, 1994). A clear-cut expression of IL-1β mRNA occurred in thecal cells of PMSG-primed immature rat ovary 6 h after hCG treatment (Hurwitz *et al.*, 1991). Moreover, IL-1 induced IL-1β message expression *in vitro* in rat whole ovarian dispersate (Hurwitz *et al.*, 1991).

Both IL and TNF-α profoundly influence follicular steroidogenesis (Adashi, 1990). TNF-α inhibited estradiol production in explanted hamster preovulatory follicles (Roby and Terranova, 1988) and cultured rat granulosa cells (Adashi *et al.*, 1989) whereas IL-1 inhibited hCG-induced androstenedione production by whole ovarian dispersate from immature rats (Hurwitz *et al.*, 1991). A detail study by Kasson and Gorospe (1989) demonstrated that whereas IL-1α and -1β inhibited FSH-induced synthesis of progesterone, estrogen and LH receptors, both forms stimulated 20α-OH progesterone production by rat granulosa cells *in vitro.* Interestingly, IL-2 and IL-3 augmented progesterone and 20α-OH-progesterone production by rat granulosa cells without altering estradiol production (Kasson and Gorospe, 1989). IL-1β synergizes with hCG to augment prostaglandin F2α and estradiol production by human granulosa cells *in vitro* (Watanabe *et al.*, 1993). A 92 kDa gelatinase protein in rat thecal cells was induced by IL-1 and the induction was significantly attenuated by TGF-β1 (Hurwitz *et al.*, 1993), indicating a critical interaction of interleukin in the ovulatory process. Apparently, ovarian action of IL-1 is regulated by LH/hCG (Piquette *et al.*, 1994), IL-1 (Hurwitz *et al.*, 1991), and a naturally occurring IL-receptor antagonist (Kokia *et al.*, 1993). Although the mechanism of IL action on ovarian cells is not clear, nitric oxide has been implicated to be an intracellular mediator (Ben-Shlomo, *et al.*, 1994).

Ovarian GnRH

The role of GnRH peptide in follicular cells has been reviewed (Leung *et al.*, 1989). GnRH-like peptide was detected in cow, human, rat and ovine ovaries (Aten *et al.*, 1987; Aten *et al.*, 1987; Ireland *et al.*, 1988); however, no high affinity GnRH binding sites had been detected in these species (Brown and Reeves, 1983; Asch *et al.*, 1981; Clayton and Hutaniemi, 1982), hence questioning the physiological role of GnRH in follicular development in these species. Nevertheless, evidence has accumulated to suggest a positive as well as negative impact of GnRH on rat follicular functions. Birnbaumer *et al.* (1985) demonstrated that GnRH antagonists potentiated FSH action on follicular development. GnRH stimulates the production of progesterone, estrogen and prostaglandin production by rat granulosa cells *in vitro* (Jones and Hsueh, 1982; Clark, *et al.*, 1980; Clark, 1982; Sheela *et al.*, 1983). One of the distinct effect of GnRH was the induction of ovulation in perifused rabbit and rat ovaries (Koos and LeMaire,1985)—the effect might be mediated by prostaglandins (Leung, *et al.*, 1989; Clark, 1982) and plasminogen activator (Hsueh *et al.*, 1988). GnRH induced rat oocyte maturation *in vitro* (Hillensjo and LeMaire, 1980; Magnusson and LeMaire, 1981; Dekel *et al.*, 1985), perhaps via protein kinase C pathway (Yoshimura *et al.*, 1992).

An equal number of evidence is available to suggest that GnRH action on the ovary is basically inhibitory (Rippel and Johnson, 1976) and perhaps, may be related to follicular atresia (Tsafriri and Braw, 1984). GnRH inhibited FSH-induction of cyclic AMP (Massicoite *et al.*, 1980), aromatase (Hsueh *et al.*, 1980), 3β-ol-dehydrogenase (Jones and Hsueh, 1982), and LH and prolactin receptors (Hsueh and Ling, 1979; Navickis *et al.*, 1982) in granulosa cells.

For the sake of brevity, it is not feasible to include all the pertinent studies on various intraovarian factors. A number of other factors, such as neurotransmitters, adenosine, eicosanoids, renin-angiotensin system (Sealey, *et al.*, 1990), histamine, etc., have been shown to influence ovarian cell functions in one way or another (see for ref., Tsafriri, 1994; Greenwald and Roy, 1994; Espey, 1994; Niswender and Nett, 1994), but their exact physiological roles in folliculogenesis and their relationship with gonadotropins are yet to be established. Nevertheless, the present review conceptualizes a regulatory process for folliculogenesis that extends far beyond the action of gonadotropins. Undoubtedly, follicular development at the local level is regulated by a complex process involving critical interaction between growth factors and cytokines; the latter originate from follicular cells as well as from resident macrophages. A constantly changing gonadotropin, hence, steroid milieu in the ovary may present a receptor environment that is crucial to intraovarian factors. Therefore, regulatory steps constitute not only the expression of ligands but also of respective receptors and the interaction between opposing factors. A precise cooperation between different follicular cell types, and a well-coordinated autocrine/paracrine action of multitude of intrafollicular faqctors is the key to successful development of healthy follicles.

Acknowledgements

The work of S.K. Roy mentioned here was supported by a grant HD 28165 from the National Institute of Health (Bethesda, MD) and the Olson Center for Women's Health (Omaha, NE). Gonadotropins used in the studies were generously provided by the National Pituitary Program, NIH.

References

Adashi, E.Y., Resnick, C.R., D 'Ercole, J.A., Svoboda, M.E. and Van Wyk, J.J. (1985) Endocrine Rev. **6:** 400–420.

Adashi, E.Y., Resnick, C.E. and Twardzik, D.R. (1987) J. Cell. Biochem. **33:** 1–13.

Adashi, E.Y., Resnick, C.E., Hernandez, E.R., Svoboda, M.E. and Van Wyk, J.J. (1988) Endocrinology **122:** 1383–1389.

Adashi, E.Y., Resnick, C.E. and Twardzik, D.R. (1987) J. Cell. Biochem. **33:** 1–13.

Adashi, E.Y., Resnick, C.E., Croft, C.S. and Payne, D.W. (1989) J. Biol. Chem. **264:** 11591–11597.

Adashi, E.Y., Resnick, C.E., Hernandez, E.R., Hurwitz, A. and Rosenfeld, R.G. (1990) Mol. Cell. Endocr. **74:** 175–185.

Adashi, E.Y. (1990) Endocrine Rev. **11:** 454–464.

Adashi, E.Y. (1990) Endocrinology **126:** 3089–3095.

Adashi, E.Y. (1994) Horm. Res. **42:** 44–48.

Asch, R.H., Sickle, M.V., Rettori, V. (1981) J. Clin. Endocrinol. Metab. **53:** 215–217.

Aten, R.F., Ireland, J.J., Weems, C.W. and Behman, H.R. (1987a) Endocrinology **120:** 1727–1733.

Aten, R.F., Polan, M.L., Bayless, R. and Behrman, H.R. (1987b) J. Clin. Endocrinol. **64:** 1288–1293.

Baird, A. and Hsueh, A.J.W. (1986) Regul. Pept. **16:** 243–250.

Baranao, J.L.S. and Hammond, J.M. (1984) Biochem. Biophys. Res. Commun. **124:** 484–490.

Bendell, J.J. and Lobb, D.K. (1987) Steroids **50:** 411–421.

Bendell, J.J. and Dorrington, J. (1991) Endocrinology **128:** 2663–2665.

Ben-Shlomo, I., Kokia, E., Jackson, M.J., Adashi, E.Y. and Payne, D.W. (1994) Biol. Reprod. **51:** 310–313.

Birnbaumer, L., Shahabi, N., Rivier, Y. and Vale, W. (1985) Endocrinology **16:** 1367–1370.

Bley, M.A., Simon, J.C., Estevez, A.G., De-Asua, L.J. and Baranao, J.L. (1992) Endocrinology **131:** 1223–1229.

Boland, N.I. and Gosden, R.G. (1994) J. Reprod. Fertil. **101:** 369–374.

Brown, J.L. and Reeves, J.J. (1983) Biol. Reprod. **29:** 1179–1182.

Buccione, R., Vanderhyden, B.C., Caron, P.J. and Eppig, J.J. (1990) Dev. Biol. **138:** 16–25.

Buck, P.A. and Schomberg, D.W. (1988) Endocrinology **122:** 28–33.

Campbell, B.K., Scaramuzzi, R.J. and Webb, R. (1995) J. Reprod. Fertil. Suppl. **49:** 335–350.

Chabot, J.G., St-ARnaud, R., Walker, P. and Pelletier G. (1986) Mol. Cell. Endocrinol. **44:** 99–108.

Chun, S.Y., Billing, H., Tilly, J.L., Furuta, I., Tsafriri, A. and Hsueh, A.J. (1994) Endocrinology **135:** 1845–1853.

Clark, M.R., Thibier, C., Marsh, J.M. and LeMaire, W. (1980) Endocrinology **107:** 17–23.

Clark, M.R. (1982) Endocrinology **110:** 146–152.

Clemmons, D.R. (1990) Trends Endocrinol. Metab. **1:** 412–417.

Clayton, R.N. and Hutaniemi, T. (1982) Nature **299:** 56–59.

Costrici, N., Lunenfeld, B., Pariente, C., Dor, J., Rabinovici, J., Kanety, H. and Karasik, A. (1994) Gnyecol. Endocrinol. **8:** 183–189.

Davoren, J.B., Kasson, B.G., Li, C.H. and Hsueh, A.J. (1986) Endocrinology **119:** 2155–2162.

Dekel, N., Sherizly, I., Phillips, D.M., Nimrod, A., Zilberstein, M. and Naor, Z. (1985) J. Reprod. Fertil. **75:** 461–466.

Devoto, L., Kohen, P., Castro, O., Vega, M., Troncoso, J.L. and Charreau, E. (1995) **80:** 1566–1570.

Dodson, W.C. and Schomberg, D.W. (1987) Endocrinology **120:** 512–516.

Dor, J., Ben-Shlomo, I., Lunenfeld, B., Pariente, C., Levran, D., Karasik, A., Seppala, M. and Mashiach, S. (1992) J. Clin. Endocrinol. Metab. **74:** 539–542.

Dorrington, J.J., Chuma, A.V. and Bendell, J.J. (1988) Endocrinology **123:** 353–359.

Downs, S.M. (1989) Biol. Reprod. **41:** 371–379.

Drummond, A.E. and Hearn, M.T. (1992) Exs. **61:** 219–229.

Echternkamp, S.E., Howard, H.J., Roberts, A.J., Grizzle, J. and Wise, T. (1994) Biol. Reprod. **51:** 971–981.

el-Roeiy, A., Chen, X., Roberts, V.J., Leroith, D., Roberts, C.T. Jr. and Yen, S.S. (1993) J. Clin. Endocrinol. Metab. **77:** 1411–1418.

Erickson, G.F., Garzo, V.G. and Magoffin, D.A. (1989) J. Clin. Endocri. Metab. **69:** 716–724.

Erickson, G.F., Garzo, V.G. and Magoffin, D.A. (1991) Hum. Reprod. **6:** 1074–1081.
Eppig, J.J. and O'Brien, M.J. (1996) Biol. Reprod. **54:** 197–207.
Espey, L.L. and Lipner, H. (1994) In: Physiology of Reproduction (E. Knobil and J.D. Neill, Eds), pp 725–780. Raven Press, New York.
Feng, P., Catt, K.J. and Knecht, M. (1986) J. Biol. Chem. **261:** 14167–14170.
Fisher, D.A. and Lakshmanan, J. (1990) Endocr. Rev. **11:** 418–442.
Fournet, N., Weitsman, S.R., Zachow, R.J. and Magoffin, D.A. (1996) Endocrinology **137:** 166–174.
Fujinaga, H., Yamoto, M., Shikone, T. and Nakano, R. (1994) J. Endocrinol. **140:** 171–177.
Feng, P., Knecht, M. and Catt, K. (1987) Endocrinology **120:** 1121–1126.
Fujinaga, H., Yamoto, M., Nakano, R. and Shima, K. (1992) Biol. Reprod. **46:** 705–709.
Gaetje, R. (1994) Clin. Exp. Obstet. Gynecol. **21:** 14–23.
Gangrade, B.K., Davis, J.S. and May, J.V. (1991) Endocrinology **129:** 2790–2792.
Gangrade, B.K., Gotcher, E.D., Davis, J. and May, J.V. (1993) Mol. Cell. Endocrinol. **93:** 117–123.
Geisthovel, F., Moreti-Rojas, I., Asch, R.H. and Rojas, F.J. (1989) Hum. Reprod. **4:** 899–902.
Ghersevich, S., Poutanen, M., Tapanainen, J. and Vihko, R. (1994) Endocrinology **135:** 1963–1971.
Gong, J.G., McBride, D., Bramley, T.A. and Webb, R. (1994) J. Endocrinol. **143:** 157–164.
Gospodarowicz, D., Ill, C.R. and Birdwell, C.R. (1977) Endocrinology **100:** 1108–1120.
Gospodarowicz, D. and Bialecki, H. (1978) Endocrinology **103:** 854–865.
Gospodarowicz, D. and Bialecki, H. (1979) Endocrinology **104:** 757–764.
Greenwald, G.S. and Roy, S.K. (1994) In: Physiology of Reproduction (E. Knobil and J.D. Neill, Eds), pp 629–988. Raven Press, New York.
Guthridge, M., Schmitt, J., Koos, R.T. and Olson, C.E. (1989) Mol. Endocrinol. **3:** 2041–2048.
Hansson, H.A., Nilsson, O., Isgaard, J., Billing, H., Isaksson, O., Stottner, A., Anderson, I.K. and Rozell, B. (1988) Histochem. **89:** 403–410.
Oliver, J.E., Aitman, T.J., Powell, J.F., Wilson, C.A. and Clayton, R.N. (1989) Endocrinology **124:** 2671–2677.
Hattori, M.A., Shinohara, Y., Yoshino, E., Kanzaki, M., Kojima, I. and Horiuchi, R. (1994) J. Endocrinol. **142:** 69–75.
Hernandez, E.R., Hurwitz, A., Payne, D.W., Dharmarajan, A.M., Purchio, A.F. and Adashi, E.Y. (1990) Endocrinology **127:** 2804–2811.
Hernandez, E.R. (1995) J. Steroid Biochem. Mol. Biol. **53:** 219–221.
Hillensjo, T. and LeMaire, W. (1980) Nature **287:** 145-146.
Holtorf, A.P., Furuya, K., Ivell, R. and McArdle, C.A. (1989) Endocrinology **125:** 2612–2620.
Hsu, C.J. and Hammond. J.M. (1987) Endocrinology **120:** 198–207.
Hsu, C.J., Holmes, S.D. and Hammond, J.M. (1987) Biochem. Biophys. Res. Commun. **147:** 242–247.
Hsueh, A.J.W. and Ling, N.C. (1979) Life Sci. **25:** 1223–1254.
Hsueh, A.J.W., Wang, C. and Erickson, G.F. (1980) Endocrinology **106:** 1697–1705.
Hsueh, A.J.W., Liu, Y.X, Cajander, S., Peng, X.-R., Dahl, K., Kristensen, P. and Ny, T. (1988) Endocrinology **122:** 1486–1495.
Hurwitz, A., Ricciarelli, E., Botero, L., Rohan, R.M., Hemandez, E.R. and Adashi, E.Y. (1991) Endocrinology **129:** 3427–3429.
Hurwitz, A., Payne, D.W., Packman, J.N., Andreani, C.L., Resnick, C.E., Heenandez, E.R. and Adashi, E.Y. (1991) **129:** 1250–1256.
Hurwitz, A., Dushnik, M., Solomon, H., Ben-Chetrit, A., Finci-Yeheskel, Z., Milwidsky, A., Mayer, M., Adashi, E.Y. and Yagel, S. (1993) Endocrinology **132:** 2709–2714.

Hurwitz, A., Hernandez, E.R., Resnick, C.E., Packman, J.N., Payne, D.W. and Adashi, E.Y. (1990) Endocrinology **126:** 3089–3095.
Kasson, B.G. and Gorospe, W.C. (1989) Mol. Cell Endocrinol. **62:** 103–111.
Ireland, J.J., Aten, R.F., Behrman, H.R. (1988) Biol. Reprod. **38:** 544–550.
Iwashita, M., Adachi, T., Katayana, E., Kudo, Y. and Takeda, Y. (1994) Horm. Res. **41:** Suppl. 1: 22–28.
Johnson, G.R., Saeki, T., Auersberg, N., Gordon, A.W., Shoyab, M., Salomon, D.S. and Stomley, K. (1991) Biochem. Biophys. Res. Commun. **180:** 481–488.
Jones, P.B.C. and Hsueh, A.J.W. (1982) Endocrinology **111:** 713–721.
Kokia, E., Ben-Shlomo, I. and Adashi, E.Y. (1995) Fertil. Steril. **63:** 176–181.
Keel, B.A., Hildebrandt, J.M., May, J.V. and Davis, J.S. (1995) Endocrinology **136:** 1197–1204.
Koos, R.D. (1989) Semin. Reprod. Endocrinol. **7:** 29–40.
Koos, R.D. and LeMaire, W. (1985) Endocrinology **116:** 628–632.
Kubota, T., Kamada, S., Taguchi, M. and Aso, T. (1995) Hum. Reprod. **9:** 2118–2122.
Kudlow, J.E., Kobrin, M.S., Purchio, A.F., Twardzik, D.R., Hernandez, E.R., Asa, S.L. and Adashi, E.Y. (1987) Endocrinology **121:** 1577–1579.
LaPolt, P.S., Piquette, G.N., Soto, D., Sincich, C. and Hsueh, A.J. (1990) Endocrinology **127:** 823–831.
Leung, P.C.K., Wang, J. and Baimbridge, K.G. (1989) Can. J. Physiol. Pharmacol. **67:** 962–967.
Liang, S.G., Taketani, Y. and Mizuno, M. (1990) Endocrinol. Jpn. **37:** 893–899.
Magnusson, C. and LeMaire, W. (1981) Acta. Physiol. Scand. **111:** 377–379.
Magoffin, D.A., Gancedo, B. and Erickson, G.F. (1989) Endocrinology **125:** 1951–1958.
Massague, J. (1990) Annu. Rev. Cell Biol. **6:** 597–641.
Massicoite, J., Veilleux, R., Lavoie, M. and Labrie, F. (1980) Biochem. Biophys. Res. Commun. **94:** 1362–1366.
Mauro, T., Ladines-Llave, C.A., Samoto, T., Matsuo, H., Manalo, A.S., Ito, H. and Mochizuki, M. (1993) Endocrinology **132:** 924–931.
McAllister, J.M., Byrd, W. and Simpson, E.R. (1994) J. Clin. Endocr. Metab. **79:** 106–112.
May, J.V., Gotcher, E.D., Gangrade, B.K. and Mau, Y-H. (1994) Endocrine **2:** 1045–1054.
May, J.V., Stephenson, L.A., Turzcynski, C.J., Fong, H.W., Mau, Y.H.L. and Davis, J.S. (1996) Biol. Reprod. **54:** 485–496.
May, J.V., Frost, J.P. and Schomberg, D.W. (1988) Endocrinology **123:** 168–179.
May, J.V. and Schomberg, D.W. (1989) Semin. Reprod. Endocrinol. **7:** 1–11.
McAllister, J.M., Byrd, W. and Simpson, E.R. (1994) J. Clin. Endocrinol. Metab. **79:** 106–112.
Meidan, R., Altstein, M. and Girsh, E. (1992) Biol. Reprod. **46:** 715–720.
Monget, P., Monniaux, D. and Durand, P. (1989) Endocrinology **125:** 2586–2593.
Monniaux, D. and Pisselet, C. (1992) Biol. Reprod. **46:** 109–119.
Morbeck, D.E., Flowers, W.L. and Britt, J.H. (1993) J. Reprod. Fertil. **99:** 577–584.
Mulheron, G.W. and Schomberg, D.W. (1990) Endocrinology **126:** 1777–1779.
Mulheron, G.W., Mulheron. J.G., Danielpour, D. and Schomberg, D.W. (1992) Endocrinology **131:** 2609–2614.
Navickis, R.J., Jones, P.B.C. and Hsueh, A.J.W. (1982) Mol. Cell. Endocrinol. **27:** 77–88.
Niswender, G.D. and Nett, T.M. (1994) In: Physiology of Reproduction (E. Knobil and J.D. Neill, Eds), pp 781–816. Raven Press, New York.
Oliver, J.E., Aitman, T.J., Powell, J.F., Wilson, C.A. and Clayton R.N. (1989) Endocrinology **124:** 2671–2677.
Olsson, J.H., Carlsson, B. and Hillensjo, T. (1990) Fertil. Steril. **54:** 1052–1057.
Oury, F. and Darbon, J.M. (1988) Biochem. Biophys. Res. Commun. **156:** 634–643.

Piquette, G.N., Simon, C., Danasouri, I.E., Frances, A. and Polan, M.L. (1994) Fertil. Steril. **62:** 760–769.
Polan, M.L., Loukides, J.A. and Honig, J. (1994) Am. J. Obstet. Gynecol. **170:** 1000–1006.
Rippel, R.H. and Johnson, E.S. (1976) Proc. Soc. Exp. Biol. Med. **152:** 432–436.
Robert, A.B. and Sporn, M.B. (1988) Adv. Cancer Res. **51:** 107–145.
Roby, K.F. and Terranova, P.F. (1988) Endocrinology **123:** 2952–2954.
Roussi, M., Royere, M., Guillonueau, M., Lansac, J. and Muh, J.P. (1989) Acta. Endocrinol. Copenh. **121:** 90–94.
Roy, S.K. and Greenwald, G.S. (1991) Biol. Reprod. **44:** 889–896.
Roy, S.K. (1994) Micros. Res. Tech. **27:** 83–96.
Roy, S.K. and Greenwald, G.S. (1990) Endocrinology **126:** 1309–1317.
Roy, S.K. and Harris, S.G. (1994) Mol. Endocrinol. **8:** 1175–1181.
Roy, S.K. and Greenwald, G.S. (1991) Endocrinology **129:** 1903–1908.
Roy, S.K., Ogren, C.L., Roy, C. and Lu, B. (1992) Biol. Reprod. **46:** 595–606.
Roy, S.K. (1993a) Biol. Reprod. **48:** 552–557.
Roy, S.K. (1993b) Biol. Reprod. **48:** 558–563.
Roy, S.K. and Hughes, J. (1994) Biol. Reprod. **51:** 821–830.
Roy, S.K. and Greenwald, G.S. (1996) J. Reprod. Fertil. (In Press).
Roy, S.K. and Kole, A.R. (1995) Endocrinology **136:** 4610–4620.
Rutanen, E.M. and Pekonen, F. (1990) Acta Endocrinol. **123:** 7–13.
Samaras, S.E., Guthrie, D., Barber, J.A.Q. and Hammond, J.M. (1993) Endocrinology **133:** 2395–2398.
Samaras, S.E. and Hammond. J.M. (1995) Am. J. Physiol. **268:** E1057–E1064.
Sara, V.R. and Hall, K. (1990) Physiol. Rev. **70:** 591–614.
Schams, D., Koll, R. and Li, Ch. (1988) J. Endocrinol. **116:** 97–100.
Sealey, J.E., Quimby, F.W., Itskovitz, J. and Rubattu, S.(1990) Front. Neuroendocrinol. **11:** 213–237.
Shaw, H.J., Boddy, S.E. and Hodges, J.K. (1993) Biol. Reprod. **49:** 568–576.
Sheela Rani, C.S., Ekholm, C., Billing, H., Magnusson, C. and Hillensjo, T. (1983) Biol. Reprod. **28:** 591–597.
Shifren, J.L., Osathanondh, R. and Yeh, J. (1993) Fertil. Steril. **59:** 1036–1040.
Shipley, G.D., Tucker, R.F. and Moses, H.L. (1985) Proc. Natl. Acad. Sci. USA **82:** 4147–4151.
Simon, C., Tsafriri, A., Chun, S.Y., Piquette, G.N., Dang, W. and Polan, M.L. (1994) Biol. Reprod. **51:** 662–667.
Simon, C. and Polan, M.L. (1994) West J. Med. **160:** 425–429.
Singh, B., Rutledge, J.M. and Armstrong, D.T. (1995) Mol. Reprod. Dev. **40:** 391–399.
Skinner, M.K. and Coffey, R.J. (1988) Endocrinology **123:** 2632–2638.
Spicer, L.J. and Stewart, R.E. (1996) Biol. Reprod. **54:** 255–263.
Spicer, L.J., Alpizar, E. and Echtekamp, S.E. (1993) J. Anim. Sci. **71:** 1232–1241.
Steinkampf, M.P., Mendelson, C.R. and Simpson, E.R. (1988) Mol. Cell. Endocrinol. **59:** 93–99.
Stewart, C.J., Owens, O.J., Richmond, J.A. and McNicol, A.M. (1992) Int. J. Gynecol. Pathol. **11:** 266–272.
Tilly, J.L., Billing, H., Kowalski, K.I. and Hsueh, A.J. (1992) Mol. Endocrinol. **6:** 1942–1950.
Tonetta, S.A. and Dizerega, G.S. (1989) Endocrine Rev. **10:** 205–229.
Tsafriri, A. and Braw, R. (1984) Oxford Rev. Reprod. Biol. **6:** 226–265.
Tsafriri, A. and Adashi, E.Y. (1994) In: Physiology of Reproduction (E. Knobil and J.D. Neill, Eds), pp 817–860. Raven Press, New York.
Trzeciak, W.H., Duda, T., Waterman, M.R. and Simpson, E.R. (1987) Mol. Cell. Endocrinol. **52:** 43–50.

Veldhuis, J.D., Furlanetto, R.W., Juchter, D.J.G. and Veldhuis, P. (1985) Endocrinology **116:** 1235–1242.
Watanabe, H., Nagai, K., Yamaguchi, M., Ikenoue, T. and Mori, N. (1993) Prost. Leu. Essn. Fatty Acids **49:** 963–967.
Wood, A.M., Lambert, A., Hooper, M.A., Mitchell, G.G. and Robertson, W.R. (1994) Hum. Reprod. **9:** 19–23.
Yamoto, M., Shikone, T. and Nakano, R. (1993) Endocr. J. **40:** 691–697.
Yeh, J., Lee, G.Y. and Anderson, E. (1993) Biol. Reprod. **48:** 1071–1081.
Yoshimura,Y., Nakamura, Y. and Ando, M. (1992) Biol. Reprod. **47:** 118–125.

Comparative Endocrinology and Reproduction
K.P. Joy, A. Krishna and C. Haldar (Eds)

20. Cytokines and Ovarian Function: The Case of TNFα

P.F. Terranova
Departments of Physiology and Obstetrics and Gynecology,
University of Kansas Medical Center, Kansas City, Kansas 66160–7401, USA

Introduction

The number of published studies on TNF and ovarian function has increased significantly in the past few years (Terranova *et al.*, 1995). TNF a prominent, multi-functional hormone-like polypeptide, is among the cytokines that are produced by many types of cells in the ovary. TNF was first described as a product of activated macrophages that lysed selected tumor cell lines (Sugarman *et al.*, 1985). However, TNF is now known to either stimulate or inhibit cell growth depending on its concentration and target cell. In the normal ovary TNF appears to be present in the absence of inflammation although TNF has been implicated in ovarian inflammation (ovulation) and pathophysiology.

This chapter describes TNF gene expression and its regulation in the ovary in an attempt to provide insight into its functional role. Working models will be proposed for the involvement of TNF in several processes, including oocyte and follicle development, ovulation, and luteal regression.

TNF Gene Family and Its Receptors

TNF is a member of the TNF family that includes nine members. Eight additional members have been identified (Baker and Reddy, 1996). The TNF gene is located in the class III region of the major histocompatibility complex. The gene codes for a 26 kDa membrane-bound pro-cytokine form that releases a 17.3 kDa soluble form after cleavage of pro-TNF (Beutler *et al.*, 1989; Camussi *et al.*, 1991; Vassalli, 1992). Physiologically active, soluble TNF circulates as a homotrimer. These trimers bind to and facilitate multimerization of two distinct cellular receptors, termed p55/p60 (Type I, b) and p75/p80 (Type II, a). The receptor genes have been cloned and sequenced in several species (Pfizenmaier *et al.*, 1992; Tartaglia and Goeddel, 1992). Soluble TNF receptors (TNF-R) inhibit TNF action by neutralizing soluble or bound TNF. These soluble inhibitors are present in serum and urine and provide protection against over expression of TNF (Fernandez-Botran, 1991).

The extracellular domains of the TNF receptors exhibit some sequence homology but the intracellular domains are different, indicating that different intracellular signaling pathways are activated by binding of TNF (Baker and Reddy, 1996).

Evidence for TNF in the Ovary

Several studies have reported the presence of TNF in the ovary using immunohistochemistry, bioassay, Northern and *in situ* hybridization and reverse transcriptase polymerase chain reaction.

Mouse Ovarian cells transcribing the gene for TNF have been identified in the adult cyclic mouse using *in situ* hybridization and immunohistochemistry (Chen *et al.*, 1993). Hybridization signals for TNF mRNA and TNF immunoreactivity were observed in oocytes of healthy follicles with two or more layers of granulosa cells. TNF mRNA and immunoreactivity were not observed in oocytes of primordial follicles, follicles with a single layer of granulosa cells and early atretic follicles. In late stages of atresia immunoreactive TNF was observed in the oocytes. Theca and/or granulosa cells from all stages of follicular development exhibited TNF mRNA. However, in some follicles only theca or granulosa were positive. In atretic follicles both theca and granulosa exhibited TNF mRNA hybridization signals. The interstitial cells were negative but macrophage-like cells within this compartment were positive for TNF mRNA and protein. Corpora lutea contained hybridization signals for TNF mRNA in luteal cells and in macrophage like cells; only macrophages contained immunoreactive TNF. TNF hybridization signals and immunoreactivity were of greater intensities in older corpora lutea than in corpora lutea of the current cycle. Northern blots revealed a 2.2 kb TNF mRNA in the ovary and the relative proportions of TNF mRNA to 28S rRNA (constitutive RNA) did not change throughout the estrous cycle. TNF hybridization signals and immunoreactivity did not change throughout the cycle. These results indicated that TNF gene transcription in the oocyte coincided with the synthesis of immunoreactive TNF and that these processes occurred at very distinct steps of follicular development. The formation of the second layer of granulosa cells in the type 4 follicle was coupled with the initial phase of TNF gene transcription in the oocyte since TNF mRNA and protein appeared in synchrony at this stage. The onset of TNF gene expression correlated with growth of the oocyte and thus, may be related to that event.

Rat Immunohistochemistry in adult ovaries revealed that the ooplasm of the oocyte was a primary site of TNF localization within the follicle (Marcinkiewicz *et al.*, 1994) similar to that observed in the mouse. However, in contrast to the mouse, immunostaining was present in primordial follicles, those with one layer of granulosa cells and larger follicles. The initial appearance of oocytic TNF in the ovary occurred 2 days after birth. Fetal oocytes one day before birth did not contain immunoreactive TNF. However, fetal and neonatal ovaries contained TNF mRNA. Thus, it appears that as soon as the pup (ovary) was no longer under the hormonal influence of the mother, TNF synthesis was initiated. Previous studies have shown that corticosterone increases in maternal circulation toward the end of pregnancy (Voogt *et al.*, 1969) and that corticoids inhibit transcription and translation of TNF (Beutler *et al.*, 1992). Thus, it is likely that maternal corticoids are causal in the inhibition of TNF synthesis during the fetal period. Therefore, removal of the pup from the high corticoid environment may allow translation of TNF into protein during the early neonatal period.

In isolated cases, the granulosa cells near the oocytes also contained immunoreactive TNF in some follicles (Marcinkiewicz *et al.*, 1994) which agreed with previous studies (Roby and Terranova, 1989). However, it is unclear if granulosa cells synthesized TNF. The granulosa cell TNF may have been sequestered from the oocyte or another cell type, such as a mononuclear cell. Reverse transcriptase polymerase chain reaction (RT-PCR) analysis of oocytes collected from preovulatory follicles before the LH surge revealed a cDNA band of 500 bp that corresponded to the predicted size for the amplified TNF cDNA (Sancho-Tello *et al.*, 1992). Oocytes contained TNF bioactivity which was similar to recombinant murine TNF (Marcinkiewicz *et al.*, 1994). These results provided firm evidence for the identification of immunoreactive and bioactive TNF within oocytes which was further supported by the presence of TNF mRNA within the oocyte. The onset of appearance of immunoreactive TNF in the oocyte at the time of birth indicated the possibility that TNF may be involved in early follicular organization.

The perfused ovary of the equine chorionic gonadotropin treated rat has been shown to release TNF into the perfusate during the periovulatory period (Brännström, *et al.*, 1994). The source of the TNF in the perfused ovary is unknown but likely to be macrophage/leukocyte in origin. LPS does stimulate TNF release from purified theca (Taylor and Terranova, 1995) and granulosa (Taylor and Terranova, unpublished). However, LPS does stimulate TNF release from whole ovarian dispersates (Li and Terranova, unpublished).

Rabbit Media obtained from cultures of luteal tissue on day 5, and days 17/19 of pseudopregnancy exhibited TNF bioactivity using the LM cell bioassay (Bagavandoss *et al.*, 1988; Bagavandoss *et al.*, 1990). *In vitro* LPS increased TNF bioactivity on day 17/19 of pseudopregnancy. The major targets of LPS were macrophages within the luteal tissue. Bioactivity of TNF was neutralized by anti-TNF. Macrophages were immunohistochemically localized to regressing corpora lutea (CL) on day 17. Macrophages were not observed in day 5 CL but significant TNF bioactivity, albeit low, was detected in conditioned media at that time. Therefore, cells other than macrophages in the CL may have produced the bioactive TNF on day 5. On day 17 numerous macrophages correlated well with high levels of LPS-induced TNF release into the culture media indicating that macrophages were the major source of bioactive TNF at that time.

Sheep, Cow and Pig TNF mRNA was undetectable by Northern blot (Ji *et al.*, 1991) in sheep CL during various stages of regression. However, TNF bioactivity increased in response to PGF2α after the CL regressed and after progesterone levels in serum had declined. These results indicated that leukocytes from circulation and/or cells from other parts of the ovary infiltrated the CL late in luteal regression and contributed to the TNF bioactivity. It was hypothesized that the mononuclear cells enter the CL after physiologic regression in order to remove cellular debris (Ji *et al.*, 1991) probably associated with apoptosis in the CL at that time.

Immunoreactive TNF (Roby and Terranova, 1989) and TNF mRNA (unpublished Marcinkiewicz, Pate and Terranova) have been detected in bovine CL. In fact, bovine CL secrete TNF but the source of secreted TNF is unknown (Shaw and Britt, 1995). Immunoreactive TNF was observed in atretic and antral follicles and in mononuclear macrophage-like cells throughout the ovary including the interstitium, theca and CL (Roby and Terranova, 1989). Immunoreactive TNF was localized to cells in the thecal cords that penetrated the CL. These cells were either macrophage-like cells, thecal cells and/or endothelial cells. A study by Zolti *et al.*, 1990) reported TNF in follicular fluid of cows using a TNF bioassay. Concentrations of bioactive TNF increased from days 3–20 of the estrous cycle with highest levels observed on the day of ovulation in corpora hemorrhagica. However, currently there are no clear insights into a role of increasing TNF in the follicles during the bovine estrous cycle.

Two studies have provided insight into the localization of TNF in pigs (Wuttke *et al.*, 1993; Hehnke-Vagnoni *et al.*, 1995). Wuttke *et al.* (1993) has the presence of TNF mRNA in large luteal cells of the pig (Wuttke, 1993 # 1684). More recently, Hehnke-Vagnoni *et al.* (1995) had immunolocalized TNF to endothelial cells in corpora lutea. The latter observation correlates well with the observance of immunoreactive TNF in bovine theca cords coarsing through the corpus luteum; it is known that these cords contain blood vessels.

Human Several studies have reported TNF in the follicular and luteal compartments of the human ovary (Barak *et al.*, 1992; Jasper and Norman, 1995; Naylor *et al.*, 1993; Punnonen *et al.*, 1992; Roby *et al.*, 1990; Wang *et al.*, 1992; Zolti *et al.*, 1992; Zolti *et al.*, 1990). Similar to the rat and cow (Roby *et al.*, 1989), healthy antral follicles exhibited immunoreactive TNF in the antral layer of granulosa cells (Roby and Terranova, 1990). Immunoreactive TNF was observed in the follicular fluid surrounding granulosa cells. Atretic follicles also exhibited TNF in the follicular fluid and in some granulosa cells lining the antral cavity. In the CL large lutein-like cells contained immunoreactive TNF (Roby *et al.*, 1990); in addition, cells within the thecal layer of the CL (paraluteal and possibly endothelial cells) contained immunoreactive TNF (Roby *et al.*, 1990); hybridization signals for TNF mRNA have also been detected in theca (Naylor *et al.*, 1993).

In vitro studies have implicated the granulosa-luteal cell as a source of TNF (Roby *et al.*, 1990; Zolti *et al.*, 1990). However, the observance of TNF in cells, follicular fluid and/or media derived from follicular aspirates may be related to the presence of contaninating leukocytes. Granulosa cells collected from patients undergoing *in vitro* fertilization (IVF) released TNF into culture medium during a 15 hour incubation (Zolti *et al.*, 1990). FSH (not hCG, colony stimulating factor (CSF), PGF2α or 8 Br-cAMP) increased TNF in the culture media (Zolti *et al.*, 1990). A synergism between CSF and hCG was observed on TNF release since neither alone increased TNF but together they increased TNF as much as FSH alone. The physiological significance of such a synergism is unknown. CSF was unable to further increase the release of TNF induced by FSH. Since granulosa cell aspirates from the IVF patients contain polymorphonuclear cells, the effects of FSH on monocytic release of TNF was investigated. No effect of FSH was observed on the monocytes. However, it is possible that FSH might cause granulosa cells to release substance(s) which induce the monocytes to release TNF.

An LPS binding protein has been observed on human granulosa cells using antibodies to LPS (Sancho-Tello *et al.*, 1992). Ninety four per cent of the cells exhibiting immunofluorescent LPS-binding were also positive for 3β-hydroxysteroid dehydrogenase, indicating that these steroidogenic (granulosa-lutein) cells exhibit LPS binding sites. In order to detect a response to LPS the enriched population of granulosa-luteal cells were cultured in the presence or absence of LPS. After 16 h of culture, the expression of TNF mRNA was enhanced when the cells were cultured in the presence of LPS, which also significantly enhanced TNF secretion into the media during the 16 h period. Since macrophages and leukocytes may have contaminated the enriched preparation of granulosa cells, additional studies are needed to confirm TNF production by granulosa cells. TNF mRNA has also been detected in human granulosa-luteal cells by Northern blot (Loy *et al.*, 1992).

Assays of TNF have been conducted on conditioned medium of oocytes, granulosa cells, cumulus cells and one to eight cell embryos from women undergoing *in vitro* fertilization (Zolti *et al.*, 1991). After a 24 h incubation period, conditioned medium only from the oocyte corona-cumulus complex contained very low levels of TNF. TNF increased further in the media of 6–8 cell embryos. TNF levels were low in oocyte conditioned medium and increased stepwise in conditioned media of 2–4 and 6–8 cell embryos.

Regulation of TNF

Factors regulating expression of ovarian TNF are unknown. Measurements of ovarian TNF mRNA during the estrous cycle of mice (Chen *et al.*, 1993) and after equine chorionic gonadotropin (eCG) treatment of immature rats revealed no significant changes (Sancho-Tello *et al.*, 1993). Qualitative immunohistochemistry in adult cyclic mice and rats and eCG-treated immature rats also revealed no significant changes (Chen *et al.*, 1993; Marcinkiewicz *et al.*, 1994; Sancho-Tello *et al.*, 1993). LPS-treated immature rats exhibited large influxes of TNF immunoreactive cells into the ovarian stroma, theca and interstitium (Sancho-Tello *et al.*, 1993). There were no changes in ovarian TNF mRNA after LPS. In the rat ovary, oocytes appear to be the major source of ovarian TNF and thus basal levels of TNF mRNA are high. Any further increase in TNF mRNA from cells migrating into the ovary may contribute little to the total amount of TNF mRNA already present.

Three reports indicate the possibility that ovarian TNF may be hormonally regulated. Zolti *et al.* (1990) has shown that bioactive TNF increased in follicular fluid of growing follicles from days 3–20 of the bovine estrous cycle. Whether this increase in TNF is hormonally regulated is unknown but the growth of follicles is regulated by gonadotropins. Possibly increased blood flow to the growing follicles delivers more macrophage-like cells that contain TNF bioactivity. Polan and her colleagues have shown that *in vitro* progesterone and estradiol reduced expression of TNF mRNA in peripheral blood monocytes taken during the luteal phase of the menstrual cycle (Loy *et al.*, 1992).

Dexamethasone, synthetic corticoid, and inhibitor of TNF transcription and translation, reduced follicular fluid levels of TNF in patients with PCOD (Zolti *et al.*, 1992). These results indicate that corticoids may regulate TNF.

In cows, TNF was evident in the dialysate during either spontaneous or prostaglandin F induced luteolysis only after the onset of progesterone decline (Shaw *et al.*, 1995). Whether the increased TNF was a result of declining progesterone is unknown. Neverthless, TNF is implicated in luteolysis because of the inverse relationship between these 2 factors. Interestingly the serum concentrations of TNF did not change during this period (Shaw and Britt, 1995).

Effects of TNF

Rat Granulosa Cell TNF inhibits FSH-stimulated aromatase activity in cultured granulosa cells from immature diethylstilbestrol (DES)-treated rats (Emoto and Baird, 1988). The inhibition was dependent on the dose of TNF and the dose was in the range of the dissociation constant for TNF receptors in porcine granulosa cells (Veldhuis *et al.*, 1991). Time course studies of the effects of TNF on FSH-stimulated aromatase activity indicated that TNF did not reduce aromatase activity but it prevented aromatase induction by FSH. Thus, existing aromatase activity in the cells was not altered by TNF. TNF also inhibited the induction of aromatase induced by TGFβ and decreased progesterone accumulation induced by FSH and TGFβ. Adashi's laboratory (Adashi *et al.*, 1989) also reported that TNF inhibited FSH-induced aromatase in granulosa cells of the DES-treated rat. A minimum of 48 hours exposure to TNF was required to inhibit the FSH-induced aromatase. Those studies clearly showed that FSH-stimulated cAMP production was reduced by TNF. TNF did not inhibit FSH binding to the granulosa. PGE2 and vasoactive intestinal peptide, and the adenyl cyclase stimulator, choleragen increased production of cAMP (that subsequently stimulated aromatase activity), were also inhibited by TNF. TNF also inhibited forskolin induced accumulation of cAMP and thus, TNF reduced adenylate cyclase activity. TNF at 1ng/ml inhibits within 24 hours the stimulatory action of FSH on formation of the LH receptor *in vitro*. The action of TNF was not due to its cytotoxicity since cell numbers were not altered by TNF. Darbon confirmed earlier reports that TNF inhibited FSH stimulated cAMP production (Darbon *et al.*, 1989). He reported that TNF could inhibit the effects of 8-Br-cAMP in stimulating LH receptor formation. Therefore, he concluded that TNF could act at a site distal to cAMP and at the level of adenylate cyclase as reported earlier.

In vitro FSH caused granulosa cells to undergo steroidogenic and morphologic differentiation indicative of luteinization (Adashi *et al.*, 1990). Treatment of granulosa cells for 72 hours with FSH followed by hCG stimulates extensive luteinization and progesterone secretion. TNF inhibited the action of hCG in stimulating progesterone secretion by the luteinizing granulosa cells.

Rat Theca-interstitial Cells *In vitro* TNF inhibited dose-dependently LH-stimulated androsterone production by theca-interstitial cells (TIC) from ovaries of immature rats (Adashi *et al.*, 1990; Zachow *et al.*, 1992). TNF's inhibitory effect was within a physiological range of the Kd for the TNF receptor, reversed by removal of the TNF from the culture media, and its action was neutralized by an antiserum to TNF. TNF inhibited the production of the second messenger, cAMP. Morphologic changes in TIC were induced by TNF. Clustering of the TIC occurred between 72 and 96 hours of culture and was dependent on the dose of TNF administered *in vitro* (Zachow *et al.*, 1992); clustering has also been observed in granulosa (Gordon and Terranova, unpublished). TIC also clustered in response to epidermal growth factor (EGF) *in vitro* (Zachow *et al.*, 1993). The physiological relevance of such a discovery is unknown but may be related to the presence of immunoreactive TNF in the oocyte of the primordial follicle. Possibly oocytic TNF acts as an organzing factor by attracting theca (and granulosa) cells to the oocyte during early stages of follicular development. Thus, TNF's action in modulating steroidogenesis may be timely to prevent the developing follicle from responding prematurely to periodic high levels of gonadotropin.

TNF inhibits LH-induced androgen and this is followed by clustering of the TIC *in vitro*. These two processes do not appear to be totally dependent on each other because inhibition of clustering using protein kinase C inhibitors does not prevent TNF from inhibiting steroidogenesis (Zachow *et al.*, 1993). Specific targets of TNF in modulating steroidogenesis are through reduction of LH receptor number (not affinity) (Zachow *et al.*, 1993), reduction of LH-stimulated cAMP (Zachow and Terranova, 1993), reduction of protein kinase A activity (Zachow *et al.*, 1993), and inhibition of hydroxylase-lyase activity (Zachow and Terranova, 1994a, b). Interestingly, phorbol esters mimic the effects of TNF on clustering and steroidogenesis and TNF increases protein kinase C activity (Zachow *et al.*, 1993). TNF induced clustering by activation of EGF-related protein tyrosine kinase since inhibitors of the EGF-related kinase blocked clustering induced by TNF. TNF-induced protein kinase C activity was also blocked by those inhibitors; however, phorbol ester-induced clustering and protein kinase C activity was not blocked by the EGF-related kinase inhibitors (Zachow *et al.*, 1993). Thus, it was proposed that TNF activates EGF-related protein tyrosine kinase which subsequently activates protein kinase C and causes induction of TIC clustering. None of these inhibitors blocked TNF's effects on inhibition of LH-stimulated androgen production by TIC (Zachow and Terranova, 1993). It may be possible that TNF induces the synthesis of EGF or a related molecule such as TGFα in theca; the EGF related molecules then mediate clustering.

Effects of TNF on Intact Follicles and Ovaries

In vitro TNF increased progesterone production by whole preovulatory rat follicles taken on the morning of proestrus (Roby and Terranova, 1988; Roby *et al.*, 1990); the follicles were cultured in 5% CO_2 in air (low O_2). At low doses TNF inhibited androstenedione accumulation but at higher doses it was stimulated. Estradiol accumulation was not altered by TNF during the 24 hour incubation but when compared to another study the basal estradiol production was very low (Braw and Tsafriri, 1980). The histology of the follicles revealed that atresia had occurred within the 24 hour culture period; this was evidenced by apoptotic granulosa cells. Thus, the effects of TNF were actually on follicles that were undergoing atresia *in vitro*. Therefore, the oxygen concentration was increased to 95% in an attempt to prevent follicular atresia *in vitro*. The 95% O_2 preserved the appearance of healthy follicles based on the absence of apoptotic granulosa cells at 24 hours of culture and estradiol accumulation increased significantly in response to TNF *in vitro*. Separation of follicle into theca and granulosa revealed that theca was the major target of TNF in this model since progesterone from theca could account for the increased progesterone observed in the intact healthy and atretic follicles. Since TNF did not alter the histology of the follicles *in vitro* then it is unlikely that TNF causes atresia under these conditions.

In vitro TNF slightly inhibited basal progesterone accumulation in granulosa cells from adult preovulatory follicles. TNF also inhibited FSH-stimulated progesterone accumulation *in vitro*. Therefore, TNF stimulated theca progesterone accumulation in the preovulatory follicle of the adult cyclic rat but inhibited granulosa cell progesterone accumulation, actions which are opposing within the preovulatory follicle.

Studies using whole preovulatory follicles from equine chorionic gonadotropin (eCG)-treated immature rats have observed that TNF stimulated the synthesis of PGE2, PGF2α, prostacyclin and progesterone (Brannstrom *et al.*, 1993). Although no inhibitory effect of TNF was observed on basal prgesterone by isolated granulosa cells as described above, it was confirmed that TNF could stimulate progesterone by isolated theca (Brannstrom *et al.*, 1993). Interestingly, the increase in progesterone accumulation induced by TNF was not dependent on the synthesis of prostaglandin since indomethacin did not block TNF induction of progesterone accumulation (Brannstrom *et al.*, 1993). Shakil and Whitehead, using granulosa cells from adult preovulatory follicles, did not observe an effect of TNF on LH-stimulated progesterone secretion (Shakil and Whitehead, 1994).

Pig Three reports have been published describing the effects of TNF on ovarian cell function in the pig (Pitzel *et al.*, 1993; Tekpety *et al.*, 1993; Veldhuis *et al.*, 1991). Veldhuis *et al.* (Veldhuis *et al.*, 1991) described an inhibitory effect of TNF on FSH and insulin stimulated progesterone accumulation by granulosa cells taken from immature ovaries. TNF's inhibitory action was at the level of adenyl cyclase since cAMP levels were reduced by TNF in the presence of gonadotropin. TNF acted a sites distal to the synthesis of cAMP since addition of cAMP analogs in the presence of TNF also reduced steroidogenesis. The inhibitory action of TNF was evident on P450scc mRNA. Tekpetey *et al.* (1993) reported that TNF inhibited LH-stimulated thecal progesterone accumulation from medium-sized prepubertal gilt follicles.

Using pig corpora lutea from days 3–10 of the cycle, Pitzel *et al.* (1993) have shown that TNF is quite inhibitory to hCG stimulated progesterone secretion and to estradiol action. Interestingly, Tekpetey using pig corpora lutea from days 10–14 of the cycle did not observe any effect of TNF on small and large luteal cells. Collectively, these studies indicate the possibility that corpora lutea may lose sensitivity to TNF as it approaches luteal regression. However, further studies are needed to clarify this point.

Cow TNF has significant effects on luteal cells of the cow during the luteal phase on days 9–12 (Benyo and Pate, 1992). TNF induced dose dependently an increase in PGF2α and 6-keto-PGF1α accumulation without altering accumulation of progesterone; although progesterone slightly decreased with 100 mg TNF/ml. Interferon-γ, which up-regulates TNF receptors, had no effect on luteal prostaglandin and progesterone accumulation; however, interferon-α reduced accumulation of PGF2α induced by TNF. Interleukin-1β increased PGF2α, and in combination with TNF synergistically increased PGF2α.

Human There are several reports describing the effects of TNF on human ovarian cells (Clinton *et al.*, 1991; Fukuoka *et al.*, 1992; Fukuoka *et al.*, 1992a, b; Terranova *et al.*, 1991a; Wang *et al.*, 1992; Wang *et al.*, 1992; Yan *et al.*, 1993; Zolti *et al.*, 1990). TNF induced an increase in PGF2α and PGE2 in media of granulosa-luteal cells (Wang *et al.*, 1992b) and also caused the cells to proliferate (Yan *et al.*, 1993). Interestingly, TNF induced significant increases in granulosa-luteal cell estradiol secretion after 4–6 days of culture (Yan *et al.*, 1993). Progesterone accumulation in the media of TNF-treated granulosa luteal cells was higher than controls; however, when values were adjusted as progesterone accumulation per cell there was no significant increase by TNF treatment (Yan *et al.*, 1993). When hCG was combined with TNF *in vitro* progesterone secretion was higher than with each treatment alone (Terranova *et al.*, 1991b). Further investigations revealed that TNF increased hCG binding on day 10 of culture. Thus, it appeared that TNF synergized with hCG by increasing the number of hCG binding sites. Scatchard analysis will be needed to determine if the increased hCG binding was due to increased number of binding sites per cell or increased affinity of hCG for its receptor on the cells. In addition, since TNF increased the number of cells *in vitro,* the data should be expressed per cell to exclude the possibility that the increased hCG binding was not due to an increase in cell number. TNF has also been shown to increase prolactin-stimulated progesterone production by human granulosa luteal cells (Clinton *et al.*, 1991).

The effects of interferon-α and -γ and TNF on progesterone and estradiol production by human luteal cells and granulosa-luteal cells have been investigated (Fukuoka *et al.*, 1992b; Wang *et al.*, 1992b). Although interferon-α had no effect on progesterone secretion *in vitro,* interferon-γ dose dependently inhibited progesterone secretion. TNF in combination with interferon-γ further suppressed progesterone production whereas TNF alone had no significant effect. There was no effect of these treatments on estradiol secretion. These studies indicate that TNF, in concert with interferon-γ, may have a role in the demise of the corpus luteum.

Functions of TNF (Table 2)

Growth Regulation TNF's originally described role in cell and tissue growth was to induce necrosis of tumors (Carswell *et al.*, 1975). Since TNF could induce apoptosis in tumors a role in luteolysis and follicular atresia appeared likely. In contrast, studies using divergent cell lineages indicated that TNF promoted growth (Hori *et al.*, 1989; Mackay *et al.*, 1993; Mutch *et al.*, 1990; Mutch *et al.*, 1992; Selmaj *et al.*, 1990; Vilcek *et al.*, 1986). Thus, it appears that TNF stimulates and inhibits growth of cells.

Table 1. Sites of TNF protein and mRNA localization in the ovary

Species	Cell/Tissue	Method of Detection	Reference
Mouse	oocytes macrophage	IHC/ISH	Chen *et al.*, 1993
	theca/gc/CL	ISH	
Rat	oocytes	IHC/RT-PCR/BA	Marcinkiewicz *et al.*, 1994
	gc	IHC	Roby *et al.*, 1989 Marcinkiewicz *et al.*, 1994
	mononuclear cells	IHC	Roby *et al.*,1989
	perfused ovary	BA	Brannstrom *et al.*, 1994
Rabbit	CL media	BA	Bagavandoss *et al.*, 1988, 1990
Sheep	CL regressing	BA	Ji *et al.*, 1991
Cow	CL (thecal cords) mononuclear cells atretic follicles	IHC	Roby *et al.*, 1989
	Follicular fluid	BA	Zolti *et al.*, 1990
	CL	BA	Shaw *et al.*, 1995
Pig	CL (endothelial cells)	IHC	Hehnke-Vagnoni *et al.*, 1995
	CL (large luteal cells)	RT-PCR	Wuttke *et al.*, 1993
Human	gc, lutein cells paraluteal cells	IHC	Roby *et al.*, 1990
	oocyte-cumulus	BA	Zolti *et al.*, 1991
	gc medium	BA	Zolti *et al.*, 1990, 1992
	FF	ELISA	Barak *et al.*, 1992
		ELISA	Punnonen *et al.*, 1992
		ELISA	Wang *et al.*, 1992
		BA/ELISA	Jasper *et al.*, 1995
	theca layer	ISH	Naylor *et al.*, 1993

**Abbreviations:* IHC, immunhistochemistry; BA, bioactivity; ELISA, enzyme linked immunosorbent assay; ISH, *in situ* hybridization; RT-PCR, reverse transcriptase polymerase chain reaction; FF, follicular fluid; gc, granulosa cell; CL, corpus luteum.

In vitro TNF had a cytotoxic effect on bovine luteal cells (Benyo and Pate, 1992). TNF and interferon-γ, each alone *in vitro,* did not alter the number of luteal cells. In combination, TNF and interferon-γ, reduced luteal cell numbers by ~ 80%. An effect of interferon-γ on human luteal cells

Table 2. Potential functions of TNFα in regulation of ovarian function

Growth Regulation (Cancer)
Follicular Development
Ovulation
Luteal Development/Regression

growth *in vitro* was not observed and its effects in combination with TNF were not studied (Wang *et al.*, 1992a). In contrast, TNF stimulated human granulosa-luteal cell division *in vitro* (Wang *et al.*, 1992b; Yan *et al.*, 1993). Granulosa-luteal cells were collected from women undergoing IVF. TNF (0.1–10ng TNF/ml) increased cell number in a dose-dependent manner (Yan *et al.*, 1993). Studies using serum free cultures of granulosa and theca cells from immature rats have not observed effects of TNF on cell number, DNA content, plating efficiency and viability (Adashi *et al.*, 1989; Andreani *et al.*, 1991; Darbon *et al.*, 1989; Yan *et al.*, 1993; Zachow *et al.*, 1992). Thus, it appears that established luteal cells are sensitive to the cytolytic actions of TNF only when combined with interferon-γ. Whereas cells in the beginning stages of luteal formation proliferate in response to TNF *in vitro.*

TNF effects on several epithelial ovarian cancer cell lines have been tested *in vitro* (Mutch *et al.*, 1990; Mutch *et al.*1992; Wu *et al.*,1993). Some are inhibited and others are stimulated by TNF. TNF and interferon-γ, each alone, stimulated growth of Caov-3 cells at 24 hours and had no effect on two other ovarian cell lines (SK-OV-3 and NIH:OVCAR 3). The two cytokines, in combination, did not have a synergistic effect on proliferation of the Caov-3 cells; however, the combined treatment reduced growth of the SK-OV-3 cells. By 72 hours, TNF increased the number of Caov-3 and NIH:OVCAR-3 cells; but the combination of interferon-γ and TNF inhibited their growth. Low doses of these two cytokines stimulatd growth whereas high doses, in combination, reduced growth. TNF stimulation of growth of other epithelial ovarian cancer cell lines has been reported (Wu *et al.*, 1993; Wu *et al.*, 1992). Interleukin-1β induced proliferation of epithelial ovarian cancer cells by increasing expression of TNF. An antibody to TNF or soluble receptors of TNF, given simultaneously with interleukin-1β prevented induction of proliferation. Thus, TNF was implicated as a promoter of epithelial ovarian cancer (Wu *et al.*, 1993). Malik and Balkwill reported that ovarian cancer may be a cytokine propelled disease (Malik and Balkwill, 1991), especially since TNF is present in the ovary.

Folicular Development The presence of immunoctive TNF in the ooplasm of the rat and mouse, coupled with the observation that TNF induced clustering of rat theca-interstitial cells and granulsoa cells *in vitro* provide basis for the hypothesis that TNF from the oocyte may be an organizing factor that participates in follicular development. Oocytic TNF, acting as a chemoattractant, may signal pregranulosa and theca cells to migrate and cluster around the oocyte. Currently, only granulosa cells have been shown to exhibit TNF receptor (Veldhuis *et al.*, 1991) (to our knowledge, no one has tested theca cells for specific TNF binding). Effects of TNF, in the range of 0.1–10 ng/ml, on theca-interstitial cells indicate that TNF receptors are present since these concentrations of TNF elicit physiological responses and are with the range of the Kd observed on the granulosa cells.

TNF from oocytes and/or macrophages may have a modulatory role in regulating steroidogenesis by granulosa and theca cells; this is evident since TNF inhibits gonadotropin-stimulated androgen production by theca and aromatase in granulosa cells (Fig. 1). In addition, TNF may regulate development of follicles. The inhibitory action of TNF must be abrogated to allow follicles to enter the growing pool. During early stages of follicular development, oocytic TNF would diffuse to nearby granulosa cells and thecal cells and modulate their functions as the follicle grows. It is possible that granulosa cells furthest away from the oocyte would escape TNF's modulatory actions since TNF would be in lower concentrations further away from the production center (oocyte). Cumulus granulosa cells, closest to the oocyte, exhibit reduced steroidogenic capacity when compared with mural granulosa (Hillensjo *et al.*, 1981a, b; Zoller and Weisz, 1979a, b). Thus, it is possible that TNF imparts the cumulus phenotype to these cells, i.e., low aromatase, low hCG(LH) binding and low steroidogenic potential. The function of the cumulus phenotype, as a result of oocytic TNF, might

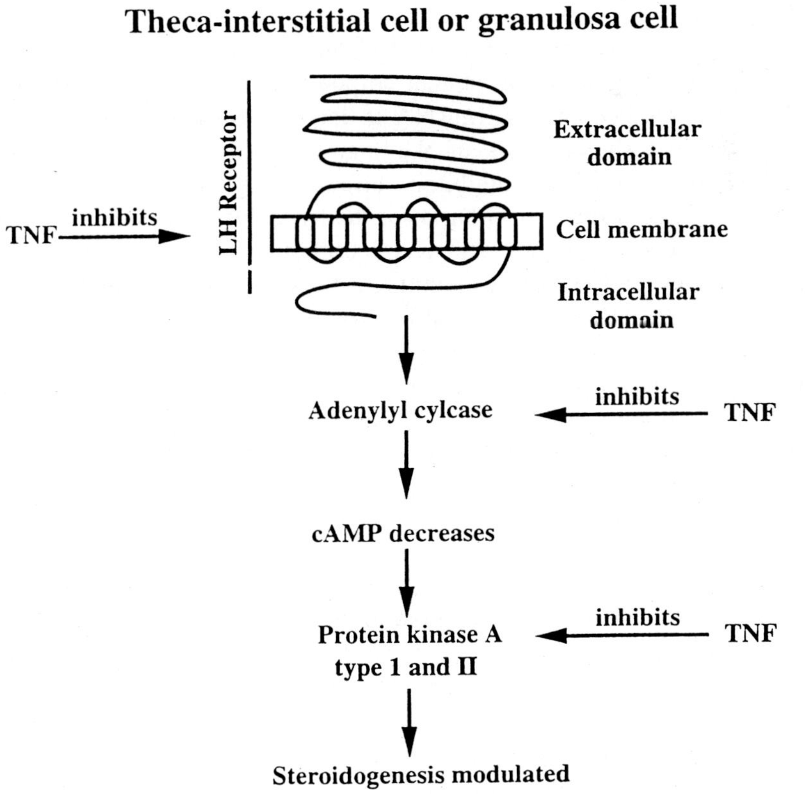

Fig. 1. Effects of TNFα on thecal-interstitial and granulosal cell steroidogenesis. TNF modulates steroidogenesis by inhibition at multiple levels in the signal transduction pathway.

be to protect the oocyte from elevated estrogen. A previous study has shown that delayed ovulation, which results in prolonged exposure of the oocyte to estrogen, increases anomalies associated with the oocyte (Butcher *et al.*, 1975). In fact, it was clearly shown that antisera to estradiol could prevent these developmental changes in the oocyte and the effects of the antisera could be reversed by diethylstilbestrol, a potent estrogen (Butcher and Pope, 1979).

TNF inhibits aromatase activity of granulosa cells and reduces estradiol that is necessary for follicular development in rats. In rats, estrogen and FSH synergize in regulation of follicular development by stimulating LH receptor formation and granulosa cell division (Richards, 1980). Thus, TNF during early stages of follicular growth could regulate follicle entry into the growing pool by modulating gonadotropin action. TNF would also retard follicles from producing androgen, which is known to be highly atretogenic. Exposure of the follicle to excessive TNF during its growth phase may also induce atresia by preventing further induction of aromatase (Adashi *et al.*, 1989; Emoto and Baird, 1988) and theca hydroxylase (Zachow *et al.*, 1994).

The presence of immunoreactive TNF in the large granulosa-lutein and small paraluteal cells of the human corpus luteum (Roby and Terranova, 1990), the presence of TNF mRNA in large luteal cells (Wuttke *et al.*, 1993) and endothelial cells of the pig, coupled with the observation that TNF has significant inhibitory effects on granulosal cell estradiol accumulation provides a firm basis for the hypothesis that luteal TNF may modulate follicular development and estradiol secretion during the luteal phase of the cycle. Recently, estradiol receptors have been located on human granulosal cells (Hurst *et al.*, 1995) providing them as potential targets of estradiol as observed in rats (Richards, 1980). It is well documented in rats that estradiol and FSH regulate follicular development by increasing the size of growing follicles and the number of LH receptors (Richards, 1980). Thus, during early stages of follicular growth TNF could regulate follicle entry into the growing pool by modulating FSH action on the secretion of estradiol.

Ovulation A recent study has determined the effects of TNF on ovulation rate, ovarian secretion of steroids, prostaglandins and on plasminogen activator in rat ovaries perfused *in vitro* for 20 h (Brannstrom *et al.*, 1995), Unstimulated ovaries did not ovulate whereas LH induced ovulations in all ovaries (~ 2.6 ova/ovary). TNF induced ovulations in 3 of 5 ovaries (~0.8 ova). However, when TNF was added simultaneously with LH, the ovulation rate was increased approximately 4 fold (10.6 ova). This effect was not observed when TNF was added 5 h after LH. Interestingly, more progesterone was released into the medium when TNF was combined with LH when compared with LH alone. Tissue concentrations of PGF2α and PGE were increased by the addition of TNF over controls (but not above LH or LH + TNF). Interestingly, plasminogen activator activity was unaffected by TNF and TNF inhibited the LH-induced increase in activity. The mechanism by which TNF promotes ovulation is unknown but possibly progesterone may be a secondary mediator.

Luteal Development Macrophages, paraluteal cells, endothelial cells and granulosa-luteal cells are potential sources of TNF in the corpus luteum. Macrophages are major sources of TNF in later stages of the luteal life span (Bagavandoss *et al.*, 1988; Bagavandoss *et al.*, 1990). TNF may have two roles dependent on the phase of luteal development. The first phase is the developmental phase of the corpus luteum and the other is regressive phase of the corpus luteum. Macrophages were originally proposed to be luteotropic since progesterone accumulation increased in media when they were in contact with mouse and human luteal cells *in vitro* (Halme *et al.*, 1985; Kirsch *et al.*, 1981; Kirsch *et al.*, 1983). Since macrophages contain TNF, it may be one of the factors that mediated the luteotropic effect. TNF increases total progesterone accumulation (Yan *et al.*, 1993) hCG binding (Terranova *et al.*, 1991a) and proliferation in human granulosa-luteal cells (Wang *et al.*, 1992; Yan *et al.*, 1993). All the latter increase during early luteal development. Therefore, it seems plausible to hypothesize that luteal macrophages and other sources of TNF may have a role in stimulating luteal development during the early phase.

Macrophages are present in the later phases of the life span of the corpus luteum (Bulmer, 1964) and the secretion of TNF by these macrophages could stimulate release of luteal prostaglandins (especially PGF2α) that induce luteal regression (Benyo and Pate, 1992). TNF might inhibit the responsiveness of the corpus luteum to LH. TNF inhibits the response of luteinized granulosa cells to gonadotropin *in vitro* (Adashi, 1990; Wang, *et al.*, 1992) and this effect of TNF is apparently mediated by inhibition of adenyl cyclase. Thus, it seems plausible to hypothesize that TNF could initiate luteal regression by inhibition of adenyl cyclase. TNF in corpora lutea can be released by LPS treatment *in vitro* (Bagavandoss *et al.*, 1988; Bagvandoss *et al.*, 1990). There is significant correlation between the number of macrophages in the corpora lutea and the amount of TNF released (Bagavandoss *et al.*,1988; Bagavandoss *et al.*, 1990). Physiologic regulation of TNF production by the corpus luteum has recently been shown in cows whereby TNF secretion increased during luteal regression (Shaw and Britt, 1995). TNF inhibits hCG(LH)-stimulated progesterone secretion *in vitro* in luteinized cells of rats, cows and humans (Adashi *et al.*, 1990; Benyo and Pate, 1992; Zachow and Terranova, 1994a, b). An increase in the number of TNF receptor or a sudden release of TNF might inhibit the luteotropic action of LH and induce luteal regression. It is well known that TNF stimulates the release of PGF2α *in vitro* (Benyo and Pate, 1992; Wang *et al.*, 1992; Zolti *et al.*, 1990) and that PGF2α is luteolytic in rats, cows and humans (Niswender and Nett, 1988). In sheep, changes in TNF mRNA were not detected in the corpus luteum during PGF2α–induced luteal regression. Nevertheless, TNF bioactivity was detected after progesterone secretion declined indicating that TNF may have a role in the later stages of luteal regression such as removal of degenerating luteal cells and debris (Ji *et al.*, 1991). The entry of TNF producing cells (leukocytes) into the corpus luteum (Ji *et al.*, 1991) may account for the increased secretion of TNF observed during luteal regression (Shaw and Britt, 1995).

Acknowledgements

The studies of the author were supported by a grant from the National Cancer Institute (CA 50616).

References

Adashi, E.Y., Resnick, C.E., Croft, C.S. and Payne, D.W. (1989). J. Biol. Chem. **264:** 11591–11597.

Adashi, E.Y., Resnick, C.E., Packman, J.N., Hurwitz, A. and Payne, D.W. (1990). Am. J. Obstet. Gynecol. **162:** 889–899.

Andreani, C.L., Payne, D.W., Packman, J.N., Resnick, C.E., Hurwitz, A. and Adashi, E.Y. (1991). J. Biol. Chem. **266:** 6761–6766.

Bagavandoss, P., Kunkel, S.L., Wiggins, R.C. and Keyes, P.L. (1988). Endocrinology. **122:** 1185–1187.

Bagavandoss, P., Wiggins, R.C., Kunkel, S.L., Remick, D.G. and Keyes, P.L. (1990). Biol. Reprod: **42:** 367–376.

Baker, T. and Reddy, E. (1996). Oncogene. **12:** 1–9.

Barak, V., Mordel, N., Holzer, H., Zajicek, G., Treves, A. and Laufer, N. (1992). Human Reproduction. **7:** 462–464.

Benyo, D.F. and Pate, J.L. (1992). Endocrinology. **130:** 854–860.

Beutler, B. and Cerami, A. (1989). Ann. Rev. Immunol. **7:** 625–655.

Beutler, B., Han, J., Kruys, V. and Giroir, B.P. (1992). Coordinate regulation of TNF biosynthesis at the levels of transcription and translation. In "Tumor Necrosis Factors: The Molecules and Their Emerging Role in Medicine" (B. Beutler), 561–574. Raven Press, New York.

Brannstrom, M., Bonello, N., Wang, L.J. and Norman, R. (1995). Reprod. Fertil. Dev. **7:** 67–73.

Brannstrom, M., Norman, R., Seamark, R. and Robertson, S. (1994). Biol. Reprod. **50:** 88–94.

Brannstrom, M., Wang, L. and Norman, R. (1993). Biol. Reprod. **48:** 165–171.

Braw, R. and Tsafriri, A. (1980). J. Reprod. Fertil. **59:** 259–268.

Bulmer, D. (1964). J. Anat. (London) **98:** 313–319.

Butcher, R.L., Collins, W.E. and Fugo, N.W. (1975). Endocrinology. **96:** 576–586.

Butcher, R.L. and Pope, R.S. (1979). Biology of Reproduction. **21:** 491–495.

Camussi, G., Albino, E., Tetta, C. and Bussolino, F. (1991). Eur. J. Biochem. **202:** 3–14.

Carswell, E.A., Old, L.J., Kassel, R.L., Green, S., Fiore, N. and Williamson, B. (1975). Proc. Natl. Acad. Sci. **72:** 3666–3670.

Chen, H.L., Marcinkiewicz, J.L., Sancho-Tello, M., Hunt, J.S. and Terrranova, P.F. (1993). Biol. Reprod. **43:** 707–714.

Clinton, T.K., Roby, K.F., Lyles, R. and Terranova, P.F. (1991). Tumor necrosis factor alpha and prolactin increase CG binding in human granulosa cells: differential interaction with hCG-stimulated progesterone secretion. In "Signaling Mechanisms and Gene Expression in the Ovary" (G. Gibori), 459–466. Springer-Verlag, New York.

Darbon, J.M., Oury, F., Laredo, J. and Bayard, F. (1989). Biochem. Biophys. Res. Commun. **163:** 1038–1046.

Emoto, N. and Baird, A. (1988). Biochem. Biophys. Res. Commun. **153:** 729–798.

Fernandez-Botran, R. (1991). FASEB J. **5:** 2567–2574.

Fukuoka, M., Yasuda, K., Emi, N., Fujiwara, H., Iwai, M., Takakura, K., Kazaki, H. and Mori, T. (1992a). Clin. Endocrinol. Metab. **75:** 254–258.

Fukuoka, M., Yasuda, K., Fujiwara, H., Kazaki, H. and Mori, T. (1992b). Hum. Reprod. **7:** 1361–1364.

Halme, J., Hammond, M.G., Syrop, C.H. and Talbert, L.M. (1985). J. Clin. Endocrinol. Metab. **61:** 912–916.

Hehnke-Vagnoni, K.E., Clark, C.L., Taylor, M.J. and Ford, S.P. (1995). Biol. Reprod. **53:** 1339–1444.

Hillensjo, T., Magnusson, C., Svensson, U. and Thelander, H. (1981a). Effect of LH and FSH on steroid synthesis by cultured rat cumulus cells. In "Dynamics of ovarian function" (N. Schwartz and M. Hunzicker-Dunn), 105–110. Raven Press, New York.

Hillensjo, T., Magnusson, C., Svensson, U. and Thelander, H. (1981b). Endocrinology. **108:** 1920–1924.

Hori, T., Kashiyama, S., Hayakawa, M., Shibamoto, S., Tsujimoto, M., Oku, N. and Ito, F. (1989). Exp. Cell Res. **185:** 41–49.

Hurst, B.S., Zilberstein, M., Chou J.Y., Litman, B., Stephens, J. and Leslie, K.K. (1995). J. Clin. Endocrinol. Metab. **80:** 229–232.

Jasper, M. and Norman, R. (1995). Hum. Reprod. **10:** 1352–1354.

Ji, I., Slaughter, R.G., Ellis, J.A., Ji, T.H. and Murdoch, W.J. (1991). Mol. Cell. Endocrinol. **81:** 77–80.

Kirsch, T., Friedman, A., Vogel, R. and Flickinger, G. (1981). Biol. Reprod. **25:** 629–632.

Kirsch, T., Vogel, R. and Flickinger, G. (1983). Endocrinology. **113:** 1910–1912.

Loy, R., Loukides, J. and Polan, M. (1992). Fertil. Stertil. **58:** 733–730

Mackay, R., Loetscher, H., Stueber, D., Gehr, G. and Lesslauer, W. (1993). J. Exp. Med. **177:** 1277–1286.

Malik, S. and Balkwill, F. (1991). Br. J. Cancer. **64:** 617–620.

Marcinkiewicz, J., Krishna, A., Cheung, C. and Terranova, P. (1994). Biol. Reprod. **50:** 1251–1260.

Mutch, D.G., Massad, L.S., Kao, M.-S. and Collins, J.L. (1990). Am. J. Obstet. Gynecol. **163:** 1920–1924.

Mutch, D.G., Powell, C.B., Kao, M.-S. and Collins, J.L. (1992). Cancer Res. **52:** 866–872.

Naylor, M.S., Stamp, G.W.H., Foulkes, W.D., Eccles, D. and Balkwill, F.R. (1993). J. Clin. Invest. **91:** 2194–2206.

Niswender, G.D. and Nett, T.M. (1988). The corpus luteum and its control. In "The Physiology of Reproduction"(E. Knobil and J. D. Neill), 489–525. Raven Press, New York.

Pfizenmaier, K., Himmler, A., Schutze, S., Scheurich, P. and Kronke, M. (1992). TNF receptors and TNF signal transduction. In "Tumor Necrosis Factors: The Molecules and Their Emerging Role in Medicine" (E.B. Beutler), 439–472. Raven Press, New York.

Pitzel, L., Jarry, H. and Wuttke, W. (1993). Endocrinology. **132:** 751–756.

Punnonen, J., Heinonen, P.K., Teisala, K., Kujansuu, E., Jansen, C.T. and Punnonen, R. (1992). Gynecol. Obstet. Invest. **33:** 80–84.

Richards, J. (1980). Physiol. Rev. **60:** 51–89.

Roby, K.F. and Terranova, P.F. (1989). Localization of tumor necrosis factor (TNF) in rat and bovine ovary using immunocytochemistry and cell blot: Evidence for granulosal production. In "Growth Factors and the Ovary" (A. N. Hirshfield), 273–278. Plenum Press, New York.

Roby, K.F. and Terranova, P.F. (1988). Endocrinology. **123:** 2952–2954.

Roby, K.F. and Terranova, P.F. (1990). Endocrinology. **126:** 2711–2718.

Roby, K.F., Weed, J., Lyles, R. and Terranova, P.F. (1990). J. Clin. Endocrinol. Metab. **71:** 1096–1102.

Sancho-Tello, M., Chen, T.-Y., Clinton, T.K., Lyles, R., Moreno, R.F., Tilzer, L., Imakawa, K. and Terranova, P.F. (1992). J. Endocr. **135:** 571–578.

Sancho-Tello, M., Pérez-Roger, I., Imakawa, K., Tilzer, L. and Terranova, P.F. (1992). Endocrinology. **130:** 1359–1364.

Sancho-Tello, M., Tash, J.S., Roby, K.F. and Terranova, P.F. (1993). Endocrine J. **1:** 503–512.

Selmaj, K., Farooq, M., Norton, W., Raine, C. and Brosnan, C. (1990). J. Biol. Chem. **144:** 129–135.

Shakil, T. and Whitehead, S. (1994). Biol. Reprod. **50:** 1183–1189.

Shaw, D. and Britt, J. (1995). Biol. Reprod. **53:** 847–854.

Sugarman, B.J., Aggarwal, B.B., Hass, P.E., Figari, I.S., Palladino, M.A. and Shepard, H.M. (1985). Science. **230:** 943–945.

Tartaglia, L.A. and Goeddel, D.V. (1992). Immunology Today. **13:** 151–153.

Taylor, C. and Terranova, P.F. (1995). Endocrinology. **136:** 5527–5532.

Tekpety, F., Engelhardt, H. and Armstrong, D. (1993). Biology of Reproduction. **48:** 936–943.

Terranova, P.F., Hunter, V., Roby, K. and Hunt, J. (1995). Proc. Soc. Exp. Biol. Med. **209:** 325–342.

Terranova, P.F., Roby, K.F., Sancho-Tello, M., Weed, J. and Lyles, R. (1991a). Tumor necrosis factor alpha: localization and actions within the preovulatory follicle. In "Growth Factors in Reproduction" (D. Schomberg), 63–78. Springer-Verlag, New York.

Terranova, P.F., Roby, K.F., Sancho-Tello, M., Weed, J. and Lyles, R. (1991b). Tumor necrosis facetor alters thecal and granulosal steroidogenesis. In "Signalling Mechanisms and Gene Expression in the Ovary" (G. Gibori), 178–189. Serono Press.

Vassalli, P. (1992). Annu. Rev. Immunol. **10:** 411–452.

Veldhuis, J.D., Garmey, J.C., Urban, R.J., Demers, L.M. and Aggarwal, B.B. (1991). Endocrinology. **129:** 641–648.

Vilcek, J., Palombella, V.J., Henryksen-DeStafano, D., Swenson, C., Feinman, R., Hirai, M. and Tsujimoto, M. (1986). J. Exp. Med. **163:** 632–643.

Voogt, J., Sar, M. and Meites, J. (1969). Am. J. Physiol. **216:** 655–658.

Wang H.Z., Sheng, W.X., Lu, S.H., Sun, Z.D., Han, X.J., Gong, Y.T. and Zhou, W. (1992a). Fertil. Steril. **58:** 941–945.

Wang, L.J., Brannstrom, M., Robertson, S.A. and Norman, R.J. (1992b). Fert. Steril. **58:** 934–940.

Wu, S., Boyer, C.M., Whitaker, R.S., Berchuck, A., Wiener, J.R., Weinberg, J.B. and Bast Jr, R.C. (1993). Cancer Res. **53:** 1939–1944.

Wu, S., Rodabaugh, K., Martinez-Maza, O., Watson, J., Silberstein, D., Boyer, C., Peters, W., Weinberg, J., Berek, J. and Bast, R. (1992). Am J. Obstet Gynecol. **166:** 997–1007.

Wuttke, W., Jarry, H., Pitzel, L., Knoke, I. and Spies, S. (1993). Hum. Reprod. **8:** 141–146.

Yan, Z., Hunter, V., Weed, J., Hutchison, S., Lyles, R. and Terranova, P.F. (1993). Fert. Steril. **53:** 332–338.

Zachow, R., Tash, J. and Terranova, P. (1993). Endocrinology. **133:** 2269–2276.

Zachow, R. and Terranova, P.F. (1993). Mol. Cell. Endocrinol. **97:** 37–49.

Zachow, R. and Terranova, P.F. (1994a). Endocrine. **2:** 1145–1150.
Zachow, R. and Terranova, P.F. (1994b). Endocrine. **2:** 625–631.
Zachow, R.J., Tash, J.S. and Terranova, P.F.(1992). Endocrinology. **131:** 2503–2513.
Zoller, L. and Weisz, J. (1979a). Endocrinology. **103:** 310–313.
Zoller, L. and Weisz, J. (1979b). Histochemistry. **62:** 125–135.
Zolti, M., Ben-Rafacl, Z., Meirom, R., Shemesh, M., Bider, D., Mashiach, S. and Apte, R.N. (1991). Fertil. Steril. **56:** 265–272.
Zolti, M., Bider, D., Seidman, D.S., Mashiach, S. and Ben-Rafael, Z. (1992). Fertil. Steril. **57**: 501–504.
Zolti, M., Meirom, R., Shemesh, M., Wollach, D., Mashiach, S., Shore, L. and Ben Rafael, Z. (1990). FEBS LETTERS. **261:** 253–255.

Comparative Endocrinology and Reproduction
K.P. Joy, A. Krishna and C. Haldar (Eds)

21. The Biology of Blastocyst Implantation

Jayasree Sengupta and Debabrata Ghosh
Department of Physiology, All India Institute of Medical Sciences,
New Delhi 110 029, India

Introduction

After fertilization in the ampullary region of the oviduct, the mammalian embryo is transported into the uterus. The timing of this event is highly species-specific. In most marsupials this occurs by 24 hours and in some carnivore species it may take one week. In most other mammalian species studied so far the transit time of embryo from oviduct to uterus takes about 2 to 4 days. In rodents it is essential that embryos enter into the uterine environment at post-morula stage of development because the uterine milieu does not support pre-morula stage growth in this species. In primates the uterine environment is not hostile to early stage embryos, since uterine transfer of one- or two-cell embryos leads to successful implantation and pregnancy in both human and non-human species. In polytocous species where more than one egg is ovulated from each ovary, the migration and spacing of embryos in the uterus appear important for fetal survivability. Uterine muscular action is important for equidistant spacing of embryos in the uterus to prevent overcrowding and embryonic death. Marula to blastocyst transition is an important landmark in embryonic life as it initiates for the first time overtly distinctive differentiation of embryonic cells. This is followed by blastocyst implantation and it represents a critical step in the establishment of pregnancy in mammalian species. At implantation an embryo assumes a fixed relationship in the uterus and establishes a close morphological and functional relationship with the mother . This review summarizes the current understanding of embryo development and differentiation, embryonic and endocrine synchrony for implantation, endometrial receptivity, signals required for implantation, types of implantation, embryo-uterus interaction and decidualization.

Embryo Development and Differentiation

The mammalian embryo is small compared to other vertebrate species. In all eutherian mammals and in marsupials it is about 0.1 mm in diameter. It is surrounded by a thin acellular membrane envelope, the zona pellucida. In most mammalian species, implantation of the embryo occurs at the blastocyst stage of development. However, there are exceptions. In ungulates such as horse, deer and antelopes where the embryo resides in the reproductive tract in the free, non-attached state for a prolonged period of time, there is considerable increase in the size of embryo and differentiation of the conceptus progresses to chorionic vesicle stage when it implants in the uterus. On the other hand, in the elephant shrew, at the 4-cell pre-blastocyst stage implantation is initiated. In anthropoidea species and in guinea-pig, the blastocyst is unilaminar at implantation; bilaminar in rabbit, insectivores and in several prosimian species. Blastocyst is at an intermediate stage of development in mouse, rat and some bats. Another feature of difference in embryonic development prior to implantation is the outer covering of the embryo. In all mammalian embryos the zona pellucida forms the outer covering from the oocyte stage onwards. In the rabbit an additional thick mucoidal layer is deposited on the zona as the embryo traverses down the oviduct and uterus. In the marsupial, outside an albuminous layer a dense shell membrane is deposited as the embryo passes through the tract.

The time taken for fertilization and the initial cleavage stages of development are remarkably slower in mammals than in other species. In the mouse, the time taken for sperm-egg fusion and second polar body extrusion into perivitelline space is about 1 to 2 hours after insemination, and pronuclei formation occurs 6 hours after insemination. Cleavage of *Drosophila* embryos on the other hand occur at intervals of 10 minutes and amphibian embryos cleave at 30 minutes interval. Mouse embryos undergo their first and second cleavages at 14 to 26 hours intervals *in vivo.* Embryo development in culture is slightly retarded and the cell cycles proceed more slowly *in vitro* than *in vivo,* so that the time required for blastocyst formation *in vitro* may be longer by at least one day than that *in vivo.* It has been suggested that the timing of events during sperm entry, fertilization, cell cleavages are intrinsically regulated by an inbuilt 'clock' which is dictated by several cellular factors regulating 'housekeeping' activities. The involvement of maturation promotion factor, cell calcium level, membrane phospholipids, phosphatidyl inositol and diacylglycerol, cellular pH, activity of oncogene related protein kinases are considered to play significant roles in the regulation of cell cycle.

The development of one-cell fertilized embryo to a blastocyst involves two distinctive processes: (i) a proliferative process by which cell numbers increase to about 100 to 120 cells, and (ii) a differentiative process by which phenotypic diversity is attained in genotypically similar cells. Intrinsic and extrinsic factors are necessary for supporting preimplantation embryo development and differentiation. In order to understand cellular differentiation during preimplantation embryo development, studies have been performed to delineate biochemical, structural, physiological and behavioural characteristics of embryonic cells. Studies aimed to understand mammalian embryogenesis during early pregnancy have been performed mainly using mouse preimplantation stage embryos. However, studies have also been performed using embryos from few non-human primate species such as rhesus monkey, baboon, marmoset and also human embryos made available to investigators as 'spare embryos' from *in vitro* fertilization (IVF) programmes.

The development of preimplantation stage embryo involves two distinctive differentiative events. (i) At the 8-cell stage of development mouse embryos undergo a process of compaction. This occurs with a change in cell shape. Cells assume columnar shape from the existing spherical shape, and they attain polarity with microtubules seen on apical cell surfaces and rearrangement of the internal cell organelles. Polarization is evident in the formation of permanent microvilli on outward surfaces of trophoblast cells, labile microvilli, membranal carbohydrates, alkaline phosphatase, and 5′–nucleotidase in areas of cell contact. Endosomes and polysomal endocytotic vesicles become distributed apically in blastomeres at 8-cell stage in mouse embryos. (ii) The second major differentiative event in the preimplantation embryo is the establishment of two distinctive cell lineages, that of inner cell mass and trophectoderm. There are both temporal and spatial coordinates to this event and it occurs in the post-compaction period. The cells located outside recognize their spatial position and develop trophectodermal lineage and cells located inside differentiate into inner cell mass (ICM) cells. This early morphogenetic event is marked by junctional differentiation in which gap junctions and incipient tight junctions are formed between outside blastomeres, and thus the process of epithelial differentiation is initiated. Membrane proteins and various antigens are formed on apical cell surfaces of differentiating trophoblast cells. Polarization is maintained in these outside cells and they lose their wide developmental potential. In contrast ICM cells remain non-polarised and totipotent. Interestingly, in metatherian species such as marsupials, embryonic compaction is not observed and thus there is no inner cell mass. The blastocyst appears as a hollow ball of cells with no apparent distinction.

Experimental manipulations using combinations of genetically or physically marked cleavage stage blastomeres have been used to form aggregation chimeras and such embryos provide evidence to support the positional hypothesis in the initiation of cellular differentiation in the preimplantation embryo. Such early differentiation is however not irreversible since cells of ICM retain until the expanding blastocyst stage the capacity to form trophectoderm; in chimera formed on aggregation with morulae, the inner cells contribute to both embryonic and trophoectodermal components of the conceptus. Totipotency is a feature of early stages of preimplantation embryo development. Two- and four-cell embryo blastomeres are each capable of developing a full individual after their injection into an oocyte. Separated blastomeres of 2-cell mouse embryo can develop into identical twins. In sheep and in rabit the time period during which cells remain totipotent is the 8-cell stage. As the outside cells become polarised in the 8-16 cell stage, the embryo no longer remains totipotent. Human embryos can retain their totipotency till the 8-cell stage as seen in one case where 7 out of 8 blastomeres failed to survive the cryopreservation procedure but the embryo after its transfer into the uterus gave rise to a normal child. A single blastomere injected into a recipient embryo can, however, colonize all tissues of the chimera.

Initial differentiation of outside cells to form trophectoderm and inside cells the inner cell mass of the embryo leads to another distinctive morphogenetic event in mammalian development: the formation of blastocyst cavity. This begins with differential fluid transport mechanism across trophectodermal cells and with the expression of Na^+, K^+-ATPase observed in these cells from the morula stage of development. Mural trophectoderm surrounds the blastocyst and polar trophectoderm overlies the inner cell mass. The cell number at the time of blastocyst formation varies among different species. In the hamster it occurs after 16-cell stage, the mouse at 32-cell stage, in the sheep it is 64-cell stage, and 128-cell stage in the rabbit embryo. In *Elephantulus* embryos, blastocoelic fluid accumulates as early as the 4-cell stage. As one trophectoderm cell divides it can form two daughter cells which are also similarly polarised; or it can also form one polarised and one non-polarised cell. The non-polarised cell can contribute to forming the inner cell mass. Thus, cells of the inner cell mass are formed during compaction and later as contributions from the outer cells. Embryogenesis during the preimplantation stages involves the expression of a temporal programme within cells and a positional programme of interaction between cells. The directional control executed by positional cues most likely operates by activation and/or repression of distinct subset of genes. Tissue-specific synthesis of polypeptides by ICM and trophectodermal cells have been detected in the early morula. Positional disparity has also been detected in radiation sensitivity of cells, ultrastructural characteristics and cellular enzymic patterns.

The preimplantation stage is a vulnerable period for embryo growth and differentiation. In the human and also in non-human primate species, significant proportion of embryos grow abnormally during this periods, and this could arise from inadequate follicular environment and from intrinsic defects in the embryo. These defects usually occur due to inherited disorders affecting chromosomal constitution or cellular metabolism. There are two sources of genetic information in embryos: (i) maternal mRNAs inherited from oocyte, and (ii) embryonic mRNAs which are synthesized after fertilization. Maternal poly (A+) mRNAs encode for many factors including zona glycoproteins, *mos* proteins, products of oncogene c-*kit,* several enzymes, and connexon (a precursor protein of junctional complexes). In addition, several maternal proteins are degraded after fertilization. The embryonic genome is activated in 4- to 8- cell human embryos and by the 2-cell stage in mouse and rabbit embryos. A 70 kDa heat shock protein mRNA is one of the earliest protein synthesized by 2-cell mouse embryo encoded by embryonic mRNA. The activation of embryonic genes also results in the expression of paternal characteristics in embryos which include the paternal variants of $\beta 2$ microglobulin and other proteins. It results in the synthesis of transcripts regulating blastocyst formation. βhCG (human chorionic gonadotrophin) transcripts in 8-cell human embryos are considered to be encoded embryonically although hCG expression occurs at the blastocyst stage. Genes located on the X-chromosomes are regulated in a different manner compared to autosomal genes. Each X-chromosome is active in female embryos between early cleavage and morulae, and then one X-chromosome is inactivated. The paternal X-chromosome is inactivated in cells of trophectoderm and primitive endoderm in mouse embryos, and in proximal endoderm and extra-embryonic ectoderm in human embryos. In embryonic cell lineages, there is random inactivation of paternal and maternal derived X-chromosomes. The inactive X-chromosome is reactivated only in oogonia. The rate of growth of mammalian embryos is influenced by their content of sex chromosomes. There is evidence that male embryos grow more rapidly than female embryos in several species. This may be due to transcription from the sex determining genes *Sry* and *Zfy*. A group of regulatory genes, oncogenes, direct embryonic differentiation. Many of these genes contain a highly conserved DNA homeobox sequence of 60 amino acids referred to as the homeobox domain. They code for proteins and polypeptides binding to DNA which can activate or repress other genes such as *oct*-3 gene which functions in early embryogenesis in regulating totipotency or multipotency in embryonic cells.

Studies on embryo metabolism have been mostly performed using mouse embryos. Mammalian embryo contains glycogen stored in oocytes and it obtains energy from its surrounding milieu of the oviduct and the uterus. All mammalian embryos require continuing energy supply from the maternal environment. There is little increase in the rate of metabolism in early cleavage stage embryos but between morula and blastocyst stages it increases sharply as indicated from studies from their oxygen uptake and carbon dioxide output. At this time there is a shift from the hexose monophosphate shunt pathway to the Embden-Meyerhof pathway and the tricarboxylic acid cycle. Such biochemical differentiation can be associated with an alteration in the ultrastructure of mitochondria. At 1- and 2-cell stage of development mitochondria are round in shape with very few cristae, and 8-cell stage onwards mitochondria assume a more adult-like shape with abundant cristae. Pyruvate is a major energy source in the early stages of development and in some species lactate is also used. Glucose metabolism and glycolytic pathway become active at later stages of cleavage. The transport of glucose in cleaving mouse embryos increase rapidly and is facilitated by GLUT1-a member of the sodium independent glucose transport system. Lipids such as phospholipids and cholesterol are synthesized in early mouse embryos and they are most likely required for extensive membrane reorganization and surface changes associated with compaction and cellular defferentiation. Mammalian embryos differ markedly from other embryos such as those of amphibians, teleost fish, echinoderm, insects in not only requiring an external energy source, but in the early onset of RNA synthesis. RNA synthesis is initiated at the 2-cell stage in mouse embryos and at 4- to 8-cell stage in human embryos. Protein synthesis increases markedly in early embryos in morula and blastocyst stages of development. Maternal proteins such as oviduct secreted glycoproteins enter into the perivitelline space of embryos and some maternal proteins have also been shown to be incorporated into embryonic cells by micropinocytotic vesicles. Indeed such information and that the ovidutal fluid contains high levels of pyruvate, lactate and chloride ions which favour early embryo development may explain the higher growth rates observed when embryos are co-cultured with oviductal epithelial cells *in vitro.* Preimplantation embryos can synthesize reactive oxygen species which may be harmful to embryo growth *in vitro.* The H_2O_2 formed in embryos growing *in vivo* may result in peroxidation of lipid and protein. Catalases and peroxidases such as glutathione peroxidase can convert H_2O_2 to H_2O in the presence of reduced glutathione. Embryos growing in culture medium can be protected against the production of excess amounts of reactive oxygen species by removing iron and copper from the medium, and by adding metal chelators to prevent the production of highly toxic hydroxyl radicals, and by avoiding excess illumination and high oxygen tension.

Embryonic cells respond to several growth factors and cytokines which in actuality may control many aspects of embryonic growth. Growth factors can be derived from embryonic cells and thus function in autocrine loop, or they may be available from the maternal environment. Genes for insulin like growth factors (IGFs), transforming growth factors (TGFs), fibroblast growth factor (FGF), platelet-derived growth factor (PDGF) and receptors for insulin, IGF, PDGF, TGFα and epidermal growth factor (EGF) are expressed by early embryos of a number of species including mouse, rat, cow and sheep. IGF-I and IGF-II mediate growth in early mouse embryos and increase cell numbers in blastocysts. EGF stimulates growth in 8-cell embryo and it also promotes the synthesis of hCG by syncytiotrophoblast cells.

Factors in Implantation

Blastocyst

The organization of blastocyst is characterized by: (i) an outer layer of trophectodemal cells which will give rise to ectoderm of fetal membranes, (ii) an eccentrically placed inner cell mass of cells (ICM) which will give rise to the embryo proper, and (iii) a fluid filled space of blastocoel which will give rise to the yolk sac activity in whole or in part.

Blastocyst undergoes a series of changes as it approaches implantation. In nearly most mammalian species the blastocyst must shed its zona pellucida before it can attach to uterine epithelium. In the mouse it has been shown that an enzyme, strypsin is synthesized in mural trophectoderm and it is responsible for digesting the zona surface. In the human and in the non-human mammals zona hatching has been observed *in vitro.* A small hole or slit first appears in the zona at one end and through it the embryo is seen to squeeze out. At this time the zona usually becomes very thin and the rupture of zona could also be assisted by increasing pressure from the expanding blastocyst. Time-lapse photography of mouse and rabbit blastocysts have shown that at this stage blastocyst undergoes a series of periodic contractions and this process, if it also occurs *in vivo,* may help in zona hatching. The hatched blastocyst initiates embryonic differentiation in preparation for implantation. Hatched human blastocysts double in volume and elongate by 2 to 3 days after hatching. The ICM differentiates to form the embryonic disc surrounded by trophectoderm. The extra-embryonic endoderm differentiaties and migrates around the blastocoelic cavity. Trophoblast cells develop prominent microvilli, large cytotrophoblast cells can be distinguished and intercellular junctions develop between trophoblast cells and glycogen is found to be synthesized by endoderm.

In several mammalian species delayed implantation is seen as a unique feature of early pregnancy. As early as mid-nineteenth century, German anatomists recognized that in the roe-deer blastocysts can remain in a state of delay or diapause prior to their implantation in the uterus. This was recognized from the hiatus between matings which occurred in August and the first overt signs of pregnancy being established only in late December. It was then discovered that in July or August ovulation and fertilization occur and the ovum undergoes cleavage stages of its development to reach to morula stage as it enters the uterus. It then develops into blastocyst and sheds its zona pellucida and then enters into a 4 to 5 month period of dormancy which is terminated in late December with activation of the blastocyst. Traphoblast cells show rapid elongation as they interact with uterine cells in initiating implantation. Delayed implantation has been observed in at least eight other mammalian orders. These are *Marsupilia, Chiroptera, Edentata, Rodentia, Carnivora, Insectivora* and *Pinnipedia.* Two forms of delayed implantation have now been documented: (i) facultative or lactational and (ii) obligatory or seasonal. In facultative delay, blastocyst undergoes a short or variable period of delay of implantation following ovulation and fertilization in the post-partum period while the mother suckles her litter. Such type of delay has been seen in a number of species which include rat, mouse, bank vole, Mongolian gerbil, Indian gerbil, several macropod marsupials. In the obligatory delay of implantation as seen in roe-deer, mink, badger the period of delay is variable. It can extend even from 4 to 10 months. The obligatory form of delay of implantation allows the animal to have mating and birth of young at times favourable to their survival. While in facultative delay, blastocyst diapause helps to enhance fecundity by ensuring rapid replacement of litter.

Delayed implantation can also be experimentally induced in species such as rat and mouse by performing ovariectomy, and embryos can be maintained in a state of diapause with progesterone administration. Embryonic diapause and delay of implantation in primate species has not been documented, however, there are few case reports in which delay of few days in the onset of hCG secretion and in the time of onset of implantation in the women were observed in IVF-ET laboratories.

As blastocysts enter into diapause, they exhibit certain changes which include a gradual cessation of cell division and a state of metabolic inactivation. Protein synthesis, RNA and DNA synthesis declines, and there is a decrease in carbon dioxide ouput. In species showing facultative delay, diapausing blastocysts remain in zona-shed condition, however, in few species with obligatory delay, the zona remains as in macropod marsupials, fur seals develop an additional sub-zonal layer which appears to be deposited by the blastocyst itself. In roe-deer and armadillo which also experience obligatory delay, blastocysts however shed their zonae. Termination of diapause is associated with metabolic reactivation of blastocyst with the resumption of mitotic activity, increases in RNA, DNA and protein syntheses, and an increase in the size of the blastocyst.

Uterus

Embryo must reach to blastocyst stage for implantation to occur. Uterus must also be synchronous to blastocyst for initiation of implantation. Thus, both partners must be in a state of developmental maturation and differentiation to allow for their mutual interaction at the time of implantation. The notion of endometrial receptivity arose with embryo transfer experiments initially performed using rabbit embryos. It was observed that a chronological relationship exists between the age of embryo and developmental status of endometrium, which becomes a determinant factor in the establishment of implantation and pregnancy. Endometrium undergoing such developmental changes is considered to be in the phase of receptivity. In species such as rat, mouse, hamster and rabbit it has been shown that there is a definite 'window' or time period of uterine receptivity during which blastocyst can implant, beyond this time period the uterus becomes non-receptive or refractory to implantation. The time frame of this window is narrow in case of rodents and the small laboratory mammals and is considered to be around 24 to 48 hours in rodents, while in primates the window is significantly wider and can span 5 to 7 days.

What is the nature of receptive endometrium? A common universal feature seen in all mammalian species is increased capillary permeability and decreased vascular impedance. In polytocous species such as rodents the entire uterine horn bearing preimplantation stage blastocysts appears hyperemic. A few hours preceding the implantation reaction further increase in local capillary permeability around the implanting blastocyst is observed, which permits location of the implantation site.

There exists considerable species variation in the nature of endometrial responses as it prepares for implantation. A significant degree of our understanding is based on studies performed in laboratory mammalian species. Characteristic and well defined morphological and biochemical changes are seen in epithelial and stromal compartments. Coincidental with such cellular changes are the vascular changes to meet the increased metabolic needs of the tissue. The cellular changes involve cell proliferation followed by differentiation, and if implantation fails to occur such differentiated cells undergo regression and cell death. In menstruating primates species where the duration of endometrial cycle is longer than the short estrous cycles seen in non-primate species, cellular proliferation dominates during the preovulatory stage and during the post-ovulatory period endometrial cellular differentiation occurs. If there is no implantation endometrial regression is associated with shedding of the upper superficial layer of endometrial lining wall associated with menstrual bleeding.

In rats and mice the pattern of endometrial cell proliferation in the post-ovulatory stage in preparation for implantation is well characterized: an initial wave of epithelial mitosis in luminal and glandular epithelial compartments and this is followed by a wave of stromal mitosis (Fig. 1). Epithelial mitosis is not prominent in glands but in luminal cells, and it is initiated two days prior to the initiation of implantation reaction. Epithelial mitosis ceases within two days and by this time stromal mitosis is initiated and within 24 hours implantation is initiated if a blastocyst is present in the uterine lumen. A characteristic feature of preimplantation stage rodent uterus is the closure of the uterine lumen so that it exhibits a slit-like appearance. With cessation of proliferation, cells begin to differentiate. Apical surfaces of luminal epithelium show bleb-like protrusions and these are responsible for the removal of uterine luminal fluid by pinocytosis and closer apposition of the uterine surfaces leading to uterine closure. Epithelial cells show typical apical vesicles and basally situated lipid droplets. At the time of embryo-uterus interaction a second stage of closure is evident by closer apposition of apical microvilli which then become short, stubby and more irregular in appearance as they intertwin with each other at the time of embryo attachment reaction. In rodents, glandular differentiation is not marked till the attachment phase when the cells begin to show secretory functions. Stromal differentiation in laboratory mammalian species occurs primarily in the post-attachment phase of implantation reaction.

As endometrium prepares for implantation metabolic parameters show changes, which correlate with the cell proliferation and differentiation profiles. Biphasic patterns of increases in RNA are observed in uterine cells, peak when luminal and stromal mitosis also exhibit maximal proliferation during the preimplantation stage. Thus the first peak is seen on the third day and the second peak on the fifth day of pregnancy in mice and rats. Ornithine decarboxylase activity increases coincident to increased RNA synthesis. The pattern of uterine protein synthesis is also biphasic in nature in the peri-implantation period. Increased rate of protein synthesis occurs at the time of maximal epithelial mitosis and then decline to increase two days later, just prior to implantation. Increased protein, RNA and DNA syntheses continue at implantation site but not in inter-implantation region of uterine horn. Interestingly the patern of RNA synthesis is greater in pregnant horns than in pseudopregnant horns of uteri even prior to the time of the attachment reaction.

Furthermore, the pattern of protein synthesis in pseudopregnant horn is different from that in the pregnant horn, in its failure to elicit a second rise in uterine protein synthesis on the fifth day of pregnancy. This supports the suggestion that embryos can stimulate uterine metabolism even prior to their attachment to uterus. Preimplantation stage endometrial receptivity as seen from the pattern of cellular and biochemical changes appear to be influenced by maternal hormonal signals and chemical signals of embryonic origin.

The relative contribution of maternal hormonal factors and embryonic factors in the process of pre-and peri-implantation endometrial synchronization can be studied by using an experimental pseudopregnancy animal model. Pseudopregnancy is a condition in which, through hormonal stimulation, uterus undergoes morphologic and biochemical changes in preparation for implantation even in the absence of fertilized ova in uterine lumen. In pseudopregnant uteri, implantation associated changes can be induced by providing an artificial stimulus at the time when implantation would normally have occurred. The nature of the artificial stimulus can vary. Placement of an air bubble or a silk thread into the uterine lumen, or scratching the luminal surface with a needle can be effective in eliciting physical interaction with luminal epithelial cells for decidualization. The nature of hormonal priming required for rendering uterus receptive to the artificial stimulus is species-specific, and in any one species it is similar to the hormonal needs as seen for normal implantation. If to a hormonally primed uterus, artificial stimulation is not provided at, or around the time of expected implantation, the uterus reverts back to its normal cyclical physiology.

Our understanding of endometrial receptivity for implantation in primate species is highly limited. In primates, uterine endometrium thickens during the proliferative phase with typical profiles of cell proliferation in epithelial and stromal cell compartments. During the post-ovulatory stage, stromal cell mitosis continues and stromal fibroblast cells adjoining spiral arterioles differentiate to form predecidual cells. There is stromal edema, and glandular differentiation is induced with increased glandular secretory functions.

Luminal epithelium expresses microvilli with apical protrusions known as pinopodes. Expression of apical pinopodes has been suggested to be highest during the mid-luteal stage at a time when implantation generally is expected to occur in the presence of fertilized ovum, that is between day 20 and day 24 of the menstrual cycle. There are extensive biochemical changes in epithelial cells during the secretory phase of the cycle. These cells secrete electronegative periodic acid-schiff staining secretory product, a layer of colloidal iron, fucose and mannosyl residues have been observed. Histological examination of endometrial biopsy tissue samples collected during the menstrual cycle has led to chronological staging of endometrial development. The cellular and extracellular features examined include epithelial and stromal cell mitosis, glandular tortuoisity, nature and location of vacuoles in epithelial cells, degree of glandular secretion, tissue edema, degree of stromal infiltration by leucocytes, presence of spiral arterioles and predecidual cells. The typical sequence of such changes in normal ovulatory menstrual cycle in the human is shown in Fig. 2.

Morphologic and biochemical distinctions between post-ovulatory stages of non-fecund and fecund cycles have been documented in rhesus monkey. Fecundity is established from the presence of stage and age-matched preimplanation embryo recovered from the reproductive trace. Morphological distinctions during preimplantation stage of gestation as compared to post-ovulatory stage endometrium from non-fecund cycles, include increased epithelial cell mitosis, changes in glandular parameters which commensurate with enhanced glandular functions, greater degree of stromal edema and increased capillary permeability. Metabolic changes in preimplantation stage rhesus endometrium have also been noted with higher retention of nuclear estrogen and progesterone receptors, high level of alkaline phosphatase and lower levels of lysosomal enzymes. These findings suggest that diffusible signals are most likely released by preimplantation stage embryo to endometrium in preparation for implantation. Embryonic signals will be discussed later.

Endocrine and Associated Factors

Significant species variations exist in the endocrine requirements for hormonal conditioning of the uterus for implantation. Bilateral overiectomy of mammals within a few days after fertilization results in the termination of pregnancy. The maintenance of corpora luteal function has been implicated as a primary factor responsible for pregnancy maintenance. The armadillo is an exception in that diapausing blastocysts are activated for implantation by the removal of both ovaries.

The maternal endocrine profiles during ovulation and the post-ovulatory stage of early pregnancy are well documented in several species. The ovulatory estrogen allows for endometrial cell proliferation. Progesterone subsequently produced by the newly formed corpus luteum renders endometrial receptivity. Estrogen has a priming action and it affects the timing and extent of stromal celi proliferation. Estrogen is obligatory for inducing implantation in species such as rat and mouse. Therefore, in these two species at least, it is now well documented that ovulatory estrogen as well as nidatory estrogen secreted during the preimplantation state are both essential for inducing endometrial receptivity for implantation. Figure 1 shows the plasma profiles of estrogen, progesterone and uterine receptors for these hormones during the pre- and peri-implantation stages of gestation. Minute doses of estrogen are required to terminate the delay state and induce implantation in either lactational delay or in experimentally induced delay of implantation. High doses of estrogen however cause inhibition of pregnancy by expelling embryos out of the reproductive tract through increasing uterine motility. Blastocysts maintained on progesterone alone will only implant if estrogen is provided to them exogenously. During lactation, secretion of gonadotrophins are inhibited and this reduces the estrogen level inducing thereby the delayed state. Induction of implantation, however, occurs in lactating mice exposed to the odours of a strange male mouse. This is due to the rise in gonadotrophin secretion and consequent estrogen secretion following exposure to strange male mouse. This is known as the Bruce effect. In rodents there is a rise in the plasma level of estrogen during the preimplantation period and ovariectomy performed on the afternoon of day 3 (mice) or on day 4 (rats) prevents implantation, which is reinitiated by the administration of estrogen.

In species like hamster and rabbit, there appears to be no obligatory requirement for estrogen derived from ovarian source for implantation since ovariectomy, as well as, adrenalectomy followed by progesterone supplementation results in implantation. In domestic animals estrogen is not required for implantation; there is no peri-implantation rise in the estrogen level in systemic circulation. But in animals not showing any need for estrogen from maternal source, a significant role may be played by steroids of embryonic origin. There is evidence to support that preimplantation blastocysts from a number of species which include human, pig, cow, sheep, camel, rabbit, hamster, and also from rat, and mouse can synthesize and secrete estrogen and progesterone during the peri-implantation stage. It is likely that embryonic steriod could be responsible for local vasodilatory effect seen as a primary endometrial response at the site of implantation, and may also help to precipitate a cascade of growth factors secreted during embryo-uterus interaction at implantation. In the species in which facultative or obligatory delay of implantation occurs, it is seen that additional requirement of estrogen from maternal ovarian source is required for timing the onset of the implantation reaction. Figure 2 shows the plasma profiles of estrogen and progesterone during normal ovulatory cycle in human and non-human primates. The luteal phase of the cycle is associated with increases in progesterone and estrogen levels in circulation. The need for luteal phase

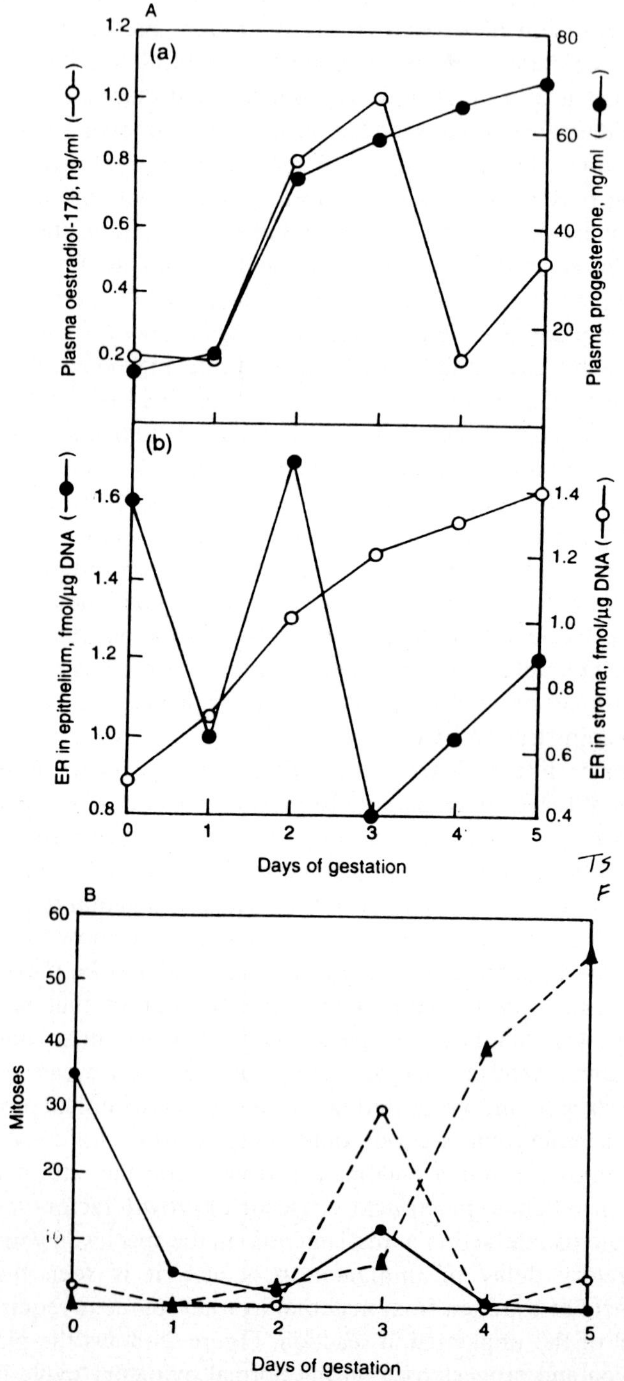

Fig. 1. **[A] Plasma concentrations of estradiol-17β and the progesterone (a), concentrations of estrogen receptor (ER) in uterine epithelium and stroma (b), and [B] degree of mitoses observed in luminal epithelium (-●-), glandular epithelium (--○--) and stromal cells (-▲-) observed during pre- and peri-implantation period of gestation in the rat.**

estrogen in the implantation reaction in human and in non-human primates has now been critically examined. In studies performed using rhesus monkey it has been shown that embryos transferred to hormonally primed recipient uteri underwent normal implantation and resulted in pregnancy even in the absence of circulatory source of estrogen during the post-transfer period. Progesterone remains obligatory, however. Similar results are now available from human embryo transfers performed in subjects having ovarian failure and supplemented with hormones exogenously. Luteal phase support with

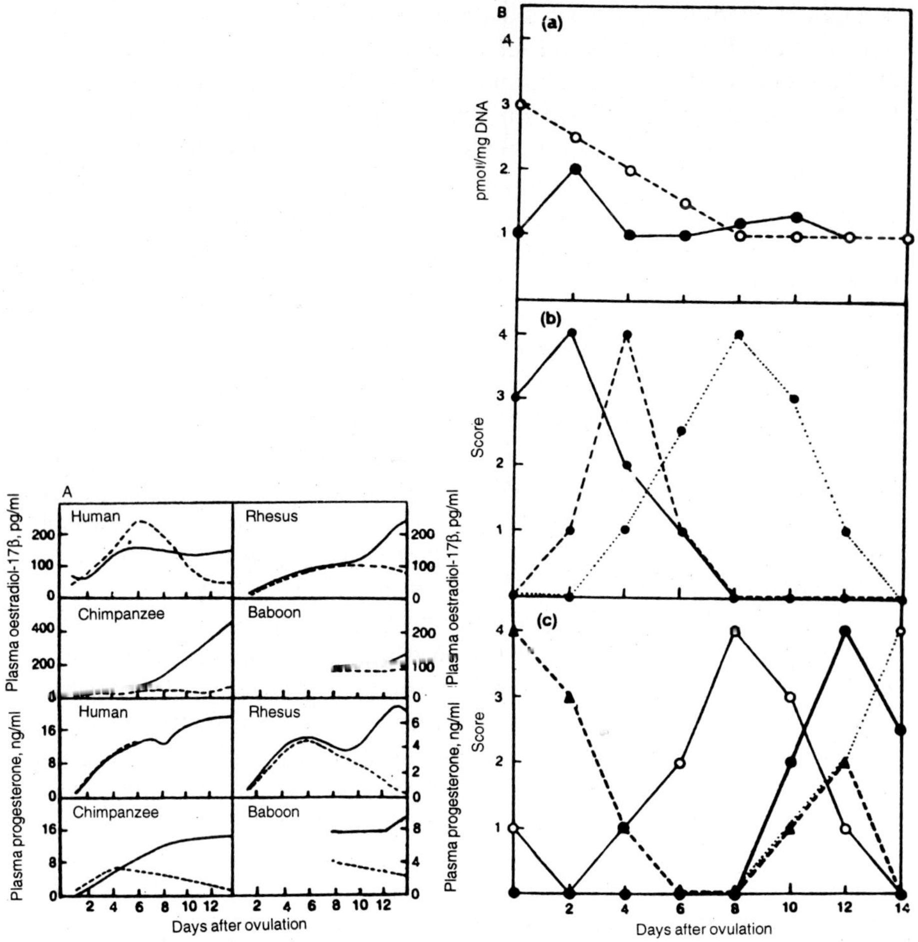

Fig. 2. [A] Peripheral estradiol and progesterone levels around the time of implantation in fecund cycles (———) compared with non-fecund menstrual cycles (------) in four primate species. [B] Endometrial concentrations of estrogen receptor (ER, —●—) and progesterone receptor (PR, --○--) (a), the degree of epithelial mitoses (—●—), secretion (···●···), and basal vacuolation (---●---) in glandular epithelium (b), and mitoses (--▲--) oedema (—○—), leukocytic emigration (--●--) and predecidual cells (—●—) in stromal compartment (c) during the luteal phase of normal menstrual cycle in the human.

only progesterone results in embryo implantation. Whether there is a critical need of estrogen available from local sources such as embryo and/or endometrium cannot, however, be ruled out. Use of competitive inhibitors which can block estrogen action at the receptor level have been shown to interfere with implantation in non-primate and primate species.

In most species studied with the exception of armadillo, progesterone is critical for implantation. Progesterone is also required for maintaining the viability of embryos during the period of delay: a dose dependent increase in embryo viability has been observed in rodents undergoing delay of implantation. This effect is considered to be mediated by an indirect action on the uterus rather than the hormone acting directly on embryonic cells. Progesterone has been shown to promote the synthesis of a number of uterine proteins specific to implantation such as blastokinin (also known as uteroglobin). This protein was initially isolated from rabbit uterus and later a similar protein was found to be present in small amounts in a number of species including human. Protein such as uteroglobin, transferin, and retinol-binding protein secreted by endometrial cells are considered to play a role in transferring substances from maternal to embryonic cells and thus contribute to the nutritional requirements of the embryo. Studies in the human and in the monkey have now shown that typical progestational maturation of endometrium can occur with progesterone alone, without any estrogen exposure in properly primed uterus. Endometrial characteristics seen at the approximate time of blastocyst implantation like expression of pinopodes, specific glycoproteins and mucin are also primarily regulated by progesterone. Inhibition of progesterone action in the peri-implantation stage of gestation with the use of progesterone antibody can lead to implantation failure. Similarly, RU486 which is a steroidal competitive inhibitor of progesterone acting at the progesterone receptor level can also inhibit implantation in a wide variety of species studied, and these include rat, mouse, rabbit, rhesus monkey and man. In the lower mammalian species studied, RU486 given at a low dose during the post-ovulatory period induced a delay of implantation which ranged from 1–2 days to even 7 days as seen in the rabbit. At a higher dose, RU486 was found to interefere with embryonic growth and viability. Endometrial asynchrony and inhibition of receptivity for implantation are found to be major factors leading to implantation failure after RU486 administration. In the human and also in non-human primate species that have been studied, a low dose single administration of RU486 in the early luteal phase of the cycle resulted in endometrial dyssynchrony characterized by inhibition of glandular functions and the altered profile of endometrial protein secretions during the pre-implantation stage of gestation; pre-implantation embryonic growth and viability were also inhibited in the rhesus monkey.

The uterus produces several prostaglandins which may be essential to support the implantation reaction. The synthesis of uterine prostaglandins is regulated by ovarian steroids. Prostaglandins (PGs) of endometrial origin are considered to play a role in regulating luteal function. In domestic animals such as sheep, cow and pig endometrial $PGF_2\alpha$ is luteolytic and the secretion of trophoblast-derived protein (interferon-*tau*) or in species such as pig, estrogen of embryonic origin induces the prolongation of corpora lutea through inhibition of endometrial synthesis of prostaglandins. In women $PGF_2\alpha$ may also be a luteolysin and its anomalous turnover in endometrium may lead to infertility and abnormal uterine bleeding. PGE_2, however, is required for endometrial decidualization at implantation. Increased concentrations of PGE_2 and $PGF_2\alpha$ have been shown in mid-luteal phase endometrial samples in the human and the binding sites for PGE_2 have been shown in endometrial epithelial cells at the time period coinciding with the implantation window in rodents and in primate species. The appearance of such binding sites in rodent species is dependent upon progesterone and administration of progesterone antagonist, RU486, inhibits their timed appearance in peri-implantation stage endometrium. Culture of rat endometrial cells *in vitro* has revealed that polarised epithelium secretes PGE_2 and $PGF_2\alpha$ from their basal domain rather than from apical end of the cells. The significance of such observation remains to be elucidated, but accumulation of PGs in the stroma suggests that their presumed site of action lies in inducing stromal cell decidualization and vascular permeability.

Prostaglandin production in uterine cells is influenced by the platelet activating factor (PAF). Chemically it is 1-*o*-alkyl-2-acetyl-sn-glycero-3-phosphorylcholine and named as PAF-acether. PAF-acether production by different cells such as neutrophils, monocytes and mast cells has been observed in response to a variety of stimuli. PAF causes platelet aggregation and is considered to be a potent mediator of inflammatory reaction. PAF has been detected in rat uteri, rabbit endometrium and in human uterine tissue. In rabbit uterus PAF secretion increases from day 3 of pregnancy or pseudopregnancy, peaks on day 4 of pseudopregnancy and day 5 of pregnancy. PAF is produced by stromal cells and it induces the secretion of PGE_2 by epithelial cells. In turn PGE_2 increases stromal PAF production. Progesterone, but not estrodiol also increases PAF production by stromal cells. PAF-acether increases the production of PGE_2 by secretory phase human endometrial epithelial cells. This effect of PAF on glandular epithelial cells is amplified by estradiol but not progesterone. The significance of endometrial PAF-prostaglandin paracrine loop may serve to amplify embryonic signal(s) and involve blastocyst derived PAF in the amplification process.

Many actions of steriod hormones are mediated by paracrine factors at the tissue level. The IGFs (IGF-I and IGF-II) are found in endometrium and their functions are mediated by membrane receptors for IGF-I and IGF-II. In addition endometrial cells also express IGF binding protein, IGFBP. Six IGFBPs have so far been identified. IGFBP1 is a major product of decidualized stroma. IGFBP1 modulates receptor-IGF binding as it has strong affinity (similar to IGF-I receptors) for the IGF peptide. Estrogen has been shown to stimulate the synthesis of a number of cytokines in the uterus during the peri-implantation stage of gestation. These include members of the EGF family, leukemia inhibitory factor (LIF) and IGFs. Mouse uterus during the peri-implantation stage synthesizes peptides of the EGF family (TGFα, heparin bind EGF amphiregulin) and interleukin (IL) family. The significance of several of these cytokines in the implantation reaction remains to be delineated. For example, it is known that mice lacking the TGFα gene can also undergo normal implantation. Therefore, the need for a specifically timed appearance of TGFα in luminal epithelium prior to blastocyst attachment reaction remains as yet highly speculative. But based on recent studies done mainly using the mouse model the possible role played by few cytokines such as LIF and IL in the induction of implantation reaction has been highlighted.

Leukemia inhibitory factor is a 45 to 56 kDa glycoprotein with multiple effects in many systems and cell types. It is known to regulate differentiation and proliferation of hemopoietic cells and various cell lines and from this it has its name. LIF transcripts are expressed by embryonic and adult tissues. LIF mRNA expression is maximal in glandular epithelial cells of mouse uterus at the time of implantation. Experiments with LIF knock-out mice (the gene expression for LIF is experimentally deleted in such animals) have shown that mated, LIF-deficient females are fertile but unable to suport implantation. Blastocysts obtained from LIF-deficient homozygous females when transferred to normal wild-type mouse uteri underwent normal implantation. Blastocysts from normal wild-type females when transferred to LIF-deficient mothers failed to implant. These results suggest that absence of maternal LIF expression by uterine cells failed to support implantation even by LIF-positive blastocysts and that at least in the mouse, maternal expression of LIF at the time of implantation is essential for implantation. LIF expression is considered to be hormonally regulated. It is suggested that estrogen priming is necessary for inducing LIF expression in uterine cells and rendering uterus responsive for implantation. Our understanding of the role played by LIF in blastocyst implantation in the human is not known for certain. Evidence is available to show that LIF is expressed by endometrial epithelial cells during the mid-luteal phase of menstrual cycle at the time when implantation is expected to occur in a fecund cycle.

Interleukin-1 (IL-1) consists of two related molecules, IL-1α and IL-1β. The IL-1α receptor is an immunomodulator expressed in T cells and in other cells. IL-1 stimulates proliferative and secretory endometria to secrete IL-6. IL-6 is a signal peptide. IL-1, CSF-1 and LIF may act via a common pathway, via the IL-1 receptor type 1 mRNA. The level of this receptor is highest during early and late luteal phases of menstrual cycle. It leads to the production of PGE_2, a prominent mediator of vascular permeability seen as an early response of endometrium to trophoblast invasion. Recombinant human IL-1 receptor antagonist inhibits implantation in mice.

Embryonic Signals

Physiological recognition of the presence of embryo by the mother is important for an early recognition of pregnancy and the maintenance of luteal function to support endometrial activities during the early stages of pregnancy. It is being increasingly recognized that an embryo-endometrial dialogue allows for endometrial receptivity for implantation to occur. Signals of trophoblastic origin play an important role in prolonging the life span of corpora luteal cells in early pregnancy. Species specificity exists in the mode of chemical signalling by peri-implantation stage conceptus. In species such as rat and mouse, luteal function is maintained by post-implantation luteotrophin secretion by trophoblast cells. These signals released prior to implantation are anti-luteolytic. Luteolysis in these animals is induced by prostaglandins of endometrial origin. Signals from peri-implantation stage conceptus cause a reduction of uterine $PGF_2\alpha$ production and thereby prolongs the life span of corpus luteum. Embryonic proteins such as ovine trophoblast protein (oTP-1) and bovine trophoblast protein (bTP-1) secreted by sheep and cow embryos, respectively, have been analyzed to be partly homologous to human interferon. These proteins are required to prevent luteal regression around the time of implantation. Human embryos do not seem to release any compound resembling interferons. Schwangerschaftsprotein (pregnancy specific β1 glycoprotein, SP-1) helps to diagnose human pregnancy 16 to 25 days after ovulation. The source and functions of SP-1 remains uncertain. In the human and non-human primate an embryonic signal by blastocysts at, or prior to implantation is chorionic gonadotrophin. The function of chorionic gonadotrophin is to 'rescue' the corpus luteum for maintaining the secretion of estrogen and progesterone during early pregnancy. Human chorionic gonadotrophin (hCG) function as essential embryonic signals required to sustain corpus luteum function and the endocrinology of the mother during the early stages of gestation. Analysis of the output of CG provides a good estimate of embryonic well-being and a prognosis of implantation. The doubling of hCG values occurs between days 12 and 18 of gestation, whereas in ectopic pregnancy and in abortive pregnancy hCG levels rise very slowly. There is evidence that primate blastocyst secretes luteinizing hormone releasing hormone (LHRH) which may function to regulate hCG secretion. In human embryos, βhCG mRNA transcription is seen as early as 8-cell stage of development, hCG secretion, however, is documented at blastocyst stage prior to embryo attachment to uterine epithelium. Receptors for hCG have been detected in human endometrial cells. Thus, besides its role in luteal maintenance, CG derived from embryonic cells may also be required for initiating endometrial receptivity for implantation.

Identification of the role of embryo-derived signals in rendering endometrial receptivity during the preimplantation period is an active area of research. There are a few candidates in this category. Embryonic steroids such as estrogen may be one candidate. Human embroys *in vitro* can effectively metabolize testosterone to estrogen which is then released. In rodents (rat, mouse and hamster) and in rabbit there is histochemical and biochemical evidence to support the hypothesis that at morula and blastocyst stages of development embryonic cells are capable of metabolizing steroid precursors and can synthesize estrogen. The presence of steroid metabolizing enzyme in preimplantation stage embryo or the ability to metabolize progesterone and testosterone by embryos *in vitro* does not necessarily mean that such steroids play an essential role in embryo development or in implantation. However, anti-estrogen added to embryo culture medium prevent morula to blastocyst transition which can be reversed after adding estrogen to the culture medium. Embryos exposed to anti-estrogen *in vitro* when transferred to pseudopregnant host uteri show reduced rate of implantation. Thus, there may be a role of embryonic estrogen in these non-primate mammalian species to support embryonic differentiation and implantation. Embryo derived steroids may function on adjacent endometrial cells at the site of implantation since there is greater translocation of nuclear estrogen receptors in luminal epithelia at implantation sites.

Another candidate is histamine. Preimplantation rat and rabbit embryos have been shown to synthesize histamine and a working hypothesis proposed for the local role of histamine in the implantation process is that embryonic histamine could stimulate endometrial prostaglandin and thereby initiate implantation. A competitive inhibitor of histidine decarboxylase prevents embryogenesis and implantation in the rat which can be reversed by giving histamine. Murine embryos can synthesize prostaglandins. Embryonic prostaglandins have been suggested to play a role in embryogenesis and in embryo hatching prior to implantation. Embryonic prostaglandins have been considered to play a role in mediating the early interactions between embryo and uterus at implantation.

Platelet activating factor (PAF) is a member of the family of biologically active phospholipids. Embryo derived PAF (EDPAF) has been shown to be released by mouse and human blastocysts, and they bear structural and biochemical homology with PAF. PAF is synthesized by embryos shortly after fertilization. Soluble PAF obtained from human embryo culture supernatants have been reported to show good correlation with viability and pregnancy potential of human embryos. Indeed, PAF increases vascular permeability and initiates many events associated with an inflammatory response. Administration of PAF antagonist can inhibit implantation. Early pregnancy factor (EPF) may also be considered as an early embryonic signal to the mother. EPF is assayed by rosette inhibition assay and its synthesis could depend upon EDPAF. Plasma mononuclear cells from pregnant women express specific receptors for EPF. EPF might be immunosuppressve in modulating the mother's immune response to the presence of embryo as a foreign allograft and inhibiting the IgG expression in mononuclear and other cells. Both EDPAF and EPF could be linked to the synthesis of prostaglandin and peptide mediators such as GM-CSF and IL-1 by endometrial cells.

Stage-specific secretion of diffusible proteins by preimplantation stage embryos has been suggested to function as putative signals for early pregnancy recognition in species such as mouse. Qualitative and quantitative changes have been observed in the pattern of proteins synthsized *de novo* by mouse blastocysts. Some of these proteins may be growth factors which may have autocrine-paracrine loops in mediating their actions on embryonic and endometrial cells in the peri-implantation stages of gestation. However, categorical information is still not available in this respect.

Immunomodulation is a unique feature of mammalian reproduction. Trophoblast cells contribute to the safety of conceptus, these cells do not express the classical human leucocyte (HLA) compatibility antigens. In implanted conceptus, trophoblast cells lining chorionic villi—cytotrophoblast and syncytiotrophoblast cells—are devoid of major histocompatability (MHC) class I and class II antigens. Trophoblast cells on the other hand express a non-classical HLA class I molecule which is a product of HLA-G locus. Unlike the polymorphic nature of HLA class I antigens, HLA-G antingen is less polymorphic and is expressed exclusively by invading trophoblast cells. Progesterone might be immunosuppressive because it reduces the levels of immunomodulators. It promotes PGE_2 production which can also function as an immunosuppressant at the site of implantation.

Types of Implantation

There are few major forms of implantation depending upon the degree of penetration of uterine epithelium by implanting blastocyst. These are: central, eccentric, interstitial and secondarily interstitial. Blastocyst orientation with reference to the uterus at implantation is also highly species-specific and it involves two components: direction of the germ disc and point of attachment by trophoblast cells to uterus. In rat, mouse and guinea-pig the ICM is situated distant to uterine epithelium and trophoblastic attachment occurs via abembryonic trophoblast cells. In human, rhesus monkey and baboon the ICM is located towards the site of implantation and embryonic trophoblast cells participate in the implantation reaction. Based on the nature of trophoblastic penetration into uterine epithelium, implantation can be classified as: intrusive, displacement or fusion type. Figure 3 is a schematic representation of different types of implantation.

Embryo-Uterus Interaction and Decidualization

Blastocyst-uterus interaction during implantation is a unique event in that it allows for cell-cell interaction between two distinct cell types bearing different genetic make-up, embryonic trophoblast cells and maternal uterine luminal epithelium. The stages of blastocyst-epithelial interaction involve apposition and adhesion of blastocyst onto uterine surface and penetration of uterus by trophoblast cells. Apposition denotes an increasingly close intimacy between blastocyst and uterine luminal epithelium. In mice and rats there is closure of uterine lumen surrounding the implanting blastocyst

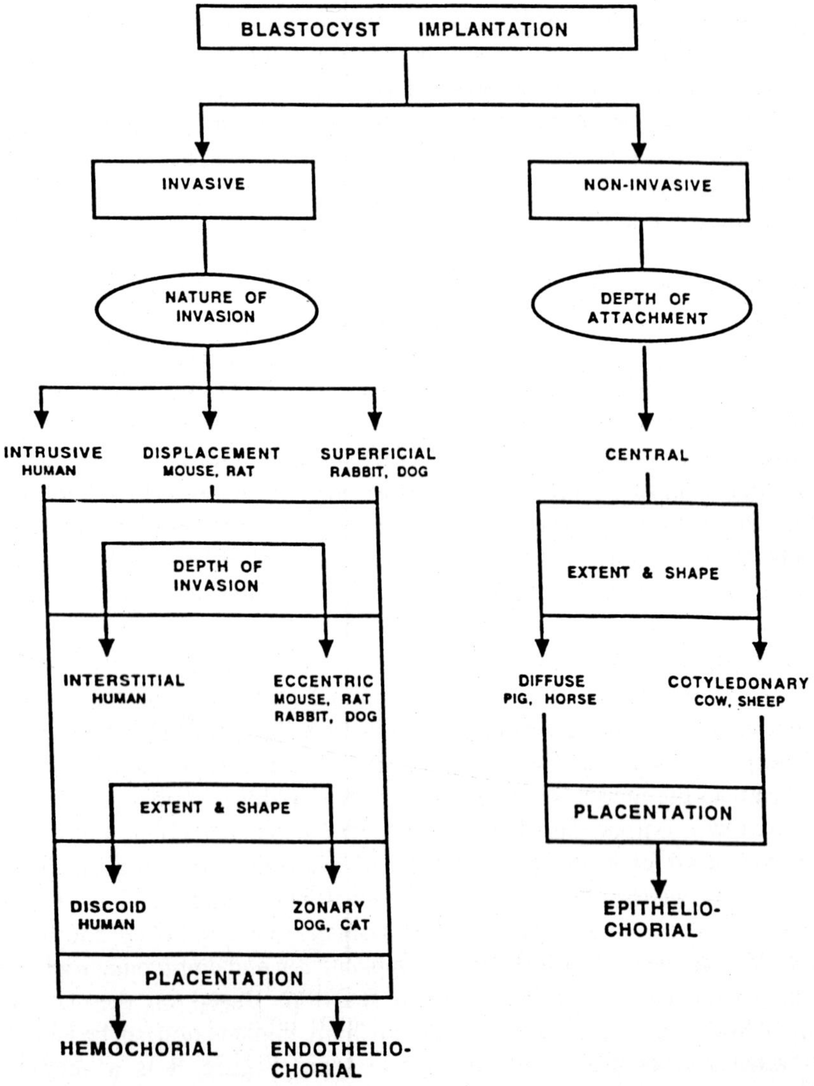

Fig. 3. Classification of implantation and placentation. Implantation can be classified based on different characteristics, namely, nature, depth, extent and shape of attachment. Placental (Grosser's) classification is based on histological features of fetal and maternal tissues at the site of implantation.

and the appearance of a 'clasping' effect of the embryo by the uterus. In species such as rabbit, however, singificant expansion of the blastocyst is seen in the preimplantation stage and as a result blastocyst apposes itself on uterine luminal surface without any significant closure of the uterine lumen. Similarly in primates, clasping type uterine luminal closure does not occur. In all species however, there is immobilization of blastocyst on uterine surface and dislodgement of the embryo at this stage can be done by gentle perfusion of the uterus. During this phase there is progressive interdigitation of microvilli on surface of luminal epithelium and on trophoblast cells. Microvilli become short and stubby as they interdigitate with each other in apposing blastocyst onto luminal surface. Luminal epithelial microvilli also show prominent bulbous cytoplasmic projections at antimesometrial end of the uterus during this stage. Close apposition of blastocyst to uterus is referred to as attachment reaction.

Prior to the attachment reaction, apical surface of luminal epithelium remains non-adhesive due to surface glycocalyx coating on the luminal apical surface. The presence of such apically distributed glycoconjugate molecules has been noted in a number of species studied. Studies done in the mouse have shown that a bulk of such glycoproteins belong to transmembrane mucin type of glycoproteins. MUC-1 is one such glycoprotein identified in mouse and in human uterus. In mouse uterus it is maximal at estrous, low during diestrous and undetectable at the time of implantation. In the human maximal loss of uterine epithelial MUC-1 is detected in mid-secretory stage at the expected time of implantation. It has been reported that there are subtle changes in the nature and character of junctional complexes in luminal surface epithelium of human endometrial biopsy samples collected during the mid-secretary phase, that is the expected time of implantation. Such changes may help to facilitate the passage of embryo derived signals or the embryo itself through the surface epithelium.

Adhesion of trophobst cells to apical ends of luminal epithelium develops as apposition progresses. The biochemical nature of the cell to cell adhesion mechanism still remains to be clearly delineated but several clues are now available based on extensive studies done using the mouse model. The initial phase of adhesion is considered to involve carbohydrate-lectin interaction and then this binding is stabilized with the binding of integrins to extracellular ligands. Mouse blastocysts express several carbohydrate structures which include selectin ligand, sialytated Lewis. They also synthesize proteoglycans such as perlecan, the basement membrane form of heparan sulphate proteoglycan and laminin on outer surface of trophectodermal cells at the time of attachment reaction. Human uterine epithelial cells also have been shown to express heparan sulphate binding proteins that may interact with trophoblast proteoglycans. Uterine epithelial cells also express H-type 1 carbohydrate at the time of implantation and abembryonic trophoblast cells specifically acquire the ability to bind H-type 1 structures at late blastocyst stage.

Integrins are cell adhesion molecules of the extracellular matrix and they are involved in regulating interactions between cells and between cell and extracellular matrix. The integrins are heterodimeric glycoprotein molecules of at least one of eleven α chains and one of six β chains. These molecules serve as receptors for collagen, fibronectin and laminin. Few integrin molecules have been shown to be continuously expressed in blastocysts and these include $\alpha_5\beta_1$, $\alpha_6\beta_1$ and $\alpha_v\beta_3$. The expression of others such as α_2, α_6A and α_7 are developmentally regulated during the peri-implantation stages of gestation. Integrin expression in uterine cells is also developmentally regulated around the time of implantation. Studies in human endometrial samples collected throughout the menstrual cycle have shown specific staining patterns for few of the integrin molecules such as $\alpha_4\beta_1$ and $\alpha_v\beta_3$ during the expected time of implantation in mid-secretory stage of the cycle. The role of such integrin molecules in the implantation reaction still remains to be delineated, but clinical studies have shown that these molecules are deficient in endometrium of women showing infertility, luteal phase deficiency or endometriosis incompatible with normal pregnancy.

Attachment of blastocyst to uterine epithelial cells is followed by penetration of these cells by trophoblast. In nearly all species studied with the exception of those species in which blastocysts exhibit superficial type of implantation, trophoblast cells penetrate uterine epithelium and underlying basal lamina to establish close structural links with maternal vasculature. Species-specific distinctions have been observed in the cytological characteristics of epithelial penetration at implantation. In species in which intrusive type of implantation is seen the blastocyst is highly invasive in nature. In human, rhesus monkey and also in ferrels intrusive type of epithelial penetration is seen. In this, syncytial trophoblst cells have been shown to penetrate between luminal epithelial cells and they share junctional complexes with surrounding epithelial cells as they penetrate to, but not through the basal lamina of epthelium. In rat and mouse, which display a displacement type of penetration, there is evidence of epithelial cell death and detachment of cells either singly or in groups in the area immediately surrounding the blastocyst. Trophoblast cells phagocytose epithelial cells undergoing apoptosis and this facilitates trophoblastic penetration. Trophoblast cells reach to the basal lamina and by this time cytoplasmic processes from underlying decidual cells breach the basal lamina. Trophoblast cells contact anti-mesometrial decidual cells, then endothelial cells of blood vessels to form the yolk-sac placenta. The mechanism responsible for cell death and autolysis of epithelial cells surrounding the implanting blastocyst is not known for certain but since it occurs only at sites of implantation and not in inter-implantation areas or in pseudopregnant uteri, it has been argued that cells undergo autolysis in response to a blastocyst signal. Histochemical and biochemical evidence suggests the involvement of lysosomal enzymes. In the rabbit, trophoblastic penetration is of fusion type. First, attachment occurs between syncytial knobs of abembryonic trophoblast cells and individaul epithelial cells. Apical membranes of epithelial cells fuse with that of trophoblastic knob and a syncyticium is formed. The syncytium appears as a 'peg' of trophoblast extending to basal lamina. There may be more than one 'peg' in each trophoblastic knob. After a delay the basal lamina is penetrated in the regions of such 'pegs' and finally trophoblastic processes penetrate endometrial blood vessels as a symplasma forms and fusion occurs in the areas between trophoblastic knobs. In ruminants, most trophoblast cells are non-invasive but specialized zones of trophoblast (cotyledons) invade into and fuse with uterine epitheluim forming binucleate trophoblasts and a synepitheliochorial placenta develops. In the pig, no invasion occurs and trophoblasts remain apposed to uterine epithelium throughout the course of pregnancy as epitheliochorial placenta is formed.

Understanding the nature of biochemical events during trophoblastic penetration of uterine epithelium and stroma during implantation and post-implantation stages of pregnancy has been achieved to some degree mainly on the basis of studies performed in mouse and from human decidual tissue obtained during 6 to 8 weeks of gestation from elective abortions. Implanting embryo penetrates through uterine epithelium by displacing epithelial cells and sending trophoblastic protrusions between epithelial cells. TNF-α of trophoblastic origin can dissociate neighbouring uterine epithelial cells by influencing the expression of cadherin and that of actin mRNA. A series of integrins expressed by invading trophoblast cells and by villous trophoblast cells may help to guide the interaction of these cells within the uterus. This repertoire of integrins include receptors for fibronectin, laminin and vitronectin. Fibronectin is a component of basement membrane and it may help to guide trophoblast cells of the implanting embryo and in the process it itself gets down graded by trophoblast. Trophoblast adheres preferentially to laminin, which is also a basement membrane component and it is also distributed around decidual cells. Trophoblast cells also secrete laminin and this may help in the mutual recognition of these molecules at implantation. Trophoblastic invasion is up-regulated by autocrine factor such as IGF II of trophoblstic origin, and by the paracrine factor such as IGFBP1 secreted by decidual cells. Uterine metalloproteinases (MMPs), serine proteinase (like plasminogen acivator, PA) and tissue inhibitor of metalloproteinase-1 (TIMP 1) are involved in maintaining a delicate balance between tissue invasion and cellular remodelling that occurs at implantation. Trophoblast produces PA and its inhibitors (PAI-1 and PAI-2). Integrins may activate MMPs in trophoblast and sustain interactions between trophoblast and various substrates as the embryo migrates deeper into the stroma. Cytotrophoblast cells secrete three classes of MMPs: (i) collagenases which can digest collagen types I, II, III, VII and X; (ii) gelatinases which can digest collagen IV and gelatin; (iii) stromelysins which can digest collagen IV, V and VII and fibronectin and laminin. Steroid hormones, growth factors and cytokines such as EGF, TGFβ and IL-1 acting locally can modulate a balance between pro MMP-1 and TIMP-1.

Following implantation polar trophoblast cells in rat and mouse proliferate to form ectoplacental cone and later the spongiotrophoblast layer. Trophoblast stem cells in ectoplacental cone remain as precursors of chorionic ectoderm. The outermost trophoblast cells differentiate into secondary giant cells and interface with maternal cells in the decidua. These cells stop proliferating but their DNA synthesis continues with endoreplication and polyploid cells are formed. These cells synthesize a number of specific proteins which include, gelatinase B, α1 intergrin, placental lactogen and pregnancy specific glycoprotein. At this time, parietal endoderm cells migrate from the enlarging ICM, now at the egg cyclinder stage, to the basal surface of trophoblast layer and deposit an extensive membrane, Reichert's membrane, between the two cells types. At this stage parietal yolk sac is responsible for the transfer of nutrients from the blood sinusoids surrounding the conceptus.

Comprehensive studies using implantation stage material have been done using several non-human Old World primate species which include rhesus monkey and baboon and the New World primate, marmoset. In the human, our information is entirely based on light microscopic histological examination of the human implantation stages available in the Carnegie collection. In all four species the inner cell mass of blastocyst is oriented towards uterine epithelium and thus initial adhesion occurs in peri-inner cell mass trophoblast cells. The initial invasion process as studied in macaque and marmoset occurs by intrusion of processes of syncytial trophoblast cells between uterine epithelial cells. This is followed by consolidation of the association with epithelium by penetration of additional trophoblastic processes to form a trophoblastic plate. This plate then spreads laterally in the plane of uterine epithelium. There is evidence to suggest that some fusion occurs between maternal epithelial cells and fetal trophoblastic cells within areas of symplasma. In human implantation sites clusters of small and large nuclei have been observed. Small nuclei have been suggested to represent areas of fusion between maternal epithelial and trophoblast cells. This stage is the trophoblastic plate stage of the implantation process and it occurs within a day of adhesion in the baboon, and one to two days after adhesion in the macaque and marmoset. There is no information available on the human.

The next state in the implantation process is the lacunar or previllous stage of implantation. Expansion of the trophoblastic plate leads to penetration of the basal lamina and underlying stroma. Species variation is observed in the extent of trophoblastic penetration. In the baboon, trophoblast cells penetrate most rapidly, subepithelial vessels dilate and these are penetrated by trophoblastic processes as soon as the basal lamina is penetrated. In the macaque this event most likely occurs a day or two later, while in the marmoset, trophoblast penetration of vessels occurs around 45 to 60 days of gestation. In the human trophoblast penetration occurs a day later than in the baboon or macaque and trophoblasts penetrate deeper into stroma before they penetrate maternal vessels. Clefts soon appear in syncytial trophoblast cells which are now differentiated from syncytial form to unilaminar absorptive type of cells. In the macaque, a secondary implantation site is formed by attachment of abembryonic trophoblast cells to opposite uterine epithelial surface. This tends to lift the implanted conceptus from the surface of the endometrium and results in superficial type of implantation. The baboon conceptus is slightly less superficial at implantation and that is because trophoblast cells invade very rapidly into the stroma and lunae are formed very near to luminal epithelial surface. In the human, conceptus is seen to be buried deep into endometrial stromal compartment since trophoblast cells invade deeper into the stroma before they tap maternal vessels and form lacunar spaces. As lacunae form, cytotrophoblast cells appear between lacunae and these are the precursors of chorionic villi.

At the end of lacunar stage, cytotrophoblast cells proliferate and arrange in columns and expand at fetal-maternal interface. Primary chorionic villi are formed. This is the villous stage of the implantation process. By 13 to 14 days of gestation, mesoderm form at fetal surfaces and convert them from primary villi to secondary villi. Vessels also form rapidly within these villi but blood circulation does not occur in these vessels till fetal heart beat stage is established.

In the human, macaque and baboon, species that display hemo-monochorial placentas, implantation occurs in undecidualized endometrium. In the macaque, by day 10 of gestation, epithelium responds to trophoblastic penetration with: (i) peripheral edema in subepithelial zone, (ii) initiation of epithelial plaque response and, (iii) hyperplasia and hypertrophy of endothelial cells of maternal vessels. The implantation site can be visualized from the endometrial surface as a pink spot (formed by the lacunar spaces being filled with maternal blood) surrounded by a translucent gelatinous area of edematous endometrium. Vascular endothelial cells at the site of implantation become nearly columnar in shape and have a gland-like appearance, with abundance of endoplasmic reticulum suggestive of increased synthetic functions. Endothelial cell hypertrophy is a feature observed only after trophoblast penetrates uterine epithelium. During the preimplantation stage, more subtle morphological changes have been detected which include increased permeability. Biochemical evidence is also available indicating increased vascular permeability at this time. In the human and in the rhesus monkey, leucocytes are seen to accumulate in the endometrium surrounding trophoblast cells prior to stromal-cell decidualization. Compared to many other species such leucocytic infiltration after trophoblast penetration is clearly more abundant in primates. Macrophages are detected in the basal plate of uterine stroma at the time of implantation in human and rhesus monkey and they have been suggested to play a critical role in maternal-fetal interactions.

A prominent feature of implantation reaction in primates is epithelial plaque response. Endometrial luminal epithelial cells immediatly adjacent to the site of trophoblast penetration respond by cellular hypertrophy and hyperplasia forming a multicellular pad or basket of cells enclosed by basal lamina. This response is known as epithelial plaque reaction and it is common in New World primates and Old World primates including macaque and baboon, but it has not been reported in the human. Plaque cells are also seen at secondary site of implantation in the rhesus monkey, where it begins a day later. Epithelial plaque cells exhibit great increases in cell size, nuclear size and polyploidy. These cells begin to store glycogen and show heightened synthetic activity with abundance of ribosome, Golgi vesicles, enlarged RER, and often secretory acinar lumina. Epithelial plaque cells have a finite lifespan. These cells begin to degenerate by day 16 of pregnancy with infiltration of leucocytes, and by day 30 the area occupied by plaque cells consist of cellular debris and lipid filled macrophages. The function of plaque cells in the implantation reaction in primates is not known. These cells may be source of nutrients and growth factors to the implanting conceptus and/or they may play a role in immunomodulation during the very early stages of gestation. Interestingly, around the time plaque cells begin to degenerate, stromal fibroblasts start to differentiate to form pregnancy specific decidual cells and there is correspondingly increase in the numbers of large endometrial granulated lymphocytes (see later) at the site of implantation.

In rat and mouse during the immediate pre-attachment stage there is localized increase in vascular permeability and increase in alkaline phosphatase in the stroma in the region adjacent to the embryo; and as the blastocyst undergoes attachment there is transduction of embryonic signal to the underlying stroma, and stromal fibroblasts then undergo morphological and biochemical transformation to form decidual cells. The decidual cell reaction is a prominent endometrial stromal cell response in species in which placentation occurs with the breach of luminal epithelium by trophoblasts and their subsequent invasion into stromal matrix. It is a complex phenomenon and like many other events in the reproductive process it is also hormone dependent. Decidualization involves both proliferation and differentation of stromal cells. Fibroblast cells become enlarged, round in shape, with increased mitotic activity and they become epitheloid with two or more polypoid nuclei. At the ultrastructural level, differentation is characterized by accumulation of glycogen, lipid, numerous bundles of microfilaments appear in cytoplasm, increase in the numbers of polyribosomes, amount and distention of rough endoplasmic reticulum. This is associated with significant increases in RNA, DNA and protein synthesis in decidualized regions of uteri and increased ornithinine decarboxylase activity has also been observed to support increased synthesis of nucleic acids. Decidua-specific proteins have been identified in rat and in hamster. Quantitative and qualitative modulation occurs in endometrial secretory protein profile during early pregnancy. Decidual cells from human, non-human primate and non-primates species secrete proteins such as prolactin, relaxin and binding protein of insulin like growth factor (IGFBP). The function and regulation of decidual proteins have been suggested to involve autocrine and paracrine type regulation of decidual cell activity and its immediate cellular environment. Endometrial decidual cells have also been shown to express the mesenchyme specific intermediate filament, vimentin and desmin. Desmin expression by decidual cells is considered to be a marker of stromal decidual cell differentiation. Decidual cells form a distinct layer surrounding the implanting blastocyst and the cells exhibit numerous junctions (adherens type and gap junctions) between cells. The concept of decidual granulation tissue providing a barrier towards restricting the extent of trophoblastic invasion of the stroma is strengthened by the observation that decidual cell generates its own basement membrane by synthesizing and secreting its components, laminin, collagen IV and heparan sulphate proteoglycan and fibronectin. Thus, such an extracellular matrix may function as a 'glue' in enabling decidual cells to form border between maternal and fetal tissues.

During pregnancy the numbers of mature T and B cells do not increase in the implantation sites confirming the absence of an immune response against the placenta. A similar response is also evident in ectopic sites of implantation. A distinctive feature of decidualized stroma is the presence of a population of lymphocytes with an eccentric nuclei and bearing acidophilic granules. These cells are known as endometrial granulated lymphocytes (eGL). These are CD56 positive cells but they do not express other cell surface markers common to the natural killer (NK) cells, such as CD57 and CD11. However, perforin which is generally elaborated by NK cells and a mediator of target cell lysis has been deteceted in granulated lumphocytes suggesting that these cells may be a type of natural killer cell. Natural killer (NK) system is evolutionarily older than the classical adaptive T/B cells. NK cells kill cells deficient in self HLA class I molecules rather than those which bear non-self molecules as is the case for T cells. This is the basis of the 'missing self' hypothesis for NK cells. Accordingly it is hypothesized that eGLs are NK cells and that these cells in direct contact with invading trophoblast cells play a role in immunomodulation during early pregnancy. The presence of HLA-G on trophoblast cells protect them against NK cells by providing adequate amount of class I signals. NK cell can lyse trophoblast cells if it is activated by IL-2 cytokine. Apparently an equilibrium is maintained between trophoblast and decidual NK cells, and this can be disrupted by a cytokine like IL-2 allowing these cells with a mechanism of regulating excessively invasive cells. Decidual NK cells express mRNAs for CSF-1, GM-CSF, TNF-α, TNF-β and LIF and these cells also secrete relaxin. This expression profile is not seen in NK cells present in blood supporting the concept that eGLs are a specialized population of lymphoid cells resident in the uterus. Although eGLs are found to be present in endometrial samples from both proliferative and secretory phases of the cycle, their numbers increase considerably during the secretory phase and then they degenerate at the end of menstrual cycle. Thus, a hormone responsive profile of eGLs is seen throughout the menstrual cycle. During decidualization, at least in part, the increase in eGL population arises from *in situ* proliferation observed from the presence of the nuclear proliferation marker Ki67 in these cells. Close temporal and spatial relationships exist between differentiating decidual cells and eGLs during early pregnancy, the significance of which remains to be determined.

Acknowledgements
The studies from the authors' laboratory reported in this review were supported by funds received from the Reckfeller Foundation, the Special Programme of Research, Development and Research training in Human Reproduction, World Health Organization, the National Institute of Health, United States India Rupee Fund and the Indian Council of Medical Research.

References

1. Blandau RJ. *The Biology of the Blastocyst.* Chicago, University of Chicago Press, 1971.
2. Finn CA, Porter DG. *The Uterus.* Blackwell Publishing Sciences Group, London, 1975.
3. Edwards RG. *Conception in the Human Female.* Academic Press, London, 1980.
4. Rossant J, Pedersen A (eds). *Experimental Approaches to Mammalian Embryonic Development.* Cambridge, Cambridge University Press, 1986.
5. O'Rahilly R Muller F. *Development Stages in Human Embryos.* Washington, DC, Carneige Institute of Washington, Publication 637, 1987.
6. Weitlauf HM. Biology of Implantation. In Knobil E, Neil JD (eds). *The Physiology of Reproduction.* New York, Raven Press, 1988, pp 231–262.
7. Wynn RM, Jollie WP (eds). *Biology of the Uterus. 2nd ed.* New York, Plenum Press, 1989.
8. Edwards RG, Brody SA. *Principles and Practice of Assisted Human Reproduction.* Philadelphia, WB Saunders Company, 1995.

Comparative Endocrinology and Reproduction
K.P. Joy, A. Krishna and C. Haldar (Eds)

22. Prolactin Family and Pregnancy: Expression of Prolactin Family Genes in Rat Decidua and Placenta

Santanu Deb*
Department of Molecular Medicine, Roswell Park Cancer Institute,
Buffalo, New York 14263, USA

Introduction

Pituitary PRL is a very versatile hormone. It is involved in broad spectrum of biological functions including reproduction, growth and development, metabolism and immunity. The classic actions of the PRL in mammals are on mammary gland and the corpus luteum. PRL directly stimulates the growth and development of the mammary glands, lactation and milk protein synthesis (Forsyth, 1986; Thordarson and Talamantes, 1987). In ovary, PRL extend the life span of the corpus luteum and estrous cycle (Morishige and Rothchild, 1974).

The pituitary PRLs from all species possess significant structural and functional similarities (Nicoll *et al.*, 1986). In rat it is composed of 191 amino acids. It has six cysteine residues, in positions 4, 10, 56, 172, 190 and 197 (Fig. 1). These cysteine residues are very important for the tertiary structure of PRL molecule. Disulfide bonds between these cysteine residues give pituitary PRL characteristic two small loops, one in N-terminal region and other in C-terminal region and a large loop in the center. Conservation of cysteine residues is an important feature of PRL-family proteins.

The gene responsible for pituitary PRL is located in chromosome 17. This gene is specifically expresed in lactotrophs in the anterior pituitary gland. Secretion of pituitary PRL show distinct changes during pregnancy. After mating, plasma PRL levels increases in the form of two daily surges (Gunnet and Freeman, 1983). One of them occurs in the late afternoon (4-6 PM) and is present up to day 8 of pregnancy; the other surge occurs in the early morning (4-6 AM) and exists until day 10 of pregnancy. These surges terminate abruptly at a time when pregnancy associated PRL-related proteins appear into the circulation. This suggests that pituitary PRL is not necessary as gestation advances and pregnancy specific functions are

*Present Address: Dept. of Physiology and Biophysics, University of Illinois at Chicago, 901 S. Wolcott Avenue Chicago, IL 60012, USA

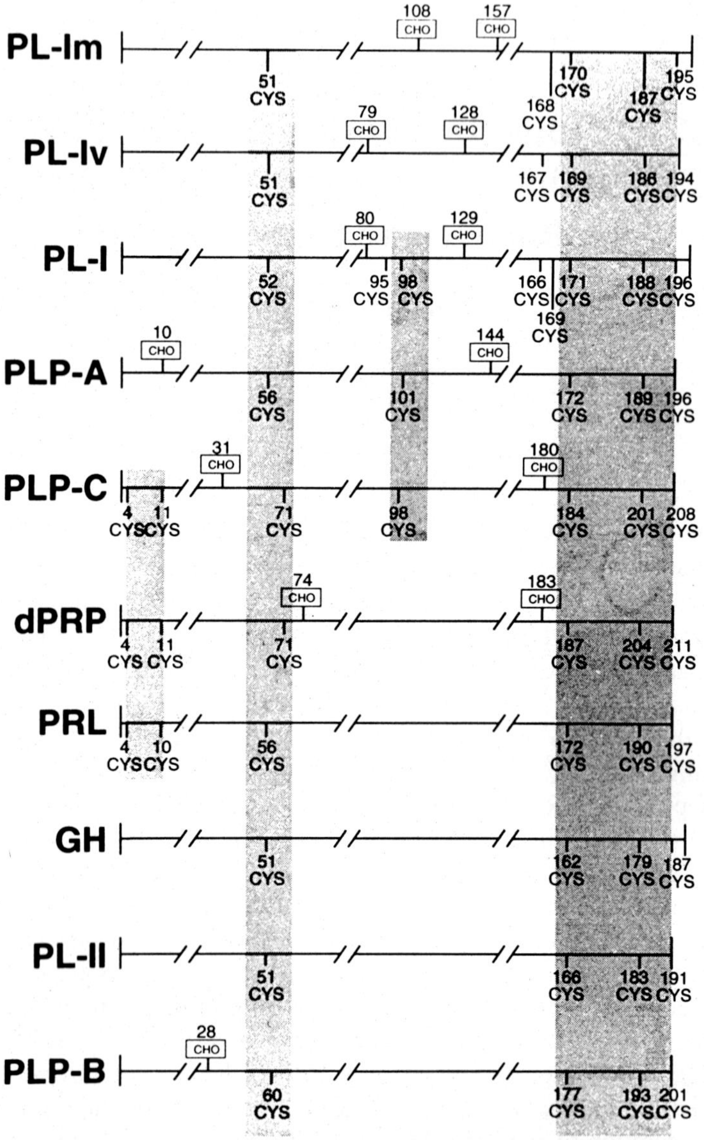

Fig. 1. Schematic comparison of the rat PRL-GH family. Sites for N-linked glycosylation are shown above each line, and cysteine residues are shown below each line. Locations of homologous cysteine residues are shaded. PRL, pituitary prolactin; GH, growth hormone; dPRP, decidual PRL-related prolactin; PL-I, placental lactogen-I; PL-Im, placental lactogen mosaic; PL-Iv, placental lactogen variant; PL-II, placental lactogen-II; PLP-A, PRL-like protein A; PLP-B, prolactin-like protein B; PLP-C, prolactin-like protein C (modified from Deb *et al.* 1991c).

performed by extrapituitary PRLs. Accumulating evidences from a variety of species suggests the roles of decidual and placental PRLs in fetal development and maternal adaptation to pregnancy.

This article discusses the expression of PRL family genes in decidua and placenta and covers following aspects: (1) biochemical and molecular characterization; (2) temporal and tissue specific expression; and (3) functions of decidual and placental PRL-related proteins. The discussion will focus primarily on the rat and where applicable on other species for comparative purposes.

In the rat, nine different PRL family genes are expressed during pregnancy. The anterior pituitary gland, uterine decidua and placenta are the site of their expression. Nomenclature and summary of biochemical properties of rat PRL-like proteins are presented in Table 1.

Expression of PRL Family Genes in Decidua

Decidua is a lining of cells in the pregnant uterus (Fig. 2). It is developed by the differentiation of endometrial stromal cells. This differentiation is induced by the contact of blastocyst to the endometrium or by artificial stimulation of the uterus. Decidualization in rat give rise to two different structures. Initially, decidualization takes place in the antimesometrial endometrium and results in the formation of decidual capsularis. Subsequently, decidualization of stromal cells in the mesometrial region of the endometrium gives rise to the decidua basalis (Fig. 2). Several functions have been described for decidual cells. It protects maternal tissue from highly invasive trophoblast cells and protect fetoplacental unit from immunological rejection by the mother. It has also been demonstrated that decidual tissue plays endocrine role in pregnancy. Its secretion prolongs the life span of corpus luteum (DeFeo, 1967; Finn, 1971). Besides prostaglandins and relaxin decidual tissues are reported to secrete PRL (Bigazzi *et al.*, 1980; Riddick *et al.*, 1978).

Biochemical and molecular characterization of decidual PRL-like proteins Characterization of decidual PRL in human revealed that it is identical to pituitary PRL (Golander *et al.*, 1979). It was Gibori, who first hypothesized that decidual tissue in rat secretes a PRL like hormone that may be responsible for the prolongation of corpus luteum functions (Gibori *et al.*, 1974). Gibori and her colleagues later performed experiments to confirm her hypothesis. They have clearly detected PRL receptor activity in decidual extract (Jayatilak *et al.*, 1985; Jayatilak and Gibori, 1986). They have further demonstrated that the PRL related substance is capable of maintaining the progesterone production by corpus luteum even in the absence of pituitary PRL (Gibori *et al.*, 1974). They named this protein as decidual luteotropin (LTH). Exact nature of decidual luteotropin is not known. It is not clear whether the LTH represent a single protein or more then one protein with PRL-like characteristics. Fractionation of decidual extracts by gel exclusion chromatography showed a major peak of PRL-like activity in the 20-25 kDa molecular weight range and several minor peaks of larger

Table 1. Characteristics of Rat PRL Family Proteins

	Structural characteristics				Functional characteristics	
Hormone	Mol. wt. of reduced forms	Cysteine residues	Similarity with Ant. Pit. PRL[a]	Glyco-protein	PRL-R binding	GH-R binding
Anterior Pituitary						
PRL	25 K	6	100%	No	Yes	No
Decidua						
dPRP	29	6	37	Yes	?	?
PLP-B	29	4	39	Yes	?	?
Trophoblast						
PL-I	30 and 36–40	8	38	Yes	Yes	No
PL-II	25	4	47	No	Yes	No
PLP-A	29 and 33	5	39	Yes	?	?
PLP-B	24–30	4	39	Yes/?	?	?
PLP-C	25 and 29	7	34	Yes/?	?	?
PL-Iv	29 and 33	5	43	Yes	Yes	?
PL-Im	?	5	44	Yes/?	?	?

Ant. Pit., Anterior pituitary; Con A, concanavalin A; PRL-R, PRL receptor; GH-R, GH receptor.
[a]Comparisons include conservative amino acid changes.

molecular weight (Jayatilak *et al.*, 1985). Decidual LTH did not cross-react with antibodies to pituitary PRL (Jayatilak *et al.*, 1989). Isolation and characterization of LTH is necessary to determine the true nature of this protein.

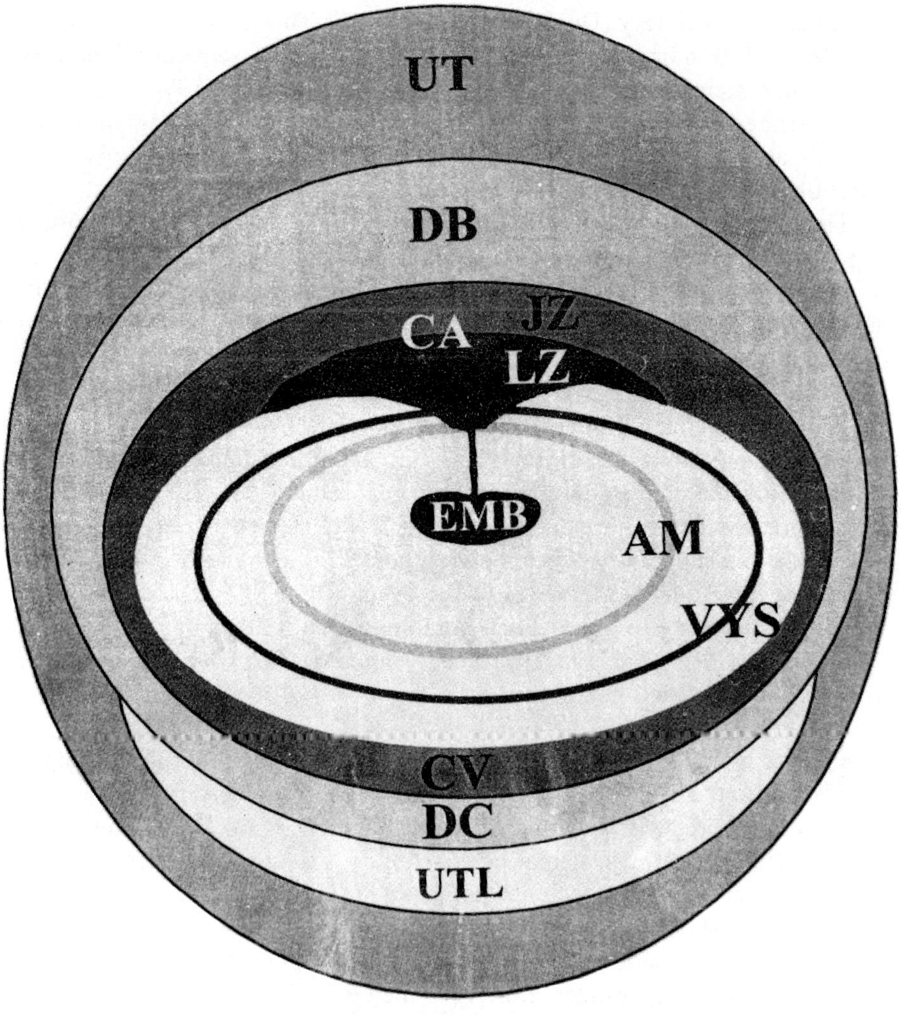

Fig. 2. Schematic drawing of a midpregnant rat conceptus. UT, uterus; DB, decidua basalis; CA, chorioallantoic placenta; JZ, junctional zone; LZ, labyrinth zone; EMB, embryo; AM, amnion, VYS, visceral yolk sac; CV, choriovitelline placenta; DC, decidua capsularis; UTL, uterine lumen. Derived from information provided by Anderson *et al.* (1959).

Recently two more PRL-like proteins were identified and characterized in rat decidua. These are prolactin like protein-B (PLP-B) and Decidual PRL-related protein (dPRP). cDNA clones of these proteins have been isolated and characterized. Considerable information regarding the biochemical nature and expressions of these protein in rat decidua are available. Whether PLP-B and dPRP represent the decidual LTH is currently being investigated.

The PLP-B Protein is an interesting exception to the other members of the rat placental PRL-like proteins in that it is expressed in both decidua and placenta. The cDNA clone of PLP-B was first isolated from a late term placental cDNA library (Duckworth *et al.*, 1988). This clone codes a protein which has 44% homology to rat PRL. This also shows sequence similarity with other members of the family. It was Croze *et al.* (1990) who first demonstrated the expression of this protein in decidua. More regarding the identification and characterization of PLP-B is discussed under placental PRL-like proteins.

The dPRP protein was purified from the medium conditioned by decidual explants. The 29 kDa protein is a glycoprotein and showed crossreactivity with antibodies to human PRL and other members of the family viz. PLP-B and PLP-C. The N-terminal sequencing of the isolated dPRP shared significant sequence identity with the N-terminal sequence of pituitary PRL and PLP-C (Roby *et al.*, 1993). The cDNA clone of dPRP was isolated from rat decidual cDNA library using the PLP-C cDNA as a probe. Nucleotide sequence analysis of the dPRP cDNA predicted a mature protein of 239 amino acids, including a 28-amino acid signal sequence (Roby *et al.*, 1993). The predicted dPRP aminoacid sequence contains two putative *N*-linked glycosylation sites and 6 cysteine residues (Fig. 1). The 6 cysteines are located in positions homologous to the cysteines of PLP-C and PRL. The dPRP also shows strong sequence homologies with other members of the family. The dPRP gene is localized in chromosome 17, which carries other members of the PRL gene family.

Temporal and tissue specific expression Croze *et al.* (1990) have demonstrated that PLP-B is specifically expressed in antimesometrial decidual tissue of the pregnant and pseudopregnant rat. PLP-B cDNA which was cloned from the term placental cDNA library was used to investigate the expression of PLP-B in decidual tissue. *In situ* hybridization study of day 10 conceptus shows that PLP-B antisense probe strongly hybridizes to the antimesometrial decidual cells surrounding the implantation site. The PLP-B antisense probe hybridizes strongly to the decidual capsularis. A temporal study of PLP-B expression during pregnancy shows that PLP-B is continued to be expressed in decidual cells till day 13 when it is concomitantly expressed in placental trophoblastic cells (Fig. 3) (Croze *et al.*, 1990). The expression of PLP-B in decidual tissue is relatively lower than term placental tissue. More about the expression of PLP-B in placental tissue is discussed below.

Examination of dPRP gene expression in pregnant and pseudopregnant rats clearly demonstrated that this gene is expressed only in decidualized uterus (Gu *et al.*, 1994). Developmental studies indicated a constant high level of dPRP mRNA in decidua tissue until day 12 of pseudopregnancy, followed by a marked decline, suggesting that dPRP is constitutively expressed in decidua. Northern blot analysis of antimesometrial and mesometrial decidual tissue revealed that dPRP gene is abundantly and solely expressed by antimesometrial tissue. Mesometrial tissue do not express this gene and showed no influence on the expression of this gene in antimesometrial tissue. dPRP is an excellent indicator of decidualization and marker of antimesometrial cell differentiation.

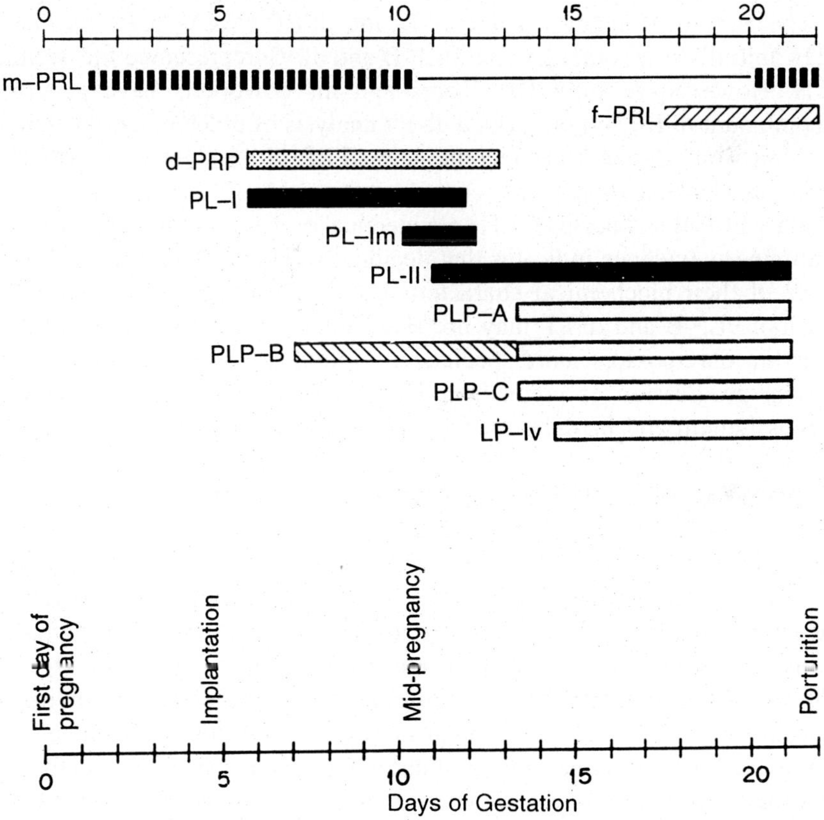

Fig. 3. Diagrammatic representation of the temporal pattern of expression of the PRL gene family during pregnancy in the rat. m-PRL, maternal anterior pituitary PRL; f-PRL, fetal anterior pituitary PRL. Other abbreviations used are explained in Figure 1. Scale at bottom and top of the diagram refers to the duration of gestation in the rat. Temporally significant events associated with pregnancy and rat PRL family are indicated in the figure (modified from Soares *et al.,* 1991).

Functions of decidual PRL-like protein Expression of PLP-B and dPRP in antimesometrial tissue is very interesting. Because these cells undergo extensive reorganization during pregnancy. Stromal cells proliferate and differentiate into decidual cells till day 11 of pregnancy (Lobel *et al.*, 1965a, b). After that growth stops, cells do not enlarge, mitoses are few and decidual begins to degenerate (Lobel *et al.*, 1965a). PLP-B and dPRP are expressed from approximately day 7 to 13 day of pregnancy in decidua. Expression of dPRP is inhibited by day 13 and PLP-B expression is switched to placental spongiotrophoblast cells. The regulation and function of dPRP and PLP-B in pregnancy remain unknown but purification of the native proteins as well as their production by recombinant techniques and search for specific receptors are necessary steps in examining its biological effects and function.

The pattern of synthesis of PLP-B and dPRP closely parallels that of LTH. Initially it was thought that PLP-B and dPRP represented the decidual LTH. However, recently Dr. Soares and his coworkers have generated recombinant dPRP and performed direct analysis of dPRP on corpus luteum and function. Unlike LTH recombinant dPRP did not interact with PRL receptors (Dr Soares, personal communication). At present it is not known whether PLP-B and decidual LTH are identical or different proteins. However, preliminary evidence indicates that decidual LTH and PLP-B may differ in some of their biochemical characteristics (Roby and Soares, unpublished results). PLP-B and dPRP may not share luteotropic function like PRL or LTH but may possess more specialized biological actions required for the normal progression of pregnancy. Their actions may extend beyond the corpus luteum and include other putative target tissues.

Expression of PRL-Family Genes in Placenta

Placenta lies at the interface between maternal and fetal tissue (Fig. 2). A very good description of morphology of rat uteroplacental unit is provided by Soares *et al.* (1991). For clear understanding the reader should refer to the above publication. In brief, the placental structures in rat are developed from the trophectoderm cells. Trophectoderm cells first differentiate to *choriovitelline* placenta then to *chlorioallantoic* placenta. Choriovitelline placenta regresses by day 14 of pregnancy. Chorioallantoic placenta develops before the complete degeneration of choriovitelline placenta. Chorioallantoic placenta is located in the mesometrial region. It is composed of two distinct zones: *junctional zone* and *labyrinth zone.* Junctional zone (sometimes called *basal zone*) is located close to the *decidua basalis.* The labyrinth zone is located closer to the embryo. Labyrinth zone is comprised of trophoblast cells (viz. *cytotrophoblasts, giant cells and syncytiotrophoblasts*) and contains fetal and maternal blood vessels. Junctional zone is also composed of fetal trophoblast cells (viz. *giant cells, spongiotrophoblast cells and glycogen cells*) and contains maternal blood vessels but no fetal blood supply.

Presence of a lactogenic factor in the placenta was first demonstrated in human (Jasimovich and Maclarren, 1962), which is now known as human placental lactogen (hPL). hPL is the most abundant protein synthesized by third trimester human placenta. Approximately 1-2 g of hPL is produced by placental tissue in the later half of pregnancy. However, cDNA cloning and sequence analysis revealed that hPL is more closely related to growth hormone (GH) than PRL. PRL and GH are very closely related and considered members of the same family. They show significant structural and functional homologies. Strong evidences are available to suggest that PRL and GH have evolved from a single ancestral gene. These PRL and GH genes then have taken different course of evolution in different species. In rodents (rat, mouse, hamster etc.) and ungulates (cow, sheep, goat etc.) PRL gene have undergone the process of gene duplication resulting in the expansion of PRL-family members. While in primates, including human, it is the GH gene which has undergone the duplication causing the expansion of GH gene family. Besides rat, investigations with mouse, cow, sheep placental tissues have resulted in the identification of several different proteins with PRL-like characteristics (Ogren and Talamantes, 1988; Schuler *et al.*, 1993; Soares *et al.*, 1991). Structurally, the placental PRLs from nonprimate species are closely elated to pituitary PRL (Nicoll *et al.*, 1986), while primate placental PRLs are closely related to pituitary growth hormone (GH) (Parks, 1989). However, the structural difference can be reconciled by the fact that primate GHs have both GH and PRL actions (Nicoll *et al.*, 1986). With the help of molecular biology techniques multiple GH related genes are identified in human which are reported to be expressed specifically in placenta (Cooke *et al.*, 1993).

Biochemical and molecular characterization of rat placental PRLs

Robertson and Firesen (1981) using a radio receptor assay (RRA) had shown that there are two peaks of lactogenic activity in pregnant rat serum: the first activity, which was designated as *placental lactogen I* (PL-I), was detected from day 8-14, the second peak was seen from day 12 to term at day 21 was designated as *placental lactogen II* (PL-II). PL-I purified and was found to be a glycoprotein, and secreted as a heterogeneous complex (36-42 kDa) (Colosi *et al.*, 1987a). PL-II is not glycosylated and is secreted as a 25 kDa protein (Deb *et al.*, 1989a). cDNAs for the mRNAs encoding PL-I and PL-II proteins have been isolated and characterized (Colosi *et al.*, 1987b; Duckworth *et al.*, 1986a; Jackson *et al.*, 1986; Robertson *et al.*, 1990). Both cDNAs and the proteins they encode show significant sequence similarity to one another and to other members of PRL-GH family (Fig. 1). Nucleotide sequence analysis of PL-I cDNA revealed that it is composed of 230 amino acids, including 30 amino acids signal peptide sequence. There is a marked conservation of cysteine residues throughout the family. Seven cysteine residues are detected in PL-I amino acid sequence. PL-I protein conserve four of the residues in homologous region of pituitary PRL (Fig. 1). Two potential glycosylation sites are detected in predicted PL-I sequence. PL-II is composed of 221 amino acid sequence of which 31 amino acids are signal sequence (Duckworth *et al.*, 1986a). Conservation of cysteine are also evident in this molecule.

Subsequent experimentation led to the identification of five more placental PRL-like proteins. During cloning of the rat PL-II cDNA, two additional members of the family were discovered and characterized (Duckworth *et al.*, 1986a, 1986b, 1988). The proteins encoded by these cDNAs have been referred as *PRL-like protein-A* (PLP-A) and *PRL-like protein-B* (PLP-B). Generation of the antibodies to predicted amino acid sequences of the PLP-A protein has permitted its identification, characterization of some of its biochemical features, and determination of its pattern of expression (Campbell *et al.*, 1989; Deb *et al.*, 1989b). PLP-A is a glycoprotein, synthesized as two distinct molecular weight species (29 and 33 kDa). The 29 kDa form, unlike the 33 kDa form is also directed to the nucleus and retained by the nuclei of the trophoblast cells (Campbell *et al.*, 1989; Deb and Soares, 1990). The amino acid structure determined for PLP-A and PLP-B by cDNA clone analysis revealed several features in these molecules which are common and shared by other members of the PRL-GH family (Fig. 1). PLP-A is composed of 227 amino acid molecule with 31 amino acid signal sequence (Duckworth *et al.*, 1986b). Characterization of PLP-B indicates that it is also a glycoprotein ranging from 24-30 kDa (Roby and Soares, 1991; Ogilvie *et al.*, 1990b). PLP-B is predicted to be composed of 234 amino acids including 33 signal sequence. PLP-A is detected in both maternal and fetal circulation (Deb and Soares, 1990). PLP-A primarily circulates as high molecular weight complexes (Deb and Soares, 1990).

The two new members, PRL-like protein-C (PLP-C) and PL-I variant (PL-Iv), were discovered during the purification of PLP-A (Deb and Soares, 1990; Deb *et al.*, 1991a, b). Identification and characterization of PL-Iv and PLP-C were also reported by others (PL-Iv, Robertson *et al.*, 1991; PLP-C, Ogilvie *et al.*, 1990a; Conlifee *et al.*, 1994). PLP-C was identified as a major contaminating protein of PLP-A with a similar molecular weight but possessing a more acidic isoelectric point and different immunoreactivities. N-terminal amino acid sequence of PLP-C showed considerable sequence homology with other members of the family, including the characteristic positioning of cysteine residues at position 4 and 11. Antisera generated to purified proteins and to PLP-C peptides specifically reacted with two major species migrating at approximately 25 and 29 kDa, respectively. The 29 kDa specifically bound to Concanavalin-A while 25 kDa failed to bind to the lectin. Furthermore, the 29 kDa could be converted to 25 kDa protein by enzymatic deglycosylation. PLP-C cDNAs have been isolated from a term placental cDNA library (Deb *et al.*, 1991c). Nucleotide sequence analysis of PLP-C cDNA clone predicate a mature protein of 238 amino acids, including a 30 amino acid signal sequence. The predicted amino acid sequence of PLP-C contains seven cysteine residues. Six of the cysteine residues in PLP-C are located in position homologous to those of pituitary PRL. Additional sequence similarities with pituitary PRL and other members of the family are evident. Northern blot analysis showed that the PLP-C cDNA clone specifically hybridized to 1.0 kb mRNA (Deb *et al.*, 1991c).

PL-Iv has been discovered after two independent observation: by protein (Deb *et al.*, 1991a) and by cDNA (Robertson *et al.*, 1991) analyses. PL-Iv protein was also isolated as a contaminant in the purification of PLP-A and was shown to possess extensive amino acid sequence identity to PL-I, 39 of 45 N-terminal amino acids were identical (Deb *et al.*, 1991a). PL-Iv crossreacted with antibodies to PL-I and was also found to be a glycoprotein. PL-Iv isolated from placental cytosol possesses two major molecular weight species with virtually identical electrophoretic mobilities as that of PLP-A (29 and 33 kDa species, Deb *et al.*, 1989b; Deb *et al.*, 1991a). Despite structural similarities between PL-I and PL-Iv, distinct differences are also evident (Deb *et al.*, 1991a). Antibodies generated to N-terminal 19 amino acids of PL-Iv specifically recognized PL-Iv, while failing to recognize PL-I. Secreted PL-Iv has an affinity for Concanavalin-A, whereas secreted PL-I lacks affinity for the lectin. PL-I is predominantly secreted as 36-40 kDa species, while PL-Iv is secreted as 33 kDa species. cDNAs to PL-Iv were isolated and sequenced revealing a sequence closely related to PL-I but are clearly distinct (85-90% sequence identity at the amino acid level). Nucleotide sequence analysis of the PL-Iv cDNA clone predicted a mature protein of 223 amino acids, including a 28-amino acids signal sequence (Robertson *et al.*, 1991; Cohick *et al.*, 1996).

Recently, Shiota and his colleagues (Hirosawa *et al.*, 1994) have reported a new member of the family, which is very closely related to PL-I and PL-Iv. It shares strong sequence homologies with both of these protein and hence named PL-I mosaic (PL-Im). Intense mRNA bands of approximately 1.0 kb were detected in the placentas obtained during mid-pregnancy (days 10 and 12), whereas the expression was not evident on the other days. They isolated a cDNA for PL-Im from a mid pregnant (D 12) rat placental cDNA library. The cDNA encode for a protein of 229 amino acids, including a signal sequence of 28 (Hirosawa *et al.*, 1994). Amino acid sequence analysis predicted it to be a glycoprotein. The cDNA was highly homologous to those of other rat PRL-family members; in particular, the homology among PL-Im, PL-I and PL-Iv was over 90%. Interestingly, the nucleotide sequence of PL-Im was also a mosaic of PL-I and PL-Iv. Southern blot analysis of PL-Im was distinct from that of PL-I and PL-Iv.

Pituitary PRL, PL-Iv, PL-II, PLP-A, PLP-B and PLP-C have each been localized to chromosome 17 of the rat genome, whereas rat pituitary GH has been localized to chromosome 10 (Deb *et al.*, 1991b; Cooke *et al.*, 1986; Levan *et al.*, 1991). Chromosomal localization of PL-I and PL-Im have not been determined.

Temporal and tissue specific expression of rat placental PRL related proteins
The seven placental PRL-like hormones are expressed during different phases of gestation and by different cell types. The temporal pattern of expression of the rat PRL family are summarized in Fig. 3.

PL-I expression is initiated shortly after implantation in mural trophoblast giant cells of the implanting blastocyst and later extends to the choriovitelline and chorioallantoic placentas (Faria *et al.*, 1990). Within a 48-h period (between days 10 and 12) of gestation in the rat PL-II expression is initiated and PL-I expression is terminated (Faria *et al.*, 1990; Soares *et al.*, 1985). At midgestation, trophoblast giant cells of the choriovitelline placenta are the primary source of PL-I and PL-II (Faria *et al.*, 1990; Soares *et al.*, 1985). Mechanism responsible for the initiation of PL-II production and termination of PL-I production are yet to be determined. The choriovitelline placenta degenerates between days 13 and 14 of gestation (Welsh and Enders, 1985), and this is accompanied by a major shift in PL-II expression to trophoblast giant cells of choriollantoic placenta. PL-II is initially expressed entirely by trophoblast giant cells of the junctional zone and then, toward the end of gestation, trophoblast giant cells of the labyrinth zone also acquire the capability of expressing PL-II (Campbell *et al.*, 1989; Deb *et al.*, 1991a; Soares, 1987). Both PL-I and PL-II hormones are present in high concentrations in maternal circulation during specific time periods of gestation (Robertson *et al.*, 1982; Soares *et al.*, 1983; Ogren *et al.*, 1989).

PLP-A, PLP-B, PLP-C and PLP-Iv, are restricted to the junctional zone of the chorioallantoic placenta (Campbell *et al.*, 1989; Ogilvie *et al.*, 1990; Deb *et al.*, 1991a, b, c; Duckworth *et al.*, 1990). Northern blot analysis, *in situ* hybridization and immunocytochemical analysis indicated that PLP-A, PLP-C, PL-Iv are primarily localized in the spongiotrophoblasts and some giant cells of junctional zone of chorioallantoic placenta. Their expression were initiated just after midgestation and continued till parturition (Fig. 3), (Campbell *et al.*, 1989; Ogilvie *et al.*, 1990; Deb *et al.*, 1991a, b, c; Duckworth *et al.*, 1990). Placental expression of PLP-B is restricted to spongiotrophoblast cells (Duckworth *et al.*, 1990). Northern blot analysis showed that PL-Im mRNA was expressed specifically during mid-pregnancy (days 10-12) in the rat placenta (Hirosawa *et al.*, 1994). It is important to note that PL-I, PL-Im and PL-Iv although structurally very similar to each other are produced at different time of pregnancy by different cell types. PL-I is produced by trophoblast giant cell during the first half of pregnancy and PL-Im is expressed by trophoblast giant cells during midgestation (Hirosawa *et al.*, 1994). While PL-Iv is produced primarily by spongiotrophoblast cells during the later half of gestation.

Factors involved in controlling the expression of the rat placental PRL-family are poorly understood. Evidences are available to suggest the role of factors from pituitary, uterus and ovary (see reviews by Ogren and Talamantes, 1988; Soares *et al.*, 1991). There are evidence to suggest that pituitary GH, ovarian estrogen and placental progesterone are responsible for the reduction of serum PL-II concentration. However, any direct effect of these hormones on PL-II expression has not been demonstrated. Calcium binding protein, *calcyclin,* has been reported to stimulate mouse PL-II secretion (Thordarson *et al.*, 1991). Calcyclin may modulate PL-II production by influencing the availability of calcium. Role of fetal factors in regulating PL-II concentration in serum is suggested by the experiments performed by Roby and Soares (1993). Factors responsible for controlling the expression of PLP-A, PLP-B, PLP-C, PL-Iv and PL-Im remain to be elucidated.

Functions of placental PRL-like proteins It is more than fifty years ago, Astwood and Greep observed that removal of the anterior pituitary in the rat or mouse after midgestation was compatible with the continuation of pregnancy, fetal growth and development (Astwood and Greep, 1938). Later on, with the advent of molecular biology techniques it has been shown that in at least three mammalian groups, the primates, rodents and ruminants, the uteroplacental unit is the chief source of PRL-like proteins in placenta which play important roles in pregnancy (see reviews, Ogren and Talamantes, 1988; Soares *et al.*, 1991).

Role of placental PRL-like protein in mammary gland development Mammary gland undergoes significant structural and functional changes during pregnancy. Pituitary PRL, GH, placental lactogens (PLs) and steroid hormones are reported to participate in this event (see review by Thordarson and Talamantes, 1987). In rodents and ruminants considerable evidences are available to suggest the role of PLs in mammary gland development. Homologous PLs are reported to stimulate the milk protein synthesis in mammary tissue *in vitro* in mice, sheep and goats (Forsyth, 1991). Talamantes and his coworkers have demonstrated differential role of mouse PLs on mammary gland development. PL-I was found to be more potent in stimulating insulin-like growth factor binding protein synthesis, while PL-II was shown to be more potent in stimulating α-lactalbumin synthesis (Fielder *et al.*, 1992). This probably suggest that PL-I effects the growth and PL-II functions of developing mammary gland. Role of rat placental PRL in mammary gland development comes from the studies by Teshima *et al.* (1983) and Faria *et al.* (1990). Teshima observed that female hosts bearing the rat choriocarcinoma showed extensive mammary gland development. Faria *et al.* (1990) demonstrated that rat choriocarcinoma predominantly secretes PL-I. Since mammary gland development continues throughout the pregnancy, even after the termination of PL-I synthesis, it is likely that other rat placental PRLs are also involved in this process.

Placental PRLs as regulators of maternal physiology It has been demonstrated that placental lactogens of various species function as fetal signals which modulate the maternal metabolism during pregnancy and thereby effect the normal growth and development of the fetus. To meet fetal needs for glucose and amino acids, maternal metabolism is altered to spare these fuels and to increase lipid mobilization. Involved in this adaptation is increased food intake, increased insulin response, hyperglycemia and increased lipid mobilization (reviewed by Ogren and Talamantes, 1988). Fasting during pregnancy leads to elevation of maternal placental lactogen in women (Kim and Felig, 1971) and in mice (Fielder *et al.*, 1987).

Placental PRLs as regulators of fetal physiology There is growing evidence that placental lactogens directly act as signals to fetal growth. Human placental lactogen has been shown to stimulate amino acid transport, DNA synthesis and mitogenesis (Hill *et al.*, 1986; Strain *et al.*, 1987; Swenne *et al.*, 1987a, b). Both human placental lactogen and human GH stimulate DNA synthesis in fetal hepatocyte. Human PL stimulates insulin-like growth factor (IGF) production from fetal liver, pancreas and fibroblasts (Hill *et al.*, 1989). Hypophysectomy studies in sheep indicates that fetal IGFs, specially IGF-I may be regulated by PL (Mesiano *et al.*, 1989). Ovine PL has a variety of anabolic and metabolic effects when tested on fetal rat tissues while Ovine GH or PRL are less potent or have no effect. Strong evidence suggest the existence of specific lactogenic receptors in fetal tissues which interact minimally with pituitary prolactin and growth hormone (Freemark and Handwerger, 1986; Freemark *et al.*, 1993). Ovine PL has been shown to stimulate ornithine decarboxylase activity, amino acid transport, glycogen synthesis, and somatomedin synthesis in fetal rat tissues (Hurley *et al.*, 1980; Freemark and Handwerger, 1983, 1984; Adams *et al.*, 1983).

Role of placental PRLs in ovarian functions The role of placental PRLs in regulating steroidogenesis has been examined in rat, mouse, sheep and human. These investigations have been focused on the effects of placental PRLs on the production of progesterones, which is required for the maintenance of pregnancy. Investigations carried out by various laboratories suggest that at least part of the luteotropic activity in rat placental extract is associated with PL-I and PL-II (Ogren and Talamantes, 1988). Effects of pure preparations of PL-I and PL-II on rat steroidogenesis have not been examined. Role of PL-Iv, PL-Im, PLP-A, PLP-B and PLP-C in ovarian functions are yet to be examined.

PRL-family proteins as immunomodulators during pregnancy Both PRL and placental lactogens have been implicated in specific interactions with the immune system. PRL is now considered as a hematopoietic growth factor which exert immunomodulatory functions. Inhibition of prolactin release causes immunosuppression (for review see Gala, 1991; Hooghe *et al.*, 1993). Recent work has demonstrated a direct effect of PRL in eliciting the proliferation of lymphocytes *in vitro* (Mukherjee *et al.*, 1990; Vaiselli *et al.*, 1991). Peripheral T lymphocytes are reported to produce PRL and to possess PRL receptors suggesting an autocrine or paracrine effect of PRL on these cells (Pellgrini *et al.*, 1992). Furthermore, the effect of PRL on lymphocytes is associated with an increase in IL-2 production and IL-2 receptor expression (Mukherjee *et al.*, 1990; Vaiselli *et al.*, 1991; Clevenger *et al.*, 1990 and 1991), suggesting that the stimulation of the IL-2 system may be one of the factors responsible for the mitogenic effects of PRL. We and others have demonstrated that like PRL, PL-I, PL-II and PL-Iv are capable of stimulating the proliferation of rat lymphoid cell lines (Robertson *et al.*, 1982; Deb *et al.*, 1991a; Cohick *et al.*, 1996). Whether these observation indicate the involvement of placental PRLs in immunological adjustment of pregnancy remains to be determined.

Role of placental PRLs in angiogenesis In mouse, four different placental PRL-like proteins are reported (Linzer *et al.*, 1985; Colosi *et al.*, 1988). These include, placental lactogen-I (PL-I), PL-II, proliferin (PLF) and proliferin related protein (PRP). Mouse PL-I and mPL-II are also reported to share the functional similarities with pituitary PRL. PLF and PRP do not appear to possess the same spectrum of biological activities the PRL possess. Recently Linzer and his colleagues provided strong evidence that PLF and PRP are involved in angiogeneis during pregnancy (Jackson *et al.*, 1994). They have demonstrated that angiogenic activity of placental secretion is associated with PLF while antiangiogenic activity is associated with PRP (Jackson *et al.*, 1994). The effects of rat placental PRLs in angiogenesis are currently being investigated.

Placental PRLs as a marker of normal placental function and fetal growth The mouse placental lactogen gene is activated early during pregnancy and is a marker of trophoblast cell differentiation. Transcription factors GATA-2, GATA-3 are reported to be responsible for the tissue specific expression of mouse PL-I gene (Ng *et al.*, 1994). GATA-3 deficient mice have several abnormalities in the central nervous system and display extensive hemorrhaging and altered hematopoiosis. Embryos homozygous for a disrupted GATA-3 gene die between day 11-12 of gestation. Interestingly, Ma *et al.* (1995) presented evidence that the GATA-3 deficient mice have lower PL-I expression then normal mice. The level of PL-I mRNA therefore provides an early molecular marker for defects in placental development and hence fetal development.

Conclusion

We have made significant progress towards understanding the role of PRL-family proteins in pregnancy. We have clearly demonstrated that multiple PRL-like proteins are synthesized in placenta and decidua during pregnancy. The expression of these genes is highly specific and developmentally regulated. Isolation and molecular characterization of cDNA clones have provided important structural information regarding these proteins and much of our current speculation regarding their function is based on this information. Structurally, these proteins represent natural mutants of pituitary PRL. They may function as agonists or antagonists of pituitary PRL. There are evidence which strongly suggests that these proteins replace pituitary PRL hormone or modulate its function during pregnancy. Currently work aiming at determining the PRL and GH like actions of the newly identified members of PRL-family is underway.

Investigation to determine whether there are unique biological actions specific to each of these proteins that are physiologically important to a given phase of pregnancy is also being carried out. Several of these PRL-family protein are being produced by recombinant means (Colosi *et al.*, 1987b; Deb *et al.*, 1993; Cohick *et al.*, 1996). These recombinant proteins provide us with promising tools to dissect the functional significance of these proteins during pregnancy. Furthermore, these advances should allow us to address the intriguing question as to why the anterior pituitary, decidua, and placenta engage in producing such highly homologous ligands during pregnancy.

Acknowledgements

I would like to express my deep sense of gratitude to Prof. Michael J. Soares, University of Kansas Medical Center, Kansas City, Kansas, for introducing me to the field and providing me all the support and encouragement to do research on pregnancy associated prolactin-like proteins.

References

Adams, S.O., Niley, S.P., Handwerger, S., and Rechler, M.M. (1983). Nature **302:** 150–153.

Anderson, J.W. (1959). Amer. J. Anat. **104:** 403–430.

Astwood E.B., Greep, R.O. (1938). Roce. Soc. Exp. Biol. **38:** 713–716.

Bigazzi, M., Nardi, E., Bruni, P., and Petrucci, F. (1980). J. Clin. Endocrinol. Metab. **51:** 939–941.

Campbell, W.J., Deb, S., Kwok, S.C.M., Joslin, J., and Soares, M.J. (1989). Endocrinol. **125:** 1565–1574.

Clevenger, C.V., Altman, S.W., and Prystowsky, M. (1991). Science **253:** 77–79.

Clevenger, C.V., Rusell, D.H., Appasamy, P.M., and Prystowsky, M. (1990). Proc. Natl. Acad. Sci. USA **87:** 6760–6764.

Cohick, C.B., Dai, G.D., Xu, L., Deb, S., Kamei, T., Levan G., Szpirer C., Kwok S.C.M., and Soares, M.J. (1996). Mol. Cell. Endocrinol. **116:** 49–58.

Colosi, P., Swiergiel J.J., Wilder, E., Oviedo, A., Linzer, D.I.H. (1988). Mol. Endocrinol. **2:** 579–586.

Colosi, P., Ogren, L., Thordarson, G., and Talamantes, F. (1987a). Endocrinol. **12:** 2500-2511.

Colosi, P., Talamantes, F., and Linzer, D.I.H. (1987a). Mol. Endocrinol. **1:** 767–776.

Conliffe, P.R., Farmerie, W.G., Charles, G.D., Buhi, W.C., Kelly, P.A., Simmen, R.C.M., and Shiverick, K.T. (1994). Mol. Cell. Endocrinol. **106:** 121–130.

Cooke, N.E., Jones, B.K., Urbanek, J.M., Misra-Press, A., Lee, A.K., Russel, J.E., Macleod, J.N., and Liebhaber, S.A. (1993). In: Trophoblast cells, pathways or maternal-embryonic communication. (M.J. Soares, S. Handwerger, and F. Talamantes, Eds). pp. 222–239. Springer Verlag, New York.

Cooke, N.E., Szpirer, C., and Levan, G. (1986). Endocrinol. **119:** 2451–2454.

Croze, F., Kennedy, T.G., Schroedter, I.C., and Friesen, H.G. (1990). Endocrinol. **127:** 2665–2672.

Deb, S., Faria, T.N., Roby, K.F., Larsen, D., Kwok, S.C.M., Talamantes, F., and Soares, M.J. (1991a). J. Biol. Chem. **266:** 1605–1610.

Deb, S., Hamlin, G.P., Roby, K.F., Kwok, S.C., and Soares, M.J. (1993). J. Biol. Chem. **268:** 3298–3305.

Deb, S., Hashizume, K., Boone, K., Southard, J.N., Talamantes, F., Rawitch, A., and Soares, M.J. (1989a). Mol. Cell. Endocrinol. **63:** 45–46.

Deb, S., Roby, K.F., Faria, T.N., Larsen, D., and Soares, M.J. (1991b). Endocrinol. **128:** 3066–3072.

Deb, S., Roby, K.F., Faria, T.N., Szpires, C., Levan, G., Kwok, S.C., and Soares, M.J. (1991c). J. Biol. Chem. **266:** 23027–23032.

Deb, S. and Soares, M.J. (1990). Mol. Cell. Endocrinol. **74:** 163–172.

Deb, S., Youngblood, T., Rawitch, A., and Soares, M.J. (1989b). J. Biol. Chem. **264:** 14348–14353.

De Feo, V.J. (1967). Decidualization. In: Wynn, R.M. (Ed.) Cellular Biology of the Uterus. Appleton-Centruy-Crofts, New York, pp. 191–290.

Duckworth, M.L., Kirk, K.L., and Friesen, H.G. (1986a). J. Biol. Chem. **261:** 10871–10878.

Duckworth, M.L., Peden, L.M., and Friesen, H.G. (1986b). J. Biol. Chem. **261:** 10879–10884.

Duckworth, M.L., Peden, L.M., and Friesen, H.G. (1988). Mol. Endocrinol. **2:** 912–920.

Duckworth, M.L., Schroedter, I.C., and Friesen, H.G. (1990). Placenta **11:** 143–155.

Faria, T.N., Deb, S., Kwok, S.C.M., Talamantes, F., and Soares, M.J. (1990). Develop. Biol. **141:** 279–291.

Fielder, P.J., Ogren, L., Edwards, D., and Talamantes, F. (1987). Am. J. Physiol. **253:** E40–E44.

Fielder, P.J., Thordarson, G., English, A., Rosenfeld, R.G., Tamalantes, F. (1992). Endocrinology **131:** 261–267.

Finn, C.A. (1971). Adv. Reprod. Physiol. **5:** 1–26.

Forsyth, I.A. (1986). J. Dairy Sci. **69:** 886–903.

Forsyth, I.A. (1991). Oxf. Rev. Reprod. Biol. **13:** 97–148.

Freemark, M. and Handwerger, S. (1983). Endocrinol. **112:** 402–404.

Freemark, M. and Handwerger, S. (1984). Am. J. Physiol. **246:** E21–E24.

Freemark, M. and Handwerger, S. (1986). Endocrinol. **118:** 613–618.

Freemark, M., Kirk, K., Pihoker, C., Robertson, M.C., Shiu, R.P., and Driscoll, P. (1993). Endocrinol. **133:** 1830–1842.

Gala, R.R. (1991). Proc. Soc. Exp. Biol. Med. **198:** 513–527.

Gibori, G., Rothchild, I., Pepe, G.J., Morishige, W.K., and Lam, P. (1974). Endocrino. **95:** 1113–1118.

Golander, A., Hurley, T., Barrett, J., and Handwerger, S. (1979). J. Endocrinol. **82:** 263–267.

Gu, Y., Soares, M.J., Srivastava, R.K., and Gibori, G. (1994). Endocrinology **135:** 1422–1427.

Gunnet, J.W. and Freeman, M.E. (1983). Endocrine Review **4:** 44–61.

Hill, D.J., Crace, C.J., Strain, A.J., and Milner, R.D. (1986). J. Clin. Endocrinol. Metabol. **62:** 753–760.

Hill, D.J., Camacho-Hubner, C., Rashid, P., Strain, A.J., and Clemmons, D.R. (1989). J. Endocrinol. **122:** 87–98.

Hirosawa, M., Miura, R., Min, K.-S., Hattori, N., Shiota, K., and Ogawa, T. (1994). Endocr. J. **41:** 387–397.

Hooghe, R., Delhase, M., Vergani, P., Malur, A., and Hooghe-Peters, E.L. (1993). Immunol. Today **14:** 212–214.

Hurley, T.W., Kuhn, C.M., Schanberg, S.M., and Handwerger, S. (1980). Life Sci. **7:** 2269–2275.

Jackson, L.L., Colosi, P., Talamantes, F., and Linzer, D.I.H. (1986). Proc. Natl. Acad. Sci. USA **83:** 8496–8500.

Jackson, D., Volpert, O.V., Bouck, N., and Linzer, D.I.H. (1994). Science **266:** 1581–1584.

Jayatilak, P.G., Glaser, L.A., Basuray, R., Kelly, P.A. and Gibori, G. (1985). Proc. Natl. Acad. Sci. USA. **82:** 217–221.

Jayatilak, P.G. and Gibori, G. (1986). J. Endocrinol. **110:** 115–121.

Jayatilak, P.G., Puryear, T.K., Herz, Z., Fazleabas, A., and Gibori, G. (1989). Endocrinol. **125:** 659–666.

Jasimovich, J.B., and Maclarren, J.A. (1962). Endocrino. **71:** 202–220.

Kim, Y.J. and Felig. P. (1971). J. Clin. Endocrinol. Metabol. **32:** 864–867.

Levan, G., Szpirer, J., Szpirer, C., Klinga, K., Hanson, C., and Islam, M.Q. (1991). Genomics **10:** 699–718.

Linzer, D.I.H., Lee, S.J., Ogren, L., Talamantes, F., Nathans, D. (1985). Proc. Natl. Acad. Sci. USA **82:** 4356–4359.

Lobel, B.L., Tic, L., and Shelesnyak, M.C. (1965a). Acta. Endocrinol. (Copenhagen). **50:** 452–468.

Lobel, B.L., Tic, L., and Shelesnyak, M.C. (1965b). Acta. Endocrinol. (Copengagen). **50:** 537–559.

Ma, G., Roth, M., Engel, J.D., and Linzer, D.I.H. (1995). Annual Meeting of the Endocrine Society, Abstract #P3–661, p. 634.

Mesiano, S., Young, I.R., Hey, A.W., Browne, C.A., and Thorburn, G. (1989). Endocrinol. **124:** 1485–1491.

Morishige, W.K. and Rothchild, I. (1974). Endocrinol. **95:** 260.

Mukherjee, P., Mastro, A.M., and Hymer, W.C. (1990). Endocrinol. **126:** 88–94.

Ng, Y.-K., George, K.M., Engel, J.D., and Linzer, D.I.H. (1994). Development **120:** 3257–3266.

Nicoll, C.S., Mayer, G.L., and Rusell, S.M. (1986). Endocrine Reviews **7:** 169–203.

Ogilvie, S., Buhi, W.C., Olson, J.A., and Shiverick, K.T. (1990a). Endocrinol. **126:** 3271–3273.

Ogilvie, S., Duckworth, M.L., Larkin, L.H., Buhi, W.C., and Hiverick, K.T. (1990b). Endocrinol. **126:** 2561–2566.

Ogren, L., Southard, J.N., Closi, P., Linzer, D.I.H., and Talamantes, F. (1989). Endocrinol. **125:** 2253–2257.

Ogren, L. and Talamantes, F. (1988). Internatl. Rev. Cytology **112:** 1–65.

Park, J.S. (1989). Acta Paediatr. Scand. (Suppl.) **349:** 127–135.

Pellgrini, I., Lebrun, J.-J., Ali, S., and Kelly, P.A. (1992). Mol. Endocrinol. **6:** 1023–1031.

Riddick, D.H., Luciano, A.A., Kusumik, W.F., and Maslar, I.A. (1978). Life Science **23:** 1913–1921.

Robertson, M.C., Croze, F., Schroedter, I.C., and Friesen, H.G. (1990). Endocrinol. **127:** 702–710.

Robertson, M.C. and Friesen, H.G. (1981). Endocrinol. **108:** 2388–2390.

Robertson, M.C., Gillispie. B., and Frieson, H.G. (1982). Endocrinol. **111:** 1862–1866.

Robertson, M.C., Gillispie, B., and Friesen, H.G. (1991). Endocrino. **129:** 2746–2756.

Roby, K.F. and Soares, M.J. (1991). Biochemical characterization of rat decidual and placental prolactinlike protein-B immunoreactive proteins. Program of the Annual Meeting of the Society for the Study of Reproduction Vancouver, B.C., Canada, p. 167 (Abstract).

Roby, K.F., Deb, S., Gibori, G., Szpirer, C., Levan, G., Kwok, S.C.M., and Soares, M.J. (1993). J. Biol. Chem. **268:** 3136–3142.

Roby, K.F. and Soares, M.J. (1993). Placenta **14:** 529–545.

Schuler, L.A. Scott, P., and Kessler, M.A. (1993). In: Trophoblast cells, pathways or maternal-embryonic communication. (M.J. Soares, S. Handwerger, and F. Talamantes, Eds). pp. 222–239. Springer Verlag, New York.

Soares, M.J. (1987). J. Reprod. Fertil. **79:** 335–341.

Soares, M.J., Colosi, P., Ogren, L., and Talamantes, F. (1983). Endocrinol. **112:** 1313–1317.

Soares, M.J., Faria, T.N., Roby, K.F., and Deb, S. (1991). Endocrine Reviews **12:** 402–423.

Soares, M.J., Julian, J.A., and Glasser, S.R. (1985). Dev. Biol. **107:** 520–526.

Strain, A.J., Hill, D.J., Swenne, I., and Milner, R.D. (1987). J. Cell. Physiol. **132:** 33–40.

Swenne, I., Hill, D.J., Strain, A.J., and Milner, R.D. (1987a). Diabetes **36:** 288–94.

Swenne, I., Hill, D.J., Strain, A.J., and Milner, R.D. (1987b). J. Endocrinol. **113:** 297–303.

Teshima, S., Sshimosato, Y., Koide, T., Kuroki, M., Kikuchi, Y., Aizawa, M. (1983). Gann **74:** 205–212.

Thordarson, G., Southard, J.N., and Talamantes, F. (1991). Endocrinol. **129:** 1257.

Thordarson, G. and Talamantes, F. (1987). In: Neville, M.C., Daniel, C. (Eds.) The Mammary Gland: Development, Regulation, and Function. pp. 459–498. Plenum Press, New York.

Vaiselli, S.M., Stanek, E.M., Mukherjee, P., Hymer, W.C., and Masro, A.M. (1991). Endocrinol. **129:** 983–990.

Welsh, A.O. and Enders, A.C. (1985). Am. J. Anat. **172:** 1–29.

Comparative Endocrinology and Reproduction
K.P. Joy, A. Krishna and C. Haldar (Eds)

23. Seasonality of Reproduction in Non-Human Primate

S.D. Kholkute and W.R. Dukelow*
Institute for Research in Reproduction (ICMR), Parel, Mumbai-400 012, India
*Endocrine Research Centre, Michigan State University, East Lansing, MI 48824, USA

Introduction

Natural selection favours adaptations that cause reproduction to occur in harmony with environmental variations. Most of the environment surrounding mammals is characterized by seasonal variation of one kind or another. Due to this, most animals exhibit at least a tendency towards reproductive seasonality while some breed during a discrete part of the year.

Reproduction in many non-human primate species has been reported to be seasonal characterized by discrete mating and birth seasons (Lindburg, 1987 for review). Although Zuckermann (1931) reported that most primates breed throughout the year, numerous studies have established temporal pattern of reproductive events in primates (Lancaster and Lee, 1965; Van Horn, 1980 for review). It must be realised that primates exhibit great variation in the length of interval between conceptions e.g. in chimpanzee, the interbirth interval encompasses 8-9 months of gestation followed by longer lactational period. On the other hand, a number of small sized, rapidly maturing prosimians and New World monkeys often produce two litters in a single year. Therefore, definition of birth season and birth peak is helpful for better understanding the seasonality of reproduction.

A birth season refers to a discrete period of the year to which all births are confined, whereas a birth peak denotes the period of the year in which high proportions of births but not all births are concentrated (Lancaster and Lee, 1965). Seasonality is a more inclusive term to denote any tendency towards temporal clustering of reproductive activity.

The seasonal components of the reproductive strategies have been shaped most strongly by the interactions of extrinsic and intrinsic variables. The major extrinsic variables are the nature, severeity and timing of the seasonal climate and dietary challenges. Furthermore, competition for resources with closely related populations and predator pressure also serve as extrinsic variables. The important intrinsic variables seem to be life span, body size, length of the female's reproductive cycle from maturation to weaning of offspring, feeding strategies, presence or absence of lactational amenorrhea, etc.

This article provides information on seasonal variations in reproduction in some of the non-human primates which are commonly used as animal models for research in human reproduction and fertility regulation. Furthermore, various proximate and ultimate factors are discussed and temporal variations in physiological functioning of males and females is also examined.

Seasonality in New World Primates

Information on seasonality in various reproductive events in New World monkeys is scanty. Even for species like squirrel monkeys and marmosets which are widely used as laboratory primates, little information is available from the wild population (Tables 1 and 2).

Callithrix: Rylands (1981) observed that in Brazil, the birth season for *C. humeralifer* was over a 7 months period (May to November) while birth peak was in September. A second birth clustering was observed between February and March. The common marmoset *C. jachhus,* is maintained in captivity in many primate facilities and appears to breed throughout the year (Epple, 1970; Hearn, 1983; Kholkute and Puri, 1988). Similarly, no distinct circannual variation in serum androgen levels was observed in adult males (Kholkute, 1984).

Saimiri: Seasonal variation in squirrel monkey is well documented. Various studies indicate that squirrel monkeys in their natural environment exhibit distinct seasonality of reproduction. Furthermore, seasonal variations in female cyclicity, birth of infants and in various male reproductive functions have been extensively studied (DuMond, 1968; Cooper, 1968). Subsequent field studies have confirmed these observations (Klein, 1971; Baldwin and Baldwin, 1981). However, a tendency towards shift in seasonality was also noted when compared to natural population. Dukelow *et al.* (1990) noted increased response to ovulation induction regimen during April-June compared to July-September. Similarly increase in *in vitro* maturation and fertilization rates were observed in April-June (Chan *et al.*, 1982).

Seasonality has also been reported for male squirrel monkey. Distinct annual cyclicity in "fatting" and sexual behaviour has been reported (Wilson *et al.*, 1978). This weight gain begins 2-3 months before mating activity. During mating season male squirrel monkeys showed higher testosterone levels (Mendoza *et al.*, 1978). Studies of Coe and Rosenblum (1978) clearly suggested that the cumulative output of higher testosterone during the long winter nights is one of the primary factors in annual variation of male squirrel monkey.

Cebus: It is generally believed that the capuchin monkeys reproduce seasonally. Classical study of Hamlett (1939) on wild *C. apela* reported birth peaks during

May-June and October-November. Klein (1971) and Pook and Pook (1981) also reported seasonal birth clustering in this species. All these studies were carried out on small sample size and infrequent observations, hence it is difficult to reach any clearcut conclusions.

Table 1. Seasonality of reproduction in female non-human primates

Species	Breeding season	Reference
Callithrix jacchus	Breeds throughout the year in captivity	Hearn, 1983 Kholkute and Puri, 1988
Callithrix humeralifer	Birth season in September-October, second birth season in February-March	Rylands, 1981
Saimiri sciureus	December-February	DuMond, 1968 Cooper, 1968
Cebus apella	Birth peak April-May. New Infants seen in September-October	Klein, 1971 Pook and Pook, 1981
Macaca mulatta	80% births March-May 20% births September-October	Prakash, 1958
	Birth peak March-May	Southwick *et al.*, 1961
	March-July	Lindburg, 1971
Macaca radiata	Births January-April	Lancaster and Lee, 1965
	January-June	Rahman and Parthasarthy, 1969
	February-May	Sugiyama, 1971
Macaca fuscata	March-July 88% births during April-June	Kawai *et al.*, 1967
Papio anubis	October-December	DeVore and Hall, 1965
	All year. Birth peak February-July	Rowell, 1969
Papio hamadryas	Major birth peak May-June. Second peak in November-December	Kummer, 1968
Papio cynocephalus	During 8 different months without clustering	Altmann, 1980
Presbytis entellus	Annual and bimodal birth clustering which was more pronounced in Himalayan populations	Bishop, 1979

To summarize, data from captive colonies suggest lack of seasonality in reproduction in common marmosets. Distinct seasonality has been observed in squirrel monkeys, however, studies carried out in capuchin monkeys are inconclusive.

Seasonality in Old World Primates

Seasonality of reproduction has been well established for macaques. In fact, these were among the first primates to be studied for discrete birth season.

Table 2. Annual variation in testicular function in male non-human primates

Species	Parameters	References
Callithrix jacchus	Breeds throughout the year in captivity	Hearn, 1983; Kholkute and Puri, 1988
	Lack of circannual variations in serum testosterone and testicular size.	Kholkute, 1984
Saimiri sciureus	Increase in body weight (fatting) during breeding season.	DuMond and Hutchinson, 1967, Mendoza *et al.*, 1978.
	Seasonal changes in seminiferous tubules and spermatogenesis; testicular volume; testosterone levels.	DuMond and Hutchinson, 1967; Coe and Rosenblum, 1978; Mendoza *et al.*, 1978.
Macaca mulatta	Annual variation in testicular volume	Sade, 1964, Wickings and Nieschlag, 1980
	Changes in seminiferous tubules and spermatogenesis	Conaway and Sade, 1965 Zamboni *et al.*, 1974
	Circannual variation in testosterone levels	Gordon *et al.*, 1976; Michael and Bonsall, 1976a, Wickings and Nieschlag, 1980.
Macaca radiata	Variation in body weight, testicular volume and testosterone during breeding and non-breeding seasons.	Glick, 1979
	Lack of circannual variations in above parameters	Kholkute *et al.*, 1993
Macaca fuscata	Annual variation in testicular volume, spermatogenesis and testosterone levels	Nigi *et al.*, 1980 Matsubayashi and Enomoto, 1983
Presbytis entellus	Lack of seasonality in spermatogenesis and testosterone levels	Lohiya *et al.*, 1984

Macaca mulatta: A breeding season for the rhesus was first proposed by Heape (1994). Most births occurred in March in his study carried out near Calcutta, India. Hartman (1931) reported presence of summer amenorrhea in captive rhesus population. However, Zuckerman (1931) reported that Old World primates breed at all times of the year and thus contradicted earlier reports. Studies conducted at Cayo Santiago island however firmly established distinct breeding seasonality characterized by mating in fall and birth peak in spring (Carpenter, 1942; Altmann, 1962; Koford, 1965).

Prakash (1958) found a major season of birth between March-May and a minor one during September-October in natural habitats in Rajasthan, India. Southwick *et al.* (1961) observed a similar pattern in Northern India. Further studies conducted by various workers confirmed a distinct seasonality of reproduction in rhesus monkey (Malik *et al.*, 1992).

Like distinct seasonality in rhesus females, several studies have also reported seasonal fluctuations in male gonadal function. Sade (1964) showed that there is a seasonal variation in the size of the testis (maximum during August to December). This change is mainly due to variations in spermatogenesis throughout the year (Conaway and Sade, 1965). Zamboni *et al.* (1974) observed that during breeding season, semen collected by electroejaculation showed highly motile and morphologically normal sperm population while during non-breeding season, semen contained low sperm density and increased number of immotile and abnormal spermatozoa. Richter *et al.* (1978) studied the annual cycle of morphological events. Total depopulation of the tubules was observed in non-breeding season, recrudescence started in July and complete spermatogenesis was again observed in September. These results are thus in accordance with the presence of distinct seasonality in male reproduction.

Macaca radiata: Field studies indicates that South Indian bonnet monkey is also a seasonal breeder, matings tend to peak during the autumn and births occur most frequently in the spring prior to the arrival of monsoon rains in native habitat (Rahman and Parthasarathy, 1969; Sugiyama, 1971).

Studiės conducted at California on a large group of bonnet monkeys living in an outdoor corral suggested that the reproductive seasonality of wild populations was not abolished by captivity (Judge and Rodman, 1976). In our primate colony amenorrhea was observed from April and lasted till July. Regular menstrual cycles were established from July-August corresponding to moist rainy climate in Bombay. In male, annual variations in testicular size, testosterone levels and body weights have been reported. This study (Glick, 1979) reported enlarged testes, increased testosterone levels and body weight gains during cool and moist months (breeding season) while reduced testicular size, decreased testosterone levels and lowered body weights during dry sunny months (non-breeding season). However, Kholkute *et al.* (1993) failed to observe any significant changes in diurnal testosterone and dihydrotestosterone levels, testicular size and body weights of 7 adult bonnet monkeys maintained in captivity under controlled photoperiod and temperature indicating lack of seasonality in this species. Thus the issue regarding seasonality of male reproduction in bonnet monkeys is open and should be resolved by future studies.

Macaca fuscata: The Japanese monkey has been extensively studied for seasonality under natural environment as well as captivity. Ovulations are restricted to the fall and winter while births occur during spring and summer (Kawai *et al.*, 1967; Nozaki and Oshima, 1987). Studies carried out at Oregon, U.S.A. on animals maintained in outdoor enclosures also revealed birth season in spring and summer months. In the same colony, Hanby *et al.* (1971) reported beginning of mating activity with occurrence of first ejaculation in the 3rd week of November. Seasonal changes in plasma levels of LH (Nozaki and Oshima, 1987), steroid hormones (Hamada *et al.*, 1989) and immunoreactive inhibin (Watanbe *et al.*, 1990) have been studied. The male monkey also show annual variations in reproductive parameters (Nigi, 1976; Matsubayashi and Enomoto, 1983).

Papio: Among African primates, various species of baboons have been extensively used as animal models for biomedical research. Although no clearcut instance of a discrete birth season has been reported, birth peaks have been observed during definite period of the year. DeVore and Hall (1965) first noted birth peaks in baboons and this was confirmed by various workers. *P. anubis* in Uganda showed significantly more births between February and July than at other times of the year. Kummer (1968) observed birth clustering for *P. hamadryas* but also noted a difference in timing of birth peaks in adjacent populations. In an outdoor colony of *P. cynocephalus* and *P. anubis* at San Antonio, U.S.A. 75% of births occurred between July and December (Goodwin and Coelho, 1982).

Presbytis: *P. entellus* (Hanuman monkey) has received attention mainly from field workers. Bishop (1979) has reviewed numerous studies on this species at various sites in India, Nepal and Sri Lanka. These studies clearly confirm some form of annual clustering of births for *P. entellus*. Poirier (1970) found infrequent births in *P. johni* in India and observed two annual birth peaks. It is possible that none of the langurs are seasonal in strict sense. Evidence to date suggests either to a birth peak or to a more or less even distribution of births throughout the year. Lack of seasonality in male reproduction has been also reported (Lohiya *et al.*, 1984).

To summarize, various macaques exhibit distinct seasonality of reproduction characterized by breeding and non breeding season. Information on langurs and baboons suggests lack of distinct birth season, although birth peaks or clustering have been observed. Clearly more detailed field studies are required to reach any conclusion in these non-human primate species.

Regulation of Seasonality

Animals are exposed to varying environment. Seasonality in reproduction helps animals to anticipate and prepare for the arrival of the young ones during optimal conditions. The selection process leading to the concentration of reproductive activity during optimal periods has been termed the ultimate cause of seasonality (Baker, 1938). At the proximate level, reproductive synchronization of individuals has been viewed as a product of the interaction between external/environmental and internal/physiological factors. Although no single environmental determinant has been identified, photoperiod, rain fall and temperature have all been implicated in regulation of reproduction in various primate species (Van Horn, 1980; Lindburg, 1987).

Photoperiod: Photoperiod is known as a primary environmental cue in a number of vertebrates (Hoffmann, 1981). Theoretically, photoperiod can regulate reproductive seasonality by three general strategies. First, for species having the capacity to reproduce more than once a year, a critical day length could be used both to initiate and to terminate gonadal activity at the appropriate times of the year. Second, in these species, a critical day length could be used either to initiate or to terminate (but not both) gonadal activity relying on an endogenous internal timer to dictate either the breeding or non breeding season. Thirdly, regardless of the number of times animal could reproduce, it's annual reproductive cycle could be driven by an endogenous circannual rhythm that is synchronized with seasonal climatic and dietary changes by photoperiod.

The role of photoperiod in stimulating gonadal recrudescence and/or regression is apparent in species living at higher altitudes where differences in daylength are greater. Precise onset or termination of mating from season to season and a reversal in timing after hemispheric transfer in primates also support a regulatory role for photoperiod. For primates living at latitudes of minor differences in day length, some other dimensions of light may function as a proximal cue.

Hemispheric Reversal: Temperate zone mammals are known to undergo an approximate 6 month shift in seasonality after hemispheric transfer (Sadlier, 1969). Among New World primates, half yearly displacement in northern hemisphere colonies is evident for the golden lion tamarin (Kleimann, 1977) and the squirrel monkey (DuMond and Hutchinson, 1967; Mendoza *et al.*, 1978). Evidence of a hemispheric shift in populations transferred to southern latitudes is available for rhesus monkeys. A reversal of about 6 months from mating and birth season reported in India was observed when the animals were shifted to southern hemisphere.

Latitudinal gradients: Photoperiod is supposed to act as a proximate stimulus to mating. If this is true, then populations distributed over a wide range of latitude may be expected to show a gradient in the timing of reproduction corresponding to latitudinal variations in day length. Kawai *et al.* (1967) examined records for 25 troops of Japanese monkey ranging from 31 to 41°N and observed no clear relationship between seasonality of reproduction and photoperiod. However, Van Horn (1980) following reanalysis of Kawai's data noted a significant relationship between latitude and conception dates underlining the importance of photoperiod. Birth records

Modifying influences of social factors on the timing of mating of various avian and mammalian species has been well documented (Vandenbergh, 1973). Social stimuli appear to act primarily as temporal modifiers in animals exhibiting seasonal gonadal recrudescence initiated by other environmental stimuli. Members of the either sex may effect the onset of estrus or may promote synchrony of estrus. Several examples of a "season" of mating in rhesus monkey groups manipulated by human have been studied. Squirrel monkeys have been reported to begin mating soon after formation of captive colonies (Wilson, 1977), inspite of distinct seasonality in wild. Similarly, Carpenter (1942) reported extensive mating in rhesus monkeys in March and April about 6 months out of phase with the mating season in India, soon after formation of colony. Studies on vervet monkeys failed to find any relationship between various environmental factors (rainfall, availability of food, photoperiod) suggesting that birth season may be regulated by social factors. In rhesus females primed with estradiol benzoate during nonmating season, births were noted to occur a month earlier than in control groups (Vandenbergh and Drickamer, 1974). In another study, ovariectomized females maintained on estradiol through mating and non-mating seasons showed sexual behaviour (Turner and Herndon, 1983). Thus females are both stimulators of, and responders to social influences and this may further help in fine tuning sexual activity. In male rhesus monkeys, ejaculatory behaviour under controlled captive conditions was maximal during December and declined in February, however, the females used in the mating tests were still sexually receptive (Michael and Keverne, 1971). Seasonal variations in sexual activity were also observed in castrated male monkeys receiving testosterone replacement (Michael and Zumpe, 1976b). Thus, the increases in testosterone accompanying testicular recrudescence are not necessary for triggering mating behaviour.

Endocrinology of Seasonality

The first evidence that the seasonal variations in testis size and the spermatogenesis in rhesus monkey is under control of cyclic testosterone production was provided by Sade (1964). Direct measurement of peripheral testosterone levels subsequently confirmed this observation (Gordon *et al.*, 1976; Wickings and Nieschlag, 1980).

The endocrinology of seasonality in male rhesus monkeys has been studied extensively. During non-breeding season (between March and May), few LH pulses of low amplitude were observed. During recrudescence, pulse frequency of 8-10 pulses per day occurring at 2 to 4 hour intervals was seen. During breeding season, LH pulses occurred more frequently during the dark phase of the 24 hr period (Plant, 1980; Wickings and Nieschlag, 1981). The capacity of the pituitary to respond to LHRH was retained throughout the year, however, during nonbreeding season the responses was much diminished (Wickings *et al.*, 1987). These results suggest that hypothalamic activity was decreased during quiescence.

Maximum serum testosterone levels were observed during the breeding season. The circannual variations in testosterone levels were present for laboratory monkeys maintained under either natural day light conditions or in controlled light-dark periods. The Leydig cells also retained their ability to respond to LH/hCG throughout the year. However, full secretory capacity was observed during breeding season. Furthermore, during breeding season even small amounts of LH was sufficient to elicit a larger increase in tesosterone (Nieschlag and Wickings, 1980). The tesosterone production rates were also lowest duing nonbreeding season and increased during breeding season (Wickings and Nieschlag, 1977). Wickings *et al.* (1981) could induce full testicular function (in terms of steroidogenesis and spermatogenesis) even in quiescent period by administration of low pulsatile LHRH suggesting that reduced hypothalamic activity during nonbreeding season is the primary controlling mechanism in regulation of seasonality in male rhesus monkey.

Conclusions

Primates are highly complex group of animals and show diverse breeding strategies. In a number of primate species, females show annual seasonality wherein breeding generally occur within the same time period from year to year. In some primates both the sexes exhibit seasonality and may respond independently to various extrinsic and intrinsic factors. Reproductive seasonality helps to concentrate births during most favourable ecological conditions and also serve to reduce predation.

From the evolutionary point of view, food availability, photoperiod, rainfall and temperature underlie variations in seasonal reproduction. Various environmental factors, however, may not always predict seasonality or lack of it, in a straight forward manner. Furthermore, which environmental cue(s) signal seasonal changes is far from clear. Several social cues may also be involved. These probably are of great importance in fine tuning reproduction during the breeding season and to promote within group synchrony rather than initiate seasonal reproduction. In the wild the reproductive state of individual at particular time is the result of interactive effects of varieties of environmental factors and social influences. However, whether this is a simple summation or a complex process or if the actions of some cues have priority over the others is not known.

From an endocrinological standpoint, very little is known about the regulation of the annual patterns of reproduction, what environmental factors shape various breeding patterns and the neuroendocrine pathways underlying these. However, the importance of the hypothalamus in regulating seasonal breeding in male rhesus monkey is established. Thus, the seasonality of reproduction is an important area involving various disciplines, e.g. evolution, social organization and behaviour and endocrinology.

Acknowledgements

This chapter is dedicated to the memory of my mentors, Prof. K.N. Udupa and Prof. P.J. Deshpande. Authors wish to thank Dr. Tarala D. Nandedkar for her suggestions and comments. The first author wishes to express his indebtedness to Profs. Thapliyal, Sathyanesan and Dominic for their guidance and encouragement during his early research career. Finally, thanks are due to all old batchmates for their support, affection and for social and cultural enrichment.

References

Abbott, D.H., George, L.M. and Barret, J. (1990). In: Socioendocrinology of Primate Reproduction (T.E. Ziegler and I.B. Bercovitch, Eds.). pp 120-141, Wiley Liss, New York.

Altmann, S.A. (1962). Ann. N.Y. Acad. Scs. **102:** 338-445.

Altmann, J. (1980). Baboon Mothers and Infants. Harvard University Press, Cambridge, MA, USA.

Baldwin, J.D. and Baldwin, J.I. (1981). In: Econolgy and Behaviour of Neotropical Primates (A.F. Coimbra-Filho and R.A. Mittermeier, Eds.). pp 277-330, Academia Brasiteira de Ciencias, Rio de Janeiro.

Baker, J.R. (1938). In: Evolution (G.R. de Beer, Eds.). pp 161-178, Clarendon Press, Oxford.

Bishop, N.H. (1979). J. Hum. Evol. **8:** 251-281.

Colishaw, G. and Dunbar, R.I.M. (1991). Anim. Behav. **41:** 1045-1056.

Carpenter, C.R. (1942). J. Comp. Psychol. **33:** 113-162.

Chan, P., Hutz, R.J. and Dukelow, W.R. (1982). Fertil. Steril. **38:** 609-615.

Chik, C.J., Almeida, O.F.X., Libre, E.A., Booth, J.D., Renquist and Merriam, G.R. (1992). J. Clin. Endocrinol. Metab. **74:** 1068-1074.

Cooper, R.W. (1968). In: The Squirrel Monkey (R.A. Rosenblum and R.W. Cooper, Eds.). pp 1-29, Academic Press, New York.

Coe C.L. and Rosenblum I.A. (1978). Folia Primatol. **29:** 19-42.

Conaway, C. and Sade, D. (1965). Folia Primatol. **3:** 1-12.

DeVore, I. and Hall, K.R.L. (1965). In: Primate Behavior-Field studies of Monkeys and Apes (I. Devore, Ed.). pp 20-52, Rinehart and Winston, New York.

Dubey, A.K., Puri, C.P., Puri, V. and Anand Kumar, T.C. (1983). Experentia **39:** 207-209.

Dukelow, W.R., Pierce, D.L., Roudebush, W.E., Jarosz, S.J. and Sengoku, K. (1990). J. Med. Primatol. **19:** 627-639.

DuMond, F.V. and Hutchinson, T.C. (1967). Science **158:** 1067-1070.

Dumond, F.V. (1968). In: The Squirrel Monkey (L.A. Rosenblum and R.W. Cooper, Eds.). Academic Press, New York.

Elvira, M.D., Herndon, J.V. and Collins, D.C. (1983). Folia Primatol. **41:** 191-203.

Epple, G. (1970). Folia Primatol. **2:** 56-76.

Epple, G. and Katz, Y. (1984). Am. J. Primatol. **6:** 215-227.

Gouzoules, H., Gouzoules, L. and Fedigan, L. (1981). Anim. Behav. **30:** 1138-1150.

Glick, B.B. (1979). Folia Primatol. **32:** 268-289.

Goodwin, W.J. and Coehlo, A.M (1982). Lab. Anim. Sci. **32:** 672-676.

Gordon, T., Rose, R. and Bernstein, I. (1976). Horm. Behav. **7:** 229-243.

Hamada, T., Watanbe, G., Kokuho T., Taya, K., Sasamoto, S., Hasegawa, Y., Miyamoto, K. and Igarashi, M. (1989). J. Endocrinol. **22:** 697-704.

Hamlett, G.W.D. (1939). Anat. Rec. **73:** 171-187.
Hanby, J.P., Robertson, L.T. and Phoenix, C.H. (1971). Folia Primatol. **16:** 123-143.
Hartman, C.G. (1931). J. Mammalogy **12:** 129-142.
Heape, W. (1994). Phil. Trans. Royal Society, London, **B 185:** 411-471.
Hearn, J.P. (1983). In: The Biology and Conservation of the Callithrichidae (D. Kleinman, Ed.). pp 163-171, Smithsonian Institution, Press, Washington, D.C.
Hoffmann, K. (1981). In: Biological Rhythms, Vol. 4 (J. Aschoff, Ed.). pp 449-473, Plenum, New York.
Judge, D.S. and Rodman, P.S. (1976). Primates **17:** 535-539.
Kawai, M., Azuma, S. and Yashiba, K. (1967). Primates **8:** 35-74.
Kholkute, S.D. (1984). Int. J. Androl. **7:** 431-438.
Kholkute, S.D., Jayaraman, S., Ashok Kumar, R. and Puri, C.P. (1987). Int. J. Androl. **10:** 635-642.
Kholkute, S.D. and Puri, C.P. (1988). In "Proceedings of the 17th Annual Meeting of the Endocrine Society of India (C.P. Puri and T.C. Anand Kumar, Eds.). pp 66-76, Good Print, Bombay.
Kholkute, S.D., Katkam, R.R. and Puri, C.P. (1993). Primates **34:** 71-76.
Kleinman, D.G. (1977). In: The Biology and Conservation of Callithrichidae (D.G. Kleinman, Ed.). pp 181-192, Smithsonian Institution Press, Washington, D.C.
Klein, L.L. (1971). Folia Primatol. **15:** 233-248.
Koford, C.B. (1965). In: Primate Social Behavior (C.H. Southwick, Ed.) Van Nostrand, Princeton.
Kummer, H. (1968). Social Organization of Hamadryas Baboons. University of Chicago Press, Chicago.
Lancaster, J.B. and Lee, R.B. (1965). In: Primate Behavior-Field studies of Monkeys and Apes (I. DeVore, Ed.). pp 486-513, Rinehart and Winston, New York.
Lehman, S.W., Taylor, L.L. and Easley, S.P. (1994). Int. J. Primatol. **15:** 115-128.
Lindburg, D.G. (1987). In: Comparative Primate Biology Vol. 2B, pp. 167-218 (J. Erwin and W.R. Dukelow, Eds.), Alan Liss, New York.
Lohiya, N.K., Sharma R.S., Sehgal, A., Pruthi, J.S., David, G.F.X. and Kumar, T.C. (1984). In: Current Primate Researches, pp. 219, (M.L. Roonwal, S.M. Manhot, N.S. Rathor, Eds.) S.K. Enterprises, Jodhpur.
Malik, I., Johnson, R.L. and Berman, C.M. (1992). Am. J. Primatol. **6:** 89-95.
Matsubayashi, K. and Enomoto, T. (1983). Primates **24:** 521-529.
Mendoza, S.P., Lowe, E.L., Resko, J.A. and Levine, S. (1978). Physiol. Behav. **20:** 515-522.
Michael, R. and Keverne, B. (1971). J. Reprod. Fertil. **25:** 95-98.
Michael, R. and Zumpe, D. (1976). Psychoneuroendocrinology **1:** 303-313.
Michael, R. and Bonsall, R.C. (1976). J. Reprod. Fertil. **49:** 129-131.
Mukku, V., Prahalada, S. and Moudgal, N.R. (1976). Nature **260:** 778-780.
Nieschlag, E. and Wickings J. (1980). In: Non-human Primate Models for study of Human Reproduction. pp. 136-147, (T.C. Anand Kumar, Ed.) Karger Verlag, Basel.
Nigi, H. (1976). Primates **17:** 81-87.
Nigi, H., Tiba, T., Yamamoto, S., Floescheim, Y. and Obsawa, N. (1980). Primates **21:** 230-240.
Nozaki, M. and Oshima, K. (1987). In: Progress in Biometerology, Vol. 5, pp. 41-49 (T. Miura, Ed.), Academia Publishing, Hague.
Plant, T.M. (1980) In: Testicular Development Structure and Function. pp. 419-429 (A. Steinberger and E. Steinberger, Eds.) Raven Press, New York.
Poirier, E.E. (1970). In: Primate Behaviour, Vol. 1, pp. 251-383 (L.A. Rosenblum, Ed.), Academic Press, New York.

Pook, A.G. and Pook, G.A. (1981). Folia Primatol. **35:** 288-312.
Prakash, I. (1958). J. Bombay Nat. Hist. Soc. **55:** 154-180.
Rahman, H. and Parthasarathy, M.D. (1969). Primates **10:** 149-162.
Rawlins, R. and Kessler, M. (1985). Amer. J. Primatol. **9:** 87-99.
Richter, K.D., Karte, R. and Senge, T. (1978). In: Physiology and Patho physiology, pp. 41-46 (T. Senge, F. Neumann and U.W. Tunn, Eds.), Thieme-Verlag, Stutgart.
Rowell, T.E. (1979). Folia Primatol. **1:** 241-251.
Rylands, A.B. (1981). Primates **22:** 46-59.
Sade, D. (1964). Folia Primatol. **2:** 171-180.
Sadlier, R.M. (1969). The Ecology of Reproduction in Wild and Domestic Mammals, Methuen, London.
Southwick, C.H., Beg, M.A. and Siddiqui, M.R. (1961). Ecology **42:** 698-710.
Sugiyama, Y. (1971). Primates **12:** 247-266.
Turner, J. and Herndon, J.C. (1983). Amer. J. Primatol. **4:** 171-177.
Vandenbergh, J. (1973). In: Primate Reproductive Behaviour, Vol. 2 (C.H. Phoenix, Ed.), pp 1-19, Karger, Basel.
Vandenbergh, J. and Drickamer, L. (1974). Physiol. Behav. **13:** 373-376.
Vandenbergh, J. and Vessey, S. (1968). J. Reprod. Fertil. **15:** 71-79.
Van Horn, R.N. (1980) In: Progress in Reproductive Biology Vol. 5, pp. 181-221, (P.O. Hubinont, Ed.), Karger, Basel.
Watanabe, G., Nozaki, M., Taya, K., Katakai, Y. and Sasamoto, S. (1990). Biol. Reprod. **43:** 196-201.
Wehrenberg, W.B. and Dyrenfurth, I. (1983). J. Reprod. Fertil. **68:** 119-122.
Wickings, E.J. and Nieschlag, E. (1977). Int. J. Fert. **22:** 56-59.
Wickings, E.J. and Nieschlag, E. (1980). Int. J. Androl. **3:** 87-104.
Wickings, E.J. and Nieschlag, E. (1981). Biol. Reprod. 24, (Suppl 1), Ab. 161.
Wickings, E.J., Zaidi, P., Brabant, G. and Nieschlag, E. (1981). J. Reprod. Fertil. **63:** 129-136.
Wickings, E.J., Marshall, G.R. and Nieschlag, E. (1987). In: Comparative Primate Biology, Vol. 3, pp. 149-170. (J. Erwin and W.R. Dukelow, Eds.), Alan Liss, New York.
Wilson, M.I. (1977). J. Reprod. Fertil. **5:** 57-63.
Wilson, M.I., Brown, G.M. and Wilson, D. (1978). Acta Endocrinol. **87:** 424-433.
Zamboni, L., Conaway, C.H. and Van Felt, L. (1974). Biol. Reprod. **11:** 251-267.
Zuckerman, S. (1931). Proc. Zool. Soc. London, 325-343.
Zuckerman, S. (1932). Proc. Zool. Soc. London, 1059-1075.

Comparative Endocrinology and Reproduction
K.P. Joy, A. Krishna and C. Haldar (Eds)

24. Reproductive Delays in Chiropterans

Amitabh Krishna
Department of Zoology, Banaras Hindu University, Varanasi-221 005, India

Introduction

Delays between copulation and parturition have been intensively studied as model systems for timing and control of reproduction. Bats are unique mammalian model for studies on reproductive delay for various reasons. First, the duration of pregnancy in chiropterans is in general much longer than one would predict on the basis of their body size. The prolonged gestation in bats is in part due to an arrest or retardation of embryonic development that can last from a few days to several months, depending on the species. Second, chiroptera is the only mammalian order in which a variety of reproductive delay has been described. Four different types of reproductive delay have been demonstrated in female bats: delayed ovulation, delayed fertilization, delayed implantation and delayed development. Delayed ovulation and delayed fertilization are uncommon and are the norm in the majority of temperate zone bats (Oxberry, 1979; Racey, 1982). It has also been recorded in several species of subtropical and tropical bats (Krutzsch, 1979; Dominic and Krishna, 1989; Singh and Krishna, 1997). Delayed implantation is the most common form that has been reported among mammals in several species belonging to 7 different Orders including Chiroptera (Mead, 1993). Post-implantation delay in development is the least common form in mammals and is known from only a few species of bats (Heideman *et al.*, 1993). The physiological mechanisms which control these interesting processes are varied. This chapter reviews what is known about the various forms of reproductive delay in bats and their endocrine and other physiological control mechanisms.

Delayed Ovulation

In certain bats final maturation and ovulation of Graafian follicles are postponed for several months. During this period they persist in the ovary. This phenomenon is termed as delayed ovulation. The remarkable longevity of the follicles and prolonged delay in ovulation are unique features in the reproductive cycle of hibernating bats. Delayed ovulation was first described in hibernating members of rhinolophid (Matthews, 1937; Gaisler, 1966) and vespertilionid bats (Courrier, 1924) from Europe. This has since been reported in numerous species from temperate latitudes throughout the world

(Wimsatt, 1944; Sluiter and Bels, 1951; Pearson *et al.*, 1952; Uchida and Mori, 1977; Oh *et al.*, 1985; Son *et al.*, 1987).

In most of the temperate zone vespertilionid bat during delayed ovulation, the ovarian follicle elicits morphological and chemical specialization not observed in other mammals (Singh and Krishna, 1996a). These follicles are characterised by a reduced antral cavity and ovum being surrounded by a large discus proligerus, the cells of which are considerably hypertrophied and vesicular (Oxberry, 1979). Cumulus cells store either glycogen (*Myotis lucifugus,* Wimsatt and Kallen, 1957; Wimsatt and Parks, 1966; *Pipistrellus abramus*, Uchida and Mori, 1977; Nakano, 1928) or lipids (*Rhinolophus ferrum-equinum nippon,* Oh *et al.*, 1985) as nutrient sources during the prolonged survival of the Graafian follicles. Glycogen as an energy source for follicle cells has been regarded as an important aspect of the prolonged survival of the Graafian follicles during hibernation (Oxberry, 1979; Son *et al.*, 1987). The ultrastructural studies of Wimsatt and Parks (1966) on the Graafian follicles of *Myotix lucifugus* have shown that glycogen stored in the granulosa cells is transferred to oocytes by follicle cell processes that pass through the zona pellucida and penetrate the oocyte surface. These authors consider the modifications as an adaptation to meet the energy requirement of the ovum-follicle complex during winter dormancy in the temperate zone vespertilionid bats.

Wimsatt (1960) suggested that the mechanism of ovulatory delay involves the depressing effect of hibernation on pituitary gonadotrophic function. It is suggested that hibernation suppresses the synthesis and release of pituitary luteinizing hormone (LH) as well as LH-responsiveness. Cytological studies demonstrated insufficient LH but not follicle stimulating hormone (FSH) in the pituitary especially during the first half of hibernation (Herland, 1956, 1968; Wimsatt, 1960). On the basis of these studies Racey (1982) speculated that the obstacle to ovulation in hibernating bats did not appear to originate in the ovary but was probably due to low pituitary secretion of LH. Our study, on the contrary, showed intense LH and FSH immunocreativity during both delayed ovulation and breeding periods in *S. heathi* (Singh and Krishna, 1996b). Only preliminary studies are undertaken on seasonal changes of LHRH system in any of the temperate zone vespertilionid bat in relation to delayed ovulation (Anthony *et al.*, 1987). Studies of Anthony *et al.* (1989) in *Myotis lucifugus* have shown enhanced tissue contents but low secretion of LHRH during hibernating pre-ovulatory period but increased LHRH secretion during the post-ovulatory period. Parallel reductions in hypothalamic and neural lobe LHRH content during peri-ovulatory period support the hypothesis that the neural lobe component of the system contributes to control of gonadotrophin secretion in this species. It may be interesting to note that as compared to other mammalian species, *Myotis lucifugus* showed more complex interactions among LHRH immunopositive elements. Whether it is related to the phenomenon of delayed ovulation is not clear.

Specialized Graafian follicles similar to that described in temperate zone vespertilionid bats have also been reported in some of the tropical/subtropical vespertilionids (Gopalakrishna *et al.*, 1974; Krishna and Dominic, 1980; Krishna and Singh, 1992). Although none of these tropical vespertilionids undergoes prolonged hibernation, they all seem to resemble the hibernating bats in having a reproductive cycle characterized by slow follicular maturation, unique specialization in the Graafian follicle and a prolonged oestrus. Thus it appears that prolonged hibernation which characterizes temperate zone species is not a necessary preadaptation for delayed ovulation.

The phenomenon of delayed ovulation is studied in great detail in my laboratory on *Scotophilus heathi*, a subtropical/tropical vespertilionid bat. In *S. heathi*, antral follicles are formed during the month of November but ovulation is delayed until early March (Krishna and Singh, 1992). All the attempts to induce ovulation during this period were unsuccessful (Singh and Krishna, 1992). A recent study on *S. heathi* has demonstrated unusually high circulating concentration of androstenedione during the period of delayed ovulation (Abhilasha and Krishna, 1996). This increase in androstenedione is shown to be correlated with seasonal increase in body weight due to accumulation of adipose tissue (Abhilasha and Krishna, 1997). The major source of this high androstenedione in this species was found to be the ovary. The most striking cytological feature of ovary is the thecal-interstitial cell dominance. In fact, entire inter-follicular stromal area of the ovary of *S. heathi* is shown to be occupied with luteinized thecal cells, similar to hyperthecosis of the human ovary as demonstrated during polycystic ovarian syndrome (PCOS) (Barbieri and Hornstein, 1988). Immunocytochemical localization of steroidogenic enzymes in the ovary of *S. heathi* showed luteinized interstitial cells as the major site of androgen production (Abhilasha and Krishna, 1996). So increased androgen production by the ovary of *S. heathi* may be the result of increased differentiation of the interstitial cells into luteinized thecal cells. Insulin is well known to induce hyperthecosis in the ovary of woman with PCOS (Poretsky and Kalin, 1987). Interestingly in *S. heathi,* the period of increased interstitial cell differentiation coincided with the period of weight gain and increased serum insulin concentration (Doval and Krishna, 1996). Increased androgen produced by thecal-interstitial cells, in turn, may have caused a number of intra-ovarian effects; as a result the ovary failed to produce dominant follicles and therefore ovulation was delayed in *S. heathi*. Interestingly, in *S. heathi* ovulation occurs during the period when the bats are having the lowest circulating androstenedione concentration and basal body weight. *S. heathi* during the period of delayed ovulation thus exhibits features such as obesity, hyperandrogenism and ovarian hyperthecosis as found in woman with PCOS. It appears that the mechanism for suppression of ovulation in *S. heathi* may be similar to the pathways through which PCOS and hyperandrogenism are caused in woman.

Delayed Fertilization

Prolonged sperm-storage is an adaptation to the reproductive cycle that is most highly developed in hibernating Microchiroptera from temperate latitudes, where sperm may be stored up to 7 months and still retained their fertilizing capacity (Racey, 1975, 1982). In temperate zone bats, mating activity commences in the autumn, spermatozoa are then stored in the female genital tract throughout the hibernating period. These sperm are used in the fertilization of ova released at ovulation after arousal in the spring (Wimsatt, 1969; Racey, 1979). The nature of specialization that has evolved in some bats to permit sperm survival for prolonged period is still poorly understood.

For most of the bat species displaying this phenomenon, storage largely coincides with hibernation, but limited storage of spermatozoa has also been described in some tropical heterothermic species that do not hibernate (Gopalakrishna and Madhavan, 1971, 1978; Medway, 1972; Myers, 1977; Krishna and Dominic, 1978; Krishna, 1984). Thus, low temperature cannot be solely responsible for the unusual longevity of bat spermatozoa. Either the surviving spermatozoa or their storage organs must be modified in some way to provide unique conditions for sperm-storage.

Spermatozoa may be stored in the female genital tract for periods varying from 16 days as in *Pipistrellus ceylonicus* (Gopalakrishna and Madhavan, 1971) to 198 days as in *Nyctalus noctula* (Racey, 1979). In general, tropical bats store spermatozoa for much shorter periods as compared with hibernating temperate zone bats. The interval between copulation and ovulation is up to five weeks in *S. heathi* (Krishna and Dominic, 1978; Gopalakrishna and Madhavan, 1978) and 30-35 days in *P. mimus* (Krishna, 1984, 1985).

The sites of sperm-storage in female bats vary with species. In many vespertilionid bats, the uterus may play an important role as a storage site for spermatozoa (Hiraiwa and Uchida, 1980; Racey and Potts, 1970). In some vespertilionid bats, viz. *Chalinolobus gouldii* (Kitchener, 1975) and *M. schreibersii fuliginosus*, the storage site is limited to the utero-tubal junction where intact spermatozoa are found in diverticulae formed by the mucosal folds (Mori and Uchida, 1980). The sperm are stored in utero-tubal junction in *M. daubentoni* (Racey, 1975), *Myotis lucifugus* and *Myotis velifer* (Krutzsch *et al.*, 1982). In *P. ceylonicus* and *S. heathi*, surviving spermatozoa are stored in both the uterus, and oviduct (Gopalakrishna and Madhavan, 1971; Krishna and Dominic, 1978). In *P. mimus*, comparatively few spermatozoa are found in the uterus and most occur in the oviduct. In *Rhinolophus ferrum-equinum, R. hipposideros, R. ferrum equinum nippon* and *Myotis nattereri*, the principal site of sperm-storage is the oviduct (Racey, 1975; Mori *et al.*, 1982).

Researches have been made to understand the biology of sperm longevity in bats by morphological (Fawcett and Ito, 1965; Uchida and Mori, 1972; Racey, 1975; Krutzsch *et al.*, 1982; Mori *et al.*, 1982; Uchida *et al.*, 1984;

Krishna, 1997); biochemical (Crichton *et al.*, 1981, 1982, 1993; Krutzsch *et al.*, 1984); and experimental (Lambert, 1981; Uchida and Mori, 1987) approaches. However, although, several theories have been offered to explain sperm storage, none has been unequivocally substantiated.

Special relationships exist between spermatozoa and female storage organs in all sperm storing bat species so far investigated, suggesting that such relationships are an integral part of the mechanism of prolonged storage of sperm (Racey, 1979; Uchida and Mori, 1987). There are several studies from different laboratories as well as from our laboratory to show the significance of close association between sperm and female genital tract during prolonged storage (Racey, 1975; Krishna, 1984). During storage, the sperm heads are either embedded in the uterine and oviductal epithelia or remain in close contact with microvilli of the host cells. According to Racey (1975) the epithelial cells, of the storage site actively provide the spermatozoa with nutrients for their prolonged survival. For example, glycogen in the uterine epithelium of *P. abramus* (Nakano, 1928) and *P. pipistrellus* (Racey, 1975) and *Myotis lucifugus* and *M. velifer* (Crichton *et al.*, 1981). Furthermore, association of spermatozoa with the epithelium of the storage organ probably protects them from various disintegrative factors (e.g. muscular movements of the uterus). Certain enzymes of glycolysis have been located in the sperm as well as in the uterus (Racey, 1975). Beside nutritional support, other factors required for long term maintenance of spermatozoa might be a mechanism that acts to suppress sperm metabolism, thereby reducing demand for available substrates. In this regard, Hunter *et al.* (1971a, b) observed that the seminal fluid of *M. lucifugus* contains a protein that is inhibitory to uterine motility and phagocytic system. Zinc has been shown to temporarily inhibit the mobility of both sperm and polymorphonuclear neutrophilic leucocytes. It has been suggested that such factors would also favour prolonged sperm-storage (Crichton *et al.*, 1982).

Krutzsch and Crichton (1991) suggested that bat spermatozoa resist capacitation and the acrosome reaction when cultured under conditions that are known to induce these events in other mammals. Furthermore, they have failed to reveal decondensing bat sperm heads within zona-free hamster ooplasm using *in vitro* fertilization procedure, even after exposure to calcium ionophore A23187, and various proteases. Thus, it was suggested that the unusual longevity, stability and resistance of bat spermatozoa to *in vitro* fertilization could be due to unique modification of their plasma membrane. A recent study in a hibernating bat (*Myotis velifer*), however, has failed to demonstrate any modification in plasma membrane lipid composition of the stored spermatozoa (Crichton *et al.*, 1993). Further studies are needed to find out the unique features of prolonged sperm longevity of bats.

Delayed Implantation

Two types of delayed implantation are recognized: facultative and obligatory. Facultative delay of implantation is restricted to species which exhibit a post-partum estrus and the retardation in embryonic development is lactation-induced. This form of embryonic diapause is known to occur in three Orders of mammals: Marsupialia, Insectivora and Rodentia. Obligate delay of implantation differs from facultative delay in that retardation of embryonic development occurs regardless of whether the female is lactating. This type of delayed implantation occurs in five Orders of mammals (Chiroptera, Insectivora, Edentata, Carnivora, Artiodactyla), but the physiological mechanisms which control this fascinating process are exceedingly different in the diverse groups which employ this reproductive strategy. Among bats, this type of delay has been demonstrated and studied in detail in long-fingered bat, *Miniopterus schreibersii* of Europe, Japan and Australia (Dwyer, 1963; Richardson, 1977; Crichton *et al.*, 1989). Delayed implantation has also been demonstrated in *Natalus stramineus* (Mitchell, 1965), *Eidolon helvum* (Mutere, 1965, 1967), *Rhinolophus rouxi* (Ramakrishna and Rao, 1977), *M. schreibersii natalensis* (van der Merwe, 1975, 1980) and *M. australis* (Dwyer, 1968; Richardson, 1977). *M. schreibersii* differs from other hibernating vespertilionids in that copulation, ovulation and fertilization occur in quick succession. The female bat enters hibernation in the pregnant stage but implantation is delayed until spring arousal. The period of delayed implantation almost coincides with that of winter hibernation (Planel *et al.*, 1961; Richardson, 1977; Kimura and Uchida, 1983). Wimsatt (1969) proposed that the delay could be due to lowered cellular metabolism during hibernation. However, Peyre and Herlant (1963a, b; 1967) suggested that an arrest in LH secretion following ovulation coupled with insufficient prolactin is responsible for delayed implantation. Both LH and prolactin secreting cells of the pituitary are enlarged when the corpus luteum assumes an active appearance (Peyre and Herlant, 1963a).

The factors causing delayed implantation in *M. schreibersii* have recently been studied in detail (Bernard *et al.*, 1991; Bernard and Bojarski, 1994). They have shown that in *M. schreibersii* torpor may be responsible for suppression of luteal activity and consequently delay in implantation. During this period plasma progesterone and LH concentrations were also found to be low. Treatment of pregnant bats with exogenous prolactin for 10 days in early delayed implantation resulted in activation of luteal activity, high plasma progesterone concentration and the initiation of implantation. Treatment of bats with hCG during this period also induces similar changes in luteal activity and plasma progesterone concentration but failed to initiate implantation. Treatment with exogenous progesterone had no influence on the luteal activity as well as the initiation of implantation. Therefore, they concluded that prolactin is the most important pituitary hormone in the control of delayed implantation in *M. schreibersii*. Interestingly, in this bat period of delayed implantation lengthens with increasing latitude (Crichton *et al.*, 1989). In all the bats that exhibit delayed implantation, it occurs as day length increases, suggesting that increasing day length may be the environmental cue that triggers implantation (Racey, 1982).

Delayed Development

Post-implantation delay in development is the least common form of delay in mammals and is seen only in a few mammals, belonging to the Order Chiroptera (Bradshaw, 1962; Fleming, 1971; Bernard and Meester, 1982; Heideman, 1988, 1989). Delayed development may be broadly categorized into two types: temperature-dependent and temperature-independent. In some species, such as *Pipistrellus pipistrellus, Myotis myotis* and *Corynorhinus rafinesquei,* reduced environmental temperature significantly lengthens gestation by slow embryonic development (Eisentraut, 1937; Pearson *et al.*, 1952; Racey, 1969). During such periods the female hibernates and gestation is prolonged approximately to the duration of hibernation (Racey, 1969; Racey and Swift, 1981). Since this form of embryonic diapause may or may not occur depending upon environmental conditions, it is referred to as facultative delayed development. Delayed development lasting about 4-5 months always occurs in the California leaf nosed bat, *Macrotus californicus* and is independent of temperature and food conditions. This is referred to as obligate form of delayed development. The fruit bat, *Haplonycteris fischeri,* from the Philippines exhibits a 9 month delay in embryonic development which is presumably of the obligate type (Heideman, 1989). Delayed embryonic development has also been observed in *Artibeus jamaicensis* (Family: Vespertilionidae, Medway, 1971) and *Hipposideros caffer* (Family: Hipposideridae, Bernard and Meester, 1982). Delays have also been implicated in several other microchiropteran species, *Rhinolophus rouxi* (Ramakrishna and Rao, 1977), *Taphozous longimanus* (Krishna and Dominic, 1982), *Cynopterus sphinx* (Krishna and Dominic, 1983) and *Pipistrellus pipisterellus* (Racey and Swift, 1981).

The mechanism and cause(s) of delayed development were mainly studied in *Macrotus californicus* (Bradshaw, 1962; Burns *et al.*, 1972; Bleir, 1975; Burns and Wallace, 1975; Burns and Easley, 1977) and *M. schreibereii* (Crichton *et al.*, 1989, 1990). Earlier study of Bradshaw (1962) found that neither food provided in excess nor elevated temperature affected the embryonic growth rate. They did find lowered plasma thyroxine concentration during part of the delay and suggested that hypothyroidism may be the cause of delay. Later, Burns and Wallace (1975) failed to increase the rate of embryonic development in this animal by treating with thyroxine up to 40 days. Burns and Wallace (1975) described biphasic peaks in plasma estrone and estradiol during development, the first peak corresponding to the early part of the delay and the second to the period after delay. Burns and Easley (1977) found a similar biphasic pattern of plasma progesterone.

They speculated that the elevated hormone concentrations might be required for implantation and suggested that the subsequently lowered hormone concentrations might be responsible for the delay. A recent study of Crichton *et al.* (1990) suggested that delayed development in *M. californicus* might be due to delayed luteal cell development and resultant nutritional incompetence of the uterus through inadequate steroid stimulation. Circulating level of plasma progesterone suggested that the corpus luteum of embryonic diapause in *Macrotus* secretes less steroid than after the onset of active embryogenesis (Crichton *et al.*, 1990). Ultrastructural observations further support this conclusion. In this species, luteal cells increase in size as transition from delayed to normal development occurs (Crichton and Krutzsch, 1985). The role of the corpus luteum in developmental delay has also been examined in *M. schreibersii* (Crichton *et al.*, 1989). In this species during the delay period the luteal cells were found to be less active and the number of steroidogenic organelles in luteal cytoplasm was greatly reduced compared with normal embryogenesis. At the end of developmental delay, mitochondria, smooth endoplasmic reticulum, and lipid droplets increase markedly resulting in hypertrophy of the luteal cells (Crichton *et al.*, 1989). The low concentration of plasma progesterone recorded by van der Merwe and Aarde (1989) in the *M. schreibersii natalensis* during the period of relatively slower embryonic development is in agreement with the depressed values reported for the Japanese long-fingered bat and a variety of other mammals experiencing embryonic diapause (Kimura *et al.*, 1987). Insufficient prolactin and gonadotrophins may be responsible for suboptimal luteal cell development and consequently failure of luteal cells to secrete sufficient hormone during delay (Kimura *et al.*, 1987; Crichton *et al.*, 1990). Studies on the pituitary during delayed development in *M. californicus* (Richardson, 1979) and during delayed implantation in *M. schreibersii* in Japan (Mikami *et al.*, 1988) and France (Peyre and Herlant, 1963b) arrived at a similar conclusion. Despite, these findings, the nature of the environmental and neuroendocrine factors controlling the onset and end of the delay is still unknown. The results from a recent study has suggested that a decline in food availability (insects) may be responsible for slow foetal growth rate and extension of gestation in *Miniopterus schreibersii* from the East Cape region of South Africa (Bernard, 1994).

Conclusion

Most of the bat species of temperate latitude are monoestrous. In these bats during spring and summer, when they are actively foraging, the energetically most expensive processes associated with reproduction take place (Genoud, 1990). In contrast during the autumn-winter period when the ambient temperature is lower and the insect preys much less, they undergo extended periods of torpor or hibernation. The effects of hibernation on life history of temperate zone bats have resulted in reproductive delay pattern. As in the temperate zone, the ultimate cause of most seasonal reproduction in the tropics is probably seasonal changes in food availability (Heideman and Bronson, 1994). Seasonal variability of insect abundance is apparently a limiting factor for breeding in several insectivorous bats. In these species birth of the young is timed to coincide with the annual insect boom (Wilson, 1973; Mutere, 1973, Krishna, 1982). In frugivorous bats the seasonal abundance of fruit and flower may be the limiting factor for breeding (Krishna and Dominic, 1983). A review of the breeding pattern in Indian bats reveals that in most species whether monoestrous or bimodal polyoestrous, the period of birth of the young is timed to coincide with the period of food abundance (Krishna, 1982; Dominic and Krishna, 1989). Similar correlations between diets and reproductive cycles of the bat in Africa are demonstrated by Happold and Happold (1989). It is evident from this study that these species feed opportunistically and breed opportunistically to take advantage of the peak(s) in insect abundance; and consequently show geographic variation in their reproduction. In these species, births are timed so that lactation and weaning occur in months of heavy rainfall. This is achieved by adjusting the interval between consecutive births (McWilliams, 1982). The reproductive delay phenomena recorded in bats may, therefore, be regarded as different ways of lengthening the reproductive cycle, possibly determined by the winter activity of the species or the period of food scarcity, so that they can programme their period of parturition with the time of year when the factors are optimal for survival of both mother and young. Recent findings of Ransoms and Mcowat (1994) further confirm the involvement of climate, particularly temperature, through its effect upon birth timing, and possibly also on sex ratios, as a major factor controlling bat populations and hence the distribution of bat species.

References

Abhilasha and Krishna, A. (1996). J. Reprod. Fert. **106:** 207-211.
Abhilasha and Krishna, A. (1997). Comp. Biochem. Physiol. **116C:** 97-101.
Anthony, E.L.P., Weston, P.J., Montvilo, J.A., Bruhn, T.O., Neel, K. and King, J.C. (1989). J. Reprod. Fert. **87:** 671-686.
Anthony, E.L.P., Wu, P., Bruhn, T.O. and Jackson, I.M.D. (1987). Brain Res. **424:** 258-263.
Barbieri, R.L. and Hornstein, M.D. (1988). Endocrinol. Metabol. Clin. North America **17:** 685-703.
Bernard, R.T.F. (1994). J. Zool. Lond. **232:** 485-490.
Bernard, R.T.F. and Bojarski, C. (1994). J. Reprod. Fert. **100:** 359-365.
Bernard, R.T.F., Bojarski, C. and Millar, R.P. (1991). J. Reprod. Fert. **93:** 31-42.
Bernard, R.T.F. and Meester, J.A.J. (1982). Ann. Transvaal, Mus. **33:** 131-144.
Bleir, W.J. (1975). Anat. Rec. **182:** 237-254.
Bradshaw, G.V.R. (1962). Science, **136:** 145-146.
Burns, J.M., Baker, R.J. and Bleier, W.J. (1972). Gen. Comp. Endocrinol. **18:** 54-58.
Burns, J.M. and Easley, R.G. (1977). Gen. Comp. Endocrinol. **32:** 163-166.

Burns, J.M. and Wallace, W.E. (1975). Gen. Comp. Endocrinol. **25:** 529-533.

Courrier, R. (1924). Arch. Biol. **34:** 369-477.

Crichton, E.G., Hoyer, P.B. and Krutzsch, P.H. (1990). Cell Tissue Res. **260:** 355-366.

Crichton, E.G. and Krutzsch, P.H. (1985). Am. J. Anat. **173:** 69-87.

Crichton, E.G., Krutzsch, P.H. and Chavapil, M. (1982). Comp. Biochem. Physiol. **71A:** 71-77.

Crichton, E.G., Krutzsch, P.H. and Wimsatt, W.A. (1981). Comp. Biochem. Physiol. **70A:** 387-395.

Crichton, E.G., Krutzsch, P.H. and Yanagimachi, R. (1993). J. Reprod. Fert. **97:** 1-4.

Crichton, E.G., Seamark, R.F. and Krutzsch, P.H. (1989). Cell Tissue Res. **258:** 183-201.

Dominic, C.J. and Krishna, A. (1989). In: Reproductive cycle of Indian Vertebrates Saidapur, S.K. (ed.), Allied Publishers Ltd., New Delhi, India, pp. 311-346.

Doval, J. and Krishna, A. (1996). Symp. Emerging Frontier Horm. Res. p 31 (abstract).

Dwyer, P.D. (1963). Aust. J. Sci. **25:** 435-436.

Dwyer, P.D. (1968). Aust. J. Zool. **16:** 49-68.

Eisentraut, M. (1937). Biol. Zentralblatt. **57:** 59-74.

Fawcett, D.W. and Ito, S. (1965). Am. J. Anat. **116:** 567-610.

Fleming, T.H. (1971). Science, **171:** 402-404.

Gaisler, J. (1966). Bijdr, Dierk, **35:** 45-64.

Genoud, M. (1990). Revue, Swisse Zool. **97:** 79-90.

Gopalakrishna, A. and Madhavan, A. (1971). Proc. Indian Acad. Sci. **73(B):** 3-47.

Gopalakrishna, A. and Madhavan, A. (1978). Indian J. Exp. Biol. **16:** 852-854.

Gopalakrishna, A., Madhavan, A., Thakur, R.S. and Rajgopal, G. (1974). Curr. Sci., **43:** 400-401.

Happold, D.C.D. and Happold, M. (1989). J. Reprod. Fert. **85:** 133-149.

Heideman, P.D. (1988). J. Zool., Lond., **215:** 577-595.

Heideman, P.D. (1989). J. Reprod. Fert. **85:** 363-382.

Heideman, P.D. and Bronson, F.H. (1994). Biol., Reprod. **50:** 607-614.

Heideman, P.D., Cummings, J.E. and Heaney, L.R. (1993). J. Mammal. **74:** 621-630.

Herlant, M. (1956). Arch. Biol., **67:** 89-180.

Herlant, M. (1968). In: Cycles genitaux saisseniers deo Mamiferes Sauvages, R. Canivence (ed.), Masson, Paris, pp. 111-131.

Hiraiwa, Y.K. and Uchida, T.A. (1980). Sci. Bull. Fac. Sci., Kyushu Univ., **15:** 253-266.

Hunter, A.G., Barker, L.D.S., Johnson, W.L., Fahring, M.L. and Schutz, R.H. (1971a). J. Reprod. Fert. **24:** 171-177.

Hunter, A.G., Johnson, W.L., Barker, L.D.S., Fahring, M.L. and Schutz, R.H. (1971b). J. Reprod. Fert. **241:** 179-186.

Kimura, K., Takeda, A. and Uchida, T.A. (1987). J. Reprod. Fert. **80:** 59-63.

Kimura, K. and Uchida, T.A. (1983). J. Reprod. Fert. **69:** 187-193.

Kitchener, D.J. (1975). Aust. J. Zool. **23:** 29-42.

Krishna, A. (1982). Life Sci. Adv. **1:** 113-119.

Krishna, A. (1984). Arch. Biol. **95:** 223-229.

Krishna, A. (1985). J. Zool. (London), **106:** 41-51.

Krishna, A. (1997). Proc. Nat. Sci. Council (ROC), Part B: Life Sci. **21 (1):** 31–36.

Krishna, A. and Dominic, C.J. (1978). J. Reprod. Fert. **54:** 319-321.

Krishna, A. and Dominic, C.J. (1980). Proc. Zool. Soc. Cal., **33:** 1-7.

Krishna, A. and Dominic, C.J. (1982). Biol. Reprod. **27:** 351-353.

Krishna, A. and Dominic, C.J. (1983). Period. Biol. **85:** 23-30.

Krishna, A. and Singh, U.P. (1992). Eur. Arch. Biol. **103:** 257-264.

Krutzsch, P.H. (1979). J. Reprod. Fert. **56:** 333-344.

Krutzsch, P.H. and Crichton, E.G. (1991). In: A comparative overview of mammalian

fertilization. Dunbar, B.S. and O'Rand, M. (eds.), Plenum Press, New York, pp. 137-149.
Krutzsch, P.H., Crichton, E.G., Lennon, D.L.F., Stratman, F.W. and Carter, A. (1984). Andrologia **16:** 34-37.
Krutzsch, P.H. and Crichton, E.G. and Nagle, R.B. (1982). Am. J. Anat. **165:** 421-434.
Lambert, H. (1981). Gamete Res. **12:** 183-224.
Matthews, L.H. (1937). Trans. Zool. Soc., Lond., **23:** 224-225.
McWilliam, A.N. (1982). Ph.D. thesis, University of Aberdeen. U.K.
Mead, R.A. (1993). J. Expt. Zool. **266:** 629-641.
Medway, Lord. (1971). J. Zool., Lond., **165:** 261-273.
Medway, Lord. (1972). Zool. J. Linn. Soc. **51:** 33-61.
Mikami, S., Chiba, S., Hojo, H., Taniguchi, K., Kubokawa, K. and Ishii, S. (1988). Cell Tissue Res. **251:** 291-299.
Mitchell, G.C. (1965). Masters Thesis, University of Arizona, USA.
Mori, T. and Uchida, T.A. (1980). J. Reprod. Fert. **58:** 429-433.
Mori, T., Oh, Y.K. and Uchida, T.A. (1982). J. Fac. Agr., Kyushu Univ., **27:** 47-53.
Mutere, F.A. (1965). Nature, **207:** 780.
Mutere, F.A. (1967). J. Zool. **153:** 153-161.
Mutere, F.A. (1973). East Afr. Wildl. J. **11:** 271-280.
Myers, P. (1977). Univ. Calif. Publ. Zool. **107:** 1-41.
Nakano, O. (1928). Folia Anat., Japan, **6:** 777-828.
Oh, Y.K., Mori, T. and Uchida, T.A. (1985). J. Reprod. Fert. **73:** 121-126.
Oxberry, B.A. (1979). J. Reprod. Fert. **56:** 347-358.
Pearson, O.P., Koford, M.R. and Pearson, A.K. (1952). J. Mammal. **33:** 273-320.
Peyre, A. and Herlant, M. (1963a). Gen. Comp. Endocr. **3:** 726-727.
Peyre, A. and Herlant, M. (1963b). C.r. Hebd, Seanc. Acad. Sci. Paris, D. **257:** 524-526.
Peyre, A. and Herlant, M. (1967). C.r. Hebd, Seanc. Soc. Biol. **161:** 1779-1782.
Planel, H.G., Guilhem, A. and Soleilhavoup, J.P. (1961). C.r. Ass. Anat. **47:** 620-633.
Poretsky, L. and Kalin. M.F. (1987). Endocrin. Rev. **8:** 132-141.
Racey, P.A. (1969). J. Reprod. Fert. **19:** 465-474.
Racey, P.A. (1975). In: The biology of the male gamete, Duckett, J.G. and Racey, P.A. (eds.), Academic Press, London, pp. 385-416.
Racey, P.A. (1979). J. Reprod. Fert. **56:** 391-402.
Racey, P.A. (1982). In: Ecology of Bats. Kunz, T.H. (ed.), Plenum Press, New York, pp. 57-164.
Racey, P.A. and Potts, D.M. (1970). J. Reprod. Fert. **22:** 57-63
Racey, P.A. and Swift, A.W. (1981). J. Reprod. Fert. **61:** 123-129.
Ransome, R.D. and Mcowat, T.P. (1994). Zool. J. Linn. Soc. **112:** 337-151.
Ramakrishna, P.A. and Rao, K.V.B. (1977). Curr. Sci. **46:** 270-271.
Richardson, B.A. (1979). J. Reprod. Fert. **56:** 379-389.
Richardson, E.G. (1977). J. Zool., Lond., **183:** 353-375.
Singh, U.P. and Krishna, A. (1992). Ind. J. Exp. Biol. **30:** 670-675.
Singh, U.P. and Krishna, A. (1996a). Zool. Sci. **13:** 893-897.
Singh, U.P. and Krishna, A. (1996b). Acta. Anat. **155:** 104-112.
Singh, U.P. and Krishna, A. (1997). J. Endocrinol. Reprod. **100:** 1–19.
Sluiter, J.W. and Bels, L. (1951). Proc. K. Red. Akad. Wt. C. **54:** 585-593.
Son, S.W., Yoon, M.H., Mori, T. and Uchida, T.A. (1987). J. Mamm. Soc. Japan. **12:** 1-14.
Uchida, T.A., Inoue, C. and Kimura, K. (1984). J. Reprod. Fert. **71:** 439-444.
Uchida, T.A. and Mori, T. (1972). Sci. Bull. Fac. Agr., Kyushu Univ., **28:** 177-184.
Uchida, T.A. and Mori, T. (1977). Metabolism (Tokyo), **4:** 154-156.
Uchida, T.A. and Mori, T. (1987). M.B. Fenton, P.A. Racey and J.N. Rayner, Cambridge University Press, U.K. pp. 251-365.

Van der Merwe, M. (1975). S. Afr. J. Sci. **71:** 237-241.
Van der Merwe, M. (1980). Proc. 5th Int. Bat. Res. Cont. Albuquerque, pp. 113-123.
Van der Merwe, M. and Aarde, R.J.V. (1989). J. Reprod. Fert. **87:** 665-669.
Wilson, D.E. (1973). Period. Biol. **75:** 215-217.
Wimsatt, W.A. (1944). Am. J. Anat. **74:** 129-173.
Wimsatt, W.A. (1960). Bull. Mus. Comp. Zool., Harv. Univ., **124:** 249-267.
Wimsatt, W.A. (1969). Symp. Soc. Expt. Biol. **23:** 511-559.
Wimsatt, W.A. and Kallen, F.C. (1957). Anat. Rec. **129:** 115-132.
Wimsatt, W.A. and Parks, H.F. (1966). Symp. Zool., Soc. Lond., **15:** 419-454.

Comparative Endocrinology and Reproduction
K.P. Joy, A. Krishna and C. Haldar (Eds)

25. Sperm Storage in the Female Reproductive Tract of Reptiles

T. Shivanandappa[1], S.R. Srinivas[2] and H.B.D. Sarkar[2]
[1]Central Food Technological Research Institute, Mysore-570 013, India
[2]Department of Studies in Zoology, University of Mysore, Mysore-570 006, India

Introduction

Reproduction in reptiles is seasonal usually limited to a certain part of the year so that the young ones are born at the most propitious time. In most species of reptiles, the onset of reproductive activity is generally synchronous in both sexes and mating takes place at the peak of reproductive activity usually coinciding with ovulation except in a few temperate species wherein the reproductive cycles in both sexes may not be concomitant (Licht, 1984; Sarkar and Shivanandappa, 1989). In several species belonging to a wide range of Orders, sperms are stored in the female genital tract for varying periods of time (Cuellar, 1966; Howarth, 1974; Saint Girons, 1975; Gist and Jones, 1987; Adams and Cooper, 1988; Palmer and Guillette, 1988). Stored sperm are used to fertilise the eggs without the need for subsequent matings. The adaptive value and evolutionary significance of this phenomenon is of great interest.

The Indian subcontinent is a home to a wide variety of reptilian fauna. Detailed studies on reproduction have been done mainly in the lizards, *Hemidactylus flaviviridis* (Geckonidae), *Calotes versicolor* (Agamidae), *Psammophilus dorsalis* (Agamidae), *Uromastix hardwickii* (Agamidae), *Mabuya carinata* (Scincidae), and the snake *Xenochrophis piscator* and very little on turtles and crocodiles except for some field observations (Sarkar and Shivanandappa, 1989). Surprisingly nothing was known till recently about the phenomenon of sperm storage in Indian reptiles although some species like the garden lizard (*C. versicolor)* has been the subject of study for more than 50 years.

Further, Agamidae which is not represented in the Europe and the American continents, comprises of a large number of species distributed all over the Indian subcontinent. In the course of our studies on reproduction of lizards for the last two decades, we discovered the phenomenon of sperm storage in two agamids, *C. versicolor,* the garden lizard and *P. dorsalis,* the rock lizard. This brief review will summarise our own work in the last ten years on sperm storage in the female genital tract of these two species.

Reproductive Cycles

C. versicolor is widely distributed in India whereas *P. dorsalis* inhabits the rocky mountainous terrain of peninsular India. Reproductive cycles in both species are very much similar (Sarkar and Shivanandappa, 1989). The peak reproductive activity extends from May to August. In the male, testicular regression begins after August and, from September to March, the testes are fully regressed. Recrudescence begins in late March and continues till April (Sarkar and Shivanandappa, 1989). In the female, ovarian recrudescence begins in March/April and vitellogenic follicles can be seen during April-May and the first clutch is invariably laid after the first showers in April-May (Shanthakumari *et al.,* 1990, 1992). However, the peak egg-laying season is during May-August. Mating takes place during May-June. Unlike the male, the female does not show gonadal regression soon after the first clutch of eggs but continues to show follicular development and the next batch of eggs developing. In these species three or more clutches of eggs are laid in a season from May to November or December. Therefore the second and subsequent clutches are laid when the males are already sexually quiescent with fully regressed testes showing no spermatogenesis nor steroidogenesis (Shanthakumari *et al.,* 1990, 1992; Srinivas *et al.,* 1995). Examination of the oviducts during the reproductive cycle revealed the presence of eggs during May-November. Field observations suggest the hatching of young ones. It was therefore concluded that the subsequent viable clutches could be possible by sperm obtained during prior mating, stored in the female genital tract.

The Oviduct

The morphology of oviduct in these species is very similar (Shanthakumari *et al.,* 1992; Srinivas *et al.,* 1995). It is a long and coiled tube comprising the proximal fallopian tube which collects the eggs, followed by the uterine region which forms the major portion of the oviduct ending with the vaginal region. Both oviducts end up in the common cloacal opening.

Histologically, three distinct layers constitute the oviduct wall, i.e, the outer serosa, middle lamina propria and inner mucosa. The mucosal layer is folded in the infundibulum and vaginal regions, whereas the uterus lacks elaborate foldings (Fig. 2). The uterus which comprises two-third of oviduct, is characterised by the presence of uterine glands, absence of mucosal foldings, a thin layer of longitudinal cells and ciliated cells interspersed with non-ciliated cells. The infundibulum is devoid of any glandular layer and the ciliated epithelium is highly folded. The uterine glands are greatly hypertrophied and secretory during the reproductive phase.

Sperm Receptacles

Examination of the oviduct during the reproductive cycle shows the presence of sperm in ‘pockets’ confined to the uterovaginal region. Stored sperm

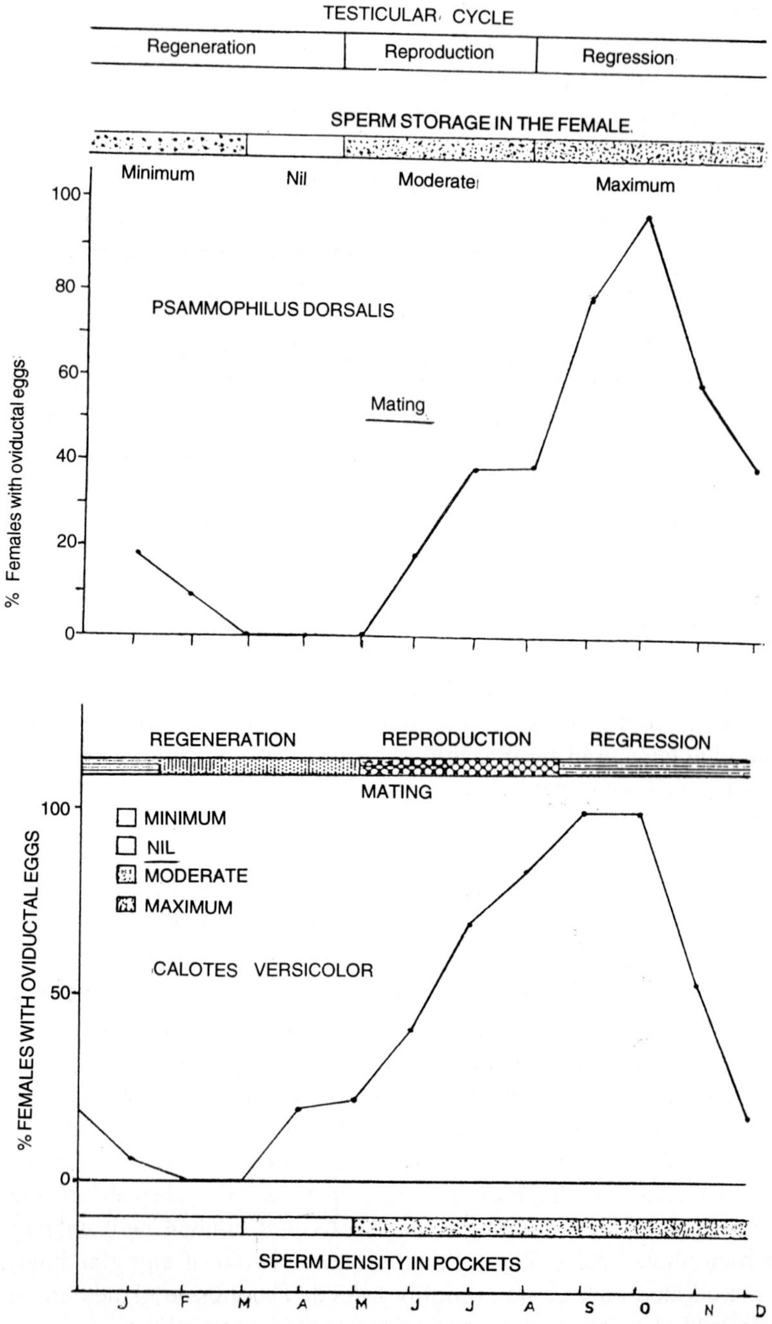

Fig. 1 **Graphs showing the percentage of gravid females in the sampled population in each month, and the relative density of stored sperm in the vaginal pockets during the annual reproductive cycle in *P. dorsalis* and *C. versicolor* (from Shanthakumari *et al.*, 1990 and Srinivas *et al.*, 1995).**

could be seen from May to December (Figs. 3 and 4). The sperm and secretory material are found as masses in the lumen of the sperm storage pockets. The sperm heads do not contact the epithelium of the sperm pockets.

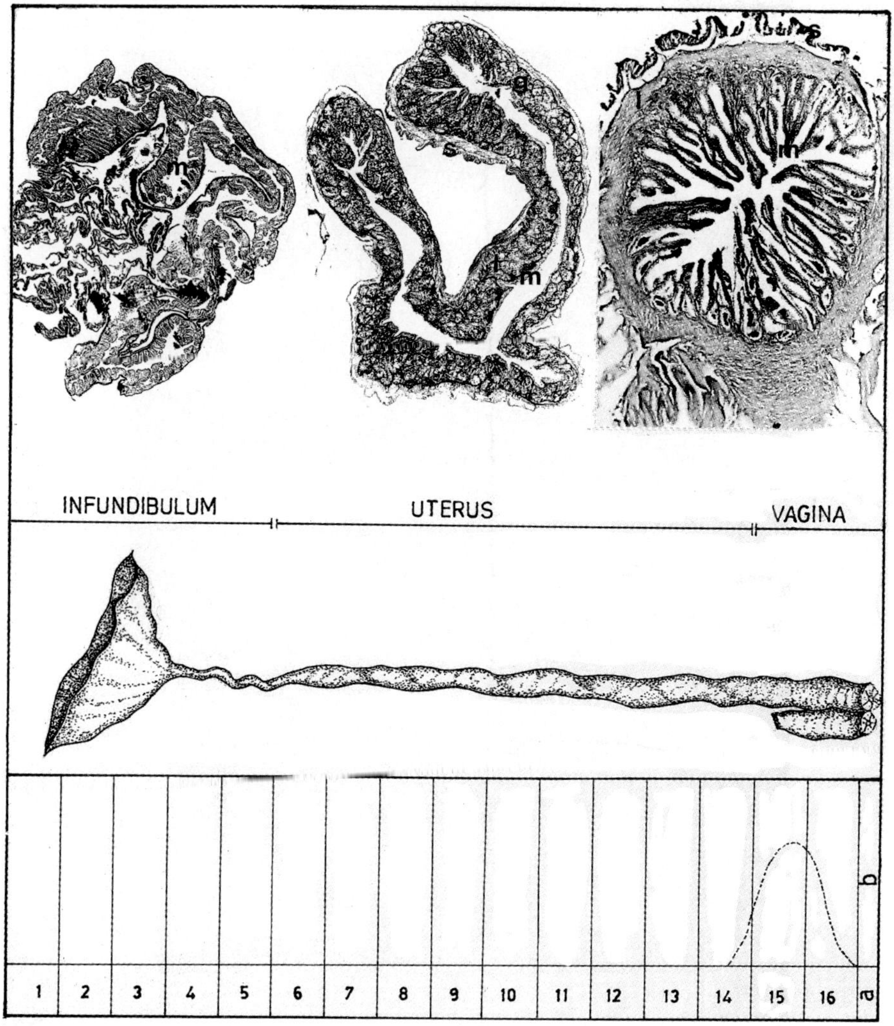

Fig. 2 Diagram showing the division of the oviduct of *C. versicolor* into three regions (infundibulum, uterus, vagina) with representative histological sections (approx. × 85). Scale at bottom (a) shows length of the oviduct in centimeters. Curve (b) shows relative distribution of sperm pockets in the vaginal region. g, uterine gland; l, lamina propria; m, mucosa; s, serosa (from Shanthakumari *et al.*, 1990).

The epithelium lining the lumen of the sperm pockets comprises of ciliated as well as non-ciliated cells. The sperm in the pockets are found with PAS +ve granular secretions. Histochemical tests indicate that these secretions also stain for proteins (Table 1). Histochemical profile of the enzymes in the sperm storage pockets shows the presence of several dehydrogenases including 3β- and 17β-hydroxysteroid dehydrogenases (HSDHs) in the epithelium. The luminal contents show strongly positive reaction for acid phosphatase.

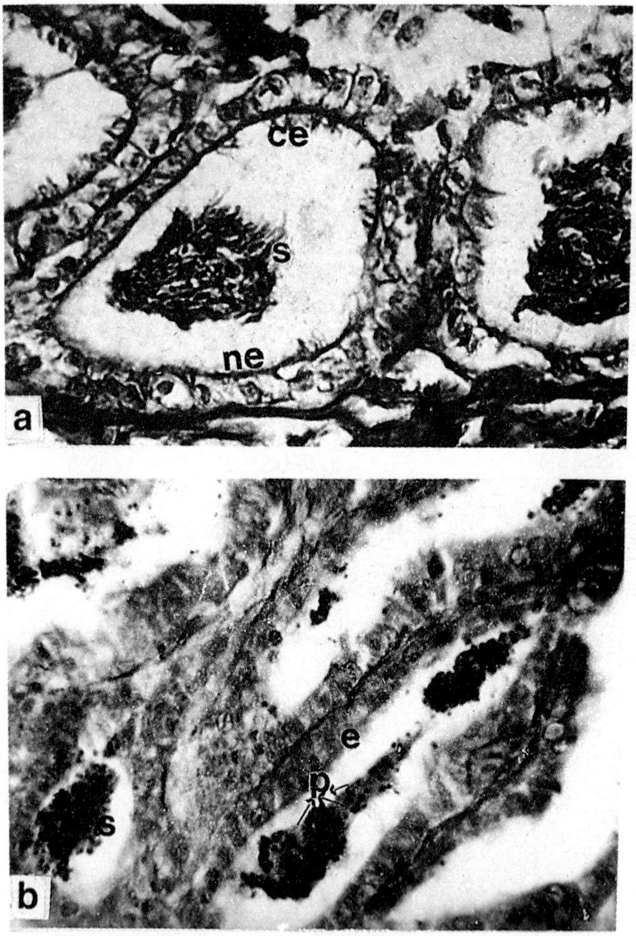

Fig. 3 **Sections of sperm pockets during September (a), consisting of ciliated and non-ciliated epithelial cells, and during the recrudescent phase (b), showing sperm distributed with PAS-positive ground substance. ce, ciliated epithelial cells; p, PAS-positive ground substance; s, sperm. a, × 600; b, × 400 (from Shanthakumari *et al.*, 1990).**

Recovery of Stored Sperm and Secretory Granules

We were successful in obtaining sperm from these pockets. The stored sperm collected from the pockets during December-January were normal and motile even after a period of 6 months in the female genital tract (Figs. 5c, d). Sperm recovery was best during the early part of the reproductive

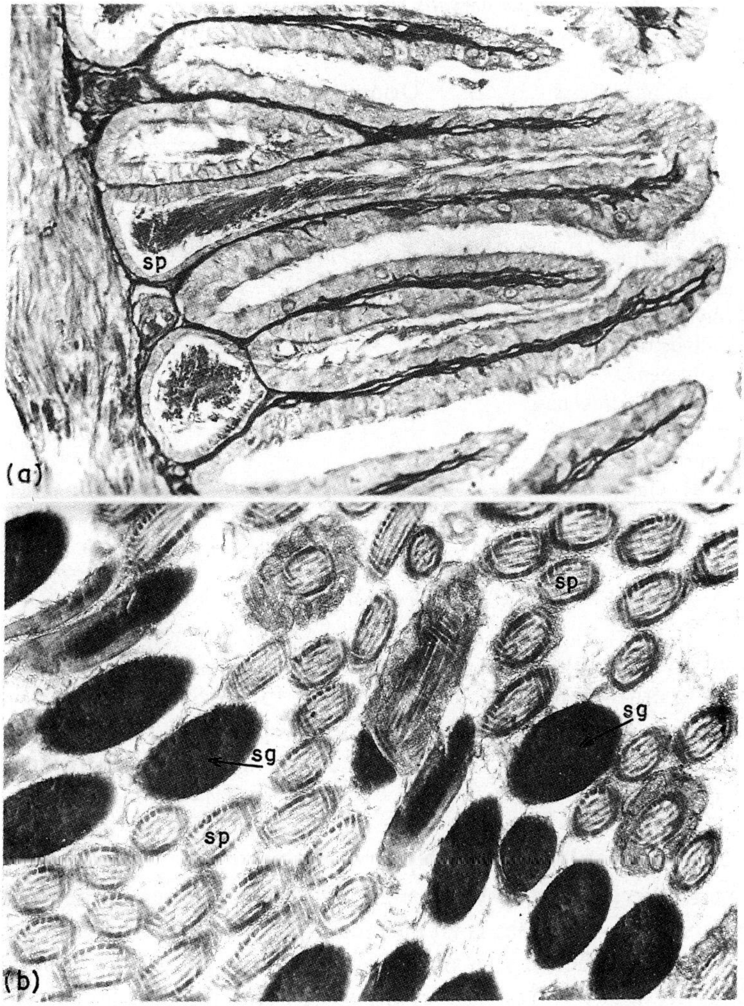

Fig. 4 (a) Sperm receptacles in the uterovaginal region of *Psammophilus dorsalis*. sp-Sperm; Mallory's Trichrome, × 400. (b) An electron micrograph of the sperm receptacle of *Psammophilus dorsalis* showing stored sperm (sp) × 10000.

phase (August-September) but somewhat poor during later months (November-January). Srinivas *et al.* (1995) showed that the flushings from the uterovaginal regions often showed sperm associated with descrete round granules (Fig. 5c, d) which on longer storage appear to become irregular and amorphous. The granules found in the sperm storage pockets stain positively to tests for protein and PAS and resemble the secretory granules isolated from the epididymis and vas deferens in size, shape and tinctorial affinities. It was further shown by us that the vas deferens of male lizards

Table 1. Histochemical reactions of the sperm storage pocket in the oviduct of *Psammophilus dorsalis* and *Calotes versicolor*

Histochemical Tests	Oviductal epithelium	Vaginal sperm storage pocket		
		Epithelium	Sperm	Secretions
Periodic acid Schiff (PAS)	+++	+++	+	++
PAS + Saliva digestion	+++	+++	+	++
PAS + Acetylation	—	—	—	—
Mercury bromophenol blue	+++	++	++	++
Sudan black B	+++	+++	±	+
3β-hydroxysteroid dehydrogenase	++	++	±	—
17β-hydroxysteroid dehydrogenase	++	++	±	—
Lactate dehydrogenase	+++	+++	+	—
Glucose-6-phosphate dehydrogenase	++++	+++	±	—
$NADH_2$ diaphorase	++++	+++	±	—
Acid phosphatase	++	++	+	++++
Esterase	++++	+++	+	+

—: nil +: Moderate ±: traces +++: High +: Low ++++: Intense

contain numerous secretory granules ranging from 4–28 μm in size mixed with sperms staining positively to tests for PAS and protein (Figs. 3a, b) (Srinivas et *al.,* 1995). This mixture of sperms and secretory granules in the sperm storage pockets are apparently derived from the males during copulation.

Physiological Basis of Sperm Storage

Sperm survival and sustenance during long periods of storage in the female genital tract of reptiles is an intriguing phenomenon. The physiological and biochemical factors that underlie sperm storage are not understood. The integrity and fertilizability of the stored sperm in the female can only be inferred from the field observations of multiple viable clutches (Sarkar and Shivanandappa, 1989). Sperm in the vaginal receptacles are found with PAS-positive secretions. Halpert *et al.* (1982), based on their studies in the garter snake, called the secretory mass "carrier matrix," which was presumed to be derived from the glycoproteinaceous secretions of the sperm pocket epithelium; they suggested that the secretions may help in sperm transport and as a source of nutrition. Our studies imply that PAS-positive glycoprotein substance, present in the storage crypts and associated with the sperm, is possibly derived from the secretions of the male reproductive tract. Our results show that the sperm in the vas deferens of male lizards are associated with innumerable PAS-positive secretory granules, apparently derived from the epididymis (Sarkar and Shivanandappa, 1989; Srinivas *et al.,* 1993, 1994). These granules could easily be recovered intact along with sperm from recently mated lizards or

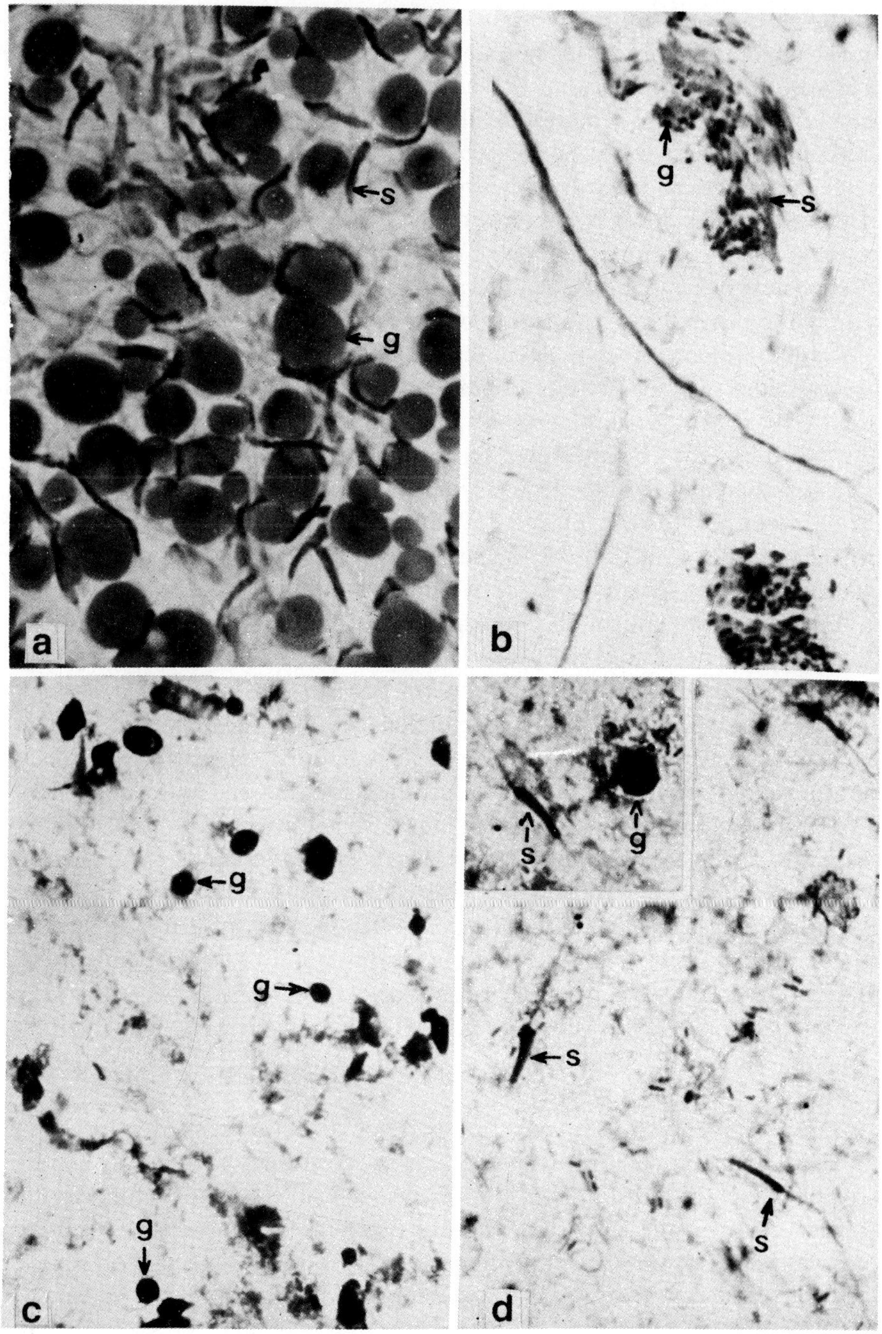

Fig. 5 ***Psammophilus dorsalis*****. Sperm and secretory granules in the male and female reproductive tract: a, Smear preparation from flushings of the vas deferens, showing numerous secretory granules (g) and sperm (s). Haematoxylin and eosin. × 1000. b, Transverse section of anterior vaginal region showing the sperm storage pocket with sperm (s) and granular secretions (g) in the lumen. PAS. × 400. c, d, Intact secretory granules (g, in figure c) and sperm (s, in figure d) recovered from vaginal sperm pockets. Inset (figure d) shows both sperm and an intact secretory granule. Heamatoxylin and eosin. × 1000 (from Srinivas** ***et al.,*** **1995).**

during early months of the reproductive phase; they seem to lose their size and shape during long periods of storage in the vaginal sperm pockets. The presence of these granules in the vas deferens and in the sperm pockets of female lizards is here reported for the first time in reptiles. The secretions present in the sperm storage pockets may also derive from the sexual segment in the kidney of male lizards. However, tinctorial and histochemical features of these granules differ from those of the epididymal granules (Srinivas *et al.*, 1994). The secretory granules of the sexual segment are PAS-negative in this species (our unpublished observations). Further direct evidence may be required to prove the exact identity of the secretions in the sperm pockets.

Another important component of the secretions present in the sperm pocket is the elevated activity of acid phosphatase. This has been reported previously in the sperm receptacles of *Calotes versicolor* (Shanthakumari *et al.*, 1990). The enzyme seems to be derived from the male accessory organs, principally from the renal sexual segment, which is known to secrete large amounts of the enzyme (Shivanandappa, 1988; Sarkar and Shivanandappa, 1989). This enzyme may have an important role in sperm sustenance in the female reproductive tract, and it is thought to be involved in the energy metabolism of an unknown substrate (Shivanandappa, 1988). It has earlier been shown by Cuellar *et al.* (1972) that sexual segment secretions increase the motility of sperm *in vitro.*

It has been proposed that the epithelial cells lining the sperm storage pockets may secrete glycogen, fructose, or a lipoidal substance which may serve as a nutritional source for the sperm (Hoffman and Wimsatt, 1972). The occurrence of sorbitol dehydrogenase in the oviductal epithelium of the garter snake, *Thamnophis sirtalis*, has been suggested as an indication for the utilisation or production of an energy substrate such as fructose. Sanyal and Prasad (1966) earlier proposed that the enzyme esterase, present in the oviductal epithelium, may be important in the metabolism of phospholipid secretions derived from the sexual segment in the wall lizard, *Hemidactylus flaviviridis.* However, almost nothing is known for certain regarding the exact biochemical basis of sperm survival in the female reproductive tract of reptiles.

There appears to be an interesting parallelism between sperm storage in the epididymis of male lizards and in sperm receptacles of females. In both situations, sperm are stored in a dormant but viable condition for varying periods of time. The activities of various enzymes of important metabolic pathways such as the Kreb's cycle, the pentose-shunt, and glycolysis are elevated in the epithelium of these storage sites but reduced in sperm (Shivanandappa and Sarkar, 1986). In birds and mammals the sperm, when ejaculated or ejected out of their site of storage, become motile and thus metabolically "activated" (Thorne and Foley, 1974). The factor(s) responsible for sperm "dormancy" during storage remains unclear. It has been hypothesized that the steroid metabolism of the epididymal epithelium has an important influence on the metabolic dormancy of sperm in mammals (McGadey *et al.*, 1966). Similar observations and postulations have been extended to the reptilian epididymis (Shivanandappa and Sarkar, 1986). It is to be noted that the sperm pocket epithelium of the lizards *C. versicolor* (Shanthakumari *et al.*, 1990) and *P. dorsalis* (Srinivas *et al.*, 1995) also shows activity for hydroxysteroid dehydrogenases, similar to that of the metabolic environment in sperm receptacle cells in the female, and the secretions derived from the male, have important roles in sperm storage and survival in the female reproductive tract.

Biological Significance

Reptiles store sperm in the female reproductive tract for periods longer than those observed in most other vertebrates (Howarth, 1974; Gist and Jones, 1987). Sperm storage in reptiles is intimately linked to the reproductive strategy of each species. This is especially true for the temperate forms in which mating and ovulation may not coincide as for instance, in the garter snake, *Thamnophis sirtalis parietalis* (Halpert *et al.*, 1982). In the anole, *Anolis carolinensis* the female ovulates an egg every 10–14 days (Crews, 1973). Although mating coincides with ovulation in this species, sperm storage in the female is uniquely advantageous; it reduces dependence on multiple matings and ensures reproductive success (Conner and Crews, 1980). Therefore, the pattern of ovulation and oviposition determines the need for sperm storage in reptiles.

In the tropical agamid *Calotes versicolor*, the reproductive cycle is of the prenuptial type, where gametogenesis in both sexes is synchronized; sperm storage helps fertilize subsequent clutches of eggs (Shanthakumari *et al.*, 1990). Sperm storage in another agamid, *Psammophilus dorsalis*, is very much like that of *C. versicolor*. Sperm storage in *C. versicolor* helps successful multiple clutches of eggs, which in effect prolongs the reproductive phase in the female for 5–6 months and thus enhances fecundity markedly (Shanthakumari *et al.*, 1992). In contrast, the sympatric skink *Mabuya carinata*, which breeds during November-December, lays only a single clutch of eggs and does not store sperm in the female reproductive tract (Sarkar and Shivanandappa, 1989). These observations in tropical species exhibiting prenuptial reproduction point out the importance of sperm storage in reptiles of diverse climatic zones, as well as the importance of adaptive strategies of reproduction in a similar habitat. It is likely that future studies will show more instances of sperm storage in several other reptiles of the Indian subcontinent.

References

Adams, C.S. and Cooper, Jr. W.E. (1988). Herpetologica **44:** 190–197.

Conner, J. and Crews, D. (1980). J. Morphol. **163:** 331–348.
Crews, D. (1973). Horm. Behav. **4:** 307–313.
Cuellar, H.S., Roth, J.J. and Fawcett, J.D. (1972). Herpetologica **288:** 53–57.
Cuellar, O. (1966). J. Morphol. **119:** 7–20.
Gist, D.H. and Jones, J.E. (1987). Scanning Microscopy **1(4):** 1839–1851.
Halpert, A.P., Garstka, W.R. and Crews, D. (1982). J. Morphol. **174:** 149–159.
Hoffman, L.H. and Wimsatt, W.A. (1972). Am. J. Anat. **134:** 71–96.
Howarth Jr, B. (1974). Sperm storage: as a function of the female reproductive tract. In: The Oviduct and its Functions. (A.D. Johnson and C.W. Foley, Eds.) pp. 238–270. Academic Press, New York.
Licht, P. (1984). Reptiles. In: Marshall's Physiology of Reproduction, Vol. I. (G.E. Lamming, Ed.) pp. 206–282. Churchill Livingstone, Edinburgh, U.K.
McGadey, J., Baillie, A.H. and Ferguson, M.M. (1966). Histochemie. **7:** 211.
Palmer, B.D. and Guillette, Jr. L.J. (1988). Amer. J. Anat., **183:** 200–211.
Saint Girons, H. (1975). In: The Biology of Spermatozoa. (E.S.E. Hafez and C.G. Thibault, Eds.) pp. 105–113. S. Karger, Basel.
Sanyal, M.K. and Prasad, M.R.N. (1966). J. Morphol. **118:** 511–528.
Sarkar, H.B.D. and Shivanandappa, T. (1989). In: Reproductive Cycles of Indian Vertebrates. (S.K. Saidapur, Ed.) pp. 225–272. Allied Publishers, New Delhi, India.
Shanthakumari, T.R., Sarkar, H.B.D. and Shivanandappa, T. (1990). J. Morphol. **203:** 97–106.
Shanthakumari, T.R., Sarkar, H.B.D. and Shivanandappa, T. (1992). J. Morphol. **211:** 295–306.
Shivanandappa, T. (1988). *XIV Intl. Congr. Biochem.*, Prague (Abst.).
Shivanandappa, T. and Sarkar, H.B.D. (1986). Curr. Sci. (India), **55:** 175–178.
Srinivas, S.R., Sarkar, H.B.D. and Shivanandappa, T. (1993). *Internatl. Symp. Recent trends Life Sci.*, Hyderabad (Abst.).
Srinivas, S.R., Sarkar, H.B.D. and Shivanandappa, T. (1994). *XII National Symp. Repr. Biol. Comp. Endocrinol.*, Pondicherry (Abst.).
Srinivas, S.R., Shivanandappa, T., Hegde, S.N. and Sarkar, H.B.D. (1995). J. Morphol. **224:** 293–301.
Thorne, J.G. and Foley, C.W. (1974). In: The oviduct and its functions. (A.D. Johnson and C.W. Foley, Eds.) pp. 221–236. Academic Press, New York.

Comparative Endocrinology and Reproduction
K.P. Joy, A. Krishna and C. Haldar (Eds)

26. Temporal Relationship Between Circannual Reproductive Cycle and Infradian Rhythms in Physiologic Functions in Lower Vertebrates

Atanu Kumar Pati and Maya Shedpure
School of Life Sciences, Pt. Ravishankar Shukla University, Raipur-492 010, India

Organisms do not exist in isolation, but live in a complex environment of many interacting components. The environmental factors in their turn are never constant from one moment of time to the next, but vary continually and most often predictably. The periodic oscillations of the environmental factors, such as day/night (solar day), lunar day, lunar month, season, year and tidal flow and ebb, etc., are determined by the rotations of the earth and the moon.

The earth, a planet inhabited by all living forms, completes one rotation on its axis in approximately 24 hours and revolves round the sun through an elliptical orbit in about 365.25 days. Organic life which began in this planet about three billion years ago must have been influenced by the rhythmical processes of the environment.

Since adaptation is one of the fundamental properties of life, it is obvious that organisms might have evolved mechanism(s) to make use of the highly predictable environment by timing their reproduction among other biological functions for successful survival. Consequently, responses of the physiologic and many other bodily functions of the organisms to periodic environment might have been essentially rhythmic. Those that did not evolve the mechanism(s) to respond and to adapt to the ever changing periodic environment probably have perished.

It is well known that most of the vertebrate species exhibit a cyclic pattern in their reproductive activities and behaviour (Pisces: Peter and Hontela, 1978; Billard, 1981; Babbar and Sareen, 1982; Mukherjee *et al.*, 1989; Amphibians: Kanamadi and Saidapur, 1982; Kanamadi *et al.*, 1983; Pramoda and Saidapur, 1984; Delgado *et al.*, 1990; Yoo *et al.*, 1995; Wake, 1995; Reptiles: Srivastava and Thapliyal, 1965; Cartland *et al.*, 1994; Lofts, 1978, 1985; Janeiro-Cinquini *et al.*, 1993; Zhang and Wu, 1994; Rostal *et al.*, 1994; Aves: Gwinner, 1973, 1987; Kumar and Tewary, 1982; Thapliyal, 1993; Mammals: Chandrakala and Devraj Sarkar, 1980; Chauhan and Saxena, 1985). Furthermore, the peak reproductive activity in the seasonally breeding vertebrates is generally confined to a particular part of the year. It would seem that in each species successful breeding requires a high degree of internal temporal compatibility with other non-reproductive, energy yielding processes for a suitable adaption to the environment (Shedpure and Pati, 1994). These eventually offer the greatest survival fitness to their offsprings.

An annual reproductive rhythm represents a chain of distinct physiologic events, namely recrudescence of the gonad, ovulation or spermiation, spawning or breeding and regression of the gonad. All these events are required to be initiated and completed within an appropriate time window. This is accomplished through the execution of two fundamental functional relationships, such as the external and the internal. The former provides adaptability whereas the latter stability to a given species with regard to the environment. A wealth of information on the environmental/exteroceptive factors such as light (Thapliyal and Lal, 1984; Gwinner, 1987), temperature (Jorgensen *et al.*, 1978), humidity (Pevet, 1987), rainfall (Rao, 1979; Saidapur, 1989), food availability (Disney and Marshall, 1956; Krishna, 1982), and social impact (Flickinger, 1961) involved in cuing of annual reproductive cycle of seasonally breeding fishes, amphibians, reptiles, birds and mammals is available.

It is indeed evident from an extensive literature that the annual reproductive cycle in vertebrates has been studied either as a mere natural phenomenon (Fitch, 1982; Tiwari *et al.*, 1988; Delgado *et al.*, 1989; Kyriakopoulou-Sklavounou and Loumbourdis, 1990; Treasurer, 1990) or has been examined as a reflection of the underlying endocrine periodicities (Rastogi *et al.*, 1976; Reiter, 1980; Haldar and Pandey, 1989; Mackenzie *et al.*, 1989; Reddy *et al.*, 1993; Malison *et al.*, 1994; Srivastava and Srivastava, 1994; Manire *et al.*, 1995). Furthermore, the relationship between the reproductive cycle and the exteroceptive factors has been validated in a number of vertebrate species (Sundararaj, 1978; Rao, 1979; Peter, 1981; Krishna, 1982; Lofts, 1985; Gwinner, 1987; Thapliyal, 1993). The main emphasis of this mini-review will be to construct peak maps for a number of physiologic functions and to establish temporal relationship between circannual rhythm of reproduction and other low frequency rhythms. It would have been too voluminous to include available data obtained on all the classes of phylum Chordata and all the species studied so far under each class. Therefore, we have selected only three different classes, viz., Pisces, Amphibian and Reptilia and furthermore we have narrowed down the scope of discussion by zooming out on the number of representative species in each class. The criteria that guided selection of species are not merely arbitrary rather the magnitude of available/documented information on a species favoured its inclusion. Eventually this review deals with the investigations performed in our as well as others' laboratories (Singh and Singh, 1979; Lal and Singh, 1987; Lal *et al.*, 1987; Singh and Singh, 1987; Joshi, 1980, 1982; Tandon and Joshi, 1974; Milone *et al.*, 1975, 1978; Rastogi, 1976; Rastogi and Chieffi,

1973; Iela *et al.*, 1979; Guarino *et al.*, 1993; Thapliyal and Sharan, 1980; Haldar-Misra, 1985; Haldar and Pandey, 1989; Goindi, 1977; Johnson and Oommen, 1982; Shanmukhamma *et al.*, 1992) on circannual/infradian rhythms in reproduction and various physiologic variables in *Clarias batrachus, Heteropneustes fossilis, Rita rita, Rana esculenta, Rana italica, Natrix piscator* and *Calotes versicolor.*

Annual or Circannual/Infradian?

The seasonal reproductive cycle, involving the morphometric and/or histomicromorphometric and/or radioimmunoassay data on gonadal state (volume or size or tinctorial property or level of relevant circulating hormones or all of these together) collected rigorously every month over a period of one or two years, is the most commonly cited/referred example of an annual endocrine cycle in ectothermic vertebrates (Van Wyk, 1994; Swain and Jones, 1994; Canterbury and Pauley, 1994; Ramirez Pinilla, 1994; Malison *et al.*, 1994). Now question arises if one could call this as a circannual endocrine/reproductive rhythm also. Technically, the answer is in the negative.

A seasonal or annual rhythm in any biological function could be called circannual only when the rhythm is endogenous and exhibits following properties: (a) Essentially rhythm should free run exhibiting a period close to 365.25 days; (b) The rhythm should be accurate and reproducible; (c) The tau of the free running rhythm should be temperature-compensated. Ideally both high (circadian) and low (circannual) frequency rhythms must be investigated in the same experiment. That would help to understand various physiologic variables in several frequency domains (Table 1). It has been rigorously demonstrated that not only the 24-h mean level but also the circadian crest time of a given physiologic variable can be the subject of a circannual rhythm (Mayersbach, 1976; Pati, 1989). However, such experiments are rare and those documenting free running circannual rhythms are very few (Chandola *et al.*, 1988; Sundararaj and Sehgal, 1970). To designate an annual rhythm endogenous and to study the above mentioned rhythm characteristics it is a prerequisite that the organism has to be put under extreme experimental manipulations (for example, isolation for more than one year and at least for 3 years, etc.). Furthermore, isolation has to be rigorous in terms of known zeitgebers, such as light (LL for day active and DD for night active organisms), temperature (its level and constancy) and humidity (high, low or moderately constant). Physiologic variables may exhibit rhythms with seemingly different frequencies and periods and it is easier to realize extreme experimental manipulations in studies involving high frequency rhythms (circadian) but not in case of low frequency rhythms (e.g., circannual etc.). However, we here prefer to designate rhythms as either circannual or infradian depending upon the frequency they exhibit. We have a hunch that various physiologic rhythms with low frequencies would also free run in organisms, when kept inside time cue less environment, by the same token as circadian rhythms behave. For the sake of clarity it is worthwhile to mention that while circannual rhythm could also be termed as infradian, the latter may not be necessarily circannual. A 6-monthly or 3-monthly rhythm could only be termed as infradian.

Table 1. Biologic time structure: period ranges for terms describing biologic rhythms.

Term	Period range
Ultradian	$\tau < 20$-h
Circadian	20-h $\leq \tau \leq$ 28-h
Dian	23.8-h $\leq \tau \leq$ 24.2-h
Infradian	$\tau >$ 28-h
Circaseptan	$\tau = 7 \pm 3$ d
Circadiseptan	$\tau = 14 \pm 3$ d
Circavigintan	$\tau = 21 \pm 3$ d
Circatrigintan	$\tau = 30 \pm 5$ d
Circannual	$\tau = 1$ y $\pm$ 2 m

τ = period; h = hour; d = day; m = month; y = year. Terms coined by analogy to usage in physics. Just as frequencies higher than those audible or visible are called *ultrasound* and *ultraviolet*, frequencies higher than 1 cycle per 20 h are designated as *ultradian*. By the same token, as frequencies lower than the audible or visible are called *infrasound* and *infrared*, rhythms with a frequency lower than 1 cycle per 28 h are designated as *infradian*. Rhythms with a frequency of 1 cycle in 24 ± 4 h are designated as *circadian* (*circa* = about, approximately; *dies* = day or 24 h) (after Halburg et al., 1977).

Conventional Practice and Current Approach

Usually in seasonal or annual studies involving one or several physiologic variables, the data are averaged per season/month and standard errors are computed, respectively, for each variable. Thereafter, a paired t-test or analysis of variance are used to validate statistically significant season and/or time effect.

However, in the present paper attempts have been made to characterize a circannual rhythm by the parameters of the best fitting cosine function approximating all data pertaining to a given variable (Figure 1). The rhythm parameters estimated by the least squares fitting of a one year cosine curve include the mesor (M, rhythm-adjusted mean, in original units), the amplitude (A, half of the difference between the minimum and the maximum of the best-fitting cosine function, in original units) and the acrophase (ϕ, time of the crest for this cosine function with local midnight of December 22 as ϕ reference in usual time units of months and days) (Figure 2). A rhythm is detected, with regard to the considered τ, when the amplitude differs from zero (non-null amplitude test) with $p \leq 0.05$. The single cosinor method was applied to all the time series for $\tau = 12$ and for 6 and 3 months (Nelson *et al.*, 1979; Gupta and Pati, 1992). An acrophase chart or a peak map was constructed for displaying relationship between annual rhythm of reproduction and other low frequency rhythms in different variables (Figure 2). The data on circannual/ infradian mesor and amplitude are not shown in this paper.

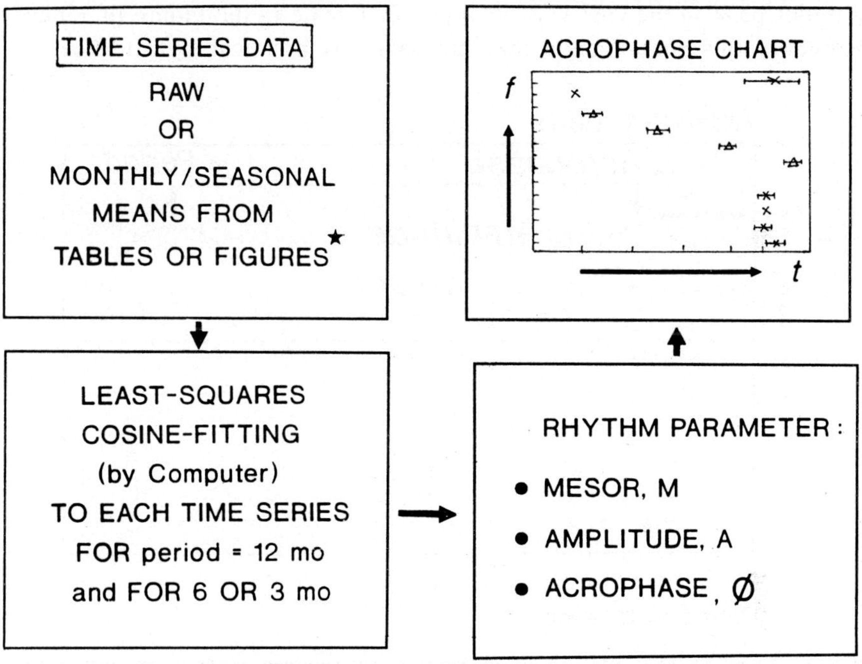

Fig. 1 Flowchart illustrating procedures of data management and presentation. *Time series data (raw) were obtained either from the author(s) directly or as monthly/seasonal means from their published papers and subjected to cosinor analyses. Rhythm parameters were obtained for each series, tabulated and presented in the form of a peak map (acrophase map).

Pisces

Clarias batrachus

A statistically significant circannual rhythm in GSI was validated for both the male and female *Clarias batrachus* (Table 2; Figures 3 and 4). The reproductive cycle of *C. batrachus* in the vicinity of Raipur (Lat.: 21°14′N; Long.: 81°38′E) can be divided into four phases i.e., preparatory phase (February-April), prespawning phase (May-June), spawning phase (July-August), and post-spawning phase (September-January) (Shedpure and Pati, 1994). It has been well established for *C. batrachus* that the gonadal cycle is cued by two important environmental factors, namely the duration of the daily photophase (photoperiod) and the external temperature (Nayak and Singh, 1988). The pattern of gonadal cycle observed at *Raipur* is comparable to that obtained at *Varanasi* (Lat.: 25°18′N; Long.: 83°1′E) (Nayak and

Singh, 1988) and *Santiniketan* (Lat.: 23°39′N; Long.: 87°47′E) (Dasgupta and Nath, 1988). Surprisingly, for female *C. batrachus* a statistically significant 6-monthly rhythm in the GSI was also obtained. Complex waveform was employed to analyze both the annual and 6-monthly rhythms. Consequently, a minor peak in the GSI was obtained for female in the month of January. However, this phenomenon may not have any practical significance.

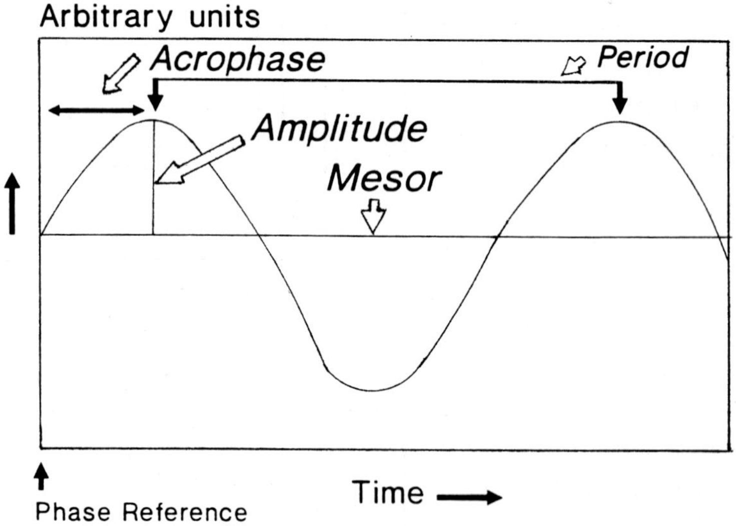

Fig. 2 Least-squares plot showing rhythm parameters, such as period, mesor, amplitude and acrophase. December 22 of the preceding year has been taken as the reference time point for estimation of acrophase.

In *Clarias batrachus,* time series for 22 variables were subjected to cosinor analyses. A statistically significant circannual rhythm was validated for 9 physiologic variables excluding those for GSI. In addition, 6-monthly and 3-monthly rhythms were validated for 11 and 3 variables, respectively. At least four variables exhibited multifrequency infradian rhythms. The acrophase map clearly shows that the peaks for almost all the variables are overlapping with the peak of the GSI except those for serum glucose, muscle protein, rate of oxygen uptake by muscle tissue, hepatic monoglyceride, plasma monoglyceride, plasma diglyceride, muscle diglyceride and ovarian testosterone (Figure 3).

In male *C. batrachus,* statistically significant circannual, 6-monthly and 3-monthly rhythms were detected in 23, 9 and 6 physiologic variables, respectively (Table 2; Figure 4). Out of these, 8 variables exhibited multifrequency infradian rhythms. It is revealed from the acrophase map that at least peaks for 13 variables are overlapping with the peak of the gonadosomatic index.

Table 2. Infradian rhythms in GSI and 22 physiologic variables of female, and 26 of male, *Clarias batrachus*

Variable[a]	τ[b]	p[c]	Acrophase (95% CL)[d]
Female			
GSI	12	0.001	Jul 19 (Jul 07, Aug 01)
	6	0.001	Jan 25 (Jan 13, Feb 07) also Jul 25
RBC	12	0.003	Nov 25 (Oct 10, Jan 10)
	6	0.001	Feb 07 (Jan 16, Feb 28) also Aug 07
Hb	12	0.09	Oct 28
	6	0.001	Feb 07 (Jan 19, Feb 25) also Aug 07
Serum			
Glucose	12	0.57	May 16
	6	0.001	Jun 10 (May 22, Jun 28) also Dec 10
Protein	12	0.001	Aug 04 (Jul 13, Aug 25)
Cholesterol	12	0.001	Jul 13 (Jul 04, Jul 22)
Muscle			
Glycogen	12	0.001	Aug 10 (Jul 01, Sep 19)
	6	0.001	Mar 01 (Feb 19, Mar 13) also Sep 01
	3	0.01	Mar 04 (Feb 19, Mar 19) also Jun, Sep & Dec 04
Protein	12	0.18	Feb 07
Oxygen consumption	12	0.06	May 28
Liver			
Glycogen	12	0.002	Jul 28 (Jun 13, Sep 13)
	6	0.001	Feb 13 (Feb 01, Feb 25) also Aug 13
Protein	12	0.19	Jul 10
	3	0.02	Feb 07 (Jan 22, Feb 22) also May, Aug & Nov 07
Oxygen consumption	12	0.009	Jun 16 (Apr 25, Aug 07)
GSI (Varanasi)	12	0.003	Aug 19 (Jul 04, Oct 04)
	6	0.001	Feb 15 (Feb 03, Feb 27) also Aug 15
Plasma			
Calcium	12	0.004	Jul 4 (May 25, Aug 13)
Monoglyceride	12	0.04	Apr 13 (Feb 16, Jun 10)
Diglyceride	12	0.1	Apr 13
Monoglyceride			
Muscle	12	0.001	Sep 19 (Aug 22, Oct 16)
Liver	12	0.11	Apr 01
Ovary	12	0.57	Jul 06
	6	0.01	Feb 25 (Feb 04, Mar 16) also Aug 25
	3	0.001	Feb 25 (Feb 19, Mar 01) also May, Aug & Nov 25
Diglyceride			
Muscle	12	0.18	Oct 22
	6	0.005	May 10 (Apr 25, May 25)
Liver	12	0.001	Aug 25 (Aug 04, Sep 16)

(Contd.)

Variable[a]	τ^b	p^c	Acrophase (95% CL)[d]
Ovary			
Testosterone	12	0.68	Apr 07
Estrone	12	0.1	Jul 07
	6	0.03	Jan 07 (Dec 01, Feb 13) also Jul 07
Estradiol	12	0.09	Jul 04
17-β	6	0.04	Dec 28 (Nov 22, Feb 04) also Jun 28
Male			
GSI	12	0.001	Jul 10 (Jun 28, Jul 22)
RBC	12	0.008	Dec 01 (Oct 17, Jan 15)
	6	0.003	Feb 19 (Jan 25, Mar 13) also Aug 19
Hb	12	0.34	Sep 10
Serum			
Glucose	12	0.01	Jun 28 (May 04, Aug 22)
	6	0.001	Jan 01 (Dec 16, Jan 16) also Jul 01
Protein	12	0.001	Aug 16 (Jul 25, Sep 07)
Cholesterol	12	0.001	Jul 13 (Jul 01, Jul 25)
Muscle			
Glycogen	12	0.001	Jul 25 (Jun 16, Sep 04)
	6	0.001	Mar 01 (Feb 22, Mar 10) also Sep 01
	3	0.009	Mar 10 (Feb 28, Mar 22) also Jun, Sep & Dec 10
Protein	12	0.04	Feb 04 (Nov 22, Apr 16)
Oxygen consumption	12	0.13	Jul 16
Liver			
Glycogen	12	0.001	Jul 28 (Jun 22, Sep 04)
	6	0.001	Feb 19 (Feb 07, Mar 01) also Aug 19
Protein	12	0.41	Jul 13
	6	0.002	May 10 (Apr 19, Jun 01) also Nov 10
Oxygen	12	0.03	May 10 (Mar 04, Jul 16)
consumption	3	0.007	Feb 28 (Feb 16, Mar 10) also May, Aug & Nov 28
Plasma			
Free cholesterol	12	0.001	Dec 19 (Nov 28, Jan 10)
Phospholipid	12	0.03	Apr 19 (Feb 28, Jun 10)
	6	0.01	May 10 (Apr 22, May 28) also Nov 10
	3	0.01	Feb 25 (Feb 13, Mar 07) alo May, Aug & Nov 25
Free fatty acid	12	0.001	Jul 25 (Jul 07, Aug 13)
Esterified cholesterol	12	0.02	Jan 01 (Oct 22, Mar 10)
Monoglyceride	12	0.006	Jul 19 (May 28, Sep 10)
Diglyceride	12	0.001	Aug 25 (Aug 13, Sep 07)
Triglyceride	12	0.06	May 25
	6	0.001	May 22 (May 19, May 25) also Nov 22
	3	0.001	Feb 22 (Feb 19, Feb 25) also May, Aug & Nov 22

Variable[a]	τ[b]	p[c]	Acrophase (95% CL)[d]
Free cholesterol			
Muscle	12	0.002	Sep 22 (Aug 22, Oct 22)
Testes	12	0.01	Apr 19 (Mar 10, May 28)
Phospholipid			
Muscle	12	0.01	Nov 13 (Oct 07, Dec 19)
Liver	12	0.001	Mar 25 (Mar 07, Apr 13)
Testes	12	0.001	Jun 01 (May 25, Jun 07)
Free fatty acid			
Muscle	12	0.001	Oct 07 (Sep 22, Oct 22)
Liver	12	0.01	Dec 10 (Oct 25, Jan 25)
	6	0.007	Mar 04 (Feb 13, Mar 25) also Sep 04
	3	0.007	Jan 01 (Dec 22, Jan 10) also Apr, Jul & Oct 01
Testes	12	0.004	Nov 13 (Oct 16, Dec 10)
	6	0.002	Feb 19 (Feb 07, Mar 01) also Aug 19
	3	0.002	Jan 07 (Jan 01, Jan 13) also Apr, Jul & Oct 07

[a]Refer Figures 3 and 4 for source and unit.
[b]Fitted period in months.
[c]From F test of null amplitude rejection hypothesis.
[d]Timing of maximum in best-fitting cosine function with 95% confidence limit. Expressed in time units of months and days.
GSI = Gonadosomatic index, RBC = Red cell number, Hb = Hemoglobin.

Rita rita

All the physiologic parameters examined for male and female *Rita rita* exhibited statistically significant circannual rhythm, except the blood glucose (Table 3; Figure 5). However, a statistically significant 6-monthly rhythm was validated for blood glucose in male *Rita rita* only (Table 3). Furthermore, interestingly the peaks of all the variables occurred within the 95% confidence limits for the peak in gonadal activity in male as well as female, *Rita rita*.

Heteropneustes fossilis

A statistically significant circannual rhythm of reproduction was validated by cosinor method (Table 4). The peak gonadal activity occurred in the early part of July with a spread from late June to late July. Only 9 physiologic variables were available for analyses and all of those exhibited statistically significant circannual rhythm. In addition, a 6-monthly rhythm was detected for ovarian cholesterol and a 3-monthly rhythm was validated for ovarian lipid, hepatic total lipid and liver cholesterol (Figure 6). Excluding hepatosomatic index, plasma glucose and plasma cholesterol, the peaks of all other variables coincided with the 95% peak spread of gonadal activity.

It is remarkable that, in general, most of the variables showing circannual periodicity exhibited peaks coinciding with the peak of gonadal activity.

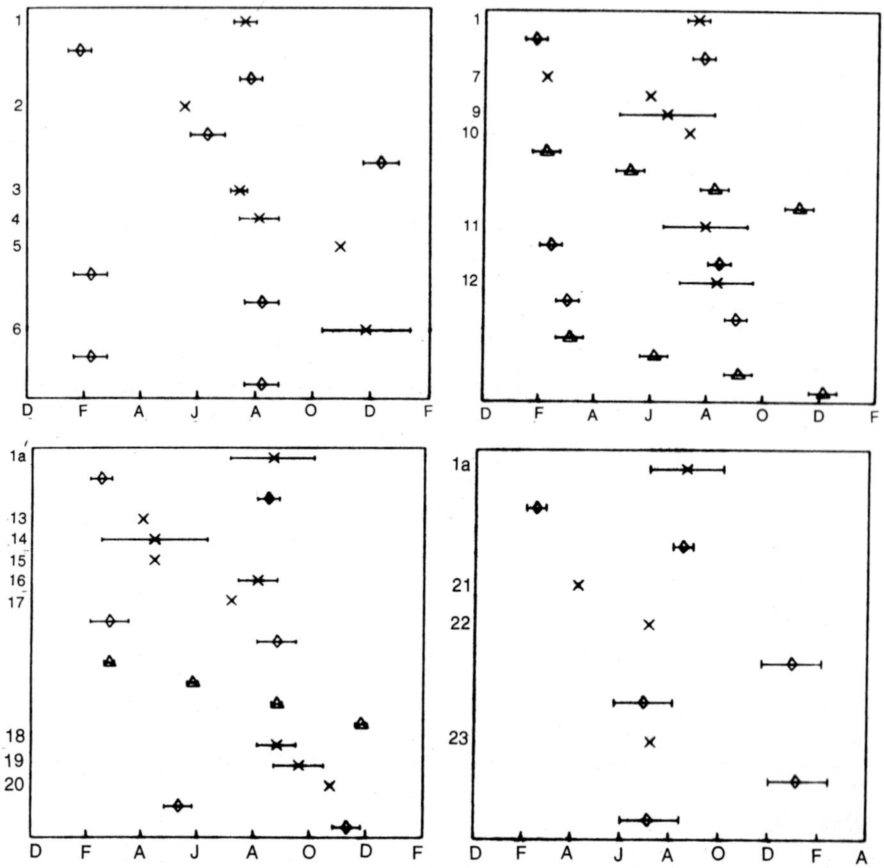

Fig. 3 Acrophase map for female *Clarias batrachus*. X-axis = Month; Y-axis = Variable code. 1, 1a = Gonadosomatic index, 2 = Serum glucose (mg/100 ml), 3 = Serum cholesterol (mg/100 ml), 4 = Serum protein (gm/100 ml), 5 = Hemoglobin (g/d1), 6 = Red cell number ($\times 10^{12}$/1), 7 = Muscle protein (mg/100 mg), 8 = Rate of oxygen consumption in muscle ($\mu 10_2$/mg/hr), 9 = Rate of oxygen consumption in liver ($\mu 10_2$/mg/hr), 10 = Liver protein (mg/100 mg), 11 = Liver glycogen (mg/100 mg), 12 = Muscle glycogen (mg/100 mg), 13 = Hepatic monoglyceride (mg/100 gm), 14 = Plasma monoglyceride (mg/100 ml), 15 = Plasma diglyceride (mg/100 ml), 16 = Plasma protein (mg/100 ml), 17 = Ovarian monoglyceride (mg/100 gm), 18 = Hepatic diglyceride (mg/100 gm), 19 = Muscle monoglyceride (mg/ 100 gm), 20 = Muscle diglyceride (mg/100 gm), 21 = Ovarian testosterone (ng/ovary), 22 = Ovarian estradiol 17–β (ng/ovary), 23 = Ovarian estrone (ng/ovary). Data source: 1–12 (Shedpure and Pati, 1994), 13–20 (Lal and Singh, 1987), 1a, 21–23 (Singh and Singh, 1987). Circannual (cross), 6-monthly (diamond) and 3-monthly (triangle) peaks obtained for each variable are shown with 95% confidence limits (horizontal line). The cross symbol without the line exhibits point estimation of the peak for those variables which do not exhibit statistically validated rhythms.

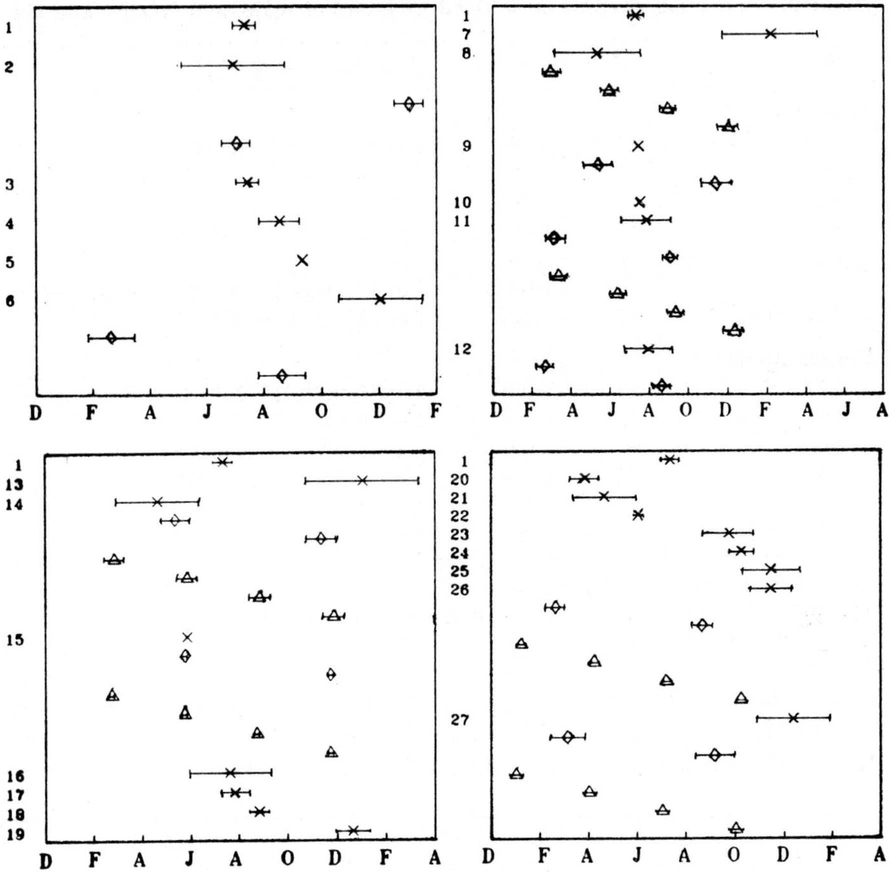

Fig. 4 Acrophase map for male *Clarias batrachus*. 1 = Gonadosomatic index, 2 = Serum glucose (mg/100 ml), 3 = Serum cholesterol (mg/100 ml), 4 = Serum protein (mg/100 ml), 5 = Hemoglobin (g/d1), 6 = Red cell number ($\times 10^{12}$/1), 7 = Muscle protein (mg/100 mg), 8 = Rate of oxygen consumption in liver (μlO_2/mg/hr), 9 = Hepatic protein (mg/100 mg), 10 = Rate of oxygen consumption in muscle (μlO_2/mg/hr), 11 = Muscle glycogen (mg/100 mg), 12 = Hepatic glycogen (mg/100 mg), 13 = Plasma esterified cholesterol (mg/100 ml), 14 = Plasma phospholipid (mg/100 ml), 15 = Plasma triglyceride (mg/100 ml), 16 = Plasma monoglyceride (mg/100 ml), 17 = Plasma free fatty acid (mg/100 ml), 18 = Plasma diglyceride (mg/100 ml), 19 = Plasma free cholesterol (mg/100 ml), 20 = Hepatic phospholipid (mg/100 gm), 21 = Testes free cholesterol (mg/100 gm), 22 = Testes phospholipid (mg/100 gm), 23 = Muscle free cholesterol (mg/100 gm), 24 = Muscle free fatty acid (mg/100 gm), 25 = Muscle phospholipid (mg/100 gm), 26 = Testes free fatty acid (mg/100 gm), 27 = Hepatic free fatty acid (mg/100 gm). Data source: 1–12 (Shedpure and Pati, 1995), 13–19, 20–27 (Lal *et al.*, 1987). See also legend to Figure 3.

Furthermore, those variables that did not show circannual periods but exhibited 6-monthly or 3-monthly periodicity showed coincidence for one of their

Table 3. Infradian rhythms in 6 blood variables of male, and female, *Rita rita*

Variable[a]	τ[b]	p[c]	Acrophase (95% CL)[d]
Male Blood			
TEC	12	0.001	Jun 25 (Jun 04, Jul 16)
Hb	12	0.001	Jun 10 (May 13, Jul 07)
PCV	12	0.001	Jul 01 (Jun 13, Jul 19)
Protein	12	0.008	May 28 (Apr 13, Jul 13)
Glucose	12	0.14	Apr 16
	6	0.03	Jun 19 (May 16, Jul 22) also Dec 19
Cholesterol	12	0.001	May 22 (Apr 22, Jun 22)
Female Blood			
TEC	12	0.001	Jun 10 (May 16, Jul 04)
Hb	12	0.001	Jun 19 (May 22, Jul 16)
PCV	12	0.001	Jun 07 (May 10, Jul 04)
Protein	12	0.004	May 25 (Apr 16, Jul 04)
Glucose	12	0.09	May 01
Cholesterol	12	0.001	May 25 (Apr 25, Jun 25)

[a]Refer Figure 5 for source and unit.
TEC = Total erythrocyte count, Hb = Hemoglobin, PCV = Packed cell volume.
See also legend to Table 2.

peaks with the peak of the circannual gonadal cycle. There are, however, couple of exceptions (Table 2; Figures 3 and 4). The spawning phase in female *C. batrachus* is energy demanding, yet during this phase energy

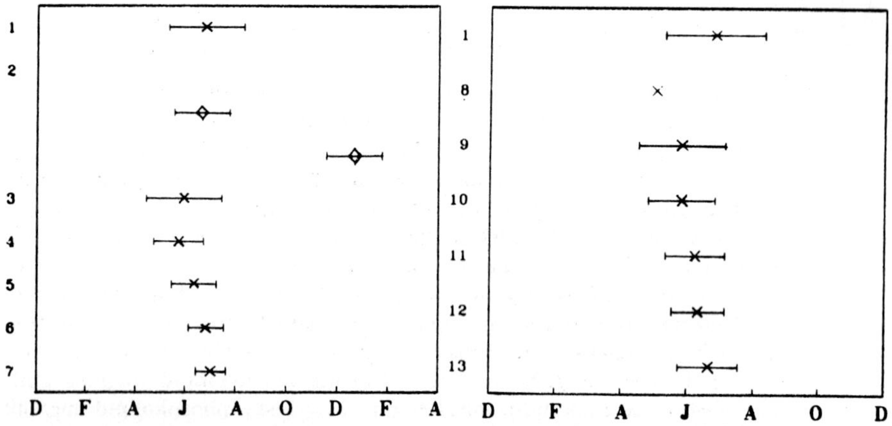

Fig. 5 **Acrophase map for *Rita rita* male: 1 = Gonadosomatic index, 2 = Blood glucose (mg/100 ml), 3 = Blood protein (gm%), 4 = Blood cholesterol (mg/100 ml), 5 = Hemoglobin (gm%), 6 = Total erythrocyte count (× 10⁶/Cmm), 7 = Packed cell volume (%); *Rita rita* female: 8 = Blood glucose (mg/100 ml), 9 = Blood protein (gm%), 10 = Blood cholesterol (mg/100 ml), 11 = Packed cell volume (%), 12 = Total erythrocyte count (× 10⁶/Cmm), 13 = Hemoglobin (gm%). Data source: 1–13 (Joshi, 1982). See also legend to Figure 3.**

input is moderately low characterized by greatly reduced food intake. The synchronization of infradian rhythms in serum cholesterol, serum protein, hemoglobin, red cell number, rate of oxygen consumption by the liver tissue, liver protein, glycogen and diglyceride, and muscle glycogen and monoglyceride with the circannual rhythm of reproduction amply suggests that various physiologic components are temporally well organized to sustain the spawning behaviour and activity (Figure 3). Comparable phenomenon is also evident in case of male, *C. batrachus* (Figure 4). In *Rita rita* and *Heteropneustes fossilis* data on large number of variables were not available.

Table 4. Infradian rhythms in GSI and 2 variables of male, and 7 of female, *Heteropneustes fossilis*

Variable[a]	τ[b]	p[c]	Acrophase (95% CL)[d]
Male			
Plasma			
Glucose	12	0.001	Jan 10 (Dec 28, Jan 22)
Cholesterol	12	0.04	Apr 28 (Feb 19, Jul 07)
Female			
GSI	12	0.001	Jul 10 (Jun 22, Jul 28)
HSI	12	0.001	Dec 28 (Dec 04, Jan 22)
Plasma			
Glucose	12	0.001	Jan 13 (Jan 04, Jan 22)
Cholesterol	12	0.005	Apr 07 (Feb 25, May 19)
Lipid			
Ovary	12	0.001	Jul 25 (Jul 13, Aug 07)
	3	0.008	Feb 04 (Jan 22, Feb 16) also May, Aug & Nov 04
Liver	12	0.001	Jan 13 (Dec 22, Feb 04)
	3	0.03	Dec 25 (Dec 10, Jan 10) also May, Jun & Sep 25
Cholesterol			
Ovary	12	0.01	Aug 04 (Jun 16, Sep 22)
	6	0.008	Feb 01 (Jan 10, Feb 22) also Aug 01
Liver	12	0.001	Jan 25 (Jan 07, Feb 13)
	3	0.02	Mar 22 (Mar 10, Apr 04) also Jun, Sep & Dec 22

[a]Refer Figure 6 for source and unit.
HSI = Hepatosomatic index. See also legend to Table 2.

However, the peak maps for *Rita rita* and *H. fossilis* present an impressive entrainment between circannual reproductive rhythm and circannual/infradian rhythms in physiologic variables (Figures 5 and 6).

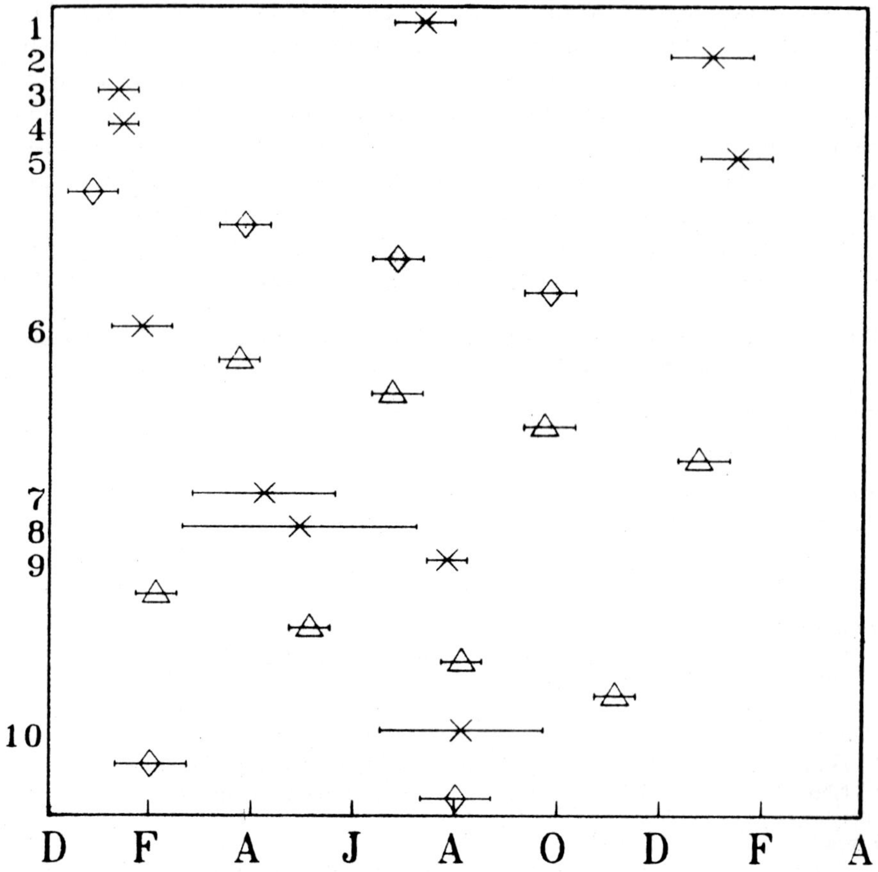

Fig. 6 **Acrophase map for *Heteropneustes fossilis*. 1 = Gonadosomatic index (female), 2 = Hepatosomatic index (female), 3 = Plasma glucose (mg/100 ml) (male), 4 = Plasma glucose (mg/100 ml) (female), 5 = Hepatic lipid (mg/100 gm) (female), 6 = Hepatic cholesterol (mg/100 gm) (female), 7 = Plasma cholesterol (mg/100 ml) (female), 8 = Plasma cholesterol (mg/100 ml) (male), 9 = Ovarian lipid (mg/100 gm), 10 = Ovarian cholesterol (mg/100 gm). Data source: 1, 2, 5, 6, 9, 10 (Singh and Singh, 1979), 3, 4, 7, 8 (Joshi, 1980). See also legend to Figure 3.**

Amphibians

Rana esculenta

Spermatogenesis in *Rana esculenta* displays a seasonal pattern. In nature spermatogenesis begins in early spring, just before the onset of breeding, continues through autumn and exhibits a brief winter stasis (Chieffi *et al.*, 1980). The ovarian cycle gauged by its weight exhibits lower values in the summer month (postnuptial period) and then a gradual and constant increase is witnessed from early autumn through the winter and reaches a peak in the period before ovulation (Milone *et al.*, 1978). The cosinor analyses revealed that GSI peaked in the month of February with a spread from December to March (Table 5; Figure 7). In male *Rana esculenta* the spermatogenetic cycle showed a statistically significant circannual rhythm with the peak activity occurring between August and October (Table 5; Figure 8).

Of the 30 variables analyzed in female *Rana esculenta,* acrophases of 20 variables overlapped with the 95% confidence limit for the peak gonadal

Table 5. Infradian rhythms in GSI and 30 physiologic variables of female, and 29 of male, *Rana esculenta*

Variable[a]	τ^b	p^c	Acrophase (95% CL)[d]
Female			
GSI	12	0.009	Feb 04 (Dec 22, Mar 16)
FBSI	12	0.002	Nov 16 (Oct 10, Dec 22)
HSI	12	0.005	Dec 22 (Nov 13, Feb 01)
Plasma			
Lipid	12	0.26	Feb 25
Protein	12	0.03	Feb 10 (Dec 10, Apr 10)
Lipid			
Ovary	12	0.009	Oct 04 (Aug 16, Nov 22)
Fat body	12	0.005	May 07 (Mar 25, Jun 19)
Liver	12	0.002	Apr 25 (Mar 19, Jun 01)
Protein			
Ovary	12	0.007	May 13 (Mar 28, Jun 28)
Fat body	12	0.01	Apr 28 (Mar 07, Jun 19)
Liver	12	0.46	Mar 10
	6	0.03	May 19 (Apr 22, Jun 16) also Nov 19
Pars distalis			
Weight	12	0.001	Apr 16 (Mar 25, May 07)
DNA content	12	0.02	Sep 25 (Jul 25, Nov 25)
Nucleic acid	12	0.41	Apr 28
	6	0.002	May 28 (May 10, Jun 16) also Nov 28
Protein	12	0.001	Apr 01 (Mar 01, May 01)
Hypothalamus			
Nucleic acid	12	0.004	Oct 28 (Sep 19, Dec 07)
Protein	12	0.003	Nov 01 (Sep 25, Dec 07)
Cerebral cortex			
Nucleic acid	12	0.18	Oct 16
Protein	12	0.68	Jan 25
Pituitary			
G6P	12	0.84	Oct 28
	6	0.001	May 04 (Apr 22, May 16) also Nov 04
	3	0.001	Feb 19 (Feb 10, Feb 28) also May, Aug & Nov 19

(Contd.)

Variable[a]	τ[b]	p[c]	Acrophase (95% CL)[d]
ACP	12	0.001	Jul 04 (Jun 10, Jul 28)
	3	0.001	Jan 22 (Jan 16, Jan 28) also Apr, Jul & Oct 22
ALP	12	0.32	Aug 22
	6	0.001	May 04 (Apr 22, May 16) also Nov 04
	3	0.007	Feb 10 (Jan 25, Feb 25) also Mar, Aug & Nov 10
G6PDH	12	0.02	Apr 01 (Feb 04, May 28)
	6	0.05	May 07 (Apr 01, Jun 13) also Nov 07
	3	0.003	Feb 07 (Jan 22, Feb 22) also May, Aug & Nov 07
GDH	12	0.19	Sep 13
	6	0.003	Apr 04 (Mar 13, Apr 25) also Oct 04
MDH	12	0.15	May 16
	6	0.005	May 10 (Apr 22, May 28) also Nov 10
	3	0.001	Feb 01 (Jan 28, Feb 04) also May, Aug & Nov 01
Hypothalamus			
G6P	12	0.01	Sep 01 (Jul 16, Oct 16)
	6	0.03	May 13 (Apr 10, Jun 16) also Nov 13
ACP	12	0.058	Jun 22
	3	0.001	Jan 22 (Jan 19, Jan 25) also Apr, Jul & Oct 22
ALP	12	0.14	May 04
	6	0.004	May 04 (Apr 19, May 19) also Nov 04
	3	0.001	Feb 01 (Jan 22, Feb 10) also May, Aug & Nov 01
G6PDH	12	0.41	Feb 16
	6	0.001	May 28 (May 16, Jun 10) also Nov 28
GDH	12	0.001	Jul 04 (Jun 04, Aug 04)
	3	0.04	Jan 28 (Jan 04, Feb 22) also Apr, Jul & Oct 28
MDH	12	0.001	Sep 13 (Aug 19, Oct 07)
Male			
GSI	12	0.81	Sep 16 (Aug 16, Oct 16)
FBSI	12	0.03	Dec 07 (Oct 01, Feb 13)
Gonadotrophin	12	0.001	Apr 01 (Mar 07, Apr 25)
Testosterone	12	0.001	Mar 04 (Jan 28, Apr 10)
Testes			
Lipid	12	0.08	Mar 13
Protein	12	0.13	Mar 01
Fat body			
Lipid	12	0.01	Apr 13 (Feb 19, Jun 07)
Protein	12	0.11	Mar 28
HSI	12	0.001	Jan 07 (Dec 13, Feb 01)

(Contd.)

Variable[a]	τ[b]	p[c]	Acrophase (95% CL)[d]
Liver			
Lipid	12	0.2	May 07
Protein	12	0.03	Feb 25 (Dec 22, Apr 28)
Blood			
Lipid	12	0.3	Feb 04
Protein	12	0.005	Jan 07 (Dec 01, Feb 13)
Pars distalis			
Weight	12	0.01	Apr 19 (Mar 01, Jun 07)
DNA	12	0.01	Sep 16 (Jul 28, Nov 04)
Nucleic acid	12	0.001	May 10 (Apr 16, Jun 04)
Protein	12	0.001	Apr 25 (Apr 04, May 16)
Cerebral cortex			
Protein	12	0.51	Aug 19
Pituitary			
G6P	12	0.36	Dec 13
ACP	12	0.02	Apr 07 (Feb 07, Jun 07)
ALP	12	0.23	Nov 25
	6	0.02	Apr 28 (Apr 07, May 19) also Oct 28
	3	0.002	Feb 25 (Feb 19, Mar 01) also May, Aug & Nov 25
G6PDH	12	0.001	Feb 07 (Jan 10, Mar 04)
GDH	12	0.35	Mar 10
	6	0.001	Mar 01 (Feb 10, Mar 22) also Sep 01
	3	0.008	Mar 07 (Feb 28, Mar 16) also Jun, Sep & Dec 07
MDH	12	0.09	Mar 19
	6	0.005	Mar 28 (Mar 04, Apr 22) also Sep 28
	3	0.009	Feb 28 (Feb 19, Mar 07) also May, Aug & Nov 28
Hypothalamus			
G6P	12	0.001	Oct 25 (Oct 07, Nov 13)
ACP	12	0.11	May 04
	6	0.001	Mar 04 (Feb 13, Mar 25) also Sep 04
ALP	12	0.07	May 04
	6	0.02	May 04 (Apr 10, May 28) also Nov 04
	3	0.001	Feb 01 (Jan 22, Feb 10) also May, Aug & Nov 01
G6PDH	12	0.002	Feb 10 (Jan 10, Mar 10)
GDH	12	0.23	Aug 04
	6	0.001	Mar 07 (Feb 22, Mar 22) also Sep 07
MDH	12	0.09	Sep 07
	6	0.004	Mar 22 (Feb 28, Apr 16) also Sep 22

[a]Refer Figures 7 and 8 for source and unit.
FBSI = Fat body somatic index, HSI = Hepatosomatic index, G6P = Glucose-6-phosphate, ACP = Acid phosphatase, ALP = Alkaline phosphatase, G6PDH = Glucose-6-phosphate dehydrogenase, GDH = Glucose dehydrogenase, MDH = Malic dehydrogenase. See also legend to Table 2.

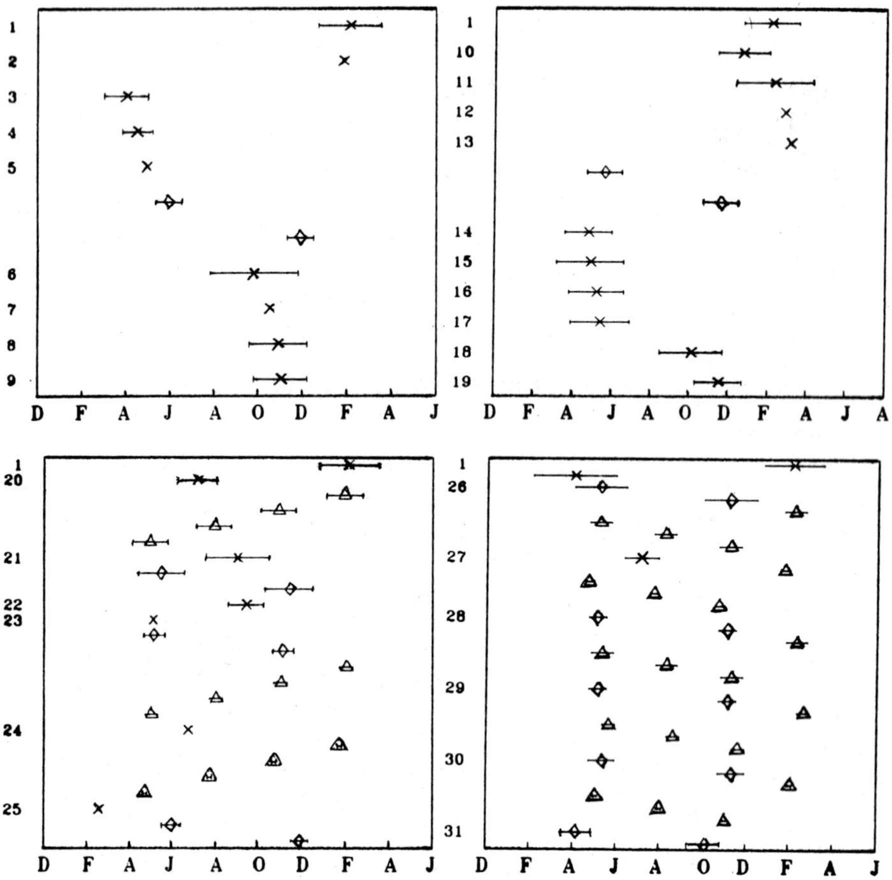

Fig. 7 Acrophase map for female *Rana esculenta.* 1 = Gonadosomatic index, 2 = Cerebral cortex protein (Protein : DNA × 10), 3 = Pars distalis protein (Protein: DNA × 10), 4 = Pars distalis weight (mg/frog), 5 = Pars distalis nucleic acid (RNA : DNA × 0.1), 6 = Pars distalis DNA (10 × μg/mg wet tissue), 7 = Cerebral cortex nucleic acid (RNA : DNA × 0.1), 8 = Hypothalamus nucleic acid (RNA : DNA × 0.1), 9 = Hypothalamus protein (Protein: DNA × 10), 10 = Hepatosomatic index, 11 = Plasma protein (μg/ml), 12 = Plasma lipid (μg/ml), 13 = Hepatic protein (μg/mg wet tissue), 14 = Hepatic lipid (μg/mg wet tissue), 15 = Fat body protein (μg/mg wet tissue), 16 = Fat body lipid (μg/mg wet tissue), 17 = Ovarian protein (μg/mg wet tissue), 18 = Ovarian lipid (μg/mg wet tissue), 19 = Fat body somatic index, 20 = Hypothalamus glutamic dehydrogenase (pm/μg DNA/min), 21 = Hypothalamus glucose-6-phosphate (pm/μg DNA/min), 22 = Hypothalamus malic dehydrogenase (nm/μg DNA/min), 23 = Hypothalamus alkaline phosphatase (nm/μg DNA/min), 24 = Hypothalamus acid phosphatase (nm/μg DNA/min), 25 = Hypothalamus glucose-6-phosphate dehydrogenase (nm/μg DNA/min), 26 = Pituitary glucose-6-phosphate dehydrogenase (nm/μg DNA/min), 27 = Pituitary acid phosphatase (nm/μg DNA/min), 28 = Pituitary alkaline phosphatase (nm/μg DNA/min), 29 = Pituitary glucose-6-phosphatase (pm/μg DNA/min), 30 = Pituitary malic dehydrogenase (nm/μg DNA/min), 31 = Pituitary glutamic dehydrogenase (nm/μg DNA/min). Data source: 1, 10–19 (Milone *et al.*, 1978), 2–9 (Rastogi and Chieffi, 1973), 20–31 (Milone *et al.*, 1975). See also legend to Figure 3.

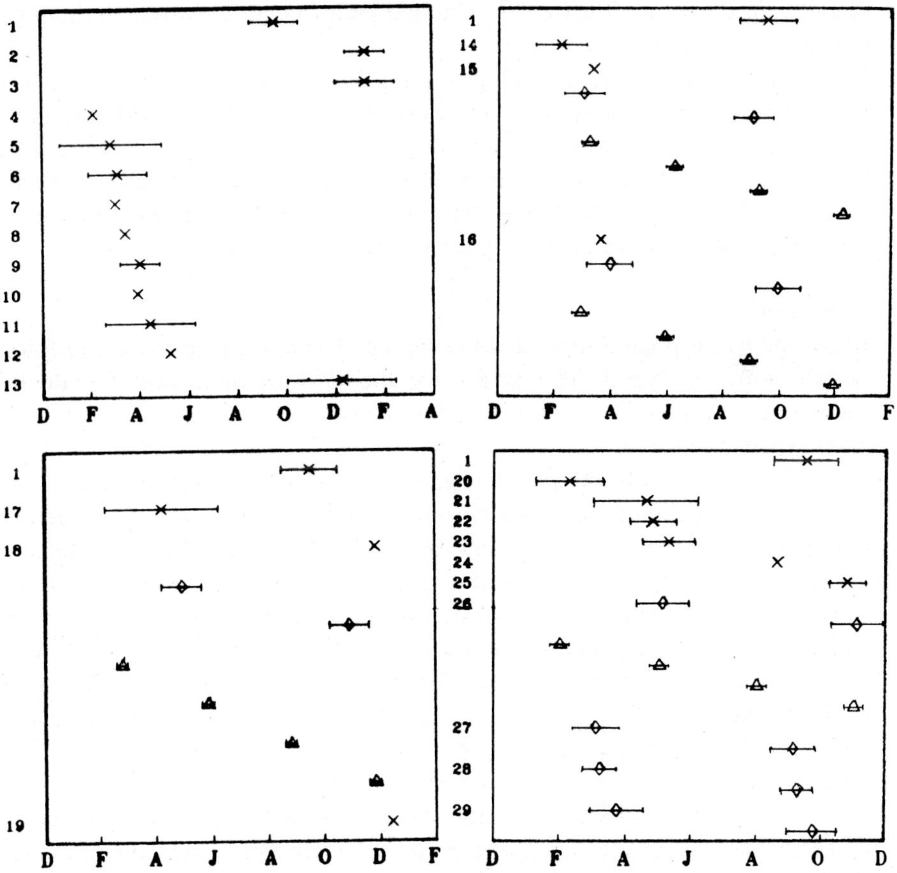

Fig. 8 **Acrophase map for male *Rana esculenta*. 1 = Gonadosomatic index, 2 = Hepatosomatic index, 3 = Blood protein (μg/ml), 4 = Blood lipid (μg/ml), 5 = Liver protein (μg/mg), 6 = Testosterone (ng/ml), 7 = Testicular protein (μg/mg), 8 = Testicular lipid (μg/mg), 9 = Gonadotrophin (ng/ml), 10 = Fat body protein (μg/mg), 11 = Fat body lipid (μg/mg), 12 = Liver lipid (μg/mg), 13 = Fat body somatic index, 14 = Pituitary glucose phosphate dehydrogenase (nm/μg DNA/min), 15 = Pituitary glucose dehydrogenase (pm/μg DNA/min), 16 = Pituitary malic dehydrogenase (nm/μg DNA/min), 17 = Pituitary acid phosphatase (nm/μg DNA/min), 18 = pituitary alkaline phosphatase (nm/μg DNA/min), 19 = Pituitary glucose-6-phosphate (pm/μg DNa/min), 20 = Hypothalamus glucose-6-phosphate dehydrogenase (nm/μg DNA/min), 21 = Pars distalis weight mg/frog), 22 = Pars distalis protein (Protein : DNA × 10), 23 = Pars distalis nucleic acid (RNA : DNA × 0.1), 24 = Cerebral cortex protein (Protein : DNA × 10), 25 = Hypothalamus glucose-6-phosphate (pm/μg DNA/min), 26 = Hypothalamus alkaline phosphatase (nm/μg DNA/min), 27 = Hypothalamus acid phosphatase (nm/μg DNA/min), 28 = Hypothalamus glutamic dehydrogenase (pm/μg DNA/min), 29 = Hypothalamus malic dehydrogenase (nm/μg DNA/min). Data source: 1–5, 7, 8, 10–13 (Iela *et al.*, 1979), 6, 9 (Rastogi, 1976), 21–24 (Rastogi and Chieffi, 1973), 14–20, 25–29 (Milone *et al.*, 1975). See also legend to Figure 3.**

activity. Furthermore, 9 variables exhibited multifrequency infradian rhythms (Figure 7; Table 5).

In male *Rana esculenta*, 29 physiologic variables were analyzed. Of those, 16 exhibited statistically significant circannual rhythms. In addition, 6-monthly and 3-monthly rhythms were validated for 7 and 4 variables, respectively. Furthermore, 4 variables exhibited multifrequency infradian rhythms. However, peaks for 10 variables showed coincidence for one of their peaks with the circannual peak of the gonadal activity.

Rana italica

In two different populations of *Rana italica*, 3 variables for male and 5 for female were analyzed. The peaks for the GSI in male and female of Montecorice population did not differ statistically significantly from its counterpart at Vesalo.

Excluding estradiol 17-β in males and FBSI in females at Vesalo, all other variables exhibited statistically significant infradian rhythms, irrespective of sex and geographic variation. FBSI and androgen exhibited multifrequency rhythms in both the populations of male *R. italica,* whereas in female, GSI, estradiol 17-β, androgen and progesterone showed multifrequency rhythms. In females, peaks of almost all the variables, excluding that of FBSI showed coincidence for one of their peaks with the peak of the circannual gonadal cycle. However, for the males above coincidence was witnessed for FBSI and androgen in the Montecorice population.

In both female and male *Rana esculenta* almost all the enzymes of the pars distalis and hypothalamus exhibit statistically validated infradian rhythms (Figures 7 and 8). In addition, the impressive synchronization of the infradian rhythm in enzyme activity of hypothalamo-hypophyseal system (HHS) with the circannual gonadal activity indicates a temporal parallelism between hormonal activities of hypothalamus and pituitary. Furthermore, in both the populations of *Rana italica,* one of the peaks of 6-monthly infradian rhythm in fat body weight shows synchronization with the circannual rhythm in gonadal activity (Figure 9). Visual inspection of the data without employment of inferential statistics, alone does not allow us to draw the above conclusion. However, in *R. italica,* unlike *R. esculenta,* limited variables have been examined over a circannual time scale.

Reptilia

Natrix piscator

This is a non-poisonous, seasonally breeding, night active ophidian. It is commonly found in paddy fields and water reservoirs (Srivastava and Thapliyal, 1965; Romulus, 1978). Its annual reproductive cycle has been divided into four phases, namely recrudescence phase (July-August), active phase (September-October), regression phase (November-February) and

Table 6. Infradian rhythms in GSI and 3 variables of male, and 5 of female, *Rana italica* at two different geographic locations in Italy

Variable[a]	τ[b]	p[c]	Acrophase (95% CL)[d]
Male			
Montecorice			
GSI	12	0.001	Nov 13 (Oct 04, Dec 22)
FBSI	12	0.001	Sep 04 (Aug 01, Oct 07)
	6	0.03	Jan 01 (Nov 25, Feb 07) also July 01
Androgen	12	0.001	Jan 07 (Dec 07, Feb 07)
	6	0.006	Jan 16 (Dec 22, Feb 10) also Jul 16
Estradiol 17–β	12	0.63	Jul 10
Vesalo			
GSI	12	0.001	Oct 13 (Sep 16, Nov 10)
FBSI	12	0.001	Nov 22 (Oct 19, Dec 25)
	6	0.04	Dec 25 (Nov 16, Feb 04) also Jun 25
Androgen	12	0.001	Jan 04 (Dec 13, Jan 25)
	6	0.005	Jan 10 (Dec 16, Feb 04) also Jul 10
Estradiol 17–β	12	0.55	Jul 10
Female			
Montecorice			
GSI	12	0.001	Dec 10 (Nov 13, Jan 07)
	6	0.03	Jan 01 (Nov 22, Feb 10) also Jul 01
FBSI	12	0.001	Jul 10 (Jun 10, Aug 10)
ODSI	12	0.001	Dec 25 (Dec 01, Jan 19)
Androgen	12	0.002	Dec 25 (Nov 19, Feb 01)
Estradiol 17–β	12	0.009	Dec 28 (Nov 13, Feb 13)
	6	0.02	Jan 22 (Dec 22, Feb 22) also Jul 22
Progesterone	12	0.002	Dec 19 (Nov 13, Jan 25)
Vesalo			
GSI	12	0.001	Dec 16 (Nov 25, Jan 07)
FBSI	12	0.11	Sep 22
	6	0.001	Jan 10 (Dec 25, Jan 25) also Jul 10
ODSI	12	0.006	Jan 13 (Dec 07, Feb 19)
Androgen	12	0.002	Dec 28 (Nov 28, Jan 28)
	6	0.04	Jan 19 (Dec 16, Feb 22) also Jul 19
Estradiol 17–β	12	0.001	Jan 01 (Dec 07, Jan 25)
Progesterone	12	0.008	Jan 04 (Nov 28, Feb 10)
	6	0.02	Jan 28 (Jan 01, Feb 25) also Jul 28

[a]Refer Figure 9 for source and unit.
ODSI = Oviductal somatic index.
See also legend to Tables 2 and 5.

inactive phase (March-June) (Haldar and Pandey, 1989). However, cosinor results reveal that its active phase extends from the month of September to early December (Table 7). The 95% confidence limits for the peaks obtained for the testicular weight, testes volume and seminiferous tubule diameter overlapped with each other.

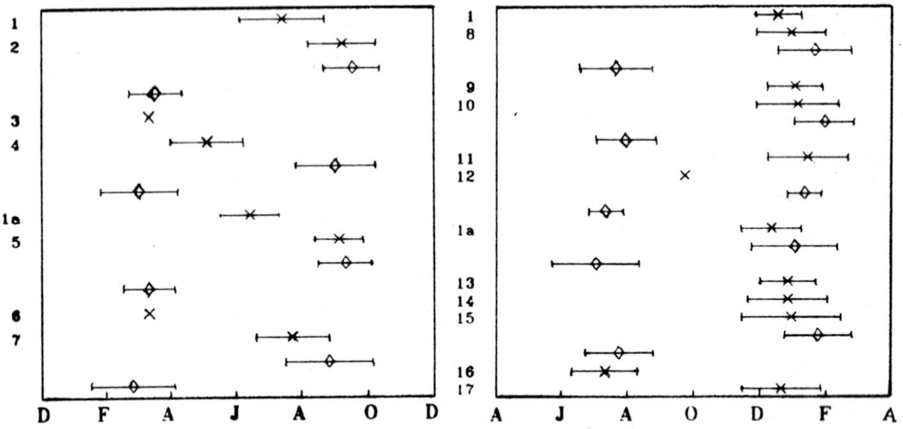

Fig. 9 **The acrophase map for *Rana italica* male: 1 = Gonadosomatic index (Montecorice), 2 = Androgen (pg/ml) (Montecorice), 3 = Estradiol 17–β (pg/ml) (Montecorice), 4 = Fat body somatic index (Montecorice), 1a = Gonadosomatic index (Vesalo), 5 = Androgen (pg/ml) (Vesalo), 6 = Estradiol 17–β (pg/ml) (Vesalo), 7 = Fat body somatic index (Vesalo); *Rana italica* female: 8 = Androgen (pg/ml) (Vesalo), 9 = Estradiol 17–β (pg/ml) (Vesalo), 10 = Progesterone (pg/ml) (Vesalo), 11 = Oviductal somatic index (Vesalo), 12 = Fat body somatic index (Vesalo), 13 = Oviductal somatic index (Montecorice), 14 = Androgen (pg/ml) (Montecorice), 15 = Estradiol 17–β (pg/ml) (Montecorice), 16 = Fat body somatic index (Montecorice), 17 = Progesterone (pg/ml) (Montecorice). Data source: 1–17 (Guarino *et al.*, 1993). See also legend to Figure 3.**

Statistically significant circannual rhythm was validated for a number of physiologic variables (Table 7; Figure 10). Circannual peaks for red cell number, hemoglobin concentration, hematocrit, plasma glucose, plasma testosterone occurred during the active phase of reproductive cycle. Furthermore, one of the peaks of 3-monthly rhythm in white blood cell number, rate of oxygen consumption by the kidney tissue also occurred during the peak reproductive phase. An inverse relationship between the testicular and pineal activity cycle has been validated (Misra nee Haldar and Thapliyal, 1980; Haldar and Pandey, 1989). The circannual peak in the concentration of circulating melatonin occurred in the month of May with a spread from April to June.

An increase in the number of red blood cells along with the levels of hemoglobin and plasma glucose clearly indicates synchronization of energy yielding cycles with that of the reproductive cycle of *Natrix piscator*.

Calotes versicolor

The Indian garden lizard, *Calotes versicolor* is a seasonal breeder and its annual gonadal cycle in the vicinity of Varanasi is divisible into five phases, such as early progressive phase (February-March), late progressive phase (April-May), peak reproductive phase (June-September), regressive phase

Table 7. Infradian rhythms in GSI and 20 physiological variables of male, *Natrix piscator*

Variable[a]	τ[b]	p[c]	Acrophase (95% CL)[d]
Testicular weight	12	0.005	Nov 01 (Sep 22, Dec 10)
RBC	12	0.001	Oct 28 (Oct 16, Nov 10)
WBC	12	0.8	Feb 16
	3	0.01	Mar 04 (Feb 22, Mar 16) also Jun, Sep & Dec 04
Eosinophil 09:00 hr	12	0.61	Mar 16
Eosinophil 21:00 hr	12	0.48	Oct 28
Hb	12	0.001	Oct 25 (Oct 13, Nov 07)
Hematocrit	12	0.001	Nov 07 (Oct 25, Nov 19)
Pineal wt.	12	0.001	Jun 07 (May 16, Jun 28)
Kidney wt.	12	0.001	Jan 04 (Dec 16, Jan 22)
Vas deferense wt.	12	0.001	Jan 22 (Dec 28, Feb 16)
Testes vol.	12	0.006	Nov 01 (Sep 19, Dec 13)
Seminiferous tubule dia.	12	0.001	Nov 07 (Oct 10, Dec 04)
Renal sex segment ht.	12	0.001	Jan 07 (Dec 22, Jan 22)
Pineal total width	12	0.001	Jun 07 (May 19, Jun 25)
Pineal nuclear dia.	12	0.001	May 28 (May 16, Jun 10)
Plasma			
Testosterone	12	0.008	Dec 19 (Nov 04, Feb 04)
Melatonin	12	0.004	May 13 (Apr 04, Jun 22)
Glucose 09:00 hr	12	0.03	Jan 10 (Nov 04, Mar 16)
Glucose 21:00 hr	12	0.03	Dec 25 (Oct 22, Feb 28)
O_2 consumption			
Brain	12	0.17	Mar 01
Kidney	12	0.61	Mar 19
	3	0.02	Feb 16 (Feb 04, Feb 28) also May, Aug & Nov 16
Liver	12	0.04	Jul 16 (May 25, Sep 07)
Muscle	12	0.13	May 07

[a]Refer Figure 10 for source and unit.
WBC = White blood corpuscles, wt. = Weight, hr. = Hour, ht. = Height.
See also legend to Table 2.

(September-December) and the quiescent phase (December-January) (Haldar-Misra and Thapliyal, 1981a, b). However, at Trivandrum, Kerala the peak reproductive phase extends till the end of October (Johnson and Oommen, 1982). The reproductive differences among populations of the same species of reptiles linked to geographic variation are not uncommon. Populations of Mexican lizard, *Sceloporus mucronatus* inhabiting high mountains and lower plateau have been shown to exhibit temporally different reproductive cycles (Mendez de la Cruz *et al.*, 1994).

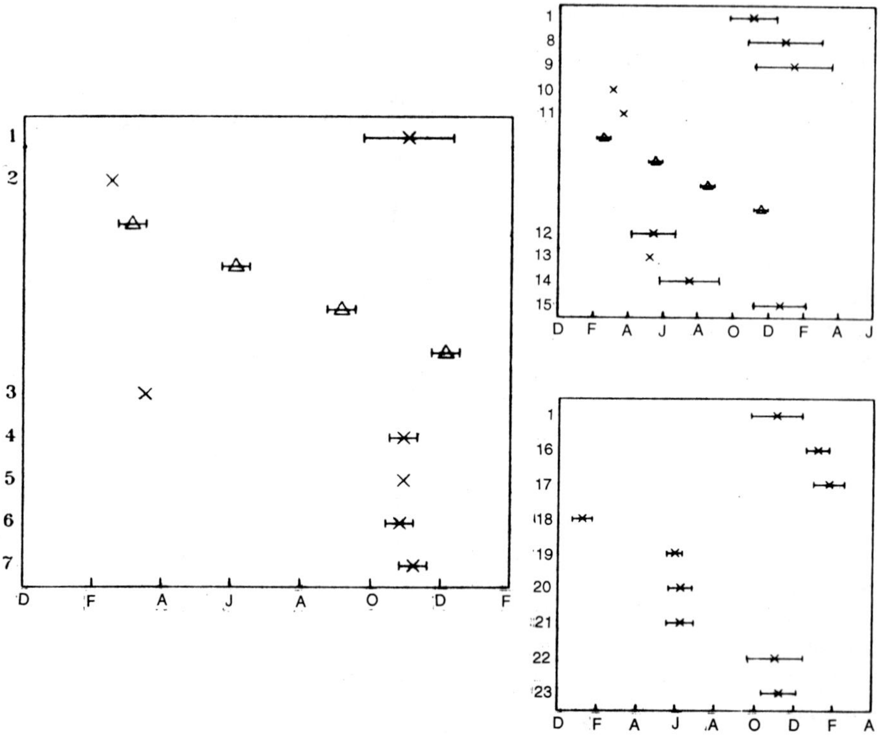

Fig. 10 **Acrophase map for male *Natrix piscator*. 1 = Testicular weight (mg/100 gm body weight), 2 = White blood corpuscles ($10^3/mm^3$), 3 = Eosinophil number ($\times 10^9/1$) 09:00 hr, 4 = Red cell number ($10^5/mm^3$), 5 = Eosinophil number ($\times 10^9/1$) 21:00 hr, 6 = Hemoglobin (g%), 7 = Hematocrit (%), 8 = Plasma glucose (mg/dl) 21:00 hr, 9 = Plasma glucose (mg/dl) 09:00 hr, 10 = Rate of oxygen consumption in brain tissue (μlO_2/mg/hr), 11 = Rate of oxygen consumption in kidney (μlO_2/mg/hr), 12 = Plasma melatonin (pg/ml), 13 = Rate of oxygen consumption in muscle (μlO_2/mg/hr), 14 = Rate of oxygen consumption in liver (μlO_2/mg/hr), 15 = Plasma testosterone (ng/ml), 16 = Kidney weight (mg/100 gm body weight), 17 = Vas deferens weight (mg/100 gm body weight), 18 = Renal sex segment cell height (μm), 19 = Pineal nucleus diameter (μm), 20 = Pineal total width (μm), 21 = Pineal weight (mg/100 gm body weight), 22 = Testes volume (mm^3), 23 = Seminiferous tubule diameter (μm). Data source: 1, 15–23 (Haldar and Pandey, 1989), 2, 4, 6, 7 (Goindi, 1977), 3, 5, 8, 9 (Pati, 1982), 10, 11, 13, 14 (Thapliyal and Sharan, 1980). See also legend to Figure 3.**

Limited data shown in the peak map (Figure 11) clearly indicate that higher levels of tissue protein during the late progressive and early part of peak reproductive span would help *C. versicolor* in gaining reproductive fitness. The lipid reserves of the muscle also remain adequately high during the peak reproductive period. The temporal distribution of the peaks in the concentration of various biochemical constituents of different tissues in *C. versicolor* appears to be meaningful from point of view of physiologic requirements of the species. Similar phenomenon could be imagined for other species of lizards (*Lacerta oxycephala*, Geller, 1976; *Anolis Carolinensis,* Dessauer, 1970; *Equrina cunninghani,* Barwick and Bryant, 1966; *Varanus oriseus,* Haggag *et al.*, 1966) so far studied for seasonal profiles in a number of physiologic variables.

Table 8. Infradian rhythms in testicular cycle and 12 physiologic variables of male *Calotes versicolor*

Variable[a]	τ^b	p^c	Acrophase (95% CL)[d]
Testicular weight	12	0.001	Sep 1 (Aug 13, Sep 19)
Lipid			
Blood	12	0.19	Apr 01
Thigh muscle	12	0.07	Sep 04
Liver	12	0.55	Nov 22
Heart	12	0.25	Nov 07
Testes	12	0.004	Dec 16 (Nov 07, Jan 25)
Fat body	12	0.001	Feb 07 (Nan 13, Mar 01)
Glycogen			
Thigh muscle	12	0.04	Oct 19 (Jul 28, Jan 10)
Liver	12	0.06	Oct 10
Heart	12	0.04	Dec 04 (Oct 01, Feb 07)
Protein			
Thigh muscle	12	0.03	Jul 04 (May 13, Aug 25)
Liver	12	0.09	Jun 19
Heart	12	0.09	Jul 10

[a]Refer Figure 11 for source and unit. See legend to Table 2.

In *N. piscator* and *C. versicolor* also infradian rhythms in various physiologic variables exhibit entrainment with the circannual reproductive cycle (Figures 10 and 11). An antiphase relationship between plasma melatonin and peak reproduction is spectacular. It has immense physiologic significance specially because higher levels of melatonin during the peak reproductive phase might upset the whole process of reproduction.

Circannual System

The circannual system appears to be paramount for seasonal breeders. This makes provision for fine temporal tuning of the machinery responsible for the initiation of reproductive process within a specified and species-specific time window. It is well known that two sets of environmental factors, namely ultimate environmental factors and proximate factors control

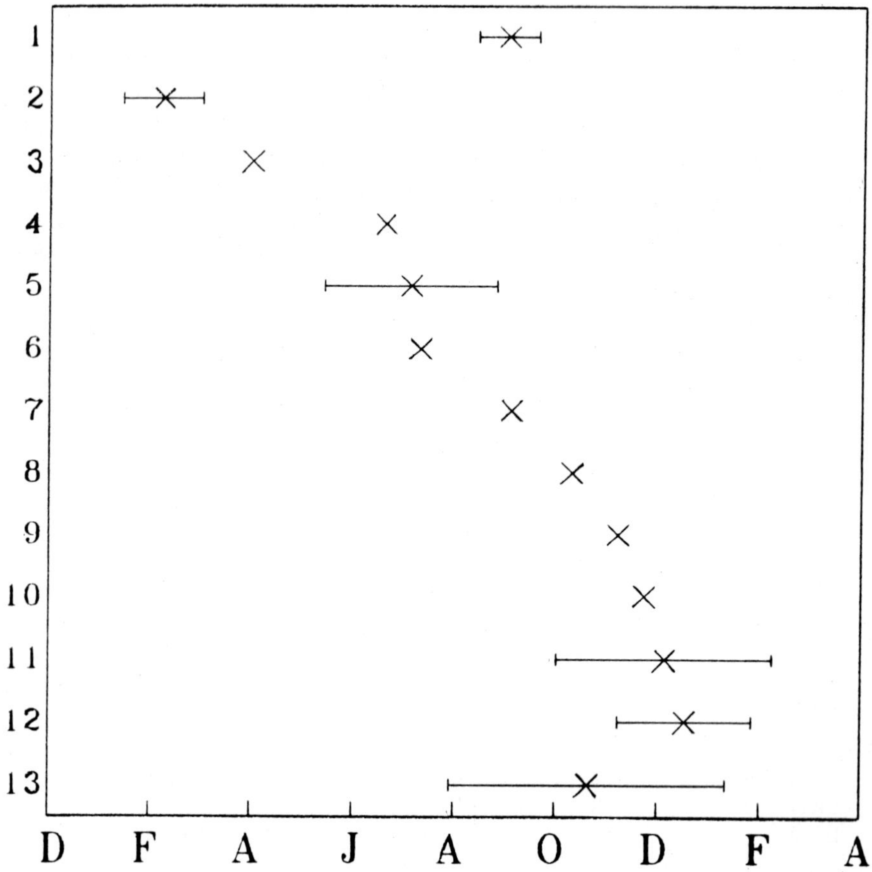

Fig. 11 **Acrophase map for *Calotes versicolor*. 1 = Testicular weight (mg/100 gm body weight), 2 = Fat body lipid (mg/g), 3 = Blood lipid (mg/ml), 4 = Hepatic protein (percentage/wet weight), 5 = Thigh muscle protein (percentage/wet weight), 6 = Heart protein (percentage/wet weight), 7 = Thigh muscle lipid (percentage/wet weight), 8 = Hepatic glycogen (μg/100 mg wet tissue), 9 = Heart lipid (percentage/wet weight), 10 = Hepatic lipid (percentage/wet weight), 11 = Heart glycogen (μg/100 mg wet tissue), 12 = Testicular lipid (mg/g). Data source: 1 (Haldar-Misra, 1985), 2, 3, 12 (Shanmukhamma *et al.*, 1992), 4–11 (Johnson and Oommen, 1982). See also legend to Figure 3.**

effectively the circannual reproductive cycles in seasonally breeding species (Farner and Wingfield, 1978). The former factors help to select the fittest individuals of a given species those would engage themselves in the act of reproduction. The latter sets of factors act as signals which upon being transmitted inform the breeder that it is the optimal time for the onset of reproduction. In this paper we propose that there could be altogether a different sets of factors those provide ideal internal milieu for successful conduct of reproduction. We prefer to designate these factors as *convenient factors.* These are necessarily internal and heighten the quality of the act of reproduction. The data presented in this paper substantiate our proposal.

In summary, the peak maps for all the species included in this paper clearly indicated that several important variables organize themselves temporally in such a way as to build an appropriate internal environment for the accomplishment of successful breeding. However, it should be borne in mind that a temporal relationship between changes of two or more variables does not prove in itself a direct causal relationship. Nonetheless it is clearly evident that all the variables are not controlled by a single circannual and/ or infradian clocks. There may be multiple clocks, yet they unite and exhibit entrainment at the appropriate time for favour of successful breeding. However, the spectrum of such relationship appears to be complex and still obscure.

Acknowledgements

We are thankful to Mr. Arvind Agrawal, Mrs. Renu Maheshwari, Mrs. Seema Gupta, Ms. Meena Balani for their help in preparing the manuscript. We are thankful to all the authors whose data we have used in this paper. We gratefully acknowledge UGC, New Delhi, for providing teacher fellowship to one of us (MS).

References

Babbar, K.K. and Sareen, M.L. (1982). Uttar Pradesh J. Zool. **2:** 97–103.

Barwick, R.D. and Bryant, C. (1966). Physiol. Zool. **39:** 1.

Billard, R. (1981). Riv. Ital. Piscic. Ittiop. **16:** 79–80.

Canterbury, R.A. and Pauley, T.K. (1994). J. Herpetol. **28:** 431–434.

Cartland, L.K., Cree, A., Sutherland, W.H.F., Grimmond, N.M. and Skeaff, C.M. (1994). New Zealand J. Zool. **21:** 399–406.

Chandola, A., Bhatt, D. and Lakhera, P. (1988). J. Comp. Physiol. **A 164:** 117–121.

Chandrakala, M.V. and Devraj Sarkar, H.B. (1980). Indian J. Exp. Biol. **18:** 806–810.

Chauhan, N.S. and Saxena, R.N. (1985). Proc. Indian Natl. Sci. Acad. **51:** 560–565.

Chieffi, G., Rastogi, R.K., Milone, M. and Iela, L. (1980). Boll. Zool. **47:** 63–70.

Dasgupta, S. and Nath, P. (1988). In: *Proc. Natn. Symp. Current Status Gen. Comp. Endocrinol.* New Delhi, November 25–27.

Delgado, M.J., Gutierrez, P. and Alonso-Bedate, M. (1989). Gen. Comp. Endocrinol. **73:** 1–11.

Delgado, M.J., Gutierrez, P. and Alonso-Bedate, M. (1990). Physiol. Zool. **63:** 373–387.

Dessauer, H.C. (1970). In: *Biology of Reptilia* (C. Gans and S. Paesons, Eds.), pp. 1. Academic Press, London, New York.

Disney, H.J. Des and Marshall, A.J. (1956). Proc. Zool. Soc. London. **127:** 379.

Farner, D.S. and Wingfield, J.C. (1978). In: *Environmental Endocrinology* (I. Assenmacher and D.S. Farner, Eds.), pp. 44–51. Springer-Verlag, Berlin.

Fitch, H.S. (1982). Univ. Kansas Mus. Nat. Hist. Occ. Pap. **96:** 1–53.

Flickinger, G.L. (1961). Gen. Comp. Endocrinol. **1:** 332.

Geller, L.I. (1976). Probl. Endocrinol. **22:** 92.

Goindi, R.J. (1977). *Hematological studies in a reptile, in relation to endocrine physiology.* Ph. D. thesis, Banaras Hindu University, Varanasi.

Guarino, F.M., Di Fiore, M.M., Caputo, V., Angelini, F., Iela, L. and Rastogi, R.K. (1993). Anim. Biol. **2:** 25–43.
Gupta, S. and Pati, A.K. (1992). J. Parasitol. Appl. Anim. Biol. **1:** 151–163.
Gwinner, E. (1973). J. Reprod. Fertil. Suppl. **19:** 51–65.
Gwinner, E. (1987). In: *Comparative Physiology of Environmental Adaptations* 3. *Adaptations to climatic changes* (P. Pevet, Ed.), pp. 30–44, Karger, Basel.
Haggag, G., Raheem, K.A. and Khalil, F. (1966). Comp. Biochem. Physiol. **17:** 341.
Halberg, F., Carandente, F., Cornelissen, G. and Katinas, S. (1977). Chronobiologia **4** (suppliment 1): 1–189.
Haldar, C. and Pandey, R. (1989). J. Reprod. Biol. Comp. Endocrinol. **1:** 14–24.
Haldar-Misra, C. (1985). J. Reprod. Briol. Comp. Endocrinol. **5:** 25–32.
Haldar-Misra, C. and Thapliyal, J.P. (1981a). Neuroendocrinology **33:** 328–332.
Haldar-Misra, C. and Thapliyal, J.P. (1981b). Gen. Comp. Endocrinol. **43:** 537–542.
Iela, L., Milone, M., Caliendo, M.F., Rastogi, R.K. and Chieffi, G. (1979). Boll. Zool. **46:** 11–16.
Janeiro Cinquini, T.R.F., Leinz, F.F. and Farias, E.C. (1993). Memorias do Instituto Butantan (Sao Paulo) **55:** 15–19.
Johnson, C. and Oommen, O.V. (1982). J. Reprod. Biol. Comp. Endocrinol. **2:** 64–68.
Jorgensen, C.B., Hede, K.-E. and Larsen, L.O. (1978). In: *Environmental Endocrinology* (I. Assenmacher and D.S. Farner, Eds.), pp. 28–36, Springer-Verlag, Berlin.
Joshi, B.D. (1980). Comp. Physiol. Ecol. **5:** 13–16.
Joshi, B.D. (1982). Uttar Pradesh J. Zool. **2:** 62–66.
Kanamadi, R.D. and Saidapur, S.K. (1982). Proc. Indian Natl. Sci. Acad. Part B. **48:** 307–316.
Kanamadi, R.D., Saidapur, S.K. and Nadkarni, V.B. (1983). Biol. Bull. India **5:** 10–19.
Krishna, A. (1982). Life Sci. Adv. **1:** 113–119.
Kumar, V. and Tewary, P.D. (1982). J. Exp. Zool. **223:** 37–40.
Kyriakopoulou-Sklavounou, P. and Loumbourdis, N. (1990). J. Herpetol. **24:** 185–191.
Lal, B., Mohan, R. and Singh, T.P. (1987). Indian J. Exp. Biol. **25:** 639–640.
Lal, B. and Singh, T.P. (1987). Fish Physiol. and Biochem. **3:** 191–201.
Lofts, B. (1978). In: *Environmental Endocrinology* (I. Assenmacher and D.S. Farner, Eds.), pp. 37–43, Springer-Verlag, Berlin.
Lofts, B. (1985). In: *The endocrine system and environment* (B.K. Follett, S. Ishii and A. Chandola, Eds.), pp. 93–103, Japan Sci. Soc. Press, Tokyo/Springer-Verlag, Berlin.
Mackenzie, D.S., Thomas, P. and Farrar, S.M. (1989). Aquaculture **78:** 63–80.
Malison, J.A., Procarione, L.S., Barry, T.P., Kapuscinski, A.R. and Kayes, T.B. (1994). Fish Physiol. and Biochem. **13:** 473–484.
Manire, C.A., Rasmussen, L.E.L., Hess, O.L. and Hueter, R.E. (1995) Gen. Comp. Endocrinol. **97:** 366–376.
Mayersbach, H.V. (1976). Arch. Toxicol. **36:** 185–216.
Mendez de la Cruz, F.R., Villagran Santa Cruz, M. and Cuellar, O. (1994). Biogeographica **70:** 59–67.
Milone, M., Iela, L., Esposito, V., Rastogi, R.K. and Chieffi, G. (1978). J. Endocr. **78:** 165–169.
Milone, M., Rastogi, R.K. and Iela, L. (1975). Gen. Comp. Endocrinol. **26:** 107–114.
Misra nee Haldar, C. and Thapliyal, J.P. (1980). Herpetologica **36:** 116–119.
Mukherjee, J., Bhattacharya, S. and Nath, P. (1989). Indian J. Exp. Biol. **27:** 764–769.
Nayak, P.K. and Singh, T.P. (1988) J. Pineal Res. **5:** 419–426.
Nelson, W., Tong, Y., Lee, J.K. and Halberg, F. (1979). Chronobiologia **6:** 305–323.
Pati, A.K. (1982). *Chronohematological and chronophysiological studies in a reptile with special reference to environment.* Ph. D. thesis, Banaras Hindu University, Varanasi.

Pati, A.K. (1989). J. Ravishankar Univ. **2B:** 18–22.
Peter, R.E. (1981). Gen. Comp. Endocr. **45:** 294.
Peter, R.E. and Hontela, A. (1978). In: *Environmental Endocrinology* (I. Assenmacher and D.S. Farner, Eds.), pp. 20–25. Springer-Verlag, Berlin.
Pevet, P. (1987). In: *Comparative Physiology of Environmental Adaptations* 3. *Adaptations to climatic changes* (P. Pevet, Ed.), pp. 82–100, Karger, Basel.
Pramoda, S. and Saidapur, S.K. (1984). Proc. Indian Natl. Sci. Acad. **50B:** 387–398.
Ramirez Pinilla, M.P. (1994). J. Herpetol. **28:** 521–524.
Rao, A.M.K.M. (1979). Indian J. Rodent. **1:** 35–45.
Rastogi, R.K. (1976). Boll. Zool. **43:** 151–172.
Rastogi, R.K. and Chieffi, G. (1973). J. Endocr. **58:** 129–130.
Rastogi, R.K., Iela, L., Saxena, P.K. and Chieffi, G. (1976). J. Exp. Zool. **196:** 151–166.
Reddy, B.C., Ramaiah, J.V., Rao, J.N., Bhaskar, M. and Govindappa, S. (1993). Indian J. Comp. Animal Physiol. **11:** 109–112.
Reiter, R.J. (1980). Endocrinol. Rev. **1:** 109–131.
Romulus, W. (1978). *Common Indian snakes, a field guide.* pp. 1–154. The Macmillan Company of India Limited, Delhi.
Rostal, D.C., Lance, V.A., Grumbles, J.S. and Alberts, A.C. (1994). Herpetological Monographs **0:** 72–82.
Saidapur, S.K., Ed. (1989). *Reproductive cycles of Indian vertebrates.* Allied Publishers Limited, New Delhi.
Shanmukhamma, H.H., Kusuma, K., Devraj Sarkar, H.B. and Rajendra, M. (1992). J. Parasit. Appl. Anim. Biol. **1:** 49–57.
Shedpure, M. and Pati, A.K. (1994). Biological Rhythm Research **25:** 451–463.
Shedpure, M. and Pati, A.K. (1995). Proc. Indian Natl. Sci. Acad. **B61:** 79–88.
Singh, A.K. and Singh, T.P. (1979). Endokrinologie. **73:** 47–54.
Singh, S. and Singh, T.P. (1987). Gen. Comp. Endocrinol. **65:** 216–224.
Srivastava, P.C. and Thapliyal, J.P. (1965). Copeia. **4:** 410–417.
Srivastavá, S.J. and Srivastava, S.K. (1994). Annales d' endocrinologie **55:** 197–202.
Sundararaj, B.I. (1978) In: *Environmental Endocrinology* (I. Assenmacher and D.S. Farner, Eds.), pp. 26–27, Springer-Verlag, Berlin.
Sundararaj, B.I. and Sehgal, A. (1970). J. Interdiscipl. Cycle Res. **1:** 219–301.
Swain, R. and Jones, S.M. (1994). Herpetologica. **50:** 502–509.
Tandon, R.S. and Joshi, B.D. (1974). Z. Tierphysiol. **33:** 108–112.
Thapliyal, J.P. and Lal, P. (1984). Gen. Comp. Endocrinol. **56:** 41–52.
Thapliyal, J.P. (1993). In: *Chronobiology* (A.K. Pati, Ed.), pp. 23–33. RSU, Raipur.
Thapliyal, J.P. and Sharan, S. (1980). Indian J. Exp. Biol. **18:** 236–239.
Tiwari, N.P., Singh, R.K. and Singh, B.N. (1988). J. Adv. Zool. **9:** 132–136.
Treasurer, J.W. (1990). J. Fish Biol. **36:** 29–46.
Van Wyk, J.H. (1994). Herpetologica. **50:** 480–493.
Wake, M.H. (1995). J. Herpetol. **29:** 119–122.
Yoo, M.S., Ra, C.H., Kim, J.Y., Kang, S.G. and Kwon, H.B. (1995). Korean J. Zool. **38:** 96–105.
Zhang, H. and Wu, M.X. (1994). Acta Zoologica Sinica. **40:** 399–404.

Comparative Endocrinology and Reproduction
K.P. Joy, A. Krishna and C. Haldar (Eds)

27. Recent Advances in the Endocrine Regulation of Reproduction and Moulting in Crustacea

T. Subramoniam[1], D. Sedlmeier[2] and R. Keller[2]
[1]Department of Zoology, Life Sciences Building, University of Madras, Guindy, Madras-600 025, India
[1]Institute für Zoophysiologie, Universität Bonn, D-5300, Bonn 1, Germany

Introduction

Crustacean endocrinology is making rapid progress in recent years, thanks to the introduction of modern microanalytical procedures in identifying a variety of molecules with hormonal functions. The ability to produce antibodies for these hormonal factors has resulted in the determination of their microquantities using radio- or nonradio-microassays such as RIA and ELISA during different physiological processes. Growth and reproduction, being the major energy demanding physiological functions in Crustacea, have been investigated in great detail with regard to their endocrine regulations. Although these two processes have been generally believed to be under the control of bihormonal systems, the concept of reproductive endocrinology is undergoing fascinating changes due to the recent discovery of many other factors with putative role in the control of reproduction. Nevertheless, many of these regulatory molecules are poorly characterized and hence their hormonal significance inadequately defined. This review analyses the recent findings on the hormonal factors that regulate molting and reproduction and discusses the interrelationship and interplay seen among them.

Endocrine Regulation of Reproduction

The most important progress made in recent years regarding the endocrine control of reproduction is the identification and elucidation of neuropeptides with specific inhibitory and stimulatory gonadal functions. The existence of a gonadal inhibitory principle in the eyestalk of decapod crustaceans was demonstrated 50 years ago by Panouse (1943). Since then, many studies involving extraction and injection of eyestalk hormones adduced further evidence to the occurrence of gonad inhibitory hormone (GIH) in a variety of crustacean species. Despite several attempts to identify and isolate this neuropeptide by conventional column chromatographic separation methods

(Subramoniam and Keller, 1993) it is only with the advent of High Pressure Liquid Chromatography, its isolation in pure form permitting elucidation of its primary structure was made possible. The regulatory role of the GIH is better understood on the vitellogenesis, rather than the other reproductive phenomena, including those in males.

The crustacean oogenesis occurs in two phases: the first growth phase upto the previtellogenesis and, the second short intense vitellogenic phase. The endocrine regulation of the first phase of oogenesis is not properly understood. However, there is ample evidence that oocyte growth during vitellogenesis is mainly controlled by the neurohormones. All the early studies concerning the gonadal inhibitory role of the eyestalk neuropeptides refer to the inhibition of vitellogenesis (Bomirsky *et al.*, 1981; Quackenbush and Herrnkind, 1983; Quackenbush and Keeley, 1988). By virtue of its specific inhibitory role in the vitellogenesis, the GIH is rechristened as Vitellogenesis Inhibiting Hormone (VIH).

The complete amino acid sequence of the VIH has been made in *Homarus americanus* (Soyez *et al.*, 1991). The primary structure of the VIH revealed the occurrence of two isoforms called VIH I and VIH II. Both forms have the same primary structure with 77 amino acid residues with six cysteines, a free N-terminus, a PI of 6.8 and a molecular weight of 9315 Daltons. Interestingly, only VIH I showed inhibitory effect on vitellogenesis. Recently, using the SMART system developed by Pharmacia/LKB, Soyez *et al.* (1994) were able to do microchromatography of the crude extract from a single sinus gland of the lobster to separate the different neuropeptides. The VIH I and II elute very closely in between two groups of another neuropeptide, the crustacean hyperglycemic hormone (CHH). The CHHs have also approximately the same primary amino acid sequence with 72 amino acid residues comprising of six cysteines and blocked *N*-terminus (Tensen *et al.*, 1991a).

Immunocytochemical localization of the VIH neuropeptide has been made in the X-Organ Sinus gland (XOSG) complex of the lobster *Homarus americanus* using polyclonal antisera raised against VIH (Meusy *et al.*, 1987). Detection of mRNA encoding VIH in the neurosecretory cells of XOSG has also been made in the lobster using nonradioactive *in situ* hybridization techniques. Recently, de Kleijn *et al.* (1992), employing complementary RNA probes, co-localized VIH mRNA with CHH mRNA in the same neurosecretory perikarya. However, these authors maintain that these two neuropeptides originate from two different precursors.

Although evidences are in favour of the inhibitory role of the eyestalk neuropeptides on vitellogenesis, the exact mode of action of VIH has not been clearly understood. The available data in this respect however suggest that the VIH is nonspecies-specific, as most of the bioassay methods have been made using heterologous crustacean species. Quackenbush and Keeley (1988) showed that the GIH from the crude eyestalk extract of the shrimp

Penaeus setiferus inhibited ^{14}C-leucine incorporation into vitellogenin by the cultured ovary of *Uca pugilator*. Jugan and Soyez (1985) demonstrated that a sinus gland extract inhibited the binding of colloidal gold labelled vitellin on oocyte microvilli of the prawn *Macrobrachium rosenbergii*. In our laboratory, we have developed a sandwich ELISA technique involving the antibodies raised against the lipovitellin of *Penaeus monodon* and *Penaeus semisulcatus* to measure the level of vitellogenin in the hemolymph of *Penaeus monodon* (Prakash *et al.*, 1996) Using this very sensitive method, it has been found that the vitellogenin levels in the hemolymph drastically declined in the VIH injected eyestalk-ablated females of *P. monodon* suggesting that the VIH inhibited the synthesis of vitellogenin in the extraovarian tissues. These experimental data are indicative of the possible multiple functions of VIH in inhibiting vitellogenesis at several stages. Clearly more evidence is required to arrive at the definitive functionality of this unique inhibitory neuropeptide.

Gonad Stimulatory Neuropeptides

A gonad stimulatory hormone (GSH), antagonistic to the VIH has been described in the brain and thoracic ganglia of some crustacean species (Otsu, 1963; Gomez, 1965; Kulkarni *et al.*, 1991; Eastman-Reks and Fingerman 1984). Recently, Yano (1992) reported the influence of thoracic ganglion on the vitellogenin secretion in the Kuruma prawn *Peneaus japonicus*. Surprisingly, the gonad stimulatory neuropeptide has neither been isolated nor characterized in any crustacean species.

Role of Biogenic Amines in the Release of Neurohormones

Biogenic amines in crustaceans function mainly as neurotransmitters and neuromodulators. In addition, some biogenic amines serve also as neuroregulators that control release of crustacean neurohormones. For example, 5-hydroxy tryptamine (5-HT) stimulates the release of the crayfish hyperglycemic hormone, red pigment dispersing hormone, neurodepressing hormone and molt inhibiting hormone (see Fingerman *et al.*, 1994). .The role of 5-HT on the ovarian development has been indicated by its stimulatory effect on the gonad stimulatory hormone in *Uca pugilator* (Richardson *et al.*, 1991). More recently, Sarojini *et al.* (1995) demonstrated the *in vitro* stimulation of GSH release under the influence of 5-HT in the crayfish *Procambarus clarkii.* However, dopamine inhibited the 5-HT stimulated ovarian maturation under *in vivo* conditions (Sarojini *et al.*, 1995). A stimulatory effect on the GSH release from the incubated thoracic ganglia has also been demonstrated by the neutrotransmitter activity of the red pigment concentrating hormone in the crayfish *P. clarkii* (Sarojini *et al.*, 1995). More interestingly, an endogenous opioid met-enkephalin has been shown to inhibit ovarian development by specific stimulation of GIH from the X-organ and/or by a concomittant inhibition of the GSH release from the brain/thoracic ganglia (Sarojini *et al.*, 1995).

Mandibular Organ and Methyl Farnesoate Secretion

Since the first description of mandibular organ in 1968 by Le Roux, several of the following studies have only confused this glandular structure with the Y-organ by virtue of its ultrastructural similarity to the vertebrate steroid producing cells (Sochasky *et al.*, 1972). However, later workers identified this as a discrete glandular organ secreting a sesquiterpenoid compound, methyl farnesoate (MF), which is the immediate precursor to the insect juvenile hormone (Laufer *et al.*, 1987). The mandibular organ (MO) has been suggested to be the crustacean analogue of the insect corpus allatum based on ultrastructural similarities and embryonic origin (Byard *et al.*, 1975, Le Roux, 1968).

The synthetic activity of mandibular organ appears to be under the inhibitory control of an eyestalk factor, as evidenced from several eyestalk ablation experiments (Le Roux, 1983; Demeusy, 1975; Hinsch, 1977; Laufer *et al.*, 1986). Subsequent studies by Tsukimura and Borst (1992) showed a distinctive elevation in the level of MF in the hemolymph and MO after uni- or bilateral eye ablation in the American lobster, *H. americanus*. The involvement of cGMP as a second messenger in the inhibitory activity of the sinus gland factor on the MF synthesis was demonstrated by Tsukimura *et al.* (1993). Surprisingly, the neuropeptide regulating the MF synthesis has not yet been identified. It is of interest to note here the findings of Landau *et al.* (1989) on the crayfish *Procambarus clarkii* that the red pigment-concentrating hormone stimulates MF synthesis, while the pigment dispersing hormone inhibits it. Such a cross functional role of the pigmentary effector neuropeptides needs further investigation.

Functional Attributes of MF

From the rather fragmentary evidences available on the physiological functions of MF in Crustacea, it may be surmised that they have a dual function, one in reproduction and the other in molting. Its functional role in vitellogenesis was first indicated by Hinsch (1980), from the observation that the MO implants stimulated ovarian growth in juvenile female spider crab, *Libinia emarginata*. By measuring the MF level in the hemolymph and MO using radio labelled methionine precursors, Laufer *et al.* (1986) demonstrated an increased synthesis and secretion of MF during vitellogenesis in the spider crab, thereby suggesting that this compound has a role in crustacean reproduction similar to that of JH in insect reproduction.

Notwithstanding the implied role of MF in the female reproduction, many workers have found the MF level in the hemolymph of males always higher than the females. For example, Sagi *et al.* (1993) found that the most active mandibular organs are in males with the highest levels of MF in their hemolymph. The varying levels of MF in the hemolymph of different male morphotypes of the spider crab, *L. emarginata* has recently been correlated with the mating behaviour (Sagi *et al.*, 1993). The polymorphic male spider crabs include type (1) that possesses unabraded exoskeleton and small claws and the type (2) that includes males with fully or partly abraded exoskeleton with large claws. The latter type has very well developed gonads and show aggressive mating behaviour. Correlated with the behaviour is the high titre of MF in the hemolymph suggesting a behavioural function to MF in Crustacea.

Perhaps, more functional evidence in favour of another important physiological process in crustacean growth and development, namely, molting is being attributed to MF in recent years. Implantation of MO resulted in the shortening of the molt cycle in the shrimp, *Penaeus setiferus* (Yudin *et al.*, 1980). Another indication for the involvement of MO in molting is the change in the ultrastructure of the cytoplasm of MO during premolt stage, signifying increased synthetic activity in the lobster, *H. americanus* (Byard *et al.*, 1975). Hormonal extract of MO from the crayfish, *Procambarus clarkii* also accelerated molting in the shrimp, *Caridina denticulata* (Taketomi *et al.*, 1989). In a more recent study, Tamone and Chang (1993) obtained direct evidence for the stimulation of ecdysteroid secretion from the Y-organ by methyl farnesoate. Y-organs of the crab, *Cancer magister,* when incubated *in vitro* along with (2E, 6E)-MF, secreted significantly more ecdysteroids into the medium after a 24 hr incubation, when compared to the controls. The MO when incubated alone secretes both farnesoic acid and MF into the medium, and hence are not themselves a source of ecdysteroids. This brings us to the fact that the secretory activity of Y-organ is positively controlled by the MF, whereas its inhibition is exercised by the molt inhibitory neuropeptide from the X-organ sinus gland. It remains to be known whether the Y-organ is additionally controlled by a PTTH-like brain hormone in crustacean too. Research towards this direction will be highly rewarding to suggest a homologous controlling mechanism in crustaceans and insects. The fact that the MO itself is negatively controlled by an eyestalk neuropeptide suggests the occurrence of a large family of inhibitory neuropeptides from the eyestalk neurosecretory centres to control diversified physiological processes in the decapod crustaceans.

Embryonic Ecdysteroids

Although the term "ecdysteroid" represents any steroid with molting hormone activity (Goodwin *et al.*, 1978), the occurrence of many species of ecdysteroids in organs, totally unassociated with molting suggests a pluripotential role for them. In insects, several studies have revealed the accumulation of ecdysteroids in the ovary to suggest a role in vitellogenesis and embryogenesis (Hoffmann and Lagueux, 1985; Hagedorn, 1989). In a similar way, many crustacean species store ecdysteroids in their eggs for their possible function in embryogenesis. The first indication of a reproductive role for ecdysteroids in Crustacea was provided by Blanchet *et al.* (1979) in an amphipod, *Orchestia gammarellus.* Here, the ovary produces the ecdysone under the influence of a brain hormone to stimulate vitellogenin synthesis in the subepidermal fat body. Although such a proposition is homologous to the hormonal regulation of vitellogenesis in the dipteran insects, supporting evidences from other crustacean species are lacking. Nevertheless, several workers have documented the accumulation of ecdysteroids in the vitellogenic ovary, implying other functions in embryogenesis. For example, in the caridian shrimp, *Palaemon serratus* the increase in the ovarian free ecdysteroids coincides with the first meiotic resumption of the oocytes, leading to the germinal vesicle breakdown (Lanot and Cledon, 1989). To confirm the effect of ecdysone in meiotic maturation of the oocytes, these authors successfully induced germinal vesicle break down by incubating oocytes with physiological doses of extraneous ecdysteroids. Interestingly, such a rise of ecdysteroid levels in the oocytes occurs during premolt stages, with the highest level reaching in D2 stage and exuviation in this shrimp. Such a temporal relationship existing among ovarian ecdysteroids, meiotic maturation and molt initiation is tempting to suggest a dual role for ovarian ecdysteroids in molting and oogenesis.

In a recent study on the giant freshwater prawn, *Macrobrachium rosenbergii*, Wilder *et al.* (1991) measured the ovarian ecdysteroids during gonad development as well as the accompanying molt stages. There is a rise in the ovarian ecdysteroids during the reproductive molt, whereas, during the non-reproductive molt the increase in the ovarian ecdysteroids is very little. This may suggest that the ovary, during its development actively synthesizes or sequesters ecdysone, as a function of oocyte maturation.

Unlike the shrimp species cited above, ovarian development and spawning occur during intermolt period in the true crabs which also accumulate significant quantities of ecdysteroids in the ovary. Lachaise and Hoffmann (1982) found the total ecdysteroids in eggs and embryos of the shore crab, *Carcinus maenas,* fluctuating during the entire period of embryonic development. The authors observed two peaks of ecdysteroids, especially Ponasterone A (PoA), during embryonic development coinciding first with the very early development and next with the secretion of the embryonic envelope (at "Naupliar" stage). The cuticle secretion occurring at the three subsequent embryonic stages ("Metanauplius", "Protozoea" and "Prezoea") also occurs in the presence of a high titre of PoA, suggesting a strong correlation for the ecdysteroids with the embryonic cuticular synthesis. In a similar way, the embryonated eggs of the mole crab, *Emerita asiatica* also have significant levels of polar, apolar and free ecdysteroids (unpublished observations). Evidently, these ecdysteroids exert their influence on several developmental programmes, including the secretion of embryonic cuticle, in the capacity of a morphogenic hormone.

An interesting aspect of embryonic ecdysteroids is the diversified forms in which they occur in the ovary. In *Carcinus maenas* the dominant embryonic ecdysteroid is PoA, the embryonic cuticle formation occurs when its level is the highest, while the level of 20-Hydroxyecdysone (20 HE) is very low, suggesting that PoA may serve as the embryonic molting hormone, whereas in all postembryonic molting processes, 20 HE participates. Interestingly, PoA is not synthesised in insects, but have significant effect in inducing molting (Lachaise, 1989). From HPLC-RIA analysis, Wilder *et al.* (1991) have recently reported several ecdysteroid species such as ecdysone, 20-HE, inokosterone, in addition to 20 hydroxy ecdysonic acid, ecdysonic acid, 20, 26-dihydroxyecdysone, PoA and 2-deoxyecdysone. Although data on the ecdysteroid species on crustacean eggs are still inadequate, it may be inferred from the available information that the embryo contains different metabolic products of ecdysone, again suggesting an active hormonal role for the embryonic ecdysteroids. In insect embryos, several conjugated ecdysteroids have been reported; but direct data from crustacean forms in this respect is wanting.

Yet another unresolved question on the crustacean embryonic ecdysteroids is their origin. The limited information available in the decapod crustaceans however indicates the existence of two patterns of ecdysteroid storage in the eggs. In the true crabs such as *Cancer magister* and *C. anthonyi,* they are presumably derived from maternal sources (Okasaki and Chang, 1991; Chaix and De Reggi, 1982). In the mole crab *E. asiatica,* the occurrence of all ecdysteroid species that are present in the egg homogenates is also present in the purified lipovitellin, suggesting their maternal derivation from the vitellogenin precursor and subsequent accumulation during vitellogenesis (Unpublished observations). On the contrary, in the penaeid shrimps, they are not stored in the freshly laid eggs, but are synthesized *de novo* by the differentiating Y-organ during the latter half of the embryonic development (Okumura *et al.*, 1989). In the first category, the derivation from the differentiating Y-organ is also not ruled out, as inferred from the fact that total free ecdysteroids register their highest level in the fully formed embryos of *E. asiatica.* Although conclusive evidences are available from the insects that the follicle cells produce ovarian ecdysteroids, no such reports are available for crustaceans. But then, the controlling mechanisms on the ecdysteroid synthesis by the follicle cells are not understood, even in insects. It will be interesting to know whether the eyestalk peptides like Moult Inhibiting Hormone (MIH) also exercise the same inhibitory effect on follicular epithelial synthesis of ecdysones.

Influence of Vertebrate-Type Steroid Hormones on Female Reproduction

In recent years, several vertebrate-type steroids such as testosterone, estradiol-17β and progesterone have been reported in ovary, hepatopancreas and hemolymph of crustaceans. Of these hormones, estradiol-17β and progesterone showed cyclical fluctuations during ovarian development suggesting their active role in vitellogenesis (Couch *et al.*, 1987; Faires *et al.*, 1990). Quinitio *et al.* (1991) demonstrated a peak in the estradiol and progesterone levels in the ovary and hepatopancreas in *Penaeus monodon* during vitellogenesis. The steroidogenic ability of the ovary has also been demonstrated by the presence of 17β-hydroxysteroid dehydrogenase. Furthermore, it has been reported that the administration of 17α-hydroxy-progesterone stimulates vitellogenin synthesis in *Penaeus japonicus* (Yano, 1987) and also maturation and spawning in *Metapenaeus ensis* (Yano, 1985). Administration of estradiol-17β stimulated lipogenesis in the freshwater prawn *Macrobrachium rosenbergii* (Ghosh and Ray, 1994) and played a prominent role at subcellular level by elevating the level of mitochondrial Na^+/K^+-ATPase in the cytosolic malate dehydrogenase and glucose-6-phosphate dehydrogenase activities in a dose-dependent manner (Ghosh and Ray, 1993). Young *et al.* (1992) showed an extensive metabolism of progesterone by vitellogenic ovary of *P. monodon.*

Although the significance of the conjugation of certain steroids is unclear till now, Faires *et al.* (1990) studied the steroid composition in *P. monodon* and reported that estradiol-17β and pregnenolone were found to be maximal in conjugated form while estrone and dehydroepiandrosterone were found to be maximal in both conjugated and unconjugated forms during II and III stages of the vitellogenic cycle. In the event of these vertebrate hormones being accumulated in the mature oocytes, there exists a possibility that they serve as morphogenic hormones during embryonic development. This hypothesis may by tested by measuring the level of both conjugated and free steroid hormone during different phases of embryogenesis. Moreover, studies on the metabolic aspects of crustacean vitellogenesis and the possible control by steroid hormones of vertebrate nature have not been investigated so far.

Control of Molting in Crustacea

Molting is a characteristic feature in the growth of all arthropods to permit periodic body growth by shedding away the old exoskeleton and forming a new cuticle in its place. Most crustaceans are generally heavy-armoured by intense calcification of the cuticle, making the process of molting most conspicuous and cumbersome. Obviously, the molting in several adult crustaceans is a protracted event in order for the animal to digest the inner cuticle and secrete a new cuticle, before the old cuticle could be shed away. The classical work of Drach and Tchernigovtzeff (1967) made it possible to recognise three major molt stages such as intermolt, premolt and post-molt which are further subdivided into several substages. Briefly, the intermolt is the interval between two successive molts, whereas, the premolt and postmolt represent the preparatory as well as post-ecdysial stages, respectively.

As in reproduction, the control of molting was first discovered by the eyestalk ablation which resulted in the increased frequency of accelerated molting (Zeleny, 1905). That a neurosecretory hormone originating from the X-organ/sinus gland was responsible for the inhibitory effect on molting was subsequently elucidated by many workers (Brown and Cunningham, 1939). Nevertheless, the actual molting gland of Crustacea, equivalent to the prothoracic gland of insects was described only later by Gabe (1953a, b; 1956) and Echalier (1955, 1959) who elegantly elucidated the role of these glands (Y-organ) in molting by their extirpation and reimplantation experiments. In the crab, *Carcinus maenas*, the surgical removal of Y-organ during early premolt stage (D0-D1), prohibits the molt preparations such as apolysis (detachment of old cuticle from the epidermis) as well as the secretion of the new cuticle. However, Y-organectomy performed during late premolt, had no effect on exuviation and the postmolt events, but subsequent molt was completely blocked at stage C4.

Y-organ: Morphology and synthetic products

Y-organs are most conspicuous as discrete glands in brachyuran crabs, usually 1-2 mm in diameter, but could grow upto 3 mm in diameter and weighing over 10 mg. In the crab, the glands may be ovoid to round in shape, yellowish to opalescent white in colour, often embedded in yellowish to brown tissue of fatty appearance (see Spaziani, 1990). Viewed from the dorsal aspect, the location of the glands in brachyurans is invariably ventral to the insertion of the lateral portion of the mandibular external adductor muscle and just above the junction of the branchiostegite with the lateral body cuticle at the anterior end of the branchial chamber (Passano, 1960). On each side, they lie against the epidermis of the ventral anterior face of the cephalothorax, lateral to the eyestalks. The glands are epidermal derivatives, but are not always attached to the epidermis, as in many brachyurans. In Caridia and Astacidea, the Y-organs are more difficult to locate, owing to less prominency and to variations in shape and position.

The structural features of Y-organ cells generally include smooth ER in much greater relative abundance than granular ER, numerous free ribosomes, polymorphic mitochondria with tubular cristae, nuclei with clumped heterochromatin in the periphery, and a relative lack of stored secretion products (Spaziani, 1990). These features however change during different molting stages with accompanying variation in the MIH titres in the hemolymph. The principal synthetic product of the Y-organ is ecdysone, the biologically functional 20 HE being produced in the target tissues from ecdysone. However, recent studies have indicated that a variety of other ecdysteroids such as 2-deoxyecdysone and 3-dehydroecdysone do occur (Lachaise *et al.*, 1993).

Regulation of Ecdysteroid Synthesis by MIH

The synthesis and secretion of ecdysone by the Y-organ is under the inhibitory control of a neurohormone (MIH) originating in the X-organ/sinus gland of the eyestalk (Soumoff and O'Connor, 1982). Purification and amino acid sequencing of the MIH for the lobster, *Homarus americanus* followed by the development of a bioassay method was provided by Bruce and Chang (1984) and Chang *et al.* (1987; 1990). MIH also shows a close structural similarity with another eyestalk neuropeptide, the crustacean hyperglycemic hormone (CHH) of this lobster (90% identity) (Chang *et al.*, 1993) as well as with CHH of a brachyuran crab, *Carcinus maenas* (61% identity) (Tensen *et al.*, 1991b). On the contrary, the CHH of *Carcinus* though showing some sequence homology with putative MIH, has an *in vitro* inhibition on Y-organ, 20 fold weaker than MIH (Webster and Keller, 1986; Webster, 1991). Another neurohormone closely related to MIH is the vitellogenesis inhibiting hormone (see Keller, 1992; Subramoniam and Keller, 1993).

Mode of Action of MIH Y-organ

The inhibitory mechanism of MIH on Y-organ activity has been investigated in great detail although the transduction mechanisms involved as well as the regulatory steps are still unclear. Early investigations of Matson and Spaziani (1985a, b) demonstrated a dose dependent but reversible effect of MIH inhibition on steroidogenesis in *Cancer antennarius.* These authors also found an increase in intracellular cAMP, preceding suppression of steroidogenesis in the Y-organ, isolated from the destalked crabs. On the contrary, in the crayfish, *Orconectes limosus,* Sedlmeier and Fenrich (1993) found evidence for the increase in cGMP levels rather than in cAMP contents in Y-organs cultured with eyestalk extract. In addition, cGMP-dependent protein kinase was only activated by eyestalk extracts (Gliscynski and Sedlmeier, 1993). In *O. limosus,* a change in the phosphorylation pattern of several proteins after treatment of the glands with eyestalk extract was demonstrated, suggesting a role of cGMP-dependent protein kinase in controlling the biosynthesis pathway of ecdysteroids from cholesterol (Böcking and Sedlmeier, 1994). In the intermoult shore crab, *Carcinus maenas,* cGMP levels increased due to application of MIH, whereas in premolt crabs an additional transient peak of cAMP developed (Saïdi *et al.*, 1994). A general scheme depicting the known mechanisms of MIH action of Y-organ is given in Fig. 1.

Stimulatory action of MF on Y-organs

Although methyl farnesoate has been shown to have reproductive functions (see above), many preliminary studies have revealed its influence on molting also (Yudin *et al.*, 1980). In a recent study on *Cancer magister,* Change *et al.* (1993) found direct evidence of the MF activity on Y-organ synthesis. By co-culturing molting glands with mandibular organs, these authors found an increased release of ecdysone into the culture medium (for more details see section on MO). Obviously, the molting gland is under the influence of a stimulatory factor from a glandular product (MF and an inhibitory neurosecretory peptide (MIH) from the eyestalk. The presence of such a dual controlling mechanism on the Y-organ activity is highly advantageous for a temporal spacing of molting from reproduction (egg production), which is also under a similar bihormonal control. Again, the seemingly complicated control mechanism operating on MO activity demonstrate how delicately the molting and reproductive events are related. More work is obviously needed to elucidate the role of MF in integrating the molting and reproductive events.

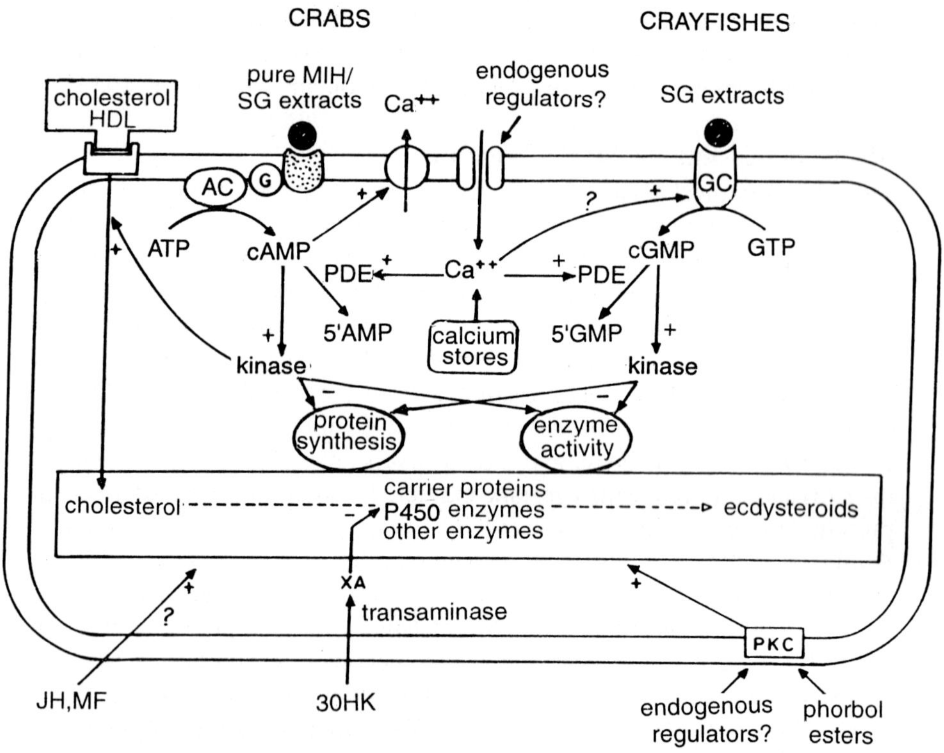

Fig. 1 **General scheme to summarize the present results concerning the mode of action of MIH or sinus gland extracts on steoidogenesis by YO. SG = sinus gland; AC = adenylate cyclase; GC = guanylate cyclase; PKC = protein kinase C; G = G protein; PDE = phosphodiesterase; JH = juvenile hormone; MF = methyl farnesoate; 3 OH K = 3-hydroxykinurenin; XA = xanthurenic acid; MIH = molt-inhibiting hormone. From Lachaise *et al.* (1993) J. Crust. Biol., 13: 198-234.**

Conclusion

A unique feature found in the crustacean endocrine system is the presence of well-defined inhibitory neurosecretory peptides synthesised in the central nervous system. For example, the two major physiological activities, namely molting and reproduction, are controlled negatively by eyestalk neuropeptides produced in the X-organ-sinus gland complex. Whereas, in the case of molting, the inhibitory neuropeptides have a direct effect on synthesis of the molt stimulating hormone by the Y-organ, the mode of action of the gonad inhibitory peptides is not clearly understood. The available information, however, indicates that the VIH inhibits the receptor mediated endocytosis of the yolk precursors by the oocytes. Its direct action on the gonad stimulatory hormone is not understood. However, the endocrine control of reproduction is much more complex than that of molting, in as much as a variety of hormonal factors are involved in the regulation of female reproduction. Unfortunately, the site at which many of these "reproductive" hormonal factors act to regulate reproductive functions remain unclear. But then, many of the reproductive processes are still not understood properly for crustaceans. For example, there is still controversy existing on the important problem of the site of vitellogenin synthesis in decapods. Another unresolved problem is the functional multiplicity of certain hormones such as methyl farnesoate whose functions have been implicated with the control of molting and reproduction.

A new perspective of neuroendocrine research in crustacean species is the application of its results to aquaculture. The regulatory role of several biogenic amines on neuropeptide secretion as well as the purported role of "vertebrate type" steroids in stimulating reproductive processes should provide new thinking on the neuroendocrine manipulation of the commercially important crustacean species. Yet again, the crustaceans are emerging as an important animal model for studying hormone action on cellular physiology.

Acknowledgment

T.S. is thankful to DAAD for a visiting fellowship to Bonn which enabled the completion of this review.

References

Blanchet, M.F., Porcheron, P. and Drey, F. (1979). Int. J. Invert. Reprod., **1:** 133–139.

Böckling, D. and Sedlemeier, D. (1994). Invert. Reprod. Devel., **26:** 237–246.

Bomirsky, A., Arendarczyk, A., Kawinska, E., Kleinholy, L.H. (1981). Int. J. Invert. Reprod., **3:** 213–219.

Brown, F.A. and Cunningham, O. (1939). Biol. Bull. (Woods Hole), **77:** 104.

Bruce, M.J. and Chang, E.S. (1984). Comp. Biochem. Physiol., **79A:** 421–424.

Byard, E.H., Shivers, R.R. and Aiken, D.E. (1975). Cell. Tiss. Res., **162:** 13–22.

Chang, E.G., Bruce, M.J. and Tamone, S.L. (1993). Amer. Zool., **33:** 324–329.

Chang, E.S., Prestwich, G.D. and Bruce, M.J. (1990). Biochem. Biophys. Res. Commun., **171:** 818–826.

Chang, E.S., Bruce, M.J. and Newcomb, R.W. (1987). Gen. Comp. Endocrinol., **65:** 56–64.

Chaix, J.C. and de Reggi, M. (1982). Gen. Comp. Endocrinol., **47:** 7–14.

Couch, E.F., Hagino, N. and Lee, J.W. (1987). Comp. Biochem. Physiol., **87:** 765–770.

De Kleijn, D.P.V., Coenen, T., Laverdure, A.M., Tenson, C.P. and Van Herp, F. (1992). Neuroscience, **51:** 121–128.

DeMeusey, N. (1975). C.R. Acad. Sci. Paris, **D281:** 1887–1889.

Drach, P. and Tchernigovtzoff, C. (1967). Vie et Milieu, **18:** 595–617.

Eastman-Reks, S. and Fingerman, F. (1984). Comp. Biochem. Physiol., **79A:** 679–684.

Echalier, G. (1959). Ann. Sci. Nat. Zool. Biol. Anim., **(12)1:** 1–59.

Echalier, G. (1955). C.R. Acad. Sci. Paris, **240:** 1581–1583.

Faires, N.J., Quinlon, P.T. and Goad, L.J. (1990). Aquaculture, **89:** 83–99.

Fingerman, M., Nagabushanam, R., Sarojini, R. and Reddy, P.S. (1994). J. Crust. Biol., **14:** 413–437.

Gabe, M. (1956). Ann. Sci. Nat. Zool. Biol. Anim., **(11) 18:** 145–152.

Gabe, M. (1953a). C.R. Acad. Sci. Paris, **237:** 1111–1113.

Gabe, M. (1953b). Bull. Microse. Appl., **3:** 153–162.

Ghosh, D. and Ray, A.K. (1993). Invert. Reprod. Develop., **25:** 43–47.

Ghosh, D. and Ray, A.K. (1994). Invert. Reprod. Develop., **25:** 43–47.

Gliscynski, V. Von, and Sedlmeier, D. (1993). J. Exp. Zool., **265:** 454–458.

Gomez, R. (1965). Naturwissenschaften, **52:** 216.

Goodwin, T.W., Horn, D.H.S., Karlson, P., Koolman, J., Nakanishi, K., Robbins, W.E., Siddall, J.B. and Takemoto, T. (1978). Nature, **272:** 122.

Hagedorn, H.H. (1989). **In:** Ecdysone: From Chemistry to Mode of Action, (Koolman, J. and G. Thieme, Eds.), pp. 279–289, Verlag, Stuttgart, New York.

Hinsch, G.W. (1977). J. Morphol., **154:** 307–316.

Hinsch, G.W. (1980). Trans. Am. Microscop. Soc., **99:** 317–322.

Hoffmann, J.A. and Lagueux, M. (1985). Endocrine aspects of embryonic development in insects. **In:** Comparative Insect Physiology and Pharmacology, Vol. I (Kerkut, G.A. and L.I. Gilbert, Eds.), pp. 435–460, Pergamon Press, New York.

Jugan, P. and Soyez, D. (1985). C.R. Acad. Sci. Paris, **300:** 705–709.

Keller, R. (1992) Experientia, **48:** 439–448.

Kulkarni, G.K., Glade, R. and Fingerman, M. (1991). J. Crust. Biol., **11:** 513–522.

Lachaie, F. (1989). **In:** Ecdysone: From Chemistry to Mode of Action, (Koolman, J. and G. Thieme, Eds.), pp. 313–318, Verlag, Stuttgart, New York.

Lachaise, F. and Hoffmann, J.A. (1982). Hoppe-Seyler's Z. Physiol. Chem., **363:** 1059–1067.

Lachaise, F., Le Roux, A., Hubert, M. and Lafont, R. (1993). J. Crust. Biol., **13:** 198–234.

Landeu, M., Laufer, H. and Homola, E. (1989). Invert. Reprod. Devl., **16:** 165–168.

Laufer, H., Landau, M., Borst, D. and Homola. E. (1986). **In:** Adv. Invert. Reprod., Vol. 4, (Porchet, M., J.C. Andries and A. Dhainaut, Eds.) pp. 135–143. Elsevier, Amsterdam and New York.

Laufer, H., Borst, V., Baker, F.C., Camesco, C., Sinkus, M., Reuter, C.C., Tsai, L.W. and Schroleg, D.A. (1987). Science, **235:** 202–205.

Lanot, R. and Cledon, P. (1989). Invert. Reprod. Develop., **16:** 169–175.

Le Roux, A. (1968). C.R. Acad. Sci. Paris, **266D:** 1414–1417.

Le Roux, A. (1983). C.R. Acad. Sci. Paris, **296D:** 697–700.

Matson, M.P. and Spaziani, E. (1985a). J. Exp. Zool., **236:** 93–101.

Matson, M.P. and Spaziani, E. (1985b). Mol. Cell. Endocrinol., **42:** 185–187.

Meusy, J.-J., Martin, G., Soyez, D., Van Deijinen, J.E. and Gallo, J.-M. (1987) Gen. Comp. Endocrinol., **67:** 333–341.

Okamura, T., Nakamura, K., Aida, K. and Hanyu, I. (1989). Nippon Suissan Gakkaishi, **55:** 2091–1098.

Okasaki, R.K. and Chang, E.S. (1991). Gen. Comp. Endocrinol., **81:** 174–186.

Otsu, T. (1963). Embryologica, **8:** 1–20.

Panouse, J.B. (1943). C.R. Acad. Sci. Paris, **217:** 553–555.

Passano, L.M. (1960). **In:** The Physiology of Crustacea, Vol. I, (Waterman, T.N., Ed.) pp. 473–536. Academic Press, New York.

Prakash, V.S.G., Keller, R. and Subramoniam, T. (1996). World Aquaculture 96 (abstract).
Quackenbush, L.S. and Herrnkind, W.H. (1983). J. Crust. Biol., **3:** 34–44.
Quackenbush, L.S. and Keeley, L.L. (1988). Biol. Bull., **175:** 321–339.
Quinitio, E.T., Yamauchi, Y. Hara, A. and Fuji, A. (1991). Gen. Comp. Endocrinol., **81:** 343–348.
Richardson, H.B., Deecaraman, M. and Fingerman, M. (1991). Comp. Biochem. Physiol., **99C:** 53–56.
Sarojini, R., Nagabushanam, R. and Fingerman, M. (1995). J. Exp. Zool., **271:** 395–400.
Saïdi, B., de Besse, N., Webster, S.G., Sedlmeier, D. and Lachaise, F. (1994). Mol. Cell Endocrinol., **102:** 53–61.
Sagi, A., Homola, E. and Laufer, H. (1993). Bio. Bull (Woods Hole), **185:** 168–173.
Sedlmeier, D. and Fenrich, R. (1993). J. Exp. Zool., **265:** 448–453.
Sochasky, J.B., Aiken, D.E. and Watson, N.H.F. (1972). Can. J. Zool., **50:** 993–997.
Soumoff, C. and O'Connor, J.D. (1982). Gen. Comp. Endocrinol., **48:** 432–439.
Soyez, D., Le Caer, J.P., Noel, P.Y. and Rossier, J. (1991). Neuropeptides, **20:** 25–32.
Soyez, D., Van Herp, F., Rossier, J., Le Caer, J.-P., Tensen, C.P. and Lafout, R. (1994) J. Biol. Chem., **269:** 18295–18298.
Spaziani, E. (1990). **In:** Morphogenetic Hormones of Arthropodes, Vol. I, Part 2 (Guptha, A.P., Ed.), pp. 233–267. Rutgers University Press, New Brunswick, New Jersey.
Subramoniam, T. and Keller, R. (1993). Curr. Sci., **65:** 619–622.
Taketomi, T., Motono, M. and Miyawaki, M. (1989). Cell Biol. Int. Rep., **13:** 463–469.
Tamone, S.L. and Chang, E.S. (1993). Gen. Comp. Endocrinol., **89:** 425–432.
Tensen, C.P., Verhoeven, A.H.M., Gaur, G., Janssen, K.P.C., Keller, R. and Van Herp, F. (1991a) Peptides, **12:** 673–681.
Tensen, C.P., De Kleijin, D.P.V. and Van Herp, F. (1991b). Eur. J. Biochem., **200:** 103–106.
Tsukimura, B., Kamamoto, F.I. and Borst, D.W. (1993). J. Exp. Zool., **265:** 427–437.
Tsukimura, B. and Borst, D.W. (1992). Gen. Comp. Endocrinol., **86:** 297–303.
Webster, S.G. (1991). Proc. Roy. Soc. London, **B244:** 247–252.
Webster, S.G. and Keller, R. (1986). J. Comp. Physiol., **B156:** 617–624.
Wilder, M.N., Okumura, T. and Aida, K. (1991). Zool. Sci., **8:** 919–927.
Yano, I. (1992). Bull. Nat. Res. Inst. Aquacult., **0:** 9–14.
Yano, I. (1987). Aquaculture, **61:** 49–57.
Yano, I. (1985). Aquaculture, **47:** 223–229.
Young, N.J., Quinlan, P.T. and Goad, L.J. (1992). Gen. Comp. Endocrinol., **87:** 300–311.
Yudin, A.I., Diener, R.A., Clarke, W.H. Jr. and Chang, E.S. (1980). Biol. Bull., **159:** 760–772.
Zeleny, C. (1905). J. Exp. Zool., **2:** 1–102.

C
Pineal Gland

Comparative Endocrinology and Reproduction
K.P. Joy, A. Krishna and C. Haldar (Eds)

28. Comparative Morphophysiology of the Pineal Organs of Vertebrates

B. Vigh and I. Vigh-Teichmann
Photoneuroendocrine Laboratory of the Department of Human Morphology and Developmental Biology, Semmelweis Medical University, Tuzolto utca 58, H-1094 Budapest, Hungary

Introduction

The pineal complex is a part of the epithalamus of the vertebrate brain. It is composed of the pineal organ accompanied by the parapineal organ in cyclostomes and teleost fishes, by the frontal organ or frontal eye in anuran amphibians and by the parietal eye in reptiles. In the following, we compare morphophysiological data about the pineal complex of the main vertebrate classes from the point of view of evolution of pineal functions. First, we discuss the question of the possible phylogenetic origin of these organs.

The *lancelets* represent recent chordates having one of the simplest central nervous systems. An organ-like pineal is not known in the lancelet, *Branchiostoma lanceolatum,* but there are numerous photosensory cells in its brain and spinal cord. In the dorsal portion of the brain vesicle and rostral spinal cord, two types of photoreceptor cells were found: numerous rhabdome-type Joseph cells supplied with microvilli, and few Satir cells, which display ciliary photoreceptor lamellae (Fig. 1). Since among the visual cells of the lancelet only the *Satir cells* are ciliary photoreceptor cells—similar to the pineal photoreceptors of vertebrates—they could be regarded as "ancestors" of the pineal organs in recent species (Eakin and Westfall, 1962; Welsch, 1968; Vigh and Vigh-Teichmann, 1982; Vigh-Teichmann and Vigh, 1983; Watanabe and Yoshida, 1986; Ruiz and Anodon, 1991).

Among recent cyclostomes, the lampreys display the "most ancient" pineal complex of chordates-vertebrates. The pineal complex of lampreys is represented by two organs, a pineal and a parapineal organ. Accordingly, the supposed ancestor of the pineal complex of vertebrates could be characterized by a *four-eyed protovertebrate* (Fig. 1), being similar to the recent lampreys, endowed with a double pineal and bilateral eyes. Petromyzontidae appeared in the upper Carbon. Earlier, more simple fishes like Anaspida and even Osteostraci—in Devon and Silur—might have had pineal organs as shown by the pineal hole present in their fossil crania (Fig. 1).

The evolution of eyes and pineal complex seems to be different. Light source of the sky is faced by the two dorsal pineal eyes measuring direct light intensity of the sky. The lateral eyes face the objects of the environment, to fulfil a locator function based on light reflection in a similar way as the bat's ear serves as a locator based on reflected sound. Locator function needs light focusing structures and highly differentiated individual photoreceptors; in contrast, the light measuring task of the pineal organs only needs suitable number of light sensitive cells.

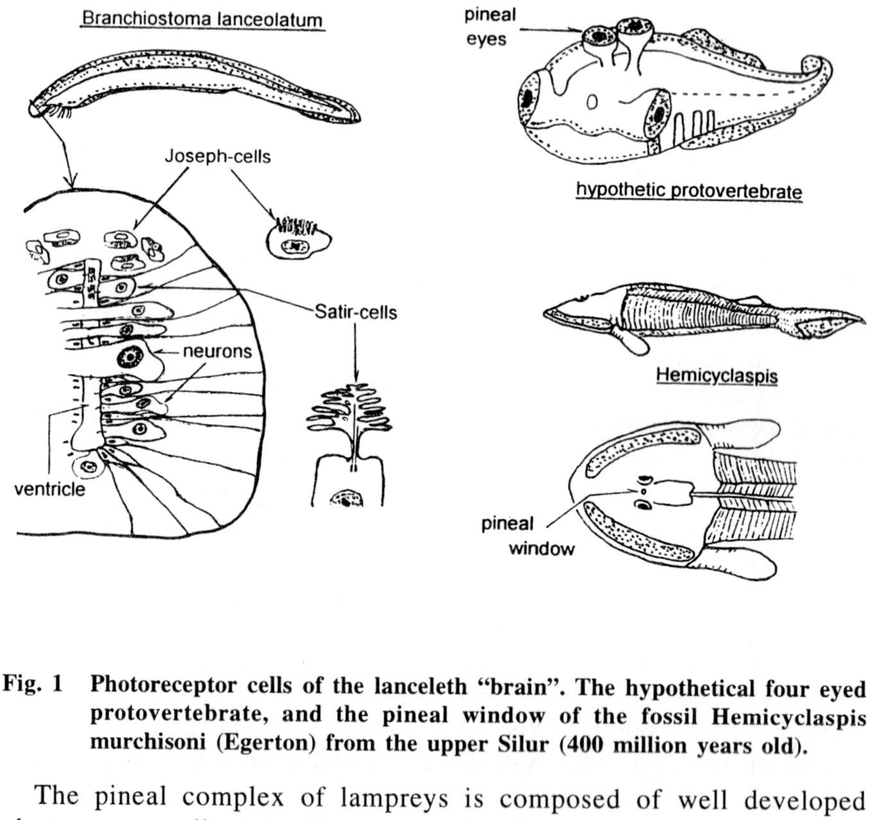

Fig. 1 Photoreceptor cells of the lanceleth "brain". The hypothetical four eyed protovertebrate, and the pineal window of the fossil Hemicyclaspis murchisoni (Egerton) from the upper Silur (400 million years old).

The pineal complex of lampreys is composed of well developed photoreceptor cells. According to several authors, the photoreceptor pineal organ of lower vertebrates transforms into the gland-like pineal of mammals during evolution (Bargmann, 1943; Oksche, 1965; Kappers, 1968; Collin, 1976; Vollrath, 1981; a.o.). In contrast to this view, secretory release sites were found in the pineal complex of the lamprey, and photoreceptor structures containing photopigments were demonstrated in higher vertebrates as well (Meiniel, 1971; Vigh and Vigh-Teichmann, 1988, 1993; Vigh-Teichmann *et al.*, 1993). Therefore, a reevaluation of comparative morphophysiological data of the pineal organs seems to be necessary in order to better understand their phylogenetic tendency.

Pineal Organs in Various Vertebrates Including Human

1. Pineal and parapineal organs of cyclostomes

As already mentioned, the pineal complex of lamprey species consists of the pineal and parapineal organs (Fig. 2). The pineal organ is a flattened vesicle. Its dorsal wall is called the pellucida and its ventral portion the retina. The pineal lumen continues toward the brain into the so-called atrium.

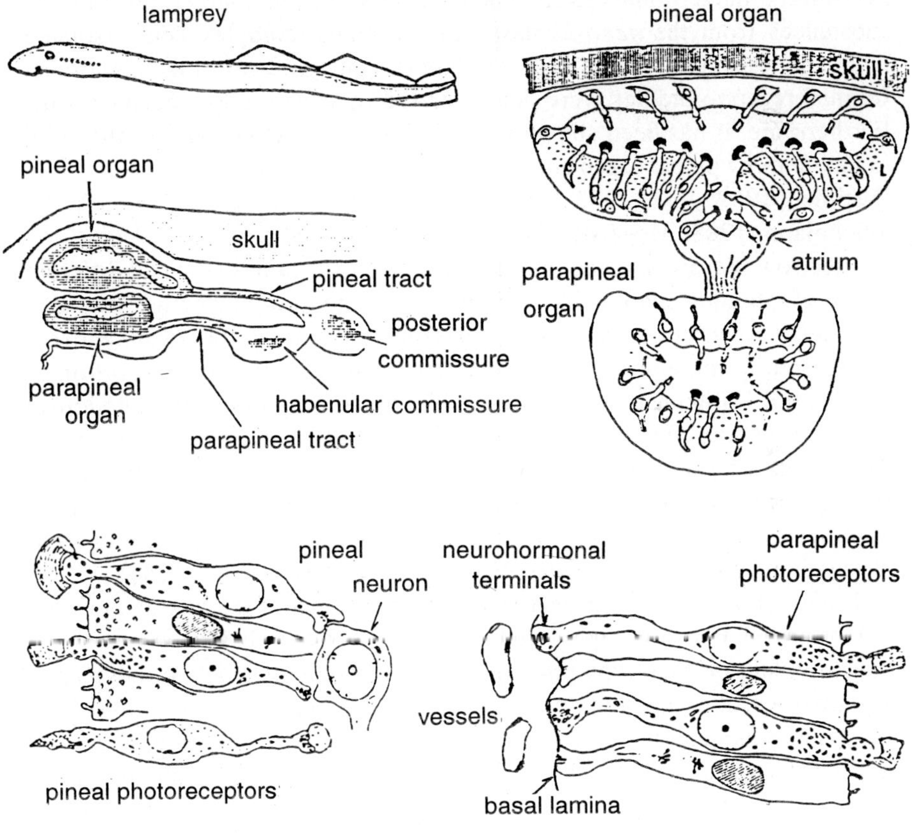

Fig. 2 The pineal and the parapineal organs of the lamprey.

The lumen of the atrium disappears in a caudal direction, and the stalk proceeds as solid tissue of the pineal tract into the brain. The **parapineal organ,** composed of a thin dorsal and a thickened ventral retina, is located below the pineal vesicle. The proximal portion of the organ is connected to the left habenular nucleus by the parapineal tract. The pineal tract projects to the habenular region, the subcommissural organ, the periventricular grey, the pretectum, the midbrain tegmentum, the posterior tubercular nucleus, the torus semicircularis, the thalamus, the hypothalamus, the optic tectum and the oculomotor nucleus (Meiniel, 1971; Cole and Yuson, 1982; Puzdrowski and Northcutt, 1989; Yanez *et al.,* 1993).

The parapineal organ exhibits morphological signs of a neuroendocrine activity. The basal processes of the photoreceptor cells of the dorsal parapineal retina give rise to axons terminating on the surface of the dorsal part of the organ as *neurohormonal endings* being similar to those of the neurohypophysis (terminal enlargement of axon-type processes attached by half-desmosomes to the basal lamina separating the nervous tissue from vascular spaces; accumulation of synaptic and granular vesicles in the terminal). Such neurohormonal terminals seem to be necessary to release larger quantities of substances from the *nervous tissue* of the pineal complex being isolated from the vascular spaces by basal lamina and glial limiting membrane. Immunoreactive indoles were demonstrated in several photoreceptor cells. Furthermore, it has been shown that melatonin is rhythmically synthesised in the lamprey pineal organ and is involved in the colour changes of the skin (Meiniel, 1969; Joss, 1973, 1977; Guerlotté *et al.*, 1986; Vigh and Vigh-Teichmann, 1988; Meyer-Rochow and Stewart, 1992).

Electron microscopic and immunocytochemical studies revealed different kinds of photoreceptor cells working at different wavelengths. Rod-like pinealocytes with photoreceptor outer segments binding rhodopsin antisera are present in the pineal retina and atrium, further in the central portion of the ventral parapineal retina. Cone-like photoreceptors being rhodopsin immunonegative were demonstrated first of all in the parapineal organ. Therefore, the parapineal organ may be regarded as a predominantly cone-type photoreceptor. The pineal organ is a predominantly rod-type photoreceptor, sensitive in the blue-green range of the light spectrum (Vigh-Teichmann *et al.*, 1983b, 1989; Vigh and Vigh-Teichmann, 1988).

S-antigen and lectins present in rods of the retina of the lateral eye were demonstrated in the rod-type pineal photoreceptors. Visinin immunoreactivity, characteristic of cone-type photoreceptors and immunoreactive iodopsin is equally present in the lamprey pineal complex. Vitamin A, the chromophore of opsin, was demonstrated immunocytochemically in the outer segments and perikarya of photoreceptors. Experimentally, the pineal photopigment can be regenerated by photo-reisomeration. Some of the perikarya and axonal processes of photoreceptors and secondary neurons contain serotonin. Further, somatostatin immunoreactive perikarya were found in the dorsal wall of the pineal organ (Meiniel and Meiniel, 1985; Kuo *et al.*, 1988; Vigh and Vigh-Teichmann, 1988; Vigh-Teichmann *et al.*, 1983a, 1989; Morita *et al.*, 1989; Cheung *et al.*, 1990; Tamotsu and Morita, 1990; Tamotsu *et al.*, 1990, 1994a, b).

Electrophysiological investigations proved the light sensitivity of the pineal organ of lamprey. Photoreceptors respond to light of 6×10^{-4} lux as a threshold with hyperpolarization, which modulates the spike discharge of the secondary pineal neurons. The secondary neurons are also hyperpolarized with concurrent suppression of spike discharge. The peak of the spectral sensitivity is at 505 nm in the larval and 525 nm in the adult animal suggesting a change of visual pigment or type of photoreceptors during metamorphosis. Electrophysiological and HPLC analysis show the presence of UV receptors having maximum sensitivity at 380 nm in the pineal organ. Light information of the environment perceived by the pineal complex is transmitted to the brainstem by axons of secondary pineal neurons. Accordingly, there is a pineal-dependent circadian rhythmicity of locamotor activity in the lamprey (Morita and Dodt, 1973; Pu and Dowling, 1981; Tamotsu and Morita, 1986; Morita *et al.,* 1985, 1989, 1991a, b, 1992; Uchida and Morita, 1990; Uchida *et al.,* 1992).

2. Pineal organ of cartilaginous fish

The cartilaginous fish *(Chondrichthyes)* comprise the *Holocephala,* the *Raja* species, and the sharks. In the literature, there are relatively few data on the pineal of these animals. Their pineal complex is represented by a rather long pineal organ exhibiting both a terminal and a proximal swelling (Fig. 3). As there are no specialized neurohormonal terminals in the pineal of these animals, the efferentation of the organ is supposed to be neural in nature (Altner, 1965; Rudeberg, 1969; Ueck, 1981; Vigh-Teichmann *et al.,* 1983a).

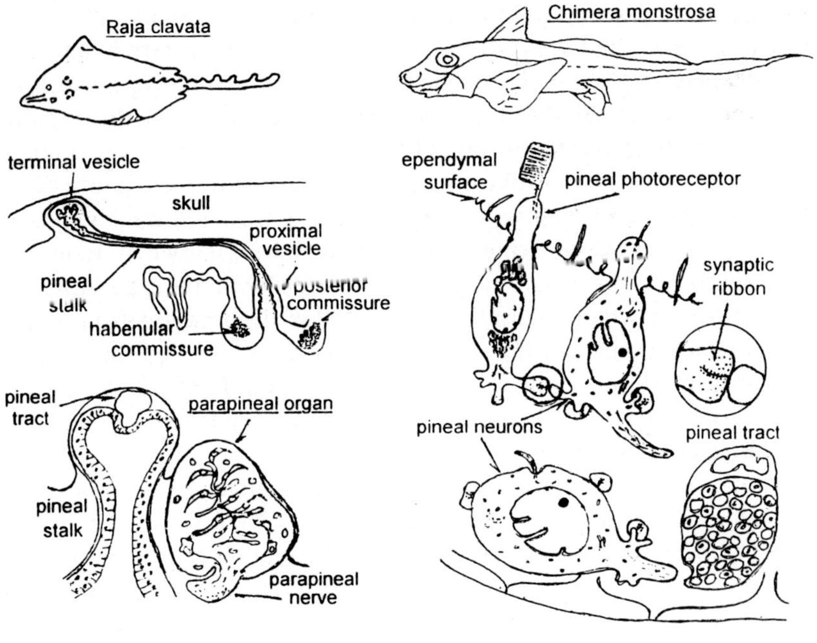

Fig. 3 The pineal organ of cartilaginous fishes (*Raja clavata, Chimaera monstrosa*). Parapineal organ of teleostean fish.

In the ray *(Raja clavata),* all three portions of the pineal organ (the proximal and the terminal vesicle as well as the intermediate narrow part) consist of photoreceptor pinealocytes, secondary neurons, and ependymal/

glial cells. Three types of photoreceptors can be distinguished: electron-lucent, electron-dense, and cells with lobulated nuclei. Most of the pineal outer segments bind rhodopsin antisera showing the predominantly rod-type character of the organ. Some of the inner segments emit two cilia, a finding rather rare among the pinealocytes of vertebrates (Vigh and Vigh-Teichmann, 1988).

The axons of the photoreceptor cells do not form neurosecretory endings, but they build up synaptic contacts with dendrites of secondary pineal neurons. Another type of axon—presumably representing fibres of intrapineal connections or of an afferentation from the brain stem—also terminates on the intrinsic neurons. Both the pinealocytes and secondary neurons exhibit glutamate and aspartate immunoreactivity. Glutamate seems to accumulate in pinealocytic terminals on secondary neurons, therefore, a role of this exciatory amino acid can be supposed in the synaptic transmission of the efferentation of the pineal (Vigh-Teichmann *et al.,* 1983a; Vigh *et al.,* 1995a, b).

The pineal organ of the ratfish *Chimaera monstrosa (Holocephala),* emerges almost vertically to the top of the skull because of the huge size of the lateral eyes. Similarly, to the pattern found in the ray, the proximal portion of the organ is broadened and continues by a thinned stalk into the sac-like terminal vesicle. Pinealocytic processes of ribbon-type terminate on the secondary neurons. Immunocytochemically, the opsin of blue-green sensitive chrysopsin—typical of the retina of deep sea fishes—and the S-antigen could not be demonstrated in the dark-adapted pineal. However, light and dark retinal rods were (chrys) opsin-immunoreactive. Interestingly, immunoreactive S-antigen was only found in retinal rod nuclei (possibly indicating start of gene expression) and in the perikarya of retinal and pineal neurons. This result may be connected with the particular light conditions under which the ratfish lived (200 m deep mesopelagic twilight zone being penetrated by blue-green light of 475-485 nm) and fed (euphotic zone down to 100 m depth also penetrated by blue light of 420 nm). The animals examined were caught at night (Vigh-Teichmann *et al.,* 1990; Vigh-Teichmann and Vigh, 1994).

Immunoreactive glutamate was found to accumulate on synaptic ribbons of pinealocytic terminals on secondary neurons. The axons of the neuronal perikarya enter the pineal tract and run to the brain stem (Vigh and Vigh-Teichmann, 1988, Vigh-Teichmann and Vigh, 1986, 1994; Vigh-Teichmann *et al.,* 1990; Vigh and Vigh-Teichmann, 1996).

There are two types of intrapineal neurons: one displaying a luminal dendrite and another exhibiting an intercellular cilium. The use of light-microscopic acetylcholinesterase reaction permits observation of neurons that send one of their processes into the pineal lumen in the shark *Triakis scyllia* as well (Ueck and Kobayashi, 1979; Ueck, 1981). These types of neurons are identical in structure to hypothalamic CSF (cerebrospinal fluid)-

contacting nerve cells that form a luminal dendrite bearing a 9x2+0 cilium. Such neurons are also known in the retina as Landolt's bipolars. Axons of pineal photoreceptor cells terminate on the pineal CSF-contacting neurons, and the axons of the latter enter the pineal tract (Vigh-Teichmann *et al.*, 1983a; Vigh, 1987; Vigh-Teichmann and Vigh; 1987, Vigh and Vigh-Teichmann, 1989a, 1996).

3. Pineal complex of teleostean fish

The pineal complex of teleostean fish consists of a pineal and a parapineal organ (Fig. 3). Numerous data are at our disposal on the morphology and physiology of the pineal organ except for the eel, which displays a proximal pineal thickening. The long pineal stalk of teleosts runs to the skull, forming a terminal vesicle there. There are day/night variations and seasonal changes in the pineal melatonin content and in the activity of enzymes (hydroxyindole-o-methyltransferase) involved in its metabolism suggesting that photoreceptors of the pineal organ translate photoperiodic light information not only into nervous but also into indolic signals. Melatonin production may be modified by temperature as well. In deep-sea fishes, both photosensory and neurendocrine functions are believed to be reduced (Vollrath, 1981; Falcon *et al.,* 1986; Birks and Ewing, 1986; McNulty *et al.,* 1988; Max and Menaker, 1992; Bolliet *et al.,* 1994).

In most teleosts, the photoreceptor cells exhibit well-developed outer segments and form ribbon-containing terminals on secondary neurons, whose axons contribute to the pineal tract, connecting the pineal organ with the brain stem (Hartwig and Pfautsch, 1973; Matsuura and Herwing, 1981; Van Veen, 1982; Omura, 1984; Ekström and Korf, 1986a, b).

Studies on opsin immunocytochemistry of the pineal and parapineal organs reveal that most of the pineal outer segments are rhodopsin positive: "rod-type pinealocytes", and some of them are rhodopsin negative: "cone-type pinealocytes". Rhodopsin-immunoreactive photoreceptors can already be demonstrated in the pineal organ and retina of prehatching charr embryos, a result indicating their early capacity to detect dim light. In some adult species, photoreceptor cells were also observed in the roof of the third ventricle, around the orifice of the pineal recess. Their outer segments bind rhodopsin antisera. When the pineal organ is removed at its stalk in experimental studies, these cells remain in the brain and may cause a persisting light sensitivity, imitating the existence of deep encephalic photoreceptors. The same problem may raise in connection with the parapineal organ (Vigh *et al.,* 1986a; Vigh and Vigh-Teichmann, 1981; Vigh-Teichmann *et al.,* 1980, 1982, 1983, 1991b, 1992).

In electrophysiological experiments, the pineal organ of fishes was found to react to light stimuli directly and to operate both in dim and bright light. Light stimuli elicit intensity-graded hyperpolarization inhibiting the discharge of secondary neurons. The receptor cells produce constant amplitude responses during steady illumination and show spectral sensitivity peaks at about 495 to 530 nm. The presence of several kinds of interneurons (cholinergic, GABAergic and substance P-containing) and of an afferent (noradrenergic and peptidergic) innervation was established by means of immunocytochemistry. Consequently, an integrative neural circuitry may be involved in the elaboration of neural signals to the brain and in the control of indolamine synthesis (Dodt, 1963; Morita, 1966; Morita and Bergmann, 1971; Hanyu *et al.,* 1978; Omura and Ali, 1980; Falcon and Meissl, 1981; Ekström and Korf, 1986b; Nakamura *et al.,* 1986; Kusmic and Marchiafava, 1990; Meissl and Ekström, 1991; Meissl and Brandstatter, 1992; Marchiafava and Kusmic, 1993).

Pineal tract fibers run to habenular nuclei, subcommissural organ and pretectum. FMRFamide immunoreactive axons originating in the nucleus nervus terminalis reach the pineal and represent a direct connection between a specialized part of the chemosensory system and the pineal organ in teleost fish (Ekström, 1984, 1987; Ekström and Korf, 1985; Ekström *et al.,* 1987, 1988; Joy and Agha, 1993).

The **parapineal organ** of most teleosts (e.g., *Phoxinus phoxinus, Carassius auratus, Cyprinus carpio, Anguilla anguilla, Salmo gairdneri*) is a small round body located on the left side of the brain above the epithalamus (Fig. 4). The parapineal tract originating from the neurons of the organ connects it with the left habenular nucleus. The organ contains neurons, glial cells, and parapinealocytes of a photoreceptor type. Their outer segments bind rhodopsin antisera, indicating the presence of rhodopsin or porphyropsin and a light-percepting ability (Vigh-Teichmann *et al.,* 1980, 1982, 1983b; Van Veen, 1982).

4. Pineal and frontal organs of amphibians

In *urodelan* species (newts, salamanders), the pineal complex is represented by a pineal organ only, whereas the *anuran* frogs display an additional frontal organ (Fig. 4). The two animal groups represent stages of different adaptation to the environmental conditions in water and on dry land and to the differences in illumination in these two biotopes. In the salamander, *Hynobius dunni* the pineal complex displays an anterior and a posterior portion. The anterior body appears to be homologous to the frontal organ of anurans (Wake *et al.,* 1974; Hartwig and Baumann, 1974; Korf *et al.,* 1981; Vigh and Vigh-Teichmann, 1981, 1986; Dobt and Meissl, 1982; Vigh *et al.,* 1985; Ueck *et al.,* 1989a, b; Kemnitz *et al.,* 1990; Takahama, 1993).

The **pineal organ** of amphibians is sac-like, emitting numerous side pockets. There are photoreceptor cells, neuronal perikarya besides of ependymal/glial cells in the wall of the pineal organ. Pedicle-like processes of photoreceptor pinealocytes form synaptic ribbon-containing axon terminals on secondary pineal neurons. Pineal nerve cells were shown to be AChE-positive. Some GABA immunoreactive perikarya are thought to be interneurons. There are axodendritic synapses on dendrites of pineal neurons whose presynaptic axoplasm contains large granulated vesicles. Some of the afferent fibers are FMRF amide immunoreactive and enter the organ via the posterior commissure (Ueck and Kobayashi, 1979; Ueck *et al.,* 1989a, b; Vigh-Teichmann *et al.,* 1986; Ekström and Meissl, 1990; Ekström *et al.,* 1990).

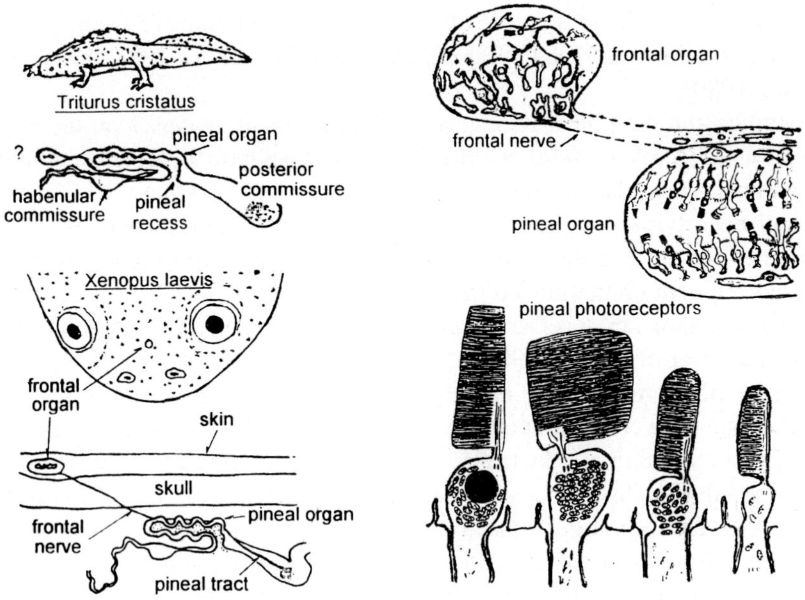

Fig. 4 The frontal and the pineal organs of amphibians.

In the *frog* pineal organ, two rod- and two cone-type photoreceptors can be found (Fig. 4). The outer segments of large and small rods bind rhodopsin antisera, and OS-2 antibodies, respectively recognizing blue-green and blue-sensitive photopigments (rhodopsin/porphyropsin and "amphibian blue-opsin"). The mitochondria of the inner segment form an ellipsoid. The cytoplasm of rods is electron dense, and their basal process terminates as a large pedicle containing synaptic vesicles of 50 to 70 nm in diameter and synaptic ribbons. Several dendrites of the secondary pineal neurons penetrate the basket-like pedicles. The large cones are represented by cells containing a lipid droplet in their inner segment. Their outer-segment is rhodopsin-immunonegative and binds monoclonal COS-1 antibodies recognizing epitopes of green- and red-sensitive pigments (iodopsin?). The outer and inner segments of the other cone-type pinealocytes are small, and this photoreceptor does not react with any of the four antiopsin antisera used. Its mitochondria are dispersed or form an ellipsoid in the dendrite of the cell. The perikaryon is located ependymofugally, and is electron-lucent similarly to its basal process and spherule-like axon terminals (Vigh-Teichmann *et al.,* 1986; Vigh-Teichmann and Vigh, 1990).

In the lumen of the pineal organ, macrophages can often be observed. In their cytoplasm, parts of outer segments exhibiting various degrees of disintegration can be detected by electron microscopy. Opsin-immunoreactive membranes also used to be present among the phagocytized elements. They indicate a shedding of photoreceptor lamellae of pineal outer segments, a phenomenon well known in the retina. Outer segments of pinealocytes of the frog seem to undergo a daily complete renewal in standard laboratory circumstances. Photoreceptors also contain immunoreactive S-antigen and vitamin A (Hartwig, 1984; Vigh-Teichmann *et al.,* 1986, 1987, 1988; Korf *et al.,* 1989).

Running in the pineal lumen, several axons form synapses on the pineal ependyma in *Pleurodeles waltlii.* Their postsynaptic cytoplasm contains myeloid bodies also present in the retinal pigment epithelium. Since the pineal ependyma contains vitamin A as detected immunocytochemically, the ependymal cells may play a role in the chromophore metabolism (like retinal pigment epithelium) of pineal photopigments. The innervating CSF-contacting axons may serve for a control of this ependymal function (Vigh-Teichmann *et al.*, 1973, 1987, 1988, 1989).

The extracranial ***frontal organ*** or frontal eye of frogs (Fig. 5) is located in the skin between and above the eyes. It is smaller than the pineal organ and has small lumina. Like the pineal organ, it is composed of photoreceptor cells, glial elements, and secondary neurons besides of nerve fibers and synapses. The organ is connected by the pineal (frontal) nerve and pineal tract to habenular, pretectal, and tegmental regions of the brain stem as well as to hypothalamic periventricular areas. The frontal organ is relatively large in young animals and presumably serves as an underwater photoreceptor in larvae (Eldred *et al.,* 1984; Vigh and Vigh-Teichmann, 1986).

The photoreceptor cells of the frontal organ are of three main kinds. The rod-type pinealocyte is characterized by a rhodopsin-immunopositive outer segment and by an electron dense cytoplasm. One of the cones has a rhodopsin-immunonegative outer segment, contains an oil droplet and is not abundant. The other cone-type photoreceptor is the most numerous. It displays a rhodopsin-immunonegative outer segment. There is no oil droplet in the inner segment, and the hyaloplasm of the cell is electron-lucent. The basal, axonal processes of the three cell types form pedicle-like terminals on the dendrites of secondary neurons of the organ. The large number of rod-like photoreceptors found in the pineal organ indicates its predominantly rod-type light-perception capacity. In contrast, the frontal organ has to be regarded as a predominately cone-type photoreceptor organ possibly most sensitive to orange and ultraviolet/blue light (Vigh-Teichmann *et al.*, 1980; Vigh *et al.*, 1982; Vigh and Vigh-Teichmann, 1986, 1988; Vigh-Teichmann and Vigh, 1990).

The ultrastructural findings are in accord with electrophysiological data demonstrating neuronal responses with spectral sensitivities at different absorption maxima, in the pineal and frontal organs. The rod-like photoreceptors seem to correspond to those photosensory elements causing inhibitory achromatic responses in secondary neurons at an absorption maximum of about 500 nm, a wavelength typical of rhodopsin. The two cone-like photoreceptors lacking rhodopsin immunoreactivity are thought to be identical with structures responsible for light perception in the orange (iodopsin) and ultraviolet-blue ranges of the spectrum. The same photopigments (rhodopsin and iodopsin) were detected in the frontal organ. Immunoreactive vitamin A was described by immonocytochemical methods in the pineal organ. HPLC analysis showed the presence of 11-cis and all-trans isomeres of retinal in both the pineal and forntal organs of the frog. These photopigments enable the animal to discern light of different wavelengths in order to adapt its behaviour to changing environmental conditions (Baumann, 1962; Dodt and Heerd, 1962; Dodt and Morita, 1964; Morita, 1965, 1969; Hamasaki, 1970; Eldred and Nolte, 1981; Korf *et al.*, 1981; Meissl and Dodt, 1981; Vigh-Teichmann and Vigh, 1990; Vigh-Teichmann *et al.*, 1987, 1988; Masuda *et al.*, 1994).

Glutamate and aspartate were shown to exert an excitatory effect on achromatic light responses of pineal neurons. Further, it is known that excitatory amino represent the synaptic mediators in the photoreceptor pathway of the retina. Using immunocytochemical methods an identical presynaptic localizations of glutamate and aspartate was found in the synapses of pineal and retinal photoreceptors on secondary neurons. These results suggest that excitatory amino acids act as synaptic mediator in the efferent photoreceptor pathway of the pineal organ. Subsequent immunoreaction on the same section showed a colocalization of both amino acids. Glutamate and aspartate were also found in some axons of pineal neurons, while other nerve fibers were immunonegative (Meissl and George, 1984; Bonaventure *et al.,* 1989; Vigh *et al.,* 1995a, b).

The ventral surface of the frog pineal organ is wavy, covered by villuslike processes of ependymal endfeet. Here, profiles can be found resembling neurosecretory terminals, the possible morphological substrate of the hormonal release of the organ. The color-change mechanism of the frog is supposed to be related to a melatonin dependent photoneuroendocrine regulation. There are several melanocytes above the pineal organ of the frog being always contracted in dark adapted animals. A supposed melatonin-secretion from the pineal may have an effect on these melanophores contracting them at night (Vigh and Vigh-Teichmann, 1986; Korf *et al.,* 1989; Vigh-Teichmann and Vigh, 1990).

5. Pineal organ and parietal eye of reptiles

Reptiles are of great evolutionary significance, since one of their branches, the sauropsides, gives rise to birds, the other, the theropsides, to mammals (Remane *et al.,* 1980). In some of the reptiles, e.g., leguan, lacertilians, the pineal complex is composed of a pineal organ and a parietal eye (Fig. 5). The **parietal eye** consists of a superficial cornea and a ventral retina. The outer- and inner segments of the photoreceptor cells of the parietal retina protrude into the lumen of the organ. The outer segments do not react with rhodopsin antisera nor with some cone-specific antibodies (Trost, 1954; Eakin *et al.,* 1961; Gundy and Wurst, 1976; Vigh-Teichmann *et al.,* 1980; Masuda *et al.*, 1994).

The reptilian **pineal organ** is a lobulated, sac-like evagination of the diencephalic roof which ascends by a long narrow stalk to the inner surface of the skull. Reptilian pinealocytes are characterized by a large amount of granulated vesicles. The presence of vesicles is considered as sign of a secretory character, and the disintegrated pinealocytic photoreceptor outer segments of most of the pinealocytes are valued as a sign of phylogenetic transformation of photoreceptor pinealocytes into secretory cells. We have to emphasize that in the relatively large pineal of reptiles, the fluid of its lumen interferes with a suitable fixation. Decomposition of outer segments may imitate phylogenetic regression (Kappers, 1968; Collin and Kappers, 1968; Oksche and Kirschstein, 1968; Vivien-Roels, 1969; Collin, 1976).

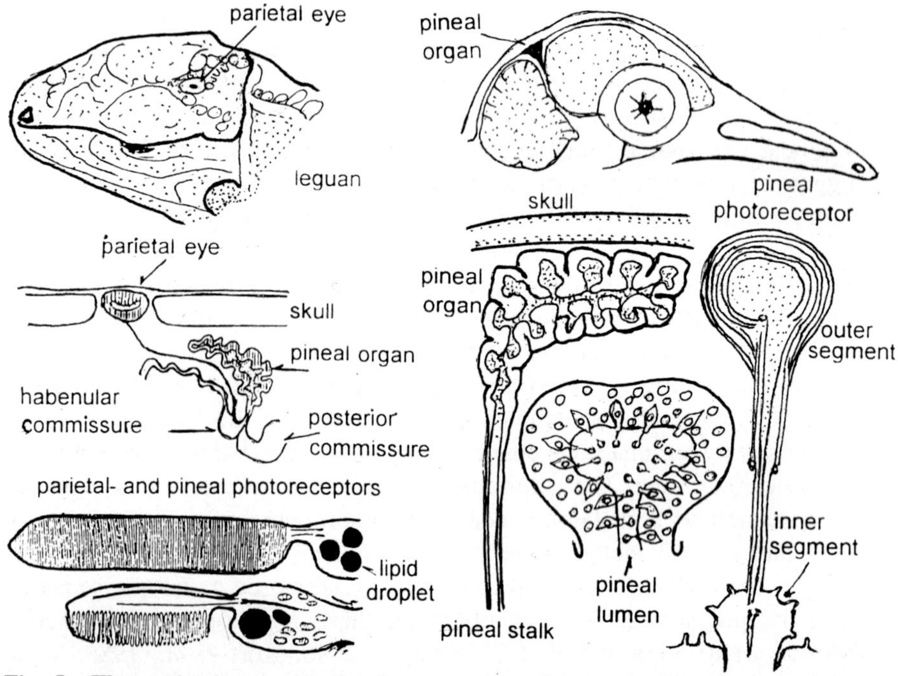

Fig. 5. The parietal eye and pineal organ of reptiles (to the left) and the pineal organ of birds (to the right).

The pineal organ has a considerable diversity in organization according to the different species. In the *turtles,* numerous photoreceptor outer segments immunoreact with rhodopsin antisera and indicate the light-percepting role of their pineal. Neurohormonal terminals are rare on the vascular surface of the pineal organ. The duration of the night peak of pineal melatonin content of the turtle depends on the daily photoperiod. The pineal melatonin content and circulating melatonin concentration is influenced by the environmental temperature as well (Vivien-Roels, 1969; Vigh *et al.*, 1975, 1982; Vivien-Roels *et al.*, 1988).

In *lacertilians,* two kinds of photoreceptor cells were described: light and dark ones. The dark one has a large outer segment with high sensitivity to the similarity of fixatives. The light one being rare, has a small outer segment and is better preserved by conventional fixation. The smaller, cone-like outer segments (or two subtypes of them) immunoreact with cone specific antisera COS1 (staining red- or green-sensitive outer segments) and/or OS2 (staining blue- or green-sensitive outer segments), the larger one is iodopsin immunoreactive. Some pinealocytes show S-antigen immunoreactivity but immunoreactive rhodopsin is absent in pinealocytes and retinal photoreceptors of lizards. The parietal eye photoreceptors are immuno-negative with all antiopsin sera used so far (Harez *et al.*, 1987; Masuda *et al.*, 1994; Vigh-Teichmann and Vigh, 1994).

The perikarya of the pineal photoreceptor of lizards contain several granular vesicles. The endings formed by the basal processes of photoreceptors terminate on secondary pineal neurons or on the surface of the pineal organ. Immunoreactive glutamate and also aspartate were found to accumulate in the presynaptic part of axodendritic and neurohormonal terminals. Therefore, excitatory amino acids may play a role in both the neural and hormonal efferentation of the pineal. The secondary pineal neurons are bipolar or rarely multipolar. In their cytoplasm, granular vesicles of 150 nm in diameter are found, their axonal processes run via the pineal stalk to the brain stem (Vigh and Vigh-Teichmann, 1988; Vigh *et al.*, 1995b).

In *Natrix* species (*Natrix natrix, Natrix tesselata*, and *Elaphe longissima*), the pineal organ strikingly differs from that of turtles and lacertilians studied so far. The pineal is spherical, small, and attached to the internal surface of the skull. The pineal lumen is rather narrow and labyrinthic and is covered by ependymal cells. The wall of the pineal organ is thin and invaginated, numerous vessels enter these foldings containing meningeal tissue. The pinealocytes give rise to double cilia. Their distal portion is thickened like a developing retinal outer segment, and may contain vesicles. Immunoreactive S-antigen, but not rhodopsin was found in the pinealocytes of snakes. On the perivascular surface of the organ, two different types of neurohormonal terminals can be found: electron-dense, large-sized terminals containing granular vesicles of 150 to 200 nm in diameter and electron-lucent ones of smaller size containing few granular and many synaptic vesicles. The large number of these endings indicates that among the vertebrates studied so far the pineal of *Natrix* represents the most highly developed secretory efferentation. The meningeal vessels of the organ are unfenestrated and accompanied by myelinated and unmyelinated nerve fibers (Vivien, 1964; Vigh and Vigh-Teichmann, 1988; Kalsow *et al.*, 1991).

In some reptiles (lacertilians), serotonin was localized to pinealocytes by histofluorescence methods. Likewise, 5-hydroxytryptophan was demonstrated autoradiographically. Only a small quantity were detected in *Tropidonotus natrix* with the same method. The two kinds of neurohormonal terminals found in *Natrix natrix* furnish morphological evidence for presence of two different substances that may be released from the pineal. Melatonin was detected in the pineal of the turtle with radioimmunoassay. Melatonin receptors of the brain are not symmetrically distributed: there is a high concentration localized to the left habenular nucleus. Several electrophysiological studies analyzed the light-percepting function of the reptilian pineal complex. In the parietal eye, an ultraviolet-sensitive mechanism was found. Experimental shielding of both the parietal eye and pineal organ positively modifies the aggressive behaviour and body temperature in lizards. The pineal was found to influence circadian locomotor activity in the iguana and ruin lizard (Miller and Wolbarsht, 1962; Quay *et al.*, 1968; Hamasaki and Dodt, 1969; Petit, 1971; Wartenberg and Baumgarten, 1969; Meissl and Ueck, 1980; Jenison and Nolte, 1980; Phillips *et al.*, 1988; Vivien-Roels, 1988; Foa, 1991; Wiechmann and Wirsing-Wiechmann, 1992; Innocenti *et al.*, 1993).

6. The pineal organ of birds

In the evolution, birds develop parallel to mammals, therefore, the avian pineal cannot be regarded as a structure transitive between reptiles and mammals. There are numerous data about the pineal of birds. Earlier, it was supposed, that pinealocytes of birds are regressive by having lost their phytoreceptor properties. By cardial perfusion, fixatives do not quickly reach the outer segments floating in the fluid of the closed lumina, and consequently, the outer segments show signs of decomposition. Intrapineal perfusion, however, results in a preservation of photoreceptor lamellae (Oksche, 1965; Oksche and Kirschstein, 1969; Bischoff, 1969; Oksche *et al.*, 1972; Kappers, 1971; Ueck and Kobayashi, 1972; Vigh and Vigh-Teichmnn, 1974; Vigh *et al.*, 1975; Collin, 1976; Omura, 1977; Vigh-Teichmann and Vigh, 1979; Vigh *et al.*, 1982, 1986a; Moller and Moller; 1990; Ohshima and Hiramatsu, 1993).

The outer segments of the avian pineal are pear-shaped, with a long connecting piece and internal granular matrix (Fig. 5). In posthatching chickens, most of the outer segments exhibit a rhodopsin immunoreaction, some of them are iodopsin immunoreactive. Visinin, a retinal cone-specific protein, further S-antigen, alpha-transducin and IRBP (interstitial retinol-binding protein) essential for the process of photochemical transduction are also present in the avian pineal. Further, vitamin A (11-cis and all-trans isomeres of retinal) was demonstrated immunocytochemically and by means of HPLC analysis (Vigh and Vigh-Teichmann, 1981, 1982; Foster *et al.*, 1987; Goto *et al.*, 1989; Araki *et al.*, 1992; Ohshima and Hiramatsu, 1993; Sun *et al.*, 1993; Masuda *et al.*, 1994; Zatz, 1994).

Using molecular biological methods, an opsin was found in the chicken which seems to be bluegreen-sensitive (lambda max.: 470 nm) and specific for the pineal ("pinopsin", Okano *et al.*,1994; Max *et al.*, 1995). Antibodies raised against a part of this molecule were found to specifically bind to pineal photoreceptor outer segments of the chicken pineal (Vigh and Vigh-Teichmann, 1998).

Neuronal perikarya demonstrated by AChE-reaction correspond to secondary neurons of the pineal light-conducting pathway. The effector pole of pinealocytes is represented by an axonal process terminating either on these secondary pineal neurons, or forming neurohormonal nerve terminals on the basal lamina of the pineal surface opposite to capillaries. In the chicken, a synaptic zone is located near the external surface of the organ. This zone is invested with synaptic ribbon-containing axon terminals of pinealocytes and regular axodendritic synapses on secondary pineal neurons. The number of the synaptic ribbons varies according to a day/night rhythmicity (Ueck and Kobayashi, 1972; Vigh *et al.*, 1982; Korf and Vigh-Teichmann, 1984; Sato and Ebisawa, 1988; Maitra and Vollrath, 1991; Masuda *et al.*, 1994).

Light-dependent changes were found in *N*-acetyltransferase activity of the organ. Light influences the pineal not only directly but also by means of the retina. In electrophysiological experiments some of the pineal cells were either activated, inhibited or unaffected by stimulation of the optic chiasm (Harbute and Bayle, 1974; Semm and Demaine, 1983; Kasal and Perez-Polo, 1980; Hamm *et al.*, 1983; Demaine and Semm, 1984; Aige-Gil and Murillo-Ferrol, 1992).

The pinealocytoes and their neurohormonal axon terminals contain a considerable amount of granular vesicles. Using electron microscopic immuno-gold method, an intense serotonin immunoreaction is detectable on these granular vesicles. The 5-methoxytryptophol and melatonin were detected in the pineal as well by radioimmuno-assay. Melatonin secretion in cultured pinealocytes of chickens and quails is directly regulated by photic variations. The terminal step of synthesis of the indolic hormone melatonin is catalysed by HIOMT (hydroxyindole-o-metyltransferase) of which the concentration shows a day/night change. Experiments using a cDNA probe encoding HIOMT demonstrated that the rhythm of HIOMT gene transcription depends on both the environmental light and an endogenous fluctuation ("oscillation"). A melatonin rhythm was observed under a light-dark regime in the pineal of chick embryos as well. Exposure of eggs to 1 h of light during the dark decreased the high melatonin concentrations in the eyes but not in pineal of the 20-day-old embryo (Ueck, 1973; Collin *et al.*, 1976; Voisin *et al.*, 1983; Guerlotte *et al.*, 1988; Vigh and Vigh-Teichmann, 1988; Pang *et al.*, 1989; Ohshima and Matsuo, 1991; Zeman *et al.*, 1992; Wiechmann and Craft, 1993; Bernard *et al.*, 1993; Ghosh *et al.*, 1994; Murakami *et al.*, 1994).

The pinealocytic perikarya also exhibit a weak-to-medium density of immunogold particles on sections reacted with glutamate and/or aspartate antisera. The reaction is stronger in the axonal processes. In the axodendritic endings, immunoreactive glutamate accumulates on synaptic ribbons and suggest a role of these structures in the synaptic release of the excitatory amino acids (Vigh *et al.,* 1995a, b).

7. The organization of the mammalian pineal organ

It is generally accepted in the literature, that the mammalian pineal is a gland secreting—among different substances—melatonin, the production of which is inhibited by environmental light. The daily fluctuation of melatonin is supposed to entrain genetically fixed circadian rhythmicity of the organism. The pineal plays a role in the seasonal timing of several functions like gonadal activity, moulting periods, pelage composition etc. Most of the physiologists deny the direct light sensitivity of the mammalian pinealocytes and suppose that light information reaches the organ via its peripheral adrenergic innervation (Wurtman *et al.,* 1964; Kappers, 1968; Vollrath, 1981; Reiter, 1981, 1988, 1993; Haldar and Pévet, 1983; Pévet, 1988; Klein, 1985, 1993; Mess and Ruzsas, 1986; Haldar *et al.,* 1992; Karasek and Reiter, 1992; Korf, 1994; Vitaterna *et al.,* 1994).

The Greatest part of our knowledge about the mammalian pineal is related to rodents. More primitive mammals, like marsupials or insectivorous species, often aid better in understanding the structural principles of the mammalian pineal organ than specialized rodents. The follicular structure of the mammalian pineal is regarded as glandular characteristic. In the marsupial **opossum** *(Didelphis Virginiana)* the small internal lumina of pineal follicles contain the inner segments of pinealocytes. This demonstrates that the lumina are homologous with the pineal photoreceptor spaces of submammalians. In the insectivorous **hedgehog** *(Erinaceus roumanicus),* the pinealocytic cilia are thickened and in addition to tubules and dense bodies they contain vesicles. In the **bat** (*Nyctalus noctula*), the cilia enlarge, and several vesicles were found in them. Similarly, in the **ferret** *(Putorius furo),* an ancient carnivorous animal, pinealocytic cilia develop photoreceptor outer segment-like structures (Fig. 6) in young animals. Studies in hereditary retinal disorder show that photoreception can be maintained by rod and cone cells without outer segments. Similarly, cilia of ascadian larvae or rhabdomeric microvilli of invertebrates are light sensitive by means of visual pigments simple located in their cell membrane. Therefore, the absence of well developed outer segments in the pineal of rodents does not contradict a direct light sensitivity of their pineal organ (Pévet *et al.,* 1977; Eakin, 1982; Vigh *et al.,* 1986; Vigh and Vigh-Teichmann, 1988, 1993; Quay, 1989; Vigh-Teichmann and Vigh, 1992, 1994; Foster *et al.,* 1993).

There are electrophysiological data on a pineal light sensitivity in the rat which was considered to partially depend on the retina (Barajas-López *et al.,* 1987). The photic evoked potential persist after superior cervical ganglionectomy and transection of the pineal stalk suggesting a direct pineal light sensitivity. The authors explain the persisting photic evoked potential by a transmission from the superior colliculus. However, there is no direct contact between the pineal and superior calliculus in this species and after interrupting the connections of the pineal with the rest of the central nervous system, two arachnoid and pial layers, further the external cerebrospinal fluid separate the organ from the brain and interfere with electrotonic transmission. Experiments showed that in bilateral enucleated neonatal rats illumination reduced pineal N-acetyltransferase activity via an extraretinal mechanism (Torres and Lytle, 1990).

Except of the 11-cis and trans retinal, several compounds known for participating in down regulation of the retinal photochemical transduction cascadè, were found in the mammalian pineal. Apparently, the cascade components that produce CGMP hydrolysis and membrane hyperpolarization are absent. However, rhodopsin immunogold labelling was demonstrated on the cell membrane of certain pinealocytes of the gerbil and ferret. Using a polymerase chain reaction, large amount of rhodopsin transcripts were detected in preparations from new-born rat pineals. Rhodopsin of the retina is a 7 helices transmembrane protein, whose carboxyl terminus faces the cytoplasm and whose NH2 terminal sequence is on the outside surface of the plasma membrane, in the intradiscal space. Rhodopsin is palmitoylated at the cysteines in its C-terminal sequence, and glycosylated at two asparagins near the NH2-terminus. Normal glycosylation is crucial for the ability of rhodopsin to bind 11-cis retinal. Since cone photopigments and related proteins of the photochemical transduction cascade are present in rather small amounts in the cone photoreceptor, their purification and elucidation of their functioning are difficult to achieve and still hindered. The same is true for the rod- and cone-like pinealocytes, the latter being a majority in the mammalian pineal (Taylor and Wilson, 1970; McClung and Dafny, 1975; Semm and Vollrath, 1978, 1980; Semm and Demaine, 1983; Luo *et al.*, 1984; Thiele and Meissl, 1985; Korf *et al.*, 1985; Ebadi and Govitrapong, 1986; Moller and Korf, 1987; Tsin *et al.*, 1989; Pazo and Gonzalez, 1991; Shi *et al.*, 1991; Araki and Taketani, 1992; Hargrave and McDowell, 1992; Lolley *et al.*, 1992; Mikkelsen, 1992; Kramm *et al.*, 1993; Vigh-Teichmann *et al.*, 1993).

Autonomic fibers reach the pineal by the so called *conarian nerves* (Fig. 6) It has to be noted that myelinated and unmyelinated fibers of the nerve run in the pineal meningeal septae separated by basal lamina and the glial limiting membrane from the pineal nervous tissue proper. Perivascular nerve fibers were demonstrated to form synapses on smooth muscle cells of pineal arterioles by us. Substance P and CGRP (calcitonin gene-related peptide) immunoreactive peripheral fibers originate from the trigeminal ganglion. VIP (Vasoactive intestinal polypeptide) containing nerves come from the pterygopalatine ganglion and NPY (neuropeptide Y) immunoreactive fibers originate from the superior cervical ganglion. These components of the conarian nerve probably represent vasoconstrictor, vasodilator and vasosensory fibers to regulate the blood supply of the organ according to its activity. They cannot mediate light-specific information, but only sympathetic or parasympathetic effects. Granulated axon profiles demonstrated in the pineal nervous tissue proper and thought to be autonomic nerves obviously represent central pinealopetal fibers, e.g. substance-P immunoreactive fibers originate from perikarya of the medial habenular nucleus, or somatostatin immunoreactive fibers presumably arrive via the stria medullaris. Further, since fibers from the retina related nuclei of the brain stem reach the pineal through its stalk, they offer a way that is neuroanatomically more evident for retino-pineal cooperation than the hitherto supposed pathway via sympathetic nerves (Finely *et al.,* 1981; Korf and Moller, 1984; Shiotani *et al.,* 1986; Moller *et al.,* 1990, 1993; Moller, 1992; Vigh and Vigh-Teichmann, 1993; Matsushima *et al.*, 1994).

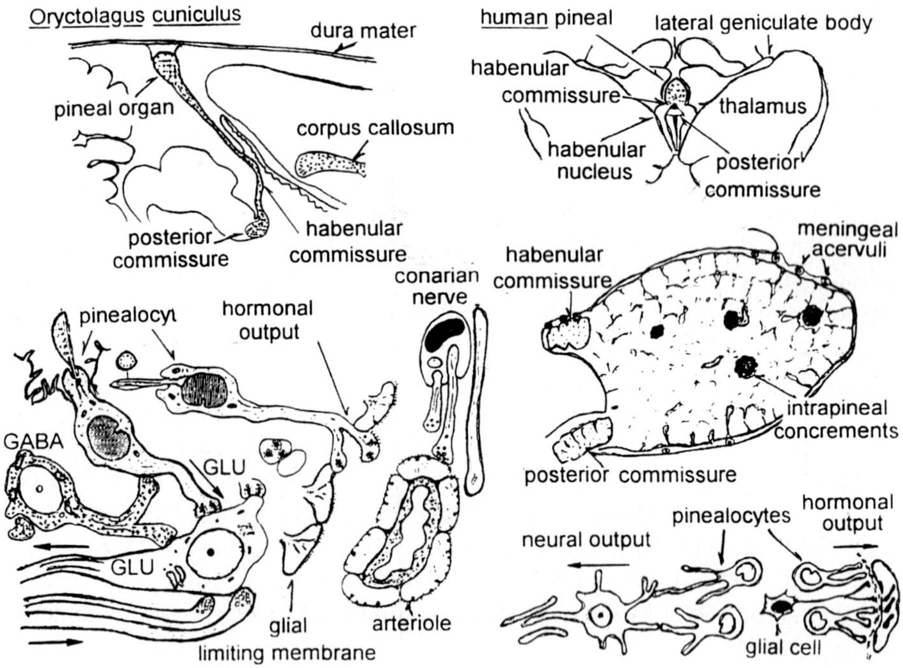

Fig. 6 The pineal organ of mammals and the human pineal organ.

Among mammals studied, the bats develop special *neurohemal release sites* composed of processes of pinealocytes and facing bilaterally the internal cerebral veins and/or their smaller intrapineal branches. These neurohormonal terminals may represent the morphological basis of pineal hormone release. Similar but diffusely arranged secretory terminals were described in the rat and rabbit. These terminals bind serotonin antibodies as demonstrated by immunogold electron microscopy. In carnivorous mammals and rodents, extended neurohemal areas have not yet been found (Vigh *et al.*, 1986b; Vigh and Vigh-Teichmann, 1988, 1986).

Finally, the presence of *nerve cells* in the mammalian pineal has to be mentioned. They are suggested to represent secondary pineal nerve cells like those in the submammalian pineal organ. Most of them are localized in the proximal part of the pineal near the habenular and posterior commissures. Synaptic ribbon-containing processes of pinealocytes as well as common axons form synapses on the dendrites of pineal neurons. Pinealocytic terminals are glutamate- and aspartate immunoreactive. Some of the neuronal perikarya show GABA and/or somatostatin, substance P immunoreactivity, while others contain excitatory amino acids, as shown by immunocytochemical studies. Axons of pineal neurons constitute a pineal tract running to the habenular nuclei and other brain stem nuclei. These morphological data show that besides of a hormonal output, there is a neural efferentation in the mammalian pineal organ which seems to use glutamate and aspartate as synaptic mediators. Summarizing the data available at present, the organization of the mammalian pineal does not differ essentially from that of submammalian pineal organs (Levin, 1938; Vigh and Vigh-Teichmann, 1975, 1988, 1992; Ichimura *et al.*, 1986; Lew and Lawson-Willey, 1987; Bhaskar and Joy, 1989; Vigh-Teichmann *et al.*, 1991a; Ichimura, 1992; Krstic and Nicolas, 1992; McNulty *et al.*, 1992; Matsushima *et al.*, 1994; Vigh *et al.*, 1995b, c; Redecker and Veh, 1994).

8. The Human Pineal Organ

Due to special aspects, data on the human pineal are incomplete in many respects. As shown by light microscopic techniques, the pineal contains neurons and glial cells besides of the sensory cell-derived pinealocytes. These elements form lobular structures separated by meningeal septae. Pinealocytes emit several processes ending in a bulb-like terminal on the vascular surface of the pineal nervous tissue. These terminals may represent the hormonal efferentation of the organ. In the proximal part of the pineal, behind the habenular and posterior commissures, several neurons were demonstrated by silver impregnation methods. The neurons are mainly multipolar and their axons form small bundles running to the medial habenular nucleus, a pattern indicating the existence of a neural efferentation. Recoverin, a retinal calcium-binding protein, rhodopsin and S-antigen were demonstrated immunocytochemically in human pinealocytes. Immunoreactive glutamate was found in intrapineal processes. In pineal parenchymal tumours, the expression of several photoreceptor-, glial- and neuronal proteins was described like that of IRBP (interphotoreceptor or retinol-binding protein), rod-opsin, cone opsin, S-antigen and cellular retinaldehyde-binding protein (Scharenberg and Liss, 1965; Parentes *et al.*, 1986; Vigh *et al.*, 1986a, 1995b; Huang *et al.*, 1992; Korf *et al.*, 1992; Lopes *et al.*, 1993).

The human pineal organ is rich in *calcified concretions* (acervuli). Two types of concrements were found: a meningeal and intrapineal one (Fig. 6). Using potassium pyroantimonate method, free calcium ions were demonstrated alongside the plasmalemma of the pinealocytes and pineal arachnoid cells. Calcium ATPase was found immunocytochemically along the plasmalemma of mammalian pinealocytes, an enzyme that uses the energy of ATP hydrolysis to pump calcium out of the cell. In the outer segments of submammalian pinealocytes, the calcium has a similar localization as known for the retina. In dark adapted rabbits and frogs, most of the calcium deposits were found to be localized inside of the retinal and pineal photoreceptor membrane, while outside in light adapted animals. Therefore, the high concentration of calcium in the pineal may be due to the receptor and effector membrane functions. The high number of tightly packed pinealocytes actively removing calcium, may be a reason for the frequent formation of corpora arenacea in mammals and human (Macpherson and Matheson, 1979; Krstic, 1986; Vigh *et al.,* 1989; Vigh and Vigh-Teichmann, 1989b, 1992; Vigh-Teichmann and Vigh, 1992).

General Conclusions

The comparison of morphophysiological data of the pineal complex in different vertebrate classes results in a conception being not fully identical with the generally accepted one about the structural organization and function of the pineal.

It seems likely, that it is the role of the pineal complex of vertebrates to measure light intensity in order to adapt functions of the organism to circadian and circannual changes of the environment. In several species, the pineal has an internal rhythm ("oscillation") and the environmental light only entrains this fluctuation. The entrainment starts in young animals or even before birth. The pineal organs influence the organism via neural (pineal tract) and hormonal (e.g. melatonin) efferentations. Hormonal signals serve the slow adaptation of the animal to the varying circadian/circannual radiation pattern of the sky. In contrast, the neural (synaptic) output may serve for a quick (locomotor) adaptation. The two types of efferentation are differently developed in various animals. The hormonal efferentation is already present in cyclostomes, therefore it does not appear as a result of the evolution of higher vertebrates, further, the hormonal output is not obligatorily connected to the regression of outer segments, as thought earlier.

In submammalians, the pineal organs have an organization and chemical composition similar to that of the retina. In addition to neurons and ependymal-glial cells, these organs contain rod- and cone-like pinealocytes according to their rhodopsin- and S-antigen immunoreactivity, though their photoreceptor membranes have a cone-like arrangement morphologically. The pinealocytes are polarized, the receptor pole bears a photoreceptor inner and outer segment protruding into the pineal lumen. These outer segments contain different cone- and cone-like pinealocytes according to their rhodopsin- and S-antigen immunoreactivity, though their photoreceptor membranes have a cone-like arrangement morphologically. The pinealocytes are polarized, the receptor pole bears a photoreceptor inner- and outer segment protruding into the pineal lumen. These outer segments contain different cone- and rod-type photopigments. Each type of photoreceptor expresses its own photopigment sensitive to light of a special range of the spectrum (UV, blue, green, red) enabling the animals to percept light of different wavelengths. A pineal-specific opsin ("pinopsin") is sensitive to blue light (470 nm). Electrophysiological experiments also suggest the presence of pineal photoreceptors being sensitive to ultraviolet light.

The efferent pole of pinealocytes is represented by a ramifying axonal process. The axons form two types of endings: neurohormonal terminals on the basal lamina of the pineal vascular spaces, or synaptic ribbon-containing axo-dendritic synapses on secondary pineal neurons. These synapses probably "use" excitatory amino acids as mediators. Some of the pineal neurons are small and GABA-ergic, and obviously represent inhibitory (inter) neurons while most of the neurons are large and contain glutamate and aspartate. The axons of glutamatergic neurons travel via the pineal stalk to the medial habenular nucleus and to different other brainstem nuclei. Pinealopetal fibers from the brain stem contain connections from retina related nuclei possibly modulating the light information percepted by the pineal.

Comparing recent species, we can state that there is no essential difference between the histologic organization of the pineal of mammals and the submammalians. Therefore, it seems not to be justified to attribute to the mammalian pineal organ a task essentially different from that of submammalians. It has to be emphasized that similarly to that of submammalians, the mammalian pineal organ is composed of neural tissue as a result of its development from the brain. It does not consist of glandular epithelium like most of the peripheral endocrine glands. Besides of pinealocytes being cytologically sensory cells, the mammalian pineal organ contains neurons, glial cells and synaptic complexes connected to the brain stem. Consequently, the expression "pineal gland" used generally in the literature is incomplete because of the integrating neural and neurohormonal nature of the mammalian pineal as a complex organ.

As the photoreceptor outer segment of mammalian pinealocytes is usually represented by a simple $9 \times 2 + 0$-type cilium, it is generally supposed that light information reaches the pineal from the retina via peripheral autonomic nerves. Visual pigments sit in the cell membrane and the outer segment is a multiplication of this membrane. Photoreception is possible without outer segments, as experiments in the retina demonstrate. Therefore, and because of the presence of rhodopsin-expressing pinealocytes, a direct light perception of the mammalian pineal cannot be excluded.

Peripheral nerves usually do not reenter the central nervous system but innervate its vessels being isolated from the brain tissue proper by a basal lamina and glial limiting membrane also present around the pineal tissue. Accordingly, in the autonomic nerves (nervus conarii) that reach the pineal, vasoconstrictor, vasodilator and vasosensory fibers have been identified. We suppose that they modify the blood supply of the pineal according to its functional state. Neuroanatomically, a retino-pineal connection would be more evident via a direct afferent projection from retina-related centres of the brain stem as already demonstrated by several authors.

References

Aige-Gil, V. and N. Murillo-Ferrol (1992). Histol. Histopathol. **7:** 1-6.

Alter, H. (1965). Progr. Brain Res. **10:** 154-171.

Araki, M. and S. Taketani (1992). Dev. Brain Res. **69:** 149-152.

Araki, M., Y. Fukada, Y. Shichida, T. Yoshizawa and F. Tokunaga (1992). Dev. Brain Res. **65:** 85-92.

Barajas-López, C., M.A. Barrientos-Martinez and C. Reyes-Vazquez (1987). J. Pineal Res. **4:** 287-294.

Bargmann, W. (1943). Die Epiphysis cerebri. In: Handbuch der mikroskopischen Anatomie des Menschen. (W. von Möllendorff, Ed.), Vol. VI/4. pp. 309-505. Springer, Berlin.

Baumann, C. H. (1962). Pflügers Arch. Gesamte Physiol. **276:** 56-65.

Bernard, M., J. Guerlotte, M. Congni, P. Greve, J.P. Collin and P. Voisin (1993). Biochem. J. **290:** 661-664.

Bhaskar, K.S. and K.P. Joy (1989). Biol. Struct. Morphog. **2:** 7-11.

Birks, E.K. and R.D. Ewing (1986). Gen. Comp. Endocrinol. **64:** 91-98.

Bischoff, M.B. (1969). J. Ultrastruct. Res. **28:** 16-26.

Bolliet, V., V. Begay, J.P. Ravault, M.A. Ali, J.P. Collin and J. Falcon (1994). J. Pineal Res. **16:** 77-84.

Bonaventure, N., B. Jardon, J. Sahel and N. Wioland, (1989). Doc. Ophthalmol. **72:** 71-82.

Cheung, R., E.M. Plisetskaya and J.H. Yuson (1990). Cell Tissue Res. **262:** 283-292.

Collin, J.P. (1976). Coll. Intern. CNRS **266:** 393-408.

Collin, J.P. and A.J. Kappers (1968). Brain Res. **11:** 85-106.

Collin, J.P., A. Calas and M.T. Juillard (1976). Exp. Brain. Res. **25:** 15-33.

Cole, W.C. and J.H. Yuson (1982). Am. J. Anat. **165:** 131-163.

Demaine, C. and P. Semm (1984). Brain Res. Bull. **13:** 629-634.

Dodt, E. (1963). Experientia **19:** 642.

Dodt, E. and E. Heerd (1962). J. Neurophysiol. **25:** 405-429.

Dodt, E. and Y. Morita (1964). Vision Res. **4:** 413-421.

Dodt, E. and H. Meissl (1982). Experientia **38:** 996-1000.

Eakin, R.M. (1982). Continuity and diversity in photoreceptors. In: Visual cells in evolution. (J.A. Westfall, Ed.), pp. 91-105. Raven Press, New York.

Eakin, R.M. and J.A. Westfall (1962). J. Ultrastruct. Res. **6:** 531-539.

Eakin, R.M., W.B. Quay and J.A. Westfall (1961). Z. Zellforsch. **53:** 449-470.

Ebadi, M. and P. Govitrapong (1986). J. Neural. Transm. **21:** 125-155.

Eldred, W.D. and J. Nolt (1981). J. Comp. Neurol. **203:** 269-295.

Eldred, W.D., T.H.E. Finger and J. Nolte (1984). Cell Tissue Res. **211:** 215-222.

Ekström, P. (1984). J. Comp. Neurol. **226:** 321-335.

Ekström, P. (1987). J. Neurosci. **7:** 987-995.

Ekström, P. and H.W. Korf (1985). Cell Tissu Res. **240:** 693-700.
Ekström, P. and H.W. Korf (1986a). Cell Tissue Res. **246:** 321-329.
Ekström, P. and H.W. Korf (1986b). Cell Tissue Res. **246:** 359-364.
Ekström, P., R.G. Foster, H.W. Korf, J.J. Schalken (1987). J. Comp. Neurol. **265:** 25-33.
Ekström, P., T. Honkanen, S.O. Ebbeson (1988). Brain Res. **460:** 68-75.
Ekström, P. and H. Meissl (1990). Vis. Neurosci. **4:** 389-397.
Ekström, P., T. Ostholm, H. Meissl, A. Bruun, J.G. Richards and H. Mohler (1990). Vis. Neurosci. **4:** 399-412.
Falcon, J. and H. Meissl (1981). J. Comp. Physiol. **144:** 127-137.
Falcon, J., P. Voisin, J. Gurerlotté and J.P. Collin (1986). Ann. Endocrinol. Paris **47:** 65-66.
Finley, J.C.V., J.L. Maderdrutt, L.J. Roger and P. Petrusz (1981). Neurosci. **6:** 2173-2192.
Foa, A. (1991). J. Comp. Physiol. **169:** 201-207.
Foster, R.G., H.W. Korf and J.J. Schalken (1987). Cell Tissue Res. **248:** 161-167.
Foster, R.G., S. Argamaso, S. Coleman, C.S. Colwell, A. Lederman and I. Provencio (1993). J. Biol. Rhythms. **8** Suppl: 17-23.
Ghosh, M., V. Csernus and B. Mess (1994). Neuroendocr. Lett. **16:** 195-200.
Goto, K., N. Miki and H. Kondo (1989). Arch. Histol. Cytol. **52** Suppl: 451-458.
Guerlotte, J., J. Falcon, P. Voisin and J.P. Collin (1986). Ann. Endocrinol. **47:** 62-65.
Guerlotte, J., P. Voisin, P. Brisson, J.P. Faure and J.P. Collin (1988). Biol. Cell. **64:** 93-96.
Gundy, G.C. and G.Z. Wurst (1976). Anat. Rec. **185:** 419-432.
Hafez, M.A., H.W. Korf and A. Oksche (1987). Cell Tissue Res. **250:** 571-578.
Haldar, C. and P. Pévet (1983). J. Neural Transm. **57:** 111-120.
Haldar D., D. Haussler and D. Gupta. (1992). J. Pineal Res. **12:** 79-83.
Hamasaki, D.I. (1970). Vision Res. **10:** 307-316.
Hamasaki, D.I. and E. Dodt (1969). Pflügers Arch. Gesamte Physiol. **313:** 9-29.
Hamm, H.E., J.S. Takahasi and M. Menaker (1983). Brain Res. **266:** 287-294.
Hanyu, L., H. Niwa and T. Tamura (1978). Comp. Biochem. Physiol. **61 A:** 49-54.
Hargrave, P.A. and J.H. McDowell (1992). FASEB J. **6:** 2323-2331.
Hartwig, H.G. (1984). Ophthalmic Res. **16:** 102-106.
Hartwig, H.G. and M. Pfautsch (1973). Z. Zellforsch. **138:** 558-589.
Hartwig, H.G. and C. Baumann (1974). Vision Res. **14:** 597-598.
Herbuté, S. and J.D. Baylé (1974). Neuroendocrinology **16:** 52-64.
Huang, S.H., D.C. Klein and H.W. Korf (1992). Cell Tissue Res. **267:** 493-498.
Ichimura, T. (1992). Microsc. Res. Techn. **21:** 124-135.
Ichimura, T., T. Arikuni and P.H. Hashimoto (1986). Cell Tissue Res. **224:** 569-576.
Innocenti, A., L. Minutini and A. Foa (1993). Physiol. Behav. **53:** 911-915.
Jenison, G. and J. Nolte (1980). Brain Res. **194:** 506-510.
Joss, J.M.P. (1973). Gen. Comp. Endocrinol. **21:** 188-195.
Joss, J.M.P. (1977). Gen. Comp. Endocrinol. **31:** 270-275.
Joy, K.P. and A.K. Agha (1993). J. Hirnforsch. **34:** 545-553.
Kalsow, C.M., S.S. Greenhouse, W. Gern, G. Adamus, P.A. Hargrave, L.S. Lang and L.A. Donoso (1991). J. Pineal Res. **11:** 49-546.
Kappers, J.A. (1968). Excerpta Med. Congr. Ser. **185:** 619-626.
Kappers, J.A. (1971). The pineal organ: An introduction. In: The pineal gland (G. E. W. Wolstenholme and J.T. Knight, Eds.). Churchill Livingstone, London-Edinburgh, pp. 3-35.
Karasek, M. and R.J. Reiter (1992). Mikrosc. Res. Techn. **21:** 136-157.

Kasal, C.A. and J.R. Perez-Polo (1980). J. Neurosci. Res. **5:** 579-585.
Kemnitz, C.P., L.M. Fox and J.A. McNulty (1990). J. Morphol. **204:** 197-208.
Klein, D.C. (1985). Photoneural regulation of the mammalian pineal gland. In: Photoperiodism, melatonin and the pineal (D. Evered and S. Clark, Eds.), pp. 38-56. Pitman Publishing Ltd., London.
Klein, D.C. (1993). The mammalian melatonin rhythm generating system. In: Light and biological rhythms in man (L. Wetterberg, Ed.), pp. 55-71. Pergamon Press, Oxford.
Korf, B., M.D. Rollag and H.W. Korf (1989). Cell Tissue Res. **258:** 319-329.
Korf, H.W. (1994). Ann. N.Y. Acad. Sci. **719:** 13-42.
Korf, H.W. and M. Moller (1984). Pineal Res. Rev. **2:** 41-86.
Korf, H.W. and I. Vigh-Teichmann (1984). Ophthalmic Res. **16:** 96-101.
Korf, H.W., R. Liesner, H. Meissl and A. Kirk (1981). Cell Tissue Res. **216:** 113-130.
Korf, H.W., R.G. Foster, P. Ekström and J.J. Schalken (1985). Cell Tissue Res. **242:** 645-648.
Korf, H.W., B.H. White, N.C. Schaad and D.C. Klein (1992). Brain Res. **595:** 57-66.
Kramm, C.M., W.J. De Grip and H.W. Korf (1993). Cell Tissue Res. **274:** 71-78.
Krstic, R. (1986). J. Neural Transm. **21:** 415-432.
Krstic, R. and D. Nicolas (1992). Acta Histochem. **93:** 382-387.
Kuo, C-H., S. Tamotsu, Y. Morita, T. Shinozawa, M. Akiyama and N. Kiki (1988). Brain Res. **442:** 147-151.
Kusmic, C. and P.L. Marchiafava (1990). Neurosci. Lett. **39:** 122.
Levin, P.M. (1938). J. Comp. Neurol. **68:** 405-409.
Lew, G.M. and A. Lawson-Willey (1987). Histochemistry **86:** 591-593.
Lolley, R.N., C.M. Craft and R.H. Lee (1992). Neurochem Res. **17:** 81-89.
Lopes, M.B., F. Gonzalez-Fernandez, B.W. Scheithauer and S.R. VandenBerg (1993). J. Neuropathol. Exp. Neurol. **52:** 516-524.
Luo, Z.R., R.L. Schultz, E.F. Whitter and L. Vollrath (1984). J. Pineal Res. **1:** 323-337.
Marchiafava, P.L. and C. Kusmic (1993). Progr. Brain Res. **95:** 3-13.
Maitra, S.K. and L. Vollrath (1991). J. Pineal Res. **11:** 140-144.
Masuda, H., T. Oishi, M. Ohtani, M. Michinomae, Y. Fukada, Y. Shichida and T. Yoshizawa (1994). Tissue Cell. **26:** 101-113.
Matsushima, S., Y. Sakai, Y. Hira, Y. Oomori and S. Daikoku (1994). Arch. Histol. Cytol. **57:** 47-58.
Matsuura, S. and H.J. Herwig (1981). Cell Tissue Res. **216:** 545-557.
Max, M. and M. Menaker (1992). J. Comp. Physiol. **170:** 479-489.
Max, M., P.J. McKinnon, K.J. Seidenman, R.K. Barrett, M.L. Applebuty, J.S. Takahashi and R.F. Margolskee (1995). Science **267:** 1502-1506.
McClung, R. and N. Dafny (1975). Life Sci. **16:** 621-628.
McNulty, J.A., M.A. Neighbors and M.H. Horn (1988). Experientia **44:** 740-742.
McNulty, J.A., L. Kus and O.P. Ottersen (1992). Cell Tissue Res. **269:** 515-523.
Meiniel, A. (1969). C.R. Acad Sci. (Paris) Ser **D268:** 2265-2268.
Meiniel, A. (1971). J. Neurovisc. Rel. **32:** 157-199.
Meiniel, A. and R. Meiniel (1985). Exp. Biol. **44:** 191-198.
Meissl, H. and M. Ueck (1980). J. Comp. Physiol. **140:** 173-179.
Meissl, H. and E. Dodt (1981). Devel. Endocrinol. **14:** 61-80.
Meissl, H. and S.R. George (1984). Ophthalmic Res. **16:** 114-118.
Meissl, H. and P. Ekström (1991). Brain Res. **562:** 71-78.
Meissl, H. and R. Brandstätter (1992). Photoreceptive function of the teleost pineal organ and its implication in biological rhythms in fishes (M.A. Ali, Ed.), pp. 235-254. Plenum Press, New York.
Macpherson, P. and M.S. Matheson (1979). Neuroradiology **18:** 67-72.

Meyer-Rochow, V.B. and D. Stewart (1992). Microsc. Electron. Biol. Celular. **16:** 69-85.

Mess, B. and C. Ruzsas (1986). Influence of pineal hormones on sexual differentiation of the rat. In: The pineal gland during development: from fetus to adult (D. Gupta and R.J. Reiter, Eds.), pp. 174-183. Croom Helm, London, Sydney.

Mikkelsen, J.D. (1992). Neurosci. **48:** 953-962.

Miller, W.H. and M.L. Wolbarsht (1962). Science **135:** 316-317.

Moller, M. (1992). Microsc. Res. Techn. **21:** 188-204.

Moller, M. and H.W. Korf (1987). Cell Tissue Res. **247:** 145-153.

Moller, M., J.D. Mikkelsen and L. Martinet (1990). Cell Tissue Res. **261:** 477-483.

Moller, M., P. Phansuwan-Pujito, P. Govitrapong and P. Smidt (1993). Brain Res. **611:** 347-351.

Moller, W. and G. Moller (1990). Cell Tissue Res. **260:** 337-348.

Morita, Y. (1965). Pflügers Arch. Gesamte Physiol. **286:** 97-108.

Morita, Y. (1966). Pflügers Arch. Gesamte Physiol. **289:** 155-167.

Morita, Y. (1969). Experientia (Basel) **25:** 1277.

Morita, Y. and G. Bergmann (1971). Z. Zellforsch. **119:** 289-294.

Morita, Y. and E. Dodt (1973). Nova Acta Leopoldina **38:** 331-339.

Morita, Y., M. Tabata and S. Tamotsu (1985). Neurosci. Res. **2** Suppl: S79-S88.

Morita, Y., M. Samejima and S. Tamotsu (1989). Arch. Histol. Cytol. **52** Suppl: 469-475.

Morita, Y., K. Uchida, S. Tamotsu and M. Samejima (1991a). Adv. Pineal Res. **5:** 97-99.

Morita, Y., K. Uchida, S. Tamotsu, M. Samejima and T. Nakamura (1991b). Adv. Pineal Res. **6:** 37-46.

Morita, Y., M. Tabata, K. Uchida and M. Samejima (1992). J. Comp. Physiol. **171:** 555-562.

Murakami, N., H. Nakamura, R. Nishi, N. Marumoto and T. Nasu (1994). Brain Res. **651:** 209-214.

Nakamura, T., G. Thiele and H. Meissl (1986). J. Comp. Physiol. **159:** 325-331.

Okano, T., T. Yoshizawa and Y. Fukuda (1994). Nature **372:** 94-97.

Oksche, A. (1965). Progr. Brain Res. **10:** 3-29.

Oksche, A. and H. Kirschstein (1968). Z. Zellforsch. **87:** 159-192.

Oksche, A. and H. Kirschstein (1969). Z. Zellforsch. **102:** 214-241.

Oksche, A., H. Kirschstein, H. Kobayashi and D.S. Farner (1972). Z. Zellforsch. **124:** 247-274.

Ohshima, K. and S. Matsuo (1991). Anat. Anz. **173:** 65-72.

Ohshima, K. and K. Hiramatsu (1993). J. Vet. Med. Sci. **55:** 945-950.

Omura, Y. (1977). Cell Tissue Res. **183:** 255-271.

Omura, Y. (1984). Cell Tissue Res. **236:** 611-617.

Omura, Y. and M. Ali (1980). Cell Tissue Res. **208:** 111-122.

Pang, S.F., L. Liu, P.P.N. Lee, A.E. Allen and N.Y.S. Woo (1989). Adv. Pineal Res. **3:** 67-76.

Parentes, E., L.J. Rubinstein, M.M. Herman and L.A. Donoso (1986). Acta Neuropathol. Berl. **71:** 224-227.

Pazo, J.H. and M. Gonzalez (1991). Neurosci. **43:** 231-235.

Petit, A. (1971). Z. Zellforsch. **120:** 246-260.

Pévet, P., J.A. Kappers and A.M. Voute (1977). Cell Tissue Res. **192:** 93-109.

Pévet, P. (1988). Reprod. Nutr. Dev. **28:** 443-458.

Phillips, J.A. and K.A. Howes (1988). Physiol. Behav. **42:** 103-108.

Pu, G.A. and J.E. Dowling (1981). J. Neurophysiol. **46:** 1018-1038.

Puzdrowski, R.L. and R.G. Northcutt (1989). Cell Tissue Res. **255:** 269-274.
Quay, W.B. (1989). Adv. Pineal Res. **3:** 55-60.
Quay, W.B., J.A. Kappers and J.F. Jongkind (1968). J. Neurovisc. Rel. **31:** 11-25.
Redecker, P. and R.W. Veh (1994). Cell Tissue Res. **278:** 579-588.
Reiter, R.J. (1981). Am. J. Anat. **162:** 287-313.
Reiter, R.J. (1988). Neuroendocrinology of melatonin. In: Melatonin-clinical perspectives (A. Miles, D.R.S. Philbrick and C. Thompson, Eds.), pp. 1-42. Oxford University Press, Oxford.
Reiter, R.J. (1993). The mammalian pineal gland as an end organ of the visual system. In: Light and biological rhythms in man (L. Wetterberg, Ed.), pp. 145-160.
Remane, A., V. Storch and U. Welsch (1980). Systematsche Zoologie. Stämme des Tierreiches. pp. 1-682. Fischer, Stuttgart.
Rudeberg, C. (1969). Z. Zellforsch. **96:** 548-581.
Ruiz, S. and R. Anadon (1991). Cell Tissue Res. **263:** 597-600.
Sato, T. and S. Ebisawa (1988). Cell Tissue Res. **252:** 287-292.
Scharengerg, K. and L. Liss (1965). Progr. Brain Res. **10:** 193-217.
Semm, P. and C. Demaine (1983). J. Neural Transm. **58:** 281-289.
Semm, P. and L. Vollrath (1978). Pflügers Arch. Ges. Physiol. (Suppl) **373:** R55.
Semm, P. and L. Vollrath (1980). J. Neural Transm. **47:** 181-190.
Shi, H.L., H.C. Furr and J.A. Olson (1991). Brain Res. Bull. **26:** 235-239.
Shiotani, Y., M. Yamano, S. Shiosaka, P.C. Emson, C.J. Hillyard, S. Girgis and I. MacIntyre (1986). Neurosci. Lett. **70:** 187-192.
Sun, J.H., R.J. Reiter, A. Hattori, K. Yaga, D.C. Herbert and A.T. Tsin (1993). J. Pineal Res. **15:** 132-137.
Takahama, H. (1993). Cell Tissue Res. **272:** 575-578.
Tamotsu, S. and Y. Morita (1986). J. Comp. Physiol. **(A) 159:** 1-5.
Tamotsu, S. and Y. Morita (1990). Comp. Biochem. Physiol. **96 (B):** 487-490.
Tamotsu, S., H.W. Korf, Y. Morita and A. Oksche (1990). Cell Tissue Res. **262:** 205-216.
Tamotsu, S., M. Samejima and Y. Morita (1994a). Adv. Pineal Res. **8:** 25-29.
Tamotsu, S., T. Oishi, K. Nakao, Y. Fukada, Y. Shichida, T. Yoshizawa and Y. Morita (1994b). Cell Tissue Res. **178:** 1-10.
Taylor, A.N. and R.W. Wilson (1970). Experiencia **26:** 267-269.
Thiele, G. and H. Meissl (1985). Pflügers Arch. Suppl. **403:** R. 64.
Torres, G. and L.D. Lytle (1990). J. Neural Tansm. **80:** 67-77.
Trost, E. (1954). Anat. Anz. **100:** 368-374.
Tsin, A.T.C., T.S. Phillips and R.J. Reiter (1989). Adv. Pineal Res. **3:** 147-150.
Uchida, K. and Y. Morita (1990). Brain Res. **534:** 237-242.
Uchida, K., T. Nakamura and Y. Morita (1992). Neurosci. **47:** 241-247.
Ueck, M. (1973). Z. Zellforsch. **137:** 37-62.
Ueck, M. (1981). Devel. Endocrinol. **14:** 151-168.
Ueck, M. and H. Kobayashi (1972). Z. Zellforsch. **129:** 140-160.
Ueck, M. and H. Kobayashi (1979). Verh. Anat. Ges. **73:** 961-963.
Ueck, M., K. Wake and H. Kobayashi (1989a). Zool. Sci. **6:** 817-831.
Ueck, M., T. Sato, S. Ohba, K. Wake and H. Kobayashi (1989b). Arch. Histol. Cytol. Suppl. **52:** 459-467.
Van Veen, T. (1982). Cell Tisse Res. **222:** 433-444.
Vigh, B. and I. Vigh-Teichmann (1974). Anat. Anz. **65:** 433-443.
Vigh, B. and I. Vigh-Teichmann (1975). Anat. Anz. **69:** 453-461.
Vigh, B. and I. Vigh-Teichmann (1981). Cell Tissue Res. **221:** 451-463.
Vigh, B. and I. Vigh-Teichmann (1982). The cerebrospinal fluid-contacting neurosecretory

cell: A protoneuron. In: Molecules, cells, systems (D.S. Farner and K. Lederis, Eds.), pp. 458-460. Plenum Press, New York.
Vigh, B. and I. Vigh-Teichmann (1986). Arch. Histol. Jap. **49:** 495-518.
Vigh, B. and I. Vigh-Teichmann (1988). Pineal Res. Rev. **6:** 1-65.
Vigh, B. and I. Vigh-Teichmann (1989a). Anat. Anz. Suppl. **164:** 915-916.
Vigh, B. and I. Vigh-Teichmann (1989b). Arch. Histol. Cytol. Suppl. **52:** 433-440.
Vigh, B. and I. Vigh-Teichmann (1992). Progr. Brain Res. **91:** 299-306.
Vigh, B. and I. Vigh-Teichmann (1993). Arch. Histol. Cytol. Suppl. **56:** 485-493.
Vigh, B. and I. Vigh-Teichmann (1998). Microsc. Res. Techn. **41:** 1-27.
Vigh, B., I. Vigh-Teichmann and B. Aros (1975). Cell Tissue Res. **158:** 409-424.
Vigh, B., I. Vigh-Teichmann and B. Aros (1989). Z. Mikrosk. Anat. Forsch. **103:** 36-45.
Vigh, B., I. Vigh-Teichmann, P. Rohlich and B. Aros (1982). Z. Mikrosk. Anat. Forsch. **96:** 113-129.
Vigh, B., I. Vigh-Teichmann, B. Aros and A. Oksche (1985). Cell Tissue Res. **240:** 143-148.
Vigh, B., I. Vigh-Teichmann, I. Reinhard, A. Szél and T. Van Veen (1986a). Opsin immunoreaction in the developing and adult pineal organ. In: The pineal gland during devlopment: from fetus to adult (D. Gupta and R.J. Reiter, Eds.), pp. 31-42.
Vigh, B., I. Vigh-Teichmann and B. Aros (1986b). Z. Mikrosk. Anat. Forsch. **100:** 745-758.
Vigh, B., I. Vigh-Teichmann, T. Heinzeller and I. Tutter (1989). Histochemistry **91:** 161-168.
Vigh, B., I. Vigh-Teichmann, K. Debreceni and J. Takács (1995a). Arch. Histol. Cytol. **58:** 37-44.
Vigh, B., Zs. Fejer and M.J. Manzano e Silva (1995b). Clin. Neurosci. Suppl. **48:** 26-27.
Vigh, B., K. Debreceni and M.J. Manzano e Silva (1995c). Acta Biol. Hung. **46:** 99-106.
Vigh-Teichmann, I. and B. Vigh (1979). Acta Biol. Hung. **30:** 1-39.
Vigh-Teichmann, I. and B. Vigh (1983). Arch. Histol. Japon. **46:** 427-468.
Vigh-Teichmann, I. and B. Vigh (1986). Cerebrospinal fluid (CSF) contacting dendrite terminals ("Landolt's clubs") in the pineal organ of Chimaera monstrosa. In: Functional morphology of neuroendocrine systems. Evolutionary and environmental aspects (B. Scharrer, H.W. Korf and H.G. Gartwig, Eds.), pp. 160. Springer, Heidelberg.
Vigh-Teichmann, I. and B. Vigh (1990). J. Pineal Res. **8:** 323-333.
Vigh-Teichmann, I. and B. Vigh (1992). Microsc. Res. Techn. **21:** 227-241.
Vigh-Teichmann, I. and B. Vigh (1994). Postembedding light and electron microscopic immunocytochemistry in pineal photoneuroendocrinology. In: Modern methods in analytical morphology (J. Gu and G.W. Hacker, Eds.), pp. 253-270. Plenum Press, New York.
Vigh-Teichmann, I., B. Vigh and B. Aros (1973). Z. Zellforsch. **144:** 139-152.
Vigh-Teichmann, I., W.J. DeGrip and B. Vigh (1993). Microscopia Elettronica **14** Suppl: 387-388.
Vigh-Teichmann, I., P. Röhlich, B. Vigh and B. Aros (1980). Z. Mikrosk. Anat. Forsch. **94:** 623-640.
Vigh-Teichmann, I., H.W. Korf, A. Oksche and B. Vigh (1982). Cell Tissue Res. **227:** 351-369.
Vigh-Teichmann, I., B. Vigh, M.J. Manzano e Silva and B. Aros (1983a). Cell Tissue Res. **228:** 139-148.
Vigh-Teichmann, I., H.W. Korf, F. Nurnberger, A. Oksche, B. Vigh and R. Olsson (1983b). Cell Tissue Res. **230:** 289-307.

Vigh-Teichmann, I., B. Vigh, I. Gery and Th. van Veen (1986). Exp. Biol. **45:** 27-43.
Vigh-Teichmann, I., B. Vigh and G.H. Wirtz (1987). Vitamin A immunocytochemistry of the retina and pineal complex in various vertebrates. In: Fundamentals and clinics in pineal research (G.-P. Trentini, C. de Gaetani and P. Pévet, Eds.), pp. 61-64.
Vigh-Teichmann, I., B. Vigh, A. Szél, P. Röhlich and G.H. Wirtz (1988). Histochemistry **88:** 533-543.
Vigh-Teichmann, I., B. Vigh and G.H. Wirtz (1989). Exp. Biol. **48:** 203-213.
Vigh-Teichmann, I., A. Szél, P. Röhlich and B. Vigh (1990). Exp. Biol. **48:** 361-371.
Vigh-Teichmann, I., H. Petter and B. Vigh (1991a). J. Pineal Res. **10:** 18-29.
Vigh-Teichmann, I., M.A. Ali, A. Szél and B. Vigh (1991b). J. Pineal Res. **10:** 196-209.
Vigh-Teichmann, I., M.A. Ali and B. Vigh (1992). Progr. Brain Res. **91:** 307-313.
Vitaterna, M.H., D.P. King, A.M. Chang, J.M. Kornhauser, P.L. Lowrey, J. FD. McDonald, W.F. Dove, L.H. Pinto, F.W. Turek and J.S. Takahashi (1994). Science **264:** 719-725.
Vivien, J.H. (1964). C.R. Acad. Sci. (Paris) **258:** 3370-3372.
Vivien-Roels, B. (1969). Z. Zellforsch. **94:** 352-390.
Vivien-Roels, B., P. Pévet and B. Claustrat (1988). Gen. Comp. Endocrinol. **69:** 163-173.
Voisin, P., M.T. Juillard and J.P. Collin (1983). Cell Tissue Res. **230:** 155-169.
Vollrath, L. (1981). The pineal organ. In: Handbuch der mikroskopischen Anatomie des Menschen, VI/7 (A. Oksche and L. Vollrath, Eds.), pp. 1-665. Springer, Berlin-Heidelberg-New York.
Wake, K., M. Ueck and A. Oksche (1974). Cell Tissue Res. **154:** 423-442.
Wartenberg, H. and H.G. Baumgarten (1969). Z. Zellforsch. **94:** 252-260.
Watanabe, T. and M. Yoshida (1986). Exp. Biol. **46:** 67-74.
Welsch, U. (1968). Z. Zellforsch. **86:** 252-261.
Wiechmann, A.F. and C.R. Wirsing-Wiechmann (1992). Brain. Res. **593:** 281-286.
Wiechmann, A.F. and C.M. Craft (1993). Neurosci. Lett. **150:** 207-211.
Wurtman, R.J., J. Axelrod, and J.E. Fischer (1964). Science **143:** 1328-1330.
Zatz, H. (1994). J. Neurochem. **62:** 2001-2011.
Zeman, M., E. Gwinner and E. Somogyiova (1992). Experientia **48:** 765-768.
Yanez, J., R. Anadon, B.I. Holmqvist and P. Ekström (1993). Neurosci. Lett. **164:** 213-216.

Comparative Endocrinology and Reproduction
K.P. Joy, A. Krishna and C. Haldar (Eds)

29. Pineal Gland and Biological Rhythms: Physiological Effects of Melatonin

P. Pevet, B. Pitrosky, V. Simonneaux, B. Vivien-Roels and M. Masson-Pevet
URA-CNRS 1332 "Neurobiologie des fonctions rythmiques et saisonnieres"
Universite L. Pasteur, Strasbourg, France

Introduction

Life on the earth is and always has been exposed to strong, rhythmic environmental changes caused by planetary movements. To ensure that certain events occur optimally just once each year (e.g. seasonal reproduction, hibernation, fur colour change, migration, etc.) or once a day (e.g. sleep) animals have to adapt themselves to these variations in the environment.

Several environmental variables such as daylength, temperature, rainfall, food availability, etc. which animals rely on to cue changes in their annual functions are presently known. However, depending on the biotope, some are more accurate indicators of the forthcoming season than others. In temperate and arctic zones, for example, drastic climatic conditions impose a very precise synchronisation for the survival of the species. In these regions for seasonal reproduction, animals use the most regularly changing phenomena, i.e. the change in daylength (photoperiod). Other environmental factors should be also considered. In small-sized herbivores plantfood has been considered to be an important "indicator" of the coming season (Bronson and Heideman, 1994) while in amphibians and reptiles thermoperiodicity is generally dominant to photoperiodicity (Lofts, 1974). In the tropical and equatorial zones of the world external cues such as rainfall are equally precise and used for timing of seasonal functions, at least in some species (Delany and Happold, 1979; Sicard *et al.*, 1993, 1994, 1995). Moreover animals living in their natural habitat are never subjected to a single factor at a time: there is always a covariation of all environmental factors. In temperate zones the decrease in photoperiod in autumn is, for example, always associated with a decrease in mean temperature and food availability. Interactions exist between the effects of all these factors (Pevet *et al.*, 1986a, b; Pevet, 1987).

We have little knowledge of the ways by which this complex of environmental stimuli once received by the exteroreceptor organs are perceived, differentiated and ultimately integrated by the central nervous system. In recent years, however, it has become clear that the functional mechanism used by mammals for this purpose is build around two structures, the suprachiasmatic nuclei of the hypothalamus (SCN) and the pineal gland whose principal hormone, melatonin (Mel) via its rhythmic pattern of secretion, seems responsible for transmitting environmental information to the central nervous system.

The SCN plays an essential role in the generation and maintenance of circadian rhythms in mammals (Rusak, 1989). The intrinsic pacemaker activity of the SCN is autonomous in nature and entrained to the daily 24 h period by environmental factors of which the light-dark cycle is the most effective. This photic information perceived by the retina is relayed directly to the SCN by the retinohypothalamic tract (RHT) and indirectly through the intergeniculate leaflet (IGL). From the SCN the rhythmic information is conveyed to the pineal through a polysynaptic pathway which includes the paraventricular nucleus of the hypothalamus, the preganglionic neurons in the intermediolateral column of the lateral funiculus of the upper thoracic cord and the superior cervical ganglia which provides a noradrenergic projection to the pineal. The SCN thus directly generates the rhythm in Mel synthesis/secretion. The synthesis induced by noradrenaline release from the nerve endings occurs during the dark phase of the light/dark cycle and appears to be tightly controlled by environmental factors which modify the temporal pattern of Mel synthesis/secretion.

The present chapter aims to evaluate the characteristics of the nocturnal Mel secretion pattern and its physiological effects and biological activities.

Changes in the Duration of Nocturnal Melatonin Peak and the Photoperiodic Response

In most of the mammalian species studied to date, the duration of the nocturnal peak of Mel secretion is positively correlated (up to a point) with the length of the dark period. These observations led Hoffmann, in 1981, to suggest that it is the duration of the nocturnal peak of Mel secretion which is the critical parameter in the transmission of photoperiodic information (now referred to as the duration hypothesis). Administration of Mel to pinealectomized animals by methods designed to produce a daily Mel peak corresponding to either a long or a short photoperiod have validated Hoffmann's concept. Carter and Goldman (1983), for example, demonstrated that 12 h Mel infusion, like a short photoperiod, inhibited testicular development in juvenile Siberian hamster whereas 4 or 6 h infusions failed to exhibit any inhibitory effect. This experimental approach has also been used in studies of different photoperiodic species (Siberian and Syrian hamsters, white-footed mice, mink, sheep) and in all situations, infusion of Mel induced the species specific gonadal response characteristic of either long (LP) or short (SP) photoperiod, depending upon the duration of the infusion (Bittman and Karsch, 1984; Dowell and Lynch, 1987; Bartness and Goldman, 1988; Karch *et al.*, 1988, Maywood *et al.*, 1990; Pitrosky *et al.*, 1991; Bonnefond *et al.*, 1990).

All these data have clearly established that Mel signal duration is an important parameter for eliciting seasonally appropriate reproductive responses but it is not necessarily the only critical feature of the Mel signal.

The pattern of Mel secretion in Syrian hamster kept under LP and SP, respectively, is shown in Fig. 1. The presence of a light-entrainable circadian phase of sensitivity to Mel starting approximately 9 to 10 h after lights off might also explain the results. Under SP, due to the increased duration of Mel secretion, the presence of Mel would coincide with this sensitive phase and so would trigger the appropriate response. This hypothesis of circadian-based variation in brain sensitivity to Mel, referred to as the "coincidence hypothesis", has been put forward primarily by Stetson and Watson-Whitmyre (1984), who with others (review Reiter, 1987) observed that in intact Syrian hamsters kept under LP, the response to exogenous melatonin was dependent upon the light/dark cycle. When injected late in the light phase, Mel was as effective, as SP, in inducing gonadal atrophy whereas Mel injections at

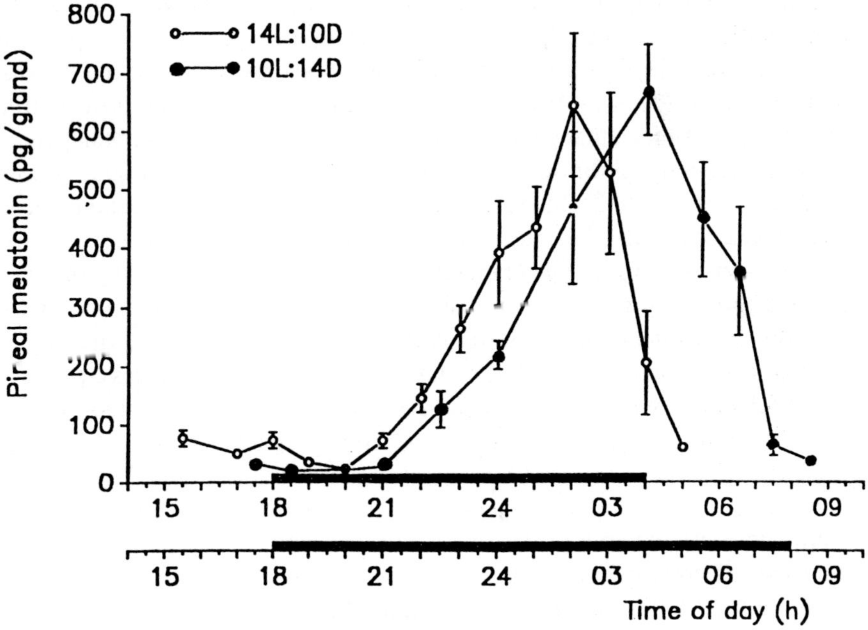

Fig. 1 **Mel content in pineal gland of adult male Syrian hamsters maintained in LP (14L: 10D) or in SP (10L: 14D) for 8 weeks (black bars indicate the period of darkness). Besides the changes in the duration of the pineal Mel content, it appears that the rise in Mel content always starts between 3 and 4 h after the onset of darkness. In SP, but not in LP, Mel concentration fell to daytime values approximately 30 min. before the end of dark period. O 14L: 10D; • 10L: 14D.**

other times (except for a brief interval before lights on) were ineffective (review Stetson and Watson-Whitmyre, 1984). The interpretation of these findings was that Mel administered in the afternoon advanced the circadian sensitive phase so that it was then coincident with the endogenous Mel peak (review Reiter, 1987).

Programmed daily infusions of Mel allow not only the duration of the Mel signal but also its phase relationships with the light/dark cycle and its frequency to be varied systematically. Using this protocol it has therefore been possible to test the relative importance of these different parameters as well as the "coincidence hypothesis". In all species studied the gonadal response to Mel infusion did not depend upon the time within the 24 h light/dark cycle at which the signal was encountered. The effect of both the long and short duration infusions were the same despite the time of day at which they were administered (Carter and Golman, 1983; Bartness and Goldman, 1988, Wayne *et al.*, 1988; Bonnefond *et al.*, 1990). The data suggested that the timing of the Mel infusion relative to the light/dark cycle was not essential in determining the effect of Mel on the reproductive system and argued against the coincidence hypothesis. This conclusion was reinforced by the observation that, when the Mel signals were delivered at periodicities of less than 24 h and, thus, during the course of the experiment scanned across the light/dark cycle, a SP-like response still occurred (Darrow and Goldman, 1986; Maywood *et al.*, 1990). Additional observations strengthened the idea that photoperiodic time measurement was not dependent upon a circadian rhythm of sensitivity to Mel in its target tissues (review of Bartness *et al.*, 1993). Interesting experiments which led to the coincidence hypothesis could also be interpreted using the duration hypothesis. Stetson and Watson-Whitmyre (1984) concluded that the gonadal atrophy observed in LP-exposed Syrian hamsters following late afternoon injections of Mel was the consequence of a Mel-induced advance of the sensitivity rhythm so that it was coincident with the endogenous Mel peak. We have observed that Mel injections given to Syrian hamsters in the late afternoon did not interfere with the endogenous pineal Mel surge (Pitrosky, 1994). The injections, however, did cause a rapid rise in circulating Mel concentrations. According to the duration hypothesis, the exogenous afternoon Mel surge summates with the endogenous Mel peak, prolongs the duration of elevated Mel and as consequence, inhibits the neuroendocrine system.

As noted by Bartness *et al.*, (1993), all these findings do not completely refute the possibility of a circadian rhythm of sensitivity of Mel. In numerous species Mel receptors have been described in the SCN, the circadian clock. It thus cannot be excluded that the Mel signal itself drives such a rhythm via its SCN receptors. However, lesioning of the SCN did not compromise the ability of the animals, at least the Syrian hamster, to exhibit an appropriate photoperiodic response to Mel infusion (Maywood *et al.*, 1990). These authors thus concluded that duration signal was not dependent upon the circadian system and that the duration is the critical parameter of the nocturnal Mel profile for the photoperiodic control of reproduction as well as for other seasonal adaptative responses.

The observation that Mel is not only involved in the photoperiodic response but is also a direct component of the circadian system requires that this concept be re-evaluated.

Melatonin Signal and Circadian System in the Induction of the Photoperiodic Response

The rhythm of Mel secretion is a circadian one which like many other circadian rhythms is directly generated by the SCN. It persists in the absence of light/dark cycles with a period slightly different from 24 h and is entrained to a 24 h period by the light/dark cycle. The Mel rhythm is an important efferent hormonal signal driven by the clock. The rhythmic secretion of Mel might thus be used as an internal circadian zeitgeber or better (in the terminology defined by Moore-Ede *et al.*, 1982) as a circadian mediator to any system than can "read" the message and so might be directly involved in the control of some circadian rhythms. In short, the SCN may use the daily Mel signal to impose circadian rhythmicity to Mel-target structures.

Clear data on the involvement of Mel in circadian rhythmicity have been obtained in non-mammalian vertebrates (for reviews see Cassone, 1990; Amstrong and Redman, 1993). In mammals experimental evidence is still fragmentary. Injection of Mel at the same time every day entrains the circadian locomotor activity of free-running male and female rats (Redman *et al.*, 1983). In contrast, Syrian hamsters are not entrained by daily injections of Mel (Hastings *et al.*, 1992). However, when administered by infusion in a way which mimics a plasma nocturnal peak, Mel is able to entrain the free running locomotor activity rhythm of pinealectomized male Siberian (Fig. 2) and Syrian hamsters. The effects corresponded to true entrainment since termination of the Mel infusion re-established the free running rhythm (Kirsch *et al.*, 1993).

All these findings strongly support the idea that in mammals, like in non-mammalian vertebrates, Mel is also involved in the control of the expression of circadian rhythmicity. The question of major importance concerning this effect of Mel is its precise site(s) and mechanism of action.

The SCN, the clock generating circadian rhythmicity in mammals, contains Mel receptors (references in Stankov and Fraschini, 1993; Masson-Pevet *et al.*, 1994). SCN metabolism (Cassone, 1990, 1991) as well as electrical activity (McArthur *et al.*, 1991) are known to be directly affected by exogenous Mel. Moreover daily injections of Mel entrain free-running locomotor activity in intact rats, but not in SCN-lesioned animals (Cassone *et al.*, 1986). Entrainment takes place when the onset of activity coincides with the time of injection, Mel inducing a phase advance (Redman *et al.*,

1983). Based on these observations, especially on the phase-advancing properties of exogenous Mel, most authors have concluded that Mel affects circadian rhythms by an action on the clock itself.

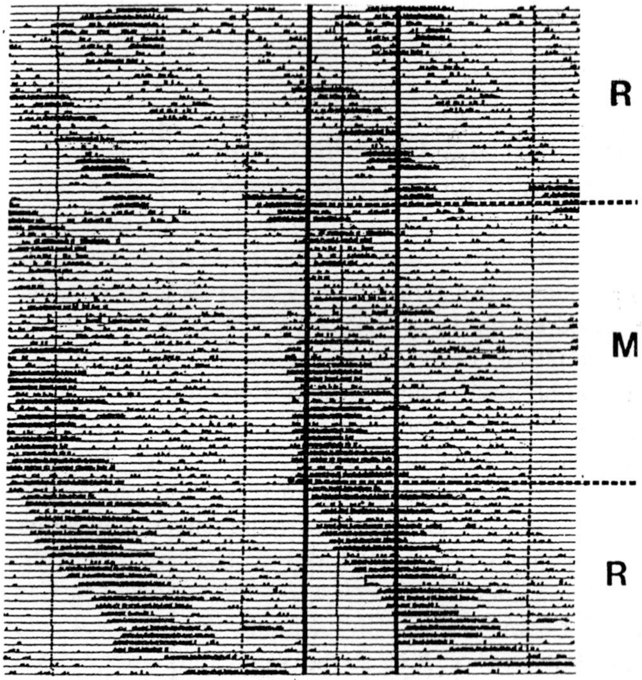

Fig. 2 **Double plotted running wheel activity record of a Syrian hamster free-running with a period smaller than 24 hours. Free-running activity is unaffected by Ringer infusion. Entrainment of activity was observed when the beginning of the Mel infusion coincided with the second part of the activity. The effect ceased when Mel was replaced by Ringer (from Kirsch *et al.*, 1993, with permission).**

The presence of Mel receptors within the SCN as well as the *in vitro* data (see above), strongly supports the idea that Mel, is involved in the regulation of SCN activity. Very probably Mel acting via the SCN-Mel receptors, is necessary to prevent internal desynchronization (the neuroendocrine loop model proposed by Cassone and Menaker, 1984; Cassone and Lu, 1994). Exogenously administered Mel may thus have an effect on clock activity but this does not mean that it will systematically shift the clock activity rhythm and so affect circadian organization. If exogenously administered Mel shifts the clock, it should act at the same time and in the same direction on all SCN-dependent circadian rhythms. To our knowledge no experimental study has specifically examined this point but results obtained in humans are interesting to consider. Exogenously administered Mel is used to manipulate human circadian rhythms (e.g. endogenous Mel production, Arendt, 1989; Lewy *et al.*, 1992; Zaidan *et al.*, 1994) and so to treat pathologically or socially induced disturbances of circadian rhythms (disturbed sleep in jet-lag, shift workers or blind people; review in Arendt, 1989). In a blind patient, Folkard *et al.*, (1990), observed that exogenously administered Mel entrained the free-running sleep-wake rhythm but not that of rectal temperature or urinary cortisol. This observation argues against Mel entrainment of the circadian oscillator and suggests that the daily administration of Mel may impose its rhythm onto the sleep-wake cycle by acting on another unidentified system.

In most nocturnal animals kept under constant conditions, the free-running period of locomotor activity is slightly longer than 24 h. Consequently the onset of activity is delayed for a few minutes every day. In order to entrain this rhythm to a 24 h period, Mel needs to induce an advance in the onset of locomotor activity every day. Based on these results, is it correct to speak of a phase-advancing effect of Mel (which implies a direct effect of the hormone on the clock)? A Few animals in each nocturnal species exhibit a free-running period shorter than 24 h. Using these animals, we have demonstrated that daily infusions of Mel were also able to entrain this locomotor activity rhythm to a 24 h period (Kirsch *et al.*, 1993) suggesting that Mel induces a daily phase delay. Thinking in terms of a direct effect of Mel on the clock, it would be necessary to accept that Mel administered at the same circadian time is able to induce, either a phase advance delay depending upon the free-running period of the animals. Although possible this is difficult to conceive. Again, these observations suggest that Mel imposes its rhythm by acting on another yet unidentified structure.

Different possibilities exist as to how such a system could operate. "Trigger circuits" initiating specific patterns of hormone secretion or behaviour (e.g. locomotor activity) that possess a circadian component and do not appear to be directly innervated by SCN neurons. Main target areas of SCN efferents are the paraventricular nuclei of the hypothalamus (PVN) and of the thalamus (PVT) (Watts and Swanson, 1987; Watts *et al.*, 1987; Kalsbeek *et al.*, 1993). The most remarkable feature of the PVT is its prominent projection to the nucleus accumbens, olfactory tubercule, central amygdala and prefontal cortex (Berendese and Groenewegen, 1990; Phillipson and Griffiths, 1985). These projections provide the SCN with an almost direct input to the mesolimbo-cortical complex considered to be the main centre in the central nervous system for the initiation and maintenance of (loco) motor behaviour. Mel binding sites have been observed in the PVT of the Siberian hamster, Syrian hamster and rat (review Masson-Pevet *et al.*, 1994). Exogenous Mel, acting on the PVT to modulate the circadian message sent by the SCN could explain the above results as well as the negative results obtained with Mel injections in SCN-lesioned animals.

Mel binding sites/receptors have not only been indentified in the SCN, the pars tuberalis and the PVT but also in a large number of brain and peripheral structures (more than 90 areas identified to date; review Stankov and Franschini, 1993; Brown *et al.*, 1994; Masson-Pevet *et al.*, 1994). Is it possible that, in some of these structures, Mel does not modulate the SCN-derived signal but directly imposes the circadian rhythm? This situation has been described in the rat pars tuberalis. The circadian rhythm of Mel receptor density in this structure disappeared after pinealectomy (Gauer *et al.*, 1994). Mel was shown to drive this circadian rhythm independently of the SCN or of the light/dark cycle. The rhythm, however, was circadian and entrained by the light/dark cycle because Mel is itself dependent upon the clock.

As Mel receptors within the pars tuberalis are thought to be involved in the photoperiodic response (Masson-Pevet and Gauer, 1994) such a concept might imply that the photoperiodic response depends upon a Mel-driven circadian phase of sensitivity to Mel. We have tested this hypothesis in the Syrian hamster and positive correlations have been obtained (Pitrosky *et al.*, 1995). When Mel was infused in pinealectomized hamsters for two 2.5 h periods separated by a 3 h non-infusion period, gonadal atrophy similar to that observed following a 10 h Mel infusion was obtained while a continuous 5 h infusion was ineffective (Fig. 3). Measurement of the plasma Mel Concentrations revealed that Mel levels reached physiological values during the infusion and fell to daytime values 1 h after the end of the infusion period. Two consecutive but separate melatonin peaks were present in the plasma. This result is not compatible with the duration hypothesis

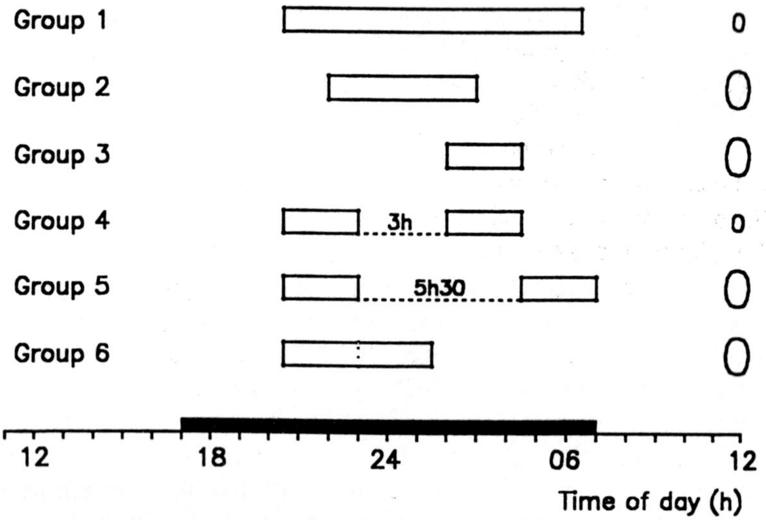

Fig. 3 **Effect of fractional infusion of Mel on the testes weight of male Syrian hamsters. Representation of infusion conditions (the black bar indicates the period of darkness; white bars indicate the infusion periods). On the right of the figure, the effect of the Mel infusion on the mean testes weight is represented graphically. A large oval represents groups in with the infusion had no effect and a small oval represents groups in with the infusion induced gonadal atrophy (from Pitrosky *et al.*, 1995, with permission).**

which implies that only variation in the duration of the Mel secretion is the important parameter of the Mel rhythm. The observed gonadal inhibition can thus only be explained by the existence of a period of sensitivity to melatonin. This result is consistent with the hypothesis of a rhythm of sensitivity to Mel driven directly by Mel itself. According to this hypothesis the first infusion signal entrains the rhythm of sensitivity to Mel, and gonadal atrophy is observed because the second signal coincides temporally with the period of sensitivity to Mel. In this case any displacement of the second signal, which would prevent the coincidence between the period of sensitivity and the presence of Mel would suppress the physiological response. Indeed, this is what we have observed by advancing or delaying the second signal. We have also demonstrated that the efficiency of this two signal paradigm was neither dependent upon the photoperiod nor on the time of day of administration (Fig. 4) (Pitrosky *et al.*, 1995).

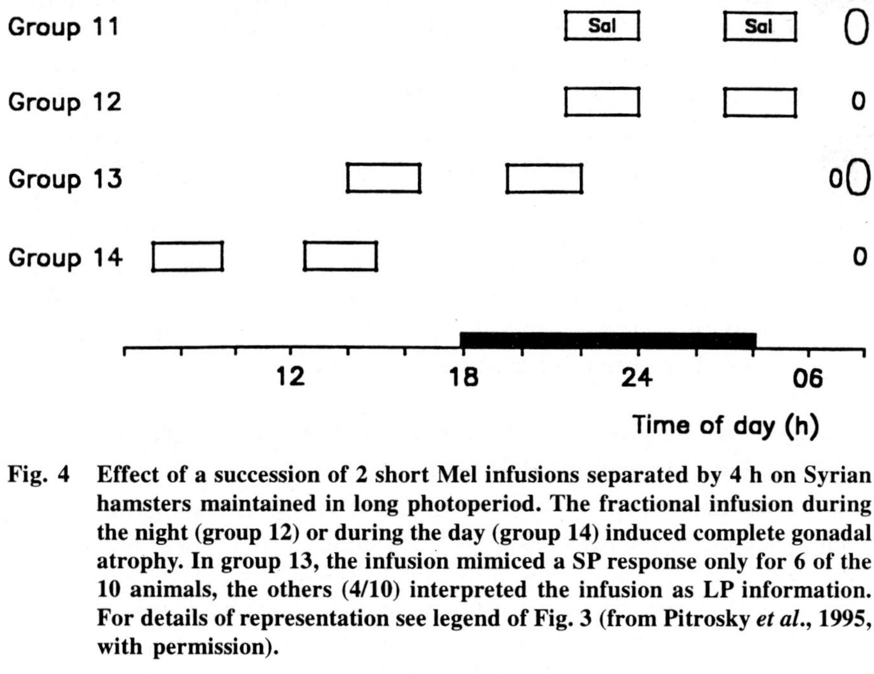

Fig. 4 Effect of a succession of 2 short Mel infusions separated by 4 h on Syrian hamsters maintained in long photoperiod. The fractional infusion during the night (group 12) or during the day (group 14) induced complete gonadal atrophy. In group 13, the infusion mimiced a SP response only for 6 of the 10 animals, the others (4/10) interpreted the infusion as LP information. For details of representation see legend of Fig. 3 (from Pitrosky *et al.*, 1995, with permission).

With respect to the photoperiodic response and its relationship with the circadian system, these data indicate that a circadian rhythm of sensitivity to Mel exists. This rhythm would be driven directly by Mel and would be circadian only because the Mel signal itself is dependent on the clock (Mel signal = circadian mediator). In normal physiological conditions, Mel synthesized and released at the beginning of the night would drive the Mel rhythm of sensitivity. A photoperiodic response would only be obtained when the increase in the duration of Mel secretion due to the increasing night length is long enough to allow high levels of circulating Mel to coincide with the Mel-driven Mel sensitivity period. This hypothesis, which is of the "coindicence" type also explains experimental data obtained by others in favour of duration hypothesis (review Bartness *et al.*, 1993). If, as suggested, the Mel signal itself directly drives the rhythm of sensitivity to Mel (thus independently of clock activity and the light/dark cycle), daily infusions of Mel during the daytime would be as effective as night-time infusions because the daytime infusion would shift the phase of Mel sensitivity into the daytime. The only condition would be that the Mel infusion time was long enough to entrain and, at the same time, to cover this sensitivity period (approximately a minimum of 7-8 h in the Syrian hamster). This has been precisely the situation in all the experimental procedures used by different authors: a SP type of response has always been obtained with a Mel infusion time of 8 h or more (review Bartness *et al.*, 1993). Similarly, when the infusion time was at least 8 h, this hypothesis could also explain the gonadal atrophy that was obtained with an alternating infusion paradigm with an infusion periodicity close to but less than 24 h.

Lesion of the SCN blocks the response to SP because it prevents all circadian organization and secretion of nocturnal Mel. Ablation of the SCN does not compromise the ability of, for example, the Syrian hamster to exhibit an appropriate photoperiodic response to Mel because daily Mel infusion would re-establish the circadian/temporal organization of the Mel sensitivity rhythm. Gonadal atrophy will be obtained if the Mel infusion time is long enough to cover the Mel sensitive period timed by the Mel infusion itself. This hypothesis also explains the photoperiodic control of gonadal activity in animal species such as the mink and ferret where no Mel receptors have been detected within the SCN.

Changes in the Amplitude of the Nocturnal Melatonin Peak and Environmental Factors

It has often been asserted that contrary to duration, the amplitude of the nocturnal pineal Mel peak does not vary significantly when animals are transferred from one photoperiod to another (Hoffmann, 1981). Reviewing the results, it appears that this is true in standard condition, that is in animal facilities where all environmental parameters, except photoperiod, remain constant. This is not true when animals are kept under natural environmental conditions. For example, in the Djungarian hamster (Steinlecher *et al.*, 1987; Stieglitz *et al.*, 1995), goat (Kanematsu *et al.*, 1989), jerboa (Lakhdar-Ghazal *et al.*, 1992), horse (Guerin *et al.*, 1995), mule (Cozzi *et al.*, 1991), sheep (Arendt, 1979; Matthews *et al.*, 1992) and the tammar (McConnel, 1986) not only the duration of nocturnal Mel secretion but also its amplitude varies with season. The nocturnal peak can even disappear. In domestic adult german Landrace sows, for example, McConnel and Ellendorf (1987) noted that a nocturnal surge in plasma melatonin concentrations was observed in animals kept under a photoperiod of 12hL:12hD but not under a photoperiod of 16hL:8hD or 8hL:16hD. This result was confirmed by Reiter *et al.* (1981), who also noted that other constituents normally associated with pineal melatonin synthesis, especially N-acetyltansferase activity, did not exhibit a nocturnal rise. Similar results have been obtained in another mammal, the European hamster (Fig. 5), (Pevet et al., 1989b) as well as in a bird, the emperor penguin (*Aptenodytes forsteri*) (Miche *et al.,* 1991). This penguin lives in Antarctica in a natural environment where photoperiod varies considerably throughout the year, ranging from almost constant darkness during the winter solstice to almost constant light during the summer solstice.

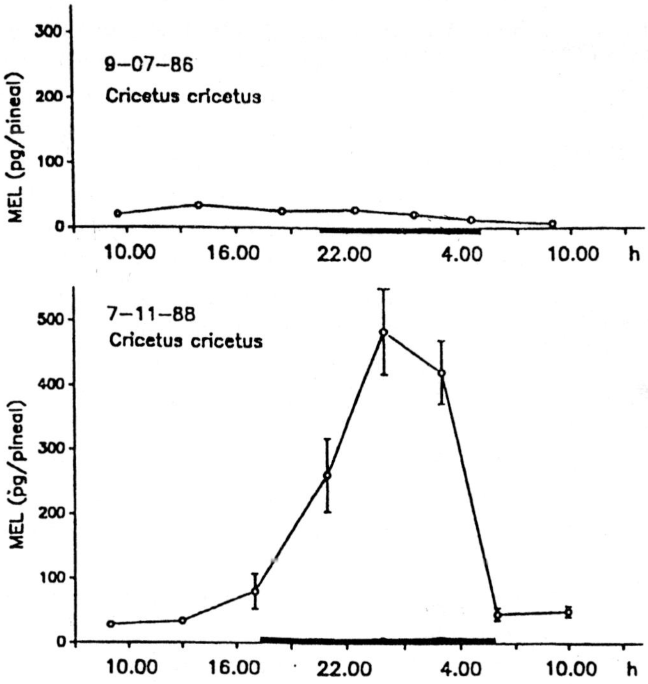

Fig. 5 Pineal Mel content over a 24 hour period in European hamsters killed either in early summer or in autumn (from Pevet *et al.*, 1991, with permission).

A strong seasonal variation in the nocturnal melatonin pattern was observed (Miche *et al.*, 1991). A clear day-night rhythm in plasma melatonin concentrations occurred in early autumn near the equinox. Subsequently an increase in Mel duration associated with a rapid decline in the amplitude of the nocturnal peak occurred and close to the winter solstice no day-night rhythm in plasma melatonin concentration was observed. Near the spring equinox the day-night variation in Mel reappeared and again in early December when approaching the summer solstice (almost constant light) the day-night variation was abolished again (Miche *et al.*, 1991). In the European hamster, a seasonal breeder living in the temperate zone, a disappearance of the day-night rhythmic Mel production has been noted in late spring-early summer (Pevet *et al.,* 1989; Vivien-Roels *et al.*, 1992). In late summer-early autumn, however, a clear day-night rhythm reappeared, its amplitude increasing regularly in autumn and early winter and then decreasing again (Vivien-Roels *et al.*, 1992).

What could the significance of these changes in Mel amplitude be? As emphasized in the introduction, animals living in their natural habitat are not only subjected to photoperiodic changes. There is always a covariation with other environmental factors and a combined type of action has been noted. For example in the Syrian hamster, low temperature which by itself has no effect on gonadal activity accelerated gonadal atrophy induced by SP (Pevet *et al.*, 1989a). We are of the opinion that these changes in the amplitude of the rhythm of Mel secretion might convey some environmental information (Pevet, 1979, 1985, 1987; Vivien-Roels and Pevet, 1983) in the same way that the duration of Mel secretion indicates daylength.

Until now very little data on this problem have been published. Nir *et al.*, (1973) have reported that exposure of rats to an ambient temperature of 33°C for 24 h induced a 17% decrease in the level of hydroxyindole-O-methyltransferase (HIOMT) activity, N-acetyltansferase (NAT) activity has also been reported to decrease after a similar exposure to high temperature both during the day and night-time (Nir and Hirschmann, 1978). In suckling rats, Ulrich *et al.*, (1973) noted that NAT activity was diminished when rats were exposed to either high (34°C) or low (7°C) temperatures. Klein and Weller (1972) studying the effect of different light-temperature regimes on pineal serotonin, HIOMT and NAT content, concluded that there could be a "dual mechanism controlling pineal function in the suckling rat, NAT being sensitive to temperature and serotonin to environmental illumination while HIOMT appeared to be influenced neither by temperature nor by light". In the collared lemming, Quay (1978) also reported evidence that both photic and thermal conditions affected pineal activity. Brainard *et al.* (1982) reported that the amplitude of the nocturnal Mel peak is higher in Syrian hamsters kept outdoors in December than in the same animals kept under laboratory SP. This result is in accordance with the observations of Stokkan *et al.*, (1993) and Tannenbaum *et al.*, (1988) that pineal activity, in the rat or hamster, is stimulated by acute night-time exposure to cold. Other environmental factors, for example, rainfall (Srivastava, 1985), food availability or quality (Brown *et al.*, 1989) or magnetic field (Bardassano, 1989) are also known to affect Mel synthesis and seasonal adaptation in mammals (Pevet, 1987).

Without doubt, changes in the pattern of Mel production are complex and it is evident, although not yet clearly demonstrated, that modifications in amplitude like those in duration can transmit some kind of environmental information. Such conclusions raise the question of the regulation of the Mel synthesis within the mammalian pineal gland.

Since 1960, when Ariens-Kappers demonstrated that the mammalian pineal gland was heavily innervated with noradrenergic fibers, numerous studies have been performed to understand the function of the sympathetic input. In short, the noradrenergic innervation was shown to be the end point of a polysynaptic pathways conveying light information from the retina to the pineal. Changes in photoperiod acting via this pathway modify the temporal pattern of Mel secretion. Noradrenaline (NA) release from the sympathetic terminals throughout the night, interacts with β and α adrenergic receptors and initiates a variety of subcellular reactions which finally increases pineal cyclic AMP and stimulates NAT activity, the rate limiting enzyme in Mel synthesis (for review see Klein, 1985; Klein *et al.*, 1981). The nocturnal activation of Mel synthesis in the pineal and its duration thus appears to be essentially under the control of noradrenergic input.

Besides noradrenergic innervation, however, an important peptidergic innervation has also been described (Moller *et al.*, 1994). Although not yet proven, the role of such innervation might be to modulate the NA-induced Mel synthesis and thus alter the melatonin signal to convey different message (Pevet *et al.*, 1994). This would explain the large annual changes in the amplitude of the daily rhythm of Mel secretion described above.

Peptidergic Innervation and Melatonin Signal

The localization of the perikarya of the peptidergic afferents to the pineal is largely unknown. In fact we face a complex problem. Considering the innervation, we know that sympathetic fibers originating from perikarya in the superior cervical ganglia dominate. The superior cervical ganglia contain numerous NPY-containing perikarya. Depending on the species immunohistochemical studies have demonstrated the presence of dense NPY innervation in the pineal with a pattern resembling the sympathetic innervation which suggests that NPY is colocalized with NA in sympathetic fibers. Although, there is evidence in the sheep (Cozzi *et al.*, 1992) as well as in rat (Zhang *et al.*, 1991), it is not true in all the species. In the mink, for example, NPY fibers are still present after superior cervical ganglionectomy (Moller *et al.*, 1990b). This indicates that the NPY innervation does not originate exclusively from the peripheral sympathetic nervous system. Evidence for fibers originating from the other peripheral ganglia, e.g. pterygopalatine or trigeminal ganglia is also available (Moller *et al.*, 1991a). Furthermore, neuronal projections from the brain directly to the pineal gland have also been identified (for details see Moller *et al.*, 1991a). These central fibers are known to project from the optic system (especially the lateral geniculate nucleus, IGL) as well as from the limbic and hypothalamic area (e.g. paraventricular nucleus, lateral hypothalamic area, habenular nuclei and the nucleus of the posterior commissure) (details in Moller *et al.*, 1991a). Peptidergic fibers are part of these neuronal projections.

Some data suggest that part of the NPY fibers in the mink as well as in the rat, Mongolian gerbil and golden hamster (Mikkelsen *et al.*, 1987) project from the IGL to the pineal gland.

Immunocytochemical studies have demonstrated the presence of vasoactive intestinal peptide (VIP) fibers in the pineal of numerous species (review in Moller *et al.*, 1987). Shiotani *et al.* (1986) have shown that at least in the Mongolian gerbil, these VIP fibers originate from neurons in the pterygopalatine ganglion. However, although VIP is classically co-localized with acetylcholine in parasympathetic neurons, it is not possible to assert that all VIP fibers are of a parasympathetic nature. Some of them may be of central origin (Moller *et al.*, 1991a).

Part of the substance P fibers observed in the pineal of the Mongolian gerbil originate from the trigeminal ganglion, a peripheral sensory ganglion (Shiotani *et al.*, 1986). Apparently from this ganglion pinealopetal GRP nerve fibers arise. In the rat, however, substance P fibers seem to be exclusively of central origin (Ronneklein and Kelly, 1984).

All the studies performed to date conclude that vasopressin (VP), oxytocin (OT) as well as luteinizing hormone releasing hormone (LHRH) fibers are of central origin (Buijs and Pevet, 1980; Nurnberger and Korf, 1981; Matsuura *et al.*, 1983). This also seems to be the case for the opioidergic fibers (Moller *et al.*, 1991b).

It is clear, that to establish the precise function of these different peptidergic fibers inervating the pineal is difficult. The same neurotransmitter/neuromodulator released from sympathetic fibers or central fibers may thus transmit different informations to the pineal. Furthermore, the mechanism of action of the same neuropeptide can differ depending upon the nature of the nervous fibers in which it is present. For example, NPY which is co-localized with NA in sympathetic fibers will be able to act at both the pre- and post-synaptic level. When released from central fibers, it will probably be able to act only at the post-synaptic level.

Very precise lesions of these different neural pathways followed by an analysis of the changes observed in pineal synthesis are needed to determine their respective roles. This also means that the effects of these neuropeptides on pineal activity must be determined.

In the last few years numerous *in vivo* as well as *in vitro* studies have been started. The results obtained have been very different depending on the neuropeptides tested as well as on the parameter studied. Some neuropeptides have been shown to directly stimulate pineal activity, others were only able to modulate an NA-induced effect. These discrepancies might also be related to the type of innervation affected.

VIP is an example of a neuropeptide mostly present in parasympathetic fibers. In incubation experiments in the rat pineal, VIP has been found to elicit an increase in cAMP (Kaku *et al.*, 1986), cGMP (Ho *et al.*, 1987), NAT activity (Kaneko *et al.*, 1980; Yuwiller, 1983) and Mel synthesis and release (Simonneaux *et al.*, 1990a) (Fig. 6). This effect, which is mediated through VIP receptors (Mikkelsen *et al.*, 1987; Simonneaux *et al.*, 1990a) is noradrenergic-independent (Simonneaux *et al.*, 1990a). Moreover, VIP

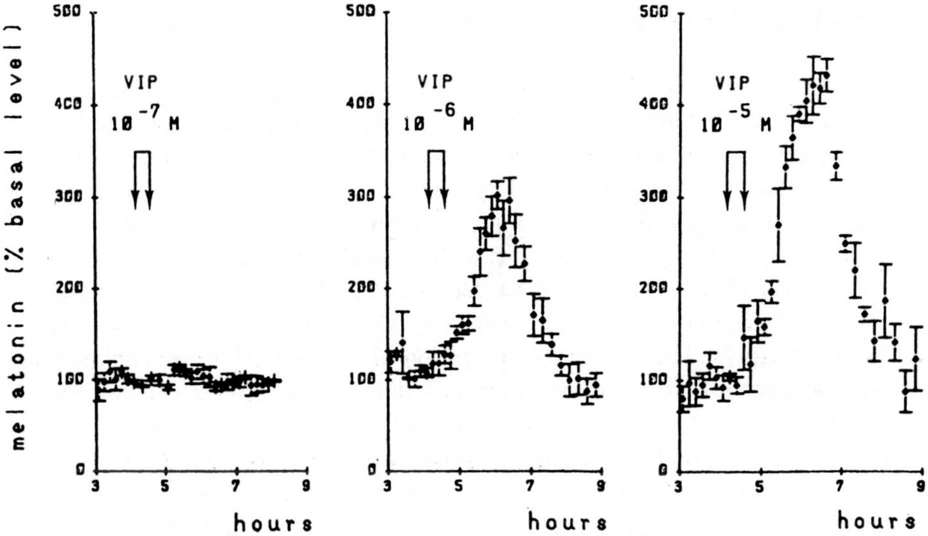

Fig. 6 Mel in the perfusate from rat pineal glands stimulated with increasing concentrations of VIP 10^{-7} to 10^{-5} M (from Simonneaux *et al.*, 1989b, with permission).

increases intracellular calcium via a mechanism involving cGMP gating of a rod type cation channel (Schaad *et al.*, 1995). VIP also stimulates cAMP production in the sheep pineal (Morgan *et al.*, 1988) but fails to activate NAT activity and Mel production in the Syrian hamster pineal (Moujir *et al.*, 1992a). Other neuropeptides of the secretin/glucagon/VIP family, histidine isoleucine (PHI) and pituitary cyclase activating polypeptide (PACAP) have also been reported to directly induce directly Mel secretion in the rat (Moujir *et al.*, 1992b; Simonneaux *et al.*, 1993). Two binding sites for VIP (Kaku *et al.*, 1983) and one binding site for PHI (Tsuchiaya *et al.*, 1987) have been characterized in the rat pineal gland. PACAP binding sites have also been demonstrated in the rat pineal by autoradiography (Masuo *et al.*, 1992). Whether the rat pineal actually has specific binding sites for these three peptides or whether some binding sites are able to bind several of them is not yet established. However, as a VIP antagonist inhibits PACAP stimulation of melatonin secretion to a less extent than VIP one (Simonneaux *et al.*, 1993), the presence of different sites is probable. VIP binding sites have also been characterized in the pineal of Mongolian gerbil (Moller *et al.*, 1994). While it is clear from several studies that VIP is a potent activator of the Mel synthesis, the information brought to the pineal by the parasympathetic fibers via this peptide has not been clarified. Since light increases pineal VIP binding sites and the cAMP response to VIP (Kaku *et al.*, 1985), and since pineal VIP content shows nycthemeral variations with night-time values being 2.5 times higher than the daytime values (Kaku *et al.*, 1986), VIP might also be involved in the transmission of light information to the pineal. Involvement in the transmission of other information, however, cannot be excluded.

Regarding the effect of the central innervation on pineal Mel synthesis, few clear results have been obtained. Most often a modulatory effect of the neuropeptide present in these central fibers has been noted. For example, a dual effect of both VP and OT on the secretion of Mel from rat pineal has been described (Simonneaux *et al.*, 1990b). VP, and to a lesser extent OT, displayed a slight inhibitory effect on basal Mel secretion. In contrast, VP and OT clearly potentiated the isoproterenol-induced increase in NAT activity (Stehle *et al.*, 1991) as well as in Mel release (Simonneaux *et al.*, 1990b) (Fig. 7). The mechanism of action of VP and OT is still unknown but it is probable that the inhibitory and potentiating effects depend upon different molecular mechanisms. Since VP receptor of the V1a subtype have been identified within the rat pineal gland (Simonneaux *et al.*, 1996) and since the VP-induced Mel potentiation is inhibited by the specific V1a receptor antagonist (Simonneaux *et al.*, 1996), it is probable that VP and OT bind to V1a and OT receptors, respectively, to induce these effects.

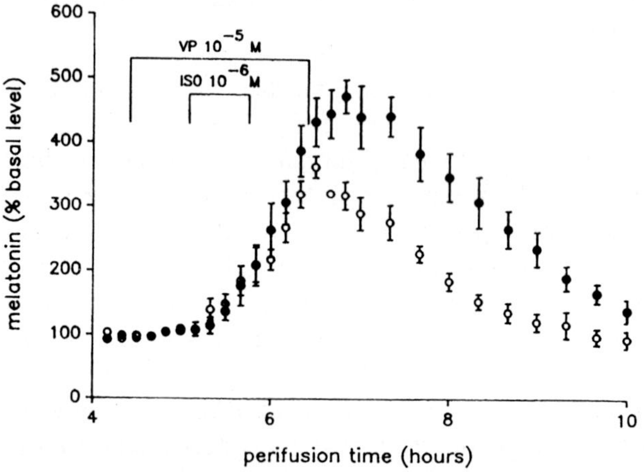

Fig. 7 **Mel in the perifusate from rat pineal glands infused either with 10^{-6} ISO for 40 min (O) or with 10^{-6} M ISO for 40 min and 10^{-5} M VP for 120 min (•). Results represent the mean+ SEM of 4 to 5 column each containing 3 pineal glands. The reference level of melatonin release (100% basal level) was calculated for each experiment as the mean of Mel output between 3 h 30 min and 4 h 10 min (4 values) after the beginning of the perfusion period (from Simonneaux *et al.*, 1989, with permission).**

Pineal VP and OT content displays temporal and physiological variations in the rat with higher levels at night-time (Gauquelin *et al.*, 1988; Liu and Burbach, 1988). Whether the nocturnal increase in VP and OT concentrations is involved with NA in the nocturnal increase in Mel synthesis has not been determined but would fit with the reported observations above. The fact that seasonal changes have been reported in pineal VP and OT (Nurnberger and Korf, 1981; Liu and Burbach, 1987) might also explain the seasonal changes in amplitude of the nocturnal Mel peak.

With regards to the effects of NPY, a neuropeptide mostly co-localized with NA in pineal sympathetic fibers, contradictory observations have been reported. NPY decreases the NA-induced cAMP accumulation in dissociated pinealocytes (Olcese 1991; Harada *et al.*, 1992; Simonneaux *et al.*, 1994a) but its effect on Mel is controversial since it was described to either potentiate (Vacas *et al.*, 1987; Mess *et al.*, 1991; Simonneaux *et al.*, 1994), inhibit (Olcese, 1991) or display no effect (William *et al.*, 1989) on NA-induced Mel release. These contradictory findings may only be a consequence of the multieffects of NPY on pineal activity.

Modulation of neurotranmitter released by NPY is a widespread phenomenon in the central and peripheral nervous system. This means that in the pineal, NPY could work simultaneously at pre- and post-synaptic levels to modulate noradrenergic effects. Vacas *et al.* (1987) and Simonneaux *et al.* (1994a) for example, have observed that NPY inhibits the release of NA from the sympathetic fibers. This inhibition which is independent but additive to the well known x2-adrenergic inhibition of NA release, involves the NPY-Y2 receptor subtype (Simonneaux *et al.*, 1994a). This NPY effect results in a decrease in NA-induced Mel synthesis, leading to a decrease in Mel secretion. On the other hand NPY at the post-synaptic level potentiates the NA-induced Mel release from organotypic (Vacas *et al.*, 1987), perfused (Mess *et al.*, 1991) or dissociated (Simonneaux *et al.*, 1994a) rat pineal gland through NPY-Y1 receptors (Olcese, 1991; Simonneaux *er al.*, 1994a). These opposite effects at the pre- and post-synaptic level are not without physiological consequences, especially if the amount of NPY released varies. Indeed a clear day-night variation in rat pineal NPY content (Olcese *et al.*, 1993) as well as a strong seasonal changes in NPY fiber density in the European hamster pineal (Moller *et al.*, 1998) has been observed.

Numerous other neuropeptides are present in the pineals of mammals (review in Simonneaux, 1995). A lot of studies are still needed to fully characterize their precise biological effects on pineal function but their presence clearly demonstrates that the regulation of pineal activity is very complex.

Conclusion

In conclusion, it appears that the pineal receives through NA and neuropeptides as well as through numerous other transmitters not considered here (e.g. serotonin, acetylcholine, GABA, glutamate, histamine) information from several areas of the nervous system as well as from the periphery. The complex seasonal changes in the daily pattern of Mel secretion are the consequence of the different action of these numerous neurotransmitters on pineal synthetic activity. It is thus easy to understand how, through these changes in Mel production, the central nervous system might integrate environmental information. To demonstrate the validity of this concept, however, the precise sites and mechanism of action of Mel for its different functions remains to be determined.

References

Arendt, J. (1979). In: The Pineal Gland of Vertebrates including Man (J. Ariens-Kappers and P. Pevet, Eds.), Progress in Brain Res., vol. 52, pp. 243-258. Elsevier/North Holland, Amsterdam.

Arendt, J. (1989). In: Biological rhythms in clinical Practice (J. Arndt, J. Minors and J.M. Waterhouse, Eds.), pp. 184-206. Wright, London.

Ariens-Kapper, J. (1990). Z. Zellforsch. **52:** 163-215.

Armstrong, S.M. and Redman, J.R. (1993). In: Melatonin: Biosynthesis, Physiological effects and clinical applications (H.S. Yu and R.J. Reiter, Eds.), pp. 187-224, CRC Press, Boca Raton, USA.

Bardassano, J.L., Cos, S. and Picazo, M.L. (1989). J. Hirnforsch. **38:** 639-643.

Bartness, T.J. and Goldman, B.D. (1988). Am. J. Physiol. **255:** R812-R822.

Bartness, T.J., Powers, J., Hasting, M.H., Bittman, E.L. and Goldman, B.D. (1993). J. Pineal Res. **15:** 161-190.

Berendse, H.W. and Groenewegen, H.J. (1990). J. Comp. Neurol. **299:** 187-228.

Bittman, E.L. and Karsch, F.J. (1984). Biol. Reprod. **30:** 585-593.

Bonnefond, C., Martinet, L. and Monnerie, R. (1990). J. Neuroendocrinol. **2:** 583-592.

Brainard, G.C., Petterbord, L.J., Richardson, B.A. and Reiter, R.J. (1982). Neuroendocrinology **35:** 342-348.

Brown, G.M., Chick, G.L., Ho, A.K., Kennedy, S. and Garfinkel, P. (1989). In: Advances in Pineal Research, vol. 3 (R.J. Reiter and O.G. Milestone, Eds.), pp. 83-92. John Libbey, London.

Brown, G.M., Pevet, P. and Pang, S.F. (1994). Biological Signals **3:** 2.

Bronson, F.H. and Heideman P.D. (1994). In: The Physiology of reproduction. 2nd Edition (E. Knobil and J.D. Neill, Eds.), pp. 541-583. Raven Press Ltd., New York.

Buijs, R.M. and Pevet, P. (1980). Cell Tissue Res. **205:** 11-17.

Carter, D.S. and Goldman, B.D. (1983) Endocrinology **113:** 1268-1273.

Cassone, V.M. (1990). T.I.N.S. **13:** 457-463.

Cassone, V.M. (1991). In: Superachiasmatic Nucleus: the Mind's Clock (D.C. Klein, R.Y. Moore and R.M. Reppert, Eds.), pp. 309-323. Oxford University Press.

Cassone, V.M., Chesworth, M.J. and Armstrong, S. (1986). Physiol. Behav. **36:** 111-112.

Cassone, V.M. and Menaker M. (1984). J. Exp. Zool. **232:** 539-549.

Cassone, V.M. and Lu J. (1994). In: Advance in Pineal Research, vol. 8 (M. Moller and P. Pevet, Eds.), pp. 31-40. John Libbey, London.

Cozzi, B., Mikkelsen, J.D., Ravault, J.P. and Moller, M. (1992). J. Comp. Neural. **316:** 238-250.

Cozzi, B., Morei, G., Ravault, J.P., Chesneau, D. and Reiter, R.J. (1991). J. Pineal Res. **10:** 130-135.

Delany, M.J. and Happold, D.C.D. (1979). Ecology of African Mammals, Longman, London.

Darrow, J.M. and Goldman, B.D. (1986). J. Biol. Rhythms **1:** 39-54.

Dowell, S.F. and Lynch, G.R. (1987). Biol. Reprod. **36:** 1095-1101.

Folkard, S., Arendt, J., Aldhous, M. and Kennett, H. (1990). Neurosci. Lett. **113:** 193-198.

Gauer, F., Masson-Pevet, M., Stehle, J. and Pevet, P. (1994). Brain Res. **641:** 92-98.
Gauquelin, G., Charib, C., Ghaemmanghami, F., Allevard, A.-M., Cherbal, F., Geelen, G., Bouzeghrane, F. and Legros, J.-J. (1988). Peptides **9:** 289-290.
Guerin, M.V., Deed, J.R., Kennaway, D.J. and Matthews, C.D. (1995). J. Pineal Res. **19:** 7-15.
Harada, Y., Okudo, M., Yaga, K. and Kaneko, T. (1992). J. Neurochem. **59:** 2178-2183.
Hastings, M.H., Mead, S.M., Vindlacheruvu, R.R., Ebling, J.F.P., Maywood, E.S. and Groose, J. (1992). Brain Res. **591:** 20-26.
Ho, A.K., Chick, C.L. and Klein, D.C. (1987) Biochem. Biophys. Res. Comm. **146:** 1478.
Hoffmann, K. (1981). In: The Pineal organ: photobiology-biochronometry-endocrinology (A. Oksche and P. Pevet, Eds.), pp. 123-138. Amsterdam, Elsevier North-Holland.
Kaku, K., Inoue, Y., Matsutani, A., Okubo, M., Hatao, K., Kaneko, T. and Yanaihara, N. (1983). Biomed. Res. **4:** 321-328.
Kaku, K., Tsuchiya, M., Matsuda, M., Inoue, Y., Kaneko, T., Yanaihara, N. and Kaneko, T. (1985). Endocrinology **117:** 2371-2375.
Kaku, K., Tsuchiya, M., Tanizawa, Y., Okuya, S., Inoue, Y., Kaneko, T. and Yanaihara, N. (1986). Peptides **7:** 193-195.
Kalsbeek, A., Teclemariam-Mesbah, R. and Pevet, P. (1993). J. Comp. Neurol. **332:** 293-314.
Keneko, T., Cheng, P.Y., Oka, H., Oda, T., Yanaihara, N. and Yanaihara, C. (1980). Biomed. Res. **1:** 84-87.
Kanamatsus, N., Mori, Y., Hayashi, S. and Hohino, K. (189). J. Pineal Res. **7:** 143-152.
Karsch, F.J., Malpaux, B., Wayne, N.L. and Robinson, J.E. (1988). Reprod. Nutr. Dev. **28:** 459-472.
Kirsch, R., Belgnaoui, S., Gourmelen, S. and Pevet, P. (1993). In: Light and Biological rhythms in man (Wetterberg L., Ed.), Wennegren International Series, vol. 63, pp. 161-185. Pergamon Press, Oxford.
Klein, D.C. (1985). In: Photoperiodism, Melatonin and the Pineal, pp. 38-56. Ciba Foundation Symposium 117. Pitman London.
Klein, D.C., Auerbach, D.A., Namboodiri, M.A.A. and Wheler, G.H.T. (1981). In: The Pineal gland, vol. I, Anatomy and Biochemistry (R.J. Reiter, Ed.), pp. 199-227. C.R.C. Press, Boca Raton, Florida.
Klein D.C. and Weller (1972). Science **177:** 532-533.
Lakhdar-Ghazal, B. Vivien-Roels and Pevet, P. (1992). J. Pineal Res. **13:** 28-35.
Lewy, A.J., Ahmed, S., Latham Jackson, J.M. and Sack, R.L. (1992). Chronobil. Int. **9:** 390-392.
Liu, B. and Burbach, J.P.H. (1988). Peptides **9:** 973-978.
Liu, B. and Burbach, J.P.H. (1987). Endocrinology **121:** 1716-1720.
Lofts, B. (1974). Physiology of the Amphibians, vol. II, Reproduction, Academic Press, New York.
Masson-Pevet, M. and Gauer, F. (1994). Biological Signals **3:** 63-70.
Masson-Pevet, M., George, D., Kalsbeek, A., Saboureau, M., Lakhdar-Ghazal, N. and P. Pevet (1994). Cell Tissue Res. **278:** 97-106.
Masuo, Y., Ohtaki, T., Masuda, Y., Tsuda, M. and Fujino, M. (1992). Brain Res. **575:** 113-123.
Matsuura, T., Kawata, M., Yamada, H., Kojima M. and Sano, Y. (1983). Arch. Hist. Jap. **46:** 373-379.
Matthews, D.C., Seamark, R.F. and Guerin, M.V. (1992). J. Reprod. Fertility **95:** 869-875.
Maywood, E.S., Buttery, R.C., Vance, G.H.S., Herbert, J. and Hastings, M.H. (1990). Biol. Reprod. **43:** 174-182.

McArthur, A.J., Gillette, M.Y. and Prosser, R.A. (1991). Brain Res. **565:** 158-161.
McConnel, S.J. (1986). J. Pineal Res. **3:** 119-125.
McConnel, S.J. and Ellendorf, F. (1987). J. of Pineal Research **4:** 201-210.
Mess, B., Csernus, V. and Rekasi, Z. (1991). In: Advances in Pineal Research, 6 (A. Foldes and R.J. Reiter, Eds.), pp. 57-66. Libbey, London.
Miche, F., Vivien-Roels, B., Pevet, P., Spehner, C., Robin, J.P. and Le Maho, Y. (1991). Gen. Comp. Endocrinol. **84:** 405-411.
Mikkelsen, J.D., Korf, H.W. and Moller, M. (1987). In: Fundamentals and Clinics in Pineal Research (G.P. Trentini, C. DeGaetani and P. Pevet, Eds.), pp. 87-89. Raven Press, New York.
Moller, M., Zhang, E. and Cozzi, M. (1994). In: Advance in Pineal Research, vol. 8 (M. Moller and P. Pevet, Eds.), pp. 135-142. J. Libbey, London.
Moller, M., Cozzi, B., Schroder, H. and Mikkelsen, J.D. (1987). In: Fundamental and Clinics in pineal research (Trentini, G.P., De Gaetani, C., Pevet, P., Eds.), pp. 71-77. Raven Press, New York.
Moller, M., Masson-Pevet, M. and Pevet, P. (1998). Cell. Tiss. Res. **291:** 415-421.
Moller, M., Mikkelsen, J.D. and Larsen, P.J. (1990a). In: Advances in pineal Research, 4 (R.J Reiter and A. Lukasyk, Eds.), pp. 1-8. John Libbey, London.
Moller, M., Mikkelsen, J.D. and Martinet, L. (1990b). Cell Tiss. Res., **261:** 477-483.
Moller, M., Mikkelsen, J.D. and Phasuwan-Pujiton, P. (1991a) In: Role of melatonin in pineal peptides in neuroimmunomodulation (F. Fraschini and R.J. Reiter, Eds.), pp. 15-26. Plenum Press.
Moller, M., Ravault, J.P., Cozzi, B., Zhang, E.-T., Pansuwan-Pujito, P., Larsen, P.J. and Mikkesen, J.D. (1991b). In: Advances in Pineal Research, 6 (A. Foldes and R.J. Reiter, Eds.), pp. 3-12. Libbey, London.
Moore-Ede, M.C., Sulzman, F.M. and Fuller, C.A. (1982). The clocks that time us. Harward Univ. Press, Cambridge, USA.
Morgan, P.J., Williams, L.M., Lawson, W. and Riddoch, G. (1988). Brain Res. **447:** 279-286.
Moujir, F., Richardson, B.A., Yage, K. and Reiter, R.J. (1992a). J. Pineal Research **12:** 35-42.
Moujir, F., Reiter, R.J., Rodriguez, C. and Yaga, K. (1992b). Endocrinology **130:** 2076-2082.
Nir, I., Hirschmann, N. and Sulman, F.G. (1973). Experientia **31:** 867-868.
Nir, I. and Hirschmann, N. (1978). Experientia **34:** 1645-1646.
Nurnberger, F. and Korf, H.H. (1981). Cell Tiss. Res. **220:** 87-97.
Olcese, J. (1991). J. Neurochem. **57:** 943-947.
Olcese, J., Sinemus, C. and Ivell, R. (1993). Mol. Cell. Neurosci. **4:** 47-54
Pevet, P. (1979). In: The pineal gland of Vertebrates including Man (J. Ariens-Kappers and P. Pevet, Eds.), pp. 149-194. Progress Brain Res. vol. 52. Elsevier/North Holland Biomedical Press, Amsterdam.
Pevet, P. (1985). In: The pineal gland: current state of pineal research (B. Mess, Ruszas Cz., Tima, L. and Pevet, P., Eds.), pp. 163-186. Akademia Kiado, Budapest and Elsevier Science Publishers, Amsterdam.
Pevet, P. (1987). In: Comparative Physiology of environmental adaptations, vol. 3 (P. Pevet, Ed.), pp. 82-100. Karger, Basel.
Pevet, P., Masson-Pevet, M., Vivien-Roels, B. and Bonn, D. (1986a). In: Advances in pineal research, vol. 1 (R.J. Reiter and M. Karasek, Eds.), pp. 185-196. John Libbey & Company, Ltd., London.
Pevet, P., Simonneaux, V. and Vivien-Roels, B. (1994). In: Advances in pineal research, 8 (M. Moller and P. Pevet, Eds.), pp. 143-152. John Libbey, London.

Pevet, P., Vivien-Roels, B. and Bonn, D. (1986b). In: Endocrine regulations as adaptive mechanisms to the environment (I. Assenmacher and J. Boissin, Eds.), pp. 201-206. Edition du CNRS, Paris.

Pevet, P., Vivien-Roels, B. and Masson-Pevet, M. (1989a). J. Neural Transm. **76:** 119-128.

Pevet, P., Vivien-Roels, B., Masson-Pevet, M., Steinlenchner, S., Skene, D. and Canguilhem, B. (1989b). J. Pineal Res. **6:** 233-242.

Philippson, O.T. and Griffiths, A.C. (1985). Neurosciences **16:** 375-296.

Pitrosky, B., Masson-Pevet, M., Kirsch, R., Vivien-Roels, B., Canguilhem, B. and Pevet, P. (1991). J. Pineal Res. **11:** 149-155.

Pitrosky, B. (1994). These de Doctorat, Universite L. Pasteur, Strasbourg.

Pitrosky B., Kirsch, R., Vivien-Roels, B., Georg-Bentz, I., Canguilhem, B. and Pevet, P. (1995). Neuroendocrinol. **7:** 889-895.

Quay (1978). Am. J. Anat. **153:** 545-562.

Redman, J.R., Armstrong, S.M. and Ng., K.T. (1983). Science **219:** 1089-1091.

Reiter, R.J. (1987). In: Comparative physiology of environmental adaptation, part 3, Adaptation in climatic changes (P. Pevet, Ed.), pp. 71-81. Karger, Basel.

Reiter, R.J., Bitt, J.H. and Armstrong, J.D. (1981). Neuroscience Letters **81:** 171-176.

Ronnekleiv, O. and Kelley, M.J. (1984). J. Pineal Res. **1:** 355-370.

Rusak, B. (1989). B. J. Biol. Rhythm **4:** 121-134.

Schaad, N., Vanecek, J., Rodriguez, I.R., D.C. Holtzclaw, L. and Russel, J.T. (1995). Mol. Pharmacol. **47:** 923-933.

Shiotani, Y., Yamano, M., Shiosaka, S., Emson, P.C., Hillyard, C.J., Grgis, S. and McIntyre, I. (1986). Neurosc. Lett. **70:** 187-192.

Sicard, B., Kylem, M., Papillon, Y., Diara, W. and Keita, M. (1995). Rongeurs nuisibles soudano-saheliens. John Libbey Eurotext, Paris.

Sicard, B., Fuminier, F., Maurel, D. and Biossin, J. (1993). Biol. Reprod. **49:** 716-722.

Sicard, B., Maurel, D., Fuminier, F. and Biossin, J. (1994). Can. J. Zool. **72:** 1-14.

Simonneaux, V. (1995). Neuroendocrinol. Lett. **17:** 115-130.

Simonneaux, V., Kozak, R., Arsenijevic, Y. and Pevet, P. (1996). Regulatory Peptides (in press).

Simonneaux, V., Ouichou, A. and Pevet, P. (1990A). J. Neural Transm. **79:** 67-79.

Simonneaux, V., Ouichou, A., Burbach, J.P.H. and Pevet, P. (1990b). Peptides **11:** 1075-1079.

Simonneaux, V., Ouichou, A. and Pevet, P. (1993). Brain Res. **603:** 148-152.

Simonneaux, V., Ouichou, A., Craft, C. and Pevet, P. (1994a). J. Neurochem. **62:** 2464-2471.

Simonneaux, V., Ouichou, A., Craft, C. and Pevet, P. (1994). In: Advance in Pineal Research, 8 (M. Moller and P. Pevet, Eds.), pp. 163-174. J. Libbey, London.

Srivastava, M. (1985). Thesis, Banaras Hindu University, Varanasi, India.

Stankov, B. and Franschini, F. (1993). Lett. **15:** 149-164.

Stehle, J., Reuss, S., Seidel, A., Riemann, R. and Vollrath, L. (1991). Neurosc. Lett., **123:** 131-134.

Steinlechner, S., Buchberger, A. and Heldmaer, G. (1987). J. Comp. Physiol., A, **160:** 593-597.

Stetson, M.H., Watson-Whitmyre, M. (1984) In: The pineal gland (R.J. Reiter, Ed), pp. 109-153. Raven Press, New York.

Stieglitz, A., Spiegelhalter, F., Klante, G. and Heldmaier, G. (1995). J. Pineal Res. **18:** 69-76.

Stokkan, K.A., Nonaka, K.O., Lerchl, A., Vaughan, M.K. and Reiter, R.J. (1991). J. Pineal Res. **10:** 43-48.

Tannenbaum, M.G., Reiter, R.J., Vaughan, M.K., Troiani, M.E. and Gonzales-Brito, A. (1988). Cryobiology **25:** 227-232.

Tsuchiya, M., Kaku, K., Matsuda, M., Kaneko, T. and Yanaihara, N. (1987). Biomed. Res. **8:** 45.

Ulrich, R., Yuwiller, A., Wetterberg, L. and Klein, D. (1973). Neuroendocrinol. **13:** 255-263.

Vacas, M.I., Keller-Sarmiento, M.I., Pereyra, E.N., Etchegoyen, G.S. and Cardinali, D.P. (1987). Cell Mol. Neurobiol. **7:** 309-315.

Vivien-Roels, B. and Pevet, P. (1983). In: Pineal Research Review (R.J. Reiter, Ed.), pp. 92-143. A.R. Liss, Inc., New York.

Vivien-Roels, B., Pevet, P., Masson-Pevet, M. and Canguilhem, B. (1992). Gen. Comp. Endocrinol. **86:** 239-247.

Watts, A.G. and Swanson, L.N. (1987). J. Comp. Neurol. **258:** 230-252.

Watts, A.G., Swanson, L.W. and Sanchez-Watts, L. (1987). J. Comp. Neurol. **258:** 204-229.

Wayne, M.L., Malpaux, B. and Karsch, F.J. (1988). Biol. Reprod. **39:** 66-75.

Williams, L.M., Morgan, P.J., Pelletier, G., Riddoch, G.I., Lawson, W. and Davidson, G.R. (1989). J. Pineal Res. **7:** 345-353.

Yuwiller, A. (1993). Science **220:** 1082-1083.

Zhang, E., Mikkelsen, J.D. and Moller, M. (1991). Cell Tiss. Res. **265:** 63-71.

Zaidan, R., Geoffriau, M., Brun, J., Taillard, J., Bureau, C., Chazot, G. and Claustrat, B. (1994). Neuroendocrinology **60:** 105-112.

Comparative Endocrinology and Reproduction
K.P. Joy, A. Krishna and C. Haldar (Eds)

30. Seasonal Variations in the Amplitude of the Daily Pattern of Melatonin Secretion in Mammalian and Non-Mammalian Vertebrates: Possible physiological consequences

B. Vivien-Roels
UMR CNRS 7518 "Neurobiologie des fonctions rythmiques et saisonnieres",
Universite Louis Pasteur, Strasbourg, France

Introduction

Most species are known to use environmental cues such as photoperiod, temperature, rainfall or food supply, to synchronize several physiological functions such as breeding, hibernation, migration or moulting.

The exact mechanisms where by animals anticipate seasons and integrate environmental information are not completely known but it is now well-established, at least in mammals living in temperate regions, that the pineal gland is implicated in the integration of photoperiodic information via the daily pattern of melatonin synthesis and secretion (Goldman and Darrow, 1983, Hoffmann, 1985, Pévet, 1988). Natural or experimental photoperiodic changes alter the duration of melatonin secretion in many species (ref. in Vivien-Roels *et al.,* 1988, Pévet *et al.,* 1991). The nocturnal peak of melatonin synthesis and secretion is proportional to the length of the dark period. Moreover, various melatonin infusions in pinealectomized hamsters or sheep can reproduce the effects of either long or short photoperiod on sexual activity (Bittman and Karsch, 1984, Bartness and Goldman, 1988, Karsch *et al.,* 1988, Bonnefond *et al.,* 1990, Maywood *et al.*, 1990, Pitrosky *et al.,* 1991). Some of these results have led to the *duration hypothesis* (Goldman and Darrow, 1983) which suggests that it is the duration of the nocturnal peak of melatonin secretion that conveys photoperiodic information to the brain.

But, when overlooking the literature, it appears that, in a large number of mammalian as well as of non-mammalian species, natural or experimentally-induced changes in the daily pattern of melatonin secretion are more complex than a simple change in the duration of the night peak of melatonin secretion. Changes in the amplitude or phase with respect to the light-dark cycle may also be affected (Reiter, 1987, Vivien-Roels *et al.,*

1988, Pévet *et al.*, 1991). Moreover, when looking at animals living in their natural habitats, most species are not only submitted to photoperiodic changes. Other environmental factors, such as environmental temperature, thermoperiod, rainfall or food supply are known to be important cues in regulating several physiological adaptations (Lofts, 1974, Delany and Happold, 1979, Sicard *et al.*, 1995). In some species these factors are also important variables implicated in the control of melatonin rhythms, especially in poikilotherm Vertebrates.

This review will be an attempt to evaluate the physiological consequences of the annual variations in the daily pattern of melatonin synthesis and secretion in vertebrates, with special reference to the seasonal variations in the amplitude of the melatonin rhythm.

Seasonal Changes in the Amplitude of the Melatonin Rhythm and Environmental Factors in Poikilotherms

It is now generally admitted that, in poikilotherms as in mammals, the pineal gland plays an important role in the transmission of photoperiodic information to the brain, via the daily rhythm of melatonin secretion. It is now also more and more evident that when animals are raised in natural conditions, where several environmental factors are interdependent and act concurrently, important seasonal variations in the daily rhythm of melatonin, affecting both duration and amplitude are observed.

Fishes

As in mammals, in fishes living in temperate regions, melatonin fluctuates diurnally, with low levels during the day and high levels during the night, and the duration of nocturnal elevated melatonin is proportional to the duration of the dark period in trouts, *Salmo gairdneri* (Duston and Bromage, 1986), pike, *Esox lucius* (Falcon *et al.*, 1994), common carp, *Cyprinus carpio* (Kezuka *et al.*, 1988) or goldfish, *Carassius auratus* (Kezuka *et al.*, 1992). Reviewing these results, it appears that this is true when fishes are maintained in standard conditions where only photoperiod is varying. When maintained in natural conditions, some species show also annual variations in the amplitude of the melatonin rhythm.

In the atlantic salmon, *Salmo salar,* maintained in natural conditions, the duration of the night peak of melatonin is related to the duration of the night, i.e. longer in winter than in summer (Randall *et al.*, 1995) but the amplitude of the melatonin rhythm is higher during the summer months, suggesting that temperature may modify circulating melatonin levels.

In the goldfish, it has recently been reported (Iigo and Aida, 1995) that, when reared under natural condition, where in addition to photoperiod, water temperature varies from 8.5°C in December to 26°C in June, there are important changes in the amplitude of the rhythm of circulating melatonin, with high levels in June and September and low levels in December and

March, showing a significant correlation with the water temperature at mid-dark (26°C in June, 25.1°C in September, 8.5°C in December and 10°C in March). The same authors report that when fishes are maintained under 16L-8D or 8L-16D in combination with temperature changes (5°C-15°C-25°C) during January-February, the duration of the night peak of melatonin is controlled by light while its amplitude clearly depends on water temperature.

Also several *in vitro* studies have demonstrated a clear effect of temperature on the rhythm of melatonin secretion from pineal glands of rainbow trout, pike and white sucker *Gastostomus commerson* (Max and Menaker, 1992, Falcon *et al.,* 1992, 1994, Zachmann *et al.,* 1992, Bolliet *et al.,* 1994).

From all these results , it clearly appears that in fishes , water temperature, which is also implicated in the control of physiological events, is implicated in the annual variations of the amplitude of the daily rhythm of melatonin. In the absence of photoperiodic changes (in constant darkness for example) temperature cycles are able to entrain the daily rhythm of melatonin secretion (Zachmann *et al.,* 1992, Falcon *et al.,* 1992).

Amphibians

In amphibians, studies reporting seasonal variations in the daily rhythm of melatonin are limited to two species, *Rana perezzi,* previously called *Rana ridibunda,* (Delgado and Vivien-Roels, 1989b) and *Rana esculenta* (d'Istria *et al.,* 1994).

In both species important changes in the amplitude of the daily rhythm of melatonin are observed in the pineal gland, retina and plasma, with higher night-time values in spring, summer and fall than in winter, where a complete lack of daily fluctuations is reported. As in fishes, in amphibians photoperiod controls the duration of the nocturnal rise of melatonin, (Gern and Norris, 1979, 1983, Pang *et al.,* 1985, Rawding and Hutchison, 1992) while the amplitude of the melatonin peak clearly depends on environmental temperature, (Gern *et al.,* 1983, Rawding and Hutchison, 1992, Delgado and Vivien-Roels, 1989a, Delgado *et al.*. 1983, Valenciano *et al.*, 1994). Contrary to fishes, where circulating melatonin originates mostly from the pineal gland, in amphibians, melatonin concentrations are always significantly higher in the retina compared to the pineal, and moreover the close parallelism observed between the seasonal variations in ocular and circulating melatonin profiles suggest that in amphibians melatonin could be released from the eyes in the general circulation (Delgado and Vivien-Roels, 1989a, b).

Reptiles

To date, annual changes in the daily rhythm of melatonin have been described in lacertilians, chelonians and ophidians.

In the australian scincid lizard, *Trachydosaurus rugosus,* important seasonal variations in the amplitude of the night peak of pineal melatonin have been observed with a rhythm of low amplitude in spring, changing to a rhythm of greater amplitude in summer and autumn (Firth *et al.,* 1979). In the tortoise, *Testudo hermanni,* the amplitude of the pineal melatonin night peak also shows important seasonal variations, with a complete lack of daily rhythm in winter and early spring, and an amplitude of the night peak increasing during late spring and early summer, being maximum in august and decreasing in autumn (Vivien-Roels *et al.,* 1979). In the black tiger snake, *Notechis ater,* Rismiller and Kennaway (1990) report that in snakes captured in the field circulating melatonin levels show a pronounced daily rhythm only at one time of the year, during the late austral summer, which corresponds to the end of the breeding season. At all other times of the year no daily rhythm of melatonin was evident. In another snake, the red sided garter snake, *Thamnophis sintalis parietalis,* Mendonca *et al.* (1995) also very recently report annual changes in daily circulating melatonin levels.

From these different studies it appears that the amplitude of the melatonin rhythm in Reptiles, varies seasonally, being highest when days are long and environmental temperature highest. Moreover several experimental studies show that either light, temperature and temperature cycles can entrain pineal and/or circulating melatonin rhythms (ref. in Underwood, 1989), temperature and temperature cycles acting more especially on the amplitude of the rhythm.

In the tortoise, *Testudo hermanni* (Vivien-Roels and Arendt, 1981, Vivien-Roels and Pévet, 1983) as well as in the box-turtle, *Terrapene carolina,* (Vivien-Roels *et al.,* 1988) we have also shown that the amplitude of the night peak of melatonin clearly depends on the environmental temperature, increasing with increasing temperature (Fig. 1). In the anole lizard, *Anolis carolinensis,* Menaker and Wisner (1983) have found that *in vitro* the pineal gland shows a rhythm of melatonin secretion, the amplitude of which being temperature dependent. Underwood (1985) in the same species, reports a clear effect of temperature cycles in the entrainment of the melatonin rhythm with higher amplitudes when temperature cycles were 12°C and lowest amplitudes when temperature cycles were 2°C. As in the tortoise and the box-turtle, a constant low temperature abolished the melatonin rhythm. Also in the scincid lizard, *Trachydosaurus rugosus,* it has been reported that the amplitude of the daily rhythm of melatonin was abolished at 15°C, high at 30°C and highest at a fluctuating temperature of 30°C during photophase and 15°C during scotophase (Firth and Kennaway, 1989). More recently, the same authors have shown that the amplitude of the night peak of melatonin in animals exposed to different thermoperiods of 33°C thermophase and 15°C cryophase, ranging from 1.5 hr to 24 hr thermophase, clearly depends on the duration of the thermophase. The amplitude of the rhythm appears to be highest when thermophase is comprised between 9 and 18 hours (Firth *et al.,* 1991). In another lizard, the gecko, *Christinus marmoratus,* Moyer *et al.* (1995) very recently reported that *in vitro* pineal glands maintained in LD 12/12 secreted melatonin rhythmically with a high amplitude of the melatonin rhythm at 30°C but not at 10 or 37°C.

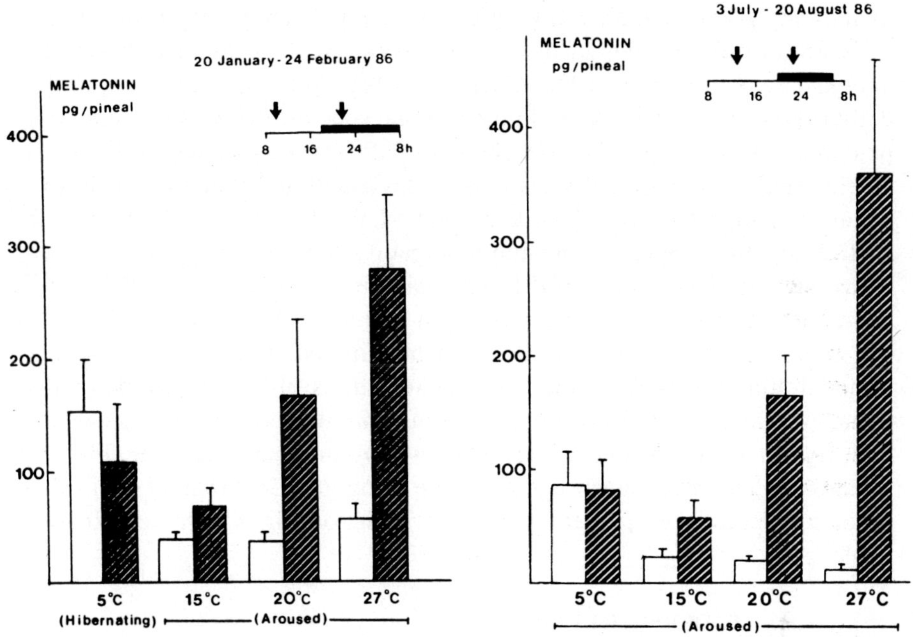

Fig. 1 Effect of environmental temperature on the amplitude of the day-night rhythm of melatonin in the pineal gland of the box-turtle, *Terrapene carolina*, in February (A) and in August (B). Arrows indicate the light and dark sampling. $p < 0.05$ and $p < .001$ as compared to day-time values (from Vivien-Roels *et al.*, 1988, with permission).

When comparing all the results obtained in the annual variations of the daily rhythm of melatonin in fishes, amphibians and reptiles, it is clear that the pineal gland of ectotherms appears to function as a transducer of both thermal and photic information and that both light and temperature show interactive effects. It is also clear that if photoperiod has a clear effect on the duration of the night peak of melatonin in poikilotherms, as in mammals, the seasonal variations in the amplitude of the day-night rhythm of melatonin are mostly controlled by annual variations in environmental temperature or temperature cycles. Whether other environmental factors, such as rainfall, humidity, food supply also affect melatonin seasonal fluctuations remains to be determined and needs further experimental investigations.

Seasonal Variations in the Amplitude of the Daily Rhythm of Melatonin in Homeotherms

Birds

If day-night variations in the melatonin content in the pineal gland, retina and plasma have been extensively studied in several bird species, chicken, *Gallus domesticus* (Pang *et al.*, 1985, Liou *et al.*, 1987), quails, *Coturnix coturnix* (Pang *et al.*, 1985, Underwood *et al.*, 1984), pigeons, *Columba livia* (Pang *et al.*, 1985), crows, *Corvus corone* (Taniguchi *et al.*, 1993), owls, *Strix uralensis* (Taniguchi *et al.*, 1993) and Adelie penguins, *Pygoscelis adelias* (Benelbaz *et al.*, 1976, Cockrem, 1991), seasonal variations in the daily rhythm of melatonin are limited to one species, the antarctic emperor penguin, *Aptenoytes forsteri* (Miché *et al.*, 1991). In this penguin living in Adelie land, where annual variations in daylength and climatic conditions are drastic, important annual variations in both the duration and the amplitude of the nocturnal peak of melatonin secretion have been reported. As in many species, the duration of the melatonin peak increases when daylength decreases and decreases when daylength increases. But when considering the amplitude of the night peak of melatonin secretion, there is a clear bimodal annual pattern in the seasonal variations of the amplitude of the melatonin rhythm with lowest values during both winter and summer solstice and highest values in late winter, early spring and early autumn (Fig. 2). In another antarctic species, the Adelie penguin, Cockrem (1991), also reports the absence of clear day-night variations in natural continuous daylength.

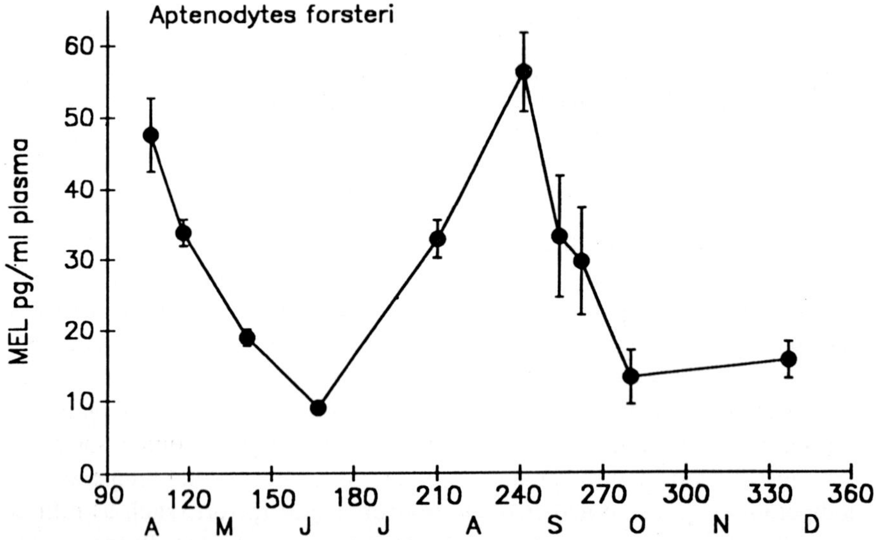

Fig. 2 Seasonal variations in the amplitude of the night peak of melatonin in the plasma of the Emperor penguin, *Aptenodytes forsteri*.

Since there are very few data reporting annual or experimental changes in the daily rhythm of melatonin in birds, it is difficult at the moment to understand the mechanisms involved in the regulation of these variations in the amplitude of the night peak of melatonin in the emperor penguin. If it is true that photoperiod varies considerably in arctic or antarctic regions, a possible effect of other factors, such as temperature, magnetic field and fasting in the regulation of these annual variations cannot be excluded and has to be considered.

Mammals

It is now well-established in most mammals living in temperate regions, that the duration of the night peak of melatonin synthesis and secretion is directly dependent on the duration of the dark period and that natural or experimental changes in the photoperiod alter this duration (ref. in Vivien-Roels *et al.,* 1988, Pévet *et al.,* 1991, Miché *et al.,* 1991). But it is now also more and more evident, when overlooking the literature, that in mammals, as in other vertebrates, seasonal or experimental changes in the pattern of melatonin secretion are more complicated than a single change in the duration of the night peak of melatonin (Pévet *et al.,* 1991, Vivien-Roels *et al.,* 1991). In many mammals also amplitude varies with seasons, especially when considering animals catched or raised in natural environments, where not only photoperiod, but also other environmental factors vary throughout the year.

To date, seasonal variations in the amplitude of the nocturnal melatonin peak and/or NAT activity have been reported in the sheep, *Ovus domesticus* (Arendt, 1979, Sheikheldin *et al.,* 1992), the horse (Guerin *et al.,* 1995), the Djungarian hamster, *Phodopus sungorus* (Steinlechner *et al.,* 1987, 1995), the golden hamster, *Mesocricetus auratus* (Brainard *et al.,* 1982), the tammar wallaby, *Macropus eugensi* (McConnel, 1986), the European hamster, *Cricetus cricetus* (Pévet *et al.,* 1989, Vivien-Roels *et al.,* 1992), the adult domestic german landrace sow, (McConnel and Ellendorf, 1987) and the deer (Paterson and Foldes, 1994).

Depending on the species, the amplitude of the daily rhythm of melatonin secretion appears to be highest in summer or in summer and autumn as for example in the tammar, the ram or the horse (McConnel, 1986, Sheikheldin *et al.,* 1992, Guerin *et al.,* 1995), in winter and spring as in golden or Djungarian hamsters (Brainard *et al.,* 1982, Steinlechner *et al.,* 1995), in autumn and winter as in the European hamster (Fig. 3) (Vivien-Roels *et al.,* 1992). The night peak of melatonin can even disappear completely in some species. In the domestic adult german landrace sow, for example, Mc-Connel and Ellendorf (1987) noted that a nocturnal surge in plasma melatonin was only observed under a photoperiod of 12L/12D. When animals are shifted to a long (16L/8D) or a short (8L/16D) photoperiod, the night peak of melatonin was suppressed. Also in the same species, Reiter (1987) has reported that in such conditions NAT activity did not show a nocturnal increase.

In the European hamster, important seasonal variations in the amplitude of the daily rhythm of melatonin are reported (Pévet *et al.,* 1989, Vivien-

Roels *et al.*, 1992). In autumn, winter and early spring, a clear day-night rhythm is present with a lack of day-night variations in spring and early summer (Fig. 3). When looking more precisely at the night pattern of melatonin it appears that a small peak of low amplitude, of one hour duration, is still present even near the summer solstice, the amplitude and the duration of which increasing rapidly when daylength decreases (Fig. 4).

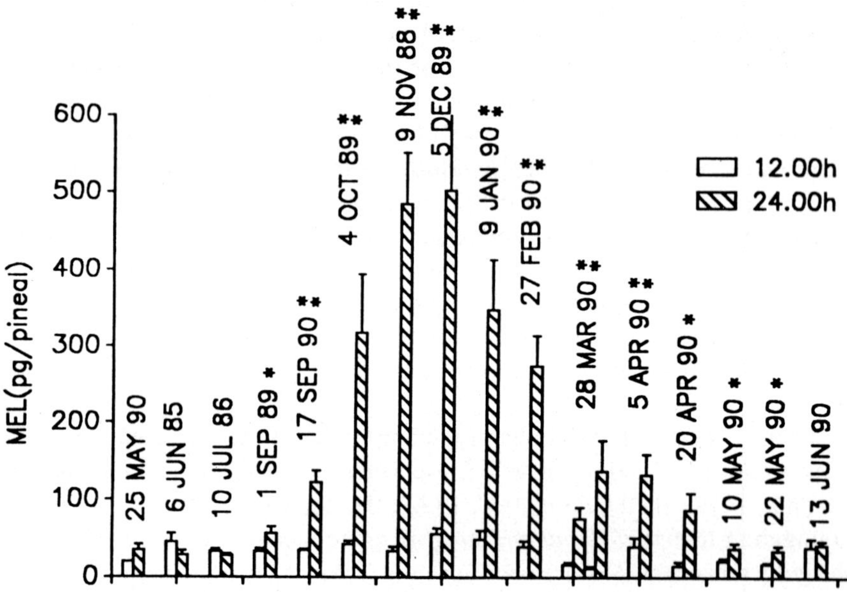

Fig. 3 Midday and midnight pineal melatonin concentrations throughout the year in the European hamster, *Cricetus cricetus*. p < .005 and p < .001 compared to daytime values (from Vivien-Roels *et al.*, 1992, with permission).

These results clearly show that, even in mammals, changes in the amplitude of the melatonin peak may be of importance in conveying the melatonin message to the brain.

To date, a few experimental studies report variations in the amplitude of the melatonin rhythm related to variations in environmental factors. When comparing the daily pattern of melatonin synthesis in the pineal gland of golden hamsters maintained either indoors under standard artificial conditions with animals maintained outdoors under natural conditions, Brainard *et al.* (1982) have shown that in addition to seasonal variations in the duration of the night peak of melatonin, in animals maintained outdoors, a clear decrease in the amplitude of the melatonin peak was observed in summer and fall, compared with winter and spring, which was not observed in standard conditions where only photoperiod varied. On the other hand, in female Djungarian hamsters, it has been shown that the amplitude of the daily rhythm of melatonin was significantly lower under long (16L/8D) than under short (8L/16D) photoperiod (Illnerova *et al.*, 1984, Lerchl *et al.*,

1993, Miguez *et al.*, 1995). In the European hamster, it has been reported that both photoperiod and environmental temperature are implicated in the regulation of the amplitude and the duration of the night peak of melatonin, (Vivien-Roels *et al.*, 1993, Vivien-Roels *et al.*, 1997), and that even the photoperiodic changes alone are able to modify the amplitude of the melatonin peak. Similar results were recently obtained in the marsh rice rat, *Oryzomys palustris* (Edmonds *et al.*, 1995); both the duration and the amplitude of the melatonin peak are lower in long photoperiod (16L/8D).

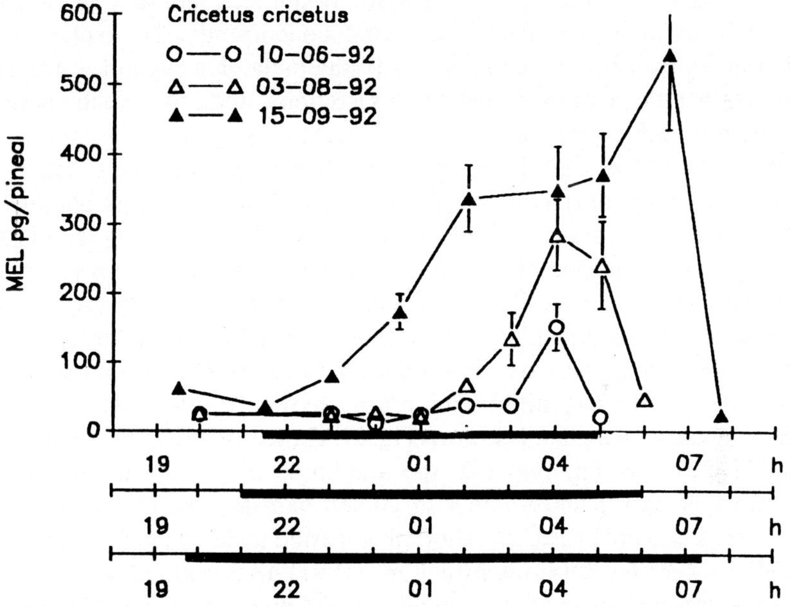

Fig. 4 **Daily variations in pineal melatonin content in the European hamster, *Cricetus cricetus*, maintained under natural conditions in June, August and September.**

Recently, Stieglitz *et al.* (1991), has shown that cold prevents the light induced inactivation of NAT in the Djungarian hamster and that circulating catecholamines may stimulate melatonin synthesis under cold load.

From these results it appears that mammals, as other Vertebrates, show seasonal variations in both the duration and the amplitude of the night peak of pineal and/or circulating melatonin, which are influenced by environmental factors.

Physiological Consequences of the Changes in the Amplitude of Nocturnal Peak of Melatonin

From this brief survey of the literature, it appears that in mammals, as well as in non-mammalian Vertebrates, seasonal variations in the daily pattern of melatonin secretion are more complicated than a single change in the duration of the night peak of melatonin. Depending on the species, the geographical location, the *modus vivendi,* other components of the melatonin message, and more especially the amplitude of the melatonin rhythm varies considerably throughout the year suggesting that amplitude may be also of physiological importance in transmitting environmental information to the brain. This raises the question of the physiological significance of these seasonal variations in the amplitude of the melatonin rhythm.

In their natural habitats, animals are submitted to various environmental cues, photoperiod, temperature, rainfall, flooding, food supply...all these factors being interdependent and acting concurrently. To explain these variations in the different components of the melatonin rhythm, it has been proposed that the duration of melatonin secretion could convey photoperiodic information, while amplitude of phase could convey other environmental informations (Pévet, 1985, Pévet *et al.,* 1989a, Vivien-Roels and Pévet 1983). Such a hypothesis is supported by the different results obtained in poikilotherm Vertebrates, where it is clear that at least temperature or thermoperiodic changes affect the amplitude of the melatonin rhythm in fishes, amphibians and reptiles. In homeotherms a clear role for environmental temperature in regulating daily variations in melatonin secretion is less clear. Nevertheless, we have seen that in the European hamster, the highest amplitude of the day-night melatonin rhythm is observed when the environmental temperature is lowest (Fig. 3). These results agree with Brainard *et al.* (1982) who reported that the amplitude of the melatonin peak in December is significantly higher in golden hamsters maintained outdoors than in animals kept in artificial short photoperiod under laboratory conditions (+ 20°C). More recently Stieglitz *et al.* (1991) also reported that 1 hr cold exposure prevents the light induced decrease of pineal NAT in the Djungarian hamster. In contrast, however, ambient temperature does not affect the daily pattern of melatonin secretion in golden hamsters kept under short photoperiod, while low temperature accelerates gonadal atrophy induced by short photoperiod (Pévet *et al.,* 1989b).

Other environmental factors, such as rainfall (Srivastva 1985), food supply (Brown *et al.,* 1989) or magnetic field (Welker *et al.,* 1983, Bardassano, 1989), are also known to affect melatonin synthesis and in some cases the amplitude of the melatonin rhythm.

On the other hand, experimental changes in daylength alone can affect the amplitude of the day-night rhythm of melatonin or NAT activity in the anole lizard (Underwood, 1985, Underwood and Calaban, 1987), the quail (Follett, 1985), the cat (Leyva *et al.,* 1984), the tammar wallaby (Mc-Connel and Tyndale-Biscoe, 1985), the Djungarian hamster (Steinlechner *et al.,* 1987) or the European hamster (Vivien-Roels *et al.,* 1993, 1997).

In some species the amplitude appears to be decreased under short photoperiod, in others, such as the European hamster, it is maximum under short photoperiod (Fig. 3). This raises the question of the regulation of the amplitude of the melatonin rhythm in the pineal gland. To date it is well-known, at least in mammals, that the nocturnal synthesis of melatonin in the pineal gland is essentially controlled by noradrenergic sympathetic innervation originating from the superior cervical ganglia. Norepinephrine (NE) released during the dark period from the sympathetic nerve endings activates the synthesis of NAT, the rate limiting enzyme in melatonin synthesis (review in Klein *et al.,* 1981).

On the other hand, since the last decade, many neuropeptides originating from various areas of the central nervous system or colocalized with NE in the sympathetic nerves, such as NPY for example, have been shown to be present in the pineal gland and to affect melatonin synthesis (ref. in Simonneaux, 1995). Some neuropeptides have been shown to stimulate pineal activity, while others are only able to modulate a NE-induced effect (ref. in Simonneaux, 1995). In the European hamster, for example, important seasonal variations in the NPY fibers (Mollu *et al.*, 1998), showing some temporal relationships with the seasonal variations in the amplitude of the melatonin rhythm, suggest a possible role of this neuropeptide in modulating the seasonal variations in the daily pattern of melatonin secretion in this species.

Conclusion

In conclusion it appears that the regulation of the seasonal variations in the daily pattern of melatonin synthesis and secretion is very complex. If, undoubtedly the seasonal variation in the duration of the night peak of melatonin is an important component of the melatonin message, other components and more especially the amplitude of the melatonin rhythm appears also very important in the transmission of the melatonin message to the brain and has seriously to be taken in account. Photoperiodic information, acting directly in the photosensitive pineal of lower vertebrates or indirectly, via the lateral eyes in mammals, is clearly implicated in the regulation of the duration of the night peak of melatonin while other environmental factors, and more especially temperature and thermoperiod appear to be implicated in the regulation of the seasonal variations in the amplitude of the melatonin rhythm, especially in Poikilotherms, corroborating the hypothesis (Pévet, 1979, 1985, Vivien-Roels and Pévet, 1983) that the pineal gland of vertebrates is primarily implicated in the integration of several environmental informations via the daily pattern of melatonin and other indoleamine secretion.

Whether endogenous information, originating from the different areas of the brain, via the neuropeptides, could modulate the seasonal variations in the daily pattern of melatonin secretion remains to be determined and needs further investigations.

References

Arendt, J. (1979). In: The pineal gland of Vertebrates including man (J. Ariens-Kappers and P. Pévet, Eds.), Progress in Brain Res., 52, 242-258, Elsevier/North Holland, Amsterdam.

Bardassano, J.L. (1989) In: The Pineal gland and magnetic fields. Instituto de Bioelectromagnetism, pp 1-167, Alonso Santa Cruz.

Bartness, T.J. and Goldman, B.D. (1988). Am. J. Physiol., 255, R812-R821.

Benellbaz, G.A., Piezzi, R.S. and Lynich, H.J. (1976), Gen. Comp. Endocrinol. 35, 43-46.

Bittman, E.L. and Karsch, F.J. (1984). Biol. Reprod., 30(3), 585-595.

Bolliet, V., Begay, V., Ravault, J.P., Ali, M.A., Collin, J.P. and Falcon, J. (1994). J. Pineal Res., 16, 77-84.

Bonnefond, C., Martinet, L. and Monnerie, R. (1990). J. Neuroendocrinol. 2, 583-592.

Brainard, G.C., Petterborg, L.J., Richardson, B.A. and Reiter, R.J. (1982). Neuroendocrinol., 35, 342-348.

Brown, G.M., Chick, G.L., Ho, A.K., Kennedy, S. and Garfinkel, P. (1989) In: Adavances in Pineal Research, vol 3 (R.J. Reiter and O.G. Milestone, Eds.), pp 83-92. John Libbey, London.

Cockrem, J.F. (1991). J. Pineal Res., 10, 2-8.

Delany, M.J. and Happold, D.C.D. (1979). In: Ecology of African Mammals, London Longman group, 1-356.

Delgado, M.J. and Vivien-Roels, B. (1989a). Gen. Comp. Endocrinol., 75, 46-53.

Delgado, M.J. and Vivien-Roels, B. (1989b). In: The Pineal Gland and Magnetic Field, Eds, J.L. Bardassano, Institudo de Bioelectromagnetismo, Alonso de Santa Cruz, 39-49.

Delgado, M.J., Alonso-Gomez, A.L., Gancedo, B., De Pedro, N., Valenciano, A.I. and Alonso-Bedate, M. (1993). Gen. Comp. Endocrinol., 92, 39-49.

D'Istria, M., Monteleone, P., Serono, I. and Chieffi, G. (1994). Gen. Comp. Endocrinol., 96, 6-11.

Duston, J. and Bromage, N. (1986). J. Endocrinol. 108, (suppl.) 77.

Edmonds, K.E., Rollag, M.D. and Stetson, M.H. (1995). J. Pineal Res. 10, 2-8.

Falcon, J., Thibault, C., Begay, V., Zachmann, A. and Collin, J.P. (1992) In: Rhythms in Fishes. Ed. Ali, A., Plenum Press, New-York, 167-198.

Falcon, J., Bolliet, V., Ravault, J.P., Chesneau, D., Ali, A. and Collin, J.P. (1994) Neuroendocrinol. 45, 479-486.

Firth, B.T., Kennaway, D.J. and Rozenbilds, M.A.M. (1979) Gen. Comp. Endocrinol. 37, 493-500.

Firth, B.T. and Kennaway, D.J. (1987) Brain Res., 404, 313-318.

Firth, B.T., Thompson, M.B., Kennaway, D.J. and Bilan I. (1991) Neurosci. Letters, 121, 139-142.

Follett, B.K., Foster, R.G. and Nicholls, D.J. (1985) In: Photoperiodism, melatonin and the pineal, Ciba Foundation Symposium 117, 93-105, Pittman, London.

Gern, W.A. and Norris, D.O. (1979) Gen. Comp. Endocrinol., 38, 393-398.

Gern, W.A., Norris, D.O. and Duvall, D. (1983) J. Herpetology, 3, 228-234.

Goldman, B.D. and Darrow, J.M. (1983) Progr. Neuroendocrinol., 37, 386-396.

Guerin, M.V., Dee, J.R., Kennaway, D.C. and Matthews, C. (1995). J. Pineal Res., 19, 7-15.

Hoffmann, K. (1985) In: The Pineal gland, current state of pineal research. B. Mess, C. Ruzsas, L. Tima and P. Pévet eds. Akad. Kiado, Budapest, and Elsevier, North Holland 211-230.

Iigo, M. and Aida, K. (1995) J. Pineal Res., 18, 62-68.
Illnerova, H., Hoffmann, K. and Vanecek, J. (1984) Neuroendocrinol., 38, 226-231.
Karsch, F.J., Malpaux, B., Wayne, N.L. and Robinson J.E. (1988) Reprod. Nutr. Dev., 28, 459-472.
Kezuka, H., Furukawa, K., Aida, K. and Hanyu, I. (1988) Gen. Comp. Endocrinol., 72, 296-302.
Kezuka, H., Iigo, M., Furukawa, K., Aida, K. and Hanyu, I. (1992) Zool. Sci., 9, 1047-1053.
Klein, D.C., Auerbach, D.A., Namboodiri, M.A.A. and Wheler, G.H.T. (1981) In: The pineal gland, vol I, Anatomy and Biochemistry (R.J. Reiter, Ed.), pp 199-277. C.R.C. Press, Boca Raton, Florida
Lerchl, A. and Schlatt, S. (1993) Neuroendocrinol. 57, 359-364.
Leyva, H., Addigo, L. and Stabenfeld, G. (1984) Endocrinol. 115, 1729-1736.
Liou, S.S., Cogburn, L.A. and Bieller, V. (1987), Gen. Comp. Endocrinol., 67, 221-226.
Lofts, B. (1974) Physiology of Amphibians, vol II, Reproduction, Academic Press, New-York, 107-218.
Max, M. and Menaker, M. (1992) J. Comp. Physiol. A, 170, 479-489.
Maywood, E.S., Buttery, R.C., Vance, G., Herbert, J. and Hastings, M. (1990) Biol. Reprod., 43, 174-182.
McConnel, S.J. (1986) J. Pineal Res., 3, 119-126.
McConnel, S.J. and Tyndale-Biscoe, C.M. (1985) J. Reprod. Fertil. 73, 529-538.
McConnel, S.J. and Ellendorf, F. (1987) J. Pineal Res., 4, 201-210.
Menaker, M. and Wisner, S. (1983) Neurobiol., 80, 6119-6121.
Mendonca, M.T., Tousignant, A.J., Crews, D. (1995) Gen. Comp. Endocrinol. 100, 226-237.
Miché, F., Vivien-Roels, B., Pévet, P., Spehner, C., Robin, J.P. and Le Maho, Y. (1991) Gen. Comp. Endocrinol., 84, 249-263.
Miguez, J.M., Recio, J., Vivien-Roels, B. and Pévet, P. (1995) J. Pineal Res., 19, 139-148.
Mollu, M., Masson-Pévet, M. and Pévet, P. (1998) Cell Tissue Res. 291, 423-431.
Moyer, R.W., Firth, B.T. and Kennaway, D.C. (1995) Brain Res., 675, 345-348.
Pang, S.F., Shiu, S.Y.W. and Tse, S.T. (1985) Gen. Comp. Endocrinol., 58, 464-470.
Paterson, A. and Foldes, A. (1994) J. Pineal Res., 16, 167-177.
Pévet, P. (1979) In: The pineal gland of vertebrates including man. J. Ariens -Kappers and P. Pévet Eds., Progr. Brain Res. 52, 149-194. Elsevier, North Holland, Amsterdam.
Pévet, P. (1985) In: The pineal gland, current state of pineal research, B. Mess, C. Ruzsas, L. Tima and P. Pévet eds. 163-186, Akad Kiado, Budapest and Elsevier, Amsterdam.
Pévet, P. (1988) Reprod. Nutr. Dev., 28, 443-458.
Pévet, P., Vivien-Roels, B., Masson-Pévet, M. and Canguilhem, B. (1989a) J. Pineal Res. 6, 233-242.
Pévet, P., Vivien-Roels, B. and Masson-Pévet, M. (1989b) J. Neural Tansm. 83, 235-241.
Pévet, P., Vivien-Roels, B. and Masson-Pévet, M. (1991) In: Advances in Pineal Research, F. Fraschini and R.J. Reiter eds 147-157.
Pitrosky, B., Masson-Pévet, M., Kirsen, R., Vivien-Roels, B., Canguilhem, B. and Pévet, P. (1991) J. Pineal Res. 11, 149-155.
Randall, C.F., Bromage, N., Thorpes, J.E., Miles, M.J. and Muir, J.J. (1995) Gen. Comp. Endocrinol., 98, 73-86.
Rawdling, R.S. and Hutchison, V.M. (1992) Gen. Comp. Endocrinol., 88, 364-374.
Reiter, R.J. (1987) In: Comparative Physiology of Environmental Aaptations, Adaptations to climatic changes, P. Pévet ed., Karger, Basel, 71-81.

Rismiller, P.D. and Kennaway, D.J. (1990) In: Abstract of the Vth colloquium of the European pineal study group, Guildford, September 2th-7th 1990, p 5.

Sheikheldin, M.A., Howland, B.E. and Palmer, W.A. (1992) J. Pineal Res., 12, 58-63.

Sicard, B., Kyelem, M., Papilon, Y., Diara, W. and Keita, M. (1995) CTA ORSTOM, Recherches, Developpement. John Libbey, 1-46.

Simonneaux, V. (1995) Neuroendocrinol. Lett. 2, 115-129.

Srivastva, M. (1985). Thesis, Banaras Hindu University, Varanasi, India.

Steinlechner, S., Buchberger, A. and Heldmaier, G. (1987), J. Comp. Physiol. A, 160, 593-597.

Steinlechner, S., Baumgarten, I., Klante, G. and Reiter, R.J. (1995), Neurochem. Int., 27, 245-251.

Stieglitz, A., Steinlechner, S., Ruf, T. and Heldmaier, G. (1991) J. Comp. Physiol., A 168, 599-603.

Taniguchi, M., Murakami, N., Nakamura, H., Nasu, T., Shinohara, S. and Etoh, T. (1993) Brain Res., 620, 297-300.

Underwood, H. (1989) Experientia, 45, 914-922.

Underwood, H. and Calaban, M. (1987) J. Biol. Rhythms, 2, 179-193.

Underwood, H., Binkley, S., Siopes, T. and Menaker, M. (1984) Gen. Comp. Endocrinol., 56, 70-81.

Valenciano, A.I., Alonso-Gómez, A.L., Alonso-Bedate, M. and Delgado, M.J. (1994) Pflugers Arch. 429, 153-159.

Vivien-Roels, B. and Arendt, J. (1981) Adv. Biosci., 29, 401-406.

Vivien-Roels, B. and Pévet, P. (1983) In: Pineal Res. Reviews, Alan R. Liss, Inc. New York, 1, 91-143.

Vivien-Roels, B., Arendt, J. and Bratdke, J. (1979) Gen. Comp. Endocrinol., 37, 197-210.

Vivien-Roels, B., Pévet, P. and Claustrat, B. (1988) Gen. Comp. Endocrinol. 86, 239-247.

Vivien-Roels, B., Masson-Pévet, M., Canguilhem, B., Bonn, D. and Pévet, P. (1993) In: Melatonin and the Pineal gland, from basic science to clinical application, Y. Touitou, J. Arendt and P. Pévet eds. Elsevier science Publ. 199-203.

Vivien-Roels, B., Pitrosky, B., Zitouri, M., Malan, A., Canguilhem, B., Bonn, D. and Pévet, P. (1997) Gen. Comp. Endocrinol. 106, 85-94.

Welker, H.A., Semm, P., Willig, R.P., Commentz, J.C., Wiltschko, W. and Vollrath, L. (1983) Exp. Brain Res. 2, 426-432.

Zachmann, A., Knijf, S., Bolliet, V. and Ali, A. (1992) Neuroendocrinol. Lett. 13, 325-330.

Comparative Endocrinology and Reproduction
K.P. Joy, A. Krishna and C. Haldar (Eds)

31. Avian Pineal Organ: Extra reproductive effects

C. Haldar
Department of Zoology, Banaras Hindu University, Varanasi-221 005, India

Introduction

Understanding of functions of avian pineal organ on the basis of the reports available on seasonally breeding laboratory birds requires both basic and detailed studies; neither is sufficient by itself. What is known of the avian pineal is only detailed effect of melatonin and pinealectomy (very less) on reproduction and behaviour of some birds. Moreover, reports on behavioral functions of birds are more common because of the pineal's function as an oscillator (Cassone, 1990; Takahashi *et al.*, 1989). In the past several years, data have accumulated mainly from tropical zone birds with particular reference to effects of pineal on seasonal reproduction and regulation of ovarian/testicular functions (Maitra and Dey, 1994; Mazi *et al.*, 1992; Ghosh and Haldar, 1994). If we go through the published literature, we find less reports on wild species of birds in relation to pineal organ (Maitra and Dey, 1993, 1994; Ghosh and Haldar, 1994); however, great extent of work exist mostly on domestic/laboratory birds (Takahashi *et al.*, 1989).

The mammalian pineal organ has already drawn full attention as a regulator of many "extra reproductive functions", while such reviews and reports are totally lacking for avian pineal. Therefore, the present review will concentrate mainly upon experimental aspects of avian pineal organ to emphasize its extra reproductive effects. Extra reproductive effects were noted while studying the inter-relationship of pineal gland with other endocrine organs, such as thyroid and adrenal known as metabolically important gland not only for body physiology, but also for reproduction of the avian group, importance of pineal gland and melatonin on metabolic functions (carbohydrate metabolism) and to some extent with Harderian gland (HG) due to its capacity of synthesizing melatonin, which has led ornithologists to search their interrelationship to immunity and reproductive functions.

Importance of the abovementioned interrelationships cannot be underrated but mechanisms by which above all control seasonality and physiology in temperate/tropical zone species are reflected faithfully in caged condition. Further, the behavioral studies by the group of Binkley on various phase changes of locomotor activity are really marvelous and well reviewed several times (Cassone, 1990; Takahashi *et al.*, 1989). Due to lack of space, I delete the same from the present article.

Little attention has been paid on pineal research in seasonally breeding birds of the tropical and sub-tropical zones and most of the studies were carried out in diurnal, herbivorous laboratory or domestic bird species. Our laboratory is engaged for quite some time with studies on pineal organ functions of a seasonally breeding bird of Coutrnix group, the Indian jungle bush quail, *P. perdicula asiatica.* Being locally migratory, its migration depends upon the movement of monsoon clouds, and restricted to green fields of paddy for shelter. With the end of monsoon these birds face major stress following harvest (lack of food and shelter) and disappear in bushes of the jungle in plain region. Our own reports and some unpublished findings along with reports in literature are being categorized as follows to explain the extra reproductive effects of avian pineal.

Pineal-thyroid Relationship

The general effect of the pineal upon the hypothalamo-hypophyseal axis was reported by Fraschini and Martin (1970). The relationship of pineal with thyroid in particular, although not very well illustrated in vertebrates, yet, reports are available (Johnson, 1980; Vriend, 1983) which allowed a well defined similarity with the effect of the pineal gland and melatonin on the neuroendocrine-gonadal axis. In the light of the current research the question of whether the pineal gland influences thyroid function can be restated as: does the pineal gland influence TSH-induced secretion of thyroid hormones? The analysis of data from literature for avian groups presents following clear interrelationship of pineal gland with thyroid function.

Annual Pineal-thyroid cyclicity

The annual cyclicity of both the glands were reported only for the tropical diurnal bird, jungle bush quail (Haldar and Ghosh, 1989). High thyroid weight was noted in April-June which coincided with the peak testicular weight. On the contrary, an inverse relationship was observed between the thyroid and pineal gland weight throughout the year. Analysis of variance of the data on thyroid weight throughout the year showed a significant variation. The positive relationship of thyroid weight with testis weight and a negative relationship with the pineal gland weight was justified by regression correlation coefficients (r_{thg} = + 0.833, ± 0.96, $P < 0.001$; r_{thp} = – 0.68, ± 0.169, $P < 0.001$). For the first time we observed a direct correlation of thyroid weight cycle with testicular cycle and an inverse correlationship of the thyroid gland weight cycle to that of the pineal weight cycle.

Effect of pinealectomy (Px)

Px in quails significantly reduced the thyroid gland weight (from the controls) in testicular recrudescence and gonadal active phases. Histologically, the activity of the thyroid increased significantly depicting less amount of colloid and increased epithelial cell height as well as nuclear diameter (Haldar and Ghosh, 1989). Thus, Px had freed the thyroid gland from its inhibition during these two phases and an inverse relation of thyroid activity with testicular activity became evident. Defronzo and Roth (1972) and Lewinski *et al.* (1984) had also observed an increased thyroid activity following Px in mammals.

Effect of different 5-methoxyindoles

Continuous release of melatonin and 5-methoxytryptamine via silastic capsule in quails showed a testicular phase dependent effect. Both the methoxyindoles increased thyroid weight significantly during all the phases except the recrudescence and active testicular phase, when a significant decrease in thyroid weight was noted. Histologically, the thyroid gland of both the groups treated with melatonin and 5-methoxytryptamine implants showed a high follicular activity (Haldar and Ghosh, 1989).

Subcutaneous injections (20 μg/day/bird) of the three methoxyindoles melatonin, 5-methoxytryptamine and 5-methoxytryptophol (MEL, 5-MT, 5-MTL) had increased thyroid weight during the inactive reproductive phase whereas, in the active reproductive phase 5-MT injections in the morning and MEL in the evening increased thyroid weight significantly from the controls. When potency of the effect of different 5-methoxindoles on thyroid weight was compared it was noted that 5-MTL is less effective than MEL and 5-MT during both the reproductive phases.

It appears that the inhibitory effects of late-afternoon melatonin injections and the counter-antithyroid influence of melatonin implants are characteristic phenomena for thyroid function. The most meaningful relationship between pineal and thyroid-gonadal function preclude precise definition of the nature of mechanism of the interaction. The concepts have been formulated by indirect modulation of the sensitivity of the gonad to its pituitary trophic hormones by thyroid and pineal hormone.

From the overall studies in avian group for pineal-thyroid interaction we can summarize that:

(i) There is a similarity of the effects of the pineal gland and melatonin on the neuroendocrine-thyroid axis and on the neuroendocrine-gonadal axis;
(ii) Under the control of environmental photoperiod (seasonal changes in daylength etc.) the pineal gland influences both reproductive hormone and hormones of thyroid, as melatonin administration inhibits both gonadal and thyroid function;
(iii) The similarity suggests a single site of action within brain, i.e. TRH-producing neurons (Vriend, 1978, 1983);

(iv) TRH can affect release of prolactin and TSH from pituitary which is once again involved in both regulation of gonadal and pineal function.

These various suggestions lead to the conclusion that the pineal gland influences the neuroendocrine-thyroid and the neuroendocrine-gonadal axis through a common controlling mechanism involving releasing factor because both pineal and thyroid poses enzyme for the synthesis of T_3, and the activity of this enzyme 5′ deiodinase is very similar to that regulates melatonin synthesis. Hence, they could be functionally dependent (Reiter *et al.*, 1993).

Pineal-adrenal Relationship

Attempts made to establish the role of the pineal gland in hypothalamo-hypophyseal-adrenal axis in rodents and reptiles have yielded inconsistent results and no conclusion can be brawn from the available results (Kitay and Altschule, 1954; Wurtman *et al.,* 1959). The pineal has been said to exert inhibitory influence on the pituitary-adrenal axis. Mess (1983) suggested that the pineal may act as a regulator influencing the function and response of adrenal gland but the mode of action was not clearly delineated. Several workers have reported that the pineal may act:

(i) directly on the adrenal cortex (Giordano *et al.,* 1970), and may
(ii) diminish adrenal function via the intermediary of pituitary ACTH (Piva *et al.,* 1973), or
(iii) perhaps both (Defronzo, 1971).

On the·other hand, importance of adrenal gland in avian reproduction is well documented (Ghosh *et al.,* 1996). In birds, no clear reports are available regarding pineal-adrenal relationship, keeping the increasing importance of the pineal in reproductive physiology of vertebrates (Wurtmen *et al.,* 1959; Reiter *et al.,* 1966; Reiter, 1975, 1991), we studied the possible relationship of pineal-adrenal-gonad in some tropical zone animals including birds. The summary of reports from our and other labs on avian groups is given below.

Annual pineal-adrenal cyclicity

Our studies with the diurnal species *P. asiatica* presented positive correlation of adrenal gland weight cycle with the testis weight cycle. Such a relationship has been reported in some other birds; for example pigeon (Riddle, 1923), starling (Burger, 1938), duck and mallard (Honh *et al.,* 1965), lal munia (Garg, 1968), baya (Bageshwar, 1969), common myna (Chaturvedi, 1977).

While correlating with pineal gland cycle, we found that the adrenal weight had a direct relation with the annual testis weight cycle, and an inverse relation with pineal weight. Therefore, high adrenal weight was noted in the month of June when the testis attained its complete maturity (Haldar and Ghosh, 1989).

Effect of pinealectomy

Px in the diurnal bird *P. asiatica* decreased the adrenal weight significantly during gonadal recrudescence and active phases; whereas, it had no effect on the adrenal in the other phases of the testicular cycle when compared with the controls. In the testicular inactive phase it increased the adrenal weight.

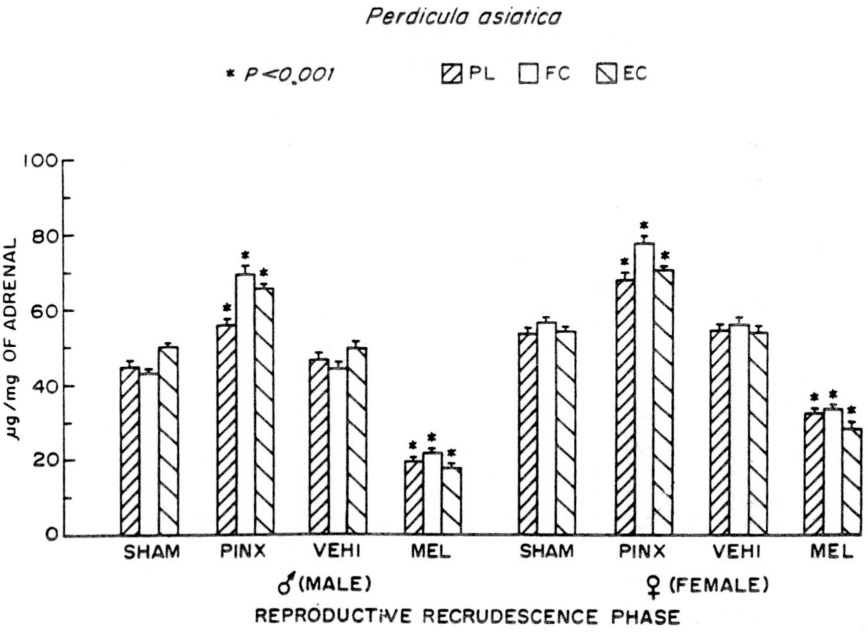

Fig. 1 **Changes in adrenal lipid fractions (µg/mg of adrenal) of *P. asiatica* during reproductive recrudescence phase after melatonin (MEL) and pinealectomy (PINX). SHAM–sham operated control; PL–phospholipids; FE–free cholesterol; EC–esterified cholesterol.**

Histologically, the adrenal gland after Px presented reduced size of the medullary cells, suggesting that only the gonadal recrudescence and active phases make adrenal sensitive to the effects of Px. Further Px, releases the adrenal from pineal inhibition only during the above two reproductive phases (Haldar and Ghosh, 1989). Px during the reproductive recrudescence phases increased the adrenal phospholipid (PL), esterified cholesterol (EC), and free cholesterol (FC) contents significantly. In the same phase, the plasma corticosterone level was also elevated significantly after Px (Figs. 1 and 2).

Effects of administration of 5-methoxyindoles

In *P. asiatica* MEL and 5-MT represent the pineal effector compounds responsible for pineal mediated adrenal function as noted by others for mammals (Defronzo and Roth , 1972) which does not include 5-MT. Injection of all the three different 5-methoxyindoles i.e. MEL, 5-MT showed a varied effect on the adrenal gland weight when examined for their effectiveness during the inactive and active phases of the testicular cycle. Melatonin in the morning hours increased the adrenal gland weight significantly during the inactive phase. However, during the active phase it had a time dependent effect, i.e. a significant decrease was noted during the morning hours and a significant increase during the evening hours (Haldar and Ghosh, 1989).

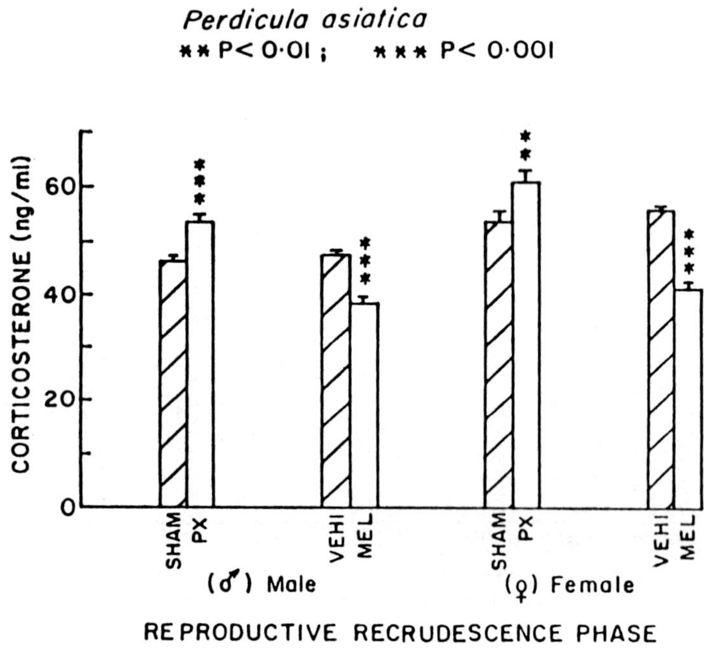

Fig. 2 Plasma corticosterone level (ng/ml) of *P. asiatica* during reproductive recrudescence phase after melatonin injection and pinealectomy.

Following the implantation of MEL and 5-MT via silastic capsule, adrenal weight of diurnal bird *P. asiatica* decreased during the gonadal recrudescence and active phases, whereas it had no effect during the other phases of the reproductive cycle. Histologically, both 5-methoxyinoles reduced the size of the medullary cells.

For adrenal-pineal interrelationship we extended our study with a nocturnal species Indian spotted owlet, *A. brama.* In the reproductive recrudescence and active phases, subcutaneous MEL administration (S.C.) for 20 days in *A. brama* significantly decreased the adrenal and gonadal weight of both the sexes. The pineal weight increased significantly after the administration of MEL in intact birds. 5-methoxytryptamine and 5-methoxytryptophol had no effect during the inactive phase but during the active phase both the methoxyindoles showed time dependent activity (Sudha, 1995). Administration of MEL in the same phase not only reduced adrenal lipid fractions, but also the plasma corticosterone level significantly. Similar type of effects was observed in the reproductive active phase. Similarly, the plasma corticosterone level was also reduced significantly in both the sexes (Figs. 3a, b; 4a, b).

Some interrelations of corticosterone and melatonin level

In general no sex-dependent activity changes in the pattern of corticosterone and MEL level were noted in birds. Therefore, its activity has no sex steroid dependent regulation, at least when 24 h cyclicity is concerned. Involvement of the pineal gland in controlling the 24 h cycle of different physiological aspects in birds has been described by Menaker and Oksche (1974) and Ebihara and Kawamura (1980). Vivien-Roels and Pevet (1986) have suggested that this gland brings about its effect by conveying photoperiodic information "via its daily pattern of MEL secretion".

During daily variation study, two peak cycles of various hormones (gonadal steroid, thyroid hormones, melatonin and corticosterone) have been noted as a common feature of tropical zone animals. In other words, the level of plasma MEL measured in the 24-hour time-scale showed a day/night rhythmicity with a distinct elevation at day time besides the night peak. Such an observation has never been reported in birds till-date, however reports are available in domestic pig, fallow deer and turtle, *Testudo hermanni* (Vivien-Roels *et al.,* 1979). This characteristic feature of plasma MEL has also been observed in our labratory for two more tropical zone vertebrates (Indian palm squirrel, Saxena, 1988; Shavali, 1992; Indian chequered snake, Pandey, 1988). It was suggested that a short peak during the day time, either at dawn or dusk may be physiologically important, which can be related to the animal's behavioral endocrinology.

However, the nocturnal bird, *A. brama* showed a different pattern. In this bird, the day time peak was much higher than the night time value (Fig. 5a, b). This is the first report of its kind, showing a strange nature of MEL rhythmicity in a nocturnal avian species. Such a presentation of high MEL level in the day time is contrary to our observation in the diurnal bird, *P. asiatica.* The day time peak in this bird was smaller than night time peak. Our observations support strongly that behavioral endocrinology is having a strong impact on the physiological processes of endocrine glands. The nocturnal *A. brama* hide themselves completely in the dark burrows and holes of the tree during day time, hence, they experience a total scotophase. It could be the reason for the high MEL level in day time in this bird. On the other hand, they expose themselves to the artificial or moonlight condition of the night time for catching their prey, and this little photoperiod may be responsible for the suppression and small peak of MEL. Since the study was performed with the animal experiencing natural condition in our lab, hence they maintained original MEL cyclicity. However, this contention requires further investigation.

The cyclicity of plasma corticosterone presented almost an inverse relationship with the daily variations of MEL in both the species of bird. In

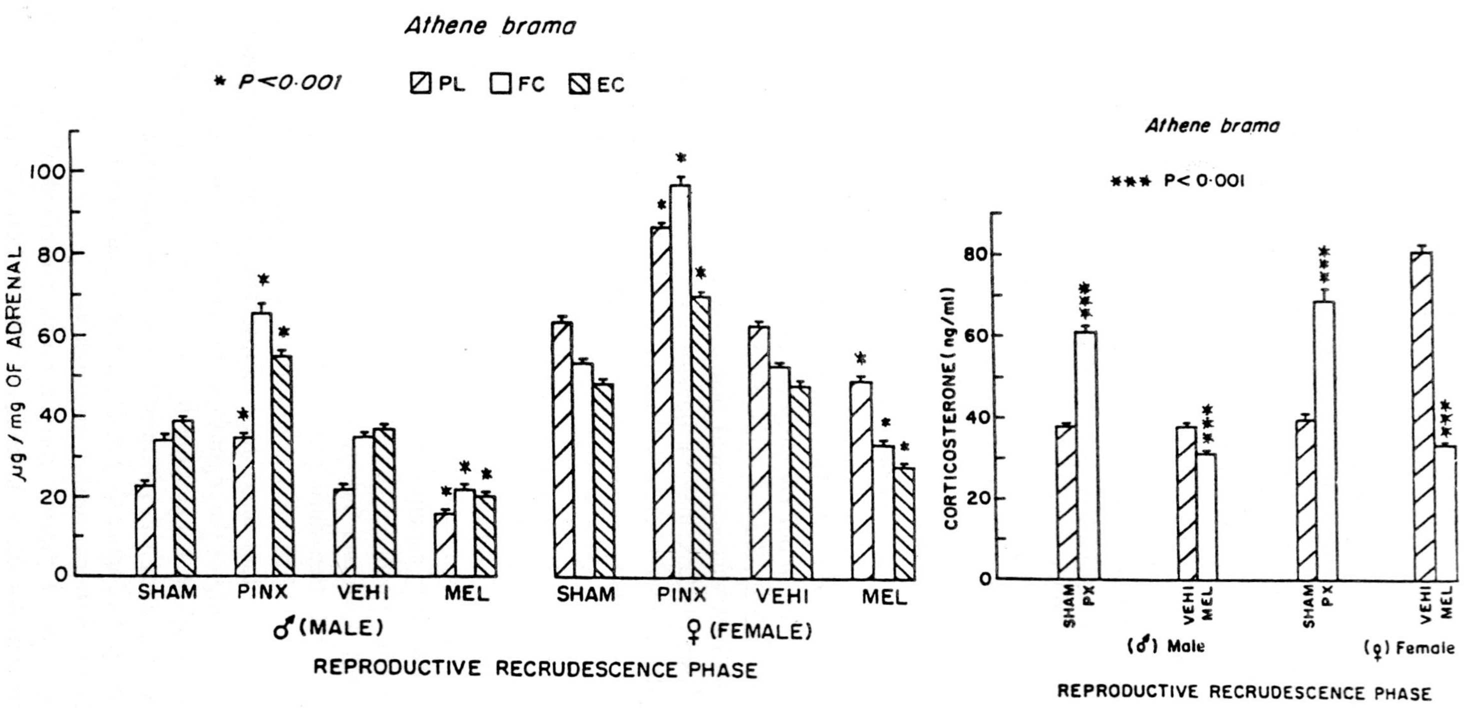

Fig. 3 Changes in adrenal lipid fractions and plasma corticosterone level of *A. brama* following melatonin evening injection and pinealectomy during recrudescence phase.

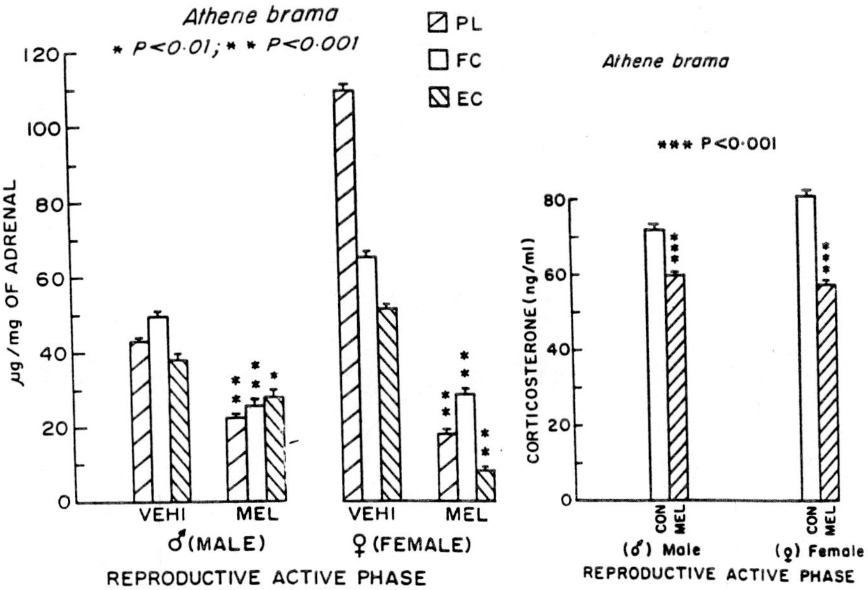

Fig. 4 Changes in adrenal lipid fractions and plasma corticosterone level of *A. brama* following melatonin evening injection during reproductive active phase.

temperate zone birds, reports on duck (Chan and Phillips, 1973) and quail, *Coturnix coturnix japonica* (Voissin and Assenamcher, 1968) suggested that corticosterone level in avian groups is influenced by waking and resting time of their habitat. However, there are no such reports available till date about the presence of diurnal pattern in corticosterone level in any of the Indian tropical zone birds. Here in both the nocturnal as well as diurnal bird, a similar effect of the habitat was noted. In the *A. brama,* plasma corticosterone level showed two peaks one at 6 pm (waking) and another at 6 am (resting), while in the diurnal bird 6 pm shows their resting peak and 6 am their waking hour. Explanation for the existence of two peaks of plasma corticosterone levels, in this circadian rhythmic observation were in accordance with earlier studies (Chan and Phillips, 1973; Voissin and Assenmacher, 1968; Dusseau and Meier, 1971). In some species, the rates of biosynthesis by adrenal tissue *in vitro* and the concentrations of corticosteroids in plasma and adrenal gland tissue may also vary according to the time of day and the season of the year (Dusseau and Meier, 1971).

Pineal-Harderian Relationship

The presence of Harderian glands (HG) were seen in several avian species, such as the chickens, quails, parrot, sparrow and rhea, and its functions to lubricate the nictitating membrane as well as the corneal surface with its secretion which might have local immune activity (Aitkin and Survashe, 1975). Support for this view has been reported from the studies of chicken (Albini *et al.,* 1974).

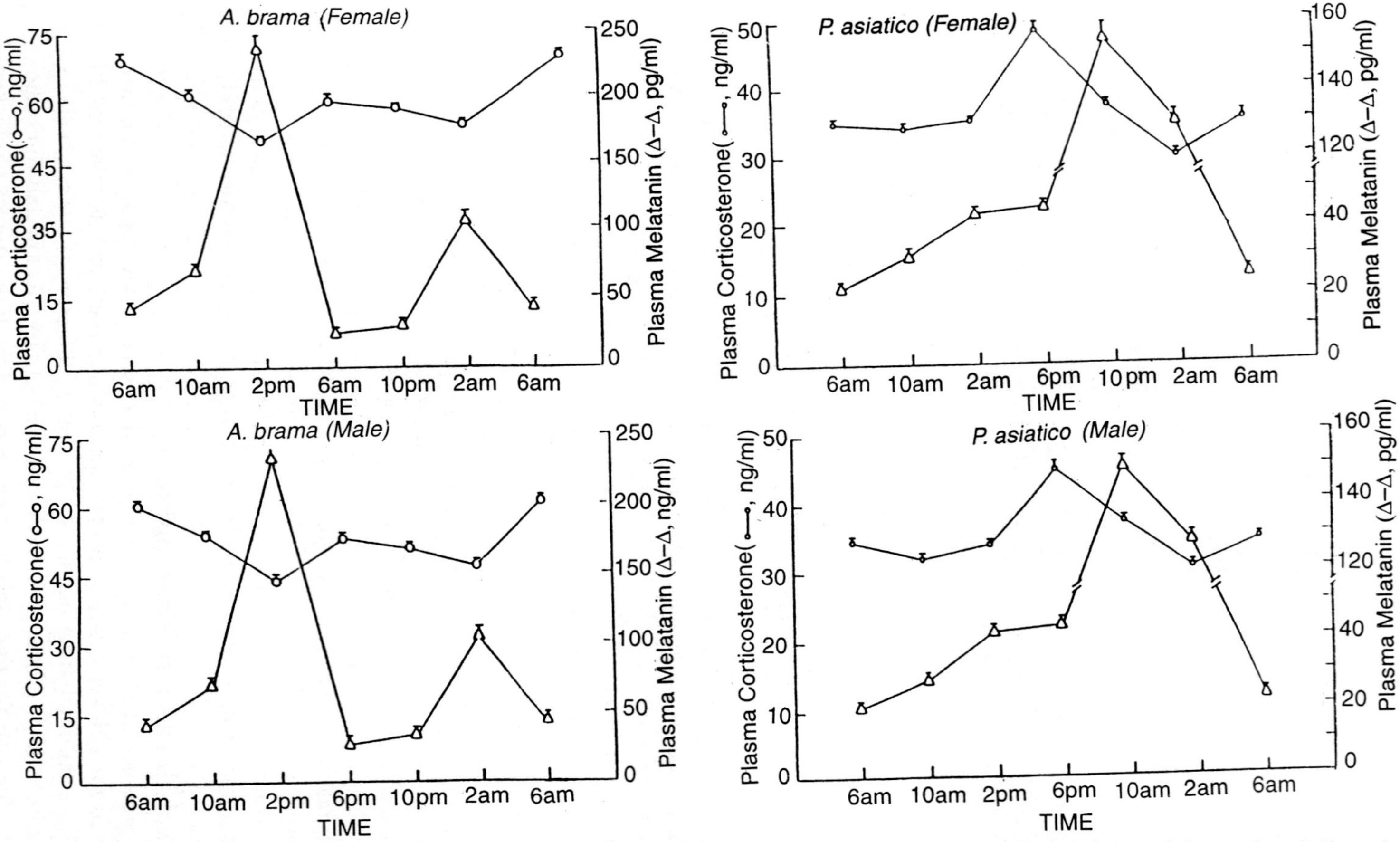

Fig. 5 **Daily variation in plasma levels of corticosterone (ng/ml) and melatonin (pg/ml) of a nocturnal bird *A. brama* (left panel) and diurnal bird *P. asiatica* (right panel). Note the drastic difference in melatonin peaks in both the species of the bird.**

Recently, after the report of the presence of MEL in the HG, the physiological aspect of the gland took momentum among pinealogists to note its functional correlation with pineal or gonad (Dubey and Haldar, 1997). Harderian gland weight exhibits a marked seasonal variation in relation to gonad and pineal. A significant decrease of the HG weight of *P. asiatica* was noted after two months of Px in the recrudescence and active phases of the testicular cycle, whereas, no difference in HG weight was noted in the other phases of the gonadal cycle. Histologically, a reduction in the height of the columnar epithelial cells and some leucocytes were noted in the lumen.

Injection of all the three 5-methoxyindoles, i.e. MEL, 5-MT, 5-MTL inhibited the HG weight during the inactive and active testicular phase. Histologically, a reduced epithelial cell height and the diameter of the HG acini were noted.

Implantation of MEL via silastic capsule in *P. asiatica* effectively reduced HG weight during the gonadal development and inactive phases, again confirming its indirect relation (via steroid hormonal level) with the testis. Therefore, melatonin could be 'the compound' involved in the control of HG function (Dubey and Haldar, 1997).

This preliminary data from the avian group suggest that there is a direct/indirect relation between pineal-harderian gland-testis but, it is too premature to give a definite suggestion about its physiological role and the nature of their relationship.

Pineal and Carbohydrate Metabolism

A relationship between the pineal and glucose homeostasis has been proposed by several workers. However, available information on the nature of the influence of putative pineal hormone MEL on the level of blood glucose in birds is scanty. Under different experimental paradigm, exogenous MEL was shown to be hyperglycemic in pigeons and hypoglycemic in parakeets. Previous studies revealed that the physiological effects of MEL administration largely depend on a number of factors including the time of administration within the daily light-dark cycle, dosage, the season or phase of the annual cycle and the age of the animals. However, the influence of these variables on the glycemic response to melatonin remains essentially unknown in a particular species of bird.

Various studies support the conjecture that in case of birds vs. mammals, the pattern of diurnal rhythmicity in the amount of glucose in blood is reverse (Hazelwood, 1986). In addition to the existing knowledge, it appears that in an avian species polysaccharide present in muscle also undergoes diurnal rhythmic changes and possibly contributes to the rhythmicity in the amount of glucose in blood.

The influence of day-time administered melatonin (20 μg/100 g body weight/day for 10 days) on blood glucose, liver glycogen and muscle glycogen was examined in sexually regressed adult male roseringed parakeets, *P. krameri.* Availability of exogenous melatonin during day time resulted in significant decrease in the values of plasma glucose and concomitant increase in the amount of hepatic and skeletal muscle glycogen in parakeets. These findings indicate that MEL has some insulin like actions in male roseringed parakeets, though, none of the pancreatic hormones has been measured in the investigation. Hypoglycemia along with an increase in the amount of glycogen in both the muscles and liver was recorded in *P. Krameri.* Therefore, it appears logical to state that exogenous MEL in parakeets, like that in mammals, possibly induces release of insulin, which may be accounted for the hypoglycemia in experimental birds. Nevertheless, studies of pancreas and its hormones in a bird after administration of melatonin and other methoxyindoles like 5-MTL and 5-MT, during morning and evening hours in the day are required to strengthen the concept of 'insulin-like' actions of MEL in birds. The study of Anam and Maitra (1992) also indicates that glycemic effects of exogenous MEL in parakeets vary not only with the dose, but also with the time in a 24 h cycle.

The increase in dose of MEL caused gradual decrease in the hypoglycemic response. It is possible that high doses of MEL have a down-regulating effect on the MEL receptors with eventual suppression of the expected effects. Nevertheless, the underlying mechanisms for dose related glycemic response in parakeets should remain speculative until further investigation (Maitra *et al.,* 1990, 1992).

An insulin-like effect of bovine pineal extract was shown by Milcou *et al.* (1957). *In vitro* study also revealed that the secretory product of the pineal organ has a stimulatory effect on insulin release by pancreatic islets suggesting a direct effect of pineal substance on hormonal secretion of pancreas. Diaz and Biazquez (1986) found an inversion of the insulin/glycogen ratio in Px rats. The ability of MEL in restoration of circulating levels of insulin was further demonstrated by the study on melatonin treated Px rats. Alternatively, in view of the studies showing altered adrenomedullary function in the Px and MEL treated animals and profound influences of adrenomedullary hormones on the blood glucose level in birds, it is not unlikely that adrenal medulla is involved in the mediation of glycemic effects of MEL. Nevertheless, it appears logical to surmise that variable glycemic response to the treatment of MEL in relation to the time of observation in a 24 h cycle may not be independent of the influence of pancreas and/or adrenal gland, the hormones of which undergo precise diurnal changes under natural light-dark cycle. A correlative study between the blood glucose and the hormonal profiles of both pancreas and adrenal in Px/MEL treatment requires further justification.

References

Aitkin, I.D. and Survashe, B.D. (1975). Comp. Biochem. Physiol., 33: 193-195.

Albini, B., Wick, G., Rose, E. and Orlans, E. (1974). Int. Arch. Allergy App. Immunol., 47: 23-34.

Anam, K.K. and Maitra, S.K. (1992). Comp. Physiol. Ecol., 17(2): 57-60.

Bageshwar, E. (1969). Avian Endocrinology. Ph.D. Thesis, Banaras Hindu University, Varanasi, India.

Burger, J.W. (1938). Amer. Natur., 22: 562.

Cassone, V.M. (1990). Trends in Neurosciences, 13: 457-463.

Chan, S.W.C. and Phillips, J.G. (1973). Gen. Comp. Endocrinol., 20: 291-296.

Chaturvedi, C.M. (1977). Compartive study of Avian adrenal gland. Ph.D. thesis, B.H.U. Varanasi, India.

Defronzo, R.A. and Roth, W.D. (1972). Acta Endocrinol., 70: 35.

Diaz, B. and Balzquez, E. (1986). Horm. Metab. Res., 18: 225-229.

Dubey, S. and Haldar, C. (1997). Gen. Comp. Endocrinol., 106: 17-22.

Dusseau, J.W. and Meier, H.M. (1971). Gen. Comp. Endocrinol. 16: 399-408.

Ebihara, S. and Kawamura, H. (1980). Neural Endocrine Aspects. Y. Tanabe, K. Tanaka and T. Okawa, (eds.), pp. 71-77, Springer-Verlag, Tokyo.

Fraschini, F. and Martin, L. (1970). The Hypothalamus. Martin, L., Motta, M. and Fraschini, F. (eds.), pp. 529-550, Academic Press, New York/London.

Garg, R.K. (1968). Avian Endocrinology. Ph.D. Thesis, Banaras Hindu University, Varanasi, India.

Ghosh, A., Ghosh, S. and Sengupta, S. (1996). Proceedings of VI Int. Avian Endocrinology, Canada, pp. 80-81.

Ghosh, M. and Haldar, C. (1994). PAVO, 32: 1-16.

Haldar, C. and Ghosh, M. (1989). J. Reprod. Biol. Comp. Endocrinol., 1: 73-81.

Hazelwood, R.L. (1986). Avian Physiology, 4th ed. Springer Verlag, New York, Berlin Heidelberg, Tokyo, pp. 123-156.

Hohn, E.O., Sarkar, A.K. and Dzubin, A. (1965). Can J. Zool., 43: 475.

Johnson, L.T. (1980). *In:* The Pineal, Extra reproductive effects. Reiter, R.J. (ed.) vol 5: 109-127, C.R.C. Press.

Kitay, J.I. and Altschule, M.D. (1954). "The Pineal Gland". Commonwealth Fund Book, Harvard University Press, Cambridge.

Lewinski, A., Webb S. and Reiter, R.J. (1984). Medical Hypothesis, 14: 141-160.

Maitra, S.K. and Dey, M. (1994). Biological Rhythm Research, 25: 1, 76-88.

Maitra, S.K. and Dey, M. (1993). Eur. Arch. Biol., 103: 157-167.

Maitra, S.K., Anam, K.K. and Dey, M. (1990). J. Reprod. Biol. Comp. Endocrinol., 2: 1, 58-61.

Maitra, S.K., Roy, T.K. and Sankar, R. (1992). J. Reprod. Biol. Comp. Endocrinol., 4: 2, 76-81.

Mazi, T.K., Some, K., Chattopadhyay, S.K. and Sarkar, A.K. (1992). J. Interdiscipl. Cycle Res., 23: (1), 72-80.

Menaker, M. and Oksche, A. (1974). *In*: Avian Biology. D.S. Farner and J.R. Kings, (eds.), pp. 79-118, Academic Press, New York.

Mess, B. (1983). The Pineal gland and its endocrine role. J. Axelrod, F. Fraschini and G.P. Velo, (eds.), pp. 477, Plenum, New York.

Milcou, I., Nanu, L. and Marcean, R. (1957). Ann. Endocrinol. (Paris), 18: 612.

Pandey, R. (1988). The pineal gland and the synchronization of reproductive cycle with environmental factors in a tropical ophidian. The Indian chequered water–snake, *Natrix piscator.* Ph.D. Thesis, Banaras Hindu University, Varanasi, India.

Piva, F., Schiaffini, O., Molta, M. and Martini, L. (1973). *In*: Hormones and Brain Functions. K. Lissak, (ed.), pp. 231, Plenum Press, New York.

Reiter, R.J., Hoffman, R.A. and Haster, R.J. (1966). J. Exp. Zool., 162: 263-268.

Reiter, R.J. (1975). Biological rhythms and endocrine function. L.W. Hedlund, J.M. Franz, A.D. Kenny (eds.), pp. 43-66, Plenum Press, New York.

Reiter, R.J. (1991). Int. J. Biomet. 35(3): 169-175.

Reiter, R.J., Guerrero, M.J. and Rutio, A. (1993). *In*: Melatonin and Pineal Gland from basic science to clinical application. Y. Touitou, J. Arendt and P. Pevet (eds.), pp. 55-58, CRC Press, New York.

Riddle, O. (1923). J. Physiol., 66: 332.

Saxena, N. (1988). Pineal in Indian palm squirrel *Funambulus pennanti,* 5-methoxyindoles and environmental control of reproduction. Ph.D. Thesis, Banaras Hindu Universty, Varanasi, India.

Shavali, S. (1992). Pineal-thyroid interrelationship in a tropical zone rodent *Funambulus pennanti.* Ph.D. Thesis, Banaras Hindu University, Varanasi, India.

Sudha, C.C. (1995). Comparative study of pineal-adrenal interrelationship in a nocturnal (*A. brama*) and a diurnal (*P. asiatica*) bird of tropical Indian origin. Ph.D. Thesis, Banaras Hindu University, Varanasi, India.

Takahashi, J.S., Murakami, N., Nikaido, S.S., Pratt, B.L. and Robertson, L.M. (1989). Recent Prog. Horm. Res., 45: 279-352.

Vivien-Roels, B. and Pevet, P. (1986) *In*: Advances in Pineal Research, Reiter, R.J. and Karasek, K. (eds.), pp. 61, John Libey & Co. (London).

Vivien-Roels, B., Arendt, J. and Bradterke, J. (1979). Gen. Comp. Hendocrinol., 37: 197-210.

Voissin, J. and Assenmacher, I. (1968). *In*: The Adrenal Cortex of Birds. W.N. Holmen and J.G. Phillips; I. Chester Jones and I.W. Enderson (eds.), vol I: pp. 319, Academic Press, London.

Vriend, J. (1978). Medical Hypothesis, 4: 376-387.

Vriend, J. (1983). *In:* Pineal Research Reviews. Reiter, R.J. (eds.), vol I: pp. 183-206, Alan R. diss. Inc., New York.

Wurtman, R.J., Altschule, M.D. and Holmgren, U. (1959). Am. J. Physiol., 197: 108-110.

D

Harmones and General Metabolism

Comparative Endocrinology and Reproduction
K.P. Joy, A. Krishna and C. Haldar (Eds)

32. Neurotransmitter-Hormonal Regulation of Diabetes and Liver Regeneration

C.S. Paulose, P.S. Padayatti and B. Sudha
Molecular Neurobiology and Cell Biology Unit, Department of Biotechnology, Cochin University of Science and Technology, Cochin-682 022, India

Introduction

Neurotransmitters like norepinephrine (NE), serotonin (5-HT), dopamine (DA), gamma-amino butyric acid (GABA), taurine, acetylcholine (ACh) as well as sphingolipids and polyamines are involved in the regulation of endocrine function through the mediation of hypothalamus and pituitary (Bilezikian, 1987; Dakshinamurti *et al.*, 1990). The release and receptor regulation of neurotransmitters are controlled by hormones secreted in response to various physiological states of the body (Halter *et al.*, 1980). The hypothalamus is the main neuroendocrine centre in the brain where the hormones of the target glands interact with the neurotransmitters (Bilezikian, 1987). The secretion by the anterior pituitary of ACTH, growth hormone, prolactin, TSH and gonadotropins (LH and FSH) is regulated by hypothalamic hormones. Regulation of the release of the stimulatory or inhibitory hormone involve complex neural circuitry in which serotonergic and catechloaminergic systems are components of the control mechanisms (Dakshinamurti *et al.*, 1986).

Neurotransmitters and Hypothyroidism

A direct relationship between hypothalamic serotonin turnover and TSH release has been reported by Smythe *et al.*, (1982). Dopaminergic neurons exert an inhibitory effect on the secretion of TSH. This effect is at the level of the pituitary, as bromocriptine blunts the stimulatory effect of thyrotropin releasing hormons (TRH) in euthyroid subjects (Brown *et al.*, 1979). The inhibitory effect of DA is abolished by DA receptor antagonists such as domperidone. The hypothalamus has high concentrations of NE , DA and 5-HT which are essentially antagonistic in their effects on pituitary hormone regulation (Krulich, 1979). The secretion of TSH is directly controlled by two factors, a negative feedback signal indicating serum thyroid status and a stimulatory factor, TRH released from the hypothalamus. The cold-induced secretion of TSH is mediated by NE. Studies using inhibitors of NE synthesis or α-adrenergic blockers have established a stimulatory role of NE in the control of TRH-mediated TSH secretion (Mannisto *et al.*, 1979). There is a difference between the role of α-adrenergic receptor subtypes, α_1- being inhibitory and α_2- being stimulatory (Krulich *et al.*, 1982). On balance, it appears that serotonergic system has a stimulatory effect on hypothalamic control of pituitary secretion of TSH in situations where the central response is natural, such as in timing of circadian rhythm and possibly in the pulsatile secretion of TSH (Mannisto, 1983). The classical formulation of hypothyroidism of hypothalamic origin is established in pyridoxine-deficient animal model. Hypothalamic hypothyroidism is due to deficient TRH secretion. Results suggest that hypothalamic type of hypothyroidism in pyridoxine-deficient rats is caused by the specific decrease in hypothalamic serotonin (Dakshinamurti *et al.,* 1986; 1990).

Neurotransmitters and Hypertension

Clinical reports (Bing *et al.*, 1980) indicate a direct correlation between hypothyroidism and hypertension. Saito *et al.* (1983) reported that the hypothyroid state accelerated the age-related increase in blood pressure and on thyroxine replacement, the blood pressure tended to normalize in the hypertensive hypothyroid patients. In view of the relationships of hypothyroidism and vitamin B6 deficiency, blood pressure regulation effect through neurotransmitter involvement using pyridoxine-deficient animal model was studied (Dakshinamurti *et al.,* 1990). Systolic blood pressure was recorded weekly by tail cuff plethysmography (Paulose *et al.*, 1988). At the end of eight-week experimental period, direct arterial blood pressure measurements were made. There was a significant increase in both systolic and diastolic blood pressure of these pyridoxine-deficient rats which was reversed by pyridoxine treatment within 24 h.

The reversible hypertension seen in the pyridoxine-deficient rat was also correlated with general sympathetic stimulation (Paulose e*t al*., 1988). Blood samples were withdrawn from conscious animal without trauma by implanting vascular access port with catheterization to the jugular vein (Paulose and Dakshinamurti, 1987). NE and epinephrine (EPI) levels in peripheral plasma of pyridoxine-deficient rats were almost three-fold higher compared with control values. Treatment of deficient rats with pyridoxine (10 mg/kg body weight, i.p.) resulted in a return of systolic blood pressure to control level. Correspondingly, the plasma catecholamine levels decreased to normal levels within 24 hrs of pyridoxine administration. The complete reversibility of the hypertension in such a short time would preclude permanent damage to the vessel wall of the deficient animal. The lesion is probably at the level of neurotransmitter regulation. The NE turnover in the heart in both deficient and control groups by estimating the decrease in the amine content due to α-methyl p-tyrosine was determined (Viswanathan *et al.*, 1990). There was no difference in myocardial NE content between the deficient and control groups. However, the NE turnover was increased significantly in the pyridoxine-deficient rats when compared to that of the control, thus supporting the observation that peripheral sympathetic activity is increased in the dificient rats. Both clonidine and α-methyl DOPA reduced the systolic blood pressure of deficient rats to normal levels (Dakshinamurti *et al.*, 1990). Clonidine like drugs exert their cardiovascular depressive effects mainly through a centrally mediated sympathetic inhibition due to the stimulation of α-adrenoreceptors (Kobinger and Pichler, 1980).

The kinetics of ligand binding to α_2-adrenoreceptors was examined using (^{3}H) p-aminoclonidine (PAC) binding to crude synaptosomal membrane preparation from the brain stem of pyridoxine-deficient and control rats (Viswanathan *et al.*, 1990). A reduced sympathetic outflow has been implicated in the mediation of the cardiovascular effects of GABA (Persson, 1980). The central mechanism of maintaining normal blood pressure in animals is regulated by the balance between sympathetic and para-sympathetic nervous system tonicities in the brain stem (De Jong *et al.*, 1975; Nomura *et al.*, 1985). The antihypertensive effect of clonidine results from its pharmacological reactivity for central α_2-adrenoreceptors in the nucleus tractus solitarri (NTS). When α_2-adrenoreceptors in NTS are stimulated, inhibitory neurons of the vasomotor centre are activated.

Neurotransmitters and Diabetes

Alterations in the neurotransmitter-hormonal regulation and sympathetic activity have been recognised as a major pathological symptom associated with diabetes mellitus. Streptozotocin (STZ)-induced diabetes is used as a model for insulin-dependent diabetes mellitus (IDDM). The major feature in this type of diabetes is a substantial reduction in insulin secreting β-cells (Gepts, 1965). Although the reason for such cellular destruction is not clearly known, genetic influences and both humoral and cell mediated immunological phenomena are involved (Eisenbarth, 1986; Lefebvre, 1988). A role for environmental factors in the etiology of IDDM has recently been indicated by epidemiological studies which have demonstrated that there is a marked increase in newly diagnosed cases of IDDM, which can only be explained by changes in environmental influences (Krowlewski *et al.*, 1987). It has been suggested that people with diabetes have a greater risk of depression compared to the general population (Lloyd and Orchard, 1993). Hypothalamic serotonin is suggested to mediate the release of catecholamines which control the insulin release and function. Yehuda and Meyer (1984) found that depletion of brain 5-HT by intraventricular injection of 5,7-dihydroxytryptamine significantly attenuated corticosterone response to insulin. Similar effect was seen in 5-HT receptor blocker, methysergide treatment studies. An increased noradrenegic and decreased serotonergic activity in the hypothalamus in diabetic rats can be a control factor in the increased hyperglycemia (Fig. 1).

The drug 8-hydroxy-2-(di-n-propylamino) tetralin (8-OHDPAT), a selective agonist for 5-HT$_{1A}$, was shown to trigger hyperglycemia and inhibition of insulin release (Chaouloff and Jeanrenaud, 1987). Pharmacological analysis showed the involvement of both 5-HT_{1A} and α_2-adrenoreceptors in the metabolic effects of this drug in elevating circulating adrenaline levels which mediated the 8-OHDPAT induced hyperglycemia (Chaouloff *et al.*, 1990). The common neuroendocrine alterations associated with diabetes mellitus include an increase in growth hormone response to exercise (Hansen, 1970), TRH and dopamine (Lorenzi *et al.,* 1980), and a decrease in GnRH release (Bowton *et al.,* 1980), TSH response to TRH and GH secretion. Most of these changes observed in diabetic state indicate an altered hypothalamic functioning in diabetes mellitus. The neuroendocrinological alterations in diabetes can be the result of either vascular insufficiency or secondary to a specific derangement in the neurotransmitter activity.

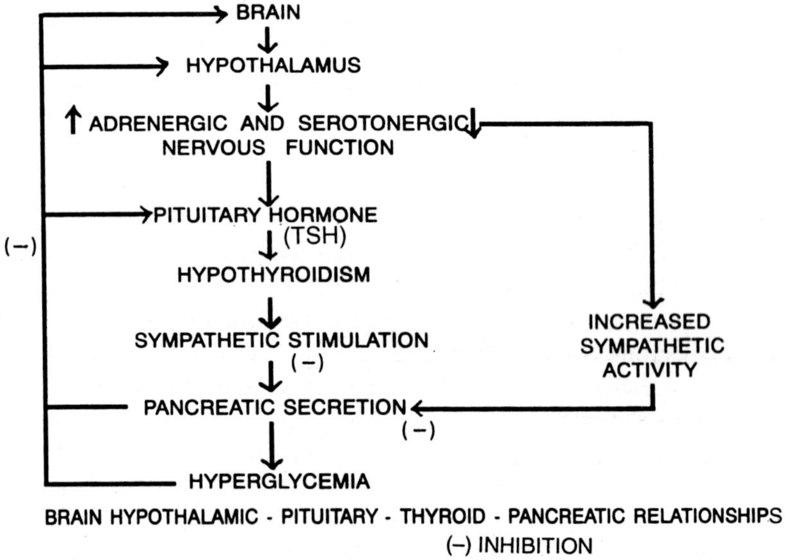

Fig. 1

During diabetes, the noradrenergic activity increased in cerebral cortex (CC), hypothalamus (Hypo) and brain stem (BS), while the cerebellum (CB) showed no change and the corpus straitum (CS) a decrease, as assessed by the monoamine content. EPI showed a significant increase in Hypo, BS and CS. Dopamine, showed a significant increase in CC, Hypo, CB, BS, and CS. The hypothalamic serotonin showed a significant decrease while that of BS and CS showed a significant increase (unpublished observation). The physiological significance of these changes is yet to be connected or delineated. The hypothalamic neurotransmitter changes are important as far as the hypothalamic endocrine regulation is considered. The increase in noradrenergic activity in the hypothalamus has been studied in detail in streptozotocin-diabetic rats by Bitar *et al.* (1987). The results of their analysis of 5-HT and 5-HIAA in hypothalamic nuclei showed difference between diabetic and control rats. 5-HT in periventricular nucleus (PVN) showed a significant increase when compared to control rats, while the 5-HIAA level was significantly decreased in suprachiasmatic nucleus (SCN), supraoptic nucleus (SON) and ventromedial nucleus (VMN). The increased noradrenergic activity and a decreased serotonergic activity together can make important decisive change on the hypothalamic hormones. The regulation of hypothalamic noradrenergic and serotonergic activity on hypothalamic corticotropic releasing factor (CRF) has been studied in stressed states (Maickel and Martel, 1983). The studies revealed that 5-HT is mediating the release of adrenal response to insulin stress. The noradrenergic neurons have a positive control of the corticotrophic response to stress. The noradrenergic activity is high in the hypothalamus and serotonergic activity is decreased in STZ diabetes. This change may be related to an increase in adrenal activity. As a consequence, the release of insulin may be inhibited causing an increased level of circulating glucose.

The most significant alteration in the diabetic state is the reduced level or absence of insulin or its function in the body. Although studies have shown about the direct cytotoxic action of the drug STZ to pancreatic β-cells, there is no study which clearly state about the extra pancreatic toxic effects of the drug. Autonomic neuropathy associated with STZ diabetes is reversed when insulin is supplemented (Sasaki and Bunag, 1983). In our study the insulin therapy was not found to fully bring back the altered brain neurotransmitter levels to normal values. These neurotransmitter changes in the brain regions may be a specific cause or associated with the direc behavioural or endocrine alteration, which have to be confirmed with receptor studies and second messenger studies at the DNA level. Also, the hormonal changes and reversal confirmatory studies have to be carried out

Hormones and Liver Regeneration

The interaction of insulin, glucagon and epidermal growth factor (EGF) showed a significant effect on liver regeneration by way of increased DNA synthesis (Chung *et al.*, 1992). A direct hepatotrophic effect of glucagon in liver regeneration was recently observed in glucagon-treated hepatectomised rats (Huang et *al.*, 1993). Measurement of various hormone levels after various intervals of partial hepatectomy showed that ACTH and corticosterone increased significantly while TSH, T4 and testosterone decreased and T3 was unchanged. This indicated an increase in pitutary-adrenal function and a decrease in pituitary-thyroid activity during regeneration (Knopp *et al.*, 1991).

Tyrosine phosphorylation of insulin receptor substrate-1 (IRS-1), a specific target for insulin receptor β subunit kinase,was strikingly enhanced prior to hepatocyte DNA synthesis. There was a parallel increase in IRS-1 protein and mRNA expression as well as the tyrosine phosphorylation of β subunit. Phosphatidyl-inositol-3-kinase (PI-3 kinase) was associated with IRS-1 following tyrosine phosphorylation (Sasaki *et al.*, 1993). The PI-3 kinase has been implicated in transmembrane signal transduction by a number of growth factor receptors (Cantley *et al.*, 1991). 3,3′,5′- triiodothyronine (T3) was implicated as a specific hepatomitogen (Short *et al.*, 1980). T3, aminoacids, glucagon and heparin were shown to stimulate DNA synthesis of unoperated rats (Short *et al.*, 1972). T3 injections stimulated the liver regeneration response that resembled in timing and magnitude, the DNA synthesis induced by 40% hepatectomy. T3 showed a dose-related response in presence of EGF in hepatocyte cultures (Francavilla *et al.*, 1994). Glucocorticoids were shown to cause an increase in EGF-induced proliferation of regenerating hepatocytes (De Juan *et al.*, 1992).

Neurotransmitters and Liver Regeneration

Neuronal stimulation of rat liver could result in changes in cellular metabolism and liver regeneration through a combination of direct innervation and intercellular communication via gap junction (Seseke *et al.*, 1992). Vagotomy was shown to cause a marked depression of cell proliferation following hepatectomy (Maros, 1970; Ohtake *et al.*, 1993). Extirpation of the brain cortex was shown to increase the rate of cell proliferation implying that the cortex exterts a normal inhibitory function on liver cell division and growth. It was reported that transection of the spinal cord above the area innervating the liver resulted in depression of thymidine incorporation into hepatic DNA (Vaptazarova *et al.*, 1973). If transection was done at 16 hrs following partial hepatectomy, the already enhanced DNA synthesis was suppressed. Similar results were obtained with mid-thoracic transection of the cord implicating a suprasegmental level of regulation of regeneration. These data suggest that an intact nervous system is an important component of the regulatory system for liver regeneration either through direct innervation of the liver or through indirect modulation of DNA synthesis resulting in cell division (Fig. 2).

The sympathetic nerve activity of the liver was reported to be suppressed at the early stage of regeneration after 1/4th hepatectomy (Iwai and Shimazu, 1992). Liver DNA synthesis following partial hepatectomy depends on extracellular and/or intracellular calcium levels and thus, one action of NE may be to function directly through modulating intracellular calcium levels (Rixon *et al.,* 1979). NE was shown to induce DNA synthesis in a dose-dependent manner in hepatocyte cultures, acting through the α_1- adrenergic receptor (Cruise *et al.*, 1985). A strong synergistic interaction between NE and EGF, a complete hepatocyte mitogen, was observed in hepatocyte cultures incubated with both EGF and NE, or pretreated with NE and then exposed to EGF, and stimulated the hepatocytes to enter the DNA synthesis (Cruise and Michalopoulos, 1985). Low affinity EGF receptors are the only type present when mitogenic response is stimulated (Vintermyr and Doskeland, 1987). It was postulated that the low affinity receptors are the true mitogenic

EGF receptors (Wollerberg *et al.*, 1989). It is hypothesised that the EGF mitogenic signal is modulated by protein kinase C which causes the switching of the EGF receptor from high affinity to low affinity states in rat liver cells (Sharma *et al.*, 1994). It was shown that NE was able to decrease the dose-dependent inhibition of TGF-β, a potent inhibitor of hepatocyte primary cultures acting through the α_1-adrenergic receptor (Houck *et al.*, 1988). Thus the results suggest that NE may trigger the initial proliferative response in liver regeneration by modulating the response of hepatocytes to growth factors such as EGF and growth inhibitors such as TGF-β.

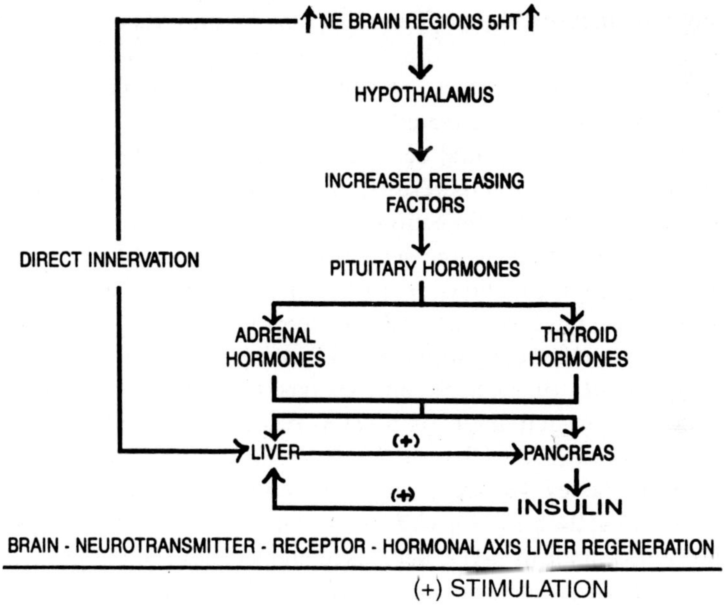

Fig. 2

Addition of NE to primary cultured adult rat hepatocytes stimulated the DNA synthesis in a dose-dependent manner, especially in the presence of insulin and EGF. The effect of EPI was also through the α_1-adrenergic receptor. However, the α_1-action of EPI was apparently not mediated by either activation of protein kinase C or Ca^{++} mobilisation (Takai *et al.*, 1988). The number of β-adrenergic receptors was found to increase with the maximal increase seen 48 hrs after partial hepatectomy. The α_1-adrenergic receptors showed a transient decrease by about 35% at 18–24 hrs (Sandnes *et al.*, 1986). An increased β-adrenergic receptor responsiveness during liver regeneration was found to result from either an increase in the receptor number or from a more efficient receptor cyclase coupling (Riles *et al.*, 1984).

In vivo, plasma catecholamines or those delivered by the synaptic stimulus may contribute to the regenerative stimulus. As a prelude to study the *in vivo* role of NE as a regulator of cell division, we have observed significant changes in the content of brain NE, especially in the cerebral cortex, brain stem and hypothalamus during 24 and 48 hrs of liver regeneration, compared to control values. Brain EPI was seen to increase significantly in the cerebral cortex during 24 and 48 hrs of liver regeneration compared to that of control suggesting that it can function synergistically with NE during the DNA replicative phase of liver regeneration. Circulating levels of NE increased significantly during 24 and 48 hrs of regeneration. We also noted that DNA synthesis was maximal during 24 and 48 hrs of the partial hepatectomy. Brain NE levels declined to near normal levels by 72 hrs of hepatectomy. This suggests that the brain NE and EPI may be important during the active phase of DNA synthesis.

Neurotransmitter receptors are usually restricted to neuronal cells, but the signalling pathways activated by these receptors are widely distributed in both neural and nonneural cells. The introduction of functional 5-HT_{1C} receptor into NIH 3T3 cells results, at a high frequency, in the generation of transformed foci. The injection of cells derived from transformed foci into nude mice results in the generation of tumors. Serotonin (5-HT_{1C}) receptor, therefore, functions as a protooncogene when expressed in NIH 3T3 fibroblasts (Julius *et al.*, 1989). Serotonin is mitogenic in vascular smooth muscle cells, acting through 5-HT_4 receptors or with 5-HT_{1D} receptors (Kavanaugh *et al.,* 1988). Serotonin synergises with mitogenic growth factors in fibroblasts via activation of 5-HT_{1B} receptor coupled to a Gi protein (Seuwens *et al.*, 1988; Seuwens and Pouyssegur, 1990). During liver regeneration, the level of serotonin increased significantly in the brain stem, CC and Hypo at 24 hrs compared to control values. 5-hydroxytrytophan (5-HTP), the precursor of 5-HT, decreased significantly in the brain stem, CC and Hypo at 24 hrs of regeneration compared to respective control values. 5-hydroxyindoleacetic acid (5-HIAA), a serotonin metabolite, increased significantly in brain stem and Hypo and 24 hrs of hepatectomy compared to control values.

It is proposed that neurotransmitters directly and/or through the hormonal pathway regulate insulin function in diabetes and DNA synthesis and cell proliferation in liver regeneration after hepatectomy.

Acknowledgement

Supported by a research grant from DST. PSP and BS thank CSIR for SRF.

References

Bilezikian, J.P. (1987) In: Adrenergic receptors in man (P.A. Insel, Ed.), pp 37–66. Marcel Dekker, Inc. New York.

Bing, R.F., Briggs, R.S.J., Burden, A.C., Russel, G.I., Swales, J.D. and Thenston, H. (1980) Clin. Endocrinol. **13:** 339–342.

Bitar, M.S., Koulu, M. and Linnoila, M. (1987) Brain Res. **400:** 236–242.

Bowton, P.A., Bryant K.R. and Whitehead, S.A. (1980) Acta Endocrinol. (Copeng) **111:** 467–501.

Brown, P.M., Buchus, R., Sachs, L., Sonsken, P.M. and Wheeler, M. (1979) Clin. Endocrinol. (Oxf) **10:** 481–488.

Cantley, L.C., Auger, K.R., Carpenter, C., Duckworth, B., Kapeller, R., Soltoff, S. (1991) Cell **64:** 281–302.

Chaouloff, F. and Jeanrenaud, B. (1987) J. Pharmacol. Exp. Ther. **243:** 1159–1166.

Chaouloff, F., Bandrie, V. and Landa, D. (1990) Br. J. Pharmacol. **99:** 39–45.

Chung, Y.C., Chen, W.J. and Chen K.M. (1992) J. Formos. Med. Assosc. **91:** 685–690.

Cruise, J.L., Houck, K.A. and Michalopoulos, G.K. (1985) Science **227:** 749–751.

Cruise, J.L., and Michalopoulos, G.K. (1985) J. Cell Physiol. **125:** 45–50

Dakshinamurti, K., Paulose, C.S. and Vriend, J. (1986) J. Endocr. **109:** 345–349.

Dakshinamurti, K., Paulose C.S., Viswanathan, M., Siow, Y. L., Sharma, S.K., Bolster, B. (1990) Annals N.Y. Acad. Sci. **585:** 128–144.

Dakshinamurti, K., Paulose, C.S., and Viswanathan, M. (1990) Annals, N.Y. Acad. Sci. **585:** 241–249.

Dejong, W., Nijkamp, F.P. and Ohns, B. (1975). Arch. Int. Pharmacodyn. **213:** 275–284.

De Juan, C., Benito, M., Alverez, A. and Fabregat, I. (1992) Exp. Cell Res. **202:** 495–500.

Eisenbarath, G.S. (1986) N. Engl. J. Med. **314:** 1360–1368.

Francavilla, A., Carr, B.I., Azzarone, A., Polimeno, L., Wang, Z., Vanthiel, D.H., Subbatin, V., Prelich, J.G. and Starzl, T.E. (1994) Hepatology **20:** 1237–1241.

Gepts, W. (1965) Diabetes **14:** 1449–1453.

Halter, I.B., Pflug, A.E. and Tolas, A.G. (1980) Metabolism **29:** 9–12.

Hansen, A.P. (1970) J. Clin. Invest. **49:** 1467–1473.

Houck, K.A., Cruise, J.L., and Michalopoulos, G.K. (1988) J. Cell Physiol. **135:** 551–555.

Huang, T.L., Chen, H.F. and Chen, T.J. (1993) J. Formos. Med. Assosc. **92:** 725–728.

Iwai, M. and Shimazu, T. (1992) J. Auton. Nerv. Syst. **41:** 309–314.

Julius, D.,'Livelli, T.J., Jessel, T.M. and Axel, R. (1989) Science **244:** 1057–1062.

Kavanaugh, W.M., William, L.T., Ives, H.E. and Coughlin, S.R. (1988). Mol. Endocrinol. **123:** 599–605.

Knopp, J., Brtko, J., Je Zova, D. and Langer, P. (1991) Horm. Metab. Res. **23:** 329–332.

Kobinger, W. and Pichler, L. (1980) Naunyn schmiedbergs Arch. Pharmacol. **315:** 21–27.

Krowlewski, A., Warram, J.H., Rand, L.I. and Kahn, C. R. (1987) N. Engl. J. Med. **317:** 1390–1398.

Krulich, L. (1979) Ann. Rev. Physiol. **41:** 603–615.

Krulich, L., Mayfield, M.A., Steele, M.K., McMillan, B.A., Mc Canne, S.M. and Korning, J.I. (1982) Endocrinology **110:** 796–804.

Lefebvre, P.J. (1988) In: The pathology of the endocrine pancreas in diabetes (P.J. Lefebvre and D.G. Pipeleers, Eds.) pp 1–16. Springer, Berlin, Heidelberg, New York.

Lloyd, C. and Orchard, T. (1993) In: Diabetes Annual/7 (S.M. Marshall, Ed.) pp 211–244, Elsevier, Amsterdam.

Lorenzi, M., Karam, J.H., McIlory, M.B., Forsham, P.H. (1980) J. Clin. Invest. **65:** 146–155.

Maickel, R.D. and Martel, R.R. (1983) Pharmacol. Biochem. Behav. **19:** 321–325.

Mannisto, P.T. (1983) Med. Biol. **61:** 92.100.

Mannisto, P., Ranta, T. and Tuomisto, J. (1979) Acta. Endocrinol. **90:** 249–258.

Maros, T. (1970) Hepatologia **16:** 21–28.

Nomura, M., Ohtsuji, M. and Nagata, Y. (1985) Neurochem. Res. **10:** 1143–1154.

Ohtake, M., Sakaguchi, T., Yoshida, K. and Muto, T. (1993) HPB Surg. **6:** 277–286.

Paulose, C.S. and Dakshinamurti, K. (1987) J. Neurosci. Methods **22:** 141–146.

Paulose, C.S. Dakshinamurti, K., Packer, C.S. and Stephens, N.L. (1988) Hypertension **11:** 387–391.
Persson, B. (1980) Naunyn schmiedebergs, Arch. Pharmacol. **313:** 224–236.
Riles, G.L., Caron M.G. and Lefkowitz, R.L. (1984) Physiol. Reviews **64:** 661–677.
Rixon, R.H., Mac Manus, J.P. and Whitfield, J.F. (1979) Mol. Cell Endocr. **15:** 79–89.
Saito, I., Ito, K. and Saruta, T. (1983) Hypertension **5:** 112–115.
Sandnes, D., Sand, T.E., Sager, G., Bronstad, G.O., Refsnes, M.R., Glandhang, P.I., Jacobsen, S. and Christofferson, T. (1986) Exp. Cell Res. **165:** 117–126.
Sasaki, S. and Bunag, R.D. (1983) Hypertension **5:** 34–40.
Sasaki, Y., Zhang, F.X., Nishiyama, H., Avruch, J., Wands, R.J. (1993) J. Biol. Chem. **268:** 3805–3808.
Seseke, F.G., Gardeman, A. and Jungermann, K. (1992). FEBS Lett. **301:** 265–270.
Seuwens, K. and Pouyssegur, J. (1990) Biochem. Pharmacol. **39:** 985–990.
Seuwens, K., Magnaldo, I. and Pouyssegur, J. (1988) Nature **335:** 254–256.
Sharma, R., Kleine, L.P. and Franks, D.J. (1994) Cell Signalling **6:** 631–643.
Short, J., Wedmore, R., Kilbert, L. and Zemel, R. (1980) Cytobios **28:** 165–177.
Short, J., Brown, R.F., Husakova, A., Gilbertson, J.R., Zemel, R. and Lieberman, I. (1972) J. Biol. Chem. **247:** 1757–1766.
Smythe, G.A., Bradshaw, J.E., Cai, W.Y. and Symons, R.G. (1982) Endocrinology **111:** 1181–1191.
Takai, S., Nakamura, T., Komi, N. and Ichihara, A. (1988) J. Biochem. **103:** 848–852.
Vaptazarova, K.I., Popop, P.G., Veseley, J. and Cihak, A. (1973) Experientia **29:** A 505–A 506.
Vintermyr, O.K. and Doskeland, S.O. (1987) J. Cell Physiol. **132:** 12–21.
Viswanathan, M., Paulose, C.S., Lal, K.J., Sharma, S.K. and Dakshinamurti, K. (1990) Neurosci. Lett. **111:** 201–205.
Woollerberg, G.K., Harris, L., Farber, E. and Hayes, M.A. (1989) Lab Invest. **60:** 254–259.
Yehuda, R. and Meyer, J.S. (1984) Neuroendocrinology **38:** 25–32.

Comparative Endocrinology and Reproduction
K.P. Joy, A. Krishna and C. Haldar (Eds)

33. Biochemistry of Thyroid and Thyroglobulin

Vijay Deshpande
Department of Biochemistry, University College of Science
Osmania University, Hyderabad-500 007, India

Introduction

Amongst the various endocrine glands, the only gland which has the unique ability of accumulating iodine in large quantities and converting it to organic bound form is "Thyroid". Thomas Wharton (1656), a London physician gave the name "Thyroid" which means shield shape in Greek. A similar function is displayed by the endostyle of protochordates, and is thought to be a precursor of the thyroid gland during evolution. Glands other than thyroid such as salivary, sub-pharyngeal and sub-maxillary are also known to concentrate iodine.

Enlargement of the gland during iodine deficiency demonstrates the importance of this essential micronutrient. The compensatory mechanism is referred to as goiter and has been recognized since ages. The advent of radioiodine has led to exhaustive research with regard to thyroid function.

The thyroid gland synthesizes and secretes two important iodine containing hormones viz., triiodothyronine (T_3) and tetraiodothyronine or thyroxine (T_4). The formation of thyroid hormones is accomplished by the iodination of tyrosine residues of the protein, thyroglobulin.

Anatomy and Histology

The thyroid is found in all vertebrates. Its origin and development is similar in various mammalian species. The thyroid, like the anterior pituitary is a derivative of the primitive buccopharyngeal cavity or the endostyle. It is of endodermal origin and arises early in foetal life as an anlage from the ventral wall of the pharynx. It first appears as an evagination from the floor of the pharynx which eventually becomes bilobed and comes to lie on the ventral surface of the trachea. The thyroid is situated a little below the larynx and consists of a lobe on either side joined by an isthmus. The weight of the gland in different animal species ranges from 0.026 to 200 g, and is influenced by environmental conditions, diet and infections.

The basic pattern of thyroid gland is similar in all vertebrates but it is of variable shape. In primitive species such as cyclostomes and teleosts, there is no glandular organization but the follicles are scattered singly or in groups. The amphibian thyroid is distinctly double and the lobes are round in shape. In snakes and turtles, it is a single discoid structure but in birds the thyroid consists of two isolated rounded lobes lying on either side of the trachea. The mammalian thyroid consists of two lobes connected by an isthmus (Gorbman, 1986).

The thyroid gland is highly vascular and is covered with a fibrous capsule. It has a rich supply of sympathetic nerves whose function is to regulate the supply of blood to the organ. The thyroid gland is an assembly of follicles. The shape of the follicles vary from species to species. Each follicle consists of a single layer of columnar epithelial cells containing a lumen or cavity filled with a yellowish viscid fluid "The colloid". The size of the follicles vary depending upon the colloid present therein, and is influenced by thyroid stimulating hormone (TSH) or thyrotropin.

Electron microscopic study of follicular cells reveals numerous well defined mitochondria, Golgi bodies, microvilli, lysosomes and intracellular vacuoles.

In addition to the follicular cells, the thyroid has parafollicular or C-cells, the derivatives of the ultimobranchial bodies which secrete calcitonin, the calcium regulating hormone.

Iodine Content of the Gland

Presence of iodine in the thyroid gland and its relation to growth and metabolism led the investigators to determine the iodine content of the gland of several animal species. Iodine contents have been determined for cattle, buffalo, sheep, pig, camel, rabbit, horse, rat, guinea pig and dogfish thyroids and the values range from traces to 0.43% (Cameron, 1930; Simpson, 1930; Dey *et al.*, 1943; Deshpande and Ramachandran, 1976; Deshpande, 1987). Various factors such as age, sex, thyroid and body weights, nutrition, environmental concentration of iodine, salinity and season which are known to influence the content of the gland have been taken into consideration.

Regulation of Thyroid Activity

Secretion of thyroid hormones (T_3 and T_4) into blood circulation from thyroid gland in vertebrates is regulated by TSH of adenohypophysis (pituitary) but its role in primitive cyclostomes is not clearly understood. The release of TSH is in turn regulated by hypothalamic factor, the thyrotropin releasing hormone (TRH), a tripeptide. The circulating thyroid hormones exert inhibitory effect by feedback mechanism primarily on the hypophysis and the hypothalamus as well. In addition to TSH regulation by TRH, another hypothalamic peptide viz., somatostatin also controls the secretory rates of TSH. Thus the adenohypophysis is under dual hypothalamic control mechanisms, positive and negative by TRH and somatostatin, respectively.

Importance in growth and metabolism

The importance of the thyroid in metamorphosis was recognized quite early. A wealth of evidence exists demonstrating that syndromes similar to myxoedema could be produced in mammals by thyroidectomy. Thyroidectomy in the immature mammal is followed by retardation of growth. Even neonatal thyroidectomy results in a marked retardation of growth and maturation in rat.

It was Murray (1891) who cured myxoedema by administration of the thyroid gland whose active principle was unaffected by drying or heating. Further, Coindet (1820) demonstrated the therapeutic value of iodine in the treatment of goiter. The presence of iodine in organic combination in sponges was reported by Vogel and later in the thyroid by Baumann (1896). The substance was identified as diiodotyrosine by Wheeler and Jamieson (1905). It was Kendall (1919) who ultimately isolated the active principle containing high iodine content viz., thyroxine from the alkaline hydrolysate of the thyroid gland.

Thyroglobulin

Thyroglobulin is the major iodoglycoprotein of thyroid with a sedimentation coefficient of 19 S and a molecular weight of 670 KDa. The protein plays an impotrant role in the synthesis, storage and secretion of thyroid hormones (T_3 and T_4). Thyroglobulin has been isolated and purified from several species such as buffalo, cattle, sheep, human, rat, chicken, whale, monkey, hog, squirrel, rabbit, guinea pig, dogfish and *Xenopus laevis* by conventional methods such as saline extraction, salting out of the protein by ammonium sulphate, DEAE-cellulose chromatography, Sepharose 4B or 6B gel-filtration and Concanavalin-A Sepharose affinity chromatography. Further, thyroglobulin from different species has been well characterised in terms of subunit structure, N-terminals, iodine, iodoamino acid, carbohydrate and amino acid contents. Even the sugar sequences of oligosaccharide chains and the amino acid sequence of thyroglobulin has been elucidated (Ui, 1974; Christophe *et al.*, 1982; Edelhoch and Robbins,1986; Deshpande,1991).

In addition to 19 S thyroglobulin, the thyroid contains other iodoproteins with sedimentation coefficient of 12 S, 27 S and 37 S. Thyroproteins with sedimentation coefficient of 3–8 S and 12 S have been reported in primitive thyroids of hagfish and lamprey species, respectively. The 27 S and 37 S iodoproteins have been purified from thyroid of various mammalian species and characterised to a limited extent (Salvatore *et al.*, 1965; Van der Walt and Van Jaarsveld, 1972).

Heterogeneity

Thyroglobulin on ion-exchange chromatography gives a complex elution pattern with multiple peaks. This behaviour has been attributed to differential iodination of thyroglobulin molecules. Variable sialic acid content, non-ionic adsorption and association-dissociation phenomena are thought to be additional factors leading to such heterogeneity.

Microheterogeneity with respect to carbohydrate in thyroglobulin has been established by Concanavalin-A Sepharose affinity chromatography and isoelectric focusing. The same has been attributed to glycosylation of thyroglobulin molecules to different extents.

Subunit Structure

The subunit structure of different thyroglobulin species has been examined by SDS–PAGE either in the absence or presence of reducing agent i.e, 2-mercaptoethanol. Thyroglobulin obtained from calf, sheep, human and pig upon SDS–PAGE in the presence of mercaptoethanol revealed two to three prominent bands. In contrast, a complex pattern with eight protein bands of apparent molecular weights ranging from 44 to 200 KDa was observed for the buffalo species. Based on these studies quaternary structure comprising of 4-8 non-identical subunits have been suggested for native thyroglobulins. However, gene cloning studies have conclusively indicated the presence of two equal sized (330 KDa) subunits in the native structure of thyroglobulin.

Iodine Content

Thyroglobulin as an iodoprotein contains variable amounts of iodine. The iodine content of thyroglobulins of different animal species including dogfish range from 0.2 to 2.0%. On the contrary, a very low iodine content of 0.0022% was observed for lamprey thyroprotein. The iodine content of thyroglobulin obtained from goitrous thyroid is lower than normal thyroid. Variation in iodine content of thyroglobulin obtained from individual or pooled glands has been noted. The variation suggests that thyroid gland of different species consists of a spectrum of thyroglobulin molecules iodinated to different extents. This property has been attributed to the elution behaviour on DEAE–cellulose either by stepwise or linear gradient.

Iodoamino Acid Composition

The iodinated amino acids of the thyroid and thyroglobulin are monoiodotyrosine (MIT), diiodotyrosine (DIT), triiodothyronine (T_3) and tetraiodothyronine or thyroxine (T_4). In addition, the thyroid and blood plasma have an analog of thyroxine i.e., reverse T_3. Amongst the iodoamino acids, only T_3 and T_4 possess potent biological activity and the potency of T_3 is ten times more than T_4. In certain thyroglobulin species, the presence of monoiodo histidine has been detected. However, its biological significance is unclear.

The iodoamino acid composition of different species is presented in Table 1. Despite species variation, the iodoamino acid composition is similar. However, a striking feature of buffalo thyroglobulin is the presence of low and high contents of DIT and T_4, respectively. The data suggests an efficient coupling mechanism.

Table 1. Total carbohydrate and iodoamino acid composition of thyroglobulin from different species

Species	Carbohydrate (%)					
	Galactose	Mannose	Fucose	Glucosamine	Sialic acid	Total
*Buffalo	1.25	3.43	0.43	2.27	1.45	8.83
+Calf	1.20	2.09	0.37	2.95	1.29	7.90
+Sheep	1.13	2.44	0.36	2.98	1.39	8.30
+Pig	1.18	2.38	0.43	3.11	1.13	8.23
+Human	1.21	3.13	0.46	3.86	1.01	9.67

Species	Iodoamino acid (moles/mole)			
	MIT	DIT	T3	T4
*Buffalo	9.02	2.09	—	15.20
†Hog	11.06	15.23	1.02	5.35
†Chicken	11.00	24.60	0.00	3.55
†Whale	7.30	7.00	0.41	2.50
±Human	10.50	14.90	—	6.90

– Not determined. *Deshpande (1991); +Spiro and Spiro (1965); †Sorimachi and Ui (1974); ±Rolland *et al.* (1970).

Analysis of iodoproteins obtained from hagfish, lamprey and dogfish thyroglobulin reveals lack of T_3. This is akin to the observation made on chicken thyroglobulin. It appears from the analysis that the composition of the iodoproteins of primitive species is not different from vertebrate thyroglobulins, although the molecular size of hagfish (3–8 S) and lamprey (12 S) iodoproteins are smaller than thyroglobulin of other species (19 S).

Carbohydrate Composition

Thyroglobulin has been characterized as a glycoprotein with galactose, mannose and fucose as neutral sugars. However, dogfish thyroglobulin has minor amounts of glucose. In addition to neutral sugars, the protein has hexosamine, glucosamine and the sugar acid viz., N-acetyl neuraminic acid or sialic acid. The total carbohydrate content of different species is presented in Table 1 and ranges from 8 to 10%. On the other hand, the total carbohydrate content of lamprey and dogfish iodoproteins are 2.49 and 5.83%, respectively and are lower than mammalian thyroglobulins. Further, the composition, sugar sequences and carbohydrate peptide linkage of the oligosaccharide chains in different thyroglobulin species have been elucidated.

Amino acid composition

The amino acid composition of thyroglobulin from several species is shown in Table 2. In general, the amino acid composition of mammals, aves, amphibians and pisces is similar. The protein is rich in acidic amino acids like glutamic and aspartic acid. The same accounts for 20% of the total weight of the protein. Comparison of amino acid composition of thyroglobulin obtained from normal and goitrous thyroids is not significantly different. There are about 220 half cystine residues in thyroglobulin. Most of them exist as inter and intra chain disulphide bridges. Very few free sulphydryl (thiol) groups are present in protein. About 5 moles/mole of the protein has been reported in different species.

Table 2. Amino acid composition of thyroglobulins

	Mole %				
Amino acid	+*Xenopus laevis*	+Chicken	+Hog	0Buffalo	+Human
Lysine	5.0	3.7	2.5	3.8	3.4
Histidine	2.3	1.1	1.3	1.0	1.4
Arginine	4.0	4.2	6.7	6.5	5.9
Aspartic acid	10.5	9.6	7.2	6.0	8.4
Threonine	5.7	5.3	5.3	3.6	5.4
Serine	9.7	8.9	8.7	5.1	9.7
Glutamic acid	11.7	12.2	12.5	13.0	13.1
Proline	5.7	9.7	7.8	7.7	7.1
Glycine	7.3	7.3	7.7	3.4	7.7
Alanine	5.7	6.3	9.7	4.6	7.6
Half cystine	2.4	2.6	3.8	2.6	2.6
Valine	6.2	5.8	5.9	4.0	6.4
Methionine	1.8	1.3	0.9	1.2	1.4
Isoleucine	4.2	4.2	2.5	2.1	2.9
Leucine	9.3	9.1	10.3	7.4	9.5
Tyrosine	2.5	3.2	2.2	3.2	2.2
Phenylalanine	5.9	5.5	4.9	5.6	5.3

0Deshpande (1991); +Hoshino and Ui (1970).

N-Terminal Analysis

The study has been employed in the evaluation of the number of polypeptide chains in native thyroglobulin. The N-terminal amino acid for various thyroglobulins has been determined by chemical methods such as of Sanger and Edman. The study suggests that the N-terminal amino acid of the mature protein for most of the species is aspartic acid or asparagine except for buffalo species which has glutamic acid. On the basis of N-terminal analysis, quaternary structures comprising of either two or four chains have been suggested for native thyroglobulin.

Biosynthesis of Thyroglobulin

The major role of the thyroid gland is to convert inorganic iodide to molecular iodine followed by iodination of the tyrosyl residues of thyroglobulin. The whole process is under the influence of TSH. The steps involved in the biosynthesis of thyroglobulin are: (a) Synthesis of protein, (b) Glycosylation and (c) Thyroid hormone synthesis.

(a) Synthesis of protein

Radiolabelled incorporation studies revealed that protein synthesis is restricted to the epithelial cells of the follicle and the mechanism is similar to that of other secretory glycoproteins. The thyroid polyribosomes synthesizing thyroglobulin are usually large with 30–60 ribosome units. However, a polyribosome containing 100 ribosome units has also been predicted for the synthesis of the 330 KDa protein subunit of thyroglobulin. A 33 S mRNA coding for this subunit has been identified (Vassart *et al.*, 1977).

Earlier investigations with radiolabelled compounds (^{3}H or ^{14}C-leucine or tyrosine) and with goitrogens such as propylthiouracil and perchlorate indicated the presence of 3–8 S and 12 S components. Thyroprotein with 3–8 S and 12 S have also been reported in primitive species such as hagfish and lamprey, respectively (Suzuki *et al.*, 1975). A progressive synthesis from 3–8 S to 12 S and finally to 19 S was proposed for the synthesis of thyroglobulin. However, this hypothesis was contradicted by subsequent studies. It has been suggested that the primary translation product of calf thyroid mRNA has a molecular weight of 660,000 daltons and the product which is observed is the cleavage product of 330,000 daltons (Eggo and Burrow, 1980).

(b) Glycosylation

The sequential addition of monosaccharide units to the polypeptide chain is termed as glycosylation and is a post-translation event. However, glycosylation of thyroglobulin occurs prior to iodination and is not under direct genetic control. Hence the mechanism is less precise than peptide chain synthesis resulting in variation in the oligosaccharide structure of thyroglobulin. The sugar sequences of the oligosaccharide chains are determined by the intracellular location and the specificity of glycosyl transferases in endoplasmic reticulum and Golgi apparatus. The mechanism of oligosaccharide chain synthesis of thyroglobulin is akin to that of other glycoproteins. Glycosylation is independent of the protein synthesis because the former process is not blocked by inhibitors of protein synthesis such as puromycin. However, when polysaccharide chain synthesis is inhibited by tunicamycin or deoxyglucose, protein synthesis and secretion are also significantly affected indicating a coordination between protein and oligosaccharide chain synthesis.

(c) Thyroid hormone synthesis

Different stages in the thyroid hormone synthesis are:

(i) *Concentration of iodide*: The thyroid accumulates iodide (I^-) against a concentration gradient and the process is energy dependent and is linked to Na^+/K^+ ATPase dependent system. A small amount of iodide enters the gland by passive diffusion. The transport of iodide across the membrane is blocked by agents such as perchlorate, pertechnetate and thiocyanate.

(ii) *Oxidation of iodide:* The next step involves the oxidation of inorganic iodide to molecular iodine and is catalysed by a peroxidase, a heme protein present at the luminal surface of the follicular cell. It is a tetrameric protein with a molecular mass of 60 KDa and has an absolute requirement for hydrogen peroxide, the oxidising agent. Regions other than thyroid such as sub-maxillary in various animal species, pharyngeal and head kidney thyroid tissue from teleosts have peroxidase activity. Anti-thyroid agents such as thiocarbamides and thioimidazoles inhibit the process of iodide oxidation and thyroid hormone synthesis.

(iii) *Iodination of tyrosine*: Iodination of the tyrosine residues in thyroglobulin is catalysed by thyroid peroxidase. Completion of the oligosaccharide chains in thyroglobulin appears to be required before iodination. *In vitro* iodination has been carried out with free tyrosine as well as proteins employing iodide and enzymes like myeloperoxidase or horseradish peroxidase or chloroperoxidase in the presence of hydrogen peroxide generating system to generate iodine *in situ* (Taurog and Howells,1966).

The process of iodination of tyrosine residues involves either an iodine free radical or the electrophile, the iodonium ion (I^+). Iodine reacts with tyrosine residues in thyroglobulin to form 3-monoiodo tyrosine (MIT). This on further iodination forms 3,5-diiodo tyrosine. Free tyrosine can also be iodinated, however, the same is not incorporated into the protein, as no tRNA codes for iodotyrosines. This has also been substantiated by radiolabelled incorporation studies using ^{14}C-tyrosine and ^{125}I-tyrosine.

(iv) *Coupling reaction*: The mechanism of thyroxine synthesis has been understood by use of tyrosine, synthetic dipeptides and polymers and co-polymers of tyrosine, as model compounds (Cahnmann *et al.*, 1977). The coupling of two molecules of DIT to form T_4 or one molecule of MIT with DIT to T_3 followed by elimination of dehydroalanine occurs within thyroglobulin. The oxidative coupling reaction appears to be catalysed by thyroid peroxidase. Dehydroalanine later decomposes to pyruvic acid and ammonia. The thyroid hormones remain as integral units of thyroglobulin and are liberated into blood circulation by a TSH dependent hydrolysis of thyroglobulin. Excess iodide is known to inhibit this process and the same has been exploited in the treatment of hyperthyroidism by use of potassium iodide.

Non-thyroidal proteins such as albumin and casein on chemical iodination *in vitro* yield iodoproteins which upon administration have thyroxine like effect. However, the formation of thyroxine in thyroglobulin is at a lower concentration of iodine.

Degradation of Thyroglobulin and Hormone Release

The release of thyroid hormones from thyroglobulin is effected by the action of proteases present in the thyroid follicular epithelial cells and is under the influence of TSH. On demand of thyroid hormones, thyroglobulin is engulfed into the follicular cells by endocytosis of colloid. The colloid droplets later fuse with lysosomes and the lytic enzymes present therein degrade thyroglobulin releasing thyroid hormones into blood circulation. The normal T_4 and T_3 levels in blood of different species are 1.51–6.22 ng/dl and 64.7–145.9 ng/dl, respectively (Reap *et al.*, 1978). Enzymes such as glycosidases required for degradation of thyroglobulin, are also present in lysosomes.

Transport of Thyroid Hormones

The thyroid hormones in blood do not exist in their free state but are protein bound. The principal serum transport protein of the thyroid hormones is thyroxine-binding globulin (TBG), a glycoprotein with a molecular mass of 55 KDa. Two other serum proteins are prealbumin (transthyretin, 55 KDa) and albumin (66 KDa). The carrier proteins vary in their concentration, binding affinity to the hormone and dissociation constants. Even species variation with regard to the degree of binding of the hormone to the carrier protein has been observed. However, in disease states, antithyroid antibodies are known to be involved in the transport of thyroid hormones.

At the target site, the carrier proteins liberate the hormone into its free state which later diffuses into the cell where it exerts its action.

The determination of protein bound iodine (PBI) is a measure of circulating thyroid hormone and is of importance in assessing the thyroid status. Circulating PBI values for a number of animal species including human have been determined, and the values range from 0.4 to 8.0 μg% (Zarrow *et al.*, 1964).

Biological Actions of Thyroid Hormones

The thyroid hormones display diverse biological effects in different vertebrates and are governed by factors such as species and tissue types, concentration and length of exposure of the hormones. The diverse functions of the thyroid hormones suggest phylogenetic adaptive mechanisms.

In all mammalian species, the thyroid hormones accelerate growth, increase in oxygen consumption, basal metabolic rate (BMR) and heat production (thermogenesis). Relation of thyroid hormones to the said factors has been established in hypo- and hyperthyroid states. In addition, its effect on growth, foetal and neonatal development has been substantiated either by thyroidectomy or in thyroxine deficiency.

Further, the normal functioning of the central nervous system depends upon the output of the thyroid hormones. In thyroxine deficiency, there is a permanent damage to the neurons impairing the adaptive behaviour.

In amphibians, the larval metamorphosis is affected in thyroxine deficiency (Oppenheimer and Surks, 1975). In addition, moulting in chicks, hibernation of warm blooded animals, reproduction, hair formation and lactation are under the influence of the thyroid hormones. That thyroxine enhances lactation in dairy cattle has been demonstrated by supplementation of thyroprotein (iodinated casein) in the feed (Sen *et al.*, 1954).

At the molecular level, the thyroid hormones increase the rate of transport of electrolytes and amino acids into the cell by modulating the pumping mechanisms of the cell membrane. The hormones also cause alterations in cellular events such as transcription and translation which are reflected in terms of increased protein synthesis, cell size, number and functions. Furthermore, the regulatory effects of the thyroid hormones on metabolism depend upon the cell type.

In the adipose tissue, there is increased lipolysis of fats followed by oxidation of fatty acids. In pituitary, the hormones modulate the activity of gonadotropins which in turn influence cell proliferation and growth. In liver, kidney and muscle tissues, the thyroid hormones stimulate the sodium pump and glycolytic mechanisms leading to increased oxidation of carbohydrates, oxygen consumption, calorigenesis, coupled to oxidative phosphorylation. On the contrary, in the brain tissue there is no calorigenic effect. However, the hormones alter the concentrations of the neurotransmitters viz., dopamine and of cholinergic and adrenergic receptors. They also modulate the activity of other hormones through receptor mediated interactions or by altering processes initiated by other hormones.

Molecular Basis of Thyroid Hormone Action

The most widely accepted model (Fig. 1) for thyroid hormone action is the nuclear mechanism. At the responsive site, the thyroid hormones diffuse freely through the plasma membrane. They bind specifically to high affinity non-histone protein receptors located in the nucleus. This is in contrast to the occurrence of cytosolic receptors for steroids which are translocated after binding. The affinity of the thyroid hormone receptor to T_3 is ten times higher than T_4. Thus thyroxine is the prohormone for the generation of T_3 at the target site, and this conversion is effected by the action of peripheral deiodinase.

The thyroid hormone (T_3) receptor has been well characterized and bears structural resemblance to that of retinoic acid receptor (Weinberger *et al.*,

1986). These receptors have a short amino terminal region and similar DNA and carboxy terminal regions in their structure. Point mutation studies have revealed carboxy terminal end of the receptor as the ligand binding domain. The percent homology of T_3 receptor with that of retinoic acid and steroid receptors is 34 and 20, respectively.

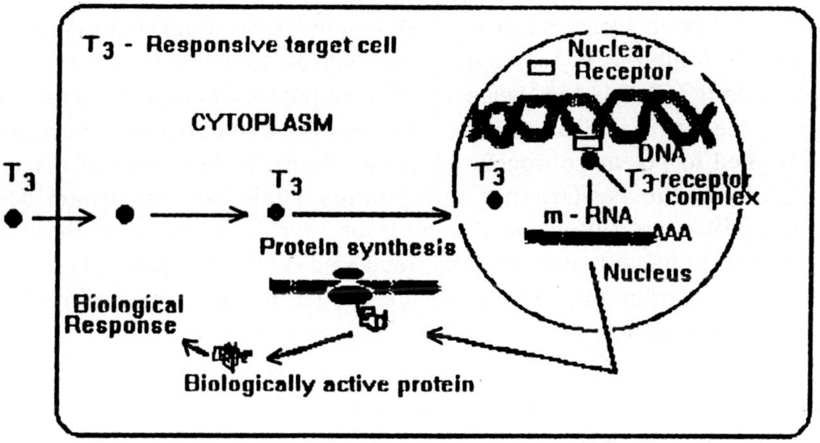

Fig. 1 Diagramatic representation of thyroid hormone action.

Binding of T_3 to its nuclear receptor is followed by increased formation of polyadenylated heterogeneous nuclear mRNA. In addition, there is an increase in activity of nuclear protein kinase. Thus T_3 stimulates the genome of the responsive cell increasing the rate of transcription and translation leading to enhanced protein/enzyme synthesis with a positive nitrogen balance.

Metabolic Fate of Thyroid Hormones

Thyroid hormones are inactivated by different metabolic reactions within the thyroid and peripheral tissues by mechanisms such as deiodination, deamination, decarboxylation and conjugation. Amongst these, deiodination of the thyroid hormones is the principle mechanism and is catalyzed by thyronine deiodinase present in the thyroid and peripheral tissues. In contrast, iodotyrosine deiodinases are restricted to the thyroid and play an important role in recycling iodine. Deiodination of T_4 yields two products T_3 and reverse T_3, the active and inactive forms, respectively. Further metabolism of thyronines by the action of deiodinases yields metabolites such as 3,5-diiodothyronine, 3,3′-diiodothyronine, 3′,5'-diiodothyronine, 3 or 3′-iodothyronine and thyronine.

Metabolities derived from deamination of T_4 and T_3 are tetraiodo- and triiodothyropyruvic acid which upon oxidative decarboxylation yield respective acids viz., tetraiodo- and triiodothyroacetic acid, the same are present as urinary metabolites. Disruption of the thyronine ring or modification of the phenolic group by conjugation (sulfation and methylation) are also additional mechanisms involved in the metabolic fate of thyroid hormones.

Antithyroid Agents

Also known as goitrogens are compounds that interfere with thyroid function by inhibiting the synthesis, release and peripheral action of thyroid hormones. The excessive consumption of foods containing goitrogens such as cabbage (Brassica) produces goiter. The same has been exploited in inducing hypothyroidism in experimental animals.

The antithyroid agents can be classified on the basis of mode of inhibition, and are: (a) Inhibitors of iodide trapping mechanism: Complex anions such as perchlorate (ClO_4^-), thiocyanate (SCN^-) are potent inhibitors of the iodide trapping mechanism. Cyanogenetic glycosides and isothiocyanates are also transformed to potent goitrogens by metabolism. (b) Inhibitors of thyroid hormone synthesis: Organic compounds such as thiocarbamides, thioimidazoles and sulfonamides inhibit the thyroid hormone synthesis at the level of iodination and coupling, mediated by thyroid peroxidase. The structures of antithyroid compounds are shown in Fig 2. (c) Inhibitors of

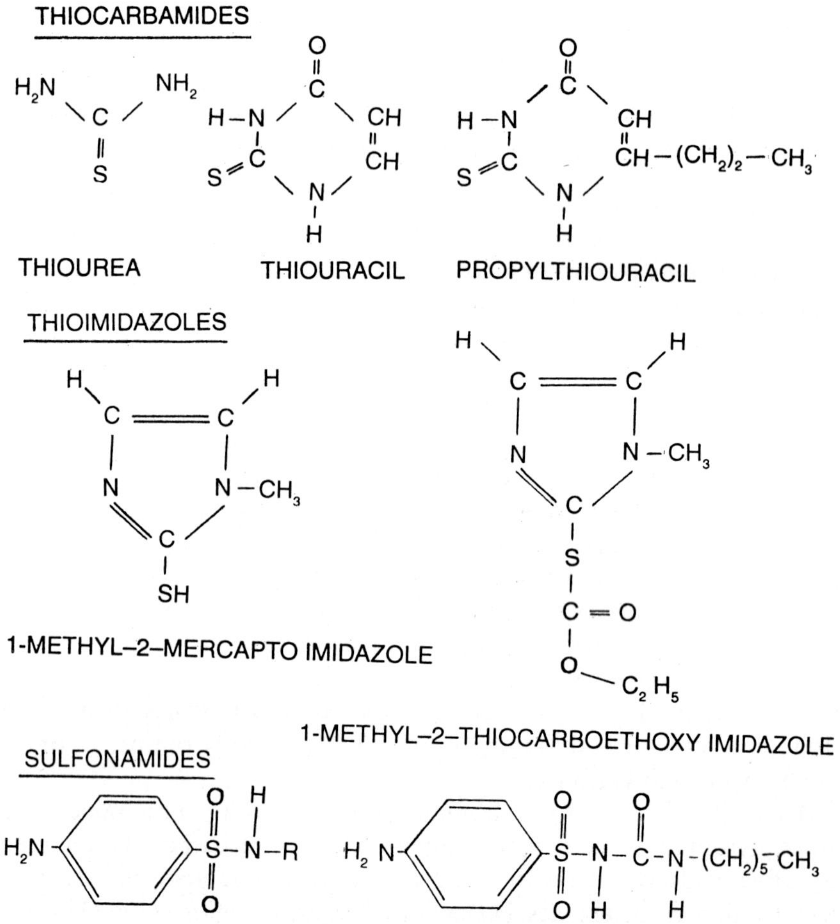

Fig. 2 Structures of antithyroid agents.

thyroid hormone secretion and deiodination: Excess iodide and lithium salts (lithium carbonate) inhibit thyroid hormone secretion whereas nitrotyrosine inhibits deiodination of iodotyrosines. (d) Inhibitors of peripheral action of thyroid hormones: Thyroxine analogs like rT_3, 3,3′-diiodothyronine, 3,3′,5–triiodothyropropionic acid and 3,3′-diiodothyropropionic acid are potent inhibitors of thyroid hormone action in peripheral tissues.

Antithyroid compounds are useful in the treatment of specific cases of hyperthyroidism. The agents may also have possible application in fattening of meat producing animals. Radioiodine (^{131}I) in large doses causes destruction of thyroid tissue. This property is advantageous in thyroid carcinoma therapy.

Disorders of Thyroid

Clinically two syndromes, hypo and hyper states, have been recognised and reflect the activity of the thyroid. Assessment of these syndromes is possible by examining the signs and symptoms. In addition, a number of diagnostic tests are available and are of value in monitoring thyroid dysfunction. Tests such as quantitation of circulating thyroid hormones (free and total) and TSH by competitive or non-competitive binding assays viz., RIA and ELISA are of significance in clinical diagnosis of thyroid disease. Further determination of circulating protein bound iodine, measurement of the capacity of the uptake of radioiodine by the gland and basal metabolic rate (BMR) are equally important in clinical diagnosis of thyroid status.

Hypothyroidism

Hypofunctioning of the thyroid due to insufficient secretion of thyroid hormones is known as hypothyroidism and the same has been observed in different animal species. The disease is characterized by impaired growth and development. If it is present at birth, the disease is known as cretinism whereas in adults the condition is called myxoedema. The disease is associated with impaired mental development and is characterized by thick scaly skin, low circulating thyroxine, hypercholesterolemia accompanied by low or normal BMR.

Hypothyroidism is of two types: primary and secondary. The primary is due to thyroid disease and has elevated TSH level because of lack of feedback operation by thyroid hormones. This elevated level leads to stimulation of gland and goiter. The insufficient secretion of TSH by adenohypophysis due to pituitary disease causes secondary hypothyroidism.

Irrespective of the type, the disease can be treated by fortification of the diet or feed by desiccated thyroid or thyroprotein or thyroxine.

Endomic Goiter

It arises due to iodine deficiency in food and water. This leads to insufficient hormone synthesis, enlargement of the thyroid, i.e. endemic goiter. Iodine deficiency can occur in any species, however it is seldom seen in modern management of livestock. Endemic goiter is prevalent in mountain regions because of deficiency of iodine in water and feed. Other factors such as presence of calcium and antithyroid compounds in diet lead to this condition. However, the use of iodized salt has been advocated as a prophylactic measure in endemic areas. In India, the state of Jammu and Kashmir, Punjab, Arunachal Pradesh and Uttar Pradesh have been identified as endemic areas. The whole of sub-Himalayan belt including Nepal, Sikkim and Bhutan are also endemic.

Hyperthyroidism

Hyperactivity of the thyroid leading to excessive secretion of the thyroid hormones is termed as hyperthyroidism or thyrotoxicosis. In domestic animal species, hyperthyroidism is less common than hypothyroidism. This state is characterized by increased BMR, cardiac irregularity and heat intolerance. Circulating thyroxine and sugar levels are elevated with hypocholesterolemia. In certain situations, the thyroxine levels are normal.

Two hyperthyroid states are recognized. The first one is known as Grave's disease. The same is also referred to as toxic diffuse or exophthalmic goiter because of abnormal protrusion of eye balls. The disease is also characterized by the presence of long acting thyroid stimulator (LATS), a thyroid stimulating immunoglobulin, in blood circulation. This mimics the action of thyrotropin and causes enlargement of the thyroid. In the other type, called toxic nodular goiter, the disease is due to thyroid tumor or adenoma. The nodular activity is not regulated by TSH. The symptoms are similar to Grave's disease except that it lacks the exophthalmic condition.

The hyperthyroid state can be cured by antithyroid agents, radioiodine and surgery.

Autoimmune Disorders

One of the well known autoimmune disorders of the thyroid is Hashimoto's disease, also referred as autoimmune thyroiditis. The disease may be described as being caused by a lesion and is characterized by atrophy of parenchymal cells, lymphocytic infiltration and fibrosis. Different variants of autoimmune thyroiditis have been recognized. A close relationship exists between Grave's disease and autoimmune thyroiditis and appears to be caused by closely related immunological factors.

It was earlier believed that thyroglobulin, the antigen, is confined to the thyroid. However, this notion has been disproved because thyroglobulin is present in normal blood circulation (human: 0–50 ng/dl). In fact, this protein is a normal blood constituent even before birth and the corresponding B-lymphocytes have been delected. Although, most of the thyroglobulin is degraded by proteolysis in the thyroid follicular cells, frequently some undegraded molecules escape into the lymphatics.

Plasma thyroglobulin levels are elevated in thyroid disorders and is generally not associated with the development of antibodies to thyroglobulin. However, thyroglobulin antibodies are characterstic of Hashimoto and Grave's disease and are found in high titers in blood circulation. The autoantibodies are mostly species specific. Additionally, antibodies to thyroid hormones (T_3 and T_4) have occasionally been observed in the said disorders only when high titers of antithyroglobulin are present. Antibodies which are capable of stimulating growth of the thyroid gland have been detected in goitrous Hashimoto's thyroiditis and are distinct from LATS of Grave's disease.

Hashimoto's disease possibly arises due to genetic defects in the immuno regulatory mechanism. Anti-microsomal antibodies bind to the thyroid cell membrane receptor leading to activation of the complement system and subsequent damage to the target organ. Alternatively, autoantibodies might behave in a similar fashion, except that they lack affinity to the complement system. Autoimmune antibodies combined with lymphocytes might also be important in explaining the cytotoxic mechanism.

Management of Hashimoto's disease is possible by immunosuppression. This approach is impractical due to side effects. However, thyroid hormone suppression or replacement therapy can be beneficial.

Acknowledgments

I thank Dr. B. Sashidhar Rao of the Department of Biochemistry, Osmania University for critically reviewing the manuscript. The whole hearted cooperation extended by my research students Dr. S. S. Shavali (PDF), Mr. S.G.Venkatesh (SRF), Ms. S. Suryakala (SRF), Ms. C. Geetha (JRF) and Ms. G. Sobha (JRF) in the preparation of the manuscript is also acknowledged.

References

Baumann, E. (1896) Munch. Med. Woehenschr. **43:** 309

Cahnmann, H.J., Pommier, J. and Nunez, J. (1977) Proc. Natl. Acad. Sci. USA **74:** 5333–5335.

Cameron, A.T. (1930) J. Biol. Chem. **16:** 465–473.

Christophe, D., Mercken, L., Brocas, H., Pohl, V. and Vassart, G. (1982) Eur. J. Biochem. **122:** 461–469.

Coindet, J.R. (1820) Ann. Chem. Phys. Paris **15:** 49.

Deshpande, V. and Ramachandran, L.K. (1976) Ind. J. Biochem. Biophys. **13:** 161–164.

Deshpande, V. (1987) Ind. J. Animal Sci. **57:** 903–904.

Deshpande, V. (1991) Comp. Biochem. Physiol. **100B:** 217–221.

Dey, B.B., Krishnan, P.B. and Giriraj, M. (1943) Curr. Sci. **12:** 272–302.

Edelhoch, H. and Robbins, J. (1986) In: Werner's The Thyroid. A fundamental and Clinical text. (S.H. Ingbar and L.E. Braverman, Eds) pp 98–115 J.B. Lippincott Company, Philadelphia.

Eggo, M.C. and Burrow, G.N. (1980) Can. J. Biochem. **58:** 446–455.

Gorbman, A. (1986) In: Werner's The Thyroid. A fundamental and Clinical text. (S. H. Ingbar and L.E. Braverman, Eds) pp 43–52 J.B. Lippincott Company, Philadelphia.

Hoshino, T. and Ui, N. (1970) Endocrinol. Japon. **17:** 521–533.
Kendall, E.C. (1919) J. Biol. Chem. **39:** 125–147.
Murray, G.R. (1891) Br. Med. J. **2:** 796.
Oppenheimer, J.H. and Surks, M.I. (1975) In: Biochemical basis of thyroid hormone action. (G. Litwack, Ed) pp 119–157 Academic Press, N.Y.
Reap, M., Cass, C. and High Tower, G. (1978) Southwestern Veterinarian **31:** 31–34.
Rolland, M., Aquaron, R. and Lissitzky, S. (1970) Anal. Biochem. **33:** 307–317.
Salvatore, G., Vecchio, G., Salvatore, M., Cahnmann, H.J. and Robbins, J. (1965) J. Biol. Chem. **240:** 2935–2943.
Sen, K.C., Premchandra, B.N., Murthy, G.K., Dastur, N.N. and Narayanan, D. (1954) Ind. J. Diary Sci. **7:** 49–63.
Simpson, B.W. (1930) N. Z. J. Agri. **41:** 302.
Sorimachi, K. and Ui, N. (1974) Gen. Comp. Endocrinol. **24:** 38–43.
Spiro, R.G. and Spiro, M.J. (1965) J. Biol. Chem. **241:** 1272–1282.
Suzuki, S., Gorbman, A., Rolland, M., Montfort, M. and Lissitzky, S. (1975) Gen. Comp. Endocrinol. **26:** 59–69.
Taurog, A. and Howells, E.N. (1966) J. Biol. Chem. **241:** 1329–1339.
Ui, N. (1974) In: Handbook of Physiology and Endocrinology. (R.O. Greep and E.B. Astwood, Eds) pp 55–79 American Physiological Society, Washington, D.C.
Van der Walt, B. and Van Jaarsveld, P. (1972) Arch. Biochem. Biophys. **150:** 786–791.
Vassart, G., Verkstreken, L. and Dinsart, C. (1977) FEBS Lett. **79:** 15–18.
Weinberger, C., Thompson, C.C., Ong, E.S., Lebo, R., Gruol, D.J. and Evans, R.M. (1986) Nature **324:** 641–646.
Wharton, T. (1656) In: Adenographia: Sive glandularum tolius corporis descriptio Chap. XVIII De glandulis thyrodaeis. pp 118–122 Londinis Typis J.G. Iempensis Authoris.
Wheeler, H.L. and Jamieson, G.S. (1905) Am. Chem. J. **33:** 365.
Zarrow, M.X., Yochim, J.M. and McCarthy, J.L. (1964) In: Experimental Endocrinology. New York Academic Press Inc.

Comparative Endocrinology and Reproduction
K.P. Joy, A. Krishna and C. Haldar (Eds)

34. Thyroid in Intermediary Metabolism

O.V. Oommen, P. Nair, J. Baby, M.A. Suresh Kumar and T.T. Valamparampil
Department of Zoology, University of Kerala, Trivandrum-695 581, India

Introduction

The thyroid gland continues to attract the attention of endocrinologists, despite the appearance of scores of papers in the last 100 years of thyroid research ever since Magnus Levey's experiments in 1895 (Magnus Levy, 1895). With the establishment, in the 1970's, of dependable and sensitive assays for the measurement of blood thyroid hormones (TH) levels, there is a reawakening of interest in the thyroid physiology (Leatherland, 1988). Consequently, a lot of information is now available on thyroidal parameters measured under a variety of environmental and physiologic conditions. Major differences still exist in the whole body responses to thyroid hormones in endothermic and ectothermic vertebrates. In poikilotherms many aspects of thyroid fuction remains to be solved, such as the role of TH in embryonic development and metamorphosis in teleost fish and amphibia. The interaction of TH, which have a wide range of metabolic effects, with other hormones, has assumed greater importance. The investigators in other areas of endocrinology, therefore, feel the need to include thyroid function in their models, and for all those reasons the thyroid continues to be an active realm of research. However, although significant advances have been made in certain areas, there is still much to be studied about the control and function of thyroid gland and its role in the normal physiology of fishes and amphibians.

There is no doubt that TH are widespread among vertebrate animals. It is believed that these hormones influence different life processes such as metabolism, growth, differentiation, reproduction and behaviour. To serve these functions TH may play roles that are permissive or synergistic with other hormones. Vertebrates have evolved a number of metabolic pathways to meet various requirements for a successful survival. The oxidative metabolism is responsible for cellular oxidation of metabolic substrates and production of energy which is conserved in the form of ATP. The entire process of biological oxidation involve a number of enzymes, metabolic substrates and utilises molecular oxygen at the mitochondrial level. Therefore, respiration or oxygen uptake acts as an index of oxidative metabolism or biological oxidation. This in turn reflects the physiological status of an individual organism.

The metabolic state of vertebrates depends on a number of external (Environmental) and internal (Physiological) factors, i.e., dose of the hormone administered, the species and age of the animal, the ambient temperature and season, the nutritional state, the type of target tissue, phylogeny etc. The neuroendocrine system plays an important role in the regulation of metabolic rate. According to Norris (1985) the major role of TH may be to maintain a state of well being in tissues which allows retention of optimal responsiveness of other regulatory factors such as hormones, neural stimulation or certain environmental stimuli.

The effects produced by TH on mammalian metabolism include the so called calorigenic or thermogenic action and specific effects related to carbohydrate, lipid and protein metabolism. They can accelerate the rate at which glucose is oxidised and thus increase the amount of metabolic heat produced in a given time. This excess heat production could be used to warm the body. Increased glucose oxidation is reflected in an increased basal metabolic rate as measured by change in the rate of oxygen consumption. TH play a major role in the regulation of BMR in birds and mammals. These hormones are also essential for the calorigenic action of catecholamines in mammalian species. Unlike in homeotherms, the calorigenic action of TH in the poikilothermic vertebrates is not fully understood.

Studies on the comparative endocrinology of thyroid in lower vertebrates, which forms the basis of this review, were initiated at Varanasi in the labratory of Prof. J.P. Thapliyal in the sixties and seventies which subsequently was extended to all over the country. The endocrine control of oxidative metabolism in fish, amphibia and reptiles has been the subject of numerous studies. The effect of thyroxine on oxygen consumption at the level of whole body or tissues *in vivo* and *in vitro* has been proved in reptiles (Thapliyal *et al.*, 1974a, b; Chandola *et al.*, 1974; Oommen, 1981). Research in lower vertebrates for the past 25 years clearly indicates the multiple control of oxidative metabolism by different hormones (Oommen, 1978, 1981; Gupta and Thapliyal, 1983, 1985, 1991).

Fish

Involvement of TH in the metabolism associated with growth of fish has been reported (Donaldson *et al.*, 1979). But, the role of TH in fish metabolism remains difficult to define. Plisetskaya *et al.* (1983) reviewed the work on the role of TH in the regulation of intermediary metabolism in cyclostomes and fish and many publications have appeared since then on the subject. There is a great deal of controversy in aspects of fish metabolism and its regulation by TH.

In teleosts, as in mammals, the primary TH are thyroxine (T4) and tri-iodothyronine (T3) that have a wide range of effects on metabolic events (Gorbman, 1978; Eales, 1979). The role of TH in the regulation of carbohydrate metabolism in fishes remains unclear (Saurez and Mommsen,

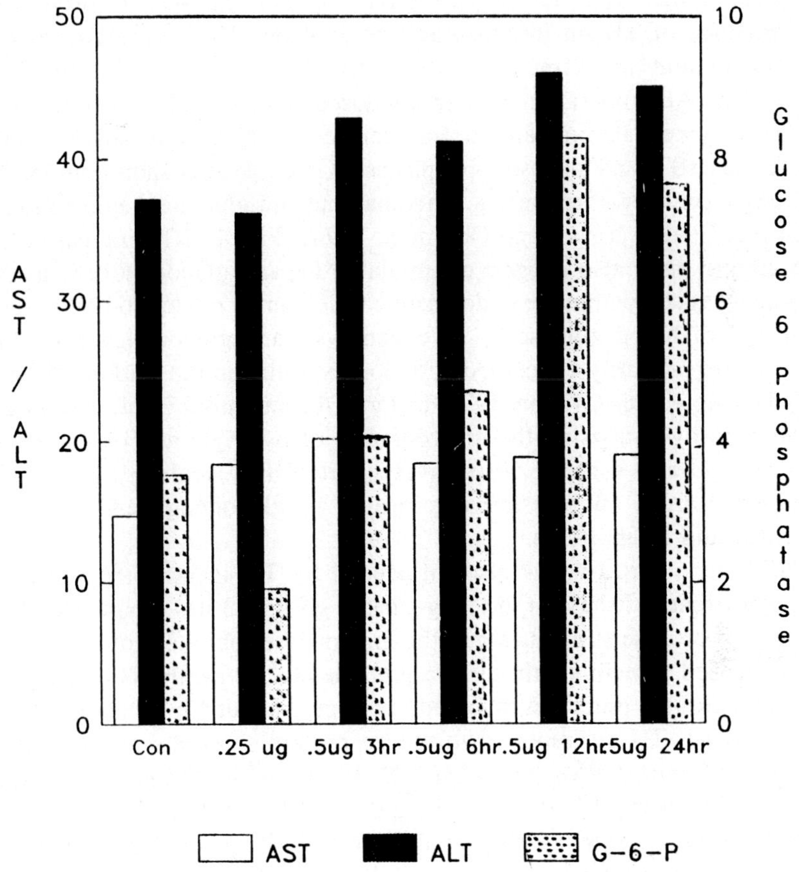

Fig. 1 Effects of T3 on intermediary metabolism in *Anabas testudineus*.

1987). Exogenous thyroxine administration causes hypoglecemia in some species and hyperglycemia in others (Plisetskaya *et al.*, 1983). Activation of gluconeogenesis from alanine has been reported in the liver of Japanese eel (Hayashi and Ooshiro, 1977). However, the mechanism by which TH exert possible gluconeogenic action and its physiological role with respect to the control of carbohydrate metabolism remains unclear. TH are known to stimulate somatic growth and metabolic function in teleost fishes (Singh, 1970; Matty *et al.*, 1982). Administration of T3 and T4 increased the plasma concentration of the same and altered the total liver mitochondrial nucleic acid, emphasizing the role of TH in teleost metabolism (Oommen and Matty, 1991). Plisetskaya *et al.* (1983) suggested that for evaluating the thyroid hormone effect on metabolism, the regulatory system has to be disturbed to some extent. Therefore, pharmacological doses of T3 and T4 may sometimes be required to elicit a biological effect.

Recent studies from our laboratory on *Anabas testudineus* clearly reveal the influence of TH on intermediary metabolism. T3 is shown to have a dose dependent and time dependent stimulatory role on intermediary metabolism. Administration of different doses of T3 stimulates the activities of the enzymes alanine amino transferase (ALT) and aspartate amino transferase (AST) and glucose-6-phophatase (G6-Pase) after short term (hours) and long term (days) treatments. This may mean a gluconeogenic action of T3 in fish (Padmaja Nair and Oommen, 1994) (Fig. 1). The increased G6-Pase activity may also suggest a stimulated hepatic glycogenolysis as this enzyme localised within the endoplasmic reticulam is related to homeostatic regulation of blood glucose by glycogenolysis and gluconeogenesis. ALT and AST are primarily concerned with non-essential amino acid metabolism and gluconeogenesis (Wilson, 1973) and interrelates carbohydrate and protein metabolism by catalysing the incorporation of pyruvate and oxalo acetate from alanine and asparate respectively (Orten and Neuhaus, 1982). The increased activity of transminases results in gluconeogenesis in fishes (Freeman and Idler, 1973).

The protein turn over is also influenced by TH in both the short and long-term treated fishes. In the long-treated group T3 has a biphasic effect i.e., the lower dose stimulates and the higher dose inhibits protein content. The increased protein synthesis is a prerequisite for synthesis of enzymes that stimulates the processes involved in bringing about the metabolic effects of T3 and T4. TH are reported to induce protein synthesis in the liver of a number of animals (Michels *et al.,* 1963; Ghosh and Medda, 1969; Tata, 1970, 1975). Thyroid hormones activates the protein synthetic machinery in the liver and muscle in fish (Thournburn and Matty, 1963; Jackin and La Roche, 1973; Narayan Singh and Eales, 1975; Medda and Ray, 1979; Matty *et al.,* 1982). Plisetskaya and Gorbman (1982) reported that TH either reduced or elevated plasma protein in lamprey and hagfishes. It was further suggested that TH deficiency decreased protein synthesis in cyclostomes (Plisetskaya *et al.,* 1983).

Administration of TH *in vivo* and *in vitro* stimulates the activities of cytochrome oxidase, a-glycerophosphate dehydrogenase and succinate dehydrogenase in teleosts (Peter and Oommen, 1989a, b, c, 1991, 1993). The increased mitochondrial respiration reported in *Anabas* may partly be due to an increase in the resporatory chain components of hepatic mitochondria. Three hours after intraperitonial injection of T3, oxidation of some substrates by mitochondria isolated from the liver of the bow fin was reduced. Similar treatment had no effect on substrate oxidation on liver mitochondria isolated from lake char. Oxidation of substrates by lake char red muscle mitochondria was stimulated by T3 injection. Citrate synthase levels were increased in red muscle suggesting that changes in enzyme activity may in part be responsible for the short-term mitochondrial responses to T3 injection (Ballantyne *et al.,* 1992). The increased oxidation of fatty acids in muscle mitochondria and the decreased lipogenic capacity of liver as suggested by the reduced malic enzyme levels (Ballantyne *et al.*, 1992) would tend to deplete lipid reserves in the fish over extended periods of time. Takashima *et al.* (1972) reported a decrease in plasma free fatty acid content and plasma lipids after treating rainbow trout with mammalian thyroid gland powder. An increase in muscle lipid was noticed after injection of T4 in coho salmon, *Onchorhynchus kisutch* (Donaldson *et al.,* 1979). Chronic treatment with T4 effectively reduces the state of hyper-cholesterolemia in hypophysectomised killifish, *Fundulus heteroclitus* (Pickford, 1973). During the period of sexual maturation of fish, the plasma fatty acids, triglyceides and cholesterol concentrations progressively fell in parallel with progressive fall in serum TH levels (Leatherland and Sonstegard, 1980, 1982). Beleznai *et al.* (1989) had reported that the cholesterol level in the mitochondrial membrane of rat liver was at its lowest in the hyperthyroid state. The regulation of fatty acid metabolism in fish by TH, however, needs careful future investigation.

Amphibia

In the life cycle of amphibians several periods of thyroidal activity have been observed. During larval metamorphosis thyroid activity is very high and the development is thyroid hormone dependent (Norris, 1983; Galton, 1988). In Xenopus laevis binding of T3 and the sensitivity to exogenous T3 begins during embryonic development much before larval metamorphosis (Tata, 1970) and prior to complete development of the circulatory system. Therefore, TH could have developmental or metabolic effects before the hormone dependent metamorphosis begins. Normal developmemt of neotenic *Ambystoma mexicanum* involves biochemical transformation of haemoglobin (Ducibella, 1974) and nitrogen excretion pathway (Schulthesis, 1977) which are similar to TH dependent transitions in metamophosing species. The biochemical transitions in *A. mexicanum* can be stimulated by low doses of thyroxine which otherwise do not cause anatomical transformation (Tomkins, 1978), suggesting that these biochemical changes are usually influenced by thyroxine. Among the various vertebrate species the thyroid gland performs many functions but none is more spectacular than its action on transformation of an aquatic fish-like tadpole into a land-dwelling frog (Dent, 1988).

In a recent study from our laboratory, we have measured the plasma concentrations of T4 and T3 empolying RIA in the tadpoles of a little known tropical amphibia, *Rana curtipes*, for the first time (Valamparampil and Oommen, 1995). Both the hormones become detectable in the late prometamorphic stages. A gradual increase in T3 and T4 is observed during this period. A sharp rise in hormone levels is apparent at the onset of metamorphic climax. Thereafter, the circulating hormones (particularly T4) decreases rapidly and reaches levels similar to prometamorphosis (Figs. 2 and 3). This kind of a variation in thyroid hormone concentration at various stages of metamorphosis is essential for the normal development of a frog. Our studies also point out to the possibility of a peripheral conversion of T4 and T3 and that T3 appears to be the most potent form of thyropid hormone in this amphibian species.

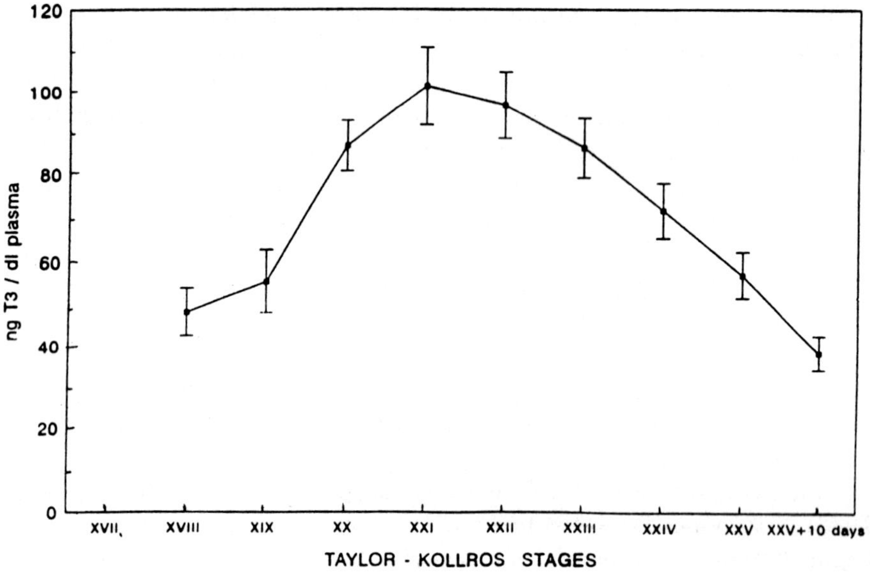

Fig. 2 Mean plasma T3 levels in tadpoles and froglets of *Rana curtipes*. Bars indicate standard deviation.

Adult amphibians have received very little attention with regard to seasonal changes of plasma concentration of thyroid hormones. The periods of maximal thyroid activity are associated with breeding or preparation for breeding in many species. Measurement of plasma concentration of thyroxine at various times of the year in *Bufo bufo* indicates that plasma T4 is slightly higher during hibernation and the spring breeding period (Rosenkilde, 1982). These data may imply that the thyroid could be involved with the preparation for breeding. There is substantial evidence that thyroid hormones have many physiological functions in adult anurans and that cellular resposnses to thyroid hormones are mediated by temperature (Packard and Packard, 1975). Attempts to correlate thyroid activity with oxidative metabolism have gained ambiguous results in amphibia (Rosenkilde, 1982). Thyroxine injections in anurans at their preferred temperature stimulates oxidative metabolism in component tissues (Packard and Packard, 1975) and in intact animals (Maher, 1967; McNabb, 1969).

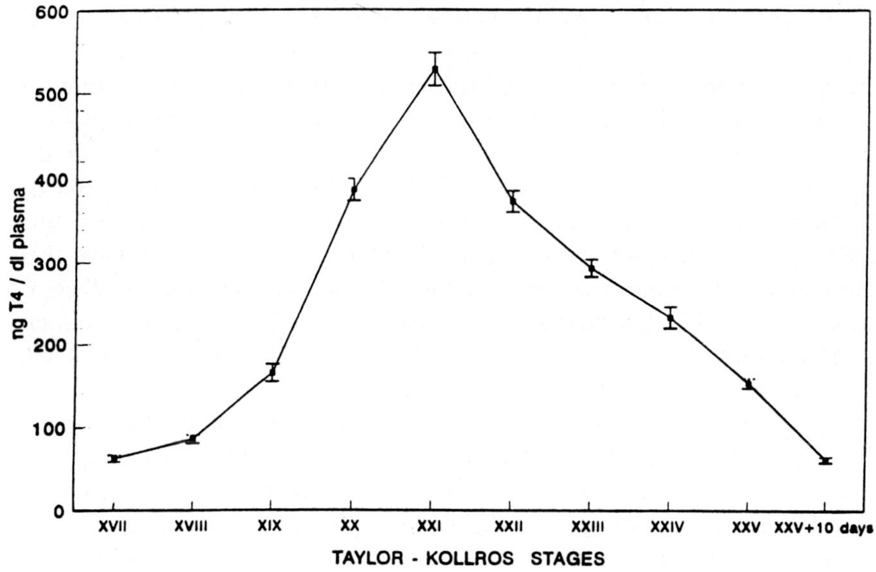

Fig. 3. Mean plasma T4 levels in tadpoles and froglets of *Rana curtipea.* Bars indicate standard deviation.

In a relatively recent report from our laboratory it has been established that TH (T4 and T3) exert a dose response effect on the activities of oxidative enzymes in an apodan amphibian, *Gegeneophis carnosus* (Sutharam *et al.*, 1990). TH administration in thiouracil treated specimens stimulates the oxidative enzyme activities. This study, for the first time, reveals that administration of TH markedly increase the oxidative capacity of liver mitochondria in an apoda (Sutharam *et al.,* 1990). In winter frog, *Rana temporaria*, Lagerstetz *et al.* (1974) have reported a decrease in the activity of SDH and oxygen consumption in skeletal muscle homogenate following a decrease in the activity of thyroid gland.

Reptiles

Energy metabolism in reptiles is influenced not only by environmental factors but also by a number of metabolic hormones (Thapliyal *et al.*, 1983). TH are well known to elicit an amazingly diverse set of developmental and physiological responses in different tissues of higher vertebrates and appear to have the same function in the maintenance and stimulation of the metabolic rate in reptiles too. The role of thyroid gland in the oxidative metabolism of reptiles is well documented by numerous investigators (Maher, 1965; Chandola *et al.,* 1974; Oommen and Sreedeviamma, 1988). The effect of thyroxine on oxygen consumption at the level of whole body or tissue has been proved in reptiles (Gupta and Thapliyal, 1985). Generally, the stimulation of oxygen consumption in the organism or tissue by TH parallels the responsiveness of its intermediary metabolism to these hormones (Gorbman *et al.*, 1983). In *Calotes versicolor*, thyroidectomy decreases and thyroxine accelerates oxygen consumption of liver and muscle *in vivo* and *in vitro* (Oommen, 1981). It has been shown that thyroid gland influences intermediary metabolism in the water snake, *Natrix piscator* (Thapliyal *et al.*, 1979; Gupta *et al.*, 1975). Menon *et al.* (1981) have reported a hyperglycemic condition in the blood of thyroidectomised lizards. But the metabolically important enzymes of reptiles have received relatively little attention in comparison with those of the other classes of vertebrates. It is clear from our recent report that thyroidectomy does not change the glycemic level, T4 or T3 administration in thyroidectomised lizards produces hyperglycemia (Jacob and Oommen, 1990) (Fig. 4). The decrease of G6-Pase activity in liver and kidney after thyroidectomy and the stimulation of the same after thyroid hormone administration with increase in blood glucose may suggest that TH increase gluconeogenesis and that hypothyroid tissues display a diminished capacity for gluconeogenesis. Evidences from our studies further suggest that the observed hyperglycemic response after thyroid hormone administration could be the result of hepatic glycogenolysis also (Fig. 5).

Thyroidectomy decreases the rate of glycogenolysis in the snake *N. piscator* (Gupta and Thapliyal, 1982). In *Calotes*, while thyroidectomy does not produce any measurable change in acid phosphatase and alkaline phosphatase activities, the administration of TH significantly increases the activity of acid phosphatase, thus generating metabolites of trophic importance for maintaining cellular integrity. The decrease in the activities of AST and ALT after thyroidectomy may be due to the probable decrease in the level of alanine and aspartic acid respectively in the tissues, thereby suggesting reduced protein breakdown in thyroidectomised lizards (Jacob and Oommen, 1990). The significant increase in the activities of these two enzymes after TH administration suggests that when the doses of T4 and T3 are increased to higher levels, protein synthesis and transamination are substantially elevated as these transaminases are concerned with non-essential amino acid metabolism and gluconeogenesis. It seems likely that the changes in transaminase activity following thyroid hormone injection are not a direct response and it is probable that the way in which the transaminases are affected is determined to some extent by the nitrogenous end product that is mostly advantageous to the animal.

In *Calotes*, there is a significant reduction in urea concentration after thyroidectomy and a stimulated production of the same after TH administration, suggesting an increased protein catabolism. The cholesterol lowering by thyroid hormones is supposed to be due to the fact that its breakdown is stimulated more than its synthesis, thus stimulating lipolysis (Jacob and Oommen, 1990). Results from our studies also suggest that thyroidal status is inversely proportional to hepatic reserves of cholesterol.

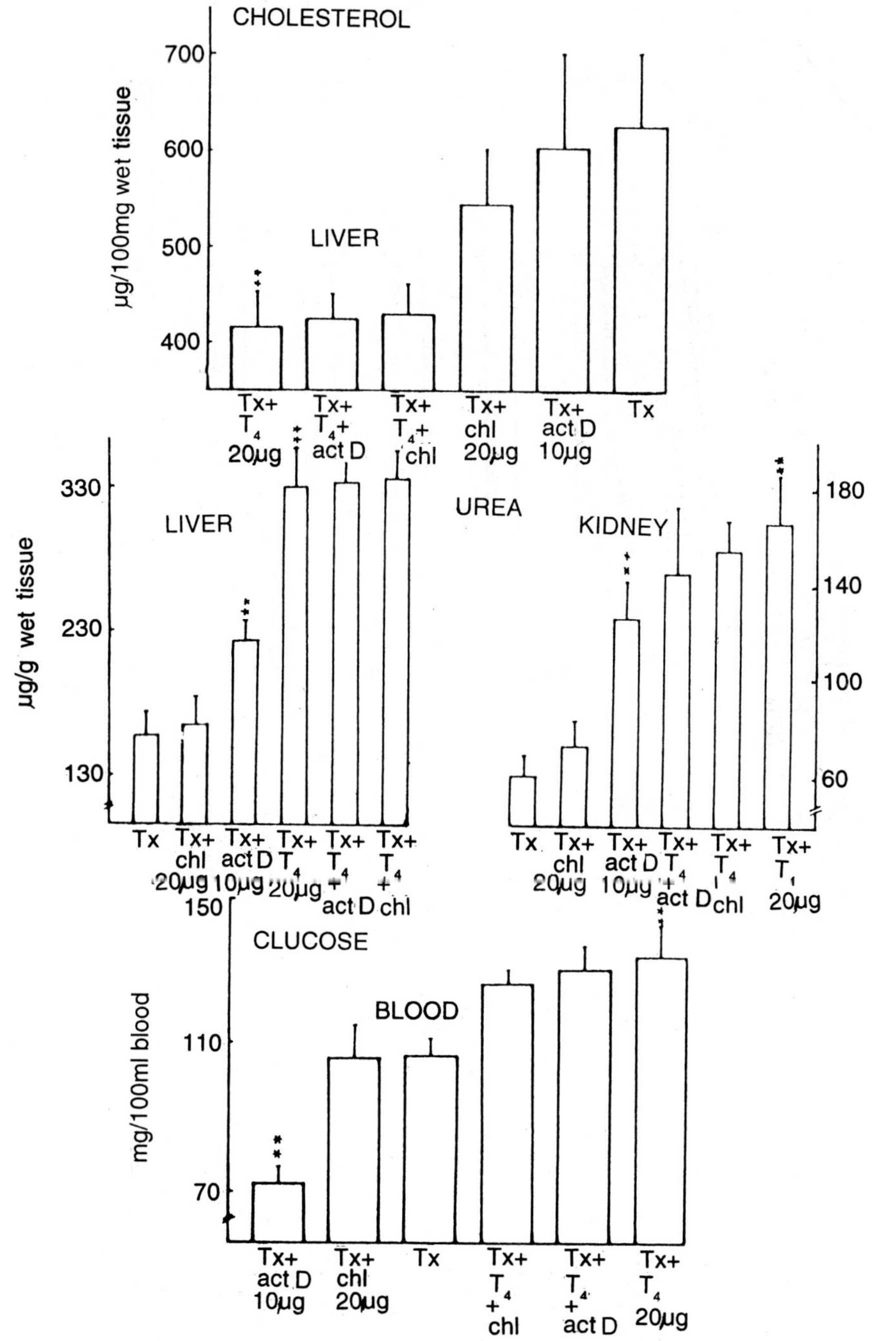

Figs. 4 Effects of administration of L-thyroxine (L-T4), actinomycin D (act D), and chloramphenicol (chl) on glucose, urea, and cholesterol levels in thyroidectomized (Tx) *C. versicolor.* Each value is CAN ± SE of five animals. **P < 0.01.

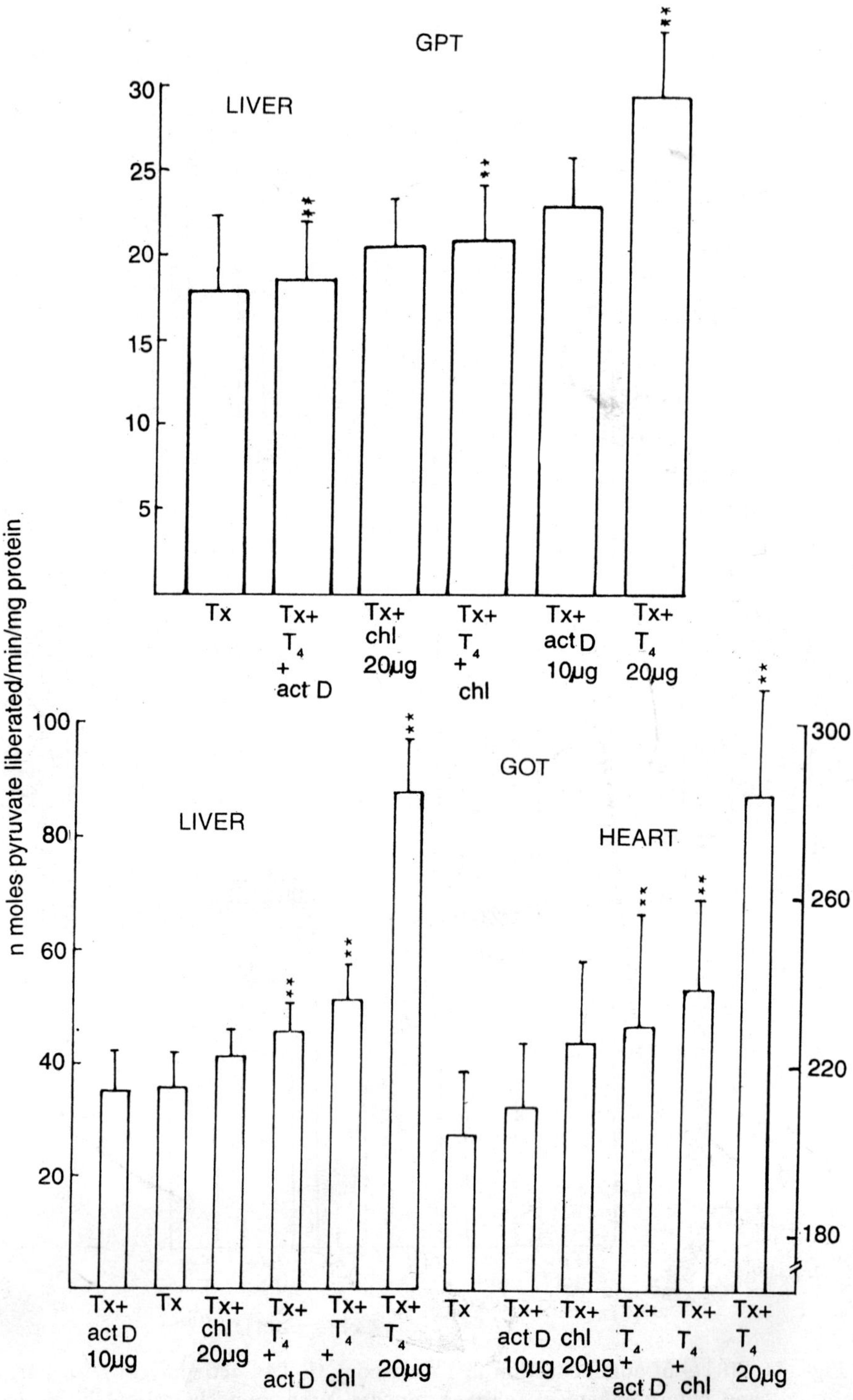

Fig. 5 Effects of administration of L-thyroxine (L-T4), actinomycin D (act d), and chloramphenicol on GOT and GPT activities in thyroidectomized (Tx) *C. versicolor*.

This view is also supported by the findings in juvenile spotted munia, *Lonchura punctulata* by Thapliyal *et al.* (1973).

In a subsequent study on the thyroid-gonad interrelationship on the intermediary metabolism in a reptile we have reported that both T4 and testosterone are effective in stimulating intermediary metabolism in lizards, irrespective of the removal of either thyroid or gonad (Peter and Oommen, 1992). It seems that thyroid and testis in *C. versicolor* exhibit a direct interrelationship which may involve gonadotropins and thyrotropin. It has been proposed that thyroid function evolved parallel with endocrine control of reproduction and that the basic function for thyroid hormones is associated primitively with maturation of gonads (Fontaine and Burzawa-Gerard, 1977). Bhattacharya *et al.* (1988) have shown that T3 has specific receptors in the *A. testudineus* testicular leydig cells and they have also reported a high affinity T3 receptor in the nuclei of human corpus luteal cells. TH are reported to elicit a stimulatory effect on the steroidogenic activity of leydig cells and hypothyroidism impairs the same in rats (Antony *et al.*, 1995).

In yet another study from our laboratory we have examined the influence of thyroxine on propanolol or alloxan treated lizards. The thyroid-catecholamine and thyroid-insulin interaction in reptiles still remains an exciting and incompletely resolved area. The results of our study indicate that thyroxine has an independent stimulatory influence on intermediary metabolism in *C. versicolor* (Jacob and Oommen, 1992). Our studies also indirectly point out that though T4 independently produces hyperglycemia, catecholamines may potentiate the effect of T4 on glycemic level. Many of the effects of TH on cellular and metabolic functions in mammals including humans occur by mechanisms totally distinct from those under adrenergic modulation (Sterling, 1979; Gelfand *et al.*, 1987). The presence of TH is a prerequisite for the response of most of the metabolic processes to adrenaline in *N. piscator* (Gupta and Thapliyal, 1982). The decrease of hepatic cholesterol in alloxan-pretreated lizards after T4 administration emphasizes that the influence of thyroid gland on lipolysis is independent of the actions of insulin (Jacob and Oommen, 1992).

Mechanism of Thyroid Hormone Action

Simultaneous administration of actinomycin D or chloramphenicol with T4 does not prevent the TH stimulated blood glucose level, suggesting that the transport of sugar into the cell by TH occurs within minutes of hormone administration and is not blocked by protein synthesis inhibitors, as has also been suggested in rats by Goldfine *et al.* (1975). Further, Segal and Gordon (1977) have reported that the 2-deoxy-D-glucose uptake affected by TH is independent from protein synthesis enhancement produced by them. Administration of actinomycin D or chloramphenicol with T4 prevents the hormone-stimulated activities of AST, ALT and acid phosphatase. This suggests that some of the TH effects on intermediary metabolism are brought about through genomic transcription in this reptile also. However, it has been understood that some of the effects of TH on metabolic processes do not require protein synthesis as has also been suggested by Gorbman *et al.* (1983).

Insects

Insects during the course of evolution, have developed their own special neural and hormonal systems to cope with the changing environment, providing clues for homeostasis, integration, differentiation and growth. Some of these systems are unique in insects. These systems, in general, are found to involve substances which are chemically or immunochemically similar to that of vertebrate hormones. Thyroxine has specific functions in the intermediary metabolism of higher as well as lower vertebrates. But nobody has demonstrated for certain any possible effect of this hormone on intermediary metabolism in insects. We conducted a preliminary study in our laboratory on the possible effect of thyroxine in an insect *Teleogryllus mitratus*, from the point of view of comparative endocrinology.

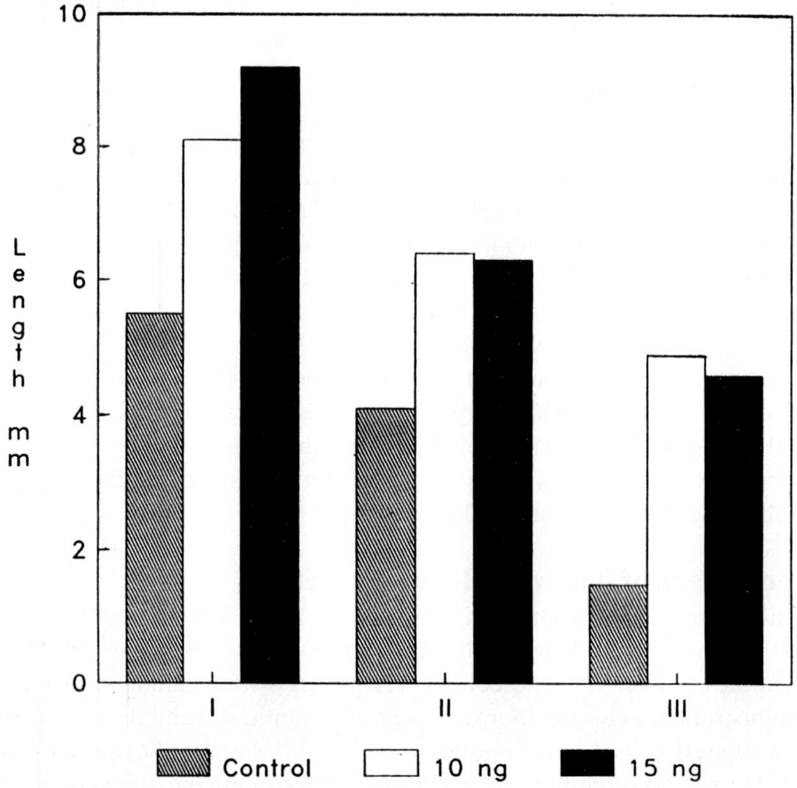

Figs. 6 Effect of L-Thyroxine on the Oocyte development of *T. mitratus*.

In our study on *Teleogryllus*, thyroxine produces marked stimulatory influence on metamorphosis and ovarian development (Suresh Kumar and

Oommen, 1994) (Fig. 6). Thyroxine treatment accelerates the development as it reduced the intermoult period in this insect. The treatment of thyroxine also accelerates the rate of growth of the oocytes and probably vitellogenesis. Thyroxine is also found to stimulate the protein turn over and gluconeogenesis as the enzymes responsible were stimulated in their activity. The enhancement of developmental activities is probably due to stimulation of the brain to release prothoracicotropin, which in turn regulates the prothoracic glands to release sufficient quantity of ecdysone resulting in early larval moulting, growth, maturity and pupation. Thyroxine-induced ecdysteroid increase has been reported in the silk worm, *Bombyx mori* (Thyagaraja *et al.*, 1991). However, a physiological role for thyroid hormones in insects remains as a topic of future investigations.

References

Antony, F.F., Aruldhas, M.M., Udayakumar, R.C.R., Maran, R.R.M. and Govindarajulu, P. (1995). J. Endocrinol. **144:** 293–300.

Ballantyne, J.S., John, T.M., Singer, T.D. and Oommen, O.V. (1992) The J. Exp. Zool. **261:** 105–109.

Beleznai, Z., Amler, E., Rauchova, H., Drahta, Z. and Jancsuk, V. (1989) FEBS Letters. **243:** 247–250.

Bhattacharya, S., Banerjee, P., Md. Jamaluddin, Banerjee, P.P. and Maitra, G. (1988). Experientia. **44:** 1005.

Chandola, A., Kumar, D.S. and Thapliyal, J.P. (1974) J. Endocrinol. **63:** 191–197.

Dent, J.N. (1988) Amer. Zool. **28:** 297–308.

Donaldson, E.M., Fagerlund, U.H.M., Higgs, D.A. and Mc Bride, J.R. (1979) Hormonal enhancement of growth. In: Fish Physiology (W.S. Hoar, D.J. Randall and J.R. Brett, Eds.) Vol 1, pp. 303–356. Academic press, New York,

Ducibella, T. (1974). Dev. Biol. **38:** 175–186.

Eales, E.G. (1979). In : Hormones and evolution (E.J.W. Barrington, Ed.) Vol. 1, pp. 341–436, Academic Press, New York.

Freeman, H.C. and Idler, D.R. (1973). Gen. Comp. Endocrinol. **20:** 69–75.

Fontaine, Y.A. and Burzawa–Gorrard, E. (1977). Gen. Comp. Endocrinol. **32:** 341–347.

Galton, V.A. (1988) Amer. Zool. **28:** 309–318.

Gelfand, R.A., Hutchinson–Williams, K.A., Bonde, A.A., Castellino, P. and Sherwin, R.R. (1987). Metabolism. **36:** 562–575.

Ghosh, R. and Medda, A.K. (1969). Endokrinologie **55:** 218–229.

Goldfine, I.D., Simons, C.G., Smith, G.J. and Ingbar, S.H. (1975). Endocrinology. **96:** 1030–1037.

Gorbman, A. (1978). Comparative anatomy and physiology. In: The Thyroid Gland (S.C.Werner and S.H. Ingbar, Eds.) pp. 22–30. Harper and Row, Hagerstown, M.D.

Gorbman, A., Dickoff, W.W., Virginia, S.R., Clark, N.B. and Ralph, C.L. (1983). In: Comparative Endocrinology. pp. 329–335. John Wiley, New York.

Gupta, B.B.P. and Thapliyal, J.P. (1983). J. Endocrinol. **99:** 211–216.

Gupta, B.B.P. and Thapliyal, J.P. (1985). Gen. Comp. Endocrinol. **58:** 20–27.

Gupta, B.B.P. and Thapliyal, J.P. (1991). Zool. Sci. **8:** 625–634.

Gupta, S.C., Thapliyal, J.P. and Garg, R.K. (1975). Gen. Comp. Endocrinol. **27:** 223–229.

Gupta, S.C. and Thapliyal, J.P. (1982). J. Endocrinol. **94:** 333–338.

Hayashi, S. and Ooshiro, Z. (1977). Mem. Fac. Fish. Kagoshima Univ. **26:** 85–89.
Jackin, E. and La Roche, G. (1973). Comp. Biochem. Physiol. **44:** 851–866.
Jacob, V. and Oommen, O.V. (1990). Gen Comp. Endocrinol. **77:** 324–336.
Jacob, V. and Oommen, O.V. (1992). The J. Exp. Zool. **262:** 263–270.
Lagerspetz, K.Y.H., Harri, H.N.E. and Okslarti, R. (1974). Gen. Comp. Endocrinol. **22:** 169–176.
Leatherland, J.F. and Sonstegard, R.A. (1980) J. Fish. Biol. **16:** 532–562.
Leatherland, J.F. (1982) Env. Biol. Fish. **7:** 83–110.
Leatherland, J.F. (1988). Amer. Zool. **28:** 319–358.
Magnus-Levy (1895) Berlin. Klin. Wochenschr. **32:** 650–652.
Maher, M.J. (1965). Gen. Comp. Endocrinol. **5:** 320–325.
Maher, M.J. (1967). Copeia. **2:** 361–365.
Matty, A.J., Chaudhry, M.A. and Lone, K.P. (1982). Gen. Comp. Endocrinol. **47:** 497–507.
McNabb, R.A.(1969). Gen. Comp. Endocrinol. **12:** 276–281.
Medda, A.K. and Ray, A.K. (1979). Gen. Comp. Endocrinol. **37:** 74–84.
Menon, J., Shah, R.V. and Hiradhar, P.K. (1981). Indian J. Exp. Biol. **19:** 1018–1021.
Michels, R.S., Cason, J. and Sokoloff L. (1963) Science **4–140:** 1417–1418.
Narayansingh, T. and Eales, J.G. (1975) Comp. Biochem. Physiol. **52B:** 399–405.
Norris, D.O. (1983). Amer. Zool. **23:** 709–718.
Norris, D.O. (1985). Thyropid gland. In: Vertebrate Endocrinology. pp. 162–210. Lea and Febiger, Philadelphia.
Oommen, O.V. (1978) J. Zool. Soc. India. **30:** 1–6.
Oommen, O.V. (1981) Comp. Physiol. Ecol. **6:** 95–98.
Oommen, O.V. and Sreedeviamma, K.K. (1988). Gen. Comp. Endocrinol. **69:** 391–398.
Oommen, O.V. and Matty, A.J. (1991). Gen. Comp. Endocrinol. **83:** 468–472.
Orten, J.M. and Neuhaus, O.W. (1982) Hormones: Chemical regulators. In: Human Biochemistry. (J.M. Orten and O.W. Neuhaus, Eds.) pp. 592–648. The C.V. Mosby Co. St. Louis.
Packard, G.C. and Packard, M.J. (975). Gen. Comp. Endocrinol. **27:** 162–168.
Padmaja, N. and Oommen, O.V. (1994). Symp. Reprod. Biol. Comp. Endocrinol. Abstract.
Pickford, G.E. (1973). Amer. Zool. **13:** 711–717.
Peter, M.C.S. and Oommen, O.V. (1989 a). Gen. Comp. Endocrinol. **73:** 96–107.
Peter, M.C.S. and Oommen, O.V. (1989 b). Current Science. **58:** 37–38.
Peter, M.C.S. and Oommen, O.V. (1989 c). Zool. Sci. **6:** 185–189.
Peter, M.C.S. and Oommen, O.V. (1991). Indian J. Exp. Biol. **29:** 481–483.
Peter, M.C.S. and Oommen, O.V. (1992). Zool. Sci. **9:** 1059–1064.
Peter, M.C.S. and Oommen, O.V. (1993). J. Exp. Zool. **266:** 85–91.
Plisetskaya, E., Woo, N.Y.S and Murrat, J. (1983). Comp. Biochem. Physiol. **74A:** 179–187.
Rosenkilde, P. (1982). In: Phylogenic aspects of thyroid hormone actions. In: (Institute of Endocrinology, Gunma University, Eds.) pp. 91–106. Centre for Academic Publications, Japan, Tokyo.
Saurez, R.K. and Mommsen, T.P. (1987). Can. J. Zool. **65:** 1869–1882.
Schulthesis, H. (1977). Gen. Comp. Endocrinol. **31:** 45–52.
Segal, J. and Gordon, A. (1977). Endocrinology. **101:** 150–156.
Singh, T.P. (1970). Indian. Biol. **2:** 1–15.
Sterling, K. (1979). New England. J. Med. **30:** 117–123.
Suresh Kumar, M.A. and Oommen, O.V. (1994). Symp. Reprod. Biol. and Comp. Endocrinol. Abstract.
Sutharam, K.K., Peter, M.C.S. and Oommen, O.V. (1990) Gen. Comp. Endocrinol. **79:** 246–252.

Tata, J.R. (1970). In: Biochemical actions of hormones. (G. Litwalk, Ed.) pp. 89–133. Academic Press, New York.

Tata, J.R. (1975) Nature. **257:** 18–23.

Takashima, F., Hebiya, T., Ngan, P-Y. and Aida, K. (1972). Bull. Japan. Soc. Sci. Fish. **38:** 43–49.

Thapliyal, J.P., Kumar, D.S. and Garg, R.K. (1974 a) Gen. Comp. Endocrinol. **22:** 308–311.

Thapliyal, J.P., Gupta, S.C. and Garg, R.K. (1974 b) J. Endocrinol. **60:** 517–524.

Thapliyal, J.P., Garg, R.K. and Murty, G.S.R.C. (1973). Gen. Comp. Endocrinol. **21:** 547–553.

Thapliyal, J.P., Lal, P., Gupta, B.B.P. and Saran, S. (1983) Gen. Comp. Endocrinol. **49:** 333–343.

Thornburn, C.C. and Matty, A.J. (1963). Comp. Biochem. Physiol. **8:** 1–12.

Tompkins, R. (1978). Amer. Zool. **18:** 313–319.

Thyagaraja, B.S., Keely, T.J., Masler, E.P. and Borkovec, A.B. (1991). J. Insect. Physiol. **37:** 153–159.

Valamparampil, T.T. and Oommen, O.V. (1995). Symp. Reprod. Biol. Comp. Endocrinol. Abstract.

Wilson, R.P. (1973). Comp. Biochem. Physiol. **46B:** 617–624.

Comparative Endocrinology and Reproduction
K.P. Joy, A. Krishna and C. Haldar (Eds)

35. Physiological Significance of Flight-Induced Changes in Certain Hormonal, Metabolite and Other Blood Values in Homing Pigeons

J.C. George
Department of Zoology, University of Guelph, Guelph, Ontario, Canada N1G 2W1

Introduction

Physiology of muscular activity in birds has been studied in a variety of ways in those forms which use their forelimbs for the aerial flight or for underwater movement and also those which use their hindlimbs for running on the ground or swimming in water. Laboratory studies using suitable avian species have employed methods such as wind tunnel flying (Tucker, 1968; Aulie, 1971, 1975; Butler *et al.,* 1977; Torre-Bueno, 1978; Torre-Bueno and Larochelle, 1978; Butler and Woakes, 1980; Butler, 1991), running on treadwheel (Rees and Harvey, 1987) and on treadmill (Pinshow *et al.*, 1976; Kiley *et al.,* 1979; Harvey and Phillips, 1982; Harvey *et al.,* 1982; Baudinette *et al.,* 1976, 1982; Bech and Nomoto, 1982; Grubb *et al.*, 1983; Kiley and Fedde, 1983; Nomoto *et al.,* 1983; Rees *et al.*, 1983; Brackenbury, 1984; Rees *et al.*, 1985; Butler, 1991). Another method used in simulating flight activity has been prolonged *in vivo* electrical stimulation of the pectoral muscles (George and Berger, 1966; Vallyathan *et al.*, 1970; Parker and George, 1974, 1975).

In bird flight several mechanisms and processes involving energetics, metabolic adaptation, thermoregulation, osmo(iono)-regulation and neuroendocrine functions, are intricately coordinated towards maintaining optimal levels of homeostasis (George *et al.,* 1992). Of the various experimental approaches to the study of bird flight, those concerned with cardiorespiratory and metabolic changes during flight have been with birds made to fly in a wind tunnel. However, flight pattern of the pigeon in a wind tunnel has been found to be different from that in natural flight (Butler *et al.,* 1977). Even though this and other experimental approaches employed have provided valuable information with regard to mobilization and utilization of energy fuel, they do have limitations when considering the overall physiology of natural flight, especially with regard to neuroendocrine and other hormonal changes associated with the control and regulation of flight as in bird migration.

In light of the above considerations, the trained homing pigeon has been found to be a highly suitable experimental model for studies on biochemical, neuroendocrine and other physiological changes during free natural flight. The present review deals with the results obtained in a series of experiments with homing pigeons. A preliminary report of the observations was presented at the 16th International Congress of Biochemistry and Molecular Biology (George, 1994).

Materials and Methods

Homing pigeons (*Columba livia)* used in all the experiments were from a colony of pigeons raised and maintained outdoors under natural photoperiod and temperature. They were given regular flight training for participation in racing competitions. Our experiments were conducted at the end of the racing season in the forenoon of a typical sunny, mid-autumn day (ambient temperature: 6°C). They did not receive their usual morning feed on the day of the experiment but water was given *ad libitum.* The control resting group of pigeons were given a 40 min car ride in order to correspond to the car ride the experimental group of birds received when transported to the release site about 48 km away from "home". The experimental group of pigeons were released between 1000 and 1030 h in batches of 2–3. All birds, took 60–80 min to return "home" except the ones used in a study in which the birds were without training for a period of 3 months prior to the day of the experiment. Birds lacking this training took 90–160 min instead. Blood samples (5 ml) were drawn from the brachial vein of each bird into heparinized EDTA coated syringes within 1–3 min of their arrival after flight and transferred to chilled polyethylene test-tubes. Blood samples from the resting control birds were obtained between 1030 and 1130 h. All blood samples were quickly transported to the laboratory and spun down in a refrigerated centrifuge to obtain plasma. The plasma was stored at −20°C for the various assays to be done. For details of the experimental procedures and of the assay methods used the appropriate references indicated against the various blood parameters in Tables 1 and 2 may be consulted.

Results

The values obtained for the various blood parameters investigated in the resting control birds and the flown birds are presented in Tables 1 and 2. Blood levels of glucose, lactate, free fatty acids (FFA), creatinine, and uric acid were significantly higher in the flown birds, but that of albumin was lower than in the resting birds (Table 1). Of the hormonal levels, growth hormone (GH) , adrenaline (A), noradrenaline (NA), arginine vastocin (AVT) and melatonin levels were significantly higher in the flown birds than in the resting ones but levels of corticosterone, thyroxine (T_4), triiodothyronine (T_3) and mesotocin in the former remained the same as in the latter birds (Table 1). No flight-induced significant changes in the levels of the enzymes

Table 1 Profile of blood parameters of resting (control) and flown homing pigeons. Values are Mean ± SEM. Number in parentheses denote number of birds. Numbers left of each parameter denotes reference from which data are compiled. *P < 0.05; **P < 0.01; ***P < 0.001; NS: not significant

Plasma Levels	Control	Flown
Metabolites and end-products		
[7]Glucose (mM/l)	15.94 ± 0.34 (12)	16.77 ± 0.22 (13)*
[7]Lactate (mM/l)	1.31 ± 0.08 (13)	2.46 ± 0.22 (14)***
[6]Free fatty acids (μEq/l)	816.2 ± 33.0 (13)	1551.6 ± 87.0 (14)***
[2]Cholesterol (mM/l)	8.43 ± 0.43 (10)	7.55 ± 0.38 (11)NS
[2]Total protein (g/l)	26.90 ± 0.78 (10)	25.55 ± 0.53 (11)NS
[2]Albumin (g/l)	10.60 ± 0.31 (10)	9.55 ± 0.34 (11)*
[2]Globulins (g/l)	16.30 ± 0.72 (10)	16.00 ± 0.38 (11)NS
[2]Creatinine (μM/l)	7.90 ± 1.11 (10)	13.27 ± 1.60 (11)**
[2]Uric acid (μM/l)	265.7 ± 60.21 (10)	450.7 ± 52.6(11)*
Hormones		
[6]Growth hormone (ng/ml)	10.10 ± 1.0 (12)	12.8 ± 0.70 (14)*
[7]Adrenaline (ng/ml)	0.16 ± 0.08 (11)	0.99 ± 0.19 (11)*
[7]Noradrenaline (ng/ml)	1.04 ± 0.09 (11)	1.47 ± 0.15 (11)**
[7]Corticosterone (ng/ml)	4.20 ± 0.54 (13)	4.88 ± 0.72 (14)NS
[6]Thyroxine (T_4) (μg/dl)	3.10 ± 0.17 (13)	3.21 ± 0.24 (14)NS
[6]Triiodothyronine (T_3) (ng/dl)	381.6 ± 13.18 (13)	352.4 ± 18.53 (14)NS
[3]Arginine vasotocin (AVT) (pg/ml)	15.7 ± 0.8 (9)	18.5 ± 0.4 (10)**
[3]Mesotocin (pg/ml)	38.6 ± 3.3 (9)	38.0 ± 2.1 (10)NS
[5]Melatonin (Serum) (pg/ml)	26.48 ± 5.38 (9)	48.22 ± 7.64 (11)*
Enzymes		
[2]Alakline phosphatase (U/l)	606.3 ± 140.9 (10)	474.8 ± 94.3 (11) NS
[2]Alanine amino-transferase (ALAT) (U/l)	7.20 ± 1.36 (10)	5.73 ± 1.21 (11)NS
[2]Aspartate amino-transferase (ASAT) (U/l)	101.9 ± 12.4 (10)	82.9 ± 6.2 (11)NS
[2]Lactate dehydrogenase (LDH) (U/l)	62.60 ± 13.69 (10)	91.91 ± 9.28 (11)NS
[2]Creatine phosphokinase (CPK) (U/l)	140.3 ± 16.04 (10)	127.8 ± 13.67 (11)NS
Electrolytes		
[2]Calcium (mM/l)	2.10 ± 0.06 (10)	2.07 ± 0.08 (11)NS
[2]Phosphorous (mM/l)	1.10 ± 0.08 (10)	1.19 ± 0.10 (11)NS
[2]Magnesium (mM/l)	0.87 ± 1.72 (10)	0.82 ± 4.20 (11)NS
[2] Sodium (mM/l)	136.38 ± 1.95 (8)	139.40 ± 1.32 (10)NS
[2]Potassium (mM/l)	4.84 ± 0.20 (8)	4.48 ± 0.12 (10)NS
[2]Chloride (mM/l)	107.70 ± 1.23 (8)	107.73 ± 0.97 (11)NS
Osmolality		
[1](mM/kg)	303.5 ± 1.1 (10)	306.1 ± 3.8 (7)NS
[2](mM/kg)	307.30 ± 3.22 (10)	307.18 ± 1.41 (11)NS
[3](mM/kg)	306.10 ± 3.40 (9)	307.20 ± 1.6 (10)NS
[5](mM/kg)	307.45 ± 3.59 (9)	307.27 ± 1.42 (11)NS

[1]George and John, 1992; [2]George and John, 1993; [3]George *et al.*, 1992; [4]George *et al.*, 1989; [5]John *et al.*, 1993; [6]John *et al.*, 1988; [7]Viswanathan *et al.*, 1987.

or electrolytes studied were seen as was also the case with osmolality (Table 1). In the second set of experiments using pigeons which were off training for 3 months (Table 2), while the blood levels of triglycerides, FFA, glucagon and reverse triiodothyronine (rT_3) were significantly higher in the flown pigeons than in the resting ones, those of T_4 and T_3 were significantly lower with the osmolality remaining unchanged (Table 2).

Table 2 Metabolite and hormonal levels in the plasma of homing pigeons which were not given flight training for 3 months prior to experiment. Values are Mean ± SEM. Numbers in parentheses denote number of birds. Number left of each parameter denotes reference from which data are compiled. *P < 0.05; **P < 0.01; ***P < 0.001; ****P < 0.000001; NS: not significant

Plasma Levels	Control	Flown
Metabolites		
[4]Triglycerides (mM/l)	1.87 ± 0.16 (10)	4.10 ± 0.75 (8)*
[4]Free fatty acids (mM/l)	0.347 ± 0.008 (10)	0.551 ± 0.016 (8)****
Hormones		
[4]Glucagon (pg/ml)	730.1 ± 56.6 (10)	1214.5 ± 160.5 (8)*
[4]Thyroxine (T_4) (ng/ml)	19.07 ± 1.54 (10)	12.54 ± 2.36 (8)*
[4]Triiodothyronine (T_3) (ng/ml)	2.29 ± 0.14 (10)	0.89 ± 0.14 (8)***
[1]Reverse triiodothyronine (rT_3) (pg/ml)	211.9 ± 16.9 (10)	493.0 ± 90.8 (7)**
Osmolality		
[1](mM/kg)	303.5 ± 1.1 (10)	306.1 ± 3.8 (7)NS

References same as in Table 1.

Discussion

In homing pigeons, flight duration is influenced by their familiarity of the release site which is referred to in literature as "release site bias" (see George and John, 1992). This bias was reflected in our experiments with pigeons which were in continuous training and those which were off training for 3 months prior to the day of the experiment. Flight duration in the former was 60–80 min whereas in the latter it was 90–160 min. Obviously, the more strenous and extended flight in the second group of birds should have had significant effects on metabolism and metabolic regulatory processes. A significant difference between the flight-induced blood values in the two groups of birds was that flight in those pigeons deprived of the 3 month period of prior training, produced a two-fold increase in plasma levels of rT_3 with concomitant decrease in those of T_4 and T_3 (Table 2). This increase in rT_3 concomitant with reductions in the levels of T_4 and T_3 with no significant change in osmolality, suggests, inhibition of the enzyme 5'-monodeiodinase which is responsible for the monodeiodination of the tyrosyl ring of the T_4 molecule in the conversion of T_4 to T_3 and the activation of the enzyme, 5-deiodinase, by which the phenolic ring is monodeiodinated to form rT_3. Thus, T_4 which is regarded as a prohormone could be deiodinated to form another inactive form rT_3 and its active form T_3 (Engler and Burger, 1984). It has been estimated that more than 80% of the circulating levels of T_3 and rT_3 in the body are the results of deiodination of T_4 in the peripheral tissues (Kühn *et al.*, 1993). Therefore, the flight-induced increase in rT_3 presumably due to the peripheral conversion of T_4 to rT_3 may be regarded as a mechanism of autoregulation of thyroid function as well as of thermoregulation.

Among the peripheral organs in which T_4 could undergo deiodination, the pineal gland has been shown to be one of them (Tanaka *et al.*, 1986; Guerrero *et al.*, 1988a). It has also been shown that the rhythmicity of T_3 formation in the rat pineal is mediated by a β-adrenergic mechanism (Tanaka *et al.*, 1987). These observation led to the testing of the hypothesis that an increase in T_3 production in the pineal at night would cause stimulation of N-acetyltransferase (NAT) activity which in turn would enhance melatonin synthesis. However, an attempt to clarify the influence of T_3 in melatonin synthesis has shown that inhibition of T_3 production has no effect on nocturnal increase of NAT activity or melatonin content in euthyroid or thyroidectomized rats (Guerrero *et al.*, 1988b). Further studies in this direction have led to the conclusion that NAT activity rhythm as well as melatonin cycle is unrelated to intracellular T_3 production (Guerrero and Reiter, 1992; Reiter and Guerrero, 1992). However, it is established that in the rat pineal the 5'-deiodinase enzyme (type II) is activated by NA via the β-adrenergic and α-adrenergic receptors in the pinealocyte membrane, thus enhancing the intrapinealocyte production of cycle AMP (cAMP) (Reiter and Guerrero, 1992). The question that arises from these observations is, what is the role of T_3 in pineal function? In the present study with homing pigeons, it has been observed that flight induced significant increase in the circulating levels of NA and A (Table 1). If in the flown pigeons, the action of NA on the adrenergic receptors in the pinealocyte membrane caused increase in intracellular cAMP which in turn enhanced 5'-deioidinase (type II) activity in the pinealocyte leading to the production of T_3 from T_4 as expected to occur in the rat, significant increase in intrapinealocyte as well as circulating levels of T_3, could be expected (see Fig. 2, Reiter and Guerrero, 1992). In the flown pigeons, however, it was found that there were no significant changes in the circulating levels of T_3 or T_4 but those of melatonin were significantly increased (Table 1). This suggests that the daytime increase in melatonin levels could have been caused by increased catecholamine stimulation of NAT and that the actual increase in T_3 levels was masked by equal hormone utilization by peripheral tissues. Exercise-induced daytime increase in melatonin levels has also been reported in humans (Strassman *et al.*, 1989). There is also evidence that NA released under stressful stimuli from distant sites such as the adrenal medulla can act like NA released from the sympathetic endings within the pineal and thus enhance melatonin production (Wu *et al.*, 1987). The relationship between pineal 5'-deiodinase activity and NAT activity, however, has remained obscure. Studies in this direction indicate that NAT rhythm does not depend on the production of T_3 (Guerrero and Reiter, 1992). Nevertheless, the fact that the potent 5'-deiodinase activity capable of turning out T_3 from T_4 does exist in the pinealocyte, suggests that the enzyme has a role in pineal metabolism. It could well be that the production of T_3 in the pineal is dependent on the production of melatonin. Should T_3 production be dependent on the level of melatonin concentration in the pineal, it would be of interest to know the effect thyroidectomy on circulating levels of melatonin. Recently it has been demonstrated in European starlings that thyroidectomy does not effect the daily or free-running rhythm of plasma melatonin (Dawson and King, 1994). However, from the Figures (1 and 2) representing the results obtained (Dawson and King, 1994), it appears that in birds exposed to long days (LD 16:8) the peak melatonin levels during the scotophase tend to be lower in the thyroidectomized birds than in those of the control birds whereas in the short day (LD 8:16) birds, the peak scotophase melatonin levels in the experimental as well as the control birds were identical (Dawson and King, 1994, Fig.1). On the other hand, in birds from the long and short days when exposed to continuous darkness (DD), the peak levels of melatonin were higher in the thyroidectomized birds (Dawson and King, 1994, Fig. 2). These observations indicate that T_3 does not alter the melatonin rhythm but may have a regulatory role in melatonin production during the scotophase.

Evidence for possible interaction between melatonin synthesis and T_3 production was indicated in our earlier studies with homing pigeons (Tables 1 and 2). In homing pigeons, after a moderate flight no significant increase in plasma levels of T_3 was observed (John *et al.*, 1988). However, after an extended flight, T_3 levels were found to be significantly reduced (George *et al.*, 1989). It was also shown that in homing pigeons flight induced marked increase in circulating levels of melatonin (John *et al.*, 1993). From these observatios it can be inferred that with the increase in melatonin production, conversion of T_4 to T_3 was significantly reduced. Since melatonin treatment has been shown to lower body temperature (T_b), breathing frequency, heart rate and metabolic rate in pigeons (John and George, 1991) and T_3 has been found to increase thermogensis and metabolism in birds (Lam and Harvey, 1990), the antagonistic relationship between these two hormones is clearly indicated. This is consistent with the observation that in hypothyroid patients, melatonin secretion has been observed to be high (Rojdmark *et al.*, 1991).

When feral pigeons acclimated to laboratory conditions were subjected to daytime sodium pentobarbital-induced sleep and exercised by electrical stimulation of the breast muscles, the circulating levels of melatonin were not any different from those of the non-exercised control pigeons (John *et al.*, 1992) unlike the results obtained with flown homing pigeons (John *et al.*, 1993). However, with regard to the circulating levels of T_4 and T_3, significant decrease in the levels of these hormones were observed in similarly exercised pigeons in an earlier investigation (Pilo *et al.*, 1984). These observations are also suggestive of the relationship between T_3 production and melatonin secretion as mentioned earlier.

The daytime post-flight increase in circulating levels of the nocturnal hormone, melatonin, in homing pigeons is indicative of its possible role in the overall homeostatic responses to muscular activity. It has been estimated in the pigeon that during sustained flight nearly eight times more heat is generated than while at rest (Aulie, 1971). In order to prevent flight-induced hyperthermia, the excess heat has to be dissipated. It has been shown that in birds, the major site for heat dissipation is the leg (Baudinette *et al.*, 1976; Martineau and Larochelle, 1988) and that in the pigeon the rate of heat loss from the legs and feet during flight is about six times that in the resting condition (Martineau and Larochelle, 1988). The hypothermic action of melatonin by facilitating peripheral heat dissipation has been demonstrated in melatonin-implanted pigeons (John *et al.*, 1978; George, 1982; John and George, 1991). The daytime increase in circulating levels of melatonin in flown homing pigeons is, therefore, a homeostatic response to flight-induced heat production.

Another hormone which has been shown to lower T_b in the pigeon is AVT which is also known to be an antidiuretic hormone (George, 1980; John and George, 1992; Hassinen *et al.*, 1994). The significant post-flight increase in plasma levels of AVT, without change in osmolality, observed in homing pigeons (Table 1), represents the body's response to possible flight-induced hyperthermia and dehydration (George *et al.,* 1992). No change in the post-flight levels of another neurohypophysial hormone, mesotocin, has been observed (Table 1). It has been suggested (George *et al.*, 1992) that mesotocin may function as a stress hormone in birds like its mammalian counterpart, oxytocin has been found to be in rats (Lang *et al.*, 1983).

It is well established in birds that in short-term muscular activity involving quick and sudden movement as in take-off and while in the air for change in direction and in avoiding predators or obstacles, carbohydrate is the preferred fuel whereas during sustained flight, lipid becomes the main fuel (George and Berger, 1966; Parker and George, 1974, 1975; Dial *et al.*, 1987). In studies with homing pigeons flight-induced increases in the circulating levels of glucose, lactate and FFA have been observed (Table 1) indicating the availability and utilization of both carbohydrate and lipid as fuels. As for cholesterol, flight produced no significant change in its concentration in the plasma (Table 1). This is consistent with studies on the effect of exercise in mammals including man (Carlson and Mossfeldt, 1964; Deshaies *et al.*, 1983). During flight, it is important that there has to be a steady flow of glucose and FFA from their storage depots to the muscle. Both GH and glucagon are known to be lipolytic as well as hyperglycemic (for review, George, 1977). The lipolytic action of glucagon has been demonstrated *in vivo* in pigeons (John and George, 1973) and also on isolated pigeon adipocytes *in vitro* (John and George, 1986). Glucagon and AVT were found to act synergistically in releasing FFA from the adipose tissue in pigeons (John and George, 1973). Both these hormones are also hyperglycemic in birds (for review, George, 1977). Circulating levels of glucagon, GH and AVT have been found to be significantly increased in flown pigeons (Table 1). The increase in circulating levels of melatonin in flown homing pigeons (Table 1) is suggestive of a possible relationship between melatonin and AVT with respect to fuel mobilization since it has been reported that melatonin increases plasma levels of GH, FFA and glucose in the pigeon (George, 1977).

The transport of FFA to the tissues that are capable of fatty acid oxidation is facilitated by FFA binding with albumin, a soluble protein having high affinity for long-chain fatty acids (Newsholme and Leech, 1983). The significant reduction in plasma levels of albumin in the flown pigeons with no change in albumin/globulin ratio or plasma osmolality, was not sufficient to lower the colloidal osmotic pressure or cause fluid loss in the circulating blood (George and John, 1993). It should also be mentioned that with lipid as the main fuel for muscular energy, substrate oxidation yields not only more than double the energy obtained from carbohydrate but also double the amount of metabolic water. With the increasing concentration of blood levels of AVT during flight and with birds being uricotelic, conservation of water in the body is ensured. In this context it is also interesting that the circulating levels of uric acid were increased in flown pigeons (Table 1). Uric acid increase in the blood has also been reported in migratory birds during migratory flight (Jenni-Eiermann and Jenni, 1991). Even in a ureotelic mammal like man, blood levels of uric acid were found to be increased with exhausting exercise (Hellsten-Westing *et al.*, 1991).

Uric acid levels in the blood may show seasonal variations since it has been reported in man that the levels tend to be higher in summer (Goldstein *et al.,* 1972). Increase in blood uric acid has also been reported in rats during hyperthermia (Frankel *et al.,* 1972). That uric acid has a role in preventing hyperthermia has been indicated in several studies (Schreiner and Maher, 1961; Kampa and Frascella, 1977; Kluger *et al.*, 1981; Gordon, 1990; George and John, 1993). Besides the thermoregulatory role of uric acid, its possible role as an antioxidant against oxidative tissue injury and its iron-ion-binding properties (Davies *et al.*, 1986) could have considerable importance in metabolic adjustments during sustained muscular activity.

The above observations suggest that even metabolic by-products such as uric acid could have a useful role in maintaining homeostasis. Another metabolic end-product that also registered a significant increase in flown homing pigeons was creatinine which is derived from the breakdown of phosphocreatine in muscle during activity (Table 1). Records of increase in blood levels of creatinine resulting from muscular exercise, are available from studies on athletes (Böhmer, 1975). Whether creatinine has a role similar to uric acid is unknown. However, it has been shown in racing (homing) pigeons that both plasma uric acid and creatinine levels are significantly increased in response to dehydration (deprived of water for 3 days) (Lumeij, 1987). Since water conservation and temperature regulation are interrelated, both uric acid and creatinine levels in blood have a common role in homeostasis.

There is no evidence in the present study to indicate that there has been any muscle tissue damage as a result of the long flight. It is, however, known that the spring migratory flight in Canada geese do produce significant degradative changes in cell structure in the breast muscle (George *et al.*, 1987). Exercise-induced structural changes in muscle have also been observed in athletes (Friden *et al.*, 1983). Circulating levels of certain muscle enzymes such as creatine phosphokinase (CPK), aspartate amino-transferase (ASAT) and lactate dehydrogenase (LDH) have been used as markers of tissue damage (Gimenez and Florentz, 1984; Jones *et al.*, 1986; Linjnen *et al.*, 1988; Van Der Meulen *et al.*, 1991). The fact that no significant changes in the blood levels of the marker enzymes, ASAT, alanine amino-transferase (ALAT), LDH and CPK in the flown pigeons (Table 1) were observed in the present study, shows that the flight of this magnitude and duration was not sufficiently strenuous or exhausting so as to produce damage to the muscle tissue that would be indicated by increase in the specific enzyme levels in the blood. The absence of any significant change in the blood levels of alkaline phosphatase and also of calcium and phosphorus (Table 1), in the flown pigeons suggest that such a flight did not affect calcium homeostasis (George and John, 1993). Similarly, the fact that flight did not cause changes in the blood levels of electrolytes and plasma osmolality (Table 1), suggests maintenance of osmo- and iono-homeostasis (George and John, 1993).

It is of considerable interest that no change in the circulating levels of corticosterone, the major glucocorticoid in birds (Frankel *et al.*, 1967) that is indicative of stress, was observed in the flown pigeons (Table 1). It is therefore clear that there was no significant amount of stress during a flight of this magnitude and duration. It has been observed in the Atlantic salmon that moderate exercise could even decrease plasma cortisol levels (Boesgaard *et al.*, 1993). In homing pigeons, however, in extended flights of 115–557 km lasting 315–561 min durations significant increases in plasma levels of corticosterone have been reported (Haase *et al.*, 1986). In a study to examine the control of the pituitary-adrenocortical system in racing pigeons it was shown that administration of ovine corticotrophin-releasing hormone and AVT acted as pituitary stimulants for corticosterone secretion (Westerhof *et al.*, 1992).

The flight-induced increase in circulating levels of catecholamines, A and NA (Table 1), is indicative of stimulation of sympathetic activity which in turn could stimulate the release of glucagon and inhibit insulin secretion. Stimulation of glucagon release and inhibition of insulin secretion has been shown in ducks by *in vivo* administration of A (Tyler *et al.*, 1972). The flight-induced increase in circulating levels of glucagon release and inhibition of insulin secretion has been shown in ducks by *in vivo* administration of A (Tyler *et al.*, 1972). The flight-induced increase in circulating levels of glucagon observed in homing pigeons may therefore be attributed to increased stimulation of glucagon release by catecholamines (Table 1). In birds, glucagon is the most powerful lipid mobilizing hormone whereas, unlike in mammals, catecholamines have only a minor role, if at all, in mobilizing fatty acids (Langslow and Hales, 1969). Hence, blood glucose levels in most birds tend to be two to three times higher than those of mammals (Heald *et al.*, 1965; Pearce, 1977). This maintenance of high levels of blood glucose in birds is also due to the relatively weak hypoglycemic action of insulin thereby rendering a low insulin/glucagon molar ratio (Hazelwood, 1971). Insulin, however, is not antilipolytic but could even act synergistically with glucagon in stimulating lipolysis (Pearce, 1977; Hazelwood, 1986). It has been well documented in humans that exercise induces rise in blood levels of glucagon and drop in insulin so as to ensure continuous supply of metabolic substrates to the exercising muscles (Felig and Wahren, 1975; Luyckx *et al.*, 1978). That exercise by prolonged electrical stimulation (1 h, 5 h) of the pectoral muscles, causes significant decrease in plasma levels of insulin has been shown in the pigeon (John *et al.*, 1989). In the same study, it was also observed that there was no significant change in the levels of avian pancreatic polypeptide (APP), an antilipolytic hormone. Since pigeon is a good flier capable of long distance flight metabolizing fatty acids as the main fuel, unlike the domestic fowl which is a very poor flier, it would be a disadvantage to have in circulation a highly potent antilipolytic hormone. The plasma levels of APP in the control and exercised (electrical stimulation of the pectoral muscles) pigeons were considerably lower than in the chicken. Levels of plasma APP in the chicken is reported to be 1.86 ng/ml plasma (Hazelwood and Langslow, 1978) against 0.63 ng/ml plasma in the pigeon (John *et al.*, 1989).

To sum up, the present study with homing pigeons, has revealed the significance of maintaining the integrity of body tissues and stability of the internal milieu with regard to plasma osmolality, electrolyte balance and plasma levels of the marker enzymes monitored, during flights of moderate distance, duration and intensity. It has also shed some light on the importance of the extent of flight training received and their familiarity with the release site. With pigeons not in regular training, "homeward" flight from the release site could become extended and the flight duration increased. In these birds flight caused a more than two-fold increase in circulating levels of rT_3 concomitant with reductions in T_4 and T_3 levels indicating the conversion of T_4 to rT_3, the inactive form of T_3 as a mechanism of autoregulation of thyroid function. The significant increases observed in the circulating levels of muscle fuels, glucose and FFA and their mobilizing hormones, glucagon, A, NA and GH in the flown pigeons are indicative of the regulatory metabolic processes in operation. However, the reduction in plasma albumin which acts as a lipid carrier, suggests that it is being lost in peripheral transport, especially to the muscles. The flight-induced increase in blood levels of melatonin suggests that melatonin by virtue of its known hypothermic effect, could prevent excessive flight-induced hyperthermia. It is postulated that the increase in plasma levels of uric acid in flown pigeons could also have a role in preventing flight-induced hyperthermia, besides uric acid excretion in birds indirectly contributing to water conservation. The flight-induced rise in blood levels of creatinine could be attributed to the breakdown of phosphocreatine for energy during the initial phase of muscular activity prior to carbohydrate and lipid being used as fuels. Whether creatinine has a role similar to the one proposed for uric acid in preventing hyperthermia is not known.

Acknowledgements

I am beholden to Professor R.C. Anderson, former Chairman of the Department of Zoology for providing me facilities for continuation of my work and to Professor P.D.N. Herbert, present Chairman, for extending the same courtesy. I wish to express my appreciation to my colleagues for their cooperation and support. I am also indebted to Mrs. Mary Anne Davis who has most diligently and willingly provided secretarial assistance over the years.

References

Aulie, A. (1971). *Comp. Biochem. Physiol.* **39A:** 173–176.

Aulie, A. (1975). *Comp. Biochem. Physiol.* **52A:** 81–84.

Baudinette, R.V., Loveridge, J.P., Wilson, K.J., Mills, C.D. and Schmidt-Nielsen, K. (1976). *Am. J. Physiol.* **230:** 920–924.

Baudinette, R.V., Tonkin, A.L., Orbach, J., Seymour, R.S. and Wheldrake, J.F. (1982). *Comp. Biochem. Physiol.* **72A:** 327–332.

Bech, C. and Nomoto, S. (1982). *J. exp. Biol.* **97:** 345–358.

Boesgaard, L., Nielsen, M.E. and Rosenkilde, P. (1993). *Comp. Biochem. Physiol.* **106A:** 641–643.

Böhmer, D. (1975). *In:* Metabolic Adaptation to Prolonged Physical Exercise. (H. Howald and J.R. Poortman, Eds), pp 229–231. Birkhaüser Verlag, Basel.

Brackenbury, J. (1984). *Biol. Rev.* **59:** 559–575.

Butler, P.J. (1991). *J. exp. Biol.* **160:** 233–262.

Butler, P.J. and Woakes, A.J. (1980). *J. exp. Biol.* **85:** 213–226.

Butler, P.J., West, N.H. and Jones, D.R. (1977). *J. exp. Biol.* **71:** 7–26.

Carlson, L.A. and Mossfeldt, F. (1964). *Acta Physiol. Scand.* **62:** 51–59.

Davies, K.J.A., Sevanian, A., Mukkassah-Kelly, S.F. and Hochstein, P. (1986). *Biochem. J.* **235:** 747–754.

Dawson, A. and King, V. (1994). *J. Biol. Rhythms.* **9(2):** 137–144.

Deshaies, Y., Vallerand, A.L. and Bukowiecki, L.J. (1983). *Life Sci.* **33:** 75–82.
Dial, K.P., Kaplan, S.R., Goslow, J.E.Jr., Jenkins, F.A.Jr. (1987). *Anat. Rec.* **218:** 284–287.
Engler, D. and Burger, A.G. (1984). *Endocr. Rev.* **5:** 151–184.
Felig, P. and Wahren, J. (1975). *N. Engl. J. Med.* **293:** 1078–1084.
Frankel, A.I., Graber, J.W. and Nalbandov, A.V. (1967). *Endocrinology.* **80:** 1013–1019.
Frankel, H.M., Yousef, M.K., Bayer, R. and Dill, D.B. (1972). *Comp. Biochem. Physiol.* **43A:** 733–738.
Friden, J., Sjostrom, M. and Ekblom, B. (1983). *Int. J. Sports. Med.* **4:** 170–176.
George, J.C. (1977). *Amer. Zool.* **17:** 787–808.
George, J.C. (1980). *In*: Avian Endocrinology. (A. Epple and M.H. Stetson, Eds), pp 85–115. Academic Press, New York.
George, J.C. (1982). *In:* The Pineal and its Hormones. (R.J. Reiter, Ed), pp 217–31. Alan R. Liss Inc., New York.
George, J.C. (1994). *Proc. 16th Inter. Congr. Biochem. Mole. Biol.* Sept. 19–22. New Delhi, India.
George, J.C. and Berger, A.J. (1966). *Avain Myology.* Academic Press, New York.
George, J.C. and John, T.M. (1992). *Zool Sci.* **9:** 675–677.
George, J.C. and John, T.M. (1993). *Comp. Biochem. Physiol.* **106A:** 707–712.
George, J.C., John, T.M. and Koike, T.I. (1992). *Biol. Singnals.* **1:** 160–166.
George, J.C., Hohn, T.M. and Mitchell, M.A. (1989). *Horm. metabol. Res.* **21:** 542–545.
George, J.C., John, T.M. and Minhas, K.J. (1987). *Cytobios* **52:** 109–126.
Gimenez, M. and Florentz, M. (1984). *Eur. J. Apple. Physiol.* **52:** 219–224.
Goldstein, R.A., Becker, K.L. and Moore, C.F. (1972). *N. Engl. J. Med.* **287:** 649–650.
Gordon, C. (1990). *J. therm. Biol.* **15:** 97–101.
Grubb, B., Jorgenson, D.A. and Conner, M. (1983). *J. exp. Biol.* **104:** 193–201.
Guerrero, J.M., Puig–Domingo, M. and Reiter, R.J. (1988a). *Endocrinology* **122:** 236–241.
Guerrero, J.M., Puig–Domingo, M., Santana, C., Menendez–Palaez, A. and Reiter, R.J. (1988b). *J. Pineal Res.* **5:** 513–520.
Guerrero, J.M. and Reiter, R.J. (1992). *Int. J. Biochem.* **24(10):** 1513–1523.
Haase, E., Rees, A. and Harvey. S. (1986). *Gen. Comp. Endocrinol.* **61:** 424–427.
Harvey, S., Klandorf, H., Foltzer, C., Strosser, M.T. and Phillips, J.G. (1982). *Gen. Comp. Endocrinol.* **48:** 415–420.
Harvey, S. and Phillips, J.G. (1982). *J. Endocrinol.* **94:** 141–146.
Hassinen, E., Pyornila, A. and Hissa, R. (1994). *Comp. Biochem. Physiol.* **107A:** 545–551.
Heald, P.J., McLaughlan, P.M. and Rookledge, K.A. (1965). *J. Endocrinol.* **33:** 83–95.
Hazelwood, R.L. (1971). *Poult. Sci.* **50:** 9–18.
Haselwood, R.L. (1986). *In*: Avian Physiology. 4th Edition. (P.D. Sturkie, Ed), pp 303–325. Springer-Verlag, New York.
Hazelwood, R.L. and Langslow, D.R. (1978). *J. Endocrinol.* **21:** 485–496.
Hellsten-Westing, Y., Sollevi, A. and Sjödin, B. (1991). Eur. *J. appl. Physiol.* **62:** 380–384.
Jenni-Eiermann, S. and Jenni, L. (1991). *J. Comp. Physiol.* **B161:** 465–474.
John, T.M. and George, J.C.(1973). *Comp. Biochem. Physiol.* **45A:** 541–547.
John, T.M. and George, J.C. (1986). *Arch. Internat. Physiol. Biochim.* **94:** 85–89.
John, T.M. and George, J.C. (1991). *In:* Photobiology. (E. Riklis, Ed), pp 597–605. Plenum Press, New York.
John, T.M. and George, J.C. (1992). *Comp. Biochem. Physiol.* **102C:** 353–359.
John, T.M., George, J.C. and Brown, G.M. (1992). *Cytobios.* **72:** 167–175.
John, T.M., George, J.C., Yie, S.M. and Brown, G.M. (1993). *Comp. Biochem. Physiol.* **106A:** 645–648.
John, T.M., Itoh, S. and George, J.C. (1978). *Horm. Res.* **9:** 41–56.
John, T.M., Pilo, B., Hazelwood, R.L. and George, J.C. (1989). *Gen. Comp. Endocrinol.* **75:** 71–77.

John, T.M., Viswanathan, M., George, J.C. and Scanes, C.G. (1988). *Horm. metabol. Res.* **20:** 271–273.

Jones, D.A., Newham, D.J., Round, J.M. and Tolfree, S.E.J. (1986). *J. Physiol. (London).* **375:** 435–448.

Kampa, I.S. and Frascella, D.W. (1977). *Life Sci.* **20:** 1373–1376.

Kiley, J.P. and Fedde, M.R. (1983). *J. appl. Physiol.* **55:** 1574–1581.

Kiley, J.P., Kuhlmann, W.D. and Fedde, M.R. (1979). *J. appl. Physiol.* **47:** 827–833.

Kluger, M.J., Turnbull, A.J., Cranston, W.I., Wing, A.J., Gross, M.P. and Rosenburg, B.A. (1981). *Am. J. Physiol.* **241:** R271–276.

Kühn, E.R., Mol, K.A. and Darras, V.M. (1993). *Zool. Sci.* **10:** 873–885.

Lam, S.K. and Harvey, S. (1990). *Comp. Biochem. Physiol.* **95A:** 435–439.

Lang, R.E., Heil, J.W.E., Ganten, D., Hermann, K., Unger, T. and Rascher, W. (1983). *Neuroendocrinology* **37:** 314–316.

Langslow, D.R. and Hales, C.N. (1969). *J. Endocrinol.* **43:** 285–294.

Linjnen, P., Hespel, P., Fagard, R., Lysens, R., Vanden Eynde, E., Goris, M., Goossens, W., Lissens, W. and Amery, A. (1988). *Int. J. Sports Med.* **9:** 108–113.

Lumeij, J.T. (1987). *Avian Pathology.* **16:** 377–382.

Luyckx, A.S., Pirnay, F. and Lefebvre, P.J. (1978). *Eur. J. Appl. Physiol.* **39:** 53–61.

Martineau, L. and Larochelle, J. (1988). *J. exp. Biol.* **136:** 193–208.

Newsholme, E.A. and Leech, A.R. (1983). *Biochemistry for Medical Sciences.* John Wiley & Sons, New York.

Nomoto, S., Bech, C., Rautenberg, W. and Johansen, K. (1983). *J. therm. Biol.* **8:** 175–177.

Parker, G.H. and George, J.C. (1974). *Int. J. Biochem.* **5:** 167–172.

Parker, G.H. and George, J.C. (1975). *Jap. J. Physiol.* **25:** 175–184.

Pearce, J. (1977). *Int. J. Biochem.* **10:** 193–199.

Pilo, B., John, T.M., George, J.C. and Etches, R.J. (1984). *Arch. internat. Physiol. Biochim.* **92:** 131–138.

Pinshow, B., Fedak, M.A., Battles, D.R. and Schmidt-Nielsen, K. (1976). *Am. J. Physiol.* **231:** 903–912.

Rees, A. and Harvey, S. (1987). *Gen. Comp. Endocrinol.* **65:** 117–120.

Rees, A., Harvey, S. and Phillips, J.G. (1983). *Gen. Comp. Endocrinol.* **49:** 485–489.

Rees, A., Harvey, S. and Phillips, J.G. (1985). *Gen. Comp. Endocrinol.* **59:** 100–104.

Reiter, R.J. and Guerrero, J.M. (1992). *Progr. Brain Res.* **91:** 315–321.

Rojdmark, S., Berg, A., Rossner, S. and Wetterberg, L. (1991). *Clin. Endocrinol.* **35:** 61–65.

Schreiner, G.E. and Maher, J.F. (1961). Uremia: Biochemistry, Pathogenesis and Treatment. Thomas, Springfield. Illinois.

Strassman, R.J., Appenzeller, O., Lewy, A.J., Qualls, C.R. and Peake, G.T. (1989). *J. Clin. Endocrinol. Metab.* **69:** 540–545.

Tanaka, K., Murakami, M. and Greer, M.A. (1986). *Biochem. Biophys. Res. Commun.* **137:** 863–868.

Tanaka, K., Murakami, M. and Greer, M.A. (1987). *Endocrinology* **121:** 74–77.

Torre-Bueno, J.R. (1978). *J. exp. Biol.* **75:** 231–236.

Torre-Bueno, J.R. and Larochelle, J. (1978). *J. exp. Biol.* **75:** 223–229.

Tucker, V.A. (1968). *J. exp. Biol.* **48:** 67–87.

Tyler, J.M., Kajinuma, H. and Mialhe, P. (1972). *Fed. Proc.* **31:** 263.

Vallyathan, N.V., Grinyer, I. and George, J.C. (1970). *Can. J. Zool.* **48:** 377–383.

Van Der Meulen, J.H., Kuipers, H. and Drukker, J. (1991). *J. Appl. Physiol.* **71:** 999–1004.

Viswanathan, M., John, T.M., George, J.C. and Etches, R.J. (1987). *Horm. metabol. Res.* **19:** 400–402.

Westerhof, I., Lumeij, J.T., Mol, J.A., van den Brom, W.E. and Rijnberk, A. (1992). *Gen. Comp. Endocrinol.* **88:** 76–82.

Wu, W., Reiter, R.J., Troiani, M.E. and Vaughan, G.M. (1987). *Life Sci.* **41:** 1473–1479.